2011 IEEE 23rd International Symposium on Power Semiconductor Devices and ICs

(ISPSD 2011)

AA002414

San Diego, California, USA
23 – 26 May 2011

IEEE Catalog Number: CFP11ISP-PRT
ISBN: 978-1-4244-8425-6

Copyright © 2011 by the Institute of Electrical and Electronic Engineers, Inc
All Rights Reserved

Copyright and Reprint Permissions: Abstracting is permitted with credit to the source. Libraries are permitted to photocopy beyond the limit of U.S. copyright law for private use of patrons those articles in this volume that carry a code at the bottom of the first page, provided the per-copy fee indicated in the code is paid through Copyright Clearance Center, 222 Rosewood Drive, Danvers, MA 01923.

For other copying, reprint or republication permission, write to IEEE Copyrights Manager, IEEE Service Center, 445 Hoes Lane, Piscataway, NJ 08854. All rights reserved.

***This publication is a representation of what appears in the IEEE Digital Libraries. Some format issues inherent in the e-media version may also appear in this print version.**

IEEE Catalog Number: CFP11ISP-PRT
ISBN 13:   978-1-4244-8425-6
ISSN: 1943-653X

Additional Copies of This Publication Are Available From:

Curran Associates, Inc
57 Morehouse Lane
Red Hook, NY 12571 USA
Phone: (845) 758-0400
Fax: (845) 758-2633
E-mail: curran@proceedings.com
Web: www.proceedings.com

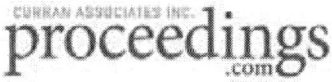

TABLE OF CONTENTS

Monday – May 23, 2011

9:00 – 11:45
Session 1 – Plenary
Chairs: Mohamed Darwish, *MaxPower Semiconductor*
Peter Moens, *On Semiconductor*
D. Disney, *Monolithic Power Systems*

9:00 – 9:45
Opportunities and Challenges with Net Zero Energy Buildings 1
Satyen Mukherjee, *Philips Research North America*

10:15 – 11:00
Trends in High-Speed Railways and the Implications on Power Electronics and Power Devices 6
Tetsuo Uzuka, *Railway Technical Research Institute*

11:15 – 12:00
SiC Power Devices – Present Status, Applications and Future Perspective 10
Mikael Östling, Reza Ghandi, Carl-Mikael Zetterling, *KTH Royal Institute of Technology*

13:30 – 15:10
Session 2 – Smart Power Technology 1
Chairs: Sameer Pendharkar, *Texas Instruments*
Tanya Trajkovic, *Cambridge Semiconductor*

13:30 – 13:55
A Novel Substrate-Assisted RESURF Technology for Small Curvature Radius Junction 16
Ming Qiao, Xi Hu, Hengjuan Wen, Meng Wang, Bo Luo, Xiao Luo, Zhuo Wang, Bo Zhang, Zhaoji Li, *University of Electronic Science and Technology of China*

13:55 – 14:20
Automotive 130 nm Smart-Power-Technology Including Embedded Flash Functionality 20
Ralf Rudolf, Cajetan Wagner, Lincoln O'Riain, Karl-Heinz Gebhardt, Barbara Kuhn-Heinrich, Birgit von Ehrenwall,
Andreas von Ehrenwall, Marc Strasser, Matthias Stecher, Ulrich Glaser, Stefano Aresu, Paul Kuepper,
Alevtina Mayerhofer, *Infineon Technologies AG*

14:20 – 14:45
Implementation of Fully Isolated Low Vgs nLDMOS with Low Specific On-Resistance 24
Choul-Joo Ko, Cheol-Ho Cho, Min-Seok Kim, Hyung-Gyun Jung, Hee-Bae Lee, Yong-Jun Lee, Min-Woo Kim,
Sung-Mo Gu, Sun-Kyung Bang, Han-Geon Kim, Sun-Kyoung Kang, Kwang-Dong Yoo, Lou Hutter, *Dongbu HiTek*

14:45 – 15:10

Wide-Voltage SOI-BiCDMOS Technology for High-Temperature Automotive Applications 28
Hidemoto Tomita, Hiroomi Eguchi, Shinya Kijima, Norihiro Honda, Tetsuya Yamada, Hideo Yamawaki,
Hirofumi Aoki, Kimimori Hamada, *Toyota Motor Corporation*

15:40 – 17:20
Session 3 – Smart Power Technology 2
Chairs: Ted Letavic, *IBM*
　　　　　Ayman Shibib, *Bournes*

15:40 – 16:05

**New Low-Resistance and Compact MOSFETs for Analog Switch ICs with
V-Groove Dielectric Isolation** 32
Kenji Hara, Junichi Sakano, Hironobu Honda, Junichi Aizawa, Taiga Arai, *Hitachi, Ltd.*

16:05 – 16:30

A Novel 0.16 μm – 300 V SOIBCD for Ultrasound Medical Applications 36
M. Sambi, D. Merlini, P. Galbiati, *STMicroelectronics TR&D;* E. Bonera, *Università degli Studi di Milano-Bicocca;*
F. Belletti, *STMicroelectronics TR&D*

16:30 – 16:55

300 V Field-MOS FETs for HV- Switching IC 40
T. Miyoshi, T. Tominari, M. Hayashi, A. Ito, M. Yoshinaga, S. Ueno, T. Oshima, S. Wada, *Hitachi, Ltd.*

16:55 – 17:20

**High Performance Pch-LDMOS Transistors in Wide Range Voltage from 35V to
200V SOI LDMOS Platform Technology** 44
Satoshi Shimamoto, Yohei Yanagida, Shinji Shirakawa, Kenji Miyakoshi, Toshinori Imai, Takayuki Oshima,
Junichi Sakano, Shinichiro Wada, *Hitachi, Ltd.*

Tuesday - May 24, 2011

8:30 – 10:10
Session 4 – IGBT 1
Chairs: Reinhard Herzer, *Semikron*
　　　　　Yasukazu Seki, *Fuji Electric*

8:30 – 8:55

**1.7kV Trench IGBT with Deep and Separate Floating p-Layer Designed for Low Loss,
Low EMI Noise, and High Reliability** 48
So Watanabe, Mutsuhiro Mori, Taiga Arai, Kohsuke Ishibashi, Yasushi Toyoda, Tetsuo Oda, Takashi Harada,
Katsuaki Saito, *Hitachi, Ltd.*

8:55 – 9:20

Development of the Next Generation 1700V Trench-Gate FS-IGBT 52
Y. Onozawa, D. Ozaki, H. Nakano, T. Yamazaki, N. Fujishima, *Fuji Electric Co. Ltd.*

9:20 – 9:45

The Radial Layout Design Concept for the Bi-Mode Insulated Gate Transistor 56
L. Storasta, M. Rahimo, M. Bellini, A. Kopta, *ABB Switzerland Ltd, Semiconductors;*
U.R. Vemulapati, N. Kaminski, *Universität Bremen*

9:45 – 10:10

Full Digital Short Circuit Protection for Advanced IGBTs 60
Takuya Tanimura, Kazufumi Yuasa, Ichiro Omura, *Kyushu Institute of Technology*

10:40 – 12:20
Session 5 – IGBT 2
Chairs: Ichiro Omura, *Kyushu Institute of Technology*
        Jean-Louis Sanchez, *LAAS-CNRS*

10:40 – 11:05

Ultrathin 400V FS IGBT for HEV Applications 64
Heike Böving, Thomas Laska, Anton Pugatschow, Waldemar Jakobi, *Infineon Technologies AG*

11:05 – 11:30

600V LPT-CSTBT™ on Advanced Thin Wafer Technology 68
Yuki Haraguchi, Shigeto Honda, Kazunari Nakata, Atsushi Narazaki, Yoshiaki Terasaki, *Mitsubishi Electric Corp.*

11:30 – 11:55

High Speed 650V IGBTs for DC-DC Conversion Up to 200 kHz 72
Hsueh-Rong Chang, Jiankang Bu, George Kong, Rachana Bou, *International Rectifier Corp.*

11:55 – 12:20

Novel High Voltage LDMOS on Partial SOI with Double-Sided Charge Trenches 76
Xiaorong Luo, Y.G. Wang, T.F. Lei, L. Lei, D.P. Fu, G.L. Yao, M. Qiao, Bo Zhang,
Zhaoji Li, *University of Electronic Science and Technology of China*

14:00 – 15:15
Session 6 – Diodes
Chairs: Dan Kinzer, *Fairchild Semiconductor*
        Stefan Linder, *ABB*

14:00 – 14:25

**Innovative Designs Enable 300-V TMBS® with Ultra-Low On-State Voltage and
Fast Switching Speed** 80
Wesley Chih-Wei Hsu, *Vishay General Semiconductor Taiwan;* Florin Udrea, *University of Cambridge;* Pai-Li Lin,
Yih-Yin Lin, Max Chen, *Vishay General Semiconductor Taiwan*

14:25 – 14:50

Ultra Low Loss Trench Gate PCI-PiN Diode with $V_F < 350mV$ 84
Motohiro Tsuda, Yasuaki Matsumoto, Ichiro Omura, *Kyushu Institute of Technology*

14:50 – 15:15

**Field Shielded Anode (FSA) Concept Enabling Higher Temperature Operation of
Fast Recovery Diodes** 88
S. Matthias, J. Vobecky, C. Corvasce, A. Kopta, M. Cammarata, *ABB Switzerland Ltd Semiconductors*

15:15 – 17:45

Posters

Chairs: Sujit Banerjee, *Power Integrations*
Ted Letavic, *IBM*

Evaluation of 1.2kV Super Junction Trench-Gate Clustered Insulated Gate Bipolar Transistor (SJ-TCIGBT) 92
N. Luther-King, M. Sweet, E.M. Shankar Narayanan, *University of Sheffield*

Relaxation of Current Filament Due to RFC Technology and Ballast Resistor for Robust FWD Operation 96
Akito Nishii, Katsumi Nakamura, Fumihito Masuoka, Tomohide Terashima, *Mitsubishi Electric Corporation*

Limits of Strongly Punch-Through Designed IGBTs 100
Thomas Raker, Hans-Peter Felsl, Franz-Josef Niedernostheide, Frank Pfirsch,
Hans-Joachim Schulze, *Infineon Technologies AG*

Filament-Induced Thermomigration of an Aluminum Drop at the Cathode-Side of High-Voltage Power Diodes 104
H.-J. Schulze, J.G. Bauer, F.-J. Niedernostheide, H.P. Felsl, J. Biermann, *Infineon Technologies AG;*
J. Lutz, R. Baburske, *Chemnitz University of Technology*

Optimization of Diodes Using the SPEED Concept and CIBH 108
Manfred Pfaffenlehner, Hans-Peter Felsl, Franz-Josef Niedernostheide, Frank Pfirsch, Hans-Joachim Schulze,
Infineon Technologies AG; Roman Baburske, Josef Lutz, *Chemnitz University of Technology*

Edge Termination Impact on Clamped Inductive Turn-Off Failure in High-Voltage IGBTs Under Overcurrent Conditions 112
X. Perpiñà, I. Cortés, J. Urresti-Ibañez, X. Jordà, J. Rebollo, J. Millán, *Instituto de Microelectrónica de Barcelona*

Hybrid Isolation Process with Deep Diffusion and V-Groove for Reverse Blocking IGBTs 116
Haruo Nakazawa, Masaaki Ogino, Hiroki Wakimoto, Tsunehiro Nakajima, Yoshikazu Takahashi,
David Hongfei Lu, *Fuji Electric Co. Ltd.*

Reduction of the Temperature Dependence of Leakage Current of IGBTs by Field-Stop Design 120
H.-J. Schulze, S. Voss, H. Huesken, F.-J. Niedernostheide, *Infineon Technologies AG*

Electro-Thermal Instability in Multi-Cellular Trench-IGBTs in Avalanche Condition: Experiments and Simulations 124
M. Riccio, A. Irace, G. Breglio, P. Spirito, E. Napoli, *Università degli Studi di Napoli Federico II;*
Y. Mizuno, *Toyota Motor Corporation*

On Chip ESD Protection of 600V Voltage Node 128
Vladislav A. Vashchenko, Antonio Gallerano, *National Semiconductor Corp.;*
Andrei Shibkov, *Angstrom Design Automation*

CSTBT™(III) Having Wide SOA Under High Temperature Condition 132
Yusuke Fukada, Kenji Suzuki, Tetsuo Takahashi, Tatsuo Harada, Hidenori Fujii, Shinichi Ishizawa,
Junichi Yamashita, *Mitsubishi Electric Corporation;* John F. Donlon, *Powerex, Inc;*
Tomohide Terashima, *Mitsubishi Electric Corporation*

Physical Analysis of Carrier Lifetime Controlled IGBT [II] 136
Chihiro Tadokoro, *Fukuryo Semicon Engineering Corp.;* M. Kaneda, *Mitsubishi Electric Corporation;*
K. Takano, *Fukuryo Semicon Engineering Corp.;* S. Kusunoki, T. Minato, *Mitsubishi Electric Corporation;*
J. Yahiro, *Fukuryo Semicon Engineering Corp.;* K. Hatade, *Mitsubishi Electric Corporation*

High Temperature Wafer Bonding Technique for the Realization of a Voltage and Current Bidirectional IGBT 140
A. Bourennane, H. Tahir, *LAAS-CNRS / Université de Toulouse;* J.-L. Sanchez, *LAAS-CNRS;* L. Pont, G. Sarrabayrouse, E. Imbernon, *LAAS-CNRS / Université de Toulouse*

Effects of Back-Side He Irradiation on MOS-GTO Performances 144
C. Ronsisvalle, V. Enea, *STMicrolectronics;* C. Abbate, G. Busatto, F. Iannuzzo, A. Sanseverino, *University of Cassino;* G.A.P. Cirrone, *Istituto Nazionale di Fisica Nucleare / LNS*

Temperature Dependence of Switching Performance in IGBT Circuits and its Compact Modeling 148
Masataka Miyake, Masaya Ueno, Junichi Nakashima, Hiroki Masuoka, Uwe Feldmann, Hans Juergen Mattausch, Mitiko Miura-Mattausch, *Hiroshima University;* Takaoki Ogawa, Takashi Ueta, *Toyota Motor Corporation*

Full Understanding of Hot-Carrier-Induced Degradation in STI-Based LDMOS Transistors in the Impact-Ionization Operating Regime 152
S. Poli, S. Reggiani, G. Baccarani, E. Gnani, A. Gnudi, *Università di Bologna;* M. Denison, S. Pendharkar, R. Wise, *Texas Instruments Inc.*

Avalanche Instability in Oxide Charge Balanced Power MOSFETS 156
J. Yedinak, R. Stokes, D. Probst, S. Kim, A. Challa, S. Sapp, *Farichild Semiconductor Corp.*

Prognostics of Power MOSFET 160
José R. Celaya, Abhinav Saxena, *NASA Ames Research Center;* Vladislav Vashchenko, *SGT Inc.;* Sankalita Saha, Kai Goebel, *NASA Ames Research Center*

Modeling the 3D Self Ballasting Behavior and Filamentation Under High Current Stressing in DeNMOS 164
Amitabh Chatterjee, *University of California, Santa Barbara;* Sameer Pendharkar, Charvaka Duvvury, *Texas Instruments Inc;* Forrest Brewer, *University of California, Santa Barbara*

Low-on-Resistance Strain-Controlled LDMOS Transistors for 0.25-μm Power ICs 168
Masafumi Miyamoto, Nobuyuki Sugii, Yukihiro Kumagai, Yoshinobu Kimura, *Hitachi, Ltd.*

0.25μm, 20V High Performance Complementary Bipolar Transistor with Dual EPI and Oxide-Filled Deep Trench Isolation for High Frequency DC-DC Converters 172
T. Kwon, S. Haynie, A. Sadovnikov, P. Allard, J. Strout, A. Strachan, *National Semiconductor Corp.*

Integration of 100V LDMOS Devices in 0.35μm CMOS Technology 176
Soon Tat Kong, Paul Stribley, Chris Lee, Michaelina Ong, *X-FAB Semiconductor Foundries*

High-Voltage Thick Layer SOI Technology for PDP Scan Driver IC 180
Ming Qiao, Lingli Jiang, Meng Wang, Yong Huang, *University of Electronic Science and Technology of China;* Hong Liao, *Changhong Electric Co., Ltd.;* Tao Liang, *University of Electronic Science and Technology of China;* Zhen Sun, *Changhong Electric Co., Ltd.;* Bo Zhang, *University of Electronic Science and Technology of China;* Zhaoji Li, *University of Electronic Science and Technology of China;* Guangzuo Huang, *Changhong Electric Co., Ltd.;* Yuanyuan Zhao, Li Lai, Xi Hu, Xiang Zhuang, Xiaorong Luo, Zhuo Wang, *University of Electronic Science and Technology of China*

Considerations on the Optimal Power Stage Segmentation Algorithm for MHz Integrated Synchronous Buck DC-DC Converters 184
Xiaopeng Wang, Alex Q. Huang, *North Carolina State University*

The ESD Failure Mechanism of Ultra-HV 700V LDMOS 188
Jian-Hsing Lee, *Independent ESD/EOS/Latch-up Consultant;* Tzu-Cheng Kao, Chien-Ling Chan, Jin-Lian Su, Hung-Der Su, Kuo-Cheng Chang, *Richtek Technology Corporation*

Techniques to Prevent Substrate Injection Induced Failure During ESD Events in Automotive Applications 192
Amaury Gendron, Chai Gill, Craig Aykroyd, Carol Zhan, *Freescale Semiconductor*

IGBT Driver Chip Set with Advanced Digital Signal Processing 196
J. Lehmann, G. Katzenberger, G. Königsmann, M. Roßberg, R. Herzer, *Semikron Elektronik GmbH & Co. KG*

Solutions to Improve Flatness of Id-Vd Curves of Rugged nLDMOS 200
S. Mouhoubi, F. Bauwens, J. Roig, P. Gassot, P. Moens, M. Tack, *ON Semiconductor*

The Vertical Voltage Termination Technique – Characterizations of Single Die Multiple 600V Power Devices 204
Kremena Vladimirova, Jean-Christophe Crébier, Christian Schaeffer, *Grenoble Electrical Engineering Lab / Grenoble University;* Delphine Constantin, *CIME Nanotech*

Investigation of Parasitic BJT Turn-On Enhanced Two-Stage Drain Saturation Current in High-Voltage NLDMOS 208
Chih-Chang Cheng, H.L. Chou, F.Y. Chu, R.S. Liou, Y.C. Lin, K.M. Wu, Y.C. Jong, C.L. Tsai, Jun Cai, H.C. Tuan, *Taiwan Semiconductor Manufacturing Company*

High Vgs MOSFET Characteristics with Thin Gate Oxide for PMIC Application 211
Jaehan Cha, Kyungho Lee, Sungoo Kim, Juho Kim, Namkyu Park, Taejong Lee, *MagnaChip Semiconductor Corporation*

Drift Design Impact on Quasi-Saturation & HCI for Scalable N-LDMOS 215
Yun Shi, Natalie Feilchenfeld, Rick Phelps, Max Levy, *IBM Microelectronics;* Martin Knaipp, Rainer Minixhofer, *Austriamicrosystems*

A Versatile 30V Analog CMOS Process in a 0.18μm Technology for Power Management Application 219
Yong-Keon Choi, Il-Yong Park, Hyun-Chol Lim, Mi-Young Kim, Chul-Jin Yoon, Nam-Joo Kim, Kwang-Dong Yoo, Lou N. Hutter, *Dongbu HiTek*

1kV AlGaN/GaN Power SBDs with Reduced on Resistances 223
Kiyeol Park, Younghwan Park, Shinwhan Hwang, Woochul Jeon, *Samsung Electro-Mechanics;* Junghee Lee, *Kyungpook National University*

3.7 mΩ-cm^2, 1500 V 4H-SiC DMOSFETs for Advanced High Power, High Frequency Applications 227
Sei-Hyung Ryu, Lin Cheng, Sarit Dhar, Craig Capell, Charlotte Jonas, Robert Callanan, Anant Agarwal, John Palmour, *Cree, Inc.;* Aivars Lelis, Charles Scozzie, Bruce Geil, *U. S. Army Research Laboratory*

High-Voltage GaN SBD on Si Substrate by Suppressing Metal Spikes 231
Min-Woo Ha, Cheong Hyun Roh, Hong Goo Choi, Jun Ho Lee, Hong Joo Song, *Korea Electronics Technology Institute;* Ogyun Seok, *Seoul National University;* Cheol-Koo Hahn, *Korea Electronics Technology Institute*

Effect of Oxygen Annealing Temperature on AlGaN/GaN HEMTs 235
Ogyun Seok, Young-Shil Kim, Jiyong Lim, Min-Koo Han, *Seoul National University*

Normally-Off High-Voltage p-GaN Gate GaN HFET with Carbon-Doped Buffer 239
O. Hilt, F. Brunner, E. Cho, A. Knauer, E. Bahat-Treidel, J. Würfl, *Ferdinand-Braun-Institut*

Safe Operating Area of AlGaAs/InGaAs/GaAs HEMT Power Transistors 243
Vipindas Pala, Mona Hella, T. Paul Chow, *Rensselaer Polytechnic Institute*

A New Vertical GaN SBD Employing in-situ Metallic Gallium Ohmic Contact 247
Jiyong Lim, Ogyun Seok, Young-Shil Kim, Min-Koo Han, *Seoul National University;*
Minki Kim, *Electronics and Telecommunication Research Institute*

High Breakdown Voltage AlGaN/GaN HEMT by Employing Selective Fluoride Plasma Treatment 251
Young-Shil Kim, Jiyong Lim, O-Gyun Seok, Min-koo Han, *Seoul National University*

Design and Characterization of a 3D Half-Bridge Semiconductor Power Module in a
DFN3x3 Package for DC-DC Buck Converter Application 256
Yi Su, Anup Bhalla, Daniel Ng, Fei Wang, Jonathan Xue, Ji Pan, *Alpha & Omega Semiconductor*

Reliability Study of Au-In Transient Liquid Phase Bonding for SiC Power Semiconductor Packaging 260
Brian Grummel, *University of Central Florida;* Habib A. Mustain, *Cree, Inc.;* Z. John Shen, *University of Central Florida;* Allen R. Hefner, *National Institute of Standards and Technology*

Thermal Impedance Spectroscopy of Power Modules During Power Cycling 264
Alexander Henlser, Daniel Wingert, Christian Herold, Josef Lutz, *Chemnitz University of Technology;*
Markus Thoben, *Infineon Technologies AG*

Application Driven Integrated Design of a Half-Bridge Power Switch 268
Adane Solomon, Alberto Castellazzi, *University of Nottingham*

Investigation on Wirebond-Less Power Module Structure with High-Density
Packaging and High Reliability 272
Yoshinari Ikeda, Yuji Iizuka, Yuichiro Hinata, Masafumi Horio, Motohito Hori,
Yoshikazu Takahashi, *Fuji Electric Co. Ltd.*

Wednesday - May 25, 2011

9:00 – 10:15
Session 7 – GaN Power Devices
Chairs: Min-Koo Han, *Seoul National University*
　　　　　Peter Moens, *On Semiconductor*

9:00 – 9:25
A Novel Normally-Off GaN Power Tunnel Junction FET 276
Li Yuan, Hongwei Chen, Qi Zhou, Chunhua Zhou, Kevin J. Chen, *Hong Kong University of Science and Technology*

9:25 – 9:50
GaN Based Super HFETs over 700V Using the Polarization Junction Concept 280
Akira Nakajima, Mahesh H. Dhyani, E.M. Sankara Narayanan, *University of Sheffield;*
Yasunobu Sumida, Hiroji Kawai, *POWDEC K.K.*

9:50 – 10:15
Over 1.7 kV Normally-Off GaN Hybrid MOS-HFETs with a Lower on-Resistance on a Si Substrate 284
Nariaki Ikeda, Ryosuke Tamura, Takuya Kokawa, Hiroshi Kambayashi, Yoshihiro Sato, Takehiko Nomura,
Sadahiro Kato, *Advanced Power Device Research Association*

10:40 – 12:20

Session 8 – SiC Power Devices

Chairs: Jose Millan, *IMB-CNM*
David Sheridan, *Semisouth Inc.*

10:40 – 11:05

Low On-Resistance 1.2 kV 4H-SiC MOSFETs Integrated with Current Sensor 288
A. Furukawa, S. Kinouchi, H. Nakatake, Y. Ebiike, Y. Kagawa, N. Miura, Y. Nakao, M. Imaizumi,
H. Sumitani, T. Oomori, *Mitsubishi Electric Corporation*

11:05 – 11:30

4H-SiC Bipolar Junction Transistors with Record Current Gains of 257 on (0001) and 335 on (000-1) 292
Hiroki Miyake, Tsunenobu Kimoto, Jun Suda, *Kyoto University*

11:30 – 11:55

**5kV Class 4H-SiC Pin Diode with Low Voltage Overshoot During Forward Recovery for
High Frequency Inverter** 296
S. Ogata, Y. Miyanagi, K. Nakayama, A. Tanaka, K. Asano, *Kansai Electric Power Co., Inc.*

11:55 – 12:20

A SiC Static Induction Transistor (SIT) Technology for Pulsed RF Power Amplifiers 300
Francis K. Chai, Bruce Odekirk, Ed Maxwell, Mar Caballero, Terri Fields, Mike Mallinger,
Dumitru Sdrulla, *Microsemi Corp.*

14:00 – 15:40

Session 9 – High Voltage MOSFET's

Chairs: Vijay Parthasarathy, *Power Integrations*
Deva Pattanayak, *Vishay-Siliconix*

14:00 – 14:25

UltiMOS : A Local Charge-Balanced Trench-Based 600V Super-Junction Device 304
P. Moens, F. Bogman, H. Ziad, H. De Vleeschouwer, J. Baele, M. Tack, G. Loechelt, G. Grivna, J. Parsey,
Y. Wu, T. Quddus, P. Zdebel, *ON Semiconductor*

14:25 – 14:50

Vertical Charge Imbalance Effect on 600 V-Class Trench-Filling Superjunction Power MOSFETs 308
T. Tamaki, Y. Nakazawa, H. Kanai, Y. Abiko, Y. Ikegami, M. Ishikawa, E. Wakimoto, T. Yasuda,
S. Eguchi, *Renesas Electronics Corporation*

14:50 – 15:15

**Energy Limits for Unclamped Inductive Switching in High-Voltage Planar and
SuperJunction Power MOSFETs** 312
J. Roig, P. Moens, J. McDonald, P. Vanmeerbeek, F. Bauwens, M. Tack, *ON Semiconductor*

15:15 – 15:40

**Improvement of Switching Trade-Off Characteristics between Noise and Loss in
High Voltage MOSFETs** 316
Wataru Saito, Satoshi Aida, Shigeo Koduki, Masaru Izumisawa, *Toshiba Corp.*

16:10 – 17:50
Session 10 – Packaging and Module Technologies
Chairs: Phil Mawby, *University of Warwick*
Dieter Silber, *University of Bremen*

16:10 – 16:35
300A 650V 70 um Thin IGBTs with Double-Sided Cooling 320
Hsueh-Rong Chang, Jiankang Bu, George Kong, Ricky Labayen, *International Rectifier Corp.*

16:35 – 17:00
SKiN: Double Side Sintering Technology for New Packages 324
Thomas Stockmeier, Peter Beckedahl, Christian Göbl, Thomas Malzer, *Semikron Elektronik GmbH & Co. KG*

17:00 – 17:25
**A Novel Power System in Package with 3D Chip on Chip Interconnections of the
Power Transistor and its Gate Driver** 328
Timothé Simonot, Nicolas Rouger, Jean-Christophe Crébier, Victor Gaude, *Grenoble Electrical Engineering Lab /
Grenoble University;* Pheng Irène, *CIME Nanotech*

17:25 – 17:50
**Innovative Heat Removal Structure for Power Devices – the Drift Region
Integrated Microchannel Cooler** 332
Kremena Vladimirova, Jean-Christophe Crébier, Yvan Avenas, Christian Schaeffer, *Grenoble Electrical Engineering
Lab / Grenoble University;* Stephane Litaudon, *CIME Nanotech*

Thursday - May 26, 2011

8:30 – 10:10
Session 11 – Device and Process Reliability
Chairs: Il-Yong Park, *Dongbu HiTech*
Rhonghu Zhu, *Maxim Integrated Products*

8:30 – 8:55
Interface Charge Trapping and Hot Carrier Reliability in High Voltage SOI SJ LDMOSFET 336
M. Antoniou, F. Udrea, *University of Cambridge;* E. Kho Ching Tee, Yang Hao, S. Pilkington, Kee Kia Yaw,
D.K. Pal, A. Hoelke, *X-FAB Sarawak Sdn. Bhd*

8:55 – 9:20
**Reliability and Performance Optimization of 42V N-Channel Drift MOS Transistor in
Advanced BCD Technology** 340
A. Molfese, P. Gattari, G. Marchesi, G. Croce, G. Pizzo, F. Alagi, F. Borella, *STMicrolectronics*

9:20 – 9:45
Practical Approaches to Improve Thermal SOA for Smart Power IC 344
T. Nitta, A. Omichi, S. Yanagi, Y. Yoshihisa, T. Kuroi, K. Hatasako, S. Maegawa, *Renesas Electronics Corp.;*
K. Furuya, *Renesas Semiconductor Engineering Corp.*

9:45 – 10:10
Hot Carrier Degradation of HV-SOI Devices Under Off-and on-State Current Conditions 348
R. van Dalen, *NXP;* S. Dhar, A. Heringa, *NXP Research;* M.J. Swanenberg, A.B. van der Wal,
P.W.M. Boos, V. Braspenning-Girault, *NXP*

10:40 – 12:20

Session 12 – Smart Power Circuit Topologies

Chairs: Sujit Banerjee, *Power Integrations*
  Wai Tung Ng, *University of Toronto*

10:40 – 11:05

A Novel Silicon-Embedded Coreless Transformer for Isolated DC-DC Converter Application 352
Rongxiang Wu, Johnny K.O. Sin, *Hong Kong University of Science and Technology;*
S.Y. (Ron) Hui, *City University of Hong Kong*

11:05 – 11:30

Integrated Low Power and High Bandwidth Optical Isolator for Monolithic Power MOSFETs Driver 356
Nicolas Rouger, Jean-Christophe Crébier, Olivier Lesaint, *Grenoble Electrical Engineering Lab / Grenoble University*

11:30 – 11:55

**Design and Characterization of a Signal Insulation Coreless Transformer Integrated in a
CMOS Gate Driver Chip** 360
Timothé Simonot, Nicolas Rouger, Jean-Christophe Crébier, *Grenoble Electrical Engineering Lab / Grenoble
University;* Jean-Daniel Arnould, *Grenoble Institute of Technology / IMEP-LAHC*

11:55 – 12:20

**Reduction of Conducted Electromagnetic Interference in SMPS Using Programmable
Gate Driving Strength** 364
A. Shorten, A.A. Fomani, W.T. Ng, *University of Toronto;* H. Nishio, Y. Takahashi, *Fuji Electric Co. Ltd.*

14:00 – 15:40

Session 13 – Low Voltage Power Devices

Chairs: Jan Sonsky, *NXP*
  Andy Strachan, *National Semiconductor*

14:00 – 14:25

Self-Heating Analysis of Power MOSFET Module During Burn-in Test 368
Evgueniy N. Stefanov, *FREESCALE Semiconductor;* Rene Escoffier, *CEA-LETI;*
Gael Blondel, Blaise Rouleau, *VALEO VES*

14:25 – 14:50

Design of an 80V-Class High-Side Capable Double-resurf JI L-IGBT 372
Hiroki Fujii, Shinichi Komatsu, Masaharu Sato, Toshihiko Ichikawa, *Renesas Electronics Corporation*

14:50 – 15:15

**150 V, 100 mΩ, SOI Power LDMOS with High Avalanche Current Capability for
MHz Frequency Power Switching Applications** 376
Patrick M. Shea, Z. John Shen, *University of Central Florida*

15:15 – 15:40

P-Type Isolated GGNMOS with a Deep Current Path for ESD Protection 380
Jae-Hyun Yoo, Jongmin Kim, Joong-Hyeok Byeon, Young-Sang Son, Jaeyoung Park, Won-Young Jung, *Dongbu HiTek*

Chairman's Message

On behalf of the conference committee, it is my great honor and pleasure to welcome you to the 23th International Symposium on Power Semiconductor Devices and ICs (ISPSD'11). The ISPSD brings together power devices and power ICs community experts to further the research and development of power electronics and its applications. The ISPSD has become the world's leading conference in the field of power devices and power ICs due to the wealth of technical work presented and we are looking forward to a great conference this year.

Following last year's very successful conference in Hiroshima, Japan, our conference once again returns to North America, this time to the beautiful seaside city of San Diego, California, one of the world's top travel destinations.

This year there were 164 submitted abstracts and a total of 95 were accepted. The demographic distribution of the submitted papers reflects the true international character of the ISPSD as 21% of the papers were from North America, 45% from Japan and the Asia Pacific region and 34% from Europe. It is noted that submissions from China, Taiwan and South Korea have significantly increased which is a strong indication of our fast growing international community.

The technical program includes three invited plenary talks, 47 contributed oral presentations, and 48 poster presentations. The three plenary talks are: Dr. Satyen Mukherjee from Philips Research North America, USA will talk about "Opportunities and Challenges with Net Zero Energy Buildings," Dr. Tetsuo Uzuka from Railway Technical Research Institute, Japan will talk about "Trends in High-Speed Railways and the Implications on Power Electronics" and Professor Mikael Östling from Royal Institute of Technology, Sweden will talk about "SiC Power Devices – Present Status, Applications and Future Perspective".

The conference will also offer an exciting short course program in six segments on Sunday, May 22: "Solid State Lighting Technology and Electronics for Displays, General Lighting and Automotive Applications" by Dr. Radu Surdeanu, "Semiconductor Opportunities in PhotoVoltaic (PV) Systems" by Dr. Henk Jan Bergveld, "Power Devices as Key Components for Photovoltaic (PV) Systems - Challenges and Solutions" Dr. Gerald Deboy, "GaN Power ICs and Design Challenges" Dr. Kevin Chen, "Physics, Challenges, and Solutions of Metal Layout Designs for Large-Area Power Devices" Maxim Ershov, and "On-Die Power Delivery" by Dr. J. Ted DiBene II.

To conclude, I would like to thank the ISPSD'11 Organizing and Technical Program Committee and in particular, Dr. Don Disney (Technical Program Chair), Professor John Shen (Publicity Chair), Dr. Sujit Banerjee (Treasurer), Dr. Gary Dolny (Publications Chair), Dr. Hsueh-Rong Chang (Local Arrangements Chair) and Professor Wai Tung Ng (Short Course Chair). It is also with great pleasure that I extend a warm welcome to all of you attending ISPSD'11 in San Diego.

Mohamed Darwish
General Chairman

Conference Organizing Committee

General Chair
M. Darwish   Maxpower Semiconductor, USA

Vice General Chair
P. Moens   On Semiconductor, Belgium
J. Sonsky   NXP, Belgium

Technical Program Chair
D. Disney   Monolithic Power Systems, USA

Program Committee Vice Chairs
A. Strachan   National Semiconductor, USA
T. Letavic   IBM, USA
J. Shen   University of Central Florida, USA
H. Yilmaz   Alpa and Omega Semiconductor, USA
D. Sheridan   Semisouth, USA

Publicity Chair
J. Shen   University of Central Florida, USA

Publications Chair
G. Dolny   Fairchild Semiconductor, USA

Short Course Chair
W. T. Ng   University of Toronto, Canada

Local Arrangements Chair
H. R. Chang   International Rectifier, USA

Treasurer
S. Banerjee   Power Integrations, USA

Past General Chair
T. P. Chow   Renessalear Polytechnic Institute, USA

Conference Advisory Members

M. Adler	IEEE, USA
G. Amaratunga	Cambridge University, UK
B. J. Baliga	N. Carolina State University, USA
T. P. Chow	Rensselaer Polytechnic Institute, USA
T. Efland	Texas Instruments, USA
W. Fichtner	Swiss Federal Institute of Technology, Switzerland
D. Kinzer	Fairchild Semiconductor, USA
L. Lorenz	Infineon Technologies, Germany
H. Ohashi	Tokyo Institute of Technology, Japan
T. Ohmi	Tohoku University, Japan
M. Okamura	Japan Advanced Institute of Science and Technology, Japan
A. Salama	University of Toronto, CA
A. Shibib	Fultec, USA
Y. Sugawara	Kansai Electric Power, Japan
Y. Uchida	Fuji Electric, Japan
R. Williams	Advanced Analogic Technologies, USA
T. Yachi	Tokyo University of Science, Japan

Technical Program Committee

Low Voltage and RF
Chair

A. Strachan	National Semiconductor, USA
G . Dolny	Fairchild Semiconductor, USA
P. Hower	Texas Instruments, USA
Y. Kawaguchui	Toshiba, Japan
V. Khemka	Maxim Integrated Products Inc, USA
K. Kobayashi	Renesas, Japan
A. Salama	University of Toronto, CA
J. Sin	HKUST, China
J. Sonsky	NXP, Belgium
F. Udrea	Cambridge University, UK
J. Zeng	Maxpower Semiconductor, USA

Integrated Power
Chair

T. Letavic	IBM, USA
S. Banerjee	Power Integrations, USA
C. Contiero	STMicroelectronics, Italy
D. Disney	Monolithic Power Systems, USA
W.T. Ng	University of Toronto, CA
I.Y. Park	Dongbu HiTek, Korea
S. Pendharkar	Texas Instruments, USA
A. Shibib	Bourns, USA
H. Tadano	Toyota, Japan
A. Tamagawa	Renesas, Japan

T. Terashima	Mitsubishi Electric, Japan
T. Trajkovic	Cambridge Semiconductor, UK
R. van Dalen	NXP, Netherlands
R. Zhu	Maxim Integrated Products, USA

High Voltage
Chair

| H. Yilmaz | Alpha and Omega Semiconductor, USA |

V. Benda	Czech Technical University, Czech Republic
R. Herzer	Semikron, Germany
M. Ishiko	Toyota Central R & D Labs, Japan
D. Kinzer	Fairchild Semiconductor, USA
L. Lorenz	Infineon Technologies, Germany
M. Mori	Hitachi, Japan
I. Omura	Kyushu Inst. Tech., Japan
V. Parthasarathy	Power Integrations, USA
D. Pattanayak	Vishay-Siliconix, USA
J.-L. Sanchez	LAAS-CNRS, France
K. Sato	Mitsubishi Electric, Japan
P. Spirito	University of Naples, Italy
Y.Seki	Fuji Electric, Japan
B. Zhang	University of Electronic Science and Technology, China

Wide Bandgap
Chair

| D. Sheridan | SemiSouth, USA |

H.R. Chang	International Rectifier, USA
T.P. Chow	Rensselaer Polytechnic Institute, USA
M.K. Han	Seoul National University, Korea
M. Hoshi	Nissan, Japan
N. Ikeda	Furukawa Electric, Japan
N. Iwamuro	AIST, Japan
T. Kimoto	Kyoto University, Japan
S. Madathil	University of Sheffield, UK
J. Millan	CNM, Spain
P. Moens	AMI on Semiconductor, Belgium
R. Rupp	Infineon Technologies, Germany
K. Sheng	Zhejiang University, China

Packaging
Chair

| J. Shen | University of Central Florida, USA |

K. Hamada	Toyota Motor, Japan
S. Linder	ABB Semiconductors, Switzerland
P. Mawby	University of Warwick, UK
D. Silber	University of Bremen, Germany

ISPSD 2010 Best Paper Award

DB (Dielectric Barrier) IGBT with Extreme Injection Enhancement

Abstract – A 1200V DB (dielectric barrier) IGBT with new surface structure and with extreme injection enhancement effect is proposed for the first time. P-base region is confined in the thin emitter layer and is almost separated from drift region by the internal buried oxide layer. The new structure is electrically equivalent to the trench gate IGBT with very narrow mesa width. The fabrication of the DB-IGBT does not need submicron process technology.

Manabu Takei

Manabu Takei was born in Nagano, Japan in 1969. He received M. Eng. Degree in nuclear engineering from the University of Tokyo, Japan, in 1994, working on electro-fluid-dynamics modeling for uranium vapor laser isotope separation. He joined Fuji Electric in Matsumoto, Japan, in 1994, where he worked on high voltage power devices. From 1999 he has led development of reverse blocking IGBT, high-voltage IGBT, DB-IGBT (this work), and SJ-MOSFET. He is now leading research on the next generation IGBT.

Tatsuya Naito

Tatsuya Naito was born in Saitama, Japan in 1976. He received B. Eng. Degree in 1999 in electronic informatics from Hosei University in Japan. In 1999 he joined Fuji Electric Research and Development in Nagano, Japan. There he was engaged in the development of the next generation Diode and IGBT. Since 2006, he has been leading the development of power devices for automotive applications.

Tomoyuki Kawashima

Tomoyuki Kawashima was born in Hyougo, Japan in 1958. He received B. Eng. Degree in engineering from the University of Kobe, Japan, in 1980. He joined Fuji Electric in 1980, where he was engaged in material research, development of semiconductor analysis technique, and development of new power devices. He is now leading the development of manufacturing technology of power devices.

Kazuo Shimoyama

Kazuo Shimoyama was born in Gunma, Japan in 1975. He received a doctoral degree in engineering from University of Tsukuba, Japan, in 2002 for a thesis on an experimental study of growth mechanism for transition metal oxide epitaxiy in ultra high vacuum. In 2002 he joined Fuji Electric in Matsumoto, Japan. He has been leeding process technology development and yield improvement for reverse blocking IGBT, especially on ultra high temperature diffusion, crystal defect analysis, device isolation, laser annealing, and thin wafer process. He has enjoyed the development of the DB-IGBT as process expert in this great team.

Shinji Fujikake

Shinji Fujikake received B. Eng. and D. Eng. Degrees in electronical engineering from Chiba University in 1986 and 2005, respectively. In 1986, he joined Fuji Electric and studied amorphous silicon materials and their application to solar cells. From 2003 to 2007, he led research on semiconductor processing technology such as CMP. He is now managing the development team of film substrate solar cells at Kumamoto factory.

Hitoshi Kuribayashi

Hitoshi Kuribayashi was born in Nagano, Japan in 1964. He received the master's degree in engineering from University of Tuskuba, Japan in 1989 and received PhD degree in engineering from Osaka University, Japan in 2007. In 1989 he joined Fuji Electric, where he has been leading the development of semiconductor manufacturing technology.

Haruo Nakazawa

Haruo Nakazawa was born in Gunma, Japan in 1964. He received M. Eng. Degree in electrical engineering from Tokyo Denki University, Japan, in 1989. He joined Fuji Electric in 1989 and worked on microactuator, micromagnetic devices, and micro machines. Since 2000, he has worked on process technology of power devices. He is now project manager of new power device development for green power electronics applications.

ISPSD 2010 Charitat Award

Self-Protected GaN Power Devices with Reverse Drain Blocking and Forward Current Limiting Capabilities

Chunhua Zhou
The Hong Kong University of Science and Technology

Chunhua Zhou received his B.S. and M.S. degrees in microelectronics from the University of Electronic Science and Technology of China, Chengdu, China, in 2005 and 2008, respectively. He is currently working toward the Ph.D. degree in the Department of Electronic & Computer Engineering, the Hong Kong University of Science and Technology, Kowloon, Hong Kong. His research interests are in the design, fabrication and modeling of GaN-based power semiconductor devices, with strong emphasis on new device concepts and device-level self-protection techniques.

Abstract—Self-protected GaN power devices were realized using AlGaN/GaN-on-Si platform, where two built-in intelligent functions were demonstrated for "Smart Discrete" applications. First, an AlGaN/GaN normally-off high electron mobility transistor (HEMT) with reverse drain blocking capability was realized, featuring a Schottky contact controlled drain barrier. Compared to the Schottky drain structures, the new design exhibits only a 0.55 V onset voltage in the forward biased "ON" state, while effectively blocks the reverse current conduction. The device fabrication is also free of extra photomask and process steps. Second, an AlGaN/GaN lateral field-effect rectifier (L-FER) with intrinsic "ON" state current limiting capability was fabricated, featuring a depletion-mode (D-mode) Schottky controlled depletion-mode channel extension (length of L_D) beyond the ohmic contact at the cathode electrode, where the on-state current of the new rectifiers are self-limited at 4.59 kA/cm^2 ($L_D = 1.3$ μm) and 3.56 kA/cm^2 ($L_D = 1.9$ μm) at room temperature. The current limiting level shows a negative temperature coefficient (TC) that is desirable for thermal stability.

Proceedings of the 23rd International Symposium on Power Semiconductor Devices & IC's
May 23-26, 2011 San Diego, CA

Opportunities and Challenges with Net Zero Energy Buildings

Satyen Mukherjee
Philips Research North America
Briarcliff Manor, NY USA
Satyen.Mukherjee@philips.com

Abstract—**Buildings represent around 41% of the total energy consumption in the US followed closely by industry (31%) and transportation (28%). One of the milestones set by the US Department of Energy is the development and deployment of net zero energy buildings defined as buildings that on a yearly average spend as much energy as they generate using renewable energy sources. Realization of net zero energy buildings require a wide ranges of technologies, systems and solutions with varying degrees of complexity and sophistication depending upon the location and surrounding environmental conditions. Lighting is a dominant load in buildings followed by heating, cooling, ventilation and various plug loads. This paper will address the roles of different technologies, devices and control strategies being developed for low energy buildings leading to net zero energy buildings. These include high efficiency lighting, daylight integration, DC power bus, solar power integration; closed loop integrated control, smart grid interface as well as emerging approaches such as chilled beams and active facades. All of these involve power conversion and controls in one form or the other where high voltage or high power integrated solutions are key to commercial viability. In addition to this, the role of whole building modeling and simulation in the development and deployment of the solutions will be addressed.**

I. INTRODUCTION

Energy consumption is receiving increasing global attention today due to several reasons including, perceived shortage of traditional resources such as fossil fuels, increasing GHG (Green House Gas) emission causing global warming and most alarming and immediate of all is increasing energy prices. The problem of energy consumption is a long term issue and is not expected to disappear by itself unless measures are taken across several areas. In this context the available data, Figure -1, indicate alarming growth projections in energy consumption and corresponding carbon dioxide emission. Furthermore, as shown in Figure – 2 for the three major fossil fuel reserves and consumption there is an imbalance between major energy supply and demand which creates additional global challenges.

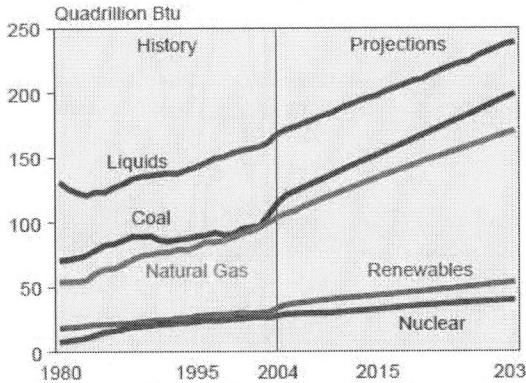

(a)

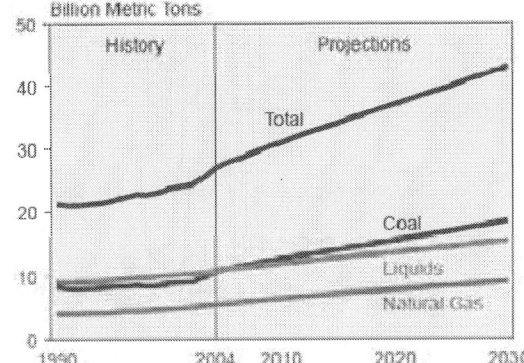

(b)

Figure 1.   (a) World Marketed Energy Use and (b) World Energy use and Carbon dioxide emission, past and projected.

978-1-4244-8425-6/11 $26.00 © 2011 IEEE

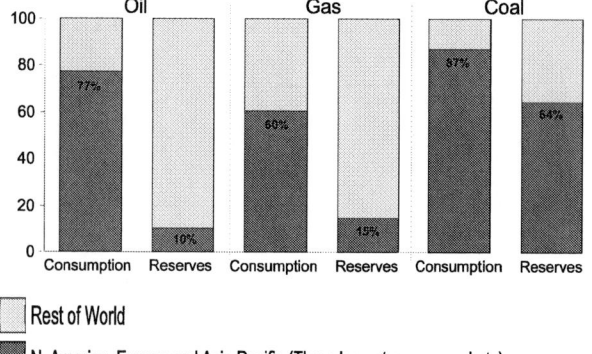

Figure 2. Fossil fuel energy consumption versus reserves 2004.

These trends raise a major concern about the sustainability of the present energy usage rate. To understand the key elements involved in the energy consumption space, Figure -3 (a) shows the world energy consumption by the major consuming sectors, buildings (44%) (residential 15% and commercial 25%); transportation(19%) and industrial (37%). Figure-3(b) shows the same information for the US and Figure-3(c) shows the breakdown of the energy consumption in commercial building by end use.

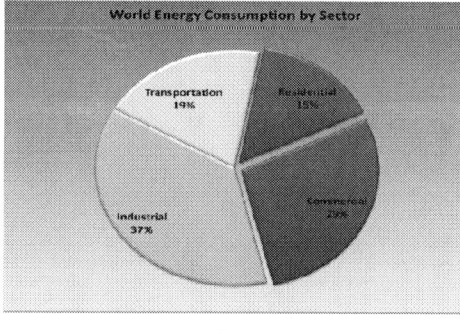

(a)

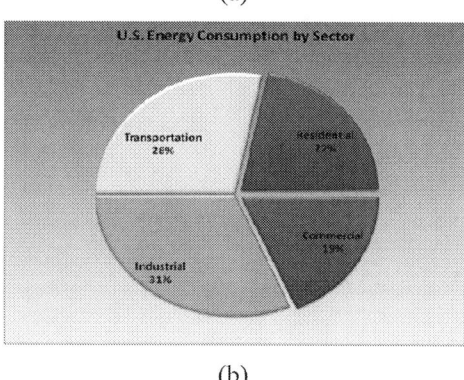

(b)

Figure 3. (a) World Energy Consumption by sectors, (b)US Energy Consumption by sectors

It is clear from Figure-3 that the largest contributor to the energy consumption is the buildings sector. Therefore, to reign in the energy spending rate, buildings have been earmarked as an area of attention across the globe. Net zero energy (NZE) or near zero energy or net zero emissions are initiatives that have been started around the globe to address this challenge. Simply stated NZE buildings (NZEBs) are those that produce as much energy on site with renewable sources such solar or wind, as they consume on a yearly basis. The building energy use may also include the embodied energy, which is the energy required to manufacture and supply to the point of use all the materials used in the buildings. In operation the buildings use electrical energy from the grid when needed and feed energy back to the grid when the production is in excess of demand. Other forms of source energy such as gas, oil, etc. need to be compensated by renewable energy sources as well.

In the US, the Department of Energy has been driving programs for several years now to develop technologies and solutions that will make NZEBs commercially viable. Roadmaps for NZE buildings deployment has been developed nationwide as well as by some states such as California which set a target of 2030 for achieving NZE for all new commercial buildings and 50% of existing commercial buildings. ASHRAE vision 2020 calls for providing tools by 2020 that enable the building community to produce market viable NZEBs by 2030.

To make NZEBs commercially and logistically viable, the energy consumption of the buildings have to be reduced by typically 60-70% over the current level and the balance generated using renewable energy sources, typically solar PV as shown in Figure – 4 [1]. These require special measures and technologies beyond established mainstream approaches. One approach is integrating several energy savings technologies with the building design to a single controllable system [2].

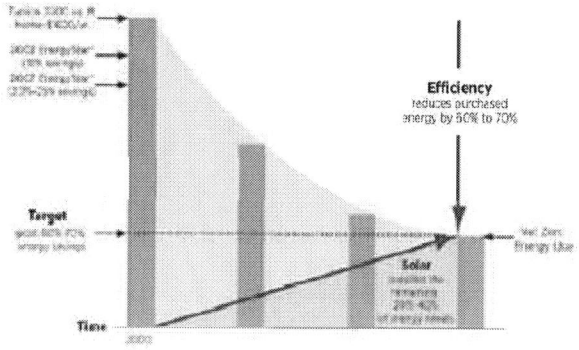

Figure 4.   Achieving Net-Zero Energy Buildings.[1]

This paper will address several key electrical energy consuming elements in buildings such as lighting, HVAC, plug loads, and describe some of the technologies being developed to address the energy efficiency challenge.

II. BUILDING ELEMENTS

The major elements/functions that are essential in modern commercial building operation include:

- Lighting
- Cooling
- Space Heating
- Ventilation
- Refrigeration
- Water Heating
- Computers
- Office equipment
- Electronics
- Mechanical systems

Independent optimization (efficiency increase) of energy use in each of these elements is the first step and the suppliers to the building industry address this on an ongoing basis. However, this does not always result in the most optimal overall building performance because several of these functions interact with one another and result in energy wastage due to competing processes. For instance, lighting systems produce heating of the environment along with illumination thus contributing to the cooling or heating load of the building. Similarly computers and office equipment contribute to the same. Therefore, intelligent control of these different elements in an integrated fashion can allow additional energy savings not possible with independent operation. This "intelligent integrated" control involves consideration of the different interactions between elements and the needs of the building environment and control at the overall building level. Ideally, the concept of integrated control needs to be deployed in the building design as well as in its operation. This is a significant challenge both technically as well as logistically considering the different businesses and industries involved. Figure -5 illustrates the concept of a commercial building employing integrated system design and operation.

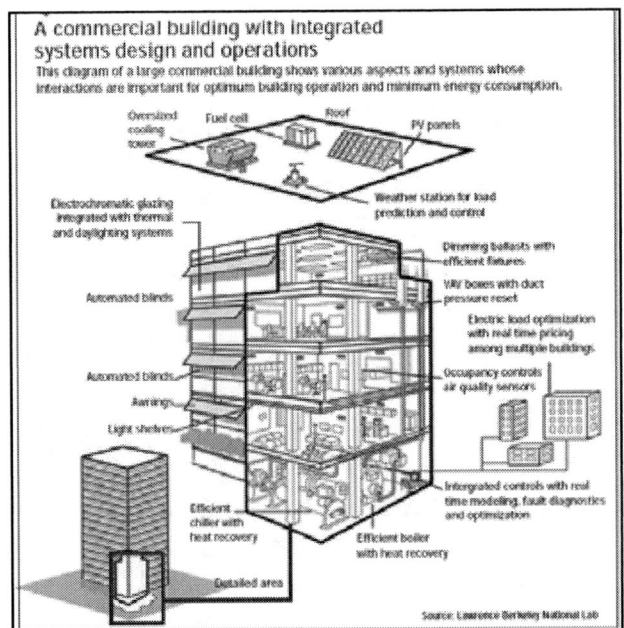

Figure 5. Illustrative Integrated System Design and Operation in Commercial Buildings.

III. LIGHTING

Lighting is the most pervasive element in all buildings and represent as much as 38% of the electrical energy consumption in commercial buildings (Figure - 6), the single largest energy consuming element. Furthermore combined with heating, cooling and air conditioning it represents as much as 70% of the electrical energy consumption. Exploiting the interaction between these systems, provides a significant opportunity for improvement in terms of energy savings.

In addition to the advancements in light source technology, from incandescent to flourescent or discharge lamps to LEDs, controlling the light to provide illumination of the right kind (in terms of spectral content), to the right place (work, recreation areas and ambience) and at the right time provide significant opportunity for energy savings. Additionally, combining the control of advanced light source technologies with daylight and HVAC systems can provide substantial energy savings and the lead the way to the commercial realization of NZEBs.

A. Lighting Controls

Lighting control systems including daylight integration are gaining prominence in low energy sustainable buildings.

Typical implementations of such systems involve two control loops, one for the dimming lights and another for the motorized blinds or the window treatments. In most applications deployed in the advanced buildings today shown in Figure -7, the two systems (lights and blinds) are controlled independently using separate sensors for each control loop to arrive at the desired light level on the work surface.

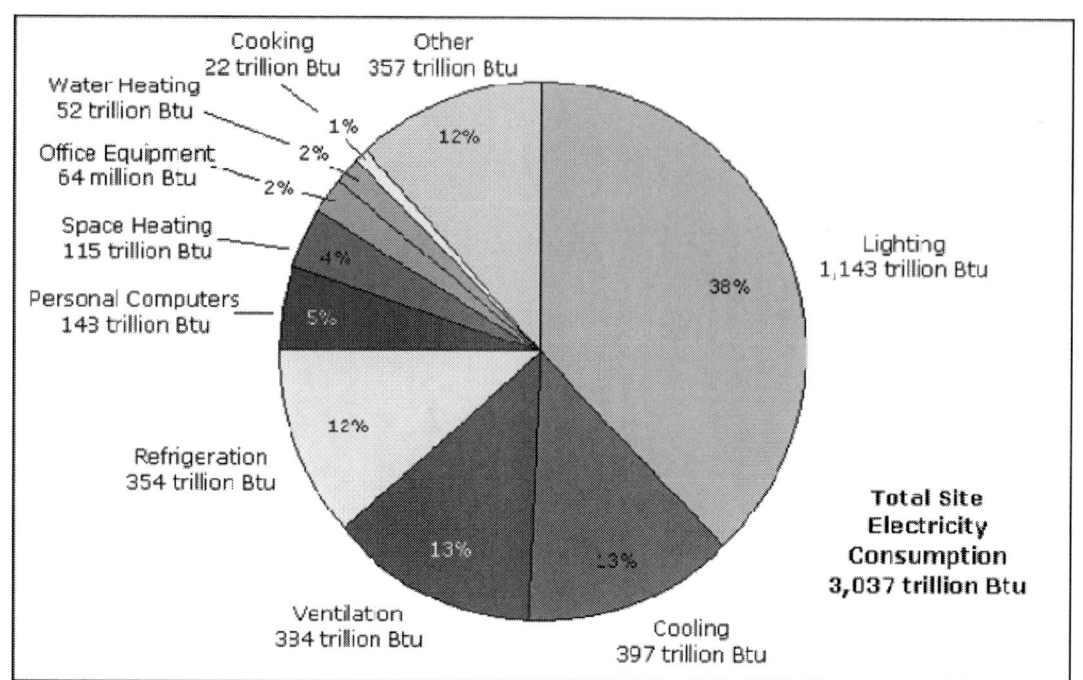

Source: Energy Information Administration, 2003 Commercial Buildings Energy Consumption Survey, Table E3.

Figure 6. Electricity Consumption in Commercial Buildings in the US

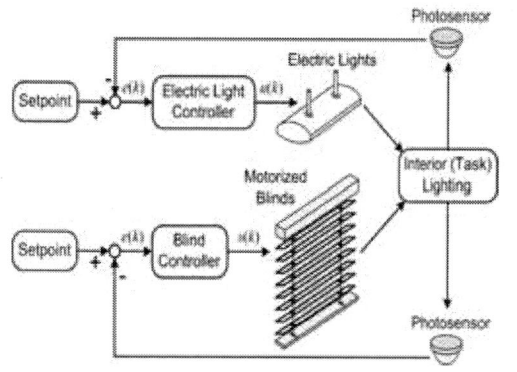

Figure 7. Independent control of closed-loop lighting system and closed-loop blind system.

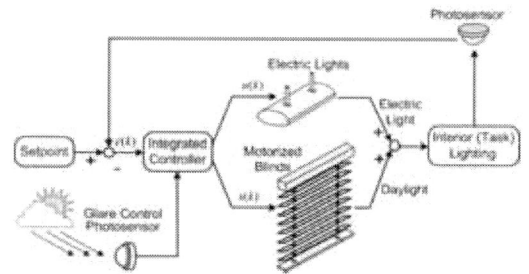

Figure 8. Integrated lighting and daylight control system.

Another strategy, shown in Figure -8 is to control the lights and the window blind slat angle in an integrated fashion using the output of a single photosensor on the work surface and deploying control algorithms to minimize the energy consumption and maximizing the use of daylight.

Analytical simulation and experimental results from a laboratory test bed have shown that improved performance in terms of energy savings and occupant comfort can be achieved by employing an integrated control strategy in comparison with independently controlled electric lighting and blinds loops [3].

Furthermore, EnergyPlus simulations of a benchmark medium sized office building have shown that integrated controls of electric lighting and blinds result in increased savings in HVAC loads compared to fixed blind position. Considerable lighting energy savings (up to 66%[3]) result from integrated controls compared to the benchmark case.

Based on our findings in [3] and published literature, we feel that achieving optimal energy savings in a real building environment is closely tied to providing occupant comfort as well. Therefore successful solutions in the market will require an integrated approach. Defining and quantifying occupant comfort comprehensively remains an open problem.

978-1-4244-8425-6/11 $26.00 © 2011 IEEE

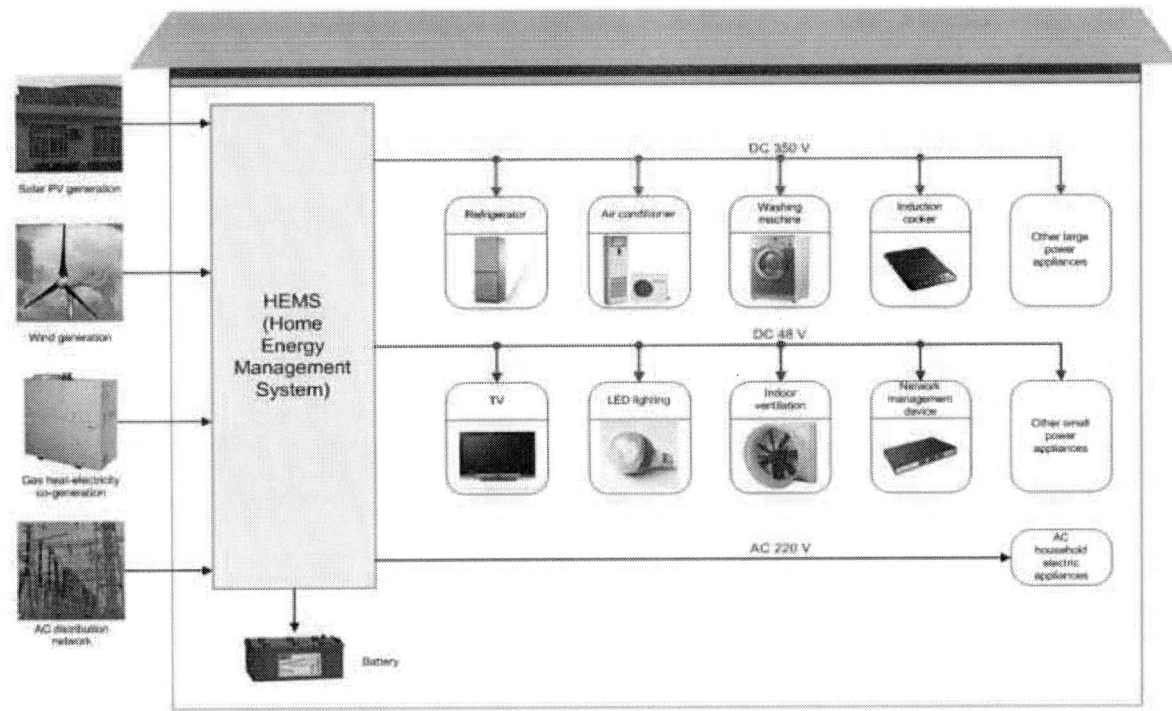

Figure 9. DC bus architecture - Guangdong White Household Appliances Innovation Alliance - China

IV. DC GRID

In the future it is envisioned that NZEB's employing Solar PV as a power source and LED lighting with a large fraction of appliances that are internally DC powered, it would be worthwhile considering a DC bus to supply the required power to the appliances in the building. This would eliminate the need for DC to AC and back from AC to DC conversions and thereby save energy on the whole. An US based industry association called Emerge Alliance has been set up to promote and develop standards for DC bus in buildings. A similar alliance called Guangdong White Household Appliances Innovation Alliance has been set up in China. Figure- 9 shows a bus architecture being explored in China.

V. CONCLUSIONS

Net Zero Energy Buildings are being developed as a means to reining in the energy consumption growth that is posing a global threat to our lifestyle. To accomplish this on a commercially viable scale, it is necessary to develop a host of technologies and system solutions. These include technologies ranging from building envelope to sensors and systems to control lighting, HVAC, appliances, and incorporating renewable energy sources such as solar PV. In the future, additional aspects such as DC grid in buildings could be of value in this context. This will provide major opportunity for the development of key components including, smart sensors, network controllers, power converters, lamp drivers, motor controllers and others to address this major challenge.

ACKNOWLEDGMENT

I would like to acknowledge the effort of the lighting controls research team at Philips Research North America for their contributions, directly and indirectly.

REFERENCES

[1] " Federal Research and Development Agenda for Net-Zero Energy, High-Performance Green Buildings", National Science and Technology Council – Committee on Technology. Report of the Subcommittee on Buildings Technology Research Program. Oct. 2008.

[2] Griffith, B; et al. "Analysis of energy performance of the Chesapeak's Bay foundation's Phillip Merril Environment Center, NREL/TP 550-34830.

[3] Mukherjee, S. et. al. "Closed Loop Integrated Lighting and Daylighting Control for Low Energy Buildings" Proceedings of 2010 ACEEE Summer Study on Energy Efficiency in Buildings, Pacific Grove, CA.

978-1-4244-8425-6/11 $26.00 © 2011 IEEE

Proceedings of the 23rd International Symposium on Power Semiconductor Devices & IC's
May 23-26, 2011 San Diego, CA

Trends in high-speed railways and the implications on power electronics and power devices

Tetsuo UZUKA,

Power Supply Technology Division, Railway Technical Research Institute (RTRI)
2-8-38 Hikari-Cho, Kokubunji City, Tokyo 185-8540, Japan

Abstract—**High Speed Rail (HSR) are expanding rapidly in the whole world in this decade. Almost all the high-speed trains are fed by high-voltage and are equipped with several large motors. In addition, High-speed trains have a strict restriction for both mass and size. Thus, HSR needs power semiconductors that can handle high-voltage and giant current. In addition, EMC problems become larger in these days, thus higher speed of switching is expected.**

From simple silicon diodes in 1960s, thyristors, GTO thyristors, IGBTs and until new wide gap devices such like SiC and GaN, the progress of power semiconductor and cooling system directly pulls the performance of high-speed rolling stock.

In some cases, fixed installations for HSR are equipped with flexible AC transmission systems (FACTS) such as static VAR compensators (SVC), also.

I.   TREND OF THE HIGH SPEED RAILWAYS

The first electric railway began commercial running in 1881 in Germany. Now electric railways convey over 90% person trips of whole railway: that has about 30% of person trips in Japan (rest of 54% car, 7% bus, and 7% air). Thus, we can consider the electric railway as a one of the backbone technologies for human life.

In 1964, Japanese National Railways started the era of high-speed railway by "Shinkansen" bullet train for 515km line with a speed of 210km/h (Figure 1). Since then, high-speed rail (HSR) means a passenger train with speed of 200km/h (164 miles per hour) or over. Almost all the high-speed trains in the world are electric railway.

Japanese Shinkansen made a great success for technical, economical, and safety aspect. Then France, Germany and Italy designed their high-speed railway called "TGV", "ICE" and "ETR" respectively in 1980-1990s. Their systems are faster (260km/h or more) than Shinkansen and more economical. Japan updated his system and these 3 countries and Japan extended their high speed lines in their lands and neighborhood.

In 1990s, the climate changing of the earth became one of the most important problems for human being. Thus, EU chose HSR as the right solution for sustainable mobility. Since then European countries began to expand their high-speed

network with a sustained momentum. Meanwhile, economic growth of Asia such as Korea, Taiwan, and China let these countries to build their own HSRs.

Figure 2, 3 and 4 [1] show growths in number of miles, in top speed, and number of HSR train sets, respectively. In 2000, total length of high-speed line was just 4,600km. In May 2010, there is 13,000km in operation, 11,000km under construction, and planned 18,000km. Maximum speed of high-speed railway in operation began with 210km/h in 1964 and became 350km/h in 2010. 2,228 train sets of HSR for 200km/h and 1,667 for 250km/h (or more) are running in April 2010.

Figure 1.   First Shinkansen departs from Tokyo station in 1964.

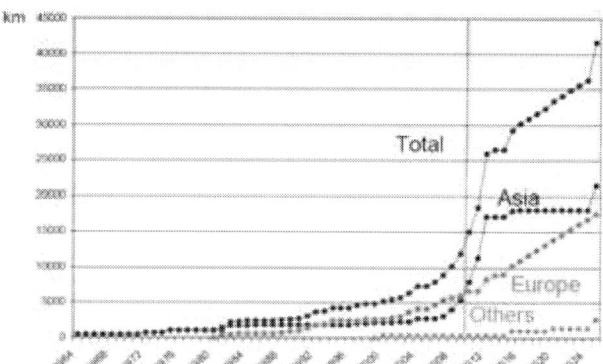

Figure 2.   Expected evolution of the world HSR network.

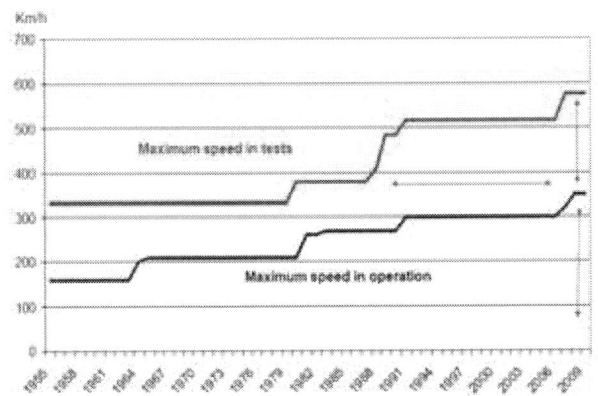

Figure 3. Evolution of maximum speed on rails.

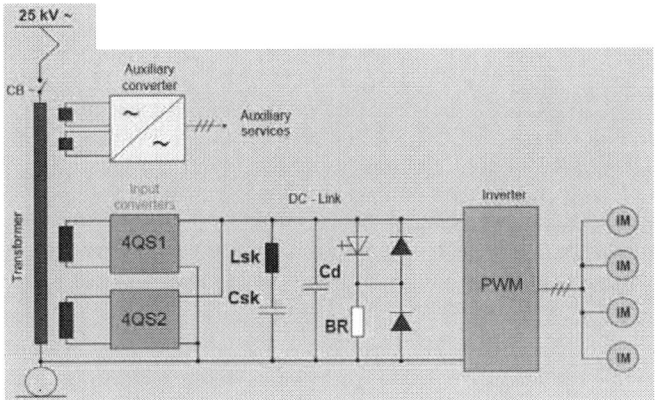

Figure 5. Typical installation of HSR main circuit.

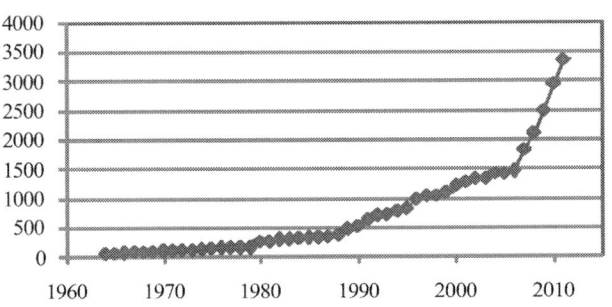

Figure 4. Evolution of HSR trainsets.

II. POWER SEMICONDUCTORS IN HSR

A. Topology of power semiconductors onboard the HSR

Electric railway includes wide variety of technologies such as motor driving, signaling, dispatcher systems, ticketing and vendor machines, illuminating, distributed information networks on the car, networks on the ground, current collecting, and fixed power supply installations on the ground. For each technology, power semiconductors occupy a center position.

Figure 5 shows the typical topology of HSR main circuit onboard the rolling stocks. Cars derive AC high voltage power from catenary and convert into DC by PWM (Pulse Width Modulation) control method of rectifying with IGBT (Insulated Gate Bipolar Transistor) power device. Then DC will convert into AC again to drive asynchronous motors, also with PWM and IGBT combination.

First Shinkansen in 1964 was driven by 200kW DC motors onboard and fed by AC 25kV 60Hz catenary. There were also silicon diode rectifiers onboard the train (Figure 6(a)). Total power for one train set (12 cars with four motors for each) was about 20MW. In those days, nominal voltage and current of diodes were so small that many diodes were used in parallel and serially instead. Transformer tap control method was used for voltage/speed control of DC motors.

In 1980s, trains adopted thyristor Ward-Leonard control method (Figure 6(b)) with a combination of phase-controlled thyristors and DC motors. Thyristors made the main circuit very simple. In France, TGV-A type HSR trains use current type thyristor control circuit with synchronous motors.

In 1990s, GTO (Gate Turn Off) thyristor such as 4500V/3000A and microprocessor leads a PWM controlled Voltage Source Inverter (VSI) based converter and inverter system for rectifying and controlling asynchronous motors (Figure 6(c)). It was a kind of revolution for rail operator companies since DC motors have some maintenance problems with brush and commutator. In addition, power factor of feeding circuit became about 1.0 (former 0.75-0.8) so that

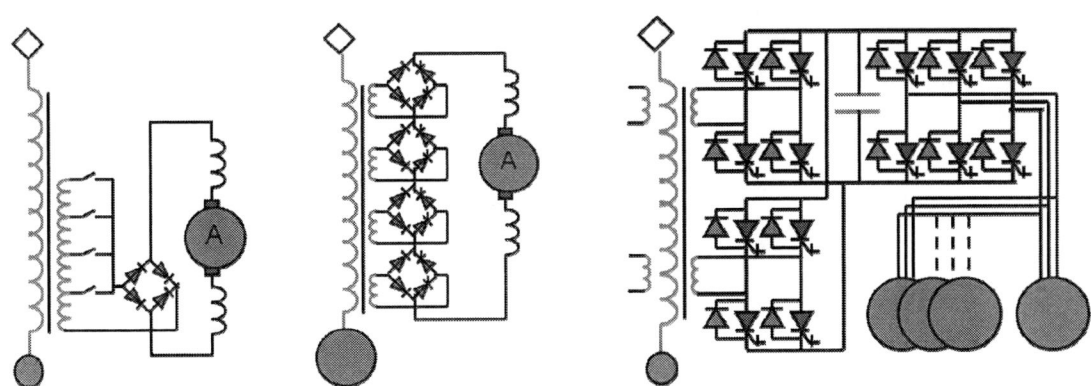

Figure 6. Topology of power circuit on HSR rolling stocks
(a)Tap control with diode rectifier, (b)Thyristor phase control (Ward-Leonard), (c)PWM control (with GTO thyristor)

feeding circuit can send more power to HSRs.

In 2000s, IGBT device occupies a major part. IGBT offered faster switching speed (1-2 kHz) than GTO (400-500Hz) that realize precise control of motors and lead an EMC problem easier. At first, withstand voltage of early generation IGBTs are not so high enough (up to 2-3kV), so that 3-level VSI converters were introduced. Then withstand voltage of latter generations of IGBT rose to 4-6kV, that enables 2-level VSI circuit. In addition, generation change of IGBTs assures less loss and more simple drive circuit so that converters are smaller than GTO or early IGBT sets.

Therefore some HSR rolling stocks of next phase shall adopt permanent magnet synchronous motors (PMSM) for higher speed. PMSM have to control with one motor by one inverter combination. Some manufactures propose power set with IEGT (Injection Enhanced Gate Transistor) or IGCT (Integrated Gate Commutated Turn-off thyristor) devices.

HSR are now waiting for new power devices with lower loss and more switching speed than silicon devices, to reduce the weight and size of power converters. Among the several developing wide-gap devices, Silicon Carbide (SiC) shot-key barrier diode (SBD) might be the closest to the application along latest papers. For example, Mitsubishi Electric reported 28% maximum improves of total loss and 18% average improves with 300kW inverter set. [2]

Besides drive systems, rolling stocks carries auxiliary power sources of several hundred kW for air conditioning and lighting for passengers and controlling and cooling. We have to count up information processing systems, telecommunication systems and signal systems. Every system is full of semiconductors.

B. Packaging and Cooling

Packaging and cooling technique of the power circuit is one of the key technologies of HSR.

French TGV and first generation of German ICE, and Italian ETR500 are concentrated power train using locomotives and trailers. In this case, power components could occupy the whole space of the floor of the locomotive; however they have to handle more power.

On the other hands, Japanese Shinkansen is distributed power train from the beginning, called as EMU (Electrical Multiple Unit). In this case, though rated power is smaller than concentrated type, every power component should be set within the limited space of under floor. Second generation of German ICE and Italian ETR adopted distributed power EMU. In addition, ALSTOM is now developing distributed power based AGV train in France.

Cooling system began with forced air driven by blower. There are several approaches available today, such as forced air, heat-pipe, ebullient cooling with perfluoro carbon, natural air, and water cooling. Figure 7 shows generations of cooling technique used in Japan [3].

In any case, all the power components are integrated in limited space. When engineers design onboard propulsion system, they should carefully consider a total balance about power flow, thermal flow, air flow, mass balance, easy to manufacturing and maintenance, and cost. Figure 8 shows a recent configuration of 1600kW / 1320kg power converter / inverter (C/I) set for N700 type Shinkansen using 3300V 1200A IGBTs with natural air flow cooling [4].

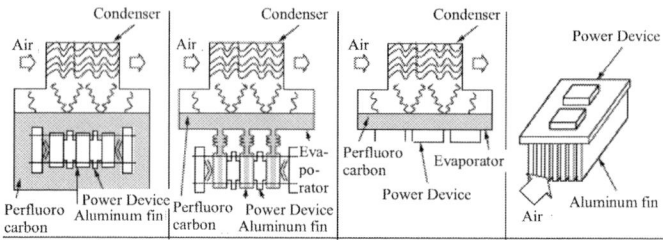

Figure 7. Cooling generation

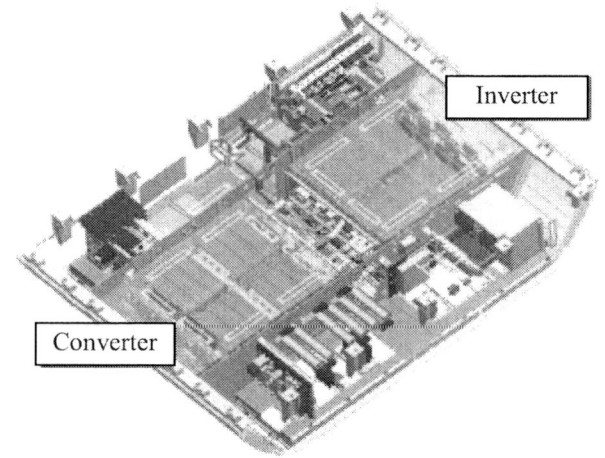

Figure 8. C/I composition of N700 type EMU.

C. Fixed Installations

Most of HSR were fed by AC 25kV 50/60 Hz electric power. HSR is a load of a single-phase, changing independently, and sometimes including harmonics. Therefore, traction load sometimes causes voltage unbalance or voltage fluctuations on the three-phase side.

As a countermeasure for this problem, some substations are equipped with flexible AC transmission systems (FACTS) such as static VAR compensators (SVC) or static compensators (STATCOM) for balancing power or suppressing voltage fluctuation of both power grid and feeding circuit since 1980s. Rated power of each HSR is about 10-20 MW, thus scale of FACTS equipments should be also 10-60 MVA to compensate HSRs. Figure 9 shows an example of a main circuit of power balancer [5].

In AC 25kV 50/60 Hz system, trains run across the phase section along the line. Shinkansen use a pair of electrical switch system to reduce no power time with VCB. Some countries are developing static switch for 25kV (Figure 10) with very high-voltage (12kV), giant current thyristors with 14 serial connected [6].

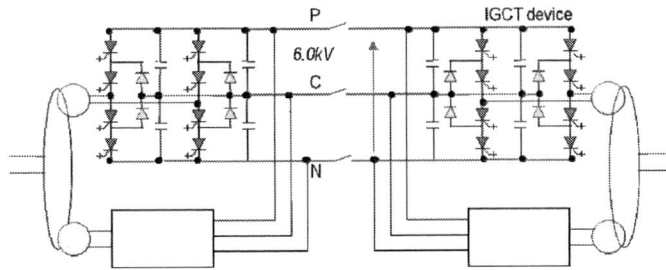

Figure 9. Power balancer for fixed installations with IGCT device.

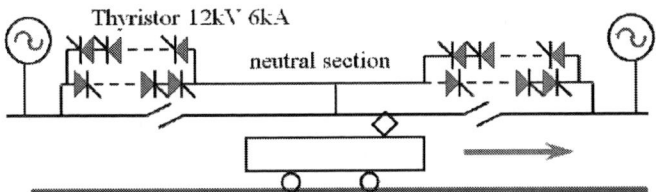

Figure 10. Static Changeover Switch with 12kV/6kA thyristor

In Sweden, German, Swiss, Austria and Norway, railway has dedicated 16.7Hz single-phase grid. There is no balancing problem with three-phase grid in these countries. Instead, they have to derive power from commercial frequency 50Hz three-phase grid to 16.7Hz single-phase grid with frequency converting. Several static frequency converters rated 15 – 100 MW are working there since 1970s [7]. North east corridor of USA between New York to Washington DC has same situation with 25Hz feeding frequency in 60Hz zone. JR central in Japan also has two static frequency converters they change 50Hz power into 60Hz to feed. Total power of static frequency converters around the world exceeds 1000MW until 2010.

III. EXPECTED POWER SEMICONDUCTORS

Expected direction of HSR might be higher speed and more comfortable. Thus, expected power devices for HSR should be higher voltage, larger current, more speed of switching and more efficiency of power devices. Of course cost is an important problem, however, the market of HSR is growing and we can hope a volume efficiency effect.

About onboard application, silicon devices already reached certain withstand voltage (6.5kV) for existing topology of circuit such as VSI. Then new property, such as reverse conductivity leads a new topology. In addition, integrated controlling circuit or protection ability within a device package is another promising approach.

Since new power semiconductors such like SiC or Gallium Nitride (GaN), etc, will improve an efficiency of onboard power converters, HSR train can reduce the weight and size of converters. For example, one of the expected applications of SiC device is a medium frequency propulsion convertor[8] which enables to omit transformer onboard, the heaviest parts of propulsion, and can reduce the total mass (Figure 11). This composition is attractive especially for 16.7Hz feeding frequency countries.

We should consider a deal between reliability, availability and redundancy of power units with new devices. New devices have to improve their reliability of course. However, for some HSR onboard application, a certain redundancy will approve to introduce new devices in early stage. In this meaning, the medium frequency propulsion convertor is one of the candidates.

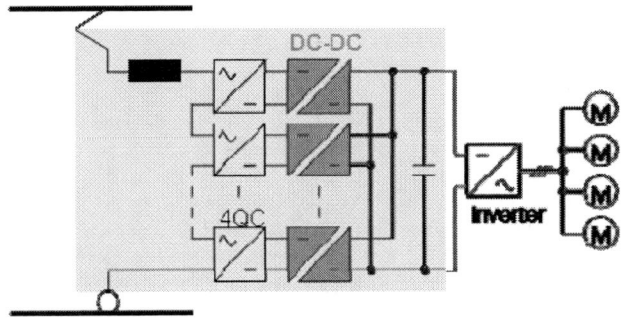

Figure 11. Medium frequency compostion

For fixed installation applications, power semiconductors have to handle higher voltage than onboard. For example, SiC active devices help to realize matrix converters without transformers for frequency conversion or power balancers as described II-C. As a bidirectional switch to replace mechanical switchgears (Figure 10), higher withstand voltage directly miniaturizes static switch composition.

IV. CONCLUSION

HSR was born and glowing up with power semiconductors with mutual influence. Thus, HSR always awaits next breakthrough of power semiconductors.

ACKNOWLEDGMENT

Special thanks to Mitsubishi Electric for offering figures and data about recent developments.

REFERENCES

[1] UIC (International Union of Railways), "High speed rail, Fast track to sustainable mobility," ISBN978-2-7461-1887-4, 2010

[2] Press release of Mitsubishi Electric, "Development of SiC diode large power module," 20 January, 2010 (in Japanese).

[3] R. Inoue, M. Oba, "Power Electronics Technology for the Railway Rolling Stock," in Fuji Electric Journal, vol. 80, No. 2, 2007, pp. 153–161 (in Japanese)

[4] Y. Yamomoto, T. Fukushima, "A Study on oprimization of cooling system of PWM power converter of Shinkansen N700 high-speed train," in JREA vol. 52, No. 5, 2009, pp.34205–34207 (in Japanese)

[5] T. Uzuka, S. Ikedo, and K. Ueda, "A Static Voltage Fluctuation Compensator for AC Electric Railway," PESC04, 2004

[6] K. Kunomura, et al., "Development of Static Changeover Switch for Shinkansen," in proceedings of IEEJ annual conference, No.5-184, 2006 (in Japanese)

[7] M. Perschbacher, "Bahnenergieversorgung der DB (Traction power supply of DB)," in elektrische bahnen, vol. 109, No. 1, 2011, pp.50–54. (in German)

[8] J. Weigel, "Medium-Frequency Traction Transformer – Outcome of Railenergy," in Railenergy Final Conference, Bruxelles, 2010

Proceedings of the 23rd International Symposium on Power Semiconductor Devices & IC's
May 23-26, 2011 San Diego, CA

SiC power devices – present status, applications and future perspective

Mikael Östling, Reza Ghandi and Carl-Mikael Zetterling

KTH Royal Institute of Technology, School of ICT, Electrum 229, SE-16440 Kista, Sweden

Email: ostling@kth.se

Abstract— **Silicon carbide (SiC) semiconductor devices for high power applications are now commercially available as discrete devices. Recently Schottky diodes are offered by both USA and Europe based companies. Active switching devices such as bipolar junction transistors (BJTs), field effect transistors (JFETs and MOSFETs) are now available on the commercial market. The interest is rapidly growing for these devices in high power and high temperature applications. The main advantages of wide bandgap semiconductors are their very high critical electric field capability. From a power device perspective the high critical field strength can be used to design switching devices with much lower losses than conventional silicon based devices both for on-state losses and reduced switching losses. This paper reviews the current state of the art in active switching device performance for both SiC and GaN. SiC material quality and epitaxy processes have greatly improved and degradation free 100 mm wafers are readily available. The SiC wafer roadmap looks very favorable as volume production takes off. For GaN materials the main application area is geared towards the lower power rating level up to 1 kV on mostly lateral FET designs. Power module demonstrations are beginning to appear in scientific reports and real applications. A short review is therefore given. Other advantages of SiC is the possibility of high temperature operation (> 300 °C) and in radiation hard environments, which could offer considerable system advantages.**

I. INTRODUCTION

Silicon carbide device technology has matured greatly over the past decades and gone from research to commercial production. Many scientific papers have been published and comprehensive text books or book chapters have been written on SiC power devices and process technology and the readers are referred to [1-4] and references therein. The device technology is greatly dependent on substrate and epitaxial material quality and the number of detrimental defects. Over the years this has been one of the limiting factors for a commercial success of the high voltage and high current device market. Today the materials quality of 4H-SiC wafers and epitaxy is at such a high quality that many companies are offering commercial SiC wafers and epitaxy on 4H-SiC with wafer diameter 100 mm. The main SiC power device products are still rectifiers based on Schottky or junction barrier diodes. A few companies are offering active power devices switches as engineering samples based on MOSFETs, JFETs and BJTs.

TABLE I. PHYSICAL PROPERTIES OF SiC AND GaN SEMICONDUCTORS REFERENCED TO Si AND GaAs [1].

Property	Si	GaAs	GaN	3C-SiC	6H-SiC	4H-SiC
Bandgap, Eg (eV at 300K)	1.12	1.43	3.4	2.4	3.0	3.2
Critical electric field, Ec (V/cm)	$2.5 \cdot 10^5$	$3 \cdot 10^5$	$3 \cdot 10^6$	$2 \cdot 10^6$	$2.5 \cdot 10^6$	$2.2 \cdot 10^6$
Thermal conductivity, λ (W/cmK at 300K)	1.5	0.5	1.3	3-4	3-4	3-4
Saturated electron drift velocity, vsat (cm/s)	$1 \cdot 10^7$	$1 \cdot 10^7$	$2.5 \cdot 10^7$	$2.5 \cdot 10^7$	$2 \cdot 10^7$	$2 \cdot 10^7$
Electron Mobility, μ_n (cm^2/V·s)	1350	8500	1000	1000	500	950
Hole Mobility, μ_p (cm^2/V·s)	480	400	30	40	80	120
Dielectric constant, ε_r	11.9	13.0	9.5	9.7	10	10

The application areas of the rectifying devices are as free-wheeling diodes in various DC-DC converters and drives. The main advantage is the absence of reverse recovery and subsequent low switching losses as well as the ability to handle large current density at elevated temperature. In Table I the key materials properties are listed for the main wide bandgap semiconductors compared with Si and GaAs.

II. HIGH VOLTAGE SiC DEVICES

The key figure of merit for power switches is the specific on-resistance R$_{on,sp}$. This parameter tells directly how much resistive loss a device generates in the forward conduction mode. For comparison between different materials, unipolar action is assumed, so that the resistance of the blocking junction dominates. The R$_{on,sp}$ for this blocking junction is usually given in mΩcm^2 and can be calculated from (1) below

$$R_{on,sp} = \frac{4V_B^2}{\varepsilon \mu_n E_c^3} \qquad (1)$$

where V$_B$ denotes the breakdown voltage and E$_c$ the critical electrical field. Since the E$_c$ of 4H-SiC is about 8-9 times higher than that of Si one can easily understand the enormous advantage of using SiC devices. Several of the published device data for diodes, JFETs and BJTs are very close to the theoretical limit, see Fig 1.

978-1-4244-8425-6/11 $26.00 © 2011 IEEE

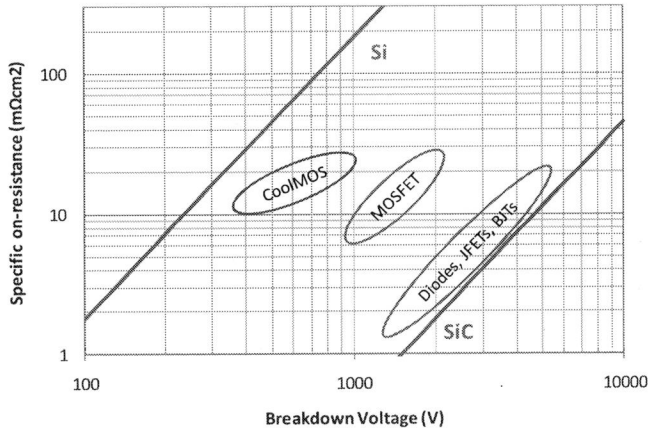

Figure 1. Comparison of unipolar limit of specific on-resistance versus blocking voltage for some device types in Si and SiC.

For MOSFETs the channel resistance in SiC dominates due to the low channel mobility, and experimental data therefore suffers. Although BJTs would be expected to perform better than the unipolar limit, high injection is seldom achieved. Therefore high injection devices such as IGBTs and GTOs are explored for the highest breakdown voltages. Note that the silicon CoolMOS actually surpasses the silicon unipolar limit and displays a different slope (proportional to V_B rather than V_B squared) due to a different scaling principle of the superjunction, see Fig. 1.

GaN actually has a higher theoretical critical field for breakdown, and would thus seem to be better than SiC. However, the GaN devices have so far shown much lower breakdown voltages than expected, possibly from the lack of free-standing GaN substrates. High injection devices are also not possible in the direct bandgap of GaN.

All device comparisons are difficult to do since the current rating varies dramatically between the data. It is encouraging to see that several of the published data by commercial companies, i.e. Cree, SemiSouth and TranSiC, refer to data for large devices. Many university results are extracted from small devices, which can be overly optimistic in on-resistance due to current spreading in the substrate.

A. SiC Schottky Diodes

Both Cree and Infineon have increased their sales substantially since 2009. The main application areas for SiC diodes are in power factor correction circuits (PFC), power supplies and recently photovoltaic (PV) inverters. The main advantage of the SBD is the absence of reverse recovery current during switching, hence it is possible to increase the switching frequency considerably. This in turn makes it possible to significantly decrease the volume, weight and cost for the system. For higher voltage (>3 kV) either pn-diode structures or merged SBD/pn designs are considered because of the superior reverse blocking and surge current capability.

B. SiC MOSFETs

The most desired power device to date is the vertical power MOSFET. It operates normally-off (enhancement mode) and with little demand on the drive circuits. The main drawbacks are the questionable reliability because of the sensitive gate dielectrics and the relatively poor channel mobility under the gate dielectrics. The low mobility gives the MOSFETs a relatively high Ron,sp for medium breakdown voltage (<2kV). Commercially available 1200V MOSFETs are recently released by Cree and a recent publication shows impressive performance [5].

C. SiC JFETs

The fabrication process of JFETs is quite straightforward. The main drawback with VJFETs is that they are usually normally-on (depletion mode) devices [6], which are considered unsafe in power applications. Recently SemiSouth has demonstrated normally-off VJFETs [7]. The operational threshold voltage margin is however limited for such a device, the on-resistance may be higher because it is limited by the pinched of region, and they may have a limited temperature of operation. JFETs exhibit a small capacitance and can thus be operated at high switching speed. From a reliability issue JFETs are considered as very promising since they rely primarily on pn-junction operation and not dependent on the quality of gate control dielectrics. SiC JFETs provide excellent high temperature operability.

D. SiC BJTs

The SiC BJT main advantages as power switch are its low conduction loss combined with fast switching. The BJT operation in the forward direction is beneficial for reaching low on-state loss since the two built-in pn-junctions cancel each other, hence the on-state loss is mostly dependent on the drift layer resistance and the substrate resistance. SiC BJTs are easy to connect in parallel since current gain decreases and on-resistance increases when the temperature increases. Increased complexity in drive circuitry and the moderate current gain are drawbacks of BJTs compared to FETs, whereas a normally-off characteristic in the BJT is an advantage over the JFET. The BJT is extremely robust with high surge current capability, high temperature performance and high cosmic-ray radiation hardness.

A main design concern is to optimize the lowly doped collector epitaxy to accommodate the high reverse voltage but not to yield any additional series resistance. The most crucial optimization from a practical point today is to increase the current gain at application temperature. Typically a current gain of 100 is desired when designing efficient drive circuitry. The base width in combination with the base doping is a key design parameter. The base-emitter junction benefits from being grown epitaxially in the same growth run to minimize the base emitter recombination current and hence lowering the current gain. A main challenge is also to ensure an optimized surface passivation in the sensitive base-emitter surface region which is very important for minimizing the

978-1-4244-8425-6/11 $26.00 © 2011 IEEE

base current recombination. In order to terminate the high electric field a sophisticated junction termination extension (JTE) needs to be employed. A field termination is needed for all the above devices [8-9].

E. Extreme High Voltage - High Injection Devices

Devices for extremely high voltage (>10 kV) require ultimate materials quality for two main reasons. Firstly, at these high voltages defects in the material will cause immature breakdown and secondly, in order to yield a reasonable low on-resistance the devices must operate during high injection which also calls for high quality SiC wafers with a minimum basal plane dislocation density to achieve long minority carrier lifetimes. Thyristor operation has recently been demonstrated for a record large 1x1 cm² sized Gate Turn-Off (GTO) thyristor with a breakdown voltage of 9 kV and a 1 ms current pulse of about 3 kA [10]. For non-pulsed applications IGBTs are preferred. Cree has demonstrated both p-IGBTs and n-IGBTs with performance already better than Si IGBTs with half the voltage rating.

III. GaN POWER DEVICES

The main interest in GaN power devices is still in applications for rf-power where the unique high electron mobility and the high critical field strength together make these devices advantageous.

The main device design of GaN is a lateral FET architecture. Since there are no available bulk wafers of GaN most technologies today are based on a GaN epitaxy on either SiC, sapphire or silicon wafers. Impressive device demonstration can be found from several research groups. In Table II a summary of recent published results from GaN based high voltage devices is given [11-19]

Recently GaN HFETs on SiC was demonstrated to yield 10,4 kV breakdown voltage[17]. Mostly lateral devices have been demonstrated for a voltage range between 600-1000 V, since higher voltages will cause surface breakdown if the gate-drain distance is too small. The best cost advantage is obtained for GaN fabricated on Si substrates and for large wafer diameter. All growth of GaN on any other substrate has to be fine tuned by a buffer layer to adapt for the lattice mismatch of GaN to SiC, sapphire or Si. A major drawback will be the loss of heat conductivity through the buffer layer and the use of less conductive Si substrate. However, the Si CoolMOS also competes in this voltage range and will be difficult to overcome in terms of price.

Smart power application may benefit in an interesting way by adopting GaN on Si to integrate both control electronics with the power devices [19].

TABLE II. GaN/AlGaN LATERAL POWER DEVICE PERFORMANCE - BREAKDOWN VOLTAGE AND ON-RESISTANCE.

GaN power device design	Vbr (V)	Ron (mΩcm²)	Ref.
AlGaN/GaN HEMTs	1050	3,4	[11]
AlGaN-GaN HEMTs, multiple field plates	900		[12]
AlGaN/GaN power HFET, AlN passivation	8300	200	[13]
AlGaN/GaN with TiO2/SiN gate insulator	1100	15	[14]
AlGaN/GaN	1900	2,2	[15]
AlGaN/GaN	1600	3,4	[16]
AlGaN/GaN	10400	200	[17]
GaN on Si by Nitronex	700	4,5	[18]

IV. APPLICATIONS

A few power system demonstrations have been made where SiC BJTs have been used in switch topologies and been compared to other SiC JFET solutions but also to reference Si IGBT (insulated gate bipolar transistor). Franke et al. [20] compared the total power loss in a switching application for 1200 V rated devices by adding the explicit switching loss, the on-state conduction loss and the loss in the driver circuitry. The study showed that the BJT has the lowest total loss of the three device types in this specific operating condition, see Fig. 2. However, if temperature, switch frequency or blocking voltage is changed, this may not be true. The main point with this figure is that the drive losses for the BJT is not necessarily the deciding factor in the comparison, and all losses have to be included.

If high operation temperatures (T) or radiation hard (RH) environments are also considered (see section VI and VII below), some new application areas are possible:

- Oil and gas drilling (T)
- Industrial motor drives (T)
- Automotive (T)
- Aviation (T, RH)
- Space exploration (T, RH)
- Nuclear energy (T, RH)

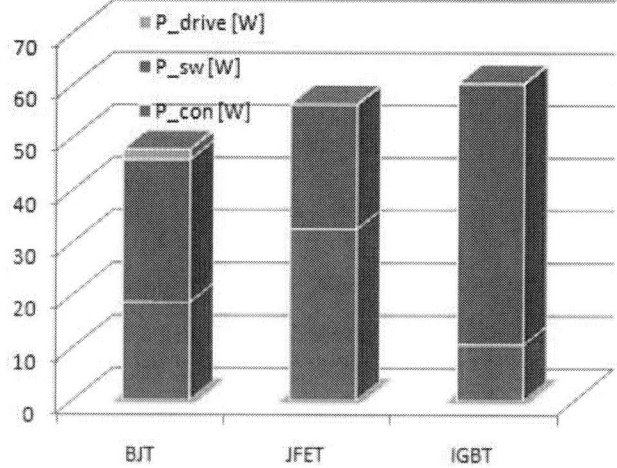

Figure 2. A total loss comparison between a BJT (1206 BitSiC TranSiC) and JFET (SiCED) in SiC vs a standard IGBT Si (IKW08T120, Infineon) configured switch topology. The BJT had the smallest total loss [20].

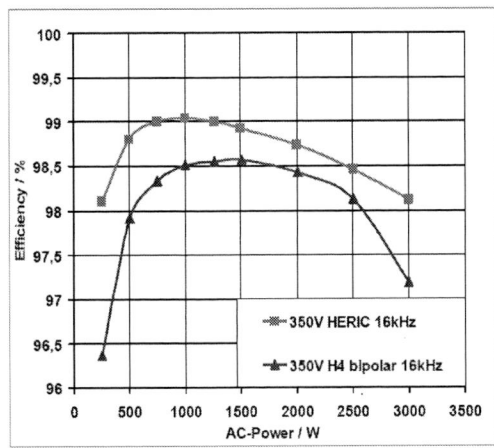

Figure 3. Power efficiency for a single phase r a DC_DC converter utilizing a SemiSouth nomally-off SiC JFET vs a Si IGBT [21].

A world record in DC-DC conversion for photovoltaic application was recently demonstrated by the Fraunhofer institute [21]. They could show 99% power conversion efficiency by using SiC JFET switch transistors from SemiSouth, see Fig. 3.

V. POWER MODULES:

A very challenging issue is to assemble and package a full power module based on SiC devices in order to take full advantage of the high potential of increasing power density. Several very promising demonstrations are recently published. In a power module demonstration recently, TranSiC presented a module consisting of 6 parallel 6A, 1200 V BJTs paired with 6 commercial SiC Schottky diodes mounted in free-wheeling configuration [22]. The demonstration showed the excellent paralleling capability of the BJTs applicable over a wide temperature range. Several other SiC modules are summarized in the references [23-28] in Table III.

Lostetter et al [24] have demonstrated that they could achieve a very good thermal match by using a substrate plate based on a powder with 80% Cu and 20% Mo. This material is easy to electroplate and to machine, with a very high thermal conductivity. Other solutions for the module packaging may be metal matrix composites based on AlSiC, CuMo, CuW and others.

TABLE III.   COMPARISON OF RECENT REPORTED SiC MODULES

Group	V_{BR} (V)	I (A)	Max T (°C)	Max freq (kHz)	Ref
TranSiC	1200	100			[22]
US Army Research Lab+CREE	1200	400	100	30	[23]
APEI		160	250	15	[24]
General Electric	600	150	150		[25]
CREE	1200	100	150		[26]
US Army Research Lab	1200	880			[26]
DARPA	10	50		20	[26]
GE	1400	200	250		[27]
SemiSouth	1200	100	150		[28]

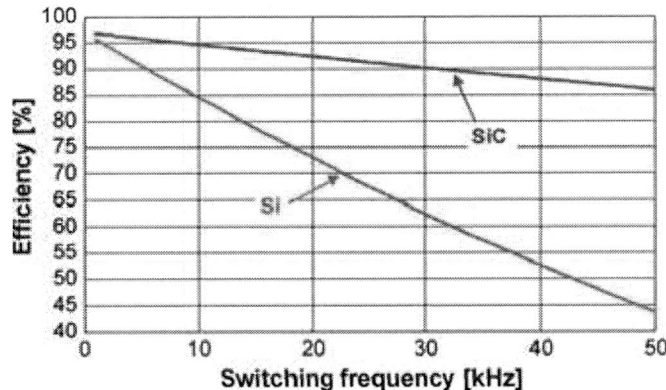

Figure 4. Wind turbine efficiency at different switching frequencies comparing Si IGBTs and SiC MOSFET converters [29].

Large power electronics based on Si IGBTs are usually limited to a few kHz switching frequency. In order to minimize the volume of the electronics a higher switching frequency is preferred. As is seen in Fig. 4 the Si IGBT efficiency is already decreased to about 73% at 20 kHz operation while the SiC MOSFET converter maintains the efficiency much better and only loses efficiency to 92%, which clearly demonstrates the potential of SiC electronics where the cost of filters and other passives are substantially reduced.

VI. HIGH TEMPERATURE OPERATION IN ALL SiC SYSTEMS

Another advantage of SiC is the operation at high temperature (> 300 °C), made possible by the wide bandgap. Even at 600 °C the intrinsic concentration is not the limiting factor, but rather the metal contacts and packaging. For a power switching system such as a motor drive, being able to operate the switches close to the motor will reduce the inductive losses. Low voltage high temperature electronics for drivers close to the switches will further reduce the system losses, see Fig. 5. An added digital interface could potentially allow optical control and galvanic isolation. If a separate cooling system for the electronics can be excluded large systems savings are achieved in most vehicular applications.

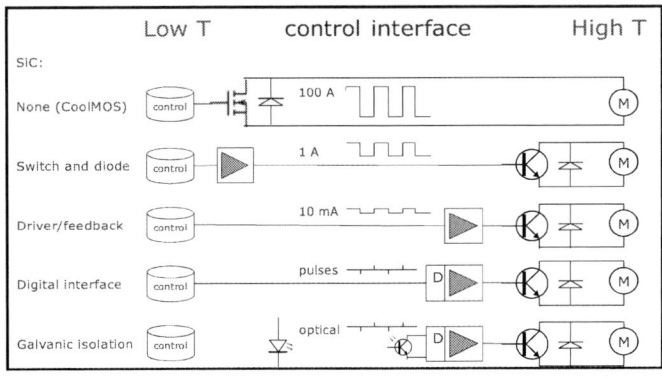

Figure 5. Power system: high T IC advantage with SiC in indicated parts.

978-1-4244-8425-6/11 $26.00 © 2011 IEEE

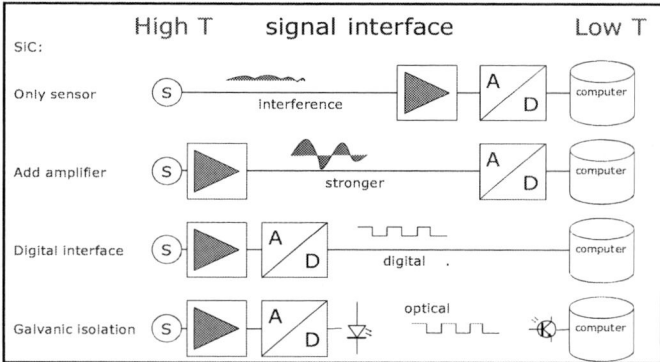

Figure 6.   Sensor system: high T IC advantage with SiC in indicated parts.

Sensors are already available in SiC for high temperature operation, for instance as gas sensors. Other potential sensor types of interest are for temperature and pressure monitoring. If amplifiers and digital circuitry is also available, more applications would open due to higher signal integrity and improved system performance. Some system advantages can be seen from Fig. 6.

VII.   RADIATION HARDNESS IN WBG

SiC and GaN have a larger tolerance to radiation than silicon. This is mainly due to the 3 times larger bandgap, which reduces the electron-hole-pair formation, and also to the threshold for atomic displacement, which is about 2-3 times larger than in Si. Although the latter effect is reduced by the higher recombination of point defects in Si, the radiation hardness of SiC devices offers an important advantage that can be further explored in aviation and space applications. JFETs and BJTs in particular are much more radiation hard than devices such as MOSFETs with a delicate gate dielectric that may be affected by radiation. In [30-31] radhard demonstration results revealed that BJTs exposed to high energy gamma rays and heavy ion bombardment can withstand 100 - 1000 times higher doses than a corresponding Si power device. The radiation hardness would also allow SiC radiation detectors with integrated amplifiers, similar to the system in Fig. 6 above.

VIII.   CONCLUSIONS

A brief review of today's technology forefront of high power device performance in SiC and GaN has been presented. Impressive device results are extensively published and several devices are commercialized. The SiC materials quality problems have been overcome to a great extent and useful device areas are now on the market. SiC offers a great variety of devices with Schottky diodes, MOSFETs and JFETs as well as BJTs. Future extreme high voltage devices will be developed on high quality thick epitaxial SiC layers including thyristor devices operating in injection mode to yield low on-resistance. For GaN the market place in high power applications is less clear. The technology is less mature but for applications below 1 kV interesting

performance has been demonstrated for GaN on Si which potentially can be cost effective, but the Si CoolMOS is also competing in this voltage range. Some other systems advantages of SiC systems at very high temperatures and radiation hard environments were also mentioned.

ACKNOWLEDGMENTS

The authors acknowledge the research funding from the Swedish Energy Agency (STEM) and the Swedish Governmental Agency for Innovation Systems (VINNOVA) and the research team at KTH.

REFERENCES

[1] C.-M. Zetterling, Ed.,"Process technology for silicon carbide devices," in EMIS processing series, IEE, 2002.

[2] M. Östling, S.-M. Koo, M. Domeij, E. Danielsson, and C.-M. Zetterling, "SiC Device Technologies," in Encyclopedia of RF and Microwave Engineering: John Wiley & Sons, Inc., 2005, pp. 4613-4619

[3] M. Östling et al "SiC Bipolar Power Transistors - Design and Technology Issues for Ultimate Performance" in *Silicon Carbide 2010 — Materials, Processing, and Devices*, edited by S.E. Saddow, E. Sanchez, F. Zhao, M. Dudley (Mater. Res. Soc. Symp. Proc. Volume 1246, Warrendale, PA, 2010, B08-01,

[4] R. Ghandi, H-S. Lee, M. Domeij, B. Buono, C-M. Zetterling, and M. Östling, IEEE Electron Device Letters, vol. 29, no. 10 (2008) pp.1135 1137.

[5] B. A. Hull, C. Jonas, S-H Ryu, M. Das, M. O'Loughlin, F. Husna, R. Callanan, J. Richmond, A. Agarwal, J. Palmour and C. Scozzie, Materials Science Forum Vols. 615-617 (2009) pp. 749-752.

[6] T. Kimoto, "SiC technologies for future energy electronics," *VLSI Technology (VLSIT), 2010 Symposium on*, pp. 9-14, 2010.

[7] D.C. Sheridan, A. Ritenour, V. Bondarenko, P. Burks, and J.B. Casady, Proceeding of 21st International Symposium on Power Semiconductor Devices & IC's, (2009) pp. 335–338.

[8] M. Domeij, C. Zaring, A.O. Konstantinov, M. Nawaz, J-O. Svedberg, K. Gumaelius, I. Keri, A.Lindgren, B. Hammarlund, M. Östling, M. Reimark, Materials Science Forum Vols. 645-648 (2010) pp 1033-1036.

[9] R. Ghandi, H-S. Lee, M. Domeij, B. Buono, C-M. Zetterling, and M..Östling, IEEE Electron Device Letters, vol. 29, no. 10 (2008) pp.1135-1137.

[10] H. O'Brien, A. Ogunniyi, Q. Jon Zhang, and A. K. Agarwal, in *Silicon Carbide 2010 — Materials, Processing, and Devices*, edited by S.E. Saddow, E. Sanchez, F. Zhao, M. Dudley (Mater. Res. Soc. Symp. Proc. Volume 1246, Warrendale, PA, 2010, B08-03

[11] N.-Q. Zhang, B. Moran, S. P. DenBaars, U. K. Mishra, X. W. Wang, and T. P. Ma, "Kilovolt AlGaN/GaN HEMTs as switching devices," *Phys.Stat. Sol. A*, vol. 188, no. 1, pp. 213–217, Nov. 16, 2001.

[12] X. Huili, Y. Dora, A. Chini, S. Heikman, S. Keller, and U. K. Mishra, "High breakdown voltage AlGaN-GaN HEMTs achieved by multiple field plates," *IEEE Electron Device Lett.*, vol. 25, no. 4, pp. 161–163, Apr. 2004.

[13] Y. Uemoto, D. Shibata, M. Yanagihara, H. Ishida, H. Matsuo, S. Nagai, N. Batta, L. Ming, T. Ueda, T. Tanaka, and D. Ueda, "8300 V blocking voltage AlGaN/GaN power HFET with thick poly-AlN passivation," in *IEDM Tech. Dig.*, Dec. 10–12, 2007, pp. 861–864.

[14] S. Yagi, M. Shimizu, H. Okumura, H. Ohashi, Y. Yano, and N. Akutsu, "High breakdown voltage AlGaN/GaN metal–insulator–semiconductor high-electron-mobility transistor with TiO2/SiN gate insulator," *Jpn. J. Appl. Phys.*, vol. 46, no. 4B, pp. 2309–2311, Apr. 2007.

[15] Y. Dora, A. Chakraborty, L. McCarthy, S. Keller, S. P. DenBaars, and U. K. Mishra, "High breakdown voltage achieved on AlGaN/GaN HEMTs with integrated slant field plates," *IEEE Electron Device Lett.*, vol. 27, no. 9, pp. 713–715, Sep. 2006.

[16] N. Tipirneni, A. Koudymov, V. Adivarahan, J. Yang, G. Simin, and M. A. Khan, "The 1.6-kV AlGaN/GaN HFETs," *IEEE Electron Device Lett.*, vol. 27, no. 9, pp. 716–718, Sep. 2006.

[17] Y. Uemoto, T. Ueda, T. Tanaka, and D. Ueda, "Recent advances of high voltage AlGaN/GaN power HFETs," in Proc. SPIE Gallium Nitride Mater. Devices IV, H. Morkoc, C. W. Litton, J. I. Chyi, Y. Nanishi, J. Piprek, and E. Yoon, Eds., San Jose, CA, 2009, vol. 7216, pp. 721 606–721 611.

[18] B. Lu, E.L. Piner, T. Palacios, "Schottky-Drain Technology for AlGaN/GaN High-Electron Mobility Transistors" *IEEE Electron Device Lett.*, vol. 31, no. 4, pp. 302–304, Apr. 2010.

[19] K.-Y. Wong, W. Chen, and K. J. Chen, "Wide Bandgap GaN Smart Power Chip Technology" CS MANTECH Conference, May 18th-21st, 2009, Tampa, Florida, USA

[20] W.-T. Franke and F.W Fuchs, 13th European Conference on Power Electronics Power Electronics and Applications, 2009. EPE '09. (p. 1-10)

[21] Bruno Burger, Dirk Kranzer, "Extreme High Efficiency PV-Power Converters," EPE, Barcelona, Spain, 8-10 September 2009

[22] M. Östling, Silicon Carbide Power Devices, IEDM Tech. Dig. p.316-319, San Francisco, December 2010 and www.transic.com

[23] D. Urciuoli, R. Green, A. Lelis, D. Ibitayo , "Performance of a dual, 1200 V, 400 A, silicon-carbide power MOSFET module," Energy Conversion Congress and Exposition (ECCE), 2010 IEEE , pp.3303-3310, 12-16 Sept. 2010

[24] A. Lostetter, J. Hornberger, B. McPherson, B. Reese, R. Shaw, M. Schupbach, B. Rowden, A. Mantooth, J. Balda, T. Otsuka, K. Okumura, M. Miura, Vehicle Power and Propulsion Conference, 2009. VPPC '09. IEEE , pp.1032-1035, 7-10 Sept. 2009

[25] L. Stevanovic, K. Matocha, Z. Stum, P. Losee, A. Gowda, J. Glaser, R. Beaupre, Control and Modeling for Power Electronics (COMPEL), 2010 IEEE 12th Workshop on , pp.1-6, 28-30 June 2010

[26] J. Richmond, S. Leslie, B. Hull, M. Das, A. Agarwal, J. Palmour, Energy Conversion Congress and Exposition, 2009. ECCE 2009. IEEE , pp.106-111, 20-24 Sept. 2009

[27] K. Matocha, P. A. Losee, A. Gowda, El. Delgado, G. Dunne, R. Beaupre, L. Stevanovic , Materials Science Forum Vols. 645-648 (2010) pp 1123-1126

[28] D. Sheridan, at the European SiC and Related Materials Conference, 2010, Oslo, Norway.

[29] H. Zhang; L. M. Tolbert.; *Industrial Electronics, IEEE Transactions on* , vol.58, no.1, pp.21-28, Jan. 2011.

[30] M. Nawaz, C. Zaring, S. Onoda, T. Ohshima and M. Östling, Proceedings of 67th Device Research Conference, The Pennsylvania State University, University Park, PA, June 22-24 2009, p. 279-280

[31] A. Hallén, M. Nawaz, C. Zaring, M. Usman, M. Domeij, and M. Östling, IEEE Electron Device Letters, Vol 31 (2010) p. 707-709

Proceedings of the 23rd International Symposium on Power Semiconductor Devices & IC's
May 23-26, 2011 San Diego, CA

A Novel Substrate-Assisted RESURF Technology for Small Curvature Radius Junction

Ming Qiao, Xi Hu, Hengjuan Wen, Meng Wang, Bo Luo, Xiaorong Luo, Zhuo Wang, Bo Zhang and Zhaoji Li
State Key Laboratory of Electronic Thin Films and Integrated Devices
University of Electronic Science and Technology of China
Chengdu, P.R.China
E-mail:qiaoming@uestc.edu.cn

Abstract—**A novel substrate-assisted (SA) RESURF technology aiming at improving off-state breakdown voltage (BV) of PN junction with small curvature radius is proposed and experimentally demonstrated in this paper. The SA RESURF technology not only realizes small curvature radius in the fingertip region, but also reduces electric field concentration in the curved metallurgical junction. Low-doped P-substrate, which increases depletion of the small curvature radius junction and reduces electric field concentration in the curved metallurgical junction, is adopted in the source fingertip region. Owing to the existence of low-doped P-substrate, the abrupt PN junction with small curvature radius is adjusted to low-doped PN junction with large curvature radius. The SA RESURF technology can be widely applied to lateral high voltage devices with small curved junction, especially to lateral super junction devices. A CBSLOP-LDMOS with the proposed SA RESURF technology has been developed. The experimental results show that the CBSLOP-LDMOS exhibits off-state BV of 700 V and specific on-resistance ($R_{on,sp}$) of 142 mΩ·cm^2.**

I. INTRODUCTION

Reduced surface field (RESURF) technology is widely used for the lateral devices with high breakdown voltage and low $R_{on,sp}$[1]-[3]. Since the depletion of vertical metallurgical junction is reinforced by the horizontal junction, the surface electric field of RESURF device is reduced at the same applied voltage. The surface electric field is far below the critical electric field and much higher applied voltage can be achieved before the avalanche breakdown occurs. The surface breakdown can be eliminated and the ideal bulk breakdown can be reached.

However, the layout of lateral high voltage semiconductor device is occlusive. Curved source or drain junction will lead to premature breakdown. The curvature radius of the source or drain fingertip region, which even affects the device area and $R_{on,sp}$, has a crucial relationship with the off-state breakdown characteristics[4]. The avalanche breakdown can be expected to occur in the fingertip region rather than in the straight edge of the conventional RESURF LDMOS, on account of the abrupt PN junction with small curvature radius strengthening the concentration of electric field. The smaller the curvature radius of the source or drain fingertip region is, the lower the breakdown voltage will be. Large curvature radius is necessary for high voltage device to gain optimal breakdown characteristics, but small device area is a significant factor to obtain low $R_{on,sp}$. Therefore, a novel SA RESURF technology for small curvature radius junction, which can improve off-state BV characteristics through substrate-assisted depletion effect and reduce $R_{on,sp}$, is proposed and experimentally demonstrated in this study.

II. SUBSTRATE-ASSISTED RESURF TECHNOLOGY

Figure 1 shows schematic surface view of the proposed SA RESURF technology. Figure 2 shows three-dimensional view of the source fingertip region with the SA RESURF technology. In the curved junction, we adopted low-doped P-substrate instead of partial high-doped N-well, by which abrupt P-well/N-well junction with small curvature radius is adjusted to low-doped P-substrate/N-well junction with large curvature radius. The depletion layer of P-substrate/N-well curved junction will sustain the applied high voltage. Due to low-doped concentration and large curvature radius of P-substrate, the electric field concentration is efficiently reduced and the breakdown voltage can be improved. Premature breakdown, which may occur at high-doped P-well/N-well junction with small curvature radius, can be avoided. Based on the low-doped P-substrate which assists depletion of small curvature radius junction and reduces surface electric field concentration of the curved metallurgical junction, the substrate-assisted RESURF (SA RESURF) technology was proposed.

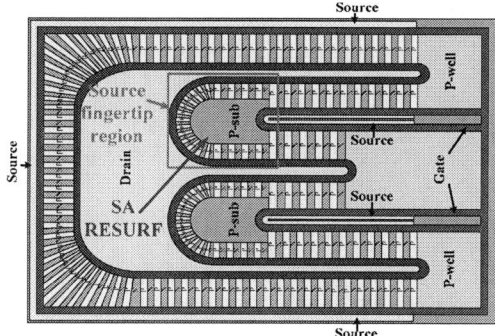

Fig.1 Schematic surface view of the CBSLOP-LDMOS with the proposed SA RESURF technology.

978-1-4244-8425-6/11 $26.00 © 2011 IEEE

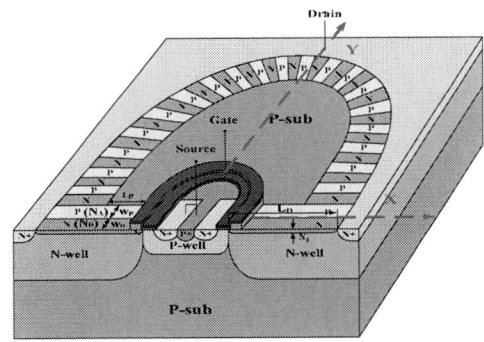

Fig.2 Three-dimensional view of the source fingertip region with the proposed SA RESURF technology.

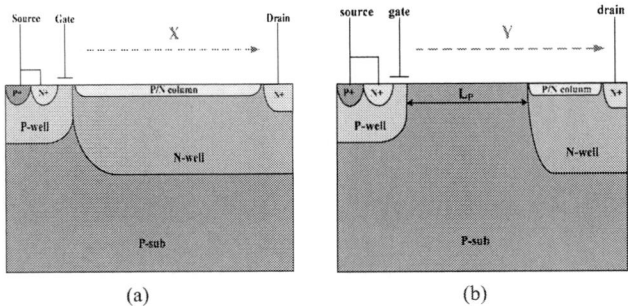

(a)   (b)

Fig.3 The cross-section of the source fingertip region along (a) X and (b) Y direction.

Figure 3 (a) shows schematic cross-sectional view along x direction. Figure 3 (b) shows schematic cross-sectional view along y direction. For straight edge part of the LDMOS, conventional RESURF structure is adopted. If we use this structure in the curved part, high-doped P-well with small curvature radius will lead to premature breakdown in curved junction. Low-doped P-substrate, which can introduce additional charges in the depletion region, can reduce the peak electric field at the P-well/N-well junction. These additional charges can create a new peak electric field at the P-substrate/N-well junction. Due to low-doped P-substrate, the curvature radius of the curved junction is remarkably increased, thus the surface electric field concentration and the breakdown voltage of the LDMOS can be improved.

Figure 4 (a) shows schematic electric field crowding effect of the source fingertip region with conventional RESURF technology. L_P is the distance between P-well and N-well in the fingertip region. When L_p is equal to 0 μm, electric field concentration occurs at high-doped P-well/N-well junction with the smallest curvature radius, leading to the lowest BV. Figure 4 (b)-(d) show schematic electric field crowding effect of source fingertip region using the proposed SA RESURF technology with various L_P. In Fig.4 (b), small L_P intensifies the electric field concentration of the curved junction and then causes premature breakdown in low-doped P-substrate/N-well junction. In Fig.4 (d), the length of p/n columns laid on N-well of the source fingertip region is cut down. Although large L_P decreases the surface electric field concentration of curved junction, the decreased p/n columns will lead to low breakdown voltage due to the reduction of depletion layer length of p/n columns. Therefore L_P can't be too small or too

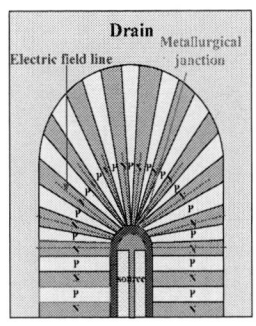 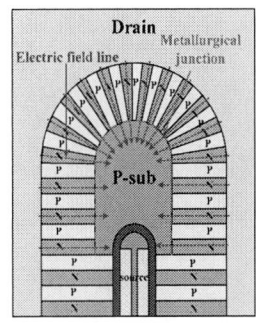

(a) The source fingertip region with conventional RESURF technology

(b) The source fingertip region with the proposed SA RESURF technology (small L_P)

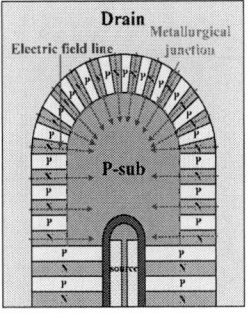 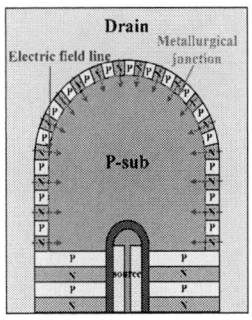

(c) The source fingertip region with the proposed SA RESURF technology (optimal L_P)

(d) The source fingertip region with the proposed SA RESURF technology (large L_P)

Fig.4 Schematic electric field crowding effect for small curvature radius junction: (a) conventional RESURF technology, and (b)-(d) SA RESURF technology with various L_P.

large. Figure 4 (c) shows the source fingertip region with the proposed SA RESURF technology when L_P is equal to an optimal value. The reduction of electric field concentration in the curved junction will eliminate the premature breakdown occurring in the low-doped P-substrate/N-well junction. At the same time, the optimal length of p/n columns will lead to a longer voltage sustain layer. Therefore, the breakdown voltage of the fingertip region with a small curvature radius can be improved.

III. RESULTS AND DISCUSSION

A LDMOS with a charge-balanced surface low on-resistance path (CBSLOP) layer is taken for example, as shown in Fig.1 [5]-[6]. A high-doped super junction region consisting of alternate P-type and N-type columns is located at the surface of N-well. High-doped CBSLOP layer supplies low on-resistance. And simultaneously, surface electric field is reduced by the assisted depletion of p/n columns, resulting in the improvement of breakdown characteristics. The CBSLOP-LDMOS based on a super junction concept can not only provide low $R_{on,sp}$ in on-state but also high breakdown voltage in off-state. By using three-dimensional and two-dimensional device simulations which are carried out by SILVACO, the CBSLOP-LDMOS with the proposed SA RESURF technology is optimized.

978-1-4244-8425-6/11 $26.00 © 2011 IEEE

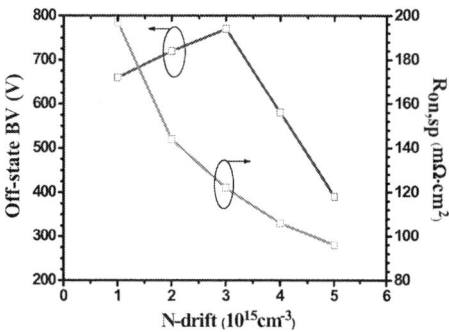

Fig.5 BV and $R_{on,sp}$ of CBSLOP-LDMOS as a function of drift concentration($W_n=W_p=1$ μm, $N_A=N_D=3E16$ cm^{-3}, $X_j=1.5$ μm).

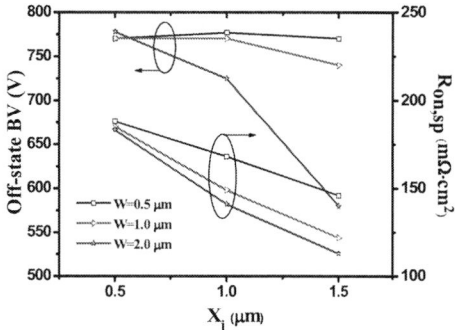

Fig.6 BV and $R_{on,sp}$ as a function of p/n column junction depth (X_j) with different p/n column widths (W) ($N_{drift}=3E15$ cm^{-3}, $N_A=N_D=3E16$ cm^{-3}).

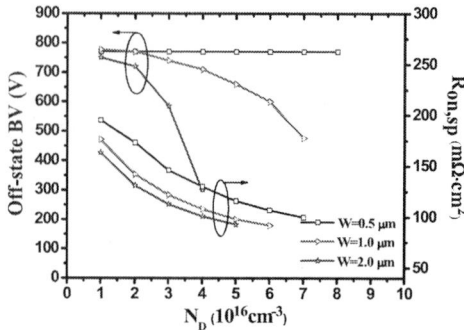

Fig.7 BV and $R_{on,sp}$ as a function of p/n column concentration with different p/n column widths ($N_{drift}=3E15$ cm^{-3}, $X_j=1.5$ μm, $N_D=N_A$).

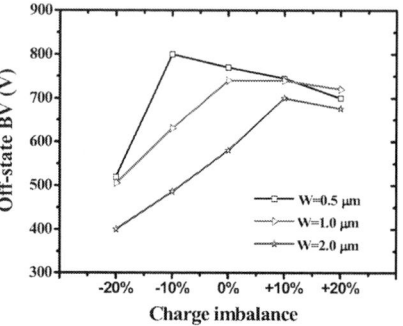

Fig.8 BV as a function of charge imbalance with different p/n column widths ($N_{drift}=3E15$ cm^{-3}, $X_j=1.5$ μm, $N_D=3E16$ cm^{-3}).

Figure 5 shows simulated off-state BV and $R_{on,sp}$ of CBSLOP-LDMOS as a function of drift concentration at $W_n=W_p=1$ μm, $N_A=N_D=3E16$ cm^{-3}, and $X_j=1.5$ μm. N_A is the concentration of p column and N_D is the concentration of n column. $R_{on,sp}$ is obviously reduced with the increase of drift concentration. However, the drift concentration can not be too large for the reduction of the off-state breakdown voltage. The peak value of the off-state BV appears when drift concentration is equal to 3E15 cm^{-3}.

Figure 6 shows simulated off-state BV and $R_{on,sp}$ as a function of p/n column junction depth (X_j) with different p/n column widths (W) at $N_{drift}=3E15$ cm^{-3} and $N_A=N_D=3E16$ cm^{-3}. Although deep depth of p/n column can reduce $R_{on,sp}$, the off-state BV is reduced as the increase of X_j at W=2 μm due to incomplete depletion of p/n columns. For X_j from 0.5 μm to 1.5 μm, high off-state breakdown voltage can be achieved at W=0.5 μm and W=1 μm.

Figure 7 shows simulated off-state BV and $R_{on,sp}$ as a function of p/n column concentration with different p/n column widths at $N_{drift}=3E15$ cm^{-3}, $X_j=1.5$ μm, and $N_D=N_A$. $R_{on,sp}$ is obviously reduced with the increase of the p/n column concentration. Due to incomplete depletion of p/n column, off-state BV is reduced at W=1 μm and W=2 μm with the rising p/n column concentration. Off-state BV is close to constant at W=0.5 μm. Therefore, small W helps to complete depletion of the p/n column.

Figure 8 shows simulated off-state BV as a function of charge imbalance with different p/n column widths at $N_{drift}=3E15$ cm^{-3}, $X_j=1.5$ μm and $N_D=3E16$ cm^{-3}. Charge imbalance is one of the main issues for the super junction concept implemented in LDMOS. The maximum BV occurs

at the condition that P-column concentration is slightly lower than N-column at W=0.5 μm. The maximum BV occurs at the condition that P-column concentration is slightly higher than N-column at W=2 μm.

In considering of the continuity of electric field, it is necessary to optimize the fingertip region with the proposed SA RESURF technology, especially the conjunction between the source fingertip region and straight edge part of the CBSLOP-LDMOS. Mutation of the structure in this conjunction will induce the radical change of surface electric field distribution at an applied voltage. Figure 9 shows surface electric field distribution of the fingertip region with the proposed SA RESURF technology. In Fig.9 (a) and Fig.9 (c), semi-CBSLOP layer is placed in N-type drift region of the source fingertip region and the difference is the p/n column sequence of smei-CBSLOP layer. The initiative column of the source fingertip region in Fig.9 (a) is N-type column. The initiative column of the source fingertip region in Fig.9 (c) is P-type column. Figure 9 (b)-(d) show the source fingertip region without smei-CBSLOP layer. Figure 9 (e) shows the surface electric field distribution along AA', BB', CC', DD' and EE' lines. With the assisted depletion of CBSLOP layer, new peaks of surface electric field between p/n columns appear and the depletion region is visibly extended. When the initiative column of the straight edge part of the CBSLOP-LDMOS is N-type column, premature breakdown occurs due to imcomplete depletion of the N-type column. When the initiative column of the straight edge part is P-type column, the breakdown voltage of the CBSLOP-LDMOS can be improved. The optimal conjunction structure for the CBSLOP-LDMOS with the proposed SA RESURF technology is shown in Fig.9 (a).

978-1-4244-8425-6/11 $26.00 © 2011 IEEE

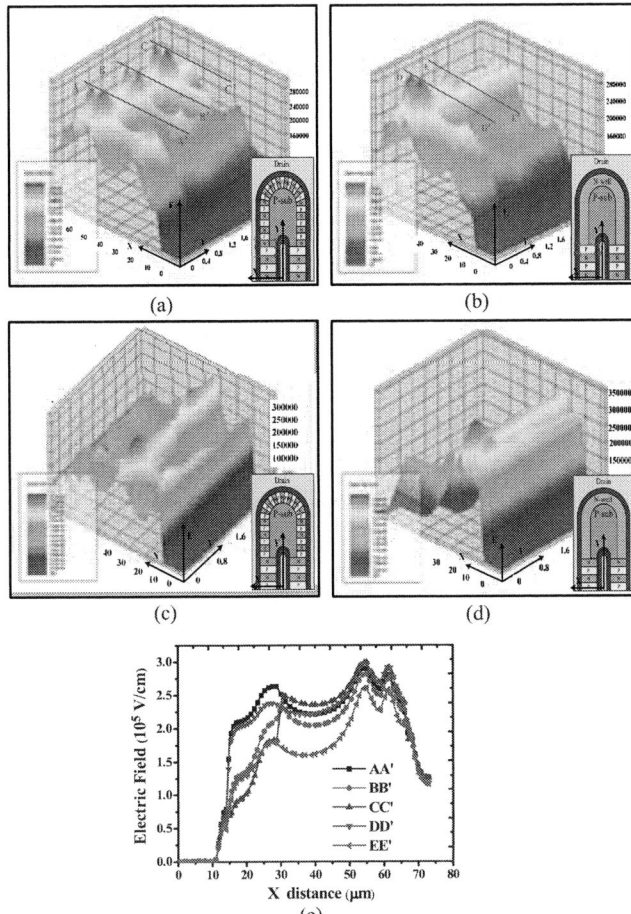

(a)

(b)

(c)

(d)

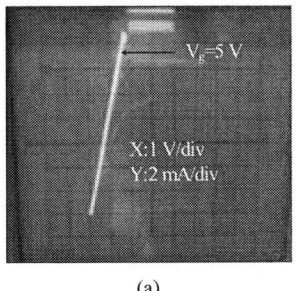

(a)

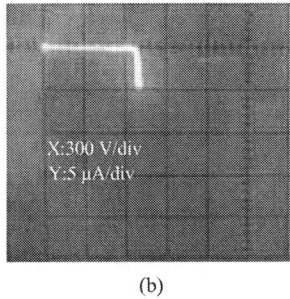

(b)

Fig.11 Experimental results of the CBSLOP-LDMOS with the proposed SA RESURF technology: (a)measured output characteristics and (b)measured off-state breakdown characteristics.

(e)

Fig.9 Surface electric field distribution of source fingertip region with the proposed SA RESURF technology.

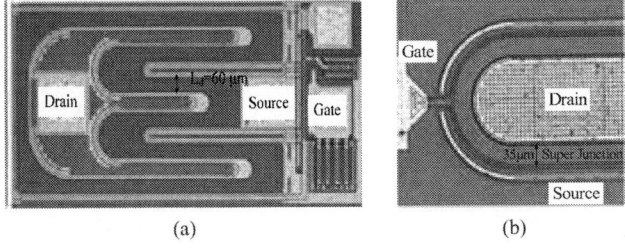

(a)

(b)

Fig.10 (a) Micrograph of the CBSLOP-LDMOS with the proposed SA RESURF technology, (b) Micrograph of the CBSLOP-LDMOS in previous work.

Figure 10 (a) shows micrograph of the CBSLOP-LDMOS with the proposed SA RESURF technology. Figure 10 (b) shows micrograph of the CBSLOP-LDMOS in previous work[6]. Compared with the previous work, the area of fingertip region is significantly reduced on account of the reduction of the curvature radius.

Figure 11 (a) shows measured output characteristics of the CBSLOP-LDMOS with the proposed SA RESURF technology. Figure 11(b) shows measured off-state breakdown characteristics of the CBSLOP-LDMOS with the proposed SA RESURF technology. The CBSLOP-LDMOS exhibits off-state BV of 700 V and $R_{on,sp}$ of 142 mΩ·cm^2, leading to power FOM, expressed as FOM=BV^2/ $R_{on,sp}$, of 3.45 MW/cm^2.

IV. CONCOLUTION

SA RESURF technology can be widely applied in lateral high voltage device, such as single RESURF, double RESURF LDMOS and super junction device, etc. The surface electric field at small curvature radius junction can be reduced and high breakdown voltage can be achieved by using the proposed SA RESURF technology. A CBSLOP-LDMOS with the proposed SA RESURF technology has been developed. Based on the simulated results, the optimal structure with the SA RESURF technology is obtained when the initiative column of the straight edge part of the CBSLOP-LDMOS is P-type column. Compared with the previous work, the area of fingertip region is reduced on account of the reduction of the curvature radius. Finally the measured results prove that the CBSLOP-LDMOS with the proposed SA RESURF technology exhibits BV of 700 V and $R_{on,sp}$ of 142 mΩ·cm^2. Low cost, small area, low $R_{on,sp}$ and high BV make the CBSLOP-LDMOS with the proposed SA RESURF technology a competitive device for high voltage power IC applications.

ACKNOWLEDGMENT

Project supported by National Natural Science Foundation of China (Grant No.60906038) and the Science-Technology Foundation for Young Scientist of University of Electronic Science and Technology of China (Grant No.L08010301JX0830).

REFERENCES

[1] J. Appels, H. Vaes and J. Verhoeven, "High Voltage Thin Layer Devices (RESURF Devices)," IEDM, Washington, DC., USA, December 1979, pp. 238-241.

[2] H. Vaes and J. Appels, "High voltage, high current lateral devices," IEDM, San Francisco, California, USA, December 1980, pp. 87-90.

[3] A.W. Ludikhuize, "A Review of RESURF Technology," in Proc. of ISPSD, Toulouse, France, May 2000, pp. 11-18.

[4] S. H. Lee, C. K. Jeon, J. W. Moon and Y. C. Choi, "700V Lateral DMOS with New Source Fingertip Design," in Proc. of ISPSD 2008, Orlando, Florida, USA, May. 2008, pp. 141-144.

[5] B. Zhang, L. Chen, J. Wu and Z. J. Li, "SLOP-LDMOS - A Novel Super-Junction Concept LDMOS and Its Experimental Demonstration," in Proc. of ICCCAS, Hong Kong, China, May 2005, pp. 1399-1402.

[6] B. Zhang, W. L. Wang, W. J. Chen, Z. H. Li and Z. J. Li, "High-Voltage LDMOS with Charge-Balanced Surface Low On-Resistance Path Layer," IEEE Electron Device Letters, vol.30, pp. 849-851, 2009.

Proceedings of the 23rd International Symposium on Power Semiconductor Devices & IC's
May 23-26, 2011 San Diego, CA

Automotive 130 nm Smart-Power-Technology including embedded Flash Functionality

Ralf Rudolf, Cajetan Wagner, Lincoln O'Riain, Karl-Heinz Gebhardt, Barbara Kuhn-Heinrich, Birgit von Ehrenwall, Andreas von Ehrenwall

Infineon Technologies AG
01099 Dresden, Germany
Ralf.Rudolf@infineon.com

Marc Strasser, Matthias Stecher, Ulrich Glaser, Stefano Aresu, Paul Kuepper, Alevtina Mayerhofer

Infineon Technologies AG
81726 Munich, Germany
Marc.Strasser@infineon.com

Abstract—In this paper a 130 nm BCD technology platform is presented. The process offers logic-devices, flash-devices and high voltage devices with rated voltages up to 60 V. There are HV analog devices with variable channel length and HV power devices with low on-resistances. To ensure the safe operation of the power devices, a superior robustness against high energetic pulses of different length and repetitions could be achieved. The isolation of the different voltage stages is ensured by deep trenches and highly doped buried layers.

I. INTRODUCTION

The suppression of parasitic diodes, bipolar-transistors, and thyristors within a BCD technology is a major challenge which becomes even more important with further shrinkage. Therefore, new isolation concepts superior to the pn-isolation concepts have to be developed. One solution to this problem is an SOI based BCD process. The drawback of SOI is its limited capability to build large power stages. Due to the buried oxide, a weak thermal connection to the substrate is the main drawback of this concept. Hence, the maximal drive current of power stages is limited in the range of 1A. Nevertheless, there are many applications in the area of the automotive electronics where larger current capabilities up to 10A are needed. If the decision is made by choosing a bulk based BCD technology, the suppression of the parasites below the flash-, logic- and analog-areas is of utmost importance. This can be realized by an extremely highly doped n-buried-layer in combination with a low-resistive sinker connection and deep-trench isolation. This allows the integration of 8bit up to 16bit micro-controllers and large power devices within the same BCD chip.

II. THE BCD-PROCESS

The presented BCD process integrates logic, analog, memory and power switching functions into one chip. Typical voltages range between 1.2 V up to 60 V and currents from some µA up to 15 A are managed. A basic ULSI-process was extended with power-transistors up to typical breakdown voltages of 75 V at room temperature. For the combination of a ULSI logic/flash process with power devices, the logic areas have to be embedded into a pseudo-substrate (Fig.1). This p-

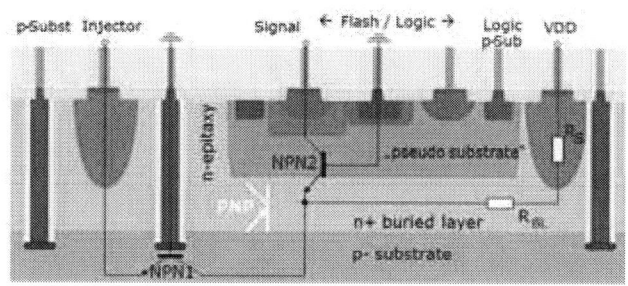

Fig. 1: Schematic cross-section of logic/flash and power areas embedded in silicon with deep trench isolation. The substrate is grounded at each device to minimize EMI.

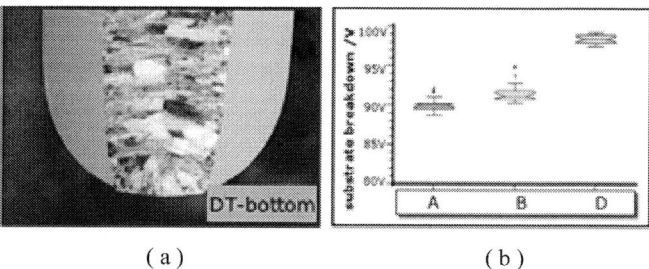

(a)   (b)

Fig. 2: (a) TEM picture of DT-bottom, (b) breakdown voltage from "buried layer" to substrate with different implant doses and thermal budgets.

well is located in the n-doped epitaxial layer. A very high doped n-buried-layer isolates this "pseudo substrate" from the real p-substrate of the wafer. This construction ensures insensitivity of logic blocks to parasitic disturbances coming from the power areas. The power devices are directly embedded into the n-epitaxial wells and are also robustly isolated from each other. One key factor for this robustness is the low gain of the vertical parasitic npn-transistor. The high base doping suppresses the current gain to less than 2%. Deep trench isolation ensures the lateral isolation of the epi-wells. This isolation technique consumes less area and makes far better physical isolation than the former approach of junction isolation achieved. Whilst the current gain of the lateral parasitic substrate-npn (NPN1) has been reduced by fifty percent, the breakdown voltage from buried-layer to substrate

978-1-4244-8425-6/11 $26.00 © 2011 IEEE

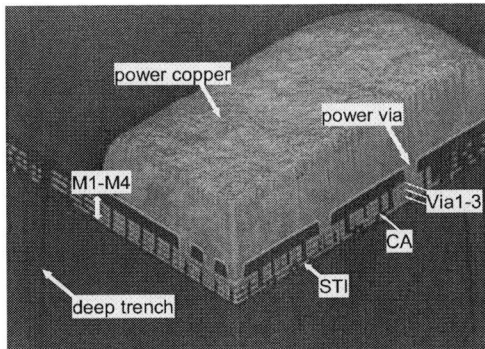

Fig. 3: Micro-image of a typical layer stack.

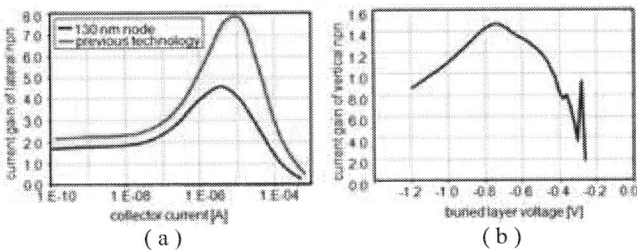

(a)   (b)

Fig. 4: current gain (a) of the lateral parasitic transistor (NPN1) and beta (b) of the vertical npn (NPN2) according to Fig.1

has been adjusted to be well above 90 V at room temperature. Additional experiments showed the direction for further improvement by adjusting the thermal budget and doping concentration at the pn-junction (Fig.2b). In contrast to fully isolating trenches, presented for different technologies [e.g. 10], a new trench isolation concept was chosen. It allows the connection of bulk substrate from the top of the wafer. The oxide liner has been removed from the deep trench bottom and the poly-fill directly connects the silicon substrate (Fig. 2a). These connections increase the EMI robustness and lead to a less complicated package since in case of isolated trenches the substrate has to be connected from the backside of the chip.

The BCD process presented also offers a three layer dual-damascene copper metallization and a single-damascene copper layer with tungsten plugs. This metal stack, suitable for high packing densities, is combined with a thick copper layer on top to enable thermal and mechanical robustness of the assembled chips (Fig. 3).

III. SUPPRESSION OF PARASITES BETWEEN POWER- AND LOGIC/ANALOG/FLASH-CIRCUIT BLOCKS

One of the major concerns in smart power circuits is the operation of large N-LDMOS stages below substrate-GND. This can appear in different applications e.g. in bridge-drivers or high-side switches during each switching cycle or in failure modes due to wire harness shorts [9]. During these operation modes, large electron currents are injected into the substrate. If they do not recombine in the substrate or at the bottom of the wafer, this electron current causes a voltage drop along "buried layers" and sinker connections. If the resistivity of these regions is not sufficiently low and the current gain of the lateral parasitic NPN-transistor is not small enough, then the "buried layer" becomes negative and the vertical parasitic NPN-transistor is turned on. Depending on the current gain of

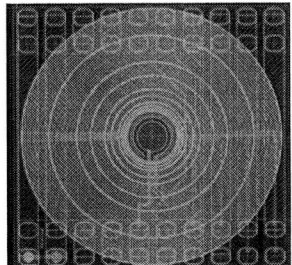
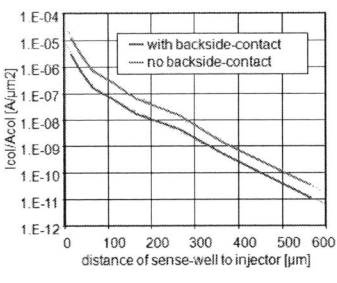

(a)   (b)

Fig. 5(a): Test structure for propagation analysis of injected electron current, if the injector in the middle is forced below ground; (b) collected current density versus distance of sense well, if a current of 1 A is injected into substrate

device	voltage class	RonA [mΩxmm^2]	Vtrigger [V]
N-LDMOS	30 V	33	50
	45 V	50	65
	60 V	70	80
P-LDMOS	30 V	75	55
	40 V	130	75
	60 V	200	85

Tab. 1: List of available power DMOS devices with respective on-resistances and SOA trigger voltages (@E_{OX}=|2.7 MV/cm|, TLP-measurement with 100ns pulses)

this transistor, a malfunction of circuit blocks could be the consequence. To analyze this effect in conjunction with our process/device concept, a special detection structure was created. Large detectors were placed circularly around a single injector (Fig. 5a). The main finding after measurement is the correlation of collected current in logic/flash-areas versus distance to the injector. The existence of a back side contact is of minor significance.

IV. DEVICES

The smart power process presented offers logic-devices (1.5 V class), flash-devices of medium voltage (5 V class) and high voltage devices (up to 60 V class). For the high voltage power devices, the lateral DMOS concept of charge compensation is used. As the underlying logic/flash process is based on STI isolation, the field plates of the power devices are now located on STI instead of field oxide as in previous technologies. Tab.1 gives the list of available power DMOS devices and shows their very low on-resistances [7][8]. As a measure of the robustness of the devices the trigger voltage is added to the table (see also Fig. 7). Due to the higher mobility of electrons N-LDMOS-transistors are preferred compared to P-LDMOS in output stages of BCD technologies. Nevertheless, the EMI requirements have been increased in recent years so drastically that high-side power stages realized with N-LDMOS stages in conjunction with charge pump circuits have become unattractive. Hence, this BCD-technology combines power N-LDMOS for different voltage classes and the corresponding P-LDMOS transistors. Also, high voltage analog MOS devices with variable channel

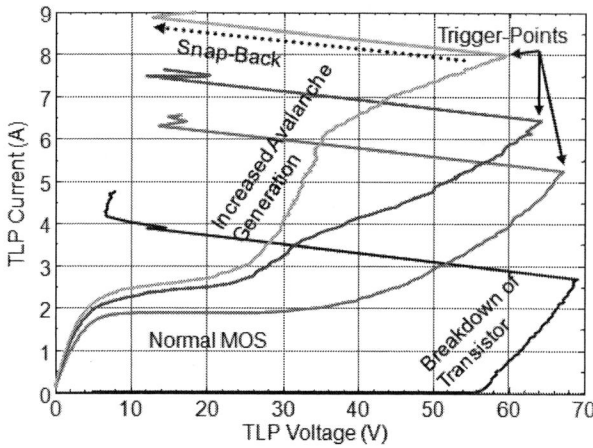

Fig. 6: Typical electrical SOA of integrated lateral 45V N-LDMOS device, measured by TLP for different gate voltages.

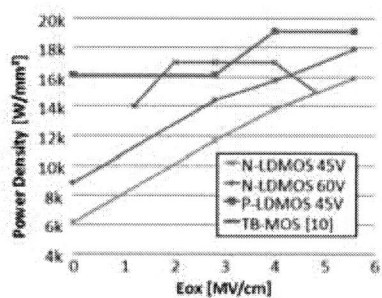

Fig. 7: e-SOA-comparison between lateral DMOS-Transistors and a Trench-Based MOS (TB-MOS) presented by [10]

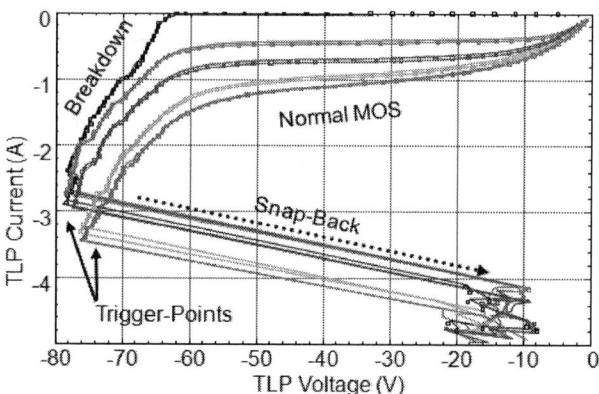

Fig. 8: Typical electrical SOA of integrated power 40V P-LDMOS device, measured by TLP for different gate voltages.

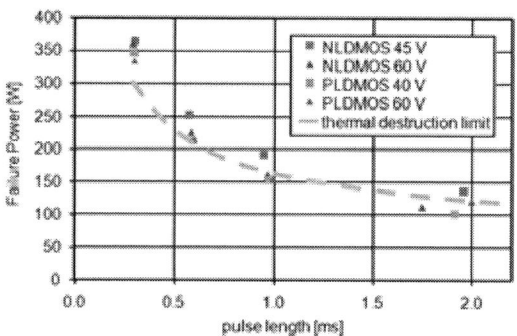

Fig.9: Power dissipation limit of the LDMOS at maximum

length, diodes, bipolar transistors, resistors and capacitors are available for the presented smart power technology platform.

V. ELECTRICAL DEVICE-ROBUSTNESS

All integrated power-transistors have to be robust against high energetic pulses. In some applications these devices may be connected directly to the cable harness within cars. Demanding ESD-requirements have to be fulfilled, not only for automotive applications. The measure of this robustness is the electrical safe-operating-area (e-SOA) shown as example of the N- and P-LDMOS-transistor (Fig. 6 and 8). The e-SOA has been analyzed with TLP pulses having a pulse width of 100ns and a rise-time of 10ns. During these events the gates were biased to substantially higher voltages than in normal operations. The higher the snap-back voltages and currents ("trigger points") the better is the e-SOA. The ESD-performance can be calculated directly from this characteristic. The calculated power density for the 45V-N-LDMOS can be given from 6kW/mm² to 16kW/mm² depending on the gate bias (Fig.8). For the 60V-N-LDMOS the power density is roughly 2kW/mm² higher for all gate voltages compared to the 45V-device. The P-LDMOS exhibits an almost constant power-density independent of the gate voltage at around 17.5kW/mm². For comparison, the lateral and vertical MOS-transistors reported in [10] show much lower power densities, only a special trench based MOS-transistor reaches the power density of 17W/mm² as a

maximum value. To obtain this excellent result for our lateral DMOS-transistors, the parasitic internal bipolar transistors of the MOS devices had to be suppressed. This could be realized by optimization of doping profiles and device layouts [4][5].

It is known that lateral DMOS transistors are strongly affected by hot carrier injection in the silicon oxide interface [9]. Moreover, with the introduction of STI instead of field oxide a potential risk arose for device reliability at extreme operative conditions (i.e. high electric field, high power, high temperature swing). However, it was possible with cautious optimized processes to have a significant low drift of the device parameters over the complete lifetime.

VI. THERMAL DEVICE-ROBUSTNESS

In case of disturbance-pulses of several micro-seconds the electro-thermal-SOA has to be evaluated to predict the robustness of the power stages against electrical overstress (EOS). The pure thermal SOA (thermal destruction limit) indicates when a destructive temperature of 500 °C is reached depending on pulse length and dissipated power (see Fig. 9). This superior thermal-SOA could be reached by using a several μm thick copper metallization layer on top of the chip (see Fig. 3). This so called Power-Copper metallization layer increases the thermal capacitance of the output stage [4]. Furthermore it improves the repetitive clamping robustness in applications like ABS and magnetic valve switching within combustion machines. Here, the electro-thermo-mechanical robustness of the signal metallization is significantly improved

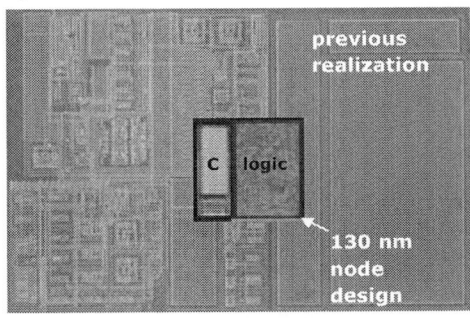

Fig. 10: Area saving by feature shrinks to 130 nm node and replacement of the analog control circuitry by an area-efficient digital based switched capacitor circuitry.

due to the cooling effect of the heat buffering Power-Copper layer [11]. Depending on the device characteristics of the transistor, i.e. its parasitic bipolar and its TCP, the electro-thermal failure behavior can deviate from the pure thermal one. Fig. 9 also shows the electro-thermal-SOA of the presented 45V N-LDMOS and 40V P-LDMOS devices. Except for long pulses, where the thermal boundary conditions of the package and introduced uncertainties become important, the devices perform better than thermally expected. In combination with an electro-thermal FEM simulator the optimum layout can be found to ensure device integrity even for the most critical pulses in application [15].

VII. SYSTEM BENEFITS

New ULSI technologies such as the BCD-process presented will have a major impact on future IC-designs and circuit concepts. The huge packing density of the logic blocks will allow new features without remarkable cost adders. Also, analog circuits may be replaced by logic blocks with an extreme area reduction. To enable the efficient replacement of analog control circuits by logic ones, lateral metal-metal capacitances are included. Fig.10 shows the shrink capability for a switched capacitor circuitry. The combination of microcontroller, flash-memory and power devices is the key for system integration with improved reliability and flexibility. Beside the pure silicon costs, the expenses for packaging and test are the main driver for the overall system costs. Higher degrees of integration and digitalization will also help to reduce these costs further.

VIII. RESULTS

The ULSI-BCD automotive power technology platform presented allows for efficient design shrinks. For power MOS device design, an optimum balance of on-resistance, breakdown voltage, HCS- and SOA-behavior was achieved. The devices show very competitive on-resistances. The use of STI instead of field oxides shows no draw-back in terms of increased HCS drift effects. The SOA of the devices is optimized to ensure sufficient headroom for ESD protection. Moreover, the electro-thermal behavior of the power devices is understood. Thus, predictive dimensioning can prevent EOS for critical pulses. Due to the copper dual damascene metallization in combination with a thick copper heat buffer layer, the mechanical robustness of the metal stack in repetitive pulsing is increased by more than one order of

magnitude in lifetime compared to conventional aluminum metallization.

ACKNOWLEDGEMENT

Many thanks to all contributors to the process development, especially Jörn Herrfurth, Marco Müller etc. and our design community for fruitful discussions. This work was sponsored by the projects MEDEA, SPOT2 (2T205) and GOLD.

REFERENCES

[1] Stecher, M.; Jensen, N.; Denison, M.; Rudolf, R.; Strzalkoswi, B.; Muenzer, M.N.; Lorenz, L.; "Key technologies for system-integration in the automotive and Industrial Applications", Power Electronics, IEEE Transactions on; Volume 20, Issue 3, May 2005 Page(s):537 - 549

[2] J. Busch, M. Denison, G. Groos, H. Gruber, R. Hofmann, N. Jensen, A. Meiser, P. Nelle, R. Weeger, W. Schwetlick, M. Stecher: "Key Features of a Smart Power Technology for Automotive Applications"; International Conference on Integrated Power Systems, 2002

[3] A. Podgaynaya, R. Rudolf, B. Elattari, D. Pogany, E. Gornik, M. Stecher, M. Strasser, "Single pulse energy capability and failure modes of n- and p-channel LDMOS with thick copper metallization", Microelectronics Reliability, 2010.

[4] A. Podgaynaya, D. Pogany, E. Gornik, and M. Stecher, "Enhancement of the Electrical Safe Operating Area of Integrated DMOS Transistors with Respect to High-Energy Short Duration Pulses", T-ED, 2010.

[5] A. Podgaynaya, D. Pogany, E. Gornik, and M. Stecher, "Improvement of the electrical safe operating area of a DMOS transistor during ESD events," in Proc. IRPS, Montreal, QC, Canada, 2009, pp. 437–442.

[6] A. Podgaynaya, R. Rudolf, D. Pogany, E. Gornik and M. Stecher, "Experimental and Theoretical Analysis of the Electrical SOA of Rugged p-Channel LDMOS", IEEE EDL, 2010.

[7] K. Shirai et al., "Ultra-low On-Resistance LDMOS Implementation in 0.13μm CD and BiCD Process Technologies for Analog Power IC's", ISPSD, 2009

[8] H. Yang et al., "Low-Leakage SMARTMOS 10W Technology At 0.13μm Node with Optimized Analog, Power and Logic Devices for SOC Design", Proc. VLSI-TSA, 2008

[9] B. Murari, F. Bertotti, G.A. Vignola, "Smart Power IC's - Technologies and Applications", Springer, 2002

[10] Peter Moens, Jaume Roig, Bart Desoete, Filip Bauwens, Angela Rinaldi, Piet Vanmeerbeek, Guillaume Jenicot, and Marnix Tack, "Safe Operating Area Considerations for Integrated Trench-Based Power Devices", IEEE T-DMR, vol. 9, no. 4, December 2009

[11] Tobias Smorodin, Peter Nelle, Jorg Busch, Jurgen Wilde, Michael Glavanovics, Matthias Stecher; "Investigation and improvement of DMOS switches under fast electro-thermal cycle stress", Solid-State Electronics, Volume 52, Issue 9; September 2008, Pages 1353-1358

[12] M. Stecher, P. Nelle, J. Busch, P. Alpern, „Interconnect Technologies for SmartPower Integrated Circuits in the area of Automotive Power Applications", to be published on IITC, 2011

[13] Y. Cao, U. Glaser, S. Frei and M. Stecher, "A Failure Levels Study of Non-Snapback ESD Devices for Automotive Applications," in IEEE International Reliability Physics Symposium (IRPS), pp. 1-8, Anaheim, USA, May 2010.

[14] Y. Cao, U. Glaser, A. Podgaynaya, J. Willemen, S. Frei and M. Stecher, "Impact of Voltage Overshoots on ESD Protection Effectiveness for High Voltage Applications," in 4th Annual International Electrostatic Discharge Workshop (IEW), Tutzing, Germany, May 2010.

[15] Pfost, M., Lachner, R., Li, H.: Simulation of Self-Heating in Advanced High-Speed SiGe Bipolar Circuits Using the Temperature Simulator TESI, Topical Meeting on Silicon Monolithic Integrated Circuits in RF Systems 2004, 2004

Proceedings of the 23rd International Symposium on Power Semiconductor Devices & IC's
May 23-26, 2011 San Diego, CA

Implementation of Fully Isolated Low Vgs nLDMOS with Low Specific On-resistance

Choul-Joo Ko, Cheol-Ho Cho, Min-Seok Kim, Hyung-Gyun Jung, Hee-Bae Lee, Yong-Jun Lee, Min-Woo Kim,
Sung-Mo Gu, Sun-Kyung Bang, Han-Geon Kim, Sun-Kyoung Kang, Kwang-Dong Yoo and Lou Hutter

Analog Foundry Business Unit
Dongbu Hitek, 222-1, Dodang-dong, Wonmi-Gu, Bucheon, Gyeonggi-Do, 420-712 Korea
Tel: +82-32-680-4133 Fax: +82-32-683-8105 Email:chouljoo.ko@dongbu.com

Abstract — **In this paper, we present a new isolated Low Vgs NLDMOS in 0.35um BCDMOS process. The proposed LDMOS is fully isolated from substrate and has very lower Rsp(specific on-resistance) than other competitors. This device can apply a negative bias to drain and it can be used in AMOLED application. The proposed LDMOS devices in 30-40V ranges have the lowest Rsp with other competitors in 0.13-0.35um BCDMOS technologies. And the Rsp of the proposed LDMOS in 40V range is 46.3% lower than Low Vgs LDMOS last reported. And the isolation efficiency of the proposed LDMOS has very good performance. Furthermore, a logic CMOS and all the other components are compatible in the proposed process.**

I. INTRODUCTION

Recently, BCDMOS(Bipolar–CMOS–LDMOS) process is widely used in variety of areas such like LED Driver, Panel Bias IC, Switching Regulator, Battery IC, Audio Amplifier, Motor Drivers, large displays (TV and monitor), small displays (handheld and mobile devices). Generally LDMOS power devices have occupied about 60% in a chip, so device engineers are trying to reduce Rsp and decrease chip size to increase die counts. We previously presented a paper on the Low Vgs LDMOS[1]. The gate voltage level of this device is same with that of CMOS so that Function Blocks(Level Shifter) are not needed. And the drift region was engineered to achieve a lower Rsp. But this device has following disadvantage because drain is shorted to n-buried layer. For outputs above supply, a vertical PNP at high-side is activated. It can be suppressed by employing n+ sinker nearby drain. For outputs below ground, a lateral NPN is activated and electrons are widely injected into the p-substrate, thus electron guard-ring is needed in this case[2]. The isolated LDMOS, which can be isolated from substrate, was reported in the past[3][4]. But, there were no report about LDMOS which can apply negative bias to drain and source/body by using

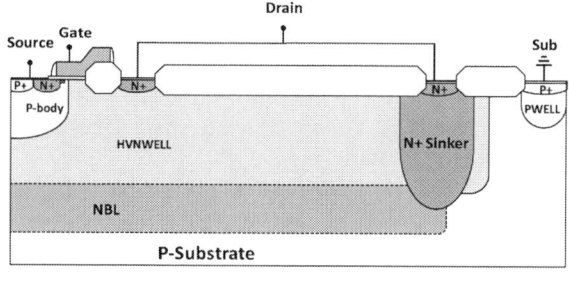

(a)

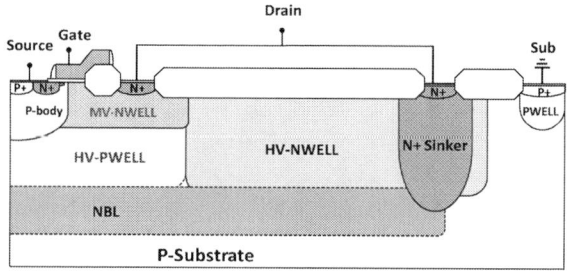

(b)

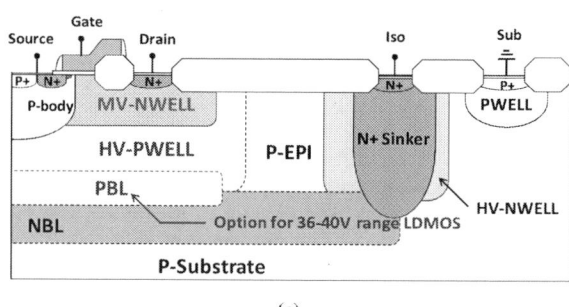

(c)

Fig. 1. Cross-sectional view for (a) conventional LDMOS high-side, (b) Low Vgs LDMOS high-side, (c) the proposed isolated Low Vgs LDMOS.

junction isolation technology. In this paper, we present a paper on the isolated Low Vgs LDMOS which can be fully isolated

978-1-4244-8425-6/11 $26.00 © 2011 IEEE

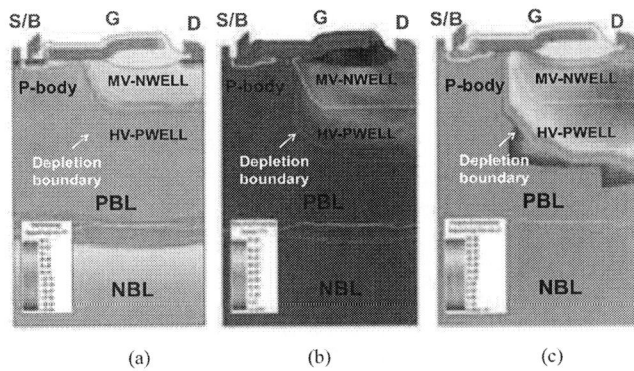

(a)   (b)   (c)

Fig. 2. Simulated device structure for the proposed 36V LDMOS (a) net doping profile, (b) potential distribution, (c) impact ionization.

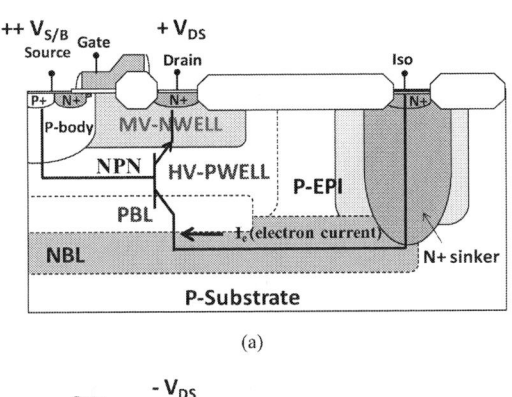

(a)

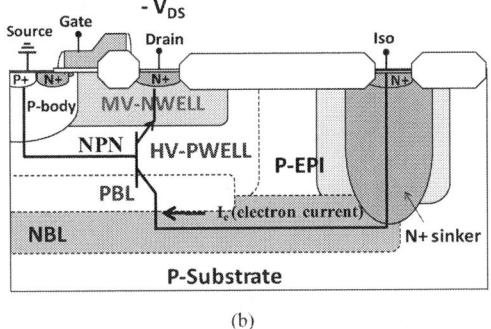

(b)

Fig. 3. Schematic cross-section of the proposed isolated Low Vgs LDMOS (a) high-side, (b) low-side.

from p-substrate and apply negative bias to drain and source, gate during switching operation. And it has very lower Rsp than other competitors.

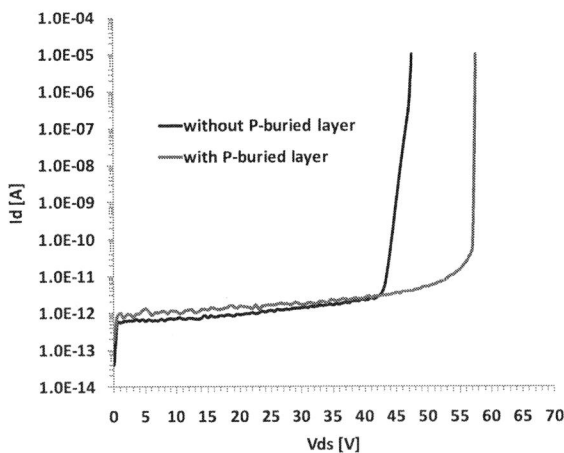

Fig. 4. The breakdown curve of the proposed LDMOS with p-buried layer or not.

II. DEVICE STRUCTURE

Fig. 1 shows the cross-sectional view of a conventional LDMOS high-side(Fig. 1a), the Low Vgs LDMOS high-side(Fig. 1b) and the proposed isolated Low Vgs LDMOS (Fig. 1c). As shown in figures, drain of the conventional LDMOS and the Low Vgs LDMOS are connected to n-buried layer. It is a major reason of electrons injection into p-substrate during a current re-circulation. In case of the isolated Low Vgs LDMOS, P-buried layer is under MV-NWELL. P-buried layer is applied only for 36-40V rated isolated Low Vgs LDMOS. Discriminatively, the 12V rated isolated Low Vgs LDMOS can be used for AMOLED application and isolation terminal of this device not be connected to source.

III. DEVICE OPERATION

Fig. 2 shows the simulated device net doping concentration, potential distribution and impact ionization rates at breakdown. The space between the potential lines is 1V. HV-PWELL releases e-field nearby gate edge and increases BVdss(breakdown voltage between drain and source) of the proposed LDMOS. And HV-PWELL and P-buried layer prevent punch-trough breakdown between MV-NWELL and n-buried layer. Therefore, the proposed structures have enhanced Rsp and ensure high breakdown voltage. Fig. 3 shows schematic cross-

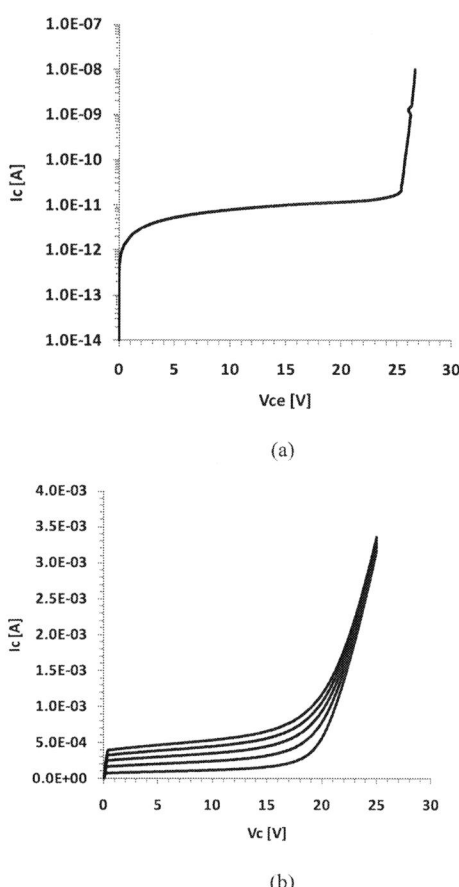

Fig. 5. Experimental I-V characteristics for the proposed LDMOS (a) I_{DS}-V_{DS} curve and (b) breakdown curve.

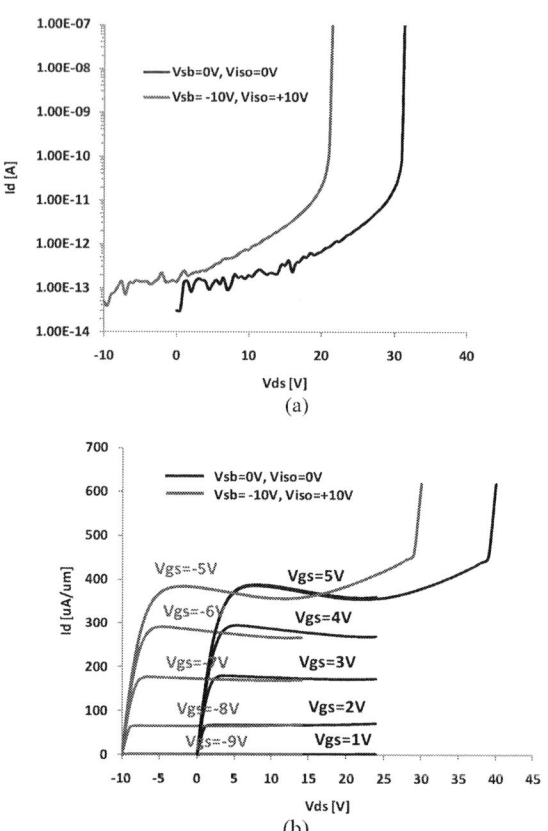

Fig. 6. I-V characteristics when +10V is applied to Iso and negative bias is applied to drain (a) BVdss characteristic, (b) I_D-V_{DS} characteristic.

section of the proposed isolated Low Vgs LDMOS when current recirculation occurs. As shown in Fig. 3(a), when $V_{S/B}$ is higher than V_{DS}, minority carriers are hard to be injected into p-substrate because electron currents flow into isolation during parasitic NPN operation. Also, minority carriers are hardly injected into p-substrate when negative bias is applied to drain in Fig. 3(b).

IV.   SILICON RESULTS

A. Electrical Characteristics

Fig. 4 shows the breakdown voltage curve of the proposed LDMOS with p-buried layer or not. We extended the half-pitch of the proposed LDMOS to check the breakdown voltage

between MV-WELL and n-buried layer. The maximum breakdown voltage between MV-NWELL and n-buried layer is 57.5V. And the breakdown voltage of 47.5V can be achieved without p-buried layer so we made the isolated 30V Low Vgs LDMOS by simply employing HV-PWELL and MV-NWELL. Fig. 5 shows that BVceo and I_C-V_C curve of internal parasitic NPN BJT in the isolated Low Vgs LDMOS without P-buried layer. The collector of parasitic NPN is NBL. As shown in figure, BVceo is above 25V so that 12V isolated Low Vgs LDMOS for AMOLED application can operate well, even when -6V is applied to source and +5V is applied to isolation. Fig. 6 shows the I-V curve of of the proposed 24V LDMOS when +10V is applied to Isolation and negative bias is applied to drain.

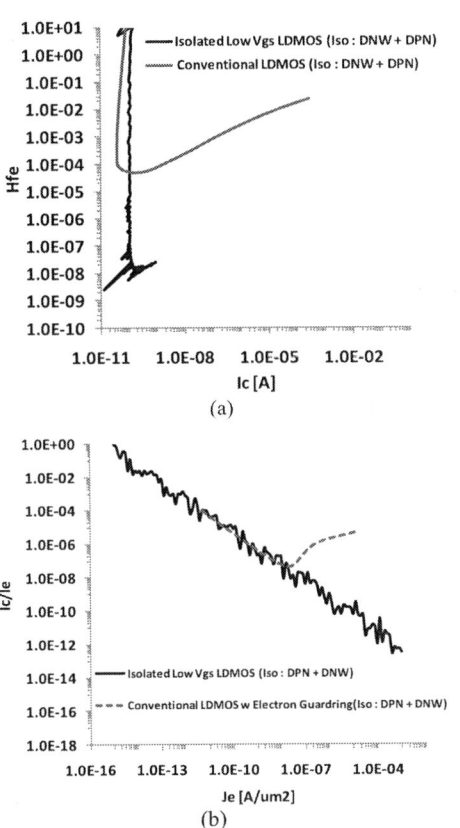

(a)

(b)

Fig. 7. The isolation efficiency of the proposed LDMOS isolation (a) Hfe characteristic when $V_{S/B}$ is higher than V_{DS}, (b) I_C/I_E characteristic when V_{DS} is lower than p-sub.

B. Isolation efficiency

Fig. 7 shows the isolation efficiency of the proposed LDMOS and conventional one. In this case, source of the isolated Low Vgs LDMOS is connected to isolation. And Fig. 8 show the isolation efficiency of the 12V isolated LDMOS for AMOLED application. The isolation efficiency of the proposed LDMOS shows very good performance.

C. Benchmark

Fig. 9 shows plots the LDMOS Rsp vs. BVdss for the proposed LDMOS and 0.15-0.25um BCDMOS technologies of other competitors. The Rps of the 40V LDMOS is lower by 46.3% than Low Vgs LDMOS at same breakdown voltage. And this plot shows that the proposed 30-40V rated LDMOS have best-in-class Rsp.

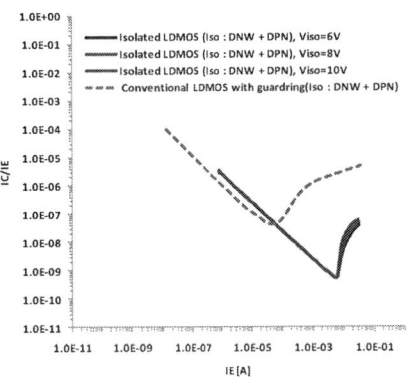

Fig. 8. The isolation efficiency of the 12V isolated Low Vgs LDMOS for AMOLED application

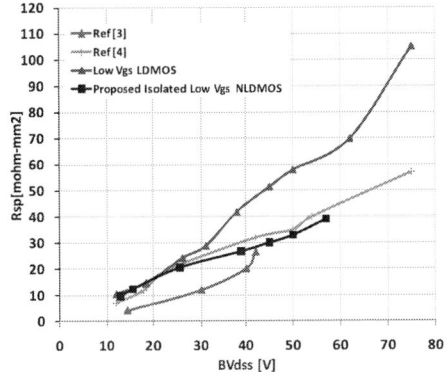

Fig. 9. The benchmark of the proposed LDMOS Rsp and BVdss for the 0.15-0.25um BCDMOS

V. SUMMARY

A new isolated Low Vgs NLDMOS in 0.35um BCDMOS process was developed. The proposed LDMOS is fully isolated from substrate and can apply a negative bias for AMOLED application. The Rsp of the 40V proposed LDMOS is 46.3% lower than Low Vgs LDMOS last reported. The isolation efficiency of the proposed LDMOS has very good performance.

REFERENCES

[1] C. J. Ko, "A New 8V – 60V rated Low Vgs NLDMOS Structure with Enhanced Specific on-Resistance" Proc. of ISPSD, p.245-248, 2010.

[2] B. Smith "Peripheral Motor Drive PIC Concerns for Integrated LDMOS Technologies", ISPSD, p.1-8, 2002.

[3] S. Pendharkar "7 to 30V state-of-art power device implementation in 0.25μm LBC7 BiCMOS-DMOS process technology" Proc. ofISPSD, p.419-422, 2004.

[4] R. Zhu, "Implementation of high-side, high-voltage RESURF LDMOS in a sub-half micron smart power technology", ISPSD, p. 403-406, 2001.

Proceedings of the 23rd International Symposium on Power Semiconductor Devices & IC's
May 23-26, 2011 San Diego, CA

Wide-Voltage SOI-BiCDMOS Technology for High-Temperature Automotive Applications

Hidemoto Tomita, Hiroomi Eguchi, Shinya Kijima, Norihiro Honda, Tetsuya Yamada,
Hideo Yamawaki, Hirofumi Aoki and Kimimori Hamada

Electronics Development Division 3
Toyota Motor Corporation
Toyota, Japan
e-mail: tomita@hidemoto.tec.toyota.co.jp

Abstract—**This paper describes a new wide-voltage SOI-BiCDMOS technology for high-temperature automotive applications. This technology is capable of integrating 35V, 60V, and 80V Nch and Pch LDMOS, 35V BJT, and 6V CMOS devices on a single chip. The devices are completely isolated dielectrically using both deep trench isolation (DTI) and a buried oxide (BOX) layer in a silicon-on-insulator (SOI) wafer for stable operation at high temperatures up to 175°C. The devices were developed using a 0.35μm process. In particular, the LDMOS devices have achieved competitive levels of low Ron*A and good SOA.**

I. INTRODUCTION

In recent years, the development of automotive technology has become focused on responding to the safety and environmental requirements of modern society while also improving comfort and usability. To accomplish these aims, vehicles are being installed with more and more electronic devices. Many of these include an electronic control unit (ECU), based on which inputs from sensors and the like are processed for driving actuators. Communication between ECUs is performed via in-vehicle local area networks (LANs). These ECUs include a circuit for processing inputs from sensors and switches, an AD conversion circuit, a microcomputer, a power source, an output processing circuit, power devices, and a circuit for communicating with other ECUs. These circuits are used over wide voltage ranges to enable compatibility with various onboard systems, including 5V and 12V input/output circuits that operate using power supplied from the battery. In a recent example, 80V blocking voltage devices have been used for level shift applications to boost the power source voltage in an electric power steering (EPS) system. Furthermore, vehicle developers are facing stronger demands to use automotive ASICs in high temperature environments. These requirements include directly installing ASICs in the engine or transmission, achieving high performance smart actuators that integrate actuators or motors with signal processing or load driving circuits, and adapting ASICs to systems with simpler or no cooling devices. At the same time,

ASICs must also be adapted to satisfy conventional requirements, such as increased integration of signal processing circuits, lower Ron*A of LDMOS devices, greater sensitivity of analog circuits, longer term stable operation under harsh environments (temperature, surge voltage, and the like), lower cost, and so on. BiCDMOS technology is one means of resolving these issues and several proposals for this technology have already been reported [1]-[6]. This paper describes the development of a cost-competitive SOI-BiCDMOS technology for automotive application ASICs that can operate at high temperatures up to 175°C and that support a wide range of voltages.

II. PROCESS TECHNOLOGY

This is the third generation of a BiCDMOS technology adopted by Toyota Motor Corporation. Fig. 1 shows a cross-sectional TEM image of a 60V Nch LDMOS with power metal that clearly depicts the outline of this process. The adopted 0.35μm process technology has a gate oxide layer thickness of 15.5nm and the maximum rated voltage of the gate electrode is set to 6V. Since the contact structure uses a W plug process, the device area can be reduced by making the contact hole diameter smaller and adopting a stacked

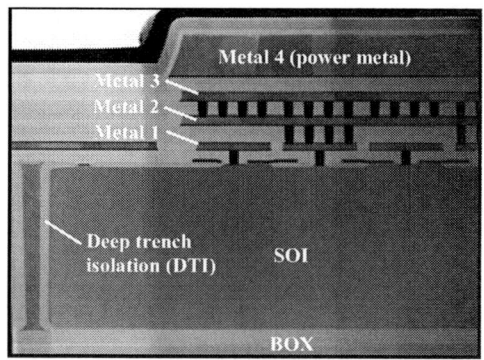

Figure 1. Cross-sectional TEM image of 0.35μm SOI-BiCDMOS (60V Nch LDMOS with power metal).

978-1-4244-8425-6/11 $26.00 © 2011 IEEE

structure. The wiring structure has a maximum of four layers: three thin metal layers and a single thick metal layer as power metal. Isolation is achieved by the same deep trench isolation (DTI) and buried oxide (BOX) layer in a silicon-on-insulator (SOI) structure as adopted in the previous generation process. This process uses a bonded 200mm SOI wafer, which has an implanted N+ layer on the BOX layer. The thickness of the BOX and N-type SOI layers is set to 1μm and 12μm, respectively. To ensure a high isolation blocking voltage, the trench is completely filled with poly-Si extremely carefully after a thermal oxidation process is applied to the trench sidewall. With this structure, DTI with a width of 2μm achieves sufficient isolation to allow application to multiple power sources and high voltages. In addition, it also helps to prevent device breakdown due to hard-to-predict parasitic current induction generated by the driving of inductive loads and the like. At the same time, since there is no PN junction in the isolation region, leak current generated at high temperatures can be reduced, thereby allowing operation at high temperatures up to 175°C.

III. DEVICE TECHNOLOGY

A. Outline of Devices

Table I lists the devices (CMOS, BJT, LDMOS, Zener, resistors, and capacitors) supported by this technology and their relevant electrical characteristics. Fig. 2 shows the overall process flow. The standard device structure (CMOS + BJT) of this BiCDMOS technology is fabricated by adding a trench process and a BJT process to the CMOS process. The BJT process can be produced by the addition of a single dedicated mask (Base P-). The 6V CMOS adopts the minimum gate length of 0.6μm. On-current performance has been improved by approximately 2.5 times compared to the previous generation, enabling application of this technology to high-speed and large-scale logic devices. In addition, this technology can also be used for analog circuit solutions. Its adaptability to BJT and passive devices including optional devices such as highly precise MIM capacitors and Poly-resistors makes it suitable for the design requirements of various analog circuits. Six types of LDMOS devices are available, which can be set optionally for each product IC. Fig. 3 shows the cross-sectional structure of a 35V Pch LDMOS. This 35V Pch LDMOS is fabricated by adding a single dedicated drift mask (35V drift P) to the two option masks (HVNW/HVPW) common to the LDMOS devices. As shown in Fig. 3, the three types of blocking voltage Pch LDMOS devices including the 35V Pch LDMOS have a conventional device structure, and the drift layer concentration can be optimized for each blocking voltage level by the adoption of a dedicated drift mask.

B. LDMOS

Fig. 4 shows the cross-sectional structure of a 35V Nch LDMOS. In the same way as the Pch LDMOS devices, the three types of blocking voltage Nch LDMOS devices including the 35V Nch LDMOS use a dedicated drift mask for each blocking voltage level. In terms of the difference between the Nch and Pch LDMOS devices, the built-in drift layer is formed so that it is an N-type (phosphorus) upward

but a P-type (boron) downward in the vertical direction by the adoption of multi-stage ion implantation using high-energy ion implantation. This gives the device a reduced surface electric field (RESURF) type structure to improve electrical properties, including achieving lower Ron*A.

TABLE I.   ELECTRICAL CHARACTERISTICS OF 0.35μM SOI-BiCDMOS.

Device	Attributes	Typical values
6V NMOS/PMOS	Vth, BVdss	0.80V, 9.6V
35V NPN/PNP	h_{FE}, BVceo	74.0, 65.0V
35V Nch LDMOS	BVdss, Ron*A	47.0V, 39m-ohm*mm²
60V Nch LDMOS	BVdss, Ron*A	75.0V, 80m-ohm*mm²
80V Nch LDMOS	BVdss, Ron*A	105.0V, 235m-ohm*mm²
35V Pch LDMOS	BVdss, Ron*A	48.0V, 110m-ohm*mm²
60V Pch LDMOS	BVdss, Ron*A	78.0V, 175m-ohm*mm²
80V Pch LDMOS	BVdss, Ron*A	100.0V, 430m-ohm*mm²
Zener diodes	Vz	5.4V, 12.2V
Diff-resistors	Sheet res.	2100ohm/sq.
Poly-resistors	Sheet res.	11.5, 1000ohm/sq.
LV capacitor	Cap./Area, BV	2140pF/mm², 16.0V
HV capacitor	Cap./Area, BV	111pF/mm², 180.0V
MIM capacitor	Cap./Area, BV	740pF/mm², 24.0V

CMOS	BJT	Nch LDMOS	Pch LDMOS
SDG		Trench	
LVNW/LVPW		HVNW/HVPW	
		35V drift N	
		60V drift N	
		80V drift N	
			35V drift P
			60V drift P
			80V drift P
	Base P-		
Gate			
LDD&SDN+/P+			
Metal 1 to 3		Metal 4 (power metal)	

Figure 2.   Overview of process flow for 0.35μm SOI-BiCDMOS.

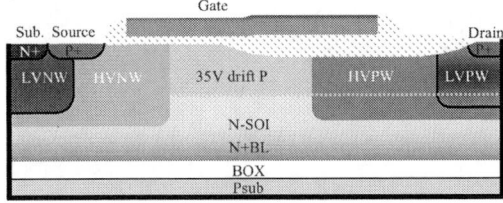

Figure 3.   Schematic cross-sectional view of 35V Pch LDMOS.

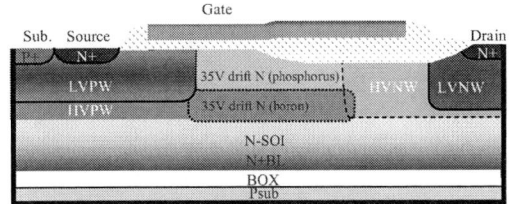

Figure 4.   Schematic cross-sectional view of 35V Nch LDMOS.

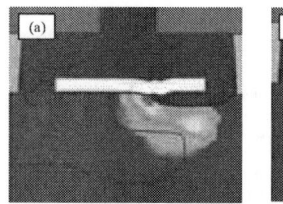 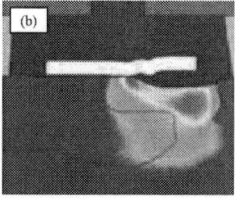

Figure 5. Simulated device structure and distribution of impact ionization in 35V Nch LDMOS with recessed LOCOS (a) and modified LOCOS (b).

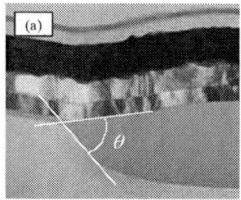 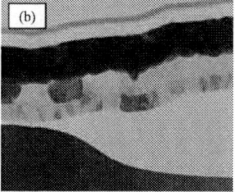

Figure 6. Cross-sectional TEM image of recessed LOCOS (a) and modified LOCOS (b). θ is defined as the angle at the LOCOS edge.

To further improve electrical properties, the shape of the local oxidation of silicon (LOCOS) was modified by omitting the recess process step in the LOCOS formation. Fig. 5a shows the distribution of impact ionization in the off-state in a 35V Nch LDMOS with a conventional recessed LOCOS. At this stage, the device shown in Fig. 5a has the following electrical characteristics: BVdss=34.9V and Ron*A=39.6m-ohm*mm^2. Its distribution indicates that the impact ionization has concentrated at one point of the Si bulk at the LOCOS edge. Based on the concept that preventing this single-point generation of impact ionization would further improve the BVdss of the device, a modified LOCOS was developed with a smaller angle at the LOCOS edge than in a recessed LOCOS. Figs. 6a and 6b show a cross-sectional TEM image of a conventional recessed and a modified LOCOS, respectively (in Fig. 6a, θ is defined as the angle at the LOCOS edge). Fig. 5b shows the distribution of impact ionization in a 35V Nch LDMOS with a modified LOCOS. The generation of impact ionization is divided and shifted from the Si bulk at the LOCOS edge to the Si bulk at the LOCOS bottom. As a result, the device shown in Fig. 5b has the following improved electrical characteristics: BVdss=49.4V and Ron*A=41.9m-ohm*mm^2. Fig. 7 shows the simulated and measured data for the trade-off characteristics between BVdss and Ron*A including the simulation data described above. Four types of simulations were carried out: three using a modified LOCOS (with the angle θ set to a typical, larger, and smaller value) and one with a conventional recessed LOCOS. It was confirmed that electrical characteristics were clearly changed by the LOCOS shape. A sample was fabricated with a drift concentration optimized in accordance with the simulation. Measurements of this sample confirmed an approximate 10% improvement in electrical characteristics with the modified LOCOS compared with the recessed LOCOS. Fig. 8 shows the Vd-Id characteristics of the developed 80V Nch LDMOS at 175°C. It was confirmed that none of the six types of LDMOS devices including an 80V Nch LDMOS exhibit snapback over the whole operation range and that all have excellent safe operating area (SOA) characteristics. In addition, it was also confirmed that these LDMOS devices satisfy the long-

term reliability requirements for automotive applications including continuous high temperature operation and hot carrier reliability. Fig. 9 compares the Ron*A data obtained in this development with the data in the literature ([2] to [6]) for recent 0.13μm to 0.8μm processes. It was confirmed that the developed LDMOS devices have strong competitiveness compared to other devices in terms of low Ron*A, especially after considering the SOA characteristics of the LDMOS devices as described above.

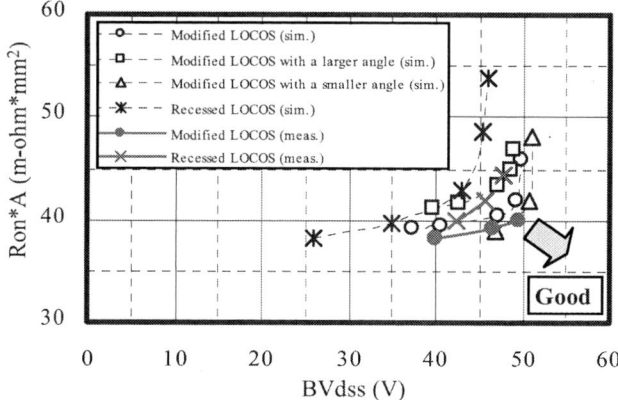

Figure 7. Simulated and measured data of trade-off characteristics between BVdss and Ron*A in 35V Nch LDMOS.

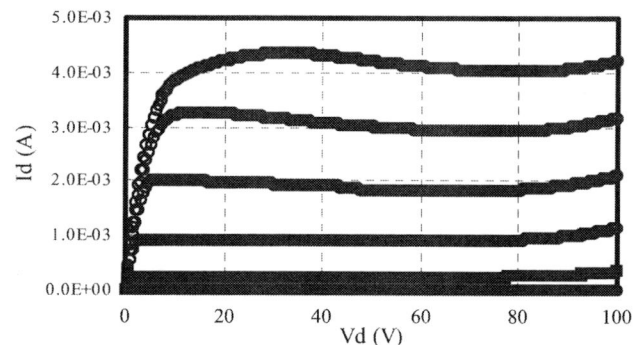

Figure 8. Vd-Id characteristics of 80V Nch LDMOS at 175°C (Vgs=0V to 6V in 1V steps).

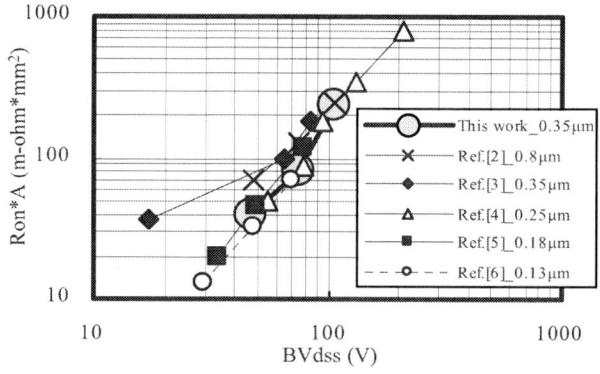

Figure 9. Ron*A comparison of Nch LDMOS devices between this work and recent 0.13μm to 0.8μm processes.

C. BJT and CMOS

Fig. 10 shows the temperature dependence of the Ic-h_{FE} characteristics of a 35V NPN. It was confirmed that h_{FE} does not fluctuate in the low collector current region at high temperatures up to 175°C. This is because the same DTI-SOI structure as the previous generation is used, which helps to prevent parasitic leak current. Fig. 11 shows the temperature dependence of the Vg-Id characteristics of a 6V NMOS. In the same way as a 6V PMOS, the structure of the 6V NMOS is carefully set to achieve a Vth of 0.5V or more at 175°C to secure the required noise margin for automotive applications. In addition, it was confirmed that these devices and process completely satisfy the reliability requirements in terms of time dependent dielectric breakdown (TDDB), positive/negative bias temperature instability (PBTI/NBTI), and electro migration (EM).

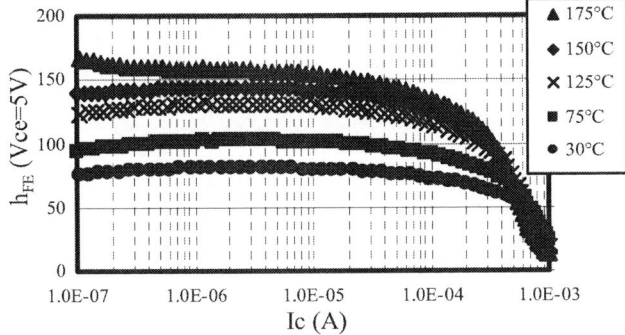

Figure 10. Temperature dependence of Ic-hFE characteristics of 35V NPN.

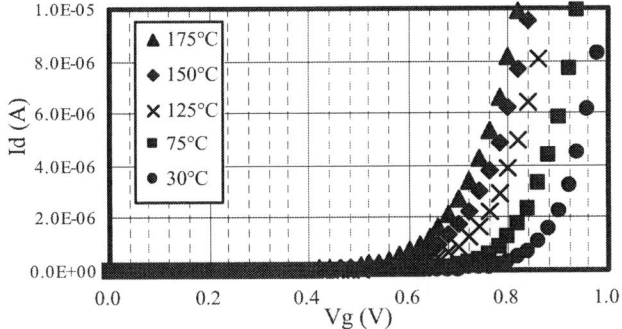

Figure 11. Temperature dependence of Vg-Id characteristics of 6V NMOS (Vd=5V).

IV. APPLICATIONS

Fig. 12 shows hybrid vehicle (HV) control ICs developed using the previous generation technology (Fig. 12a) and the new generation technology (Fig. 12b). The two ICs have virtually the same function. The newly developed IC takes advantage of the characteristics of this technology to increase output current, operating temperature, and sensitivity. At the same time, while improving performance, Fig. 13 shows that substantial reductions in chip size are also achieved compared to the previous generation technology. In specific terms, chip area is reduced by 85% for analog devices, 80% for digital devices, 54% for power devices and 52% at the chip periphery. As a result, the developed IC has a 65% smaller chip area than the previous generation IC.

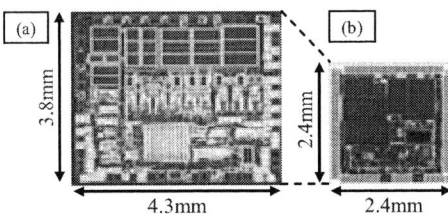

Figure 12. Comparison of product design examples of IC using the previous generation technology (a) and IC using the new generation technology (b).

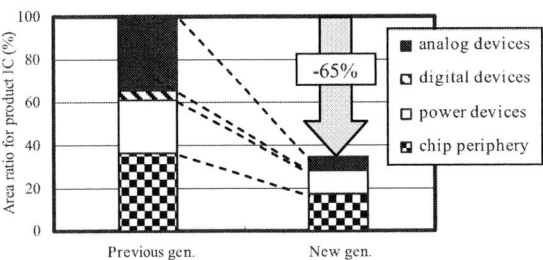

Figure 13. Comparison of chip area using the previous generation technology and the new generation technology.

Furthermore, it was confirmed that this smaller chip also satisfies the reliability requirements for automotive ASICs in terms of long-term durability, electrostatic discharge, noise resistance, and the like.

V. CONCLUSION

A new wide-voltage SOI-BiCDMOS technology has been developed for high-temperature automotive applications. This technology enables the integration of 35V, 60V, and 80V Nch and Pch LDMOS, 35V BJT, 6V CMOS, and various passive devices on a single chip. It was confirmed that these devices can be operated stably at high temperatures up to 175°C due to the adoption of a completely dielectrically isolated structure consisting of DTI and a SOI wafer. In particular, LDMOS devices with three types of blocking voltage were realized for a variety of automotive applications. These LDMOS devices have strong cost-competitiveness in terms of low Ron*A due to the adoption of a modified LOCOS shape and an excellent SOA up to 175°C.

REFERENCES

[1] K. Hamada et al., "A 60V BiCDMOS device technology for automotive applications," Conference Record of the 1995 IEEE Industry Applications Society 30th IAS Annual Meeting, 1995, pp. 986–990.

[2] F. Kawai et al., "Multi-voltage SOI-BiCDMOS for 14V & 42V automotive applications," Proc. of the ISPSD 2004, pp. 165–168.

[3] P. Moens et al., "I3T80: A 0.35μm based system-on-chip technology for 42V battery automotive applications," Proc. of the ISPSD 2002, pp. 225–228.

[4] T. Nitta et al., "Wide voltage power device implementation in 0.25μm SOI BiC-DMOS," Proc. of the ISPSD 2006, pp. 341–344.

[5] D. Riccardi et al., "BCD8 form 7V to 70V: a new 0.18μm technology platform to address the evolution of applications towards smart power ICs with high logic contents," Proc. of the ISPSD 2007, pp. 73–76.

[6] K. Shirai et al., "Ultra-low on-resistance LDMOS implementation in 0.13μm CD and BiCD process technologies for analog power ICs," Proc. of the ISPSD 2009, pp. 77–79.

978-1-4244-8425-6/11 $26.00 © 2011 IEEE

Proceedings of the 23rd International Symposium on Power Semiconductor Devices & IC's
May 23-26, 2011 San Diego, CA

New Low-Resistance and Compact MOSFETs for Analog Switch ICs with V-groove Dielectric Isolation

Kenji Hara and Junichi Sakano

Hitachi Research Laboratory
Hitachi Ltd.
7-1-1 Omika-cho Hitachi-shi Ibaraki-ken 319-1292, Japan
E-mail: kenji.hara.vf@hitachi.com

Hironobu Honda, Junichi Aizawa and Taiga Arai

Power & Industrial Systems Division
Power Systems Company, Hitachi Ltd.
Hitachi-shi, Ibaraki-ken, Japan

Abstract— **A low-resistance and compact MOSFET for analog switch ICs with Dielectric Isolation (DI) process technology is proposed. To obtain a high current density, we have developed new MOSFET with internal prominence, which reduce the drift resistance of devices with a high breakdown voltage. New N-ch and P-ch compact MOSFETs for level shifters have also been developed that can control saturation current with a low electric field under the gate region by using a junction field effect transistor structure for higher hot carrier reliability. The areas of these MOSFETs can be shrunk about 40% in 220-V devices.**

I. INTRODUCTION

High-voltage analog switch ICs are widely used in medical ultrasound imaging systems and printed circuit board tester applications [1-2]. Figure 1 shows a block diagram of a conventional high-voltage analog switch IC [3]. To reduce system size and cost, low on-resistance and high saturation current have become more important for main switch MOSFETs with thick gate oxide. Moreover, MOSFETs for level shifters that also have thick gate oxide should have to reduce the saturation current and electric field for hot carrier reliability with a small device area.

To satisfy these requirements, we propose a new concept of low-resistance and compact MOSFET for analog switch ICs with dielectric isolation. To obtain a high current density, we developed a MOSFET with internal prominences, which reduces the drift resistance of devices with a high breakdown voltage. These prominences are formed without any additional processes in a conventional V-groove DI process by using isotropic etching [4-5]. New N-ch and P-ch compact MOSFETs for level shifters have also been developed that can control saturation current with a low electric field under the gate region by using a junction field effect transistor (JFET) structure for higher hot carrier reliability.

II. CONCEPT OF COMPACT MOSFETs

A. MOSFET for Main Switches

Figure 2 shows cross sections of the conventional and proposed MOSFETs for the main switches. The conventional structure has channel regions that are arranged like a mesh. The proposed structure has channel regions that are arranged like stripes and has internal prominences, which are formed between the channel regions. These prominences reduce drift resistance in devices with a high breakdown voltage. Moreover, they improve the performance of parasitic diodes and reduce the impedance of the main switch circuit. They can be formed with narrow V-groove patterns.

Figure 3 shows the process for fabricating the V-grooves and internal prominences. The L1 indicates the resist material space for V-groove etching for the device isolation area. A prominence is formed in the device area by making L2 shorter than L1. Prominence height is controlled by adjusting L2.

The buried n+ layer is connected with the drain electrode of the MOSFET. In the conventional structure, the current flows in the horizontal and vertical directions at the edge of the device and in only the vertical direction at the center of the device. The thickness of the n- substrate depends on the edge structure of the device to maintain a high breakdown voltage. Therefore, the distance between the drain region and the

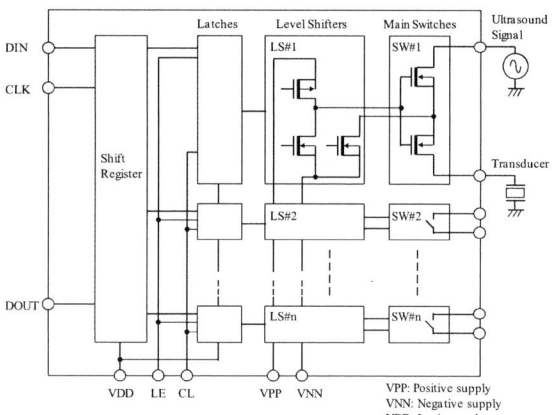

Figure 1. Block diagram of high-voltage analog switch IC.

source region is excessively long at the center of the device, which degrades drain current characteristics.

In the proposed structure, the distance between the drain region and the source region at the center of the device is shorter due to the formation of the internal prominences, and the current density is higher due to the reduction in the effective drift layer length and drift resistance. The breakdown voltage does not drop if the length between the drain and source regions at (A) in Fig. 2(b) is equal or longer than at (B). Figure 4 shows the potential distributions for the conventional and proposed MOSFETs for a main switch at the same current density. The drift resistance becomes lower for the proposed one.

B. N-ch MOSFET for Level Shifters

Figure 5 shows cross sections of the conventional and proposed N-ch MOSFETs for the level shifters. The conventional structure has a very long channel region to reduce the electric field under the gate region for higher hot carrier reliability. The proposed structure controls the saturation current by using a JFET structure instead of the conventional long channel region.

The source electrode in the proposed structure surrounds the gate electrode, and the channel region is formed along the edge of the gate electrode. The JFET resistance around the channel region increases with the voltage of the drain electrode. Therefore, the saturation current is suppressed, and the electric field under the gate region is reduced. The saturation current can be controlled by optimizing the space between adjacent channels (Lgate in Fig. 5(b)).

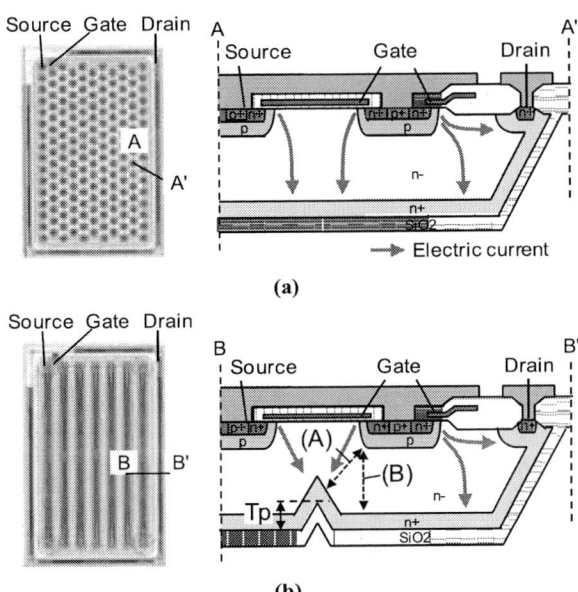

Figure 2. Cross sections of MOSFETs for main switches: (a) conventional and (b) proposed.

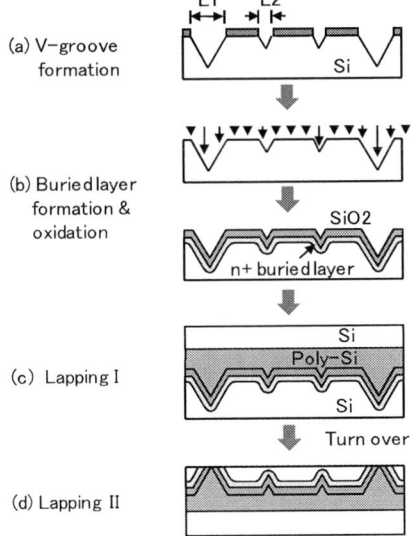

Figure 3. Process flow of DI wafer.

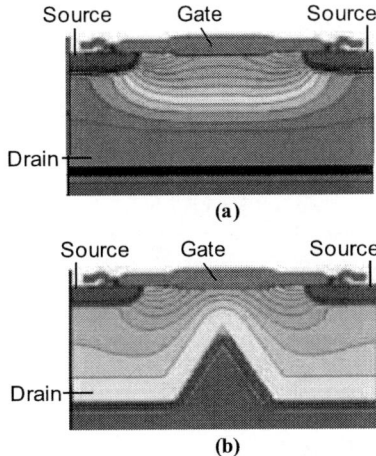

Figure 4. Calculated potential distributions for (a) conventional and (b) proposed MOSFETs at same current density. (0.2 V/line, J=0.06 mA/um)

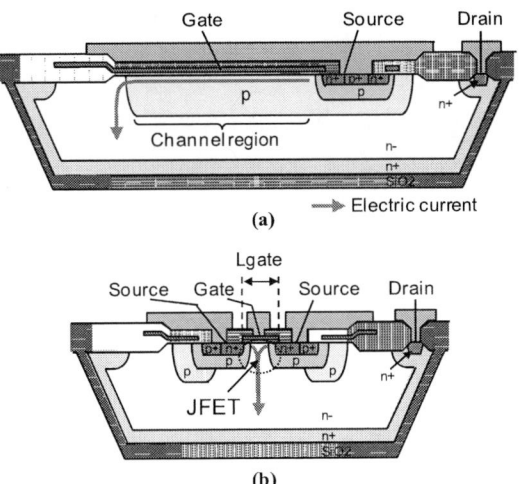

Figure 5. Cross sections of N-ch MOSFETs for level shifters: (a) conventional and (b) proposed.

C. P-ch MOSFET for Level Shifters

Figure 6 shows the proposed P-ch MOSFET device structure for the level shifters. The conventional P-ch MOSFET for the level shifters also has a very long channel region for the same reason as for the N-ch one. The proposed structure also controls the saturation current by using a JFET structure instead of the conventional long channel region. The JFET region in the P-ch MOSFET is formed adjacent to the RESURF (REduced SURface Field) layer in the plane direction.

In the proposed P-ch MOSFET, additional n layers are formed on both sides of the RESURF p layer. These n layers are connected to the source electrode with large resistance and form the JFET region. The JFET resistance in the RESURF p layer increases with the voltage of the source electrode. Therefore, the saturation current is suppressed, and the electric field under the gate region can be reduced. The saturation current can be controlled by optimizing the space between the additional n layers (Lnn in Fig. 6(c)).

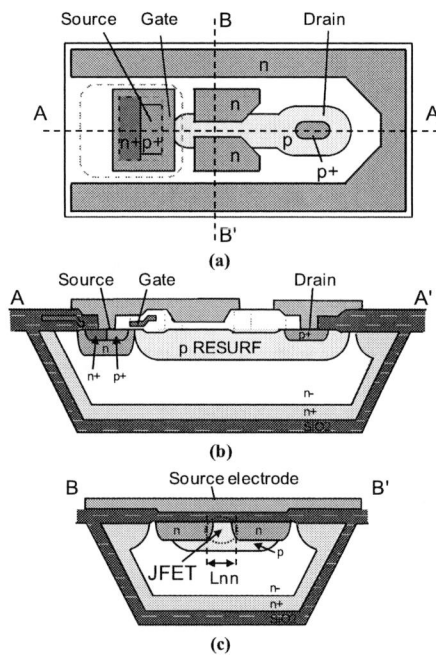

Figure 6. Proposed P-ch MOSFETs device structure for level shifters: (a) top view, (b) cross section of A-A', and (c) cross section of B-B'.

III. EXPERIMENTAL RESULTS

To verify the effectiveness of the proposed MOSFETs, we have fabricated sample devices using a 6-inch DI process.

A. MOSFET for Main Switch

Figure 7 compares the characteristics of the proposed MOSFET device for the main switches with those of the conventional one. The proposed device has 40% higher current density than the conventional one. Figure 8 shows the current density and breakdown voltage for various heights of the internal prominences (Tp, see Fig. 2(b)). The current density increases when Tp is increase. The breakdown voltage is constant up to Tp=1.0(au), and then started to deteriorate at

Tp=1.0(au). This is because the drift length between the drain and source at (A) (see Fig. 2(b)) is equal to that at (B) when Tp=1.0(au). Therefore, the prominences reduce the effective drift length without increasing the peak electric field below Tp= 1.0(au).

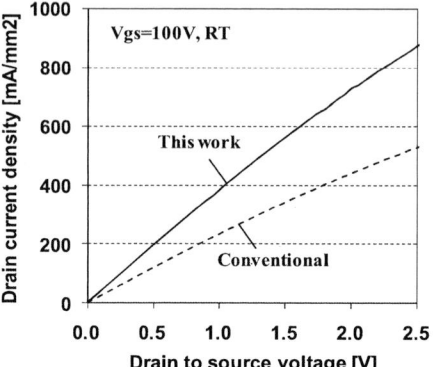

Figure 7. Current density – drain voltage characteristics of main switch devices.

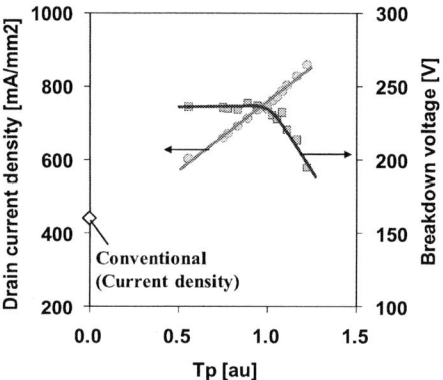

Figure 8. Current density and breakdown voltage for various heights of internal prominences.

B. MOSFET for Level Shifters

Figure 9 shows the drain current characteristics of the proposed Nch-MOSFET for the level shifters at various gate-to-source voltages (Vgs). The saturation current is almost constant when Vgs is higher than the threshold voltage. This means that the saturation current is not only controlled by the gate voltage but also suppressed by the JFET resistance. The saturation current can be controlled by adjusting the width of the gate electrode (Lgate). Figure 10 shows the relationship between the drain current and Lgate. The drain current increase as Lgate is widened. Device size compared with that of the conventional type is also plotted in Fig. 10. With the proposed structure, the device size is less than 50% that of a conventional device.

Figure 11 shows the drain current characteristics of the proposed Pch-MOSFET for the level shifters at various Vgs. The saturation current is almost constant when Vgs is higher than the threshold voltage, as it is for the N-ch one. The saturation current can be controlled by adjusting the space between the additional n layers (Lnn). Figure 12 shows the relationship between the drain current and Lnn. The drain current increase as Lnn is widened. Device size compared with that of the conventional type is also plotted in Fig. 12. With the proposed structure, the device size is less than 44% that of a conventional device.

IV. CONCLUSION

To reduce the size and cost of high-voltage analog switch ICs, we have developed a low-resistance and compact MOSFET for analog switch ICs with dielectric isolation that is fabricated using isotropic etching. To obtain high current density, we form internal prominences, which reduce the drift resistance of devices with a high breakdown voltage. The current distribution in the drift layer at the center of the device is made uniform by the internal prominence, which improves the drain current characteristics. The proposed structure has 40% higher current density than the conventional structure. N-ch and P-ch compact MOSFETs for level shifters have also been developed that can control saturation current with a low electric field under the gate region by using a JFET structure for higher hot carrier reliability. The areas of these MOSFETs can be shrunk about 40% in 220-V devices.

ACKNOWLEDGMENT

The authors would like to thank Koichi Suda, Shinichi Kurita and Takeya Ikeda for fruitful discussions and support on device fabrication.

REFERENCES

[1] Bruno Haider, "Power Drive Circuits for Diagnostic Medical Ultrasound," Proceedings of ISPSD '06, Plenary Session p. xxxiii, 2006.

[2] Richard K. Williams, Larry T. Sevilla, Eric Ruetz and James D. Plummer, "A DI/JI-Compatible Monolithic High-Voltage Multiplexer," IEEE Trans. on Elec. Dev., vol. ED-33, pp. 1977 –1984, 1986.

[3] ECN3290 Datasheet, Hitachi Ltd.

[4] Y. Sugawara, Y. Inoue, S. Ogawa and S. Kurita, "New Dielectric Isolation for High Voltage Power ICs by Single Silicon Poly Silicon Direct Bonding (SPSDB) Technique," Proceedings of ISPSD '92, pp. 316-321, 1992.

[5] B. Murari, F. Bertotti and G. A. Vignola, "Smart Power ICs," Springer-Verlag Berlin Heidelberg, 2002, pp. 114–118.

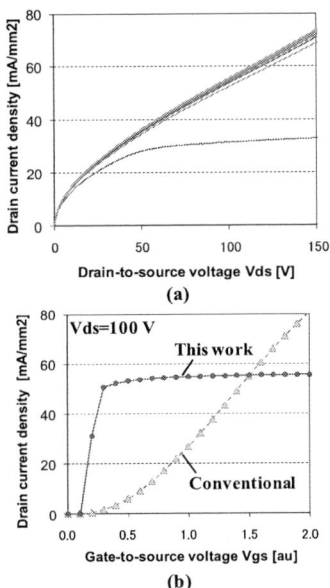

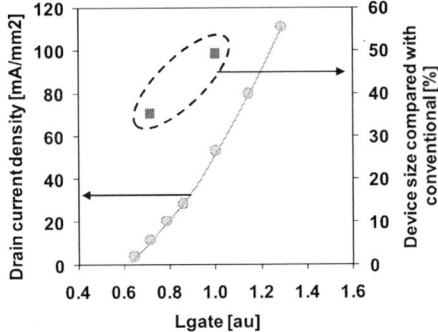

Figure 9. Drain current characteristics of proposed N-ch MOSFET for level shifters: (a) Id-Vds and (b) Id-Vgs.

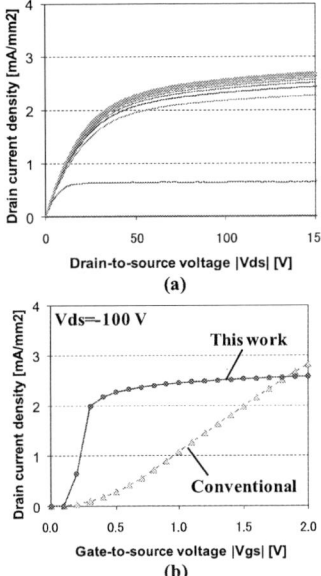

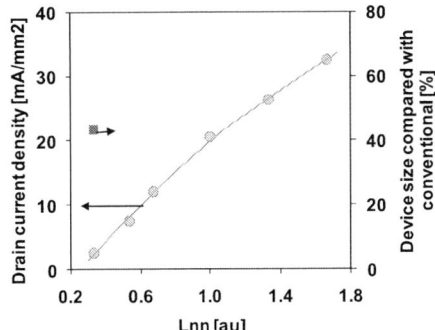

Figure 11. Drain current characteristics of proposed P-ch MOSFET for level shifters: (a) Id-Vds and (b) Id-Vgs.

Figure 10. Drain current characteristics of proposed N-ch MOSFET for level shifters and device size compared with that of conventional one.

Figure 12. Drain current characteristics of proposed P-ch MOSFET for level shifters and device size compared with that of conventional one.

Proceedings of the 23rd International Symposium on Power Semiconductor Devices & IC's
May 23-26, 2011 San Diego, CA

A novel 0.16 µm – 300 V SOIBCD for Ultrasound Medical Applications

M. Sambi[*], D. Merlini[*], P. Galbiati[*]

[*]STMicroelectronics TRD, Agrate Brianza (MI), Italy,
marco.sambi@st.com, daniele.merlini@st.com

E. Bonera[^], F. Belletti[*,+]

[+]University of Milano, Dept. of Physics, Milano, Italy
[^] Dipartimento di Scienza dei Materiali , Università degli Studi di Milano-Bicocca, Milano, Italy

Abstract — **The development of a new 0.16 µm SOIBCD technology integrating components with breakdown voltage higher than 300 V is here described. Process integration and mechanical stress due to buried oxide and lateral dielectric isolation was investigated with TCAD simulations, morphological analysis and Raman spectroscopy measurements. Component portfolio was derived from existing junction isolated (JI) 0.16 µm technologies and expanded with high voltage MOS. Stress induced by the full dielectric isolation is far from critical values. Breakdown voltages over 350 V were measured.**

I. INTRODUCTION

BCD technologies are widely known to allow integration of different kind of circuitry on the same die. SOIBCD is a BCD technology where isolation is made with dielectrics; SOIBCD8S is derived from existing 0.16 µm JI BCD and keeps all the main technology features. Target applications are the ultrasound 3D/4D medical echo-graphs probes, which require thousands of high voltage output stages, LNA and complex logic blocks in a single die. The main feature of such ASIC is the high number of I/O channels; each has to drive a capacitive load (the piezoelectric element of the probe) with hundreds of volts and has to read the small signal coming back from the body. These requirements cannot be satisfied with standard JI technologies. SOIBCD8S is the best technology for such ASIC: it is easier to make coexist high voltage MOS belonging to different voltage classes, low noise components and CMOS in a single process; moreover the superior insulating efficiency of dielectrics allows controlling the cross talk among adjacent stages with reduced area consumption.

II. SOIBCD8S OUTLINE

SOIBCD8S process is based on our 0.16 µm JI BCD8S, from which it inherits the main technology features and low voltage components [8]: STI as field oxide, double flavored columnar poly, borderless stacked vias and contacts, double gate oxide 3.5 nm and 7 nm for 1.8V and 3.3V CMOS respectively. The substrate is an N- SOI with thick buried oxide and thick active silicon, with deep trenches for lateral isolation.

High sheet-resistance poly, high linearity capacitors, n and p channel power MOS up to 300V, low noise devices complete the component portfolio. ASIC for ultrasound probes have operating voltages up to hundreds of volts. For this reason, 110 V / 220 V / 300 V power MOS, both N- and P-Channel, were integrated in SOIBCD8S.

III. PROCESS INTEGRATION AND STRESS ANALYSIS

A. Trench integration

Deep trench isolation (DTI) is realized under shallow trench isolation (STI), encapsulating the oxide at the top, which is the safer situation to ensure the trench integrity during subsequent wafer processing. After the N- type epitaxy growth, silicon is etched till reaching the buried oxide. After a dedicated sidewall and oxide filling, the trench top is made planar for optimal STI module processing (Figure 1).

Figure 1. SEM image of a trench triplet and (small insert) trench and active area used for stress measurements

Mechanical stress evolution induced by DTI during silicon processing has to be monitored to identify possible marginalities both on silicon robustness and device performances (see Figure 12).

B. Stress determination and analysis

In semiconductor devices, the mechanical local stresses are known to affect process yield and device performances and reliability. For example, stress-induced band gap narrowing increases junction leakages and modifies carrier mobility,

978-1-4244-8425-6/11 $26.00 © 2011 IEEE

otherwise stresses induce extended defect generation, i.e. dislocations. It is well know also that the peak stresses reached during the process steps may create extended defect regions, while the residual stresses at the end of the process flow determine the electrical performances of the device [5]. Simultaneously with the elementary devices development, it has been designed a matrix of geometrical structures for stress characterization. In particular, an STI array was set up near and far away from trenches. We perform 2D process simulation to determine strain fields and comparing them with direct measures of stress using the micro-Raman spectroscopy ([1], [3], [6]).

The assumption on Physics model is to have a combination of visco-elastic/elastic mechanical approach for oxide/silicon, respectively. It has been compared the final residual stress values for the σ_{xx} stress-tensor component and the shear stress values inside a designed average integration area. The analysis shows a complete view of the stressed region in xy plane section, focusing on the differences between silicon with and without the trench. The stress is compressive for concave silicon-oxide interface geometry, as expected, and tensile outside. Most significant steps by steps residual stresses are also plotted to figure out the trend during wafer diffusion (Figure 2) and unsafe mechanical stress are not present (stress is below 0.2GPa, critical point is higher than 1GPa).

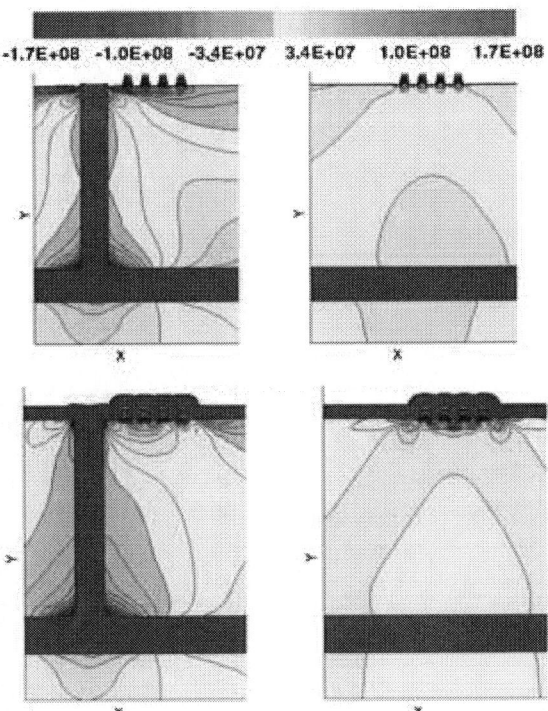

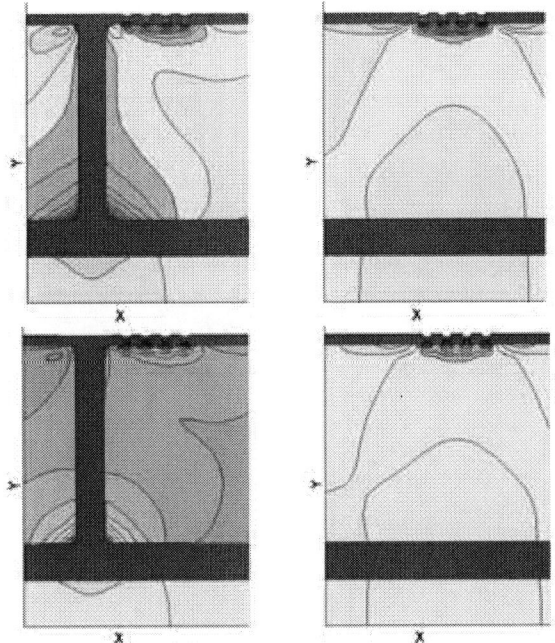

Figure 2. Evolution of mechanical stress during trench process steps and comparison with SOI silicon without lateral isolation. Mechanical critical point is higher than 1 GPa while simulations show stress well below this limit. Colorbar levels are in Pa.

A Raman direct measure of stress was performed inside the shallow trench array, near and far from the DTI. The experiment was set up using UV-light excitation Ar$^+$ laser (364 nm) in a resonant configuration [7]. The light penetration in silicon is about 10 nm, with a spot diameter on the sample around 1 µm. Assuming uniaxial stress model [4], we obtain data summarized in Figure 3. The shift between trench / no trench is detectable and well matched with simulated data. The simulation grid was averaged assuming the same condition of the Raman measurement (light penetration and spot step and spot dimensions). We detected an average wave number [7] shift of 0.35(5) cm^{-1}, which means 0.16(2) GPa, in the case of sample without trench, and of 0.55(5) cm^{-1} which means 0.25(2) GPa (Figure 3). Active area strain is significantly increased by the trench integration, respect to standard STI situation.

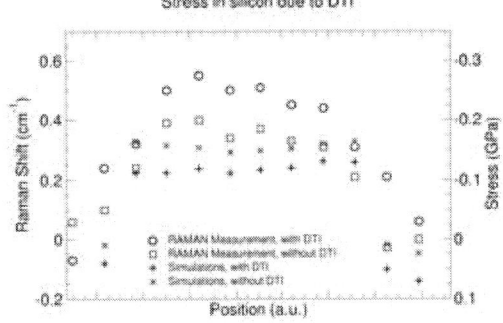

Figure 3. Comparison of simulated and measured mechanical stress in SOI substrates, with or without trench. Region with higher stress is near STI / active area transitions.

The last, it has been performed a SEM analysis on the same Raman sample in order to detect eventually crystal defects inside silicon lattice with a standard Secco d'Aragona methodology. The lattice beneath trench and SOI, and inside active, is defect-free (Figure 4).

Figure 4.   Silicon images near trench edge with Secco d'Aragona process applied. No defects were detected.

IV.   ELECTRICAL PERFORMANCES

Typical ASIC for ultrasound probes need the coexistence of different high voltage classes in the same die. Careful compromise between cost and performances of high voltage components is a figure of merit is the design of such ASIC. 220V class is the most area-consuming in our products: drain wells were tuned to optimize this class; other voltage classes are derived only acting on geometrical parameters. Double resurf [9] architecture (Figure 5) was used for both N-Channel and P-Channel; components have been simulated and characterized on silicon (Figure 6).

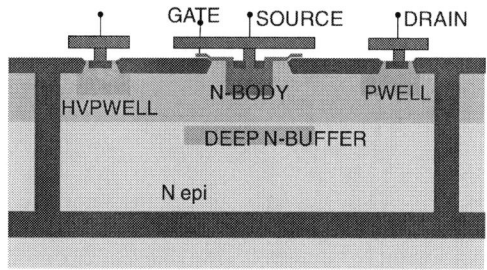

Figure 5.   Cross section of P-Channel power MOS. N-Channel devices have the same structure with opposite dopings.

Performances are well aligned with simulations and a high voltage NMOS with breakdown voltage of 254V and R_{ON} x Area of 11.5 mOhm cm^2 is available.

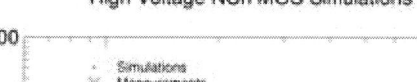

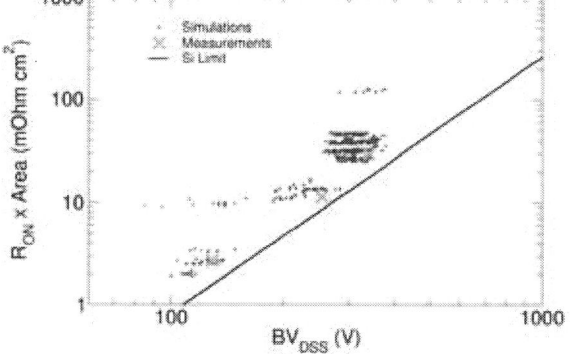

Figure 6.   Simulations of N-Channel power MOS (blue points) and optimal structure from silicon measurements (orange cross) [10]

P-channel components are equally important to be used in output bridges as high side. High voltage PMOS simplify circuit design (e.g. absence of charge pump) of these stages and their optimization is crucial. Current capability or on-state resistance is usually matched between high side and low side: HV-PMOS consequently occupy the majority of the area, due to its lower current capability or higher on-state resistance. Our 220V P-channel has a |BV$_{DSS}$| of 255V and R_{ON} x Area of 34 mOhm cm^2.

Output characteristics of high voltage MOS up to 220V were measured with short pulses (100 ns) on the gate to avoid self-heating effects. Electrical safe operating area is guaranteed in the full temperature range (-40 °C to 175 °C). 300V components were measured with a curve tracer due to limitation of previous instrument.

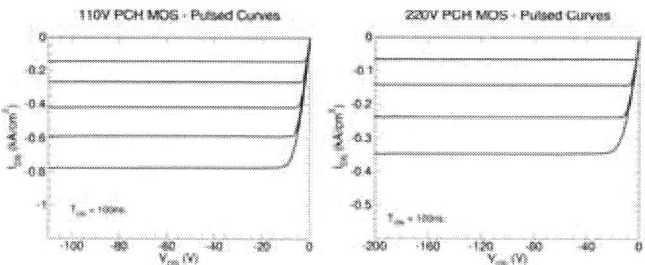

Figure 7.   Output characteristics of 110V and 220V P-Channel MOS. |V$_{DS}$| is limited to 200V due to instruments limitation. |V$_{GS}$| max is 3.3V, step 0.5V

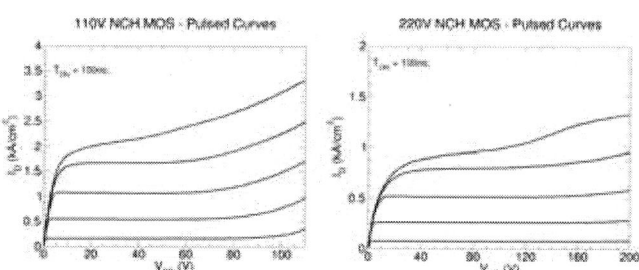

Figure 8.   Output characteristics of 110V and 220V N-Channel MOS. V$_{DS}$ is limited to 200V due to instruments limitation. V$_{GS}$ max is 3.3V, step 0.5V

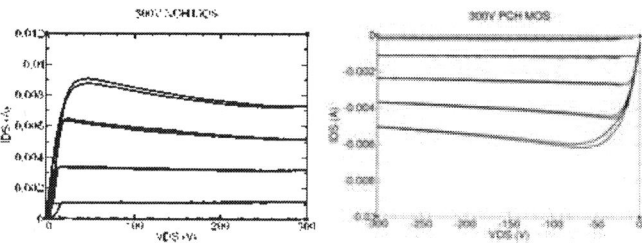

Figure 9.   Output characteristics of 300V P-Channel MOS. |V$_{GS}$| max is 3.3V, step 0.5V

Together with high voltage power MOS, an npn bipolar transistor and an n-channel JFET were integrated in the technology to address analog circuitry with low noise requirements. Such stages need low voltage, 3.3V are enough.

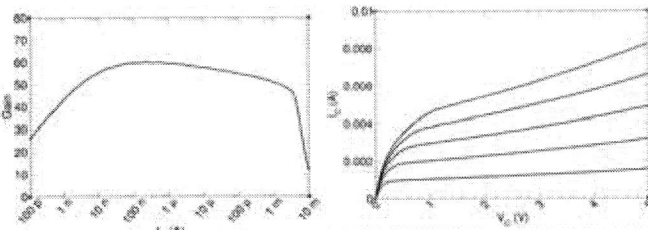

Figure 10. Electrical characteristics of low voltage npn bipolar transistor.

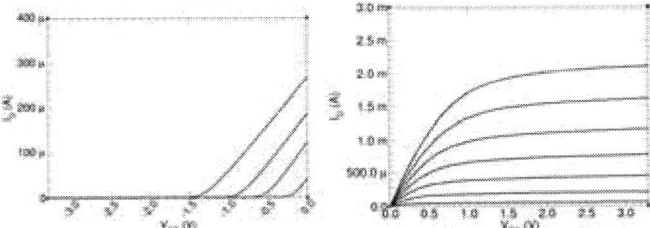

Figure 11. Low voltage N-channel JFET. Different dopings of the channel were evaluated to find the optimal pinch-off voltage (left) and output characteristics (right).

CMOS transistors (both 1.8V and 3.3V) have been ported from existing JI BCD and electrical performances are aligned. The critical point is to evaluate possible criticalities due to the presence of dielectric isolation, which modifies silicon stress and consequently carrier mobility. The analysis has been carried out measuring electrical parameters of CMOS with DTI at different distances from source/drain active area. The results are depicted in Figure 12. The impact of DTI on electrical parameters is negligible and limited to a few percent points in the worst case.

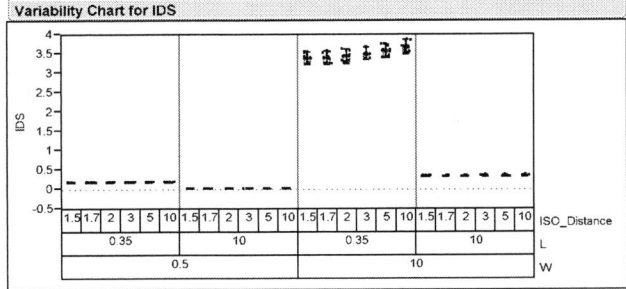

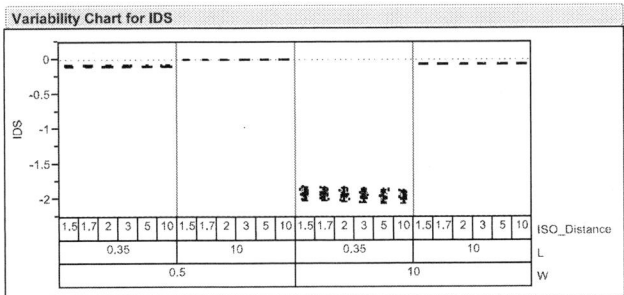

Figure 12. Measured current of 3.3V NMOS (top) and PMOS (bottom). W and L are the electrical width and length of the channel. ISO_Distance is the spacing between active area and DTI. All dimensions are in in µm.

V. CONCLUSIONS

0.16 µm SOIBCD8S technology has been developed and high voltage power MOS with BV higher than 300V have been integrated. SOIBCD8S is available for ultrasound application but also for all ASIC requiring high voltage, higher levels of integration with respect to JI BCD, low power consumption, noise immunity, such as PDP or AMOLED display drivers, automotive audio amplifiers and sensor interfaces and many others. Simulations and measurements of stress induced by fully oxide-filled trenches in a 0.16 µm show that silicon lattice is integer and no impact on electrical characteristics is detectable. Moreover, form at mechanical-only point of view the technology is safe.

VI. ACKNOWLEDGEMENTS

The authors would like to thank L. Zullino for her support throughout all the TCAD simulations activities; A. Ronchi, A. Vinay, L. Labate and S. Cozzi for their contributions in device characterization.

The research leading to these results has received funding from the ENIAC Joint Undertaking under grant agreement n° 120008 and from national programs/funding authorities German Federal Ministry of Education and Research (BMBF); the Belgian IWT-Flanders; the French Ministère de l'Economie, des Finances et de l'Industrie; Secrétariat d'Etat à l'Industrie (STSI); Enterprise Ireland; the Italian Ministero Istruzione Università Ricerca; APRE Agenzia per la Promozione della Ricerca Europea; the Dutch SenterNovem; the Research Council of Norway; the Spanish DGI-Ministerio de Educación y Ciencia; and the Swedish Vinnova.

[1] De Wolf, I., "Micro-Raman spectroscopy to study local mechanical stress in silicon integrated circuits", Semicond. Sci. Technolol. 11 (1996) 139 – 154.

[2] Senez,V., Armigliato, A. et al. "Strain determination in silicon microstructures by combined convergent beam electron diffraction, process simulation, and micro-Raman spectroscopy", Journal of Applied Physics, vol. 94 (2003) 5574 – 5583.

[3] Bonera, E. et al., "Combing high resolution and tensorial analysis in Raman stress measurement of silicon", J. Appl. Phys. 94 (4): 2729-2740, (2003).

[4] Anastassakis, E. et al., "Effect of static uniaxial stress on the Raman spectrum of silicon", Solid State Commun. 8: 133-138, (1970).

[5] I. Mica et al., "Crystal defects and junction properties in the evolution of device fabrication technology" , J. Phys. Cond. Matt. 14 13403-13410 (2002).

[6] E. Bonera, et al. "Raman stress maps from finite-element models of silicon structures", J. Appl. Phys. 100, 033516 (2006).

[7] E. Bonera, et al., "Raman spectroscopy of strain in subwavelength microelectronic devices", Appl. Phys. Lett. 87, 111913 (2005)

[8] D. Riccardi et al., "BCD8 from 7V to 70V: a new 0.18µm Technology Platform to Address the Evolution of Applications towards Smart Power Ics with High Logic Contents", Proc. ISPSD2007

[9] P.Wessel et al. "Advanced BCD technology for automotive, audio and power applications" Solid-State Electronics, 2007.

[10] T. Fujihira, "Theory of Semiconductor Superjunction Devices," Jpn. J. Appl. Phys. vol. 36 (1997), Part 1 No. 10, Oct. 1997.

Proceedings of the 23rd International Symposium on Power Semiconductor Devices & IC's
May 23-26, 2011 San Diego, CA

300 V Field-MOS FETs for HV- Switching IC

T. Miyoshi, T. Tominari, M. Hayashi, A. Ito, M. Yoshinaga, S. Ueno, T. Oshima, and S. Wada

Micro Device Division, Hitachi, Ltd.
6-16-3 Oume-shinmachi, Tokyo 198-8512, Japan
Phone: +81-428-33-2222 ext. 5383, Fax: +81-428-33-2164, E-mail: tomoyuki.miyoshi.pt@hitachi.com

Abstract— **We have developed 300 V Field-MOS FETs for High-Voltage switching IC. The breakdown voltages are 410 V/370 V with specific on-resistance of 1845/11000 mΩ·mm² for Field-NMOS/PMOS FETs, respectively. The vertical and lateral electric fields are both optimized to maximize a breakdown voltage at wide range of substrate voltages and minimize a specific on-resistance with a device layout optimization by introducing a Field Pate and a Extended-Drain layer. This technology can apply to 300 V High-Voltage switching IC with a low leak current and a low switching resistance.**

I. INTRODUCTION

The lateral High Voltage MOS FETs, like LDMOS and Field MOS FETs are suitable as power devices for Smart-Power applications thanks to their ease of integration and isolation with submicron CMOS technology. High-Voltage switching ICs, which are kind of Analog switch circuits, are applied for clinical ultrasound imaging system [1] and bare board and packaging testers. These ICs constructed with low voltage logic, level-shifting and output circuits requires the operation voltage up to 300 V. In order to implement a large amount of switching actions, simple circuit design and low leakage current are important. Field-MOS FETs, in which high voltages can be applied to the gate, are effective for this purpose. Fig. 1 shows the level-shifting and output circuits with Field-MOS FETs.

In the work covered in this paper, we focused on the development of Field-MOS FETs with a high breakdown voltage over 300V, low specific on-resistance and low leakage current on relatively thin SOI wafer. The device was fabricated by 0.35-μm SOI process technology for 300 V High-Voltage switching IC.

II. DESIGN OF FIELD-MOS FETS

Fig. 2 shows the structure of the Field-NMOS FETs in 0.35-μm SOI technology. The use of silicon on insulator (SOI) wafers is the initial measure towards high breakdown voltage characteristics. Combination of deep trench isolation and SOI substrate contributes to small area low leakage current. Moreover, parasitic bipolar action can be perfectly suppressed by the buried oxide (BOX). Fig. 3 shows the outline of the

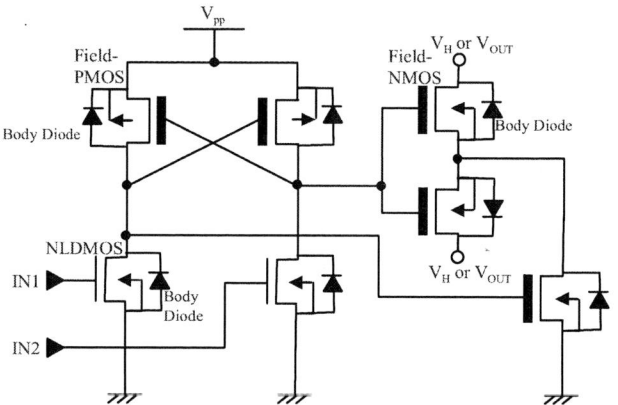

Figure 1. Output circuit of the High-Voltage switching IC.

process flow. P type SOI start wafer that has high resistivity over 200 Ωcm and high-energy phosphorus ion implantation are used to form deep N-Well layer. The channel is formed in the P-Well (PW) with the threshold voltage (V_{th}) of 13 V. The gate Poly-Si is formed on the LOCOS oxide to withstand 300 V. Since both the PW and the sources are designed self-aligned to the LOCOS, the effective gate length can be less than 1 μm, leading to a high current density.

A. Optimization for high off-state breakdown voltage

To obtain a high breakdown voltage, using the thick SOI/BOX wafers is already reported [2]. In this work, our optimized SOI thickness is less than 8 μm to simplify the complexity of deep trench isolation process and minimize the wafer cost. The calculated breakdown voltages are 320 V and 380 V on thin and thick SOI wafer respectively. Fig. 4 shows the off-state electric potential distribution near the drain region under V_{ds}=320 V on the Field-NMOS FETs on thin SOI wafer and thick SOI. In thin SOI, the vertical electric field at the drain region is strong and the impact ionization is generated under this voltage. On the other hand, the thick SOI reduces the electrical field.

978-1-4244-8425-6/11 $26.00 © 2011 IEEE

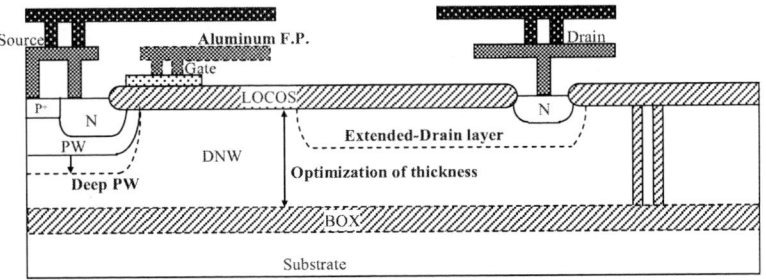

Figure 2. Cross sectional view of the Field-NMOS FET.

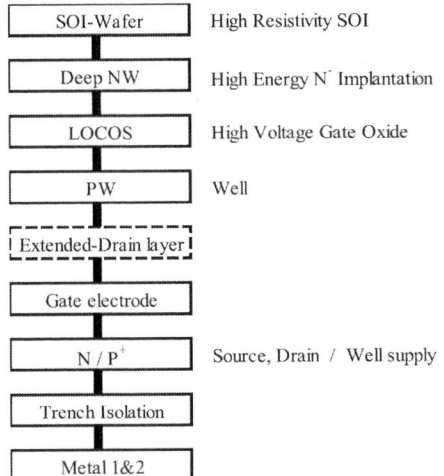

Figure 3. Field-NMOS process flow in 0.35-µm
SOI technology

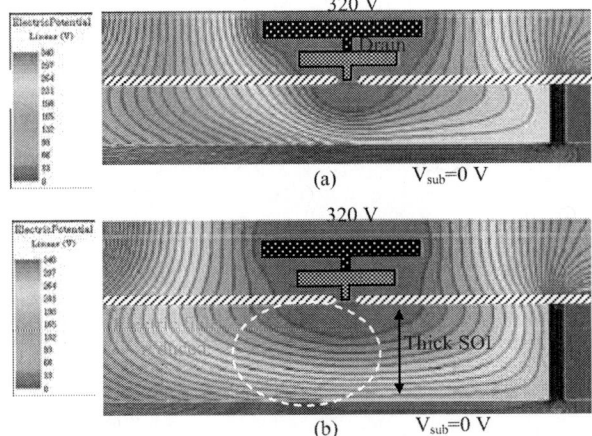

Figure 4. Electric potential distribution of off-state Field -
NMOS FETs with thin SOI (a) and thick SOI (b).
(V_{sub}=0 V, V_d=320 V, V_s=0 V)

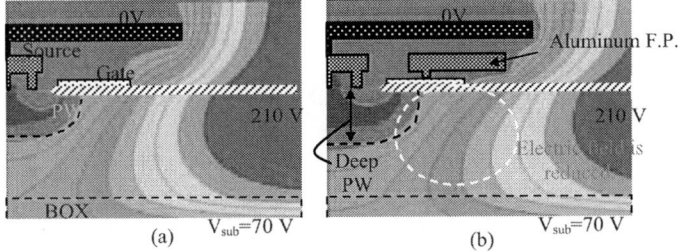

Figure 5. Electric potential distribution of off-state Field-
NMOS FETs before (a) and after (b) optimization.
(V_{sub}=70 V, V_d=210 V, V_s=0 V)

However, the substrate voltage (V_{sub}) dependence on the breakdown voltage for the thick SOI device is degraded. Fig. 6 shows the calculated and measured breakdown voltages as a function of substrate voltages. For the thick SOI device, the breakdown voltage increases with the V_{sub} when $V_{sub}<0$ V, but sharply falls when $V_{sub}\geqq0$ V, which is inferior characteristics to the thin SOI device. This indicates unstable breakdown voltage for the fluctuation of SOI thickness. With the thick SOI device, the large electric field is created under the gate region for the positive substrate voltage as indicated in Fig. 5(a), when V_{ds}=210 V and V_{sub}=70 V. In this area, the impact ionization is generated, which leads to a breakdown failure. In order to reduce the lateral electric field under the gate, the Aluminum Field Plate and the deep PW are introduced. Here, the Aluminum Field Plate is connected to the gate electrode as shown in Fig. 2. In the off-state, the Aluminum Field Plate with the same voltage to the source can reduce the electric field. The deep PW has also the same effect. Fig. 5(b) shows the calculated electric potential distribution after introducing Aluminum F.P. and deep PW under the same condition as Fig. 5(a). The electric potential distribution moves toward the drain region and the electric potential under the gate is relaxed. Fig. 6 shows the calculated and measured breakdown voltages of the devices before and after structure optimization. The calculated breakdown voltage is 400 V when V_{sub}=70 V. The

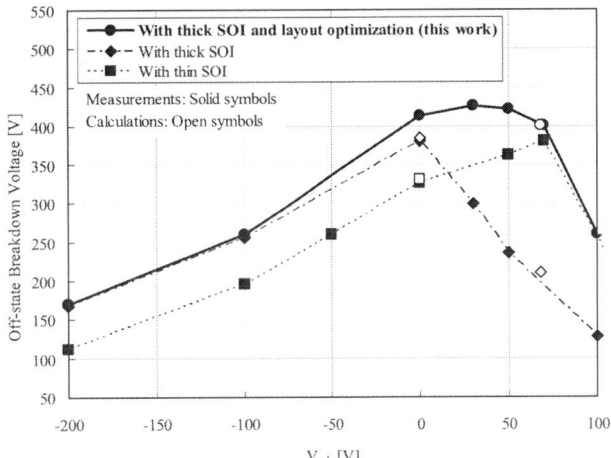

Figure 6. Measured and calculated off-state breakdown
voltage of the Field-NMOS FETs.

reduction of lateral electric field when $V_{sub}<0$ V and $V_{sub}\geqq0$ V leads to the breakdown voltage larger than 400 V, up to V_{sub} of 70 V. This also contributes to stable off-state breakdown voltage for the fluctuation of SOI thickness.

978-1-4244-8425-6/11 $26.00 © 2011 IEEE

B. Optimization for low specific on-resistance

In this work, we introduced the Extended-Drain layer to the drain region to reduce the specific on- resistance ($R_{on,sp}$) and increase the saturation current. Fig. 7 shows the electric potential distribution of on-state Field-NMOS without and with Extended-Drain layer when V_{ds}=150 V, V_{gs}=275 V and V_{sub}=0 V, respectively. Extended-Drain layer has an effect to reduce the electric field at drain region and heighten the electric potential at the drift region near the source. This leads that carrier flow from source to drain efficiently through Extended-Drain layer. Moreover, this layer also has the effect to reduce the resistance of drift region at drain. These effects reduce the $R_{on,sp}$. Fig. 8 shows the measured $R_{on,sp}$ when Vgs=275 V and off-state breakdown voltage depending on the distance between source region and Extended-Drain layer. The shorter the distance, the lower the $R_{on,sp}$ is, because the effect on carrier flow and low drift resistance with the Extended-Drain layer is more efficient. On the other hand, the breakdown voltage at off-state degraded to lower than 400V when the distance is smaller than Distance A. This is because the electric field under the gate at off-state is large, with reducing the distance between Extended-Drain layer and source. With the Distance A, a low specific on-resistance of 1845 m$\Omega \cdot$mm^2, reduced by 18% from non Extended-Drain layer, is obtained with maintaining the off-state breakdown voltage. Fig. 9(a) shows the measured V_{ds}-I_{ds} characteristics of the Field-NMOS FETs with thin SOI, non Extended-Drain layer (before optimization) and with thick SOI and Extended-Drain layer (after optimization). The saturation current under V_{ds}=100 V is also increased by 20% by Extended-Drain layer. Moreover, the on-state breakdown voltage under V_{gs}=275 V is increased, because of the reduced electric field near the drain region and introducing Extended-Drain layer on thick SOI.

These optimizations are also applied for Field-PMOS. Fig. 9(b) shows the measured V_{ds}-I_{ds} characteristics of the Field-PMOS FET. The off-state breakdown voltage of 370 V and the specific on-resistance of 11000 m$\Omega \cdot$mm^2 are achieved, respectively.

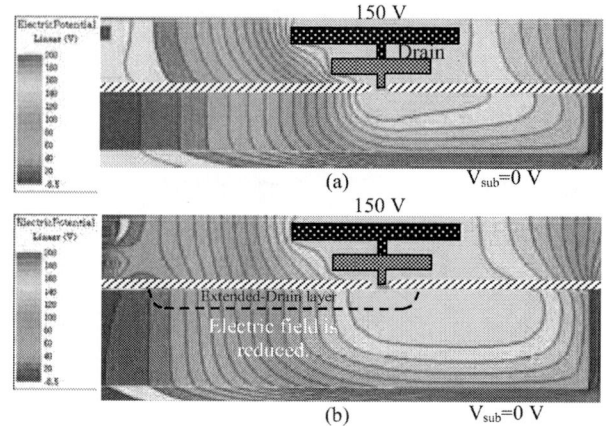

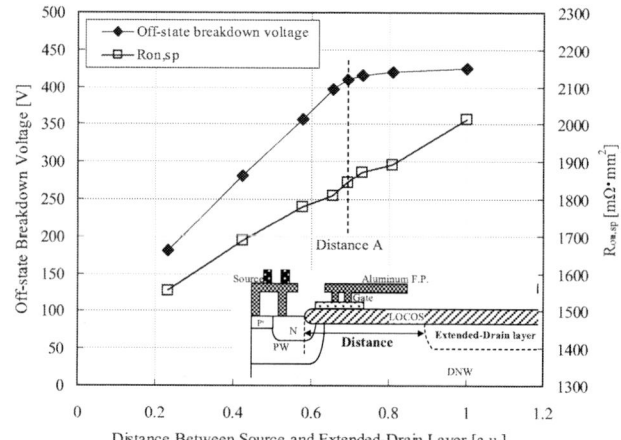

Figure 7. Electric potential distribution of on-state Field-NMOS FETs before optimization (a) and after optimization (b). (V_g=275 V, V_d=150 V, V_s=0V, V_{sub}=0 V)

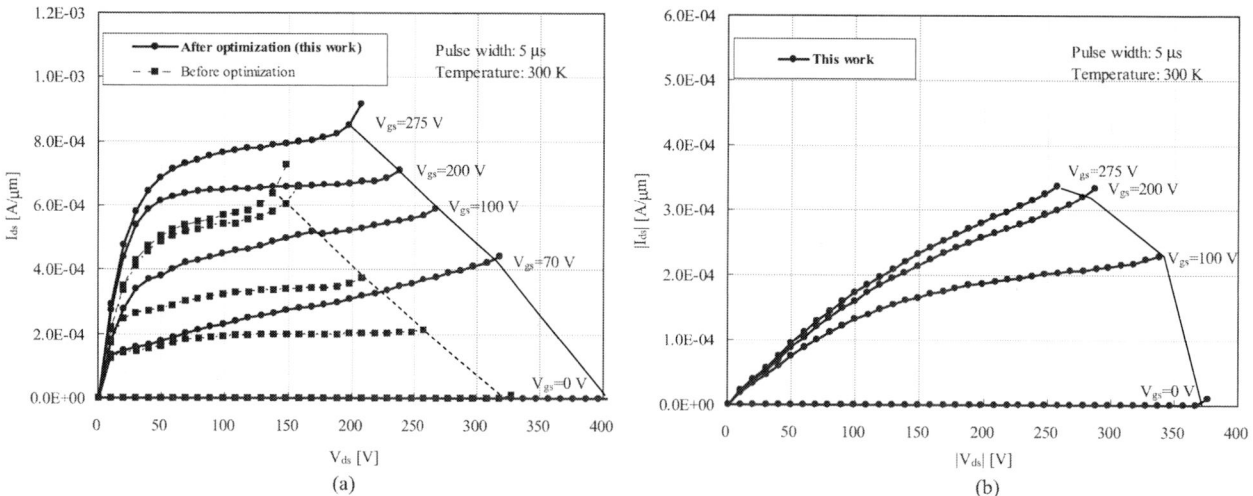

Figure 8. $R_{on,sp}$ when Vgs=275 V and breakdown voltage dependence on the distance between source region and Extended-Drain layer.

Figure9. Measured I_{ds}-V_{ds} characteristics of the Field-NMOS FET (a) and the Field-PMOS FET (b).

Table. 1 Device parameters

Device	Parameter	Unit	Value	Condition
Field-MOS FETs	Off-state breakdown voltage (N / P)	[V]	410 / 370	V_{sub}=0 V
	On-state breakdown voltage (N / P)	[V]	210 / 275	$|V_{gs}|$=275 V
	V_{th} (N / P)	[V]	13 / -17	$|V_{ds}|$=200 V
	I_{ds} (saturation current) (N / P)	[A/μm]	7.8E-04 / 2.1E-04	$|V_{gs}|$=275 V $|V_{ds}|$=130 V
	$R_{on,sp}$ (N / P)	[mΩ·mm²]	1845 / 11000	$|V_{gs}|$=275 V $|V_{ds}|$=2 V
	Off-state leakage current (N / P)	[A/μm]	2.79E-14 / 3.50E-14	$|V_{ds}|$=275 V, Ta=300 K
NLDMOS FET	Off-state breakdown voltage	[V]	380	V_{sub}=0 V
	On-state breakdown voltage	[V]	350	V_{gs}=5 V
	V_{th}	[V]	1.75	V_{ds}=10 V
	$R_{on,sp}$	[mΩ·mm²]	2125	V_{gs}=5 V V_{ds}=2 V
5 V CMOS FETs	V_{th} (N / P)	[V]	0.6 / -0.6	$|V_{ds}|$=5 V
	I_{ds} (N / P)	[μA/μm]	194 / 85	$|V_{gs}|$=5 V $|V_{ds}|$=5 V
Diode	Zener, Low-Voltage, High-Voltage			
Resistor	Diff. (720 [Ω/sq.])			
Capacitor	1.4 [fF/μm]			

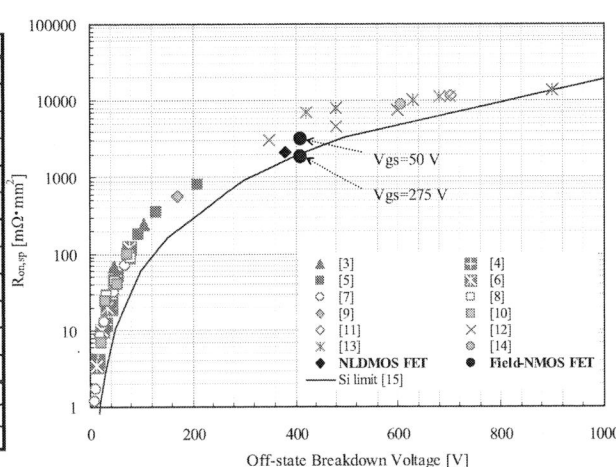

Figure 10. $R_{on,sp}$ vs. off-state breakdown voltage trade-off relation of various LDMOS and Lateral power MOS FETs including the technology reported in this work.

III. SUMMARY OF DEVICE PARAMETERS

Table 1 summarizes the device parameters of the 0.35-μm technology with Field-NMOS/PMOS FETs. A quite low off-state leak current of 0.28 fA/μm at 25°C is achieved in the NMOS FETs. The high-performance NLDMOS with on/off-states breakdown voltages of 350/380 V, 5 V CMOS FETs, Diode and passive device including Resistor and Capacitor are also integrated in our 0.35-μm process technology. In Fig. 10, we compared the $R_{on,sp}$ vs. off-state breakdown voltage trade-off relation of the fabricated Field-NMOS FET and NLDMOS FET with the reported LDMOS FET or Lateral power MOS FET [3-14]. Our NLDMOS FET is at the lowest $R_{on,sp}$ within 400 V class Lateral power MOSFET. In addition, our Field-NMOS FET's characteristic is positioned near the one dimensional Silicon limit calculated with critical electric field and mobility of silicon [15]. The Field-NMOS FETs are more suitable for High-Voltage Switching IC and Analog switch circuit than LDMOS FETs or other lateral power MOS FETs, because of simplicity of circuit simplicity.

IV. CONCLUSION

We have developed 300 V Field-MOS FETs for High-Voltage switching IC. The breakdown voltages are 410 V/370 V with specific on-resistance of 1845 /11000 mΩ·mm² for Field-NMOS/PMOS FETs, respectively. A high off-state breakdown voltage was accomplished field by optimization of SOI thickness, establishment of Aluminum F.P., and optimization of PW depth. A quite low specific on- resistance of Field-NMOS was achieved by introducing Extended-Drain layer. Our 0.35-μm SOI process technology, including Field-MOS, NLDMOS and 5V-MOS FETs, realize a 300 V High-Voltage switching IC with low resistance and leak current.

ACKNOWLEDGMENT

The authors would like to thank Mr. S. Shirakawa, Dr. J. Sakano, and Dr. M. Mori for useful discussion and support on process and device development.

REFERENCES

[1] K. Hara et al., "A New 80V 32x32ch Low Loss Multiplexer LSI for a 3D Ultrasound Imaging System" ISPSD2005, pp.359-362

[2] M. yamaji et al., "A Novel 600V-LDMOS with HV-Interconnection for HVIC on Thick SOI", ISPSD2010, pp.101-104.

[3] F. Kawai et al., "Multi-Voltage SOI-BiCDMOS for 14V & 42V Automotive Applications" ISPSD2004, pp.165-168.

[4] Sameer Pendharkar et al., "7 to 30V state-of-art power device implementation in 0.25um LBC7 BiCMOS-DMOS process technology" ISPSD2004, pp.419-422.

[5] T. Nitta et al., "Wide Voltage Power Device Implementation in 0.25um SOI BiC-DMOS" ISPSD2006

[6] Damiano Riccardi et al., "BCD8 from 7V to 70V:a new 0.18um Technology Platform to Address the Evolution of Applications toward Smart Power Ics with High Logic Contents" ISPSD2007, pp.73-76.

[7] Koji Shirai et al., "Ultra-low On-Resistance LDMOS Implementation in 0.13um CD and BiCD Process Technologies for Analog Power IC's" ISPSD2009, pp.77-79.

[8] Choul-Joo Ko et al., "A New 8V-60V rated Low Vgs NLDMOS Structure with Enhanced Specific on-Resistance" ISPSD2010, pp.245-248.

[9] Alexander Holke et al., "A 200V Partial SOI 0.18um CMOS technology" ISPSD2010, pp.257-260.

[10] Ken Chen et al., "The Foundry Perspective on Integrated Power Technologies" ISPSD2009, pp.6-8.

[11] M. H. Kim et al., "A Low On Resistance 700V Charge Balanced LDMOS with Intersected WELL Structure" ISPSD2003, pp.220-223.

[12] T.Letavic et al., "High Performance 600V Smart Power Technology Based on Thin Layer Silicon-on-Insulator" ISPSD1997, pp.49-52.

[13] K. Permthammasin et al., "New 600V Lateral Superjunction Power MOSFETs Based on Embedded Non-Uniform Column Structure" ASDAM'06, pp.263-266.

[14] M. Rub et al., "A 600V 8.7Ohmmm2 Lateral Superjunction Transistor" ISPSD2006

[15] C. Hu, "Optimum doping profile for minimum ohmic resistance and high-breakdown voltage" IEEE Trans. Electron Devices, vol. 26, pp.243-244

Proceedings of the 23rd International Symposium on Power Semiconductor Devices & IC's
May 23-26, 2011 San Diego, CA

High Performance Pch-LDMOS Transistors in Wide Range Voltage from 35V to 200V SOI LDMOS Platform Technology

Satoshi Shimamoto, Yohei Yanagida, Shinji Shirakawa[+], Kenji Miyakoshi, Toshinori Imai, Takayuki Oshima, Junichi Sakano[+], and Shinichiro Wada

Micro Device Division, Hitachi, Ltd.
6-16-3 Shinmachi, Ome, Tokoy 198-8512, Japan
Phone: +81-428-33-2222, Fax: +81-428-33-2161, E-mail: satoshi.shimamoto.xh@hitachi.com
[+]Hitachi Research Laboratory, Hitachi Ltd.

Abstract—We have developed high performance Pch-LDMOS transistors in wide range rated voltage from 35V to 200V SOI LDMOS platform technology. By applying a novel channel structure, a high saturation drain current of 172 μA/μm in the 200V Pch-LDMOS transistor was achieved, which is comparable to that of the Nch-LDMOS transistor. A low on-resistance of 3470 mΩ*mm² was obtained while maintaining high on- and off-state breakdown voltages of −240 and −284 V. The 35V to 200V LDMOS transistors with a competitive low on-resistance were also demonstrated by layout optimization such as RESURF structure and field plate.

I. INTRODUCTION

Lateral double-diffused MOS (LDMOS) transistors are actively being studied for smart-power ICs, and for automotive and industrial applications [1-6]. LDMOS transistors on SOI wafers are also actively being studied [7-9], since they show superior characteristics such as a high breakdown voltage, strong immunity to latch-up, and a smaller layout size. Moreover, a high-side LDMOS transistor with a high breakdown voltage can be easily formed on an SOI wafer, since a high-breakdown voltage between the source and substrate can be obtained. A high-side Pch-LDMOS transistor is used with an Nch-LDMOS transistor for a symmetrical bipolar output waveform, which affects on a harmonic performance in transmitter IC in medical ultrasound imaging applications. In general, a poor performance of the Pch-LDMOS transistor degrades not only the wave form performance but also chip sizes. For this reason, a high performance Pch-LDMOS transistor which was comparable to a Nch-LDMOS was developed.

In this paper, we mainly present a high performance Pch-LDMOS transistor in our SOI LDMOS platform technology covering the wide range of rated voltages from 35V to 200V.

II. NECESSITY FOR HIGH PERFORMANCE PCH-LDMOS TRANSISTOR

Harmonic imaging has become a major sub-field of medical ultrasound imaging [10], which is one of our target applications. For a symmetrical bipolar output waveform, matched P/N channel output current and impedance become particular important. In general, drain current ratio of Pch to Nch-MOS transistor is about 1/2 to 1/3. For this reason, the channel width of the Pch-MOS transistor has to be wider than that of the Nch-MOS transistor, which degrades impedance matching. Fig. 1 shows output waveforms. A good positive/negative bipolar pulse symmetry is obtained by well-controlled P/Nch matching property. For this reason, a high saturation drain current for the Pch-LDMOS transistor is important.

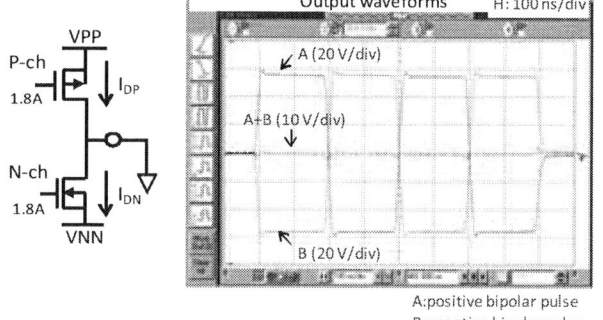

Figure 1. Output waveform in harmonic imaging.

III. EXPERIMENTAL RESULTS AND DISCUSSION

A Outline of Our SOI LDMOS Process

LDMOS transistors are fabricated in an 8-inch SOI isolation process. The shrinked process technology from our conventional 0.35μm process is used. Fig. 2 shows an outline of the process flow. P type SOI start wafer that has high resistivity over 200 Ωcm and high-energy phosphorus ion implantation are used to form deep Nwell layer. This layer is optimized to achieve a high breakdown voltage when the handle wafer is biased to the ground at the +VPP/−VNN power supply condition. The LDMOS channel layer is formed by self-aligned ion implantation using poly-Si gate electrode.

We succeeded in integrating LDMOS transistors covering a wide range of rated voltage from 35V to 200V. Fig. 3 shows a schematic view of Pch-LDMOS transistor. The high-voltage (HV) pwell layer forms the drift and drain regions and is optimized for the proper reduced surface field (RESURF) action [11]. Layout lengths such as drift layer, Poly-Si gate and metal field plate are optimized for each rated voltage LDMOS transistor without additional photo mask.

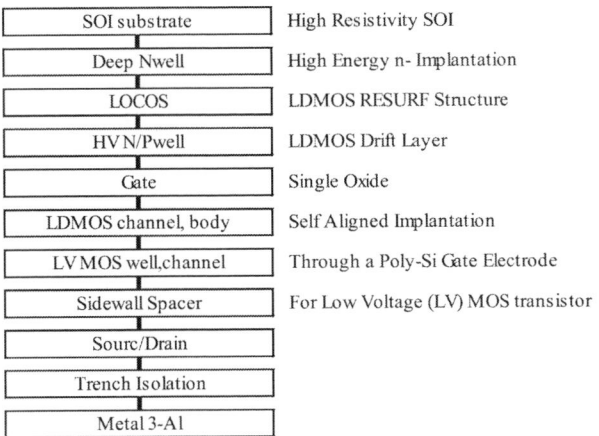

Figure 2.   Outline of process flow.

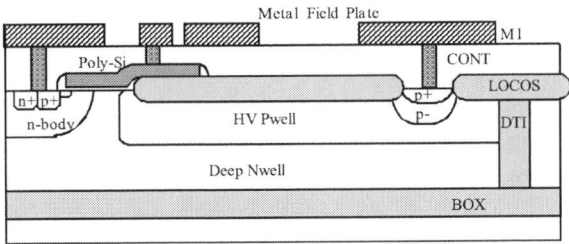

Figure 3.   Schematic view of the Pch-LDMOS transistor.

B High Drain Current Pch-LDMOS transistor

To increase the saturation drain current ($I_{ds,sat}$) of the Pch-LDMOS transistor, the effective channel length (L_{eff}) was effectively reduced while maintaining the Nwell profile to keep a high breakdown voltage. Fig. 4 shows the simulated Pch-LDMOS device structure of net doping profile. By introducing the channel profile minimization process, the L_{eff} is reduced to about 1/4 for the conventional device. A combination of self-aligned ion implantation for the channel and ion implantation through Poly-Si gate electrode enables this profile. The measured $I_{ds,sat}$ is increased by 74% as shown in Fig. 5, while the high on-state breakdown voltage (BV_{on}) of −240 V is retained. With threshold voltage optimization, an $I_{ds,sat}$ of 172 μA/μm is achieved for the 200V Pch-LDMOS transistor, which is comparable to that for the Nch-LDMOS transistor as shown in Fig. 6.

The off-state breakdown voltage (BV_{off}) of the 200V Pch-LDMOS transistor is affected by the voltage of SOI handle wafer (V_{sub}), since the electric potential field in the LDMOS transistor is changed by the V_{sub}. The 200V Pch-LDMOS transistor is optimized in its profile for a use at the supply voltage of +/−100 V. Therefore, the off-sate breakdown voltage increases with the V_{sub}, and becomes −284 V when the V_{sub} is −100 V as shown in Fig. 7.

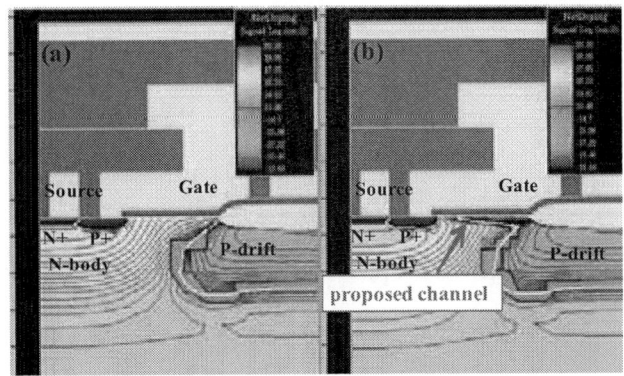

Figure 4.   Simulated device structure of net doping profile for (a) the conventional channel and (b) the proposed channel for the Pch-LDMOS transistor.

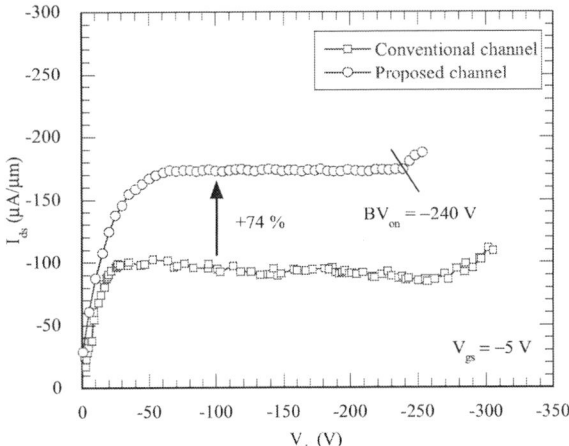

Figure 5.   Measured I_{ds} - V_{ds} characteristics of 200V Pch-LDMOS transistor comparing conventional and proposed channel profiles.

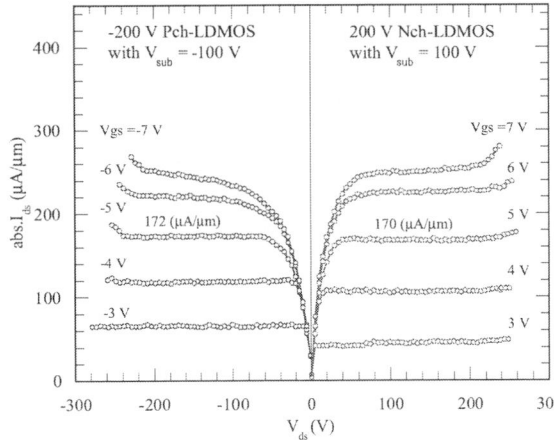

Figure 6.   Measured I_{ds} - V_{ds} characteristics of Pch and Nch-LDMOS transistors.

978-1-4244-8425-6/11 $26.00 © 2011 IEEE   45

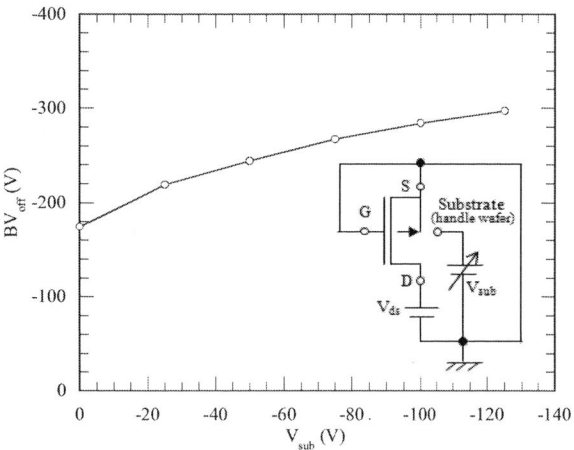

Figure 7. V_{sub} dependence of measured BV_{off} in the 200V Pch-LDMOS transistor.

Fig. 8 shows the measured $I_{ds,sat}$ - V_{ds} characteristics when V_{sub} is 0 V and −100 V. Both the $I_{ds,sat}$ and the BV_{on} are increased when the V_{sub} is −100 V. This result is because hole current flows more widely when the V_{sub} is −100 V, as shown in Fig. 9. The electrical potential from the SOI handle wafer leads to the accumulation of holes under the p-drift region.

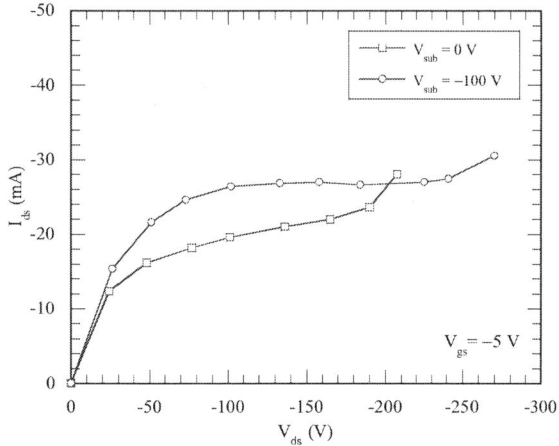

Figure 8. Measured I_{ds} - V_{ds} characteristics of the Pch-LDMOS transistor at $V_{sub} = 0$ V and −100 V.

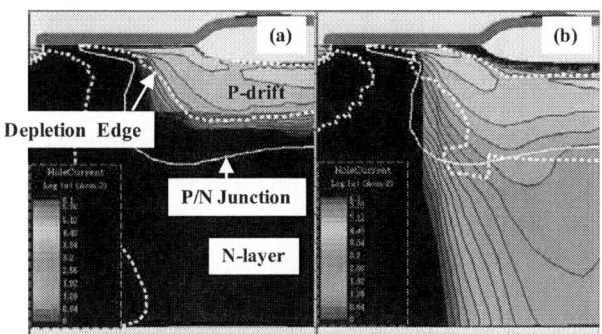

Figure 9. Simulated hole current distribution in the Pch-LDMOS transistor with (a) $V_{sub} = 0$ V and (b) $V_{sub} = -100$ V at $V_{gs} = -5$ V, $V_{ds} = -100$ V.

A high temperature reverse bias (HTRB) reliability test at T=150°C and at constant $V_{ds} = -210$ V, $V_{sub} = -100$ V was performed. As shown in Fig. 10 and 11, the shifts in both R_{on} and normalized off-state leakage current for the proposed channel remain in the same range as those for the conventional channel structure.

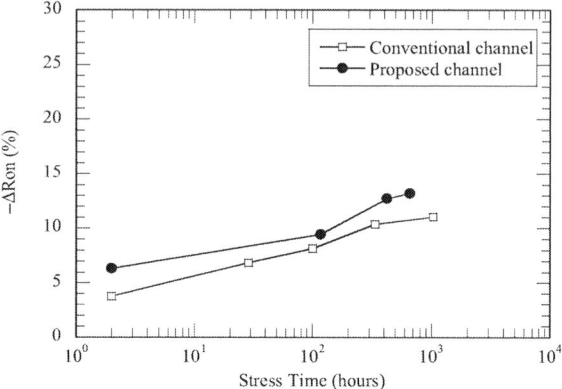

Figure 10. R_{on} shift of 200V Pch-LDMOS transistors under HTRB stress.

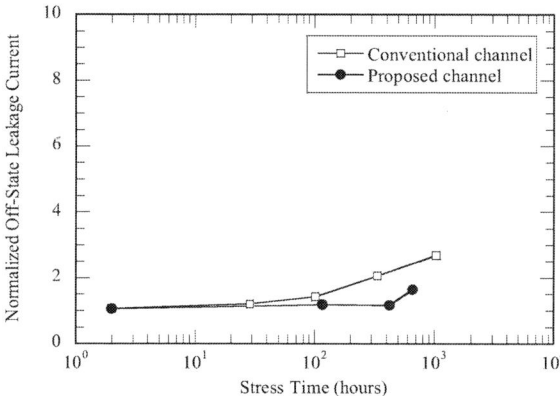

Figure 11. Normalized off-state leakage current divided by initial off-state leakage current of 200V Pch-LDMOS transistors under HTRB stress.

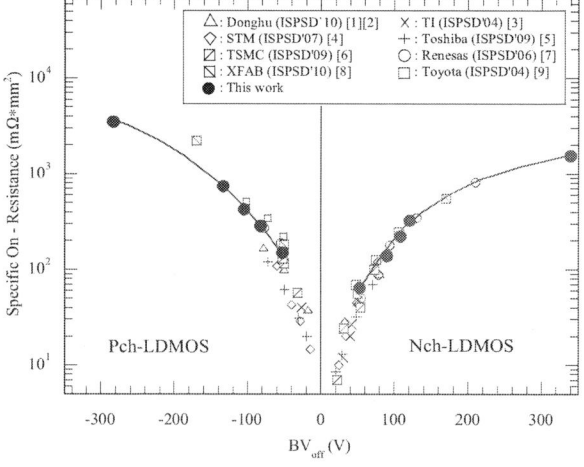

Figure 12. Relationship between $R_{on,sp}$ and BV_{off} for this work and previously reported values.

978-1-4244-8425-6/11 $26.00 © 2011 IEEE

Fig. 12 shows the relationship between specific on-resistance ($R_{on,sp}$) and BV_{off} for 35V to 200V LDMOS transistors in our platform technology compared with previously reported values [1-9]. In the 200V Pch-LDMOS transistor, a competitive low on-resistance of 3470 mΩ*mm^2 with the BV_{off} of −284 V are achieved. In wide range rated voltage, a low on-resistance characteristics of the LDMOS transistors is also achieved.

IV. DEVICE LINEUP

Table I shows the device lineup of our SOI LDMOS technology with high performance Pch-LDMOS transistors. LDMOS transistors for operation over the wide range of rated voltage from 35V to 200V are fabricated simultaneously with 5V-CMOSFETs and passive devices by using our CMOSFET platform technology. This platform technology can be applied to smart power ICs, and to industrial, medical and automotive high-voltage driver ICs.

TABLE I.   DEVICE LINEUP OF OUR SOI LDMOS TECHNOLOGY

Device	Parameters
Substrate	SOI
LV-CMOS V_{DD}	5V
HV-LDMOS	35V,60V,80V,100V,200V
Diode	Zener, LV, HV:35V~200V
BJT	NPN:h_{FE}=50, PNP:h_{FE}=37
Resistor	Poly-Si, Diff.
Capacitor	1.7 fF/μm^2 300V-MIM:0.034 fF/μm^2
Fuse	Poly-Si
Metal	3- layer

V. CONCLUSION

We have developed high performance SOI Pch-LDMOS transistors in 35V to 200V SOI LDMOS technology. By introducing a novel channel structure, an $I_{ds,sat}$ of 172 μA/μm in the 200V Pch-LDMOS transistor was achieved, which was comparable to that of Nch-LDMOS transistor. A wide range of rated voltage was efficiently accomplished for the LDMOS transistors by optimizing layout lengths such as drift layer with RESURF structure and field plate. This platform technology is applicable to a broad range of products.

ACKNOWLEDGMENT

The authors would like to thank Dr. J. Noguchi, Mr. K. Kitazawa, and Mr. T. Miyoshi for support in process and device development, and Mr. S. Hanazawa, Mr. T. Shinomiya, and Dr. H. Yoshizawa for useful discussions on harmonic imaging driver IC characteristics.

REFERENCES

[1] Choul-Joo Ko, et al., "Low Vgs P-ch LDMOS with Shallow Pwell from 8V to 60V", ISPSD'10, pp.177-180.

[2] Choul-Joo Ko, et al., "A New 8V – 60V rated Low Vgs NLDMOS Structure with Enhanced Specific on-Resistance", ISPSD2010, pp. 245-248.

[3] Sarneer Pendharkar, et al., "7 to 30V state-of-art power device implementation in 0.25μm LBC7 BiCMOS-DMOS process technology", ISPSD2004, pp. 419-422.

[4] Damiano Riccardi, et al., "BDC8 from 7V to 70V : a new 0.18mm Technology Platform to Address the Evolution of Applications towards Smart Power Ics with High Logic Contents", ISPSD2007, pp. 73-76.

[5] Koji Shirai, et al., "Ultra-low On-Resistance LDMOS Implementation in 0.13mm CD and BiCD Procss Technologies for Analog Power IC's", ISPSD2009, pp. 77-79.

[6] Ken Chen, et al., "The Foundry Perspective on Integrated Power Technologies", ISPSD2009, pp. 6-8.

[7] T. Nita, et al., "Wide Volatge Power Device implementation in 0.25mm SOI BiC-DMOS", ISPSD2006, pp.341-344.

[8] Alexander Hölke, et al., "A 200V Partial SOI 0.18mm CMOS technology", ISPSD2010, pp.257-260.

[9] Fumiaki Kawai, et al., "Multi-Voltage SOI-BiCDMOS for 14V&42V Automotive Applications", ISPSD2004, pp165-168.

[10] Torfinn Taxt et al., "Supperresolution of Ultrasound Images Using the First and Second Harmonic Signal", IEEE Trans. Ultrason., Ferroelect., Freq. Contr., vol. 51, no. 2, pp. 163-175, 2004.

[11] Adriaan W. Ludikhuize, "A Review of RESURF Technology", ISPSD2000, pp. 11-18.

Proceedings of the 23rd International Symposium on Power Semiconductor Devices & IC's
May 23-26, 2011 San Diego, CA

1.7kV Trench IGBT with Deep and Separate Floating p-Layer Designed for Low Loss, Low EMI Noise, and High Reliability

So Watanabe and Mutsuhiro Mori

Hitachi Research Laboratory
Hitachi Ltd.
Hitachi-shi, Ibaraki, Japan
so.watanabe.du@hitachi.com

Taiga Arai, Kohsuke Ishibashi, Yasushi Toyoda,
Tetsuo Oda, Takashi Harada and Katsuaki Saito

Power & Industrial Systems Division
Power Systems Company, Hitachi Ltd.
Hitachi-shi, Ibaraki, Japan

Abstract— A novel 1.7kV IGBT with deep floating-p layers separated from trench gates has been developed to realize low loss, low EMI noise, and high reliability. Separating floating-p layers from the trench gates reduces excess V_{GE} overshoot, which results in a 51% smaller reverse recovery dV_{AK}/dt than the conventional IGBT. The deep floating p-layers weaken the electric field under the trenches, which results in an avalanche breakdown voltage of 2250V. In addition, the E_{on} + E_{off} for the proposed structure can be reduced by 47% more than that of the conventional one, maintaining a low $V_{CE(sat)}$ of 2.3V at 125°C.

I. INTRODUCTION

Controllability of dV/dt during switching operation is one of the main issues in IGBT design. A high switching speed can reduce switching loss, but it also increases EMI noise caused by a high dV/dt. Therefore, the dV/dt must be controllable over a wide range so that users can make the most suitable choice for their system. Some trench-type IGBTs to reduce EMI noise have been proposed[1][2]. These IGBTs can achieve a low-recovery dV/dt of a free-wheeling diode of the opposite arm by suppressing excess V_{GE} overshoot during the turn-on operation. This paper presents a newly developed 1.7kV IGBT for low loss, low EMI noise, and high reliability that includes a new design to suppress V_{GE} overshoot.

II. DEVICE DESIGN

Fig. 1 shows the three IGBT structures described in this paper. The conventional IGBT (Fig. 1(b)) has two different intervals between trench gates. The n$^+$-sources that form the MOS channels are arrayed only in the narrower intervals and contact the emitter electrode. In the wider intervals, floating p-layers are formed to ensure a high avalanche breakdown voltage. This arrangement enables both saturation current and saturation voltage to be decreased, which results in low conduction loss and sufficient short-circuit capability. We call this concept high-conductivity IGBT (HiGT)[3][4].

The proposed structure is shown in Fig. 1(a). This structure also follows the HiGT concept and has the two different intervals between trenches. The originality in this

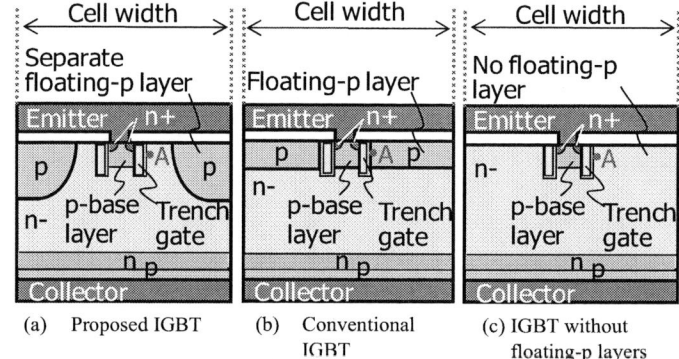

(a) Proposed IGBT   (b) Conventional IGBT   (c) IGBT without floating-p layers

Figure 1. Schematic cross-sectional structures of 1.7kV trench IGBTs (1 cell).

structure is the floating p-layers which are deeper than the trench gates and separated from the trenches. In this paper, we discuss and compare the effects of the proposed structure with both the conventional IGBT with the floating p-layers formed on the entire surface, and the IGBT without floating p-layers as an extreme structure (Fig. 1(c)).

III. CHARACTERISTICS

Fig. 2 shows the simulated cell-width dependencies of breakdown voltage BV_{CE} of the active cell. The IGBT without floating-p layers has a much lower BV_{CE} than the other IGBTs. This is because removing the floating p-layers increases the electric field under the trench gates (Fig. 3) as also described in a previous work[4]. In the proposed IGBT, the deep floating p-layers pull equi-potential lines deeper, resulting in the weaker electric field and higher BV_{CE} of 2250V.

Fig. 4 shows the simulated cell-width dependencies of maximum recovery dV_{AK}/dt at 25°C, $J_C = 0$~115A/cm^2. The proposed IGBT can be designed to have a lower dV_{AK}/dt than the conventional IGBT. It is because fewer holes are injected to the floating p-layers as the layer is separated farther from the trench gates, which suppresses the increase of the V_{FP} (the electrostatic potential at point A as shown in Fig. 1) during turn-on operation. The smaller increase of V_{FP} can reduce

978-1-4244-8425-6/11 $26.00 © 2011 IEEE

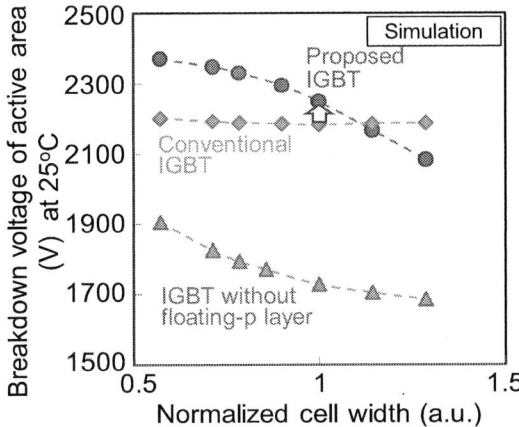

Figure 2. Simulated cell-width dependence of breakdown voltage.

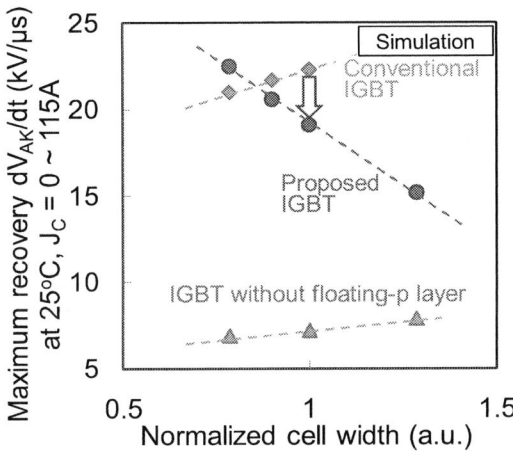

Figure 4. Simulated cell-width dependence of maximum recovery dV_{AK}/dt.

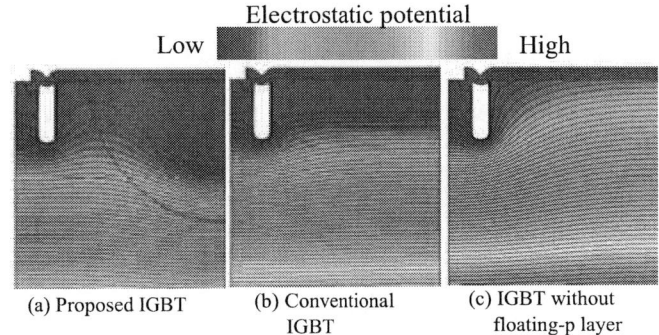

Figure 3. Simulated electrostatic potential of surface and equi-potential lines at 25°C, V_{CE}=1700V.

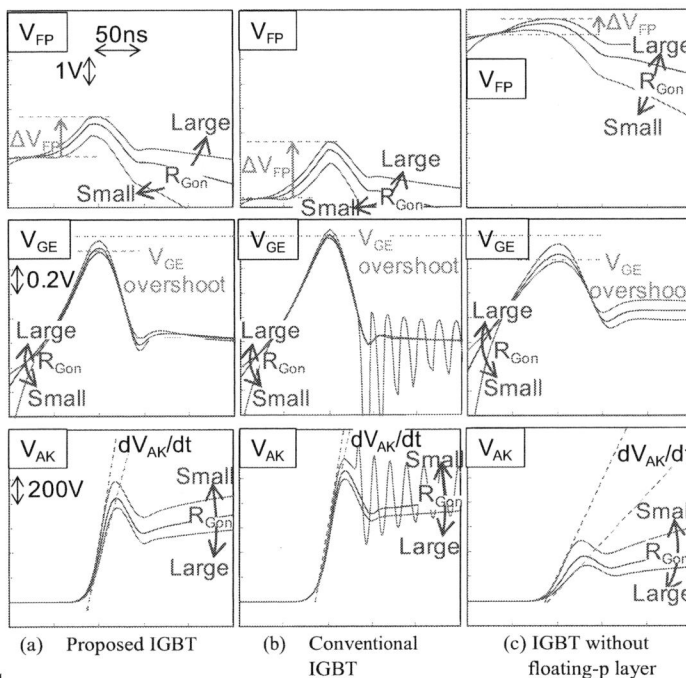

(a) Proposed IGBT (b) Conventional IGBT (c) IGBT without floating-p layer

Figure 5. Simulated waveforms of V_{FP}, V_{GE} and V_{AK} during turn-on and recovery operation.

excess V_{GE} overshoot and dV_{AK}/dt[1][2]. Therefore, the proposed IGBT can reduce dV_{AK}/dt, and the IGBT without floating p-layer can further reduce. The mechanism is discussed in detail in section IV. Fig. 5 shows the simulated waveforms of V_{FP}, V_{GE}, and V_{AK} during turn-on and recovery operation when the turn-on gate resistance R_{Gon} is changed. It has been verified that the IGBT with a lower ΔV_{FP} (hopping voltage of V_{FP}) can reduce more V_{GE} overshoot and dV_{AK}/dt. The results described above indicate that the proposed IGBT can achieve both higher reliability of the gate oxides and lower EMI noise than the conventional IGBT.

Fig. 6 shows the measured recovery dV_{AK}/dt and $E_{on}+E_{off}$ of the proposed IGBT and the conventional IGBT. The $V_{CE(sat)}$ of both structures is 2.3V at 125°C, J_C = 115A/cm^2. The parameter of dV_{AK}/dt and E_{on} is R_{Gon}. The turn-off gate resistance R_{Goff} is constant, so E_{off} is also constant. The lower limit of the proposed IGBT has been reduced by at least 51%, so users can select gate-drive conditions that are suitable for their systems. In addition, $E_{on} + E_{off}$ at dV_{AK}/dt = 15kV/μs for the proposed IGBT can be reduced by 47% more than that for the conventional IGBT.

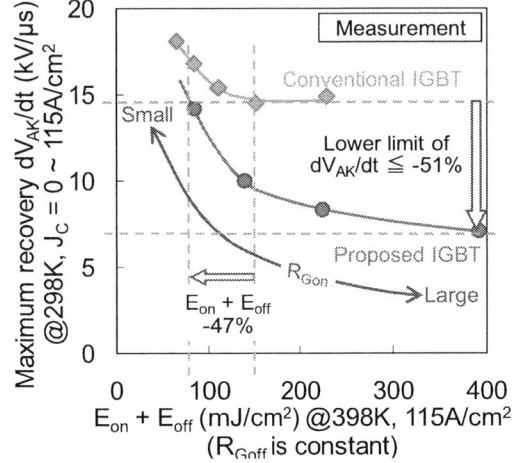

Figure 6. Measured trade-off of maximum recovery dV_{AK}/dt and $E_{on} + E_{off}$.

978-1-4244-8425-6/11 $26.00 © 2011 IEEE

IV. DISCUSSION

The mechanism that enables the proposed IGBT to reduce the V_{GE} overshoot during turn-on operation is discussed below. The effect of the hole current on ΔV_{FP} is simulated by using a circuit (shown in Fig. 7). V_{CE} is kept constant (900V) so that the V_{FP} is not affected by the V_{CE} waveform difference. The V_{GE} is decided only by a gate voltage source without the feedback from ΔV_{FP} because the gate is directly connected to the source without gate resistance. Therefore, this circuit enables that the susceptibilities of the V_{FP} to only the collector current density are compared among the structures as shown in Fig. 1.

Fig. 8 shows the simulated waveforms of the IGBTs shown in Fig. 1. The V_{GE} is linearly increased from 0 to 10V at $dV_{GE}/dt = 15V/\mu s$. The threshold voltage is adjusted to 6.5V so that the dI_C/dt of all IGBTs is the same. For all IGBTs, the V_{FP} abruptly increases after I_C starts to flow at $t = 0.47\mu s$, and then it slows down again. However, the dV_{FP}/dt during the period when V_{FP} abruptly increases is different for each IGBT. The conventional IGBT has the largest dV_{FP}/dt, the proposed IGBT has the second largest, and the IGBT without floating p-layer has the smallest. The reasons for this are considered below.

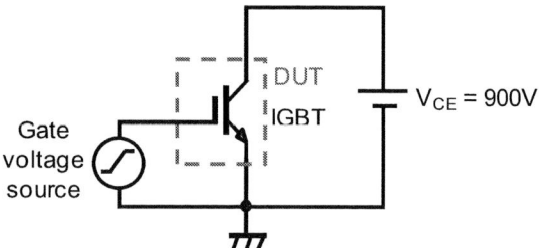

Figure 7. Circuit for simulation of low noise mechanism.

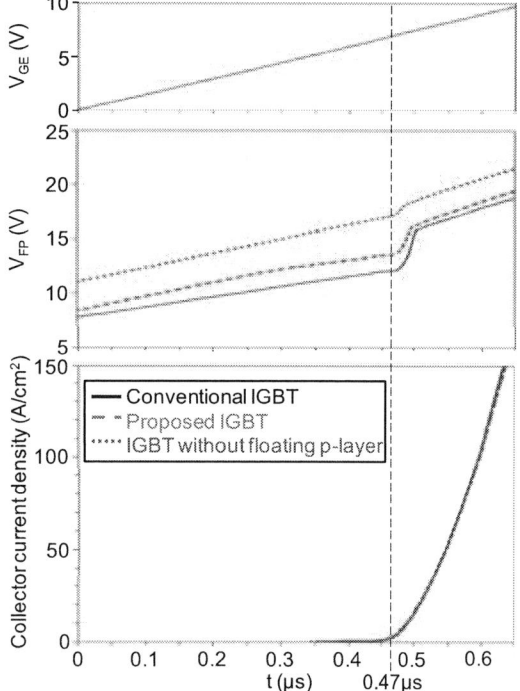

Figure 8. Simulated waveforms of circuit shown in Fig. 7.

Fig. 9 shows the enlarged waveforms focusing on the abrupt increase of the V_{FP}, and the hole density around the trench gate at $t = 0.488, 0.496$, and $0.504\mu s$. The hole density

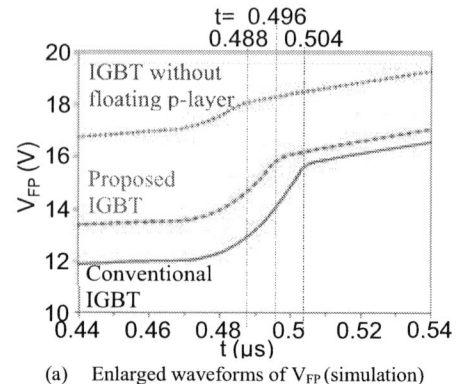

(a)   Enlarged waveforms of V_{FP} (simulation)

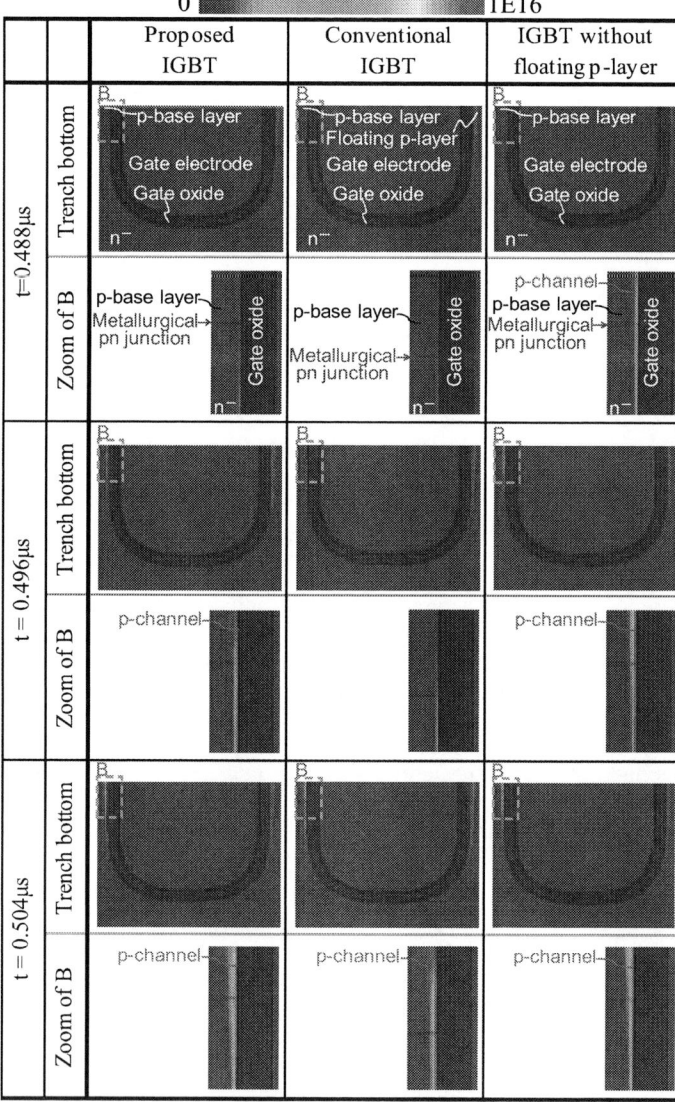

(b)   Simulated hole density around trench bottom at $t = 0.488, 0.496, 0.504\mu s$

Figure 9. Formation of p-channel on gate oxide/silicon interface.

on the gate oxide/silicon interface increases as the hole current flows. In the conventional IGBT, a p-layer with a high hole density larger than 1E15cm^{-3} is formed at t = 0.504μs. Such a transient hole accumulation has been reported as an undesirable phenomenon that can cause, for example, the negative capacitance during turn-off[5][6]. In the case of the IGBTs shown in Fig. 1, the formed p-layer has another effect: a p-channel from the floating p-layer to the p-base layer. The low resistance of the p-channel stabilizes the V_{FP} at the level close to the emitter potential. This is why the increase of V_{FP} slows down after t=0.504μs for the conventional IGBT. This analysis is consistent with the phenomenon as described in [7]: "discharge" of the holes from the floating p-layer to the p-base layer under an inductive-load turn-on operation. The V_{FP} of the proposed IGBT starts to slow down earlier than the conventional IGBT (t=0.496μs), while the IGBT without floating p-layer is the earliest (t=0.488μs). For all the IGBTs, the time at which the p-channel is formed is consistent with the time at which the increase of V_{FP} slows down.

Fig. 10 shows the simulated hole-current density around the surface of the IGBTs at t = 0.488μs. The hole current of the IGBT without floating p-layer concentrates on the bottom of the trench gate. This result indicates that the concentration of the hole current forms the p-channel as shown in Fig. 9. The concentrated hole current around the trench gate of the conventional IGBT is the smallest because the floating p-layer with low potential for holes attracts a part of the hole current. Since the floating p-layer of the proposed IGBT is separated from the trench gate, the layer attracts the hole current less than that of the conventional IGBT. Therefore, the hole concentration around the trench gate of the proposed IGBT is larger than the conventional IGBT; that is, the p-channel with low resistance is formed earlier, and the dV_{FP}/dt is reduced. This is the reduction mechanism of the V_{GE} overshoot during the turn-on operation of the proposed IGBT.

Hole current density
(A/cm^2, linear scale)

0   10

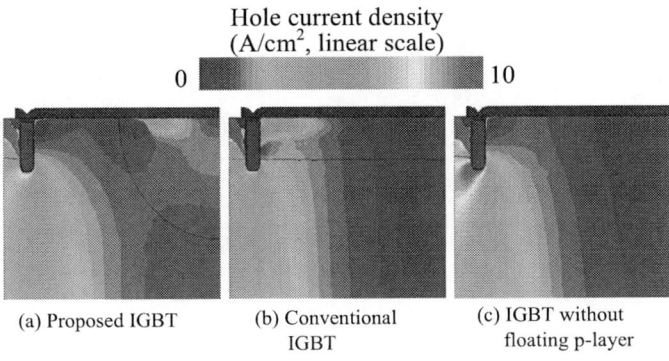

(a) Proposed IGBT   (b) Conventional IGBT   (c) IGBT without floating p-layer

Figure 10. Simulated hole current density at t=0.488μs.

V. CONCLUSION

A novel 1.7kV IGBT with deep floating p-layers separated from trench gates has been proposed and developed. The deep floating p-layers weaken the electric field under the trenches, which results in an avalanche breakdown voltage of 2250V. Separating floating-p layers from the trench gates reduces excess V_{GE} overshoot, which results in 51% smaller reverse recovery dV_{AK}/dt than the conventional IGBT. E_{on} + E_{off} at dV_{AK}/dt = 15kV/μs for the proposed IGBT can be reduced by 47% more than that for the conventional IGBT. Therefore, the proposed IGBT can achieve low loss, low EMI noise, and higher reliability.

The reduction mechanism of the V_{GE} overshoot for the proposed IGBT was investigated by simulation. Results have shown that the potential of the floating p-layer of the proposed IGBT is more stabilized than the conventional IGBT, because the hole current concentrates on the bottom of the trench gates and forms the low-resistance p-channel from the floating p-layer to the p-base layer.

REFERENCES

[1] M. Yamaguchi, I. Omura, S. Urano, S. Umekawa, M. Tanaka, T. Okuno, T. Tsunoda and T. Ogura, "IEGT design criterion for reducing EMI noise", Proc. 16th ISPSD, pp. 115-118, 2004.

[2] Y. Onozawa, H. Nakano, M. Otsuki, K. Yoshikawa, T. Miyasaka and Y. Seki, "Development of the next generation 1200V trench-gate FS-IGBT featuring lower EMI noise and lower switching loss", Proc. 19th ISPSD., pp. 13-16, 2007.

[3] K. Oyama, T. Arai, K. Saitou, K. Masuda and M. Mori, "Advanced HiGT with low-injection punch through (LiPT) structure", Proc. 16th ISPSD., pp. 111-114, 2004.

[4] M. Mori, K. Oyama, Y. Kohno, J. Sakano, J. Uruno, K. Ishizaka and D. Kawase, "A trench-gate high-conductivity IGBT (HiGT) with short-circuit capability", IEEE Trans. on Elec. Dev., pp. 2011-2016, 2007.

[5] I. Omura, T. Domon, T. Miyanagi, T. Ogura and H. Ohashi, "IEGT design concept against operation instability and its impact to application", Proc. 12th ISPSD., pp. 25-28, 2000.

[6] Y. Onozawa, M. Otsuki, N. Iwamuro, S. Miyashita, T. Miyasaka and Y. Seki, "1200V super low loss IGBT module with low noise characteristics and high dI/dt controllability", Proc. 2005 IEEE-IAS Annual Meeting., pp. 383-387.

[7] N. Tokura, "Influence of floating p-base on turn-on characteristics of trench-gate FS-IGBT", IEEJ Trans. IA, Vol. 130-D, No. 6, pp. 728-733, 2010., in Japanese.

Proceedings of the 23rd International Symposium on Power Semiconductor Devices & IC's
May 23-26, 2011 San Diego, CA

Development of the next generation 1700V trench-gate FS-IGBT

Y.Onozawa, D.Ozaki, H.Nakano, T.Yamazaki and N.Fujishima

Fuji Electric Systems Co., Ltd. 4-18-1, Tsukama, Matsumoto, Nagano 390-0821, Japan

Tel: +81-263-28-7167, Fax: +81-263-26-6945, e-mail:onozawa-yuichi@fujielectric.co.jp

Abstract— **This paper describes the next generation 1700V trench-gate FS-IGBT utilized the micro p-base structure for the first time. The new 1700V IGBT has been achieved that "better turn-on di/dt controllability", "oscillation free turn-off" and "improved Von-Eoff trade-off relationship" as well as 600V and 1200V IGBTs. Furthermore, the critical thermal runaway temperature has successfully been elevated by the newly developed field-stop layer, which leads to increase of maximum junction temperature as high as 175 deg. C.**

Keywords: micro p-base structure, backside avalanche, thermal runaway, field-Stop layer optimization

I. INTRODUCTION

Insulated Gate Bipolar Transistor (IGBT) modules are widely used in a variety of power switching applications, such as motor drives, traction control and power supplies. In recent years, the market of high power IGBT module has been expanding because large wind power and photovoltaic generation systems that use renewable energy source have recently become widespread in the world. Therefore, there is a great need for further improvement of the high blocking voltage IGBT such as 1700V device.

Trade-off relationship between the on-state voltage drop and the turn-off power dissipation of IGBT has been improved drastically by utilizing the trench gate structure and the field-stop concept [1][2]. Furthermore, we have proposed the new IGBT surface structure named micro p-base which has realized better turn-on di/dt controllability, oscillation free turn-off and improved Von-Eoff trade-off relationship [3]-[5].

This paper presents the superior characteristics of newly developed 1700V IGBT chip which is utilized the micro p-base structure and newly developed field-stop layer. The new 1700V IGBT has been successfully realized 30% lower turn-on power dissipation under the same collector current peak and 0.2V lower on-state voltage drop under the same turn-off power dissipation compared to the conventional trench-gate FS-IGBT. In addition, the critical thermal runaway temperature has been elevated which leads to increase of maximum junction temperature as high as 175 deg. C by utilizing the new field-stop layer.

II. DEVICE STRUCTURE AND DESIGN

A. Surface structure

The new 1700V trench-gate FS-IGBT has been utilized the micro p-base structure as shown in Fig.1. The concept of this structure has already been reported in previous our work [3]-[5]. The localized micro p-bases increase carrier density near the surface emitter side. Since there are no floating p-base regions, undesirable current increasing can be avoided during IGBT turn-on transient. In addition, the surface electric field distribution is uniformed due to periodical trench-gate pattern. The device thickness can be reduced without turn-off oscillation because lower resistivity drift layer can be utilized.

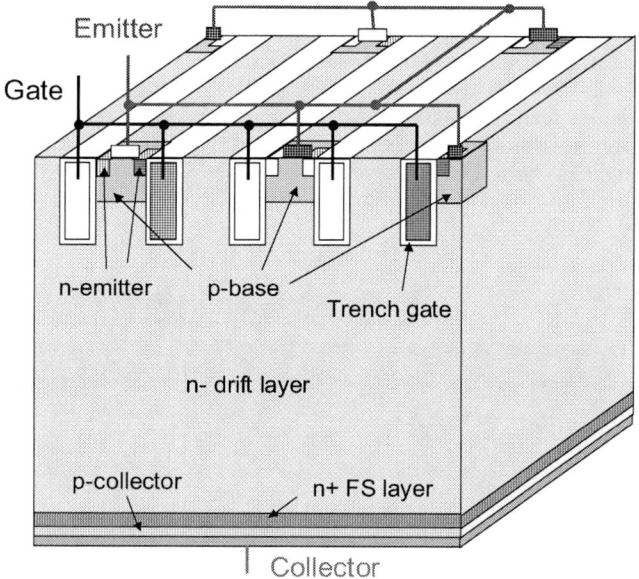

Fig. 1 Schematic viewing of the newly developed 1700V IGBT.

B. Back-side structure

Another technology for the new IGBT is newly developed field-stop layer. Especially in high blocking voltage IGBT, net positive charge of the drift layer is easy to decrease by negative charge of the injected excess electron during the short-circuit transient. As a result, the electric field peak moves toward the backside and avalanche breakdown occurs at this region, and then the IGBT will be destructed [6][7].

978-1-4244-8425-6/11 $26.00 © 2011 IEEE   52

To avoid this failure, it is essential to reduce the electron current density from the surface side or increase of hole injection from the backside during the short circuit transient.

Former can be achieved by reduction of the surface MOSFET channel width in order to decrease the saturation current, but it causes decrease of capable output current and unacceptable on-state voltage drop increasing.

Latter can be realized by increase of the parasitic PNP-transistor gain, as described

$$a_{PNP} = \alpha * \beta * \gamma \qquad (1)$$

where a_{PNP} is common-base current gain and α, β, γ are the collector efficiency, the base transfer efficiency and the emitter injection efficiency. According to eq. (1), a_{PNP} is controlled by β which is determined by neutral region in the field-stop layer [8]. As the field-stop layer is set to low concentration or shallow, the hole injection from the backside is increased with a_{PNP} increasing. As a result, the backside avalanche failure mode during the short circuit can be avoided. Figure 2 shows the calculation results of the electric field distribution in 1700V IGBTs during the short circuit transient. It is clear that increase of the electric field in the backside is prevented in higher a_{PNP} condition.

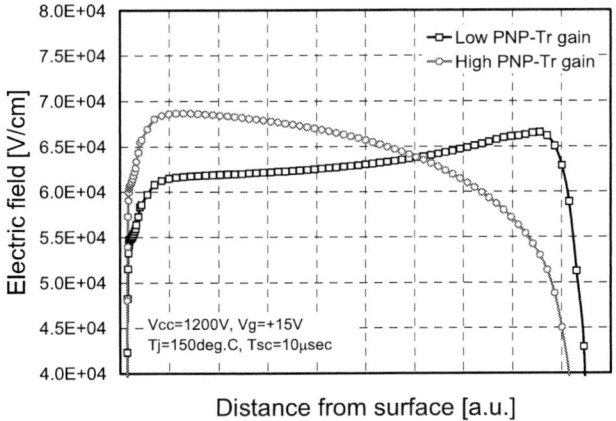

Fig. 2 Calculation results of the electric field distribution in 1700V IGBT during the short circuit.

However, the low concentration or shallow field-stop layer causes high leakage current especially in high temperature ambient. It is obvious that higher leakage current is severely disadvantage in high temperature operation.

To overcome this issue, we have proposed newly developed field-stop layer. In order to decrease the leakage current, high concentration and shallow layer is added to the original field-stop layer. The additional field-stop layer is well controlled so that high a_{PNP} is maintained. Therefore, the new field-stop layer can realize both higher hole injection during short circuit transient and lower leakage current even in high temperature ambient.

III. STATIC AND DYNAMIC CHARACTERISTICS

A. Static characteristics

Figure 3 shows forward characteristics of 1700V150A IGBT. As a result of high carrier accumulation effect by the micro p-base structure and reduction of the device thickness, the on-state voltage drop has successfully been reduced as low as 0.25V compared to the conventional trench-gate FS-IGBT under the same current density of 125A/cm^2.

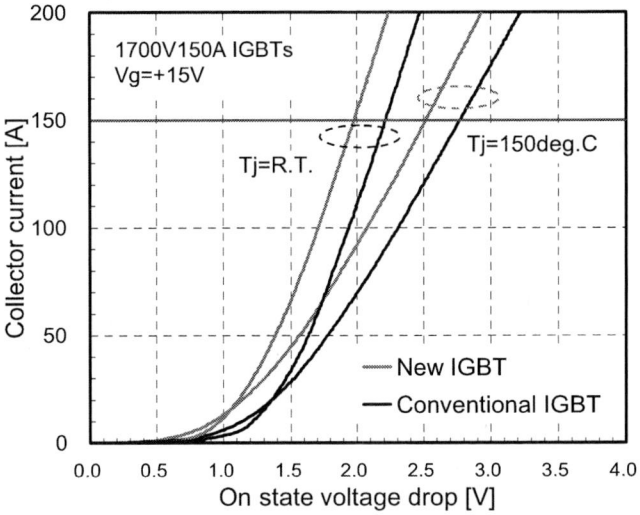

Fig. 3 Forward characteristics of 1700V150A IGBT.

The leakage currents of 1700V150A IGBT with the new field-stop layer and conventional one at 175 deg. C are shown in Fig.4. The leakage current has been reduced to one-third by utilizing the new field-stop layer, thanks to the additional shallow field-stop layer.

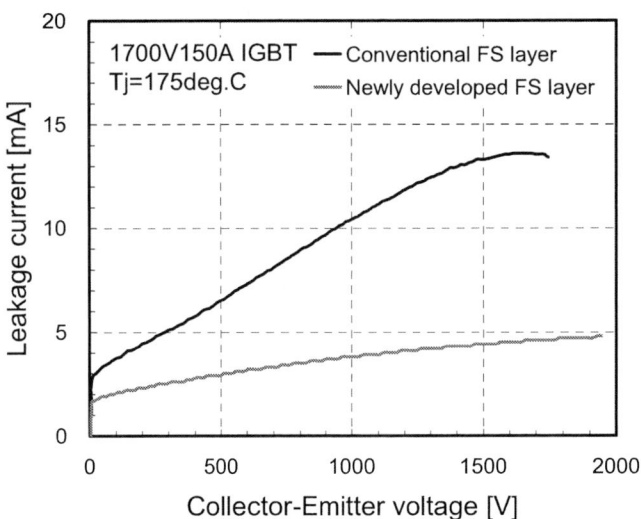

Fig. 4 Comparison of leakage current at 175 deg.C.

B. Turn-off characteristics

Figure 5 shows trade-off relationship between the on-state voltage drop and the turn-off energy. There is 0.2V on-state drop reduction compared to the conventional IGBT. In other words, 25% turn-off energy reduction has been achieved under the same on-state voltage drop.

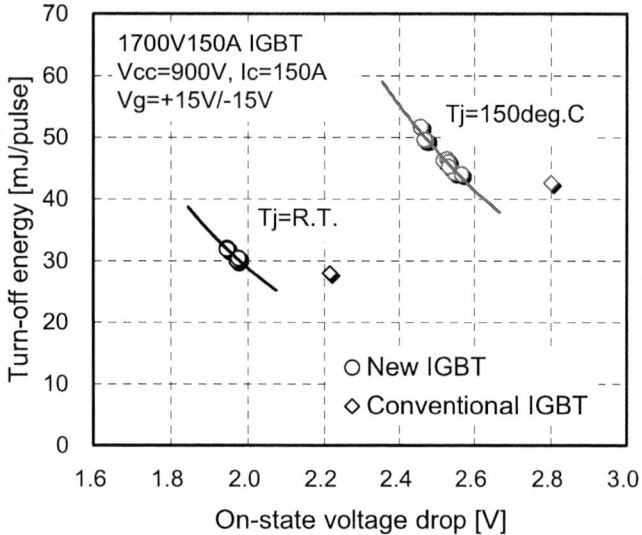

Fig. 5 Trade-off relationship between Von and Eoff.

Figure 6 shows comparison of the turn-off waveforms with 1000V DC-bus voltage and 360nH inductance. It is clear that the turn-off oscillation does not occur in the new IGBT even in this extreme condition. These results indicate that the new 1700V IGBT is capable of turn-off performance at critical condition without surge protection function.

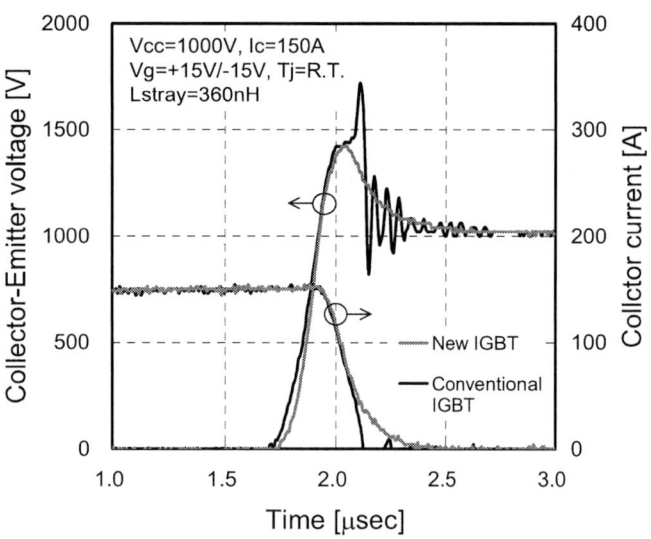

Fig. 6 Turn-off waveforms with large inductance.

C. Turn-on characteristics

Figure 7 shows trade-off relationship between turn-on energy at rated current and the collector peak current of 1700V150A IGBT. The trade-off relationship of the new IGBT is better than that of the conventional one. Under the same collector current peak at 70A, turn-on energy of the new IGBT is decreased by 30%. It means that the new IGBT can reduce the turn-on power dissipation under the same emission noise.

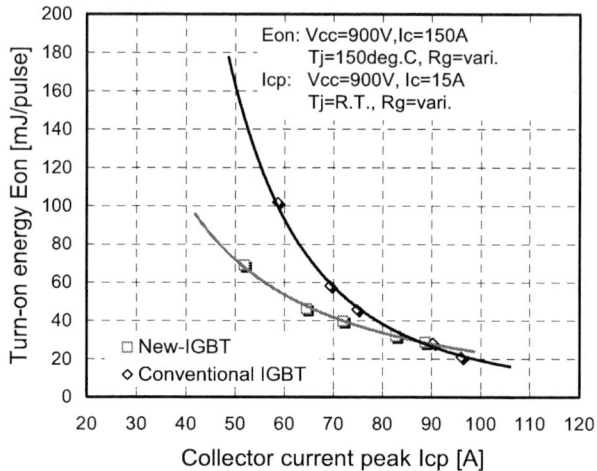

Fig. 7 Trade-off relationship between Eon and Icp.

IV. RUGGEDNESS

A. Large current short circuit withstand capability

The experimental measurement of large current short-circuit test of the new 1700V150A IGBT has been done. As shown in Fig. 8, the gate voltage was set at 29V. This condition produces large short circuit current of 2800A which is about 19 times larger than the rated current. Even in this extreme condition, the device survived after 5μsec short-circuit. Therefore, it is clear that the backside avalanche failure has been successfully prevented in the new 1700V IGBT.

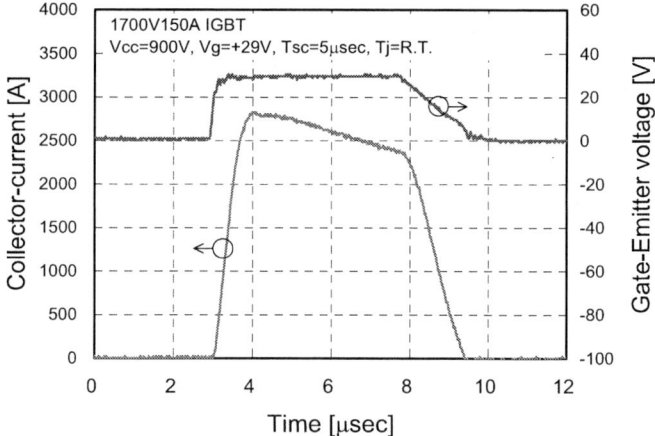

Fig. 8 Short circuit waveforms with large current.

978-1-4244-8425-6/11 $26.00 © 2011 IEEE

B. High Junction Temperature withstand capability

In order to increase a power density of the IGBT module, high temperature operation such as 175deg.C is one of the important issues of the IGBT. Therefore, we measured the critical thermal runaway temperature of the new and conventional 1700V IGBT. Figure 9 shows the test circuit for this measurement and the 1700V IGBT photograph after thermal runaway failure. The IGBT is set on a hot plate with temperature controller. After Vcc is applied to the IGBT, the IGBT temperature is increased by self-heating that caused by the product of Vcc and a leakage current. Finally, the IGBT will be destroyed due to a thermal runaway.

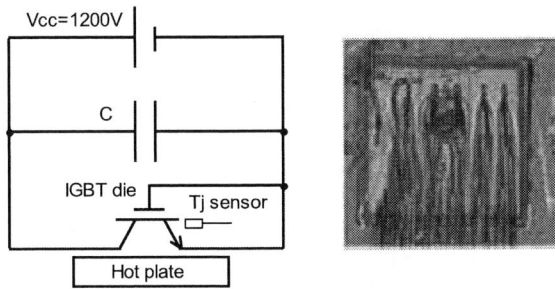

Fig. 9 Measurement circuit of thermal runaway temperature
and the 1700V IGBT photograph after failure.

Figure 10 shows comparison of the thermal runaway measurement results of the conventional and new 1700V IGBT. The new IGBT has higher critical thermal runaway temperature as high as 210 deg. C. It is obvious that the high critical thermal runaway temperature is achieved by low leakage current by utilizing the new field-stop layer. This result indicates that the new IGBT is capable of 175deg.C operation.

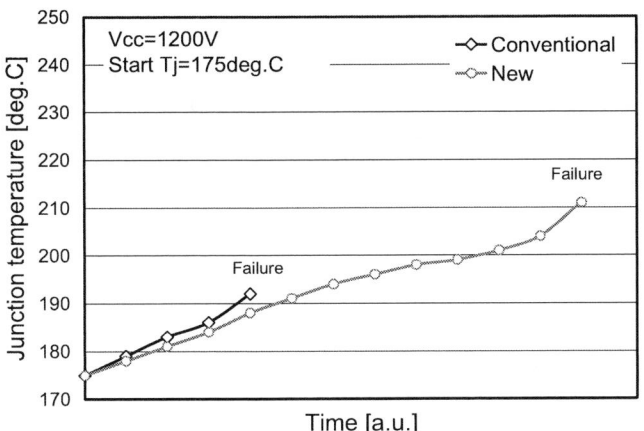

Fig. 10 Comparison of the thermal runway measurement results.

V. Conclusion

The new 1700V IGBT with low noise and low power dissipation has been successfully developed by utilizing the micro p-base structure and newly developed field-stop layer. The new IGBT has been able to realize 30% reduction in the turn-on power dissipation when compared to the conventional IGBT under the same collector peak current. The on-state voltage drop successfully reduced by 0.25V compared to the conventional trench-gate FS-IGBT under the same current density. The trade-off relationship between the on-state voltage and the turn-off power dissipation has been improved about 25% without the turn-off oscillation even in the extreme condition. Furthermore, large current short circuit withstand capability and higher critical thermal runaway temperature have been achieved by the new field-stop layer.

References

[1] M. Kitagawa, I. Omura, S. Hasegawa, T. Inoue, and A. Nakagawa, "A 4500V injection enhanced insulated gate bipolar transistor (IEGT) operating in a mode similar to a thyristor," in IEDM Tech. Dig., 1993, pp. 679–682.

[2] T.Laska, M.Münzer, F.Pfirsch, C.Shaeffer and T.Schmidt, "The Field Stop IGBT (FS IGBT) – A New Power Device Concept with a grate improvement Potential", in Proc.12th ISPSD, pp.355-358, June 2000.

[3] Y.Onozawa, H.Nakano, M.Otsuki, K.Yoshikawa, T.Miyasaka and Y.Seki, "Development of the next generation 1200V trench-gate FS-IGBT featuring lower EMI noise and lower switching loss" in Proc. 19th ISPSD, pp.13-16, June 2007.

[4] H.Nakano, Y.Onozawa, R.Kawano, T.Yamazaki and Y.Seki, "600V trnench-gate FS-IGBT with Micro-P structure" in Proc. 21st ISPSD, pp.13-16, June 2009.

[5] M.Momose, K.Kumada, H.Wakimoto, Y.Onozawa, A.Nakamori, K.Sekigawa, M.Watanabe, T.Yamazaki and N.Fujishima, "A 600V Super Low Loss IGBT with Advanced Micro-P structure for the next Generation IPM", in Proc.22nd ISPSD, pp.355-358, June 2002.

[6] M. Pfaffenlehner, T.Laska, R.Mallwitz, A. Mauder, F.Pfirsch and C.Shaeffer, "1700V IGBT3:Field Stop Technology with Optimized Trench Structure – Trend setting for the High Power Application in Industry and Traction", in Proc.14th ISPSD, pp.355-358, June 2002.

[7] A.Nakagawa, T.Matsudai, T.Matsuda, M.Yamaguchi and T.Ogura, "MOSFET-mode Ultra-Thin Wafer PTIGBTs for Soft Switching Application --- Theory and Experiments", in Proc.16th ISPSD, pp.103-106, 2004.

[8] M.Otsuki, Y.Onozawa, S.Yoshiwatari and Y.Seki, "1200V FS-IGBT module with enhanced dynamic clamping capability", in Proc.16th ISPSD, pp.339-342, 2004.

Proceedings of the 23rd International Symposium on Power Semiconductor Devices & IC's
May 23-26, 2011 San Diego, CA

The Radial Layout Design Concept
for the Bi-mode Insulated Gate Transistor

L. Storasta, M. Rahimo, M. Bellini, A. Kopta
ABB Switzerland Ltd, Semiconductors
Lenzburg, Switzerland
Email: liutauras.storasta@ch.abb.com

U. R. Vemulapati, N. Kaminski
Institute for Electrical Drives, Power Electronics and Devices (IALB)
University of Bremen
Bremen, Germany

Abstract—In this paper we present a new radial design concept for an optimized layout of anode shorts in the Bi-mode Insulating Gate Transistor (BiGT). The study shows that the arrangement of the n^+-stripes plays a key role for the on-state characteristics of the BiGT. With the aid of 3D device simulations the visualization of the plasma distribution during the on-state conduction was obtained in a 0.25×4 mm^2 large BiGT model area. The influence of the dimensioning and layout of the anode shorts was simulated and compared with measured on-state curves. A clear improvement of plasma distribution in the device when the stripes are arranged orthogonally (radially) to the pilot-IGBT boundary is observed in 3D simulations. Measurements confirm lower on-state losses as a result of better utilization of the device area.

I. INTRODUCTION

Until recently, the use of reverse conductive (RC) IGBT devices has been limited to low voltage and/or soft switching applications with reduced diode requirements. With the introduction of the Bi-mode Insulated Gate Transistor (BiGT) [1], a new target to replace the high voltage IGBT - Free wheeling diode (FWD) pair in high power applications has been set. The BiGT device is expected to outperform the state of the art IGBT and diode in both soft and hard switching conditions, and fulfil rigorous robustness standards set on power devices today.

In a reverse conducting IGBT, alternating n^+-type doped areas are introduced into the collector contact, which act as a cathode contact in the internal diode conduction mode. The area ratio between the IGBT anode (p^+-areas) and the diode cathode (n^+-areas) determines which part of the collector area is available in IGBT and diode modes, respectively. During conduction of the body diode, p^+-areas are inactive and do

not directly influence the diode performance. However, the n^+ areas act as anode shorts, strongly influencing the bipolar gain of the IGBT. During the design, trade-offs between the diode and IGBT modes must be carefully considered. One of the implications of anode shorting is the voltage snap-back, or negative resistance region in the device IGBT mode *I-V* characteristics. This effect could have a negative impact when devices are paralleled, especially at low temperature conditions. It has been shown that the initial snap-back can be controlled and eliminated by introducing a wide anode region, also called pilot-IGBT into the device [1], [2]. This resulted in a BiGT – a hybrid structure consisting of an RC-IGBT and a standard IGBT (Fig. 1). The sizing of the pilot-IGBT is an important design parameter determining the smooth on-set of the output characteristics with minimum snap-back. However, when the electron-hole plasma is built up in the pilot-IGBT area, only a small region of the BiGT is conductivity modulated. Further smooth and fast lateral expansion of the plasma towards the RC-IGBT region is crucial for strong conductivity modulation of the full device area and depends on the scaling, shape and arrangement of the anode shorts. For obtaining the largest possible diode to IGBT area ratio and widest anode areas simultaneously, stripe shaped anode shorts are utilized in the BiGT device. It has been demonstrated [2] that stripe design might lead to secondary snap-backs in the *I-V* characteristics. In this work, we analyze in detail experimentally and by device simulation the influence of the shape and arrangement of the n^+-shorts on the trade-off between conduction losses of the BiGT in both diode and IGBT modes.

II. EXPERIMENT AND SIMULATION SETUP

For the experiments we used 4500 V / 50 A Enhanced-Planar IGBT devices with anode shorts. Fig. 2 (a) shows the lithographic test-masks employed for the introduction of the n^+-shorts in the collector contact for the devices tested in this work. All the structures have identically sized pilot-IGBT regions, but different layout designs of the n^+-shorts. Structures S1, S2, S3 have stripe designs with different orientation of the stripes with respect to the pilot-IGBT area. The widths of the n^+ and p^+ regions (L_{n+} and L_{p+}, see Fig. 1) are 100 μm and 400 μm respectively, which results in an n area 25% of the total RC-IGBT collector contact area. Structures D1 and D2

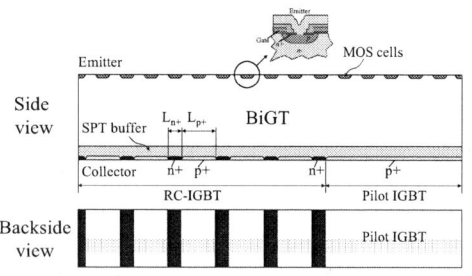

Fig. 1. Design features of the BiGT

978-1-4244-8425-6/11 $26.00 © 2011 IEEE

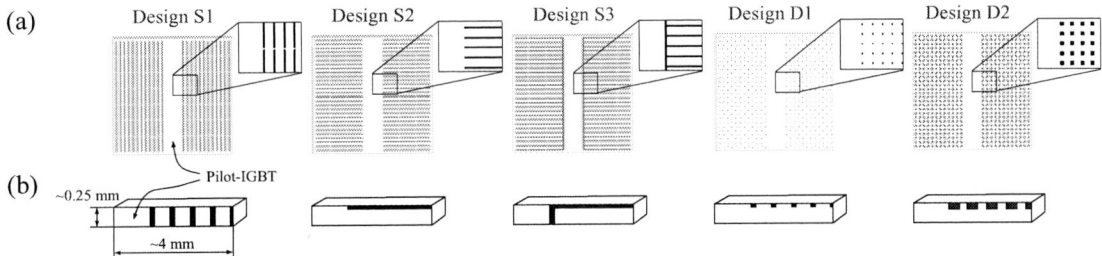

Fig. 2.   Different anode shorts layouts of BiGTs. (a) lithographic test masks used to structure the anode: Stripes radial (S1), parallel (S2) and parallel-radial (S3); Dots matching n^+-stripe width L_{n+} (D1) and matching diode area (D2). (b) structures used in 3D device simulations to represent the above experimental layouts.

are designed with square dot shaped n^+-shorts. Structure D1 has the dot dimension equal to the n^+-stripe width in structures S1, S2, S3 (100 μm), while in the structure D2 the dots are sized to match the n area ratio to total collector area of 25% . Devices with different Soft Punch Through n buffers have been manufactured for comparison. The on-state characteristics of the BiGTs were measured in diode and IGBT mode. A gate voltage of +15 V was applied during both IGBT and diode measurements.

To explain the experimental findings, device simulations were performed for the anode patterns shown in Fig. 2 (a). Except for design S1, all other variants can only be modelled in a 3D-mode. As a basis, a 3.3 kV Enhanced-Planar IGBT structure was utilized for quasi-stationary simulations in the IGBT mode. Large device structures with detailed layout of the anode and n^+-shorts measuring up to 0.25 mm × 4 mm were simulated in order to provide a realistic representation of the BiGT concept including a sufficiently large IGBT-pilot region. With the aim to reduce the number of mesh points, the MOS cell structure was replaced by a continuous n-contact to represent the forward conduction of the BiGT. On the device collector side, the n^+ diffusions (anode shorts) were introduced. Structures showing the collector side layouts used in the simulations are presented in Fig. 2 (b). When mirrored, they closely resemble the experimentally verified structures.

III. RESULTS AND DISCUSSION

A. Influence of the different n^+-stripe layouts

Initially, stripe designs S1, S2 and S3 were compared. Fig. 3 shows the measured I-V characteristics of the three above designs. The difference between the designs can be close to 1 V at nominal (50 A) current, depending on the n buffer used. When the n^+-stripes are placed parallel to the boundary between the pilot-IGBT and RC-IGBT (design S1), the on-state voltage drop is the highest for the same current flowing in the device, whereas the orthogonally (radially) arranged stripes (design S2) yield the lowest on-state voltage drop values. Also, there are small but clearly visible secondary snap-backs at low currents in the parallel stripe (S1) design while in the case of orthogonal stripe design the I-V curve is completely smooth. Diode characteristics of both structures

are overlapping and do not show any dependency on the stripe layout. The design with orthogonal stripes with an n^+-stripe running along the boundary of the pilot-IGBT area (design S3, Fig. 2) has features of both designs: high on-state voltage drop at low currents similar to the S1 design, and low losses similar to the S2 design at high currents.

To visualize the carrier plasma spread in the device, 3D device simulations were performed. The carrier density was extracted in the collector plane 30 μm above the contact. Fig. 4 shows a comparison of the simulated carrier density evolution for the S1, S2 and S3 designs. For all designs, at low current densities the electron-hole plasma is predominantly concentrated in the pilot-IGBT area. Further lateral expansion of the plasma strongly depends on the n^+-shorts layout. In the parallel stripes design (S1) case, the injection from the anode segments starts in a step-like manner as each anode segment becomes forward biased. When the stripes are placed orthogonally to the pilot-IGBT boundary, the injection from the anode stripes is initiated at a much lower current density. It starts at the position closest to the pilot-IGBT and smoothly extends towards the device periphery. It is important that for the same current, the device area filled with plasma is larger in the S2 design, giving lower on-state voltage drop and better current distribution compared to the parallel stripe (S1) design for the same n^+-shorts dimensions. Due to the low buffer doping applied in the simulations, the enhanced plasma spread

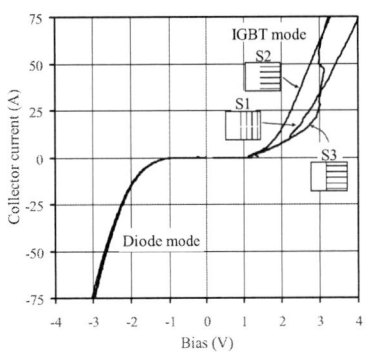

Fig. 3.   Measured on-state I-V characteristics of BiGTs with different stripe orientation.

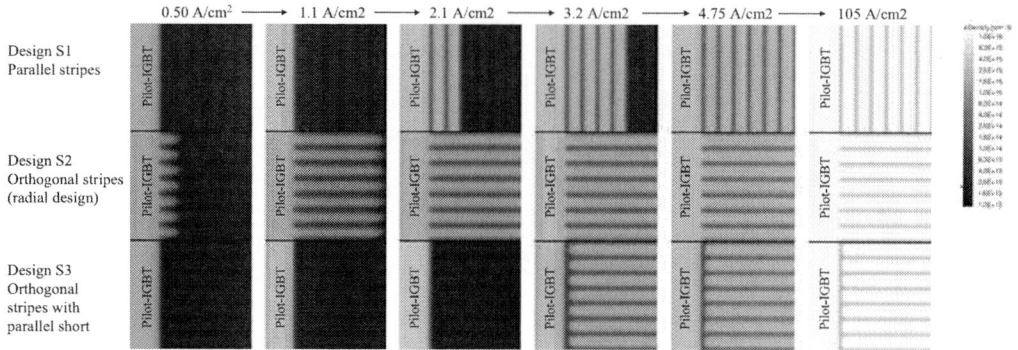

Fig. 4. Evolution of the simulated carrier density 30 μm above the anode contact during the current rise in the BiGTs with different backside n^+-stripe layouts. n^+-shorts ratio to full collector area is 25%, and pilot-IGBT size is maintained between designs.

effect is observed at significantly lower currents compared to the measurements.

In the case of design S3 the end of the anode stripe closest to the pilot-IGBT is shorted, which prevents forward biasing of the anode stripes and smooth expansion of the plasma seen in the orthogonal S2 design. The hole injection from the anode stripe starts at a higher current density in the middle of the stripe and almost instantly fills the whole device area with plasma, yielding at high current an identical *I-V* curve to the radial design S2.

We attribute the above phenomenon to the different direction of the lateral current in respect to the anode stripes. For all designs, the plasma is confined to the pilot-IGBT initially and the device current flow is directed vertically and laterally, towards the pilot-IGBT. In case when the n^+-stripes are placed parallel to the pilot-IGBT boundary, lateral current flows perpendicular to the anode segments and the voltage drop across the anode segment is small due to the limited non-shorted anode segment width L_{p+} available. Therefore, a high device current is required to forward bias the anode segments in the RC-IGBT region. Injection from each anode segment is visible in the *I-V* characteristics as a small secondary snap-back (see Fig. 3, design S1). This effect has been treated using 2D device simulations as well as analytical models [2]–[4] . Orthogonal (radial) placement of the n^+-stripes is a complex case and can no longer be treated in 2D. On the one hand, the anode stripe is now oriented along the lateral current flow and provides its full length for the forward biasing to be achieved. At the same time, the anode stripe is shorted along both sides. Evidently, this arrangement enables smooth transition from the pilot-IGBT towards the periphery of the device without negative resistance regions due to the early injection from the anode stripe at the point of connection to the pilot-IGBT (Fig. 4, compare 1.1 A/cm² plasma distribution between the designs). If this position is shorted as in design S3 (Fig. 2), injection from the anode stripe occurs in the middle of the stripe, which in this case is the point of highest potential difference. However, the required current is much higher as compared to design S2, therefore, up to 3.2 A, the current has

to concentrate in the pilot-IGBT area and then quickly spreads to the whole device area.

B. Comparison with a square dots pattern

It is interesting to see if a dot shaped n^+-shorts can bring further advantage to the best performing radial stripe design S2. Similar to the radial n^+-stripes design, dotted patterns also have long continuous p^+-doped anode regions along the lateral current direction. If the dots have the same dimension as the n^+-stripe (compare designs D1 and S2, Fig. 2) and the distance between the p^+-dots is equal to the width L_{p+} of the anode stripes in design S2, this inevitably reduces the diode contact area. To compensate for this, either the dot pitch or the dot size has to be adjusted, as done in design D2 (Fig. 2). Fig. 5 shows the comparison of the *I-V* characteristics in IGBT and diode mode for the stripe and dot designs. While the D1 design has slightly lower on-state voltage in the IGBT mode due to less shorting of the anode, the diode mode suffers from high conduction losses, as the diode contact area is reduced by 80%. Design D2 has the same diode mode conduction losses as for the stripe designs which is expected from the same diode contact area. However, the anode p^+ spacing between the dots is smaller by 25%, which is the cause for slightly higher losses in IGBT mode. 3D simulations in Fig. 6 also

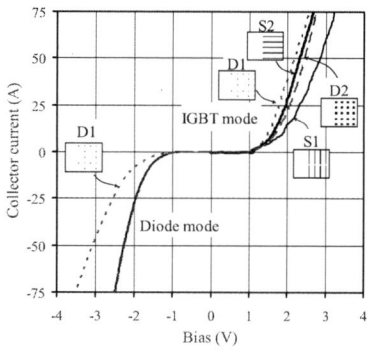

Fig. 5. Comparison of on-state *I-V* characteristics of BiGT devices with n^+-stripe and n^+-dot designs

978-1-4244-8425-6/11 $26.00 © 2011 IEEE

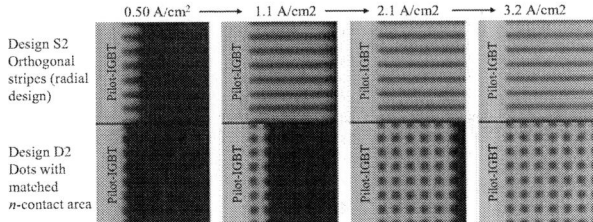

Fig. 6. Comparison of the simulated carrier density 30 μm above the collector contact during the current rise in the BiGTs with backside n^+-stripe and square n^+-dots designs.

confirm that less of the device area is filled with plasma in the D2 design, compared to S2. It is clear that radially arranged n^+-stripes achieve better trade-off between diode and IGBT on-state losses.

C. Influence of the n buffer

The resistivity of the n buffer determines the lateral voltage drop required for forward biasing the anode segments, as reported in [2]. Therefore, a higher resistivity (or lower doped) buffer is preferred to initiate the injection from the anode segments at low currents. However, adjustment of the buffer affects other design parameters such as the leakage current. Fig. 7 shows the I-V characteristics of the parallel stripe (S1) and radial stripe (S2) designs, measured on samples with different buffer doping concentrations. The change in buffer design increases the on-state losses at the nominal current from 2.8 V to 4.5 V for the parallel n^+-stripe (S1) design. In addition, secondary snap-backs become very prominent with the increase of the buffer doping. Radial n^+-stripe design (S2) has much lower sensitivity to the buffer and changes the on-state voltage drop from 2.4 V to 3.3 V for the same buffer modifications. The weaker sensitivity to the n buffer doping opens additional flexibility in the design which is important for optimizing the device for high temperature operation.

D. Switching characteristics

Switching characteristics were measured for the samples with parallel stripe (S1) and radial stripe (S2) designs at

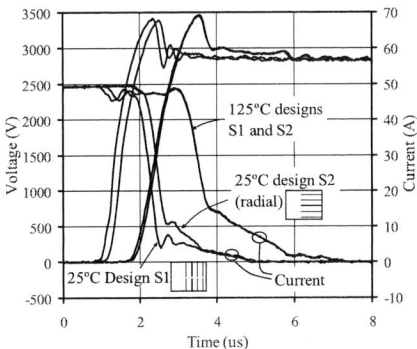

Fig. 8. Measured turn-off characteristics of BiGTs with different orientation of the stripes. Switching conditions 2800 V 50 A, $L_\sigma = 4400$ nH

nominal voltage and current (2800 V, 50 A), as shown in Fig. 8. Only a small difference in the current tail at room temperature is visible as a result of the different carrier density distribution. The higher and more evenly distributed plasma in the radial design provides additional carriers for the softer turn-off at low temperatures. At 125 °C, the waveforms become indistinguishable from each other.

IV. CONCLUSIONS

We have presented a comparison between the different layout designs of the anode n^+-shorts for the optimization of the BiGT. The investigation shows that the choice of shape and arrangement of the n^+-shorts determines the on-state conduction losses of the BiGT in IGBT mode. With the aid of 3D device simulations it has been demonstrated that the radial n^+-shorts stripes significantly improve the plasma spread in the device. Measurements confirm lower on-state losses as a result of better utilization of the device area. Square dot shaped n^+-shorts also offer good plasma spread in the device, but have a worse trade-off between diode and IGBT on-state losses. The radial n^+-stripe design of the anode shorts achieves the best diode and IGBT conduction losses trade-off and is the optimum design for the BiGT.

ACKNOWLEDGMENT

The authors wish to acknowledge the help of R. Jabrany and K. Ruef with device processing.

REFERENCES

[1] M. Rahimo, A. Kopta, U. Schlapbach, J. Vobecky, R. Schnell, S. Klaka, *The Bi-mode Insulated Gate Transistor (BiGT) A potential technology for higher power applications*, Proc. ISPSD09, p 283, 2009.

[2] L. Storasta, A. Kopta, M. Rahimo, *A comparison of charge dynamics in the Reverse-Conducting RC IGBT and Bi-mode Insulated Gate Transistor BiGT*, Proc. ISPSD10, p.283, 2010.

[3] A. Bourennane, J-L. Sanchez, F. Richardeau, E. Imbernon, M. Breil, *On the integration of a PIN diode and an IGBT for a specific application* Proc. ISPS06, Czech Rep., p 145, 2006.

[4] M. Gärtner, D. Vietzke, D. Reznik, M. Stoisiek, K.-G. Oppermann, W. Gerlach, *Bistability and hysteresis in the characteristics of segmented anode lateral IGBTs*, IEEE Trans. Electron. Dev., vol. 45, pp. 1575-1579, July 1998.

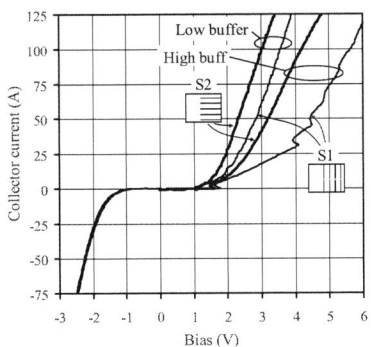

Fig. 7. Influence of n buffer doping on the on-state I-V characteristics of BiGT devices with different layout of the n^+-stripes

Proceedings of the 23rd International Symposium on Power Semiconductor Devices & IC's
May 23-26, 2011 San Diego, CA

Full Digital Short Circuit Protection
for Advanced IGBTs

Takuya Tanimura, Kazufumi Yuasa and Ichiro Omura
Department of Electrical Engineering and Electronics
Kyushu Institute of Technology
1-1 Sensui-cho, Tobata-ku, Kitakyushu-shi, Fukuoka 804-8550, JAPAN
j349529t@tobata.isc.kyutech.ac.jp, omura@ele.kyutech.ac.jp

Abstract— **A full digital short circuit protection method for advanced IGBTs has been proposed and experimentally demonstrated for the first time. The method employs combination of digital circuit, the gate charge sense instead of the conventional sense IGBT and analog circuit configuration. Digital protection scheme has significant advantages in the protection speed and flexibility.**

I. INTRODUCTION

As the increase in power density of IGBT with the reduction in chip thickness [1], the required time to protection becomes short, so that conventional protection method with sense IGBT ([2], [3]) will come to the limit of the speed. The required time to protection against short circuit for advanced IGBTs is expected to be less than 1 micro second (Fig. 1), which is much shorter than the protection speed with conventional sense IGBT method. To break the speed limit, the new approach has been introduced. The approach utilizes the gate charge as the indicator of the short circuit condition as explained in Fig. 2([4], [5]). This method is expected to improve the detection speed for the advanced IGBT.

In this paper, the new protection method is demonstrated on full digital configuration aiming to install automatic detection function of protection threshold and self–adjustment function for the deviation of device characteristics with temperature etc.. With these functions, the method can be applied to any type of IGBT without any change in the gate circuit and protection circuit. The new approach was experimentally demonstrated with a Field Programmable Gate Array (FPGA).

II. PROTECTION METHOD WITH DIGITAL CIRCUIT

A. Comparison with Proposed method and Conventional method

Conventional short circuit protection method employs sense IGBT configuration with a gate drive circuit, a sense resistor R_s and the sense IGBT in the chip, as shown in Fig. 3(a). A sense IGBT is a small IGBT imbedded in the chip with a tiny emitter with common gate and collector with main IGBT. The high current under short circuit condition is detected with the voltage drop V_{sense} across R_S, which is

proportional to the collector current. This method, however, has a drawback in the detection speed because of the noise from the high current through the main emitter under the short circuit condition.

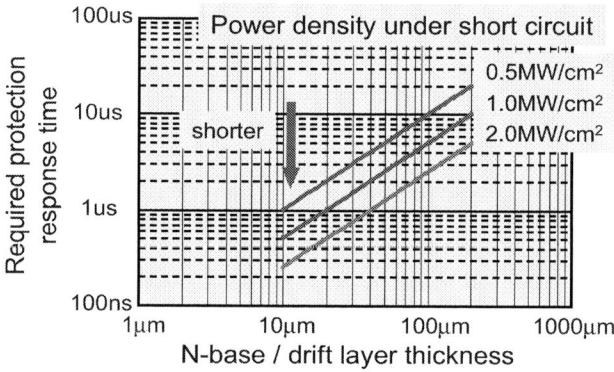

Figure. 1. Relationship between N-base thickness and the short circuit

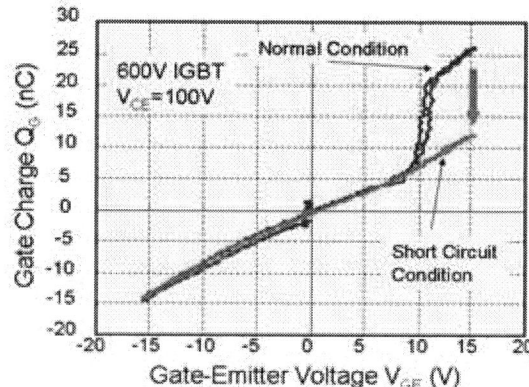

Figure 2. Difference of gate charge to the gate voltage under normal condition and short circuit condition.

To solve this problem, we propose a new method for the protection with gate charge as an indicator of the short circuit condition. Figure 2 shows the gate charge under short circuit condition in comparison with normal condition. Under short

978-1-4244-8425-6/11 $26.00 © 2011 IEEE

circuit condition, the gate charge is substantially reduced from the value of the normal condition. In the new method, the decrease of the gate charge is detected with the gate circuit and feeds back the protection signal. A schematic illustration of the new protection circuit is shown in Fig. 3(b). The circuit consists of a gate driver, a gate charge sense, a reference voltage generator and a comparator. Once the voltage from the charge sense dropped to the predetermined reference voltage, the comparator output changed to the negative state which immediately pulled down the input voltage to the gate driver to protect the IGBT.

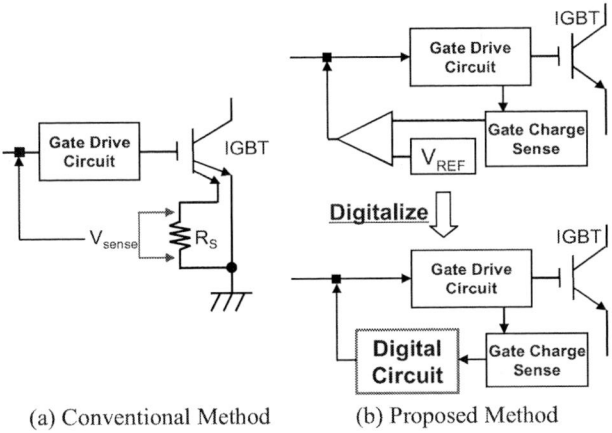

(a) Conventional Method   (b) Proposed Method

Figure 3. Conventional method and proposed method

Applying the new method to practical IGBT protection, there are a couple of problems to be solved. In the method, the reference voltage is to be determined prior to install in the gate driver in the analog circuit for a particular IGBT to be protected and, more over, the reference voltage may changed with IGBT junction temperature, which means that the reference voltage generation must be temperature sensitive.

B. Advantage of digital protection over analog circuit

We employ digital circuit instead of analog configuration. The proposed method includes digital circuit with AD / DA converters. All the required functions such as reference voltage generation and comparator are implemented into the digital logic circuit. Major advantage of the digital logic over analog for the IGBT protection is the automatic reference voltage generation by self capturing the gate charge under normal condition. This function eliminates the procedure to determine the pre-setup of parameters, such as reference voltage, prior to implement the protection circuit into the IGBT inverters and makes it possible to adjust the reference voltage to the change in IGBT characteristics with temperature even under operation.

C. Gate charge sense circuit

The charge sense circuit is integrated into the gate driver as shown in Fig. 4. The gate driver is connected to bus voltage via two current mirror circuits. The equivalent current I_G^* to the gate current I_G flow into a capacitance C_M and the voltage V_{QG} across the capacitance indicates the gate charge of the IGBT driven by the gate circuit, i.e. the gate charge can be measured by the voltage V_{QG}.

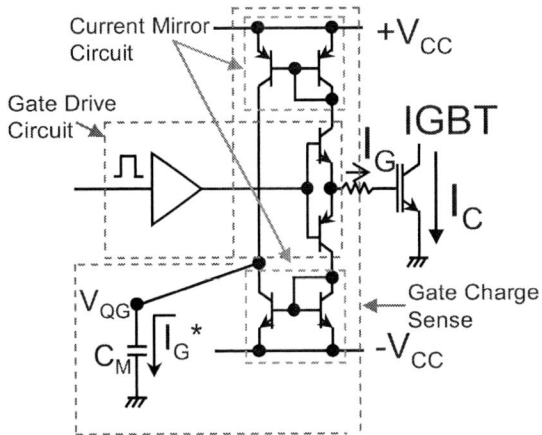

Figure 4. Configuration of gate charge sense

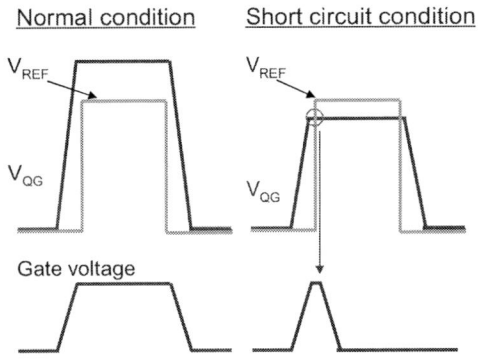

Figure 5. Schematic waveforms for reference voltage, gate charge voltage and gate voltage under normal condition and short circuit condition.

D. Protection scheme

The IGBT protection threshold is determined by the reference voltage V_{REF}, as shown in Fig. 5, and once the gate charge voltage V_{QG} drops below V_{REF}, the digital comparator outputs the protection signal.

III. EXPERIMENT AND RESULT

The digital protection with the gate charge sense circuit was demonstrated using a FPGA. All the function explained below are implemented into the FPGA (32MHz) using Hardware Description Language (VHDL). A 60MHz A/D converter and a 125MHz D/A converter are used for analog interface.

Digital filter: The voltage V_{QG} waveform is digitally filtered against the error with circuit noise. A FIR (Finite duration impulse response) filter is installed in the FPGA. The sampling frequency of about 30 MHz and the order of the filter is designed to be 4 for minimum time delay.

Peak detector: The reference voltage V_{REF} is generated in this block. V_{REF} is determined between V_{QG} values under the normal condition and short circuit condition with appropriate margin to prevent possible error caused by noise on V_{QG} waveform.

Pulse generator: Gate drive pulse for the experiment is generated in this block.

Digital Comparator: The gate charge voltage V_{QG} is digitally compared with the reference voltage V_{REF} to trigger the IGBT protection when V_{QG} decreases below V_{REF}.

Gate controller: The signal from the pulse generator is interrupted and switched to the off gate voltage when the protection scheme is triggered by the digital comparator so that the IGBT is immediately turned off.

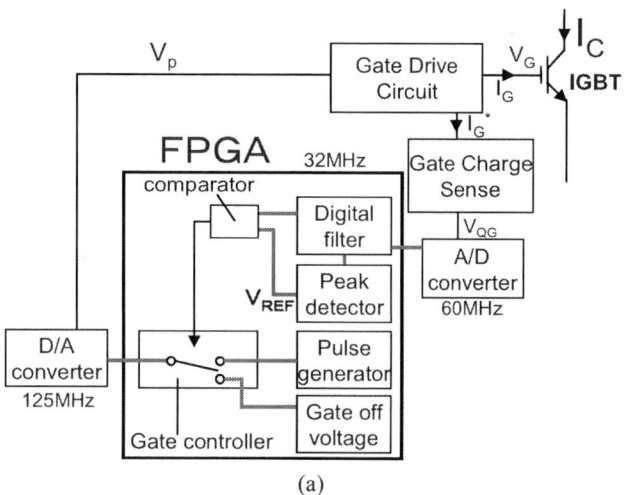

(a)

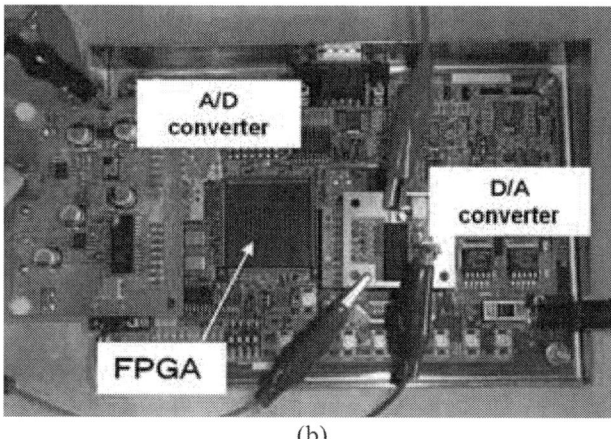

(b)

Figure 6. Block diagram for the demonstrated digital protection circuit(a) and the photograph of the FPGA board used in the experiment(b).

The short circuit protection with the digital circuit is demonstrated with the FPGA as shown in Fig 6. In this experiment, an IGBT with rated current of 10 A and the gate resistor of 27Ω. The main circuit voltage was 300 V.

(1) Under normal condition, the peak detector obtains the data V_{PEAK} and generates the V_{REF} automatically. (2) Under short circuit condition, collector current I_C was shut down successfully by detecting the V_{QG} reduction.

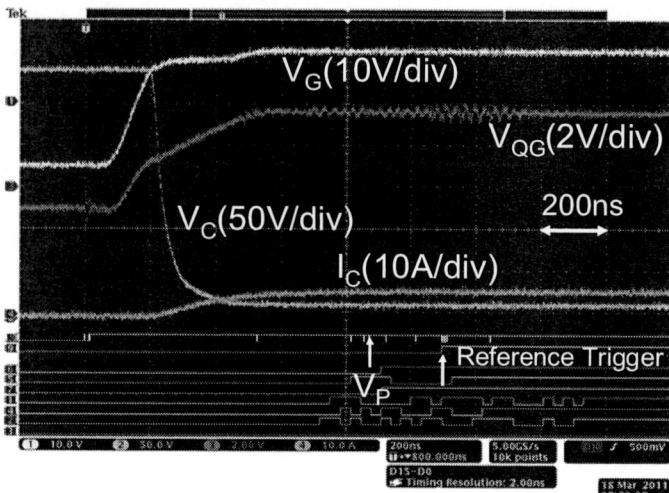

(a)Normal condition

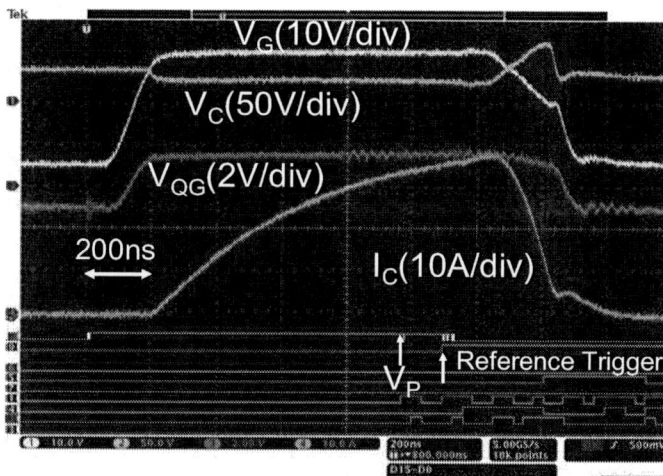

(b)Short circuit condition

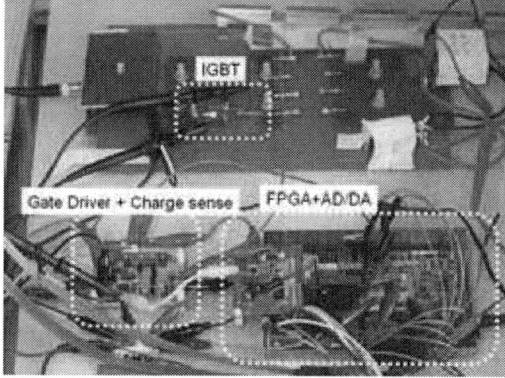

(c)Experimental setup

Figure 7. Protection waveform under normal condition (a) the short circuit condition (b) with an IGBT GT10J303 and photograph for experimental setup(c).

The high speed protection was successfully demonstrated and the required time to protect the IGBT with this setup was about 1.5 μs. The most clock consuming process was the digital filtering. The overall protection time will be reduced to less than 1μs with the increase of clock frequency.

The detail of the experiment system is listed in Table 1.

IV. SUMMARY

We demonstrated the short circuit protection method for advance IGBT. The new method features the potential protection speed less than 1μs with the automatic capturing of the protection threshold and adjustment without pre-setting of parameters prior to setup the protection circuit into IGBT inverters. This function is enabled by the digital real time processing. The concept was experimentally demonstrated with a FPGA at 32MHz with high speed AD / DA converters as analog interfaces.

REFERENCES

[1] I. Omura, Presentation at ECPE Workshop on Power Electronics Reseach & Technology Roadmaps -Copenhagen, Denmark, September 2007.

[2] E. Motto, J. Donlon, S. Ming, K. Kuriaki, T. Iwagami, H. Kawafuji and T Nakano, "Large package transfer molded DIP-IPM," Proc. of IAS'08,pp. 1-5, 2008.

[3] M. Kudoh, Y. Hohi, S. Momota, T. Fujiwara and K. Sakurai, "Current sensing IGBT for future intelligent power module," Proc. of ISPSD' 96, pp. 303-306, 1996.

[4] I. Omura, H. Ohashi and W. Fichtner, "IGBT megative gate capacitance and related instability effects", IEEE ED-letters, Vol. 18, No.12, pp. 622-624, 1997.

[5] K. Yuasa, S. Nakamichi and I. Omura, "Ultra high speed short circuit protection for IGBT with gate charge sensing" Proc. of ISPSD'10, pp.37-40, 2010.

Table 1. Protection system detail

FPGA board	
Device	XILINX Spartan XC3S400-4
Clock	32 MHz
Number of Gate	400kgates Max
AD Converter (8-Bit)	
Device	Analog Devices AD9283
Clock	60MHz
DA Converter(8bit out of 10-bit)	
Device	Analog Devcies AD9760
Clock	125MHz
Gate drive and gate charge sense	
Op-amp	National Semiconductor LM7171
NPN Bip Tr	Toshiba 2SC1815 x 3
PNP Bip Tr	Toshiba 2SA1015 x 3
IGBT	Toshiba GT10J303
	Rg=27Ω

Proceedings of the 23rd International Symposium on Power Semiconductor Devices & IC's
May 23-26, 2011 San Diego, CA

Ultrathin 400V FS IGBT for HEV applications

Heike Böving[1], Thomas Laska[1], Anton Pugatschow[2], Waldemar Jakobi[3]

[1]Infineon Technologies, Neubiberg, Germany, heike.boeving@infineon.com
[2]Infineon Technologies, Villach, Austria
[3]Infineon Technologies, Warstein, Germany

Abstract—400V IGBT and freewheeling diode as well based on 40μm thin wafer technology have been developed for electric and hybrid electric vehicles with a DC link voltage of 120V to 200V. First prototype ultrathin devices worldwide showed clearly reduced overall losses since both on state and switching losses are directly dependent on the chip thickness. The new 40μm chips also exhibited a very high dI/dt during switching resulting in high voltage overshoots exceeding the maximum allowed breakdown voltage of 400V. For this reason an overall stray inductance as small as possible is required to make use of the fast switching behavior of the new devices. Optimization of the switching behavior of both IGBT and Diode could be obtained by adapting dI/dt to an overall stray inductance of 33nH but still with reduced losses at the same time. On state voltage of both IGBT and Diode could be decreased by about 200mV. Turn off energy loss could be decreased by 10%, total losses of IGBT and Diode during turn on could be reduced by about 10% in comparison to standard 650V devices

I. INTRODUCTION

In the emerging market of electric and hybrid electric vehicles a consequent pursuit of optimum cost and energy efficiency is important. Improved cost performance can be obtained by shrink of the chip size offering the possibility to higher power density and thus decreased IGBT module and inverter sizes. Optimized energy efficiency is achievable by decrease of the chip thickness which means reduction of on state and switching losses at the same time. Consequently 400V IGBT in trench field stop technology [1, 2] and 400V Diode in the emitter controlled diode concept [3] both including a shrink has been developed based on 40μm thin wafer technology.

Reduction of the chip thickness from 70μm of 650V standard devices to 40μm means decreased losses but also reduced breakdown voltage of 400V. An IGBT with a rated blocking voltage of 400V is adapted to drive applications especially for mild hybrids with an inverter DC link voltage of 120-200V in the medium switching frequency range of some kHz. Comparing characteristics of IGBT and MOSFET makes clear why an IGBT usually is the best choice in this field of application (see Fig. 1).

As one can see the unipolar MOSFET shows a continuously resistive characteristic whereas the IGBT shows

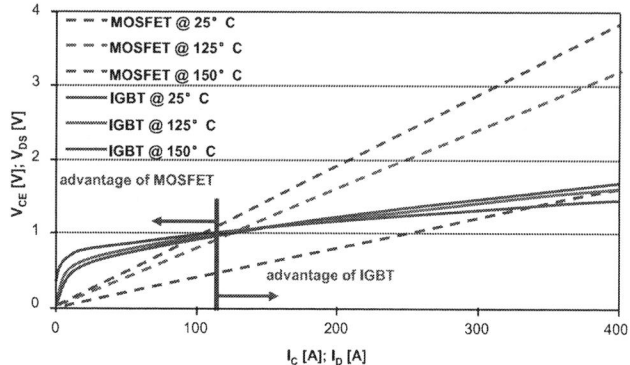

Figure 1. Characteristics of IGBT and MOSFET devices

a steeper VI-characteristic above a threshold voltage of about 0,8V due to the additional pn junction at the back [4]. Electrons and holes contribute to the current flow so that switching losses of an IGBT are significantly higher than those of a MOSFET. As a result applications with high switching frequencies (>100 kHz) are the domain of MOSFETs while IGBTs typically are used for applications with low switching frequencies (<10 kHz). Due to their non-resistive characteristic IGBT semiconductors are better suited for applications which utilize higher current densities.

II. PROCESSING OF 40μm THIN DEVICES

Looking back to the past, a continuous trend in chip thickness reduction for 1200V and 600V IGBTs and freewheeling diodes as well took place over the last 15 years [5, 6, 7]. This movement will go on in the future, with even bigger challenges to be solved due increasing wafer diameters.

So in particular the fabrication of 400V IGBT and diode based on 8" 40μm wafer technology means a huge challenge for frontend production. Equipment has to be adapted to handling of ultrathin wafers. Several unit processes have to be optimized to 40μm wafer thickness as well. Especially the wafer thinning process from 625μm to 40μm has to be improved regarding thickness distribution on a wafer and from wafer to wafer since thickness variation of ultrathin wafers has a much larger impact on electrical device characteristics compared to thicker devices of higher voltage classes.

978-1-4244-8425-6/11 $26.00 © 2011 IEEE

Additionally an ideal concept for passivation and metallization layers leading to an acceptable wafer bow has to be found. Furthermore a wafer thickness of 40μm demands an improved chip separation method to avoid chipping.

III. CHARACTERISTICS OF A 400V IGBT

A. Output characteristic

The output characteristic of a 400V IGBT including a shrink of 10% in comparison to a 650V standard device is illustrated in Fig. 2. As one can see the 400V IGBT shows an on state voltage at 150°C clearly reduced by more than 200mV compared to the reference.

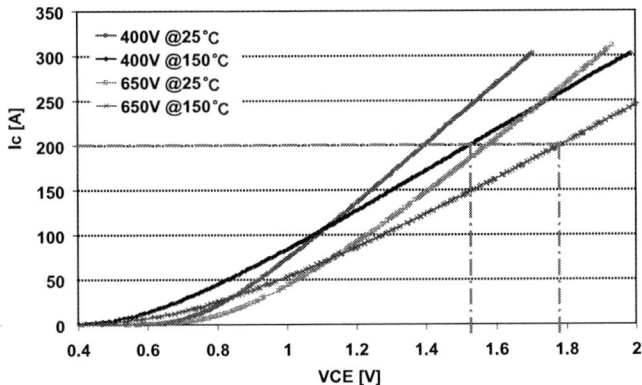

Figure 2. Output characteristic of a 200A 400V IGBT at 25°C and 150°C in comparison to a 650V IGBT

B. Turn off characteristic

Turn off characteristics of a 400V IGBT chip set and a standard 650V IGBT chip set is shown in Fig. 3 and Fig. 4 respectively. Dynamic characterization was carried out at chip sets (2 IGBT chips connected in parallel) assembled in HybridPACK™2 modules offering a stray inductance of 14nH.

During switching at a DC link voltage of only 120V the 400V IGBT exhibits a voltage overshoot of 324V in compa-

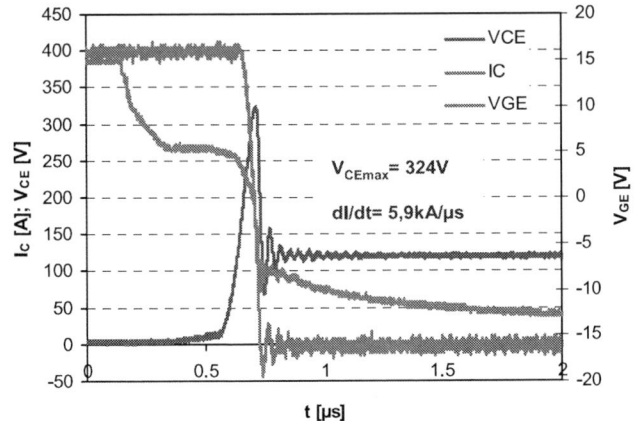

Figure 3. Turn off characteristic of first 200A 400V IGBT chip set (2 chips connected in parallel), measured at T = 25°C and Rg =3.6Ω

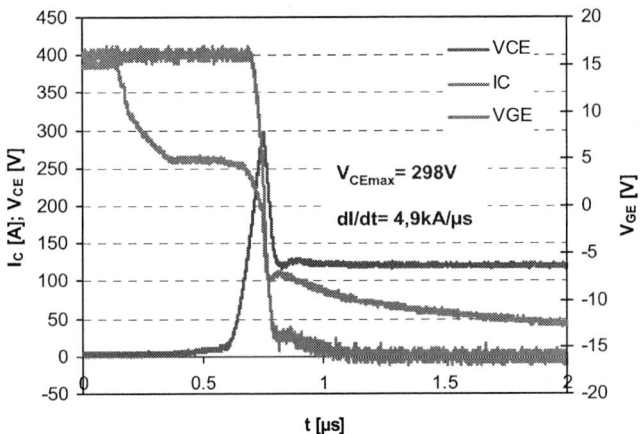

Figure 4. Turn off characteristic of a 200A 650V IGBT chip set (2 chips connected in parallel), measured at T = 25°C and Rg =3.6Ω

rison to 298V of the reference. Increasing the DC link voltage up to 200V would increase the voltage overshoot to values exceeding the maximum allowed breakdown voltage of 400V. As the voltage overshoot is given by dI/dt multiplied by the total stray inductance ($dV = L_{σtotal} \cdot dI/dt$), it is essential to compare dI/dt of the 400V IGBT chip set with the standard 650V IGBT. As illustrated in Fig. 3 and Fig. 4 the 400V IGBT exhibits a dI/dt of 5,9kA/μs in comparison to 4,9kA/μs of the 650V reference at the same gate resistance R_G.

The outcome of high dI/dt and high voltage overshoot of the ultrathin 400V IGBT is on the one hand the requirement of a stray inductance as small as possible to make use of fast switching devices. The influence of the stray inductance on the voltage overshoot will be demonstrated exemplarily in chapter V. On the other hand optimization of the switching behavior of the 40μm IGBT by limiting dI/dt was necessary. Reducing dI/dt of a field stop IGBT is possible by adjusting the back emitter efficiency. Fig. 5 shows the turn off characteristic of an optimized 400V IGBT including slightly increased back emitter efficiency. dI/dt is reduced from 5,9kA/μs (see Fig. 3) to 5,3kA/μs (see Fig. 5). The resulting voltage overshoot is decreased from 324V to 302V and is

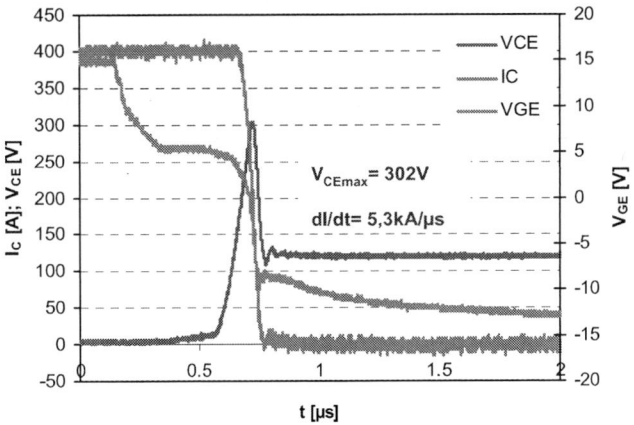

Figure 5. Turn off characteristic of a 200A 400V IGBT chip set with slightly increased back emitter efficiency (2 chips connected in parallel), measured at T = 25°C and Rg =3.6Ω

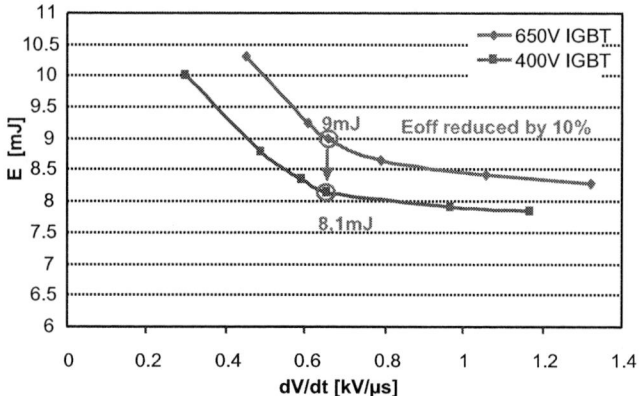

Figure 6. Turn off energy loss of a 200A 400V IGBT chip set in comparison to a 650V reference

comparable to the 650V reference (VCEmax=298V, see Fig. 4). The adjusted dI/dt allows switching up to DC link voltages of 200V. Composition of turn off energy losses of the optimized 40µm IGBT and the standard 650V IGBT is illustrated in Fig. 6. As one can see turn off energy loss of the 400V IGBT is decreased by about 10% in comparison to the 70µm reference for equal dV/dt values.

IV. CHARACTERISTICS OF A 400V DIODE

Additional to ultrathin FS IGBTs 400V diodes have been developed. First 40µm prototype diodes showed a very high di/dt during diode recovery resulting in high voltage overshoots. Improvement of the switching behavior could be obtained by an adapted field stop profile and slight increase of the chip thickness with reduced losses at the same time in comparison to standard 650V devices. On state voltage could be decreased by about 150mV to 200mV. Amount of reverse recovery energy and turn on loss of a 400V chip set could be reduced by about 10% for equal dI/dt values.

V. INFLUENCE OF STRAY INDUCTANCE

The influence of a modules stray inductance on the switching behavior of a 400V-IGBT/diode chipset was investigated at existing Infineon HybridPACK™ 1 and 2 packages. The major difference between these packages is the internal stray inductance. A HybridPACK™1 module offers a stray inductance of 30nH whereas a HybridPACK™2 shows a reduced stray inductance of only 14nH. A modules stray inductance depends on design of package and DBC.

The investigated modules were prepared with not optimized 400V chip sets consisting of 2 × 200A IGBTs and 2 × 200A Diodes per switch resulting in a nominal current of 400A per half-bridge. Turn-off characteristics of a 400V IGBT chip set assembled in modules HybridPACK™1 and HybridPACK™2 measured at varying DC link voltages are illustrated in Fig. 7. The stray inductance of the used test bench setup was Lσsetup = 19nH.

As one can see depending on the stray inductance of the module a 400V IGBT/diode chipset reaches the maximum allowed breakdown voltage of 400V during switching at different DC-link voltages (voltage overshoot

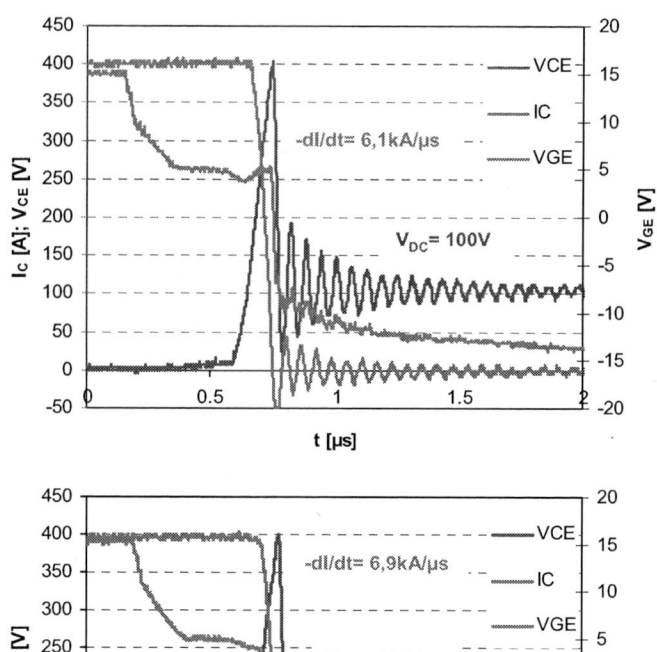

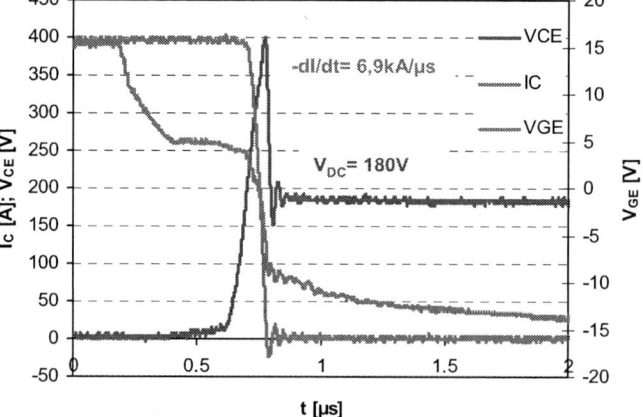

Figure 7. Turn off characteristics of a 2·200A 400V IGBT/Diode in HybridPACK™ 1 (upper curve) and HybridPACK™ 2 (lower curve) packages, measured at T = 25°C and Rg = 3.6Ω

dV=Lσtotal·dI/dt). For the chosen R_G value the HybridPACK™1 package enables switching of nominal current only up to a DC-link voltage of 100V whereas the HybridPACK™2 package allows switching up to 180V. The maximum possible DC-link voltage without destruction of the chip is limited by the module stray inductance. Using 400V IGBT/diode chipsets in applications with occurring DC link voltages up to 200V requires an optimized package with an internal stray inductance below 14nH and a total stray inductance below 33nH.

VI. SHORT CIRCUIT BEHAVIOR

Short circuit behavior of an optimized 400V IGBT was investigated. The corresponding measurements were carried out at chip sets assembled in HybridPACK™2 modules (2 × 200A IGBTs and 2 × 200A diodes per switch). Fig. 8 shows the turn-on characteristic of a 400V IGBT chip set with existing short circuit in the output path.

The measurement carried out at a DC link voltage of 240V and at a junction temperature of 150°C reveals that a 40µm thin IGBT is at least 8µs short circuit proof. This short circuit robustness is comparable to the one of a 650V standard IGBT during operation at a DC link voltage of 360V

978-1-4244-8425-6/11 $26.00 © 2011 IEEE

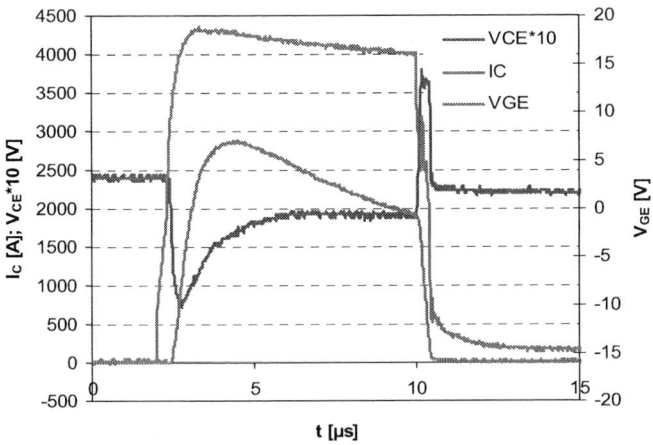

Figure 8. Short circuit in HybridPACK™ 2 package, measured at T = 150°C; Vce =240V; Rg = 3,6Ω

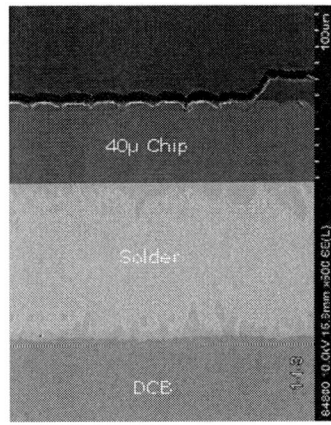 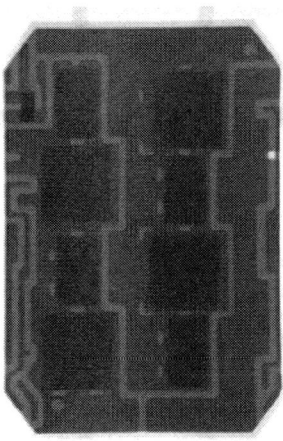

Figure 9. Cross section and X-Ray images of 40μm devices soldered on a DBC

and at a junction temperature of 150°C. Therefore short circuit robustness equal to the robustness of a 650V reference can be ensured for the ultrathin 400V IGBT.

VII. MODULE ASSEMBLY

A. Pick and place

There are several parameters with influence on a pick and place process like wafer thickness, back metallization, lamination foil, chip separation and a suitable pick and place equipment. An optimum combination of all these parameters had to be found. Several investigations were carried out until a final concept for pick and place of ultrathin chips free from breakage could be defined.

B. Soldering

In standard IGBT modules the vacuum soldering process is used for electrical and thermal connection between metallic backside of semiconductor chip and DBC. Parameters which influence the chip soldering are the metallic backside, chip dimensions, but also bending of the chip. Optimum solder layer shows no voids, which would increase the thermal resistance of the junction to the DBC and would cause higher risk of chip fail. Fig. 9 shows a cross section through a soldered 40μm chip and an x-ray image of several 40μm chips soldered on a DBC. No voids are visible.

C. Ultrasonic Bonding

Ultrasonic bonding is a typical connection between approx. 3μm thin metallic topside of a semiconductor chip and DBC. During this process a 400μm thick aluminum wire is welded to a contact pad with use of ultrasound. The results of several investigations with varying process parameters approve the feasibility of ultrasonic bonding process for 40μm thin chips resulting in bond connections reliable and free from chip breakage.

VIII. CONCLUSION

Results of static and dynamic characterization of first prototype 40μm IGBT's revealed clearly reduced overall

losses in comparison to standard 650V devices with a chip thickness of 70μm. The ultrathin devices also showed a very high dI/dt during switching resulting in voltage overshoots exceeding the maximum allowed breakdown voltage. The trend for the IGBT chip development is further loss reduction and fast switching behavior. Consequently, reduction of the stray inductance has to be the driving force for module development. On the other hand adapting dI/dt to a modules stray inductance and optimizing softness of fast switching devices respectively gets more and more important and difficult at the same time with decreasing chip thickness.

ACKNOWLEDGMENT

We would like to thank our UPD and UPS colleagues as well as our colleagues from backend assembly for their continuous technical support. Part of this work was supported by the ENIAC E³Car funding Project.

REFERENCES

[1] T.Laska et al., "The Field Stop IGBT (FS IGBT) – A New Power Device Concept with a Great Improvement Potential", Proceedings of the 12th ISPSD, pp.355-358, 2000

[2] H.Rüthing et al., "The 600V-IGBT3: Trench Field Stop Technology in 70μm Ultra Thin Wafer Technology", Proceedings of the 15th ISPSD, 2003

[3] A.Mauder.et al., "The Field Stop IGBT Concept and Emcon High Efficiency Diode", Proceedings PCIM USA, 2000.

[4] Münzer, M.; Thoben, M. & Gietzold, T. (2006), 'HEV Power Electronics - Adapting to automotive requirements', The 22nd International Battery, Hybrid and Fuel Cell Electric Vehicle Symposium & Exposition, 2312-2319

[5] D. Burns et al.: "NPT- IGBT - Optimizing for Manufacturability", Proceedings of the 8th ISPSD, 1996

[6] T. Laska et al.: "Ultrathin-Wafer Technology for a new 600V-NPT-IGBT", Proceedings of the 9th ISPSD, pp.361-364, 1997

[7] T. Matsudai et al.: "New 600V Trench Gate Punch-Through IGBT Concept with Very Thin Wafer and Low Efficiency p-emitter, having an On-state Voltage Drop lower than Diodes", Proceedings of the IPEC Tokyo, pp.292-296, 2000

Proceedings of the 23rd International Symposium on Power Semiconductor Devices & IC's
May 23-26, 2011 San Diego, CA

600V LPT-CSTBT^{TM} on Advanced Thin Wafer Technology

Yuki Haraguchi, Shigeto Honda, Kazunari Nakata[1], Atsushi Narazaki and Yoshiaki Terasaki

Power Semiconductor Device Development Department, [1]Manufacturing Engineering Center, Mitsubishi Electric Corporation

997 Miyoshi Koshi, Kumamoto, 861-1197, JAPAN

Phone: +81-96-242-5805 E-mail: Haraguchi.Yuki@ct.MitsubishiElectric.co.jp

Abstract— Electrical characteristics of the fabricated 600V class CSTBT^{TM} with a Light Punch Through (LPT) structure on an advanced thin wafer technology are presented for the first time. The electrical characteristics of LPT-CSTBT are superior to the conventional Punch Through type (PT) one, especially in low current density regions because of the inherent lower built-in potential. Furthermore, we also have evaluated the effects of the mechanical stress on the device characteristics after soldering, utilizing a novel evaluation method with a very small size sub-chip layout. The results validate the proposed tool is useful to examine the influence of the mechanical stress on the electrical characteristics.

I.   INTRODUCTION

In order to improve the electrical characteristics of IGBT, we have developed several technologies, which include an optimization of the vertical carrier distribution by introducing the design concept named CSTBT^{TM} [1] and an application of the Light Punch Though structure [2] which enables the reduction of the n-base layer. In 600V class, we believe our LPT type CSTBT^{TM} as an IGBT structure is one of the best solutions, as we demonstrated in 1200V class devices. On the other hands, efforts to fabricate 600V class thin IGBT chips [3-4] using a thin wafer technology are extremely critical from both the wafer fabrication process and the electrical characteristics points of view. Not only in comparison with the conventional PT type device in a trade-off relationship between the forward ON state voltage drop $V_{CE}(sat)$ or Von and the turn-off switching Eoff, but we could also confirm the advantage of the LPT type structure is low Von characteristics in low current density regions, because of the low doping concentration for both n-buffer and p-collector layers precisely controlled by the backside ion implantation without any carrier lifetime control like an electron beam irradiation. Lowering a knee point of a forward output I-V characteristic is welcome in consumer electronics application such as refrigerators and air-conditioners, whose one of the indices is known as the Annual Performing Factor (APF) in Asia. It is widely recognized that the electrical characteristics such as Vth and $V_{CE}(sat)$ are shifted when stress is applied to IGBT. [5] It is thought that the reason of this phenomenon is the change of carrier mobility by the

strain of crystal structure. Considering about the mechanical stress influence for the about a half of a hundred micrometer thin IGBT chip during a packaging process especially around a soldering process as a die-bonding, it is very important to understand the details of the stress distribution inside of the single chip as the first step of an improvement to reduce the packaging effect for thin power chip. By using special sub-chip arrays configuration, our approach of evaluation is successfully achieved.

II.   DEVICE CHARACTERISTICS

Figure 1 shows a cross sectional view of the Conventional PT type CSTBT and the LPT type CSTBT. The LPT type device has an almost same N-drift thickness, but the total device thickness is about a quarter of the conventional PT type one.

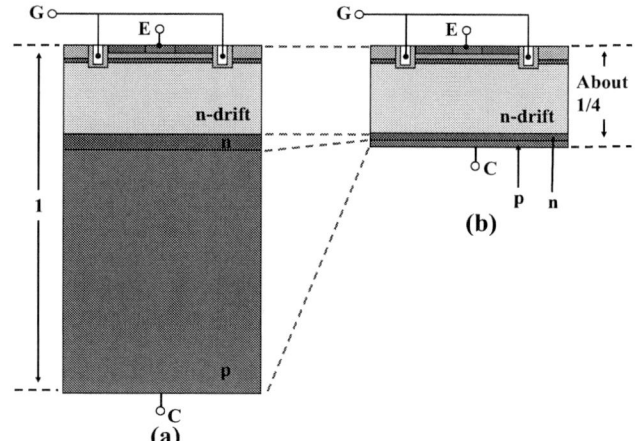

Fig.1: A cross sectional view of (a) Conventional PT type CSTBT and (b) LPT type CSTBT

Figure 2 shows Jc-V_{CE} characteristics of the fabricated LPT-CSTBT and the conventional PT-CSTBT. Maintaining low Eoff, the proposed LPT type has a lower $V_{CE}(sat)$ than the conventional one even in the high current density region, since the LPT structure with the low doping concentration for p-collector realizes not only low Eoff value but also small tail

978-1-4244-8425-6/11 $26.00 © 2011 IEEE

current value, which is caused by the low gain of parasitic pnp transistor, without any carrier lifetime control such as an EB irradiation, and maintains relatively long enough value to lowering the V_{CE}(sat) of LPT type. Moreover, in low current density regions (Fig. 3), Von of LPT type is much lower than the conventional one and very close to the built-in potential of Si pn junction of 0.55V, a half the value of energy band gap Eg(Si), because of the low doping concentration for both n-buffer and p-collector layers.

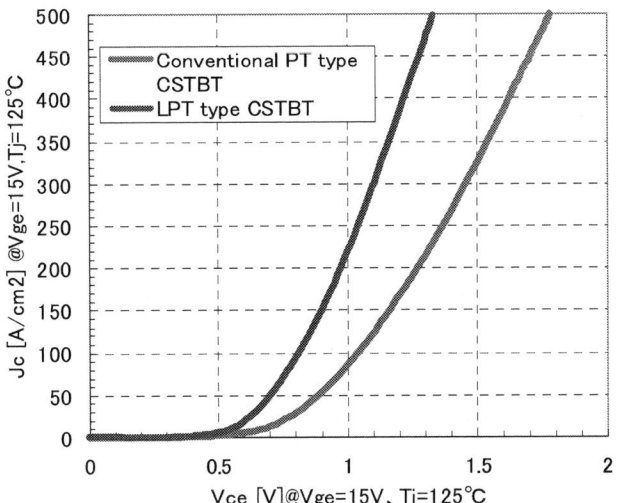

Fig.2: Jc-Vce characteristics of LPT-CSTBT and conventional PT-CSTBT (We measured up to high current density region.)

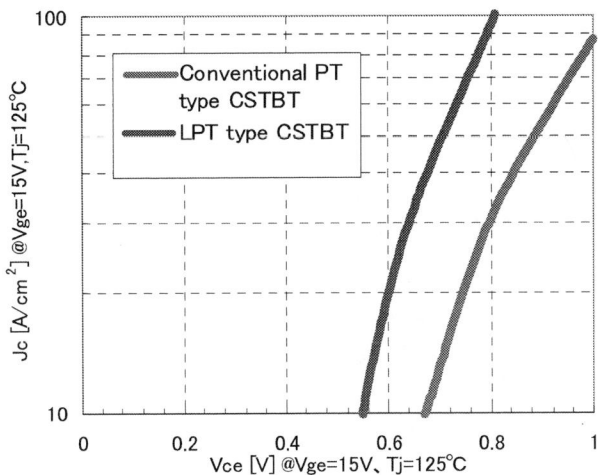

Fig.3: The enlarged graph in a low current density region of Fig.2

Figure 4 shows Eoff-V_{CE}(sat) trade-off characteristics. As a chip is thinner, the trade-off relationships between V_{CE}(sat) and Eoff of the proposed LPT types are better than our conventional PT type, and are improved by N-drift thinning, in which the breakdown characteristic of the thinnest chip is higher enough than this voltage class requirement. This is because of our newly developed advanced thin wafer process technology including the

optimization of the backside doping profile for both n-buffer and p-collector regions.

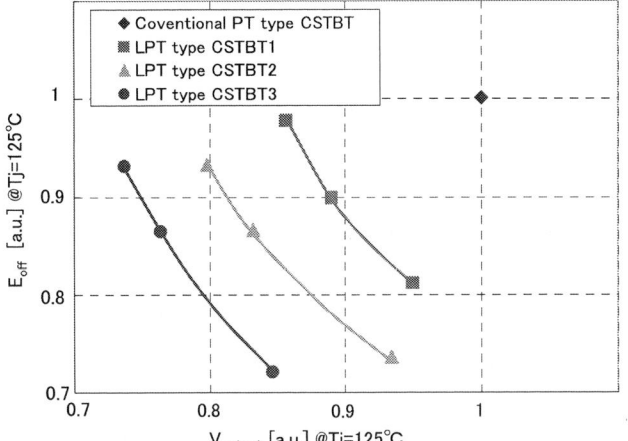

Fig.4: Eoff-V_{CE} (sat) trade-off characteristics (Wafer thickness: LPT type CSTBT1 > CSTBT2 > CSTBT3. As a chip is thinner, Eoff-V_{CE} (sat) trade-off characteristics is improved. We normalized the trade-off characteristic on a linear scale.)

III.   INFLUENCE OF STRESS

To simulate the mechanical stress during a power module assembly process, we extracted a die-bond soldering step as the one of highest stress process.

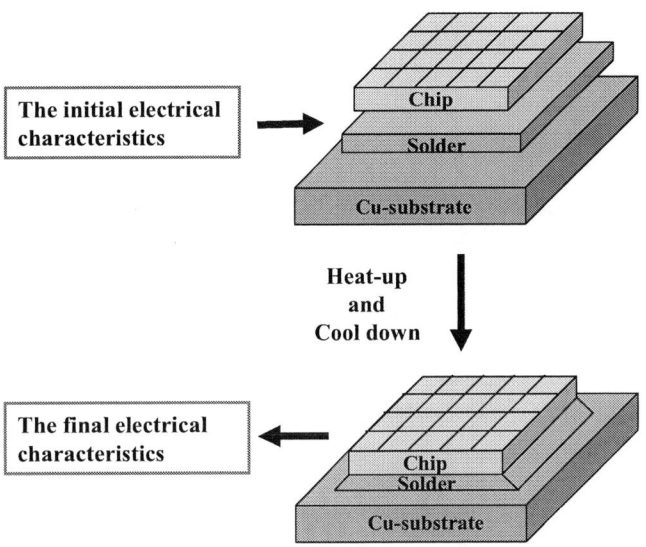

Fig. 5: The mechanical stress evaluation procedure through electrical characteristics measurement both before and after soldering process

Figure 5 shows the mechanical stress evaluation procedure. Both before and after the soldering step, fundamental electrical characteristics of IGBT, including threshold voltage Vth and ON state forward voltage drop V_{CE}(sat) or Von, are measured, at the room temperature for special sub-chip array, whose entire sizes are enlarged to a stepper's image field and the number of sub-chip is 256 as the product of 16 X 16 square arrangement . Figure 6 shows a photograph of the sub-chip that consists of 256 small IGBT.

Fig. 6: Photograph of the sub-chip that consists of 256 small IGBT (The size of the chip is same as one shot of stepper image. A small IGBT has a junction termination, gate pad)

The rates of shift in the Vth and the V_{CE}(sat) are shown in Fig.7. The center chip of the array of each thickness was measured. Relatively large shifts are measured only by Vth and V_{CE}(sat), and their shifting value is a negative coefficient, i.e. the "after" values are smaller than the initial values. And the thinner the chip is, the larger V_{CE}(sat) shift is. On the other hand, Vth is shifted but maintains a constant value. Basically Vth shift is caused by tensile stress in the direction of the wafer thickness, along which the trench gate is formed with a vertical channel region. In contrast, V_{CE}(sat) is affected by not only a channel region but also an N-drift. This is the reason why there is the different tendency between Vth and V_{CE}(sat).

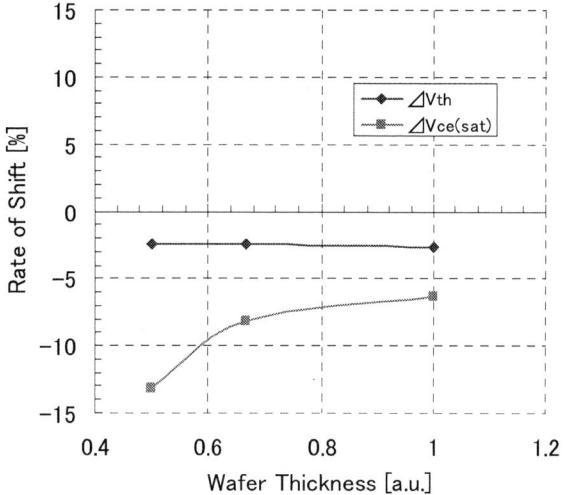

Fig.7: The rate of shift at the center chips for each thickness Δ ="final value"–"initial value". Each V_{CE}(sat) value is also normalized by the wafer thickness.

Figure 8(a) shows the distribution of the mechanical stress after soldering, and this map is simulated by using a simplified structure with the isothermal model. We estimated the distribution of the electrical characteristics (Fig.8 (b)).

The simulation and measurement regions are the second quadrants and the model structure of the simulation is Si / Solder / Cu-substrate. As a result of the simulation, it was shown that a compressive stress is generated in the chip. The red area shows higher compressive stress in Fig.8 (a). ΔVth pattern shown in Fig.8 (b) is different from a radial uniform distribution of the mechanical stress on numerical simulation values, because the real device has an anisotropic structure like a pattern of trench and Si mesa regions (Fig.9). The distribution of ΔVth in a vertical to trench is changed easily, but the distribution of ΔVth in a parallel to trench is changed hardly. It is thought that Vth is easily shifted due to the strain of crystal structure in channel region. We expect that ΔVth has the formation direction dependency of the trench because the mechanical strength of the chip is different by the direction of trench formation.

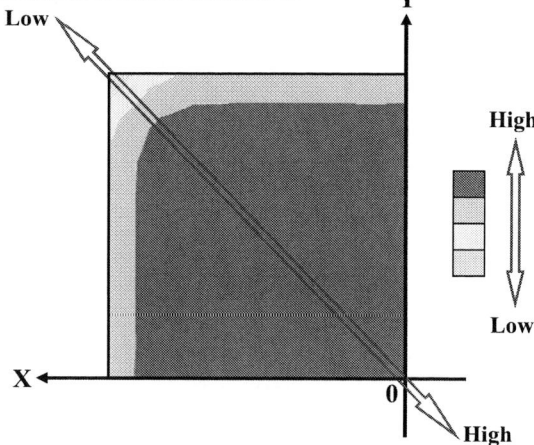

Fig.8 (a): The result of thermal stress simulation (The model structure of this simulation is Si-substrate / Solder / Cu-substrate. As a result of simulation, the compressive stress occurs to of a chip.)

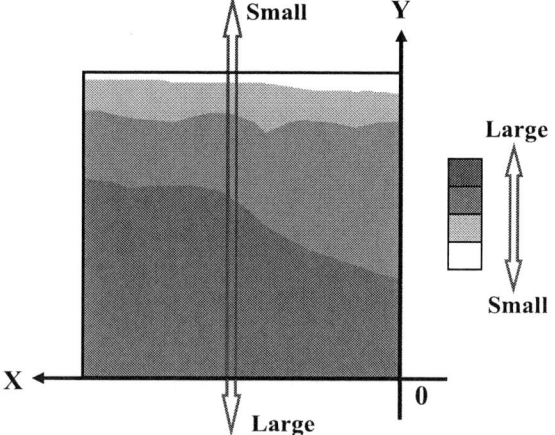

Fig.8 (b): A distribution of ΔV_{th}
The difference of the measurement values before and after soldering is shown by the monochrome coloring. The darker color shows the larger shift of the electrical characteristics.

Fig.8: A mechanical stress contour map and an electrical characteristic (Only the second quadrants are shown)

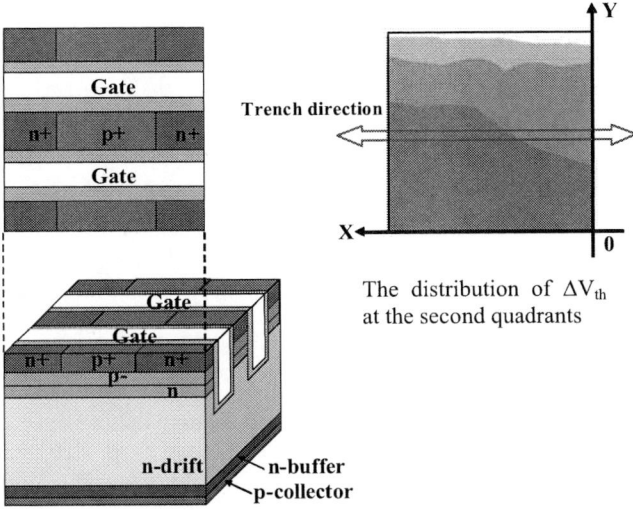

The distribution of ΔV_{th} at the second quadrants

Fig. 9: Simple 3D figure of sub-chip
(The electrical characteristic of Vth pattern is different from the distribution of mechanical stress (Fig.8).)

In contrast, ΔV_{CE}(sat) pattern is similar to the radial uniform distribution of the mechanical stress on numerical simulation. Figure 10 shows the ΔV_{CE}(sat) pattern. V_{CE}(sat) is easily shifted by the strain of crystal structure in a cannel, N-drift and backside region as mentioned above. Therefore, we expect the distribution of ΔV_{CE} (sat) became like Fig.10 because the mechanical stress like Fig. 8(a) is caused in N-drift and backside regions.

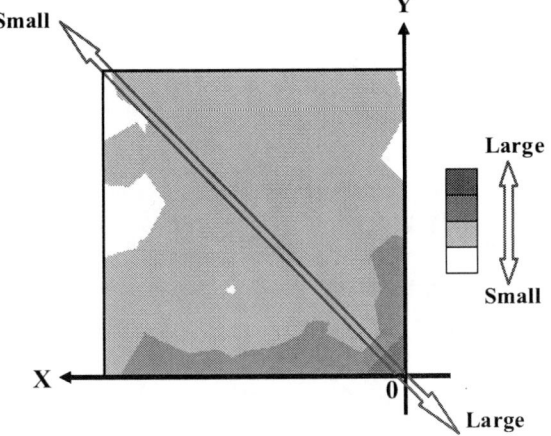

Fig.10: A distribution of ΔV_{CE}(sat)
(The difference of the measurement values before and after soldering is shown by darkness. As it is darker, it shows the larger shift of the electrical characteristics.)

IV. CONCLUSION

In this paper, we presented the electrical characteristics of the fabricated 600V class LPT type device using the advance thin wafer process technology, and the distribution of electrical shifts due to the mechanical stress. We successfully fabricated LPT type device with excellent characteristics, and the above clear distribution difference among stress, Vth and Von on the map is very good evidence that our proposed sub-chip array approach for the analysis of the mechanical stress effect for the electrical characteristics of IGBT device.

V. FUTURE WORK

This evaluation using the TEG array is very effective approach to investigate the mechanical stress effect on the chip surface and the upper side of N-drift, but is insufficient to understand the entire chip volume especially for backside structure. We continue to study the detail distribution of the mechanical stress inside the device to influence the electrical characteristics through both the device experiment and the combination usage of three dimensional stress and device simulations.

ACKNOWLEDGEMENT

The authors would like to thank Dr. K. Sato, Mr. T. Minato, Mr. Y. Omoto and Mr. M. Shinkai for their technical advice, also appreciate to Mr. T. Shitomi and Mrs. A. Horita for their intense support for the special device fabrication.

REFERENCES

[1] H.Takahashi, H.Harugichi, H.Hagino and T.Yamada, "Carrier Stored Trench-Gate Bipolar Transistor (CSTBT) -A Novel Power Device for High Voltage Application-"ISPSD'96, p349-352(1996)

[2] K. Nakamura, S. Kusunoki, H. Nakamura, Y. Ishimura, Y. Tomomatsu and T. Minato, "Advanced Wide Cell Pitch CSTBTs Having Light Punch-Through (LPT) Structures", Proc. ISPSD'02, pp.285-288, 2002

[3] T. Matsudai, H. Nozaki, S. Umekawa, M. Tanaka, M Kobayashi, H. Hattori and A. Nakagawa, "Advanced 60p.m Thin 600V Punch-Through IGBT Concept for Extremely Low Forward Voltage and Low Turn-off Loss" Proc. ISPSD'01, pp.441-444, Jun. 2001

[4] T. Laska, M. Münzer, F. Pirsch, C. Schaeffer, T. Schmidt, "The Field Stop IGBT (FS IGBT) A new device concept with a great improvement Potential", Proc. ISPSD'06, pp.355-358, Jun. 2006

[5] H. Tanaka, K. Hotta, S. Kuwano, M Usui, M. Ishiko, "Mechanical stress dependence of power device electrical characteristics" Proc. ISPSD'06, pp.1-4, Jun. 2006

Proceedings of the 23rd International Symposium on Power Semiconductor Devices & IC's
May 23-26, 2011 San Diego, CA

High Speed 650V IGBTs for DC-DC Conversion up to 200 kHz

Hsueh-Rong Chang , Jiankang Bu , George Kong [*] and Rachana Bou
Automotive Power Switches Development, [*] Temecula Manufacture Center
International Rectifier Corp. 101 N. Sepulveda Blvd, El Segundo, CA 90245
Phone: 310-726-8854, Email: hrchang1@irf.com

Abstract— The increasing demand for higher power density and lower cost in high voltage power supplies has driven semiconductor manufacturers to expand IGBT performance for high switching frequency beyond 100 kHz. An ultra thin punch-through IGBT with a blocking voltage of 650V has been developed and optimized targeting DC-DC conversion up to 200 kHz. Its high Tjmax of 175°C further enhances the converter compactness. This paper describes the feature of this ultra fast IGBT in a critical comparison with equivalent products available on the market today.

Keywords: IGBT, ultra-thin wafer, DC-DC converter, switching losses

I. INTRODUCTION

IGBTs have been available as power switches for more than 30 years, largely adopted in motor control and inverter applications which require relatively low switching frequency and high current density. As the trend in power conversion constantly increases in switching frequency, IGBTs found limited use in the switching mode power supply (SMPS) applications. Power MOSFETs have been the choice of switches for high voltage SMPS applications.

Superjunction (SJ) MOSFET can achieve a better trade-off between conduction loss and breakdown voltage than the conventional power MOSFETs, for high frequency and high voltage applications at the expense of extra masks and complex fabrication process [1-4]. Due to presence of the minority carriers in IGBTs, it has been a challenge for IGBT to achieve high performance for the high frequency, hard switching applications above 100 kHz. In this paper we report for the first time, that ultra fast IGBTs fabricated with the ultra-thin wafer technology have achieved low power dissipation similar to SJ MOSFET but with a much simpler fabrication process. It has a high Tjmax of 175°C as compared with 150°C for SJ MOSFET. This new IGBT offers a cost-competitive, high performance option for the DC-DC converters up to 200 kHz.

II Device STRUCTURE and Fabrication

650V high speed IGBTs are fabricated on 70 μm thin wafers using the Punch-Through (PT) structure as shown in Fig. 1. The use of ultra thin wafers allows a lightly doped collector which reduces stored charge, thus resulting in better switching perofrmance, especially at high temperatures.

Conventional Punch-Through IGBTs use minority-carriers lifetime killing techniques, such as electron irradiation or metal doping, to increase switching speed. One of the side effects of these processes is that the leakage current increases rapidly with increasing temperature, which limits the Tjmax to 150°C.

The high speed thin IGBTs are not processed with lifetime killing techniques. The leakage current is maintained low at 175°C which enables an operation with Tjmax of 175°C. The current carrying capability of the fast switching thin IGBTs are therefore increased, further reducing the converter size.

The design of ultra fast IGBT for 200 kHz DC-DC converter involved major changes in comparison to the standard IGBT for the motion control. The SMPS applications do not require large short-circuit SOA. IGBT with higher cell density is preferred to produce lower on-state voltage drop while a smaller gate capacitance with minimal internal gate resistance R_G is required to achieve fast switching speed. The threshold voltage is designed to the standard 3-5V range. Fig. 2 shows a photograph of the high speed thin IGBT.

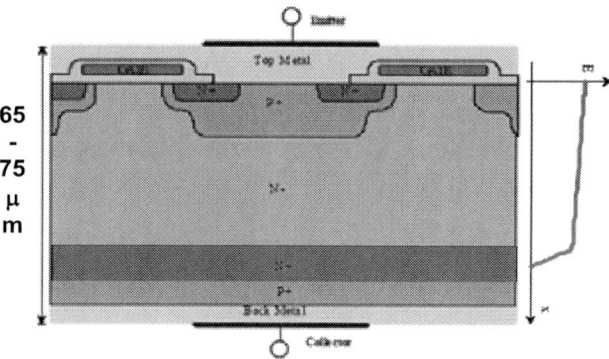

Fig. 1 Punch-Through thin IGBT structure

978-1-4244-8425-6/11 $26.00 © 2011 IEEE

Fig. 2 Photograph of a high speed thin IGBT

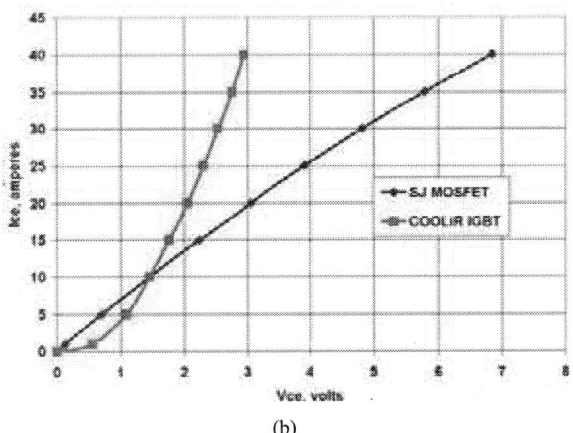

(b)

Fig. 3 Comparison of the forward conduction characteristics of COOLiRIGBT™ and SJ MOSFET: (a) 25°C and (b) 150°C.

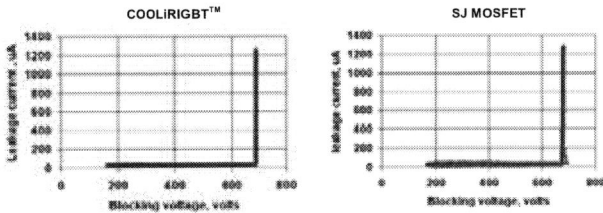

Fig. 4 Comparison of the reverse blocking characteristics of COOLiRIGBT™ and SJ MOSFET.

II. RESULTS AND DISCUSSION

Comparison between MOSFET and IGBT can be difficult sometimes because devices are rated in different ways. MOSFET shows a resistive behavior and therefore on-state voltage drop increases linearly with current. IGBT has threshold-like I-V characteristic where the V_{CEsat} does not vary linearly with current.

Fig. 3 compares the I-V curves of the fast switching IGBT and a commercially available SJ MOSFET. The die area is almost identical for these two devices. The Rdson of the SJ MOSFET is low at room temperature; it increases rapidly with increasing temperature, by 260% from 25°C to 150°C, while the high speed thin IGBT only increased by 37%. The reverse blocking characteristics of the fast switching IGBT and SJ MOSFET is shown in Fig. 4. Both devices have similar blocking voltage of 690V.

The turn-on and turn-off power losses were measured in the test circuit shown in Fig. 5. The turn-on energy of the DUT is strongly dependent on the reverse recovery characteristics of the fast diode [5]. An external diode is chosen for the test circuit to ensure a valid comparison for the turn-on of IGBT and MOSFET.

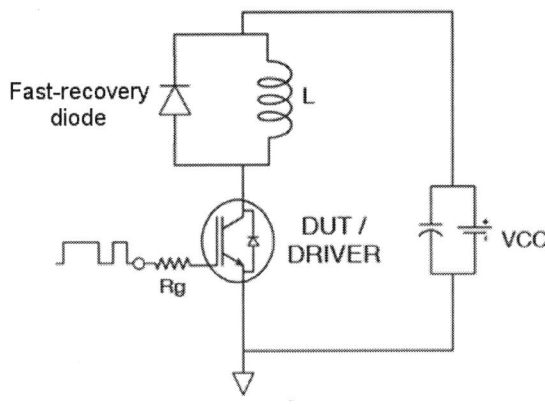

Fig. 5 Test circuit used for the measurements of the switching characteristics of IGBT and MOSFET.

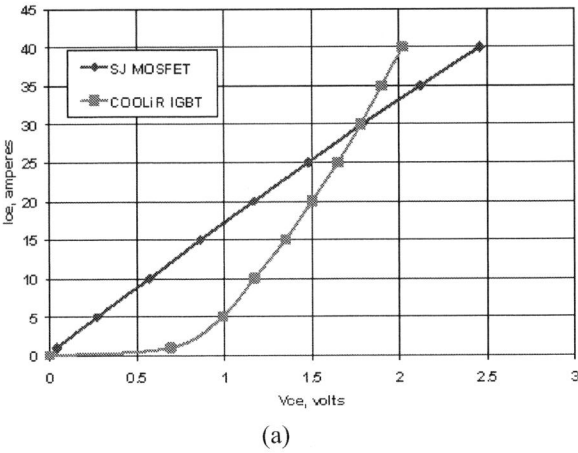

(a)

The switching waveforms of the high speed IGBT and SJ MOSFET at 150°C are shown in Fig. 6 and 7, respectively. The turn-on energy of the high speed IGBT at 10A is 156 uJ as compared with 135 uJ observed for the SJ MOSFET, which is mainly influenced by the recovery behavior of the fast-recovery diode. The turn-off energy is controlled by the DUT itself. The high speed IGBT exhibits a relatively low turn-off energy of 87 uJ while 70 uJ was recorded for the SJ MOSFET.

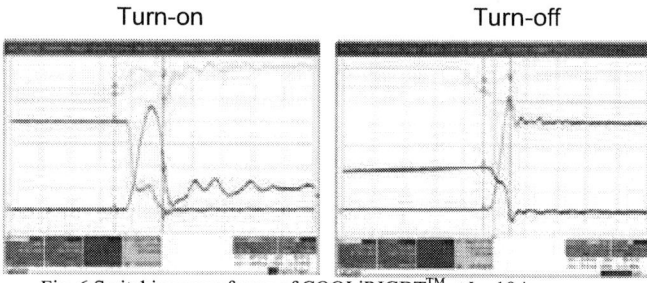

Fig. 6 Switching waveforms of COOLiRIGBT™ at Ic=10A, Vcc=400V and 150°C

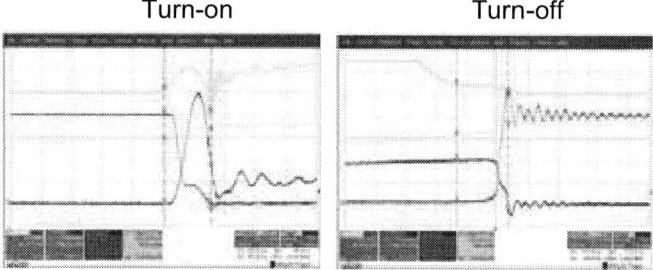

Fig. 7 Switching waveforms of a SJ MOSFET at Ic=10A, Vcc=400V and 150°C

For hard switching pulse-width-modulated (PWM) applications, the total power loss incurred in each power transistor consists of conduction loss, turn-on loss and turn-off loss. For a DC-DC converter with a switching frequency of 200 kHz, the total power losses were calculated for the high speed IGBT and SJ MOSFET, base on 50% duty cycle. The results are summarized in Table 1. The high speed IGBT and SJ MOSFET show similar on-state forward voltage drop at a low current of 10A. The switching loss of the IGBT is 14% higher than that in SJ MOSFET. Thus the total power loss for the high speed IGBT is 11% higher than SJ MOSFET.

Table 1 power losses at Ic=10A and 150°C

	COOLiRIGBT™	SJ MOSFET
Eon, uJ	156	143
Eoff, uJ	87	70
switching loss, W	24.3	21.3
conduction loss, W	7.25	7.25
total power loss, W	31.55	28.55

As the current level is increased to 15A, the on-state voltage of the high speed IGBT becomes much lower than that of SJ MOSFET (1.77 vs 2.23V as shown in Fig. 3b). The IGBT conduction loss becomes significantly lower than that of SJ MOSFETs by 26% while its switching loss is slightly higher by 8%. At this current level of 15A, the high speed IGBT and SJ MOSFET show similar power losses.

Table 2 power losses at Ic=15A and 150C

	COOLiRIGBT™	SJ MOSFET
Eon, uJ	214	195
Eoff, uJ	143	139
switching loss, W	35.7	33
conduction loss, W	13.28	16.73
total power loss, W	48.98	49.73

For soft-switching such as zero-voltage–switching (ZVS) applications, the turn-on energy is negligible because the transistors get switched to on-state when the voltage across the device reaches zero. In this case, the conduction loss and turn-off loss are the main contributors to the total power loss for each transistor. Table 3 compares the power losses for the high speed IGBT and SJ MOSFET for ZVS applications. The total power loss of the high speed IGBT is 6% lower than that in SJ MOSFET. For converters with higher output power, COOLiRIGBT™ can offer much lower conduction loss than SJ MSOFET while keeping similar turn-off power loss as SJ MOSFET. This will make COOLiRIGBT™ a better choice of transistors for DC-DC converters to achieve a higher efficiency.

Table 3 power loss at Ic-15A and 150C

	COOLiRIGBT™	SJ MOSFET
Eoff, uJ	143	139
switching loss, W	28.6	27.8
conduction loss, W	13.28	16.73
total power loss	41.86	44.53

For ZVS converts with a bus voltage of 400V and 50% duty cycle, the turn-on loss is negligible. The maximum current allowed for various switching frequencies were calculated for COOLiRIGBT™ and SJ MOSFET using Tjmax=150C. The results are shown in Fig. 8. COOLiRIGBT™ and SJ MOSFET show similar maximal input current, ~14A at 200 kHz.

978-1-4244-8425-6/11 $26.00 © 2011 IEEE

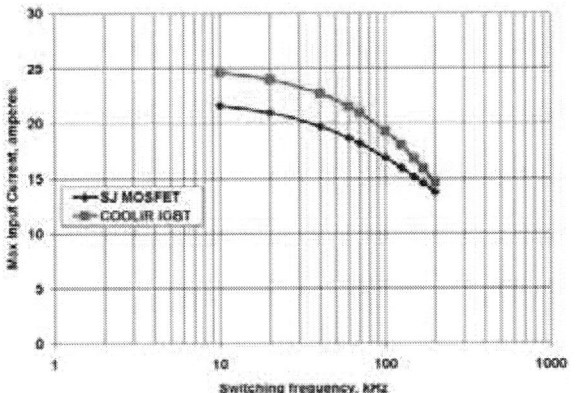

Fig. 8 Comparison of maximum input current at various switching frequencies for COOLiRIGBT™ and SJ MOSFET at 150°C for ZVS applications.

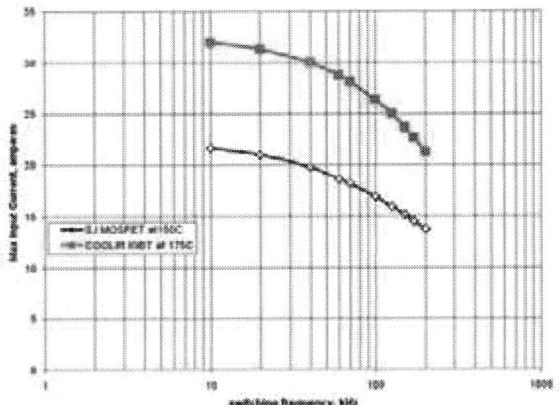

Fig. 9 Maximum input current at 200 kHz is increased by 50% for COOLiRIGBT™ with Tjmax=175°C comparing with 14A for SJ MOSFET with Tjmax=150°C.

The Tjmax of the COOLiRIGBT™ is rated at 175°C while the SJ MOSFET is limited to a Tjmax of 150°C. At 175°C, the on-state voltage drop of the COOLiRIGBT™ is increased by 8% and the turn-off energy is increased by 1.4%. Fig. 9 compares the maximum input current vs switching frequency for COOLiRIGBT™ at Tjmax=175°C and SJ MOSFET at Tjmax=150°C. The COOLiRIGBT™ has its maximum input current increased to 21A, which is 50% higher than SJ MOSFET.

In addition, COOLiRIGBT™ exhibits superior robustness with large safe-operating-area (SOA) - full square RBSOA at Vcc=480V and Ic=120A and short-circuit SOA of 10 μs at Vcc=400V and 150°C.

A simple fabrication process along with high current carrying capability and large SOA enables the COOLiRIGBT™ a low cost and high performance switch option for high frequency DC-DC converters.

III. CONCLUSION

High speed punch-through IGBT on ultra thin wafer was successfully developed for DC-DC conversion up to 200 kHz. A simple fabrication process in conjunction with high current carrying capability and large SOA make the high speed IGBT a cost-competitive and high performance switch option for the DC-DC converters.

ACKNOWLEDGMENT

We would like to thank Henning Hauenstein and Temecula management and engineering team for their support.

REFERENCES

[1] "Superjunction FETS Boost Efficiency in PWM", J. Hancock, Power Electronics Technology, 2005

[2] W. Saito, et. al. Proceeding ISPSD 2006, pp. 293-296

[3] K. Takabashi et. al., Proceeding ISPSD 2008, pp. 299-302

[4] J. Sakakibara, et. al., Proceeding ISPSD 2006, pp. 293-296

[5] B.J. Baliga, "Powe Semiductor Devices", PWS, 1995.

Proceedings of the 23rd International Symposium on Power Semiconductor Devices & IC's
May 23-26, 2011 San Diego, CA

Novel High Voltage LDMOS on Partial SOI with double-sided Charge Trenches

Xiaorong Luo, IEEE member, Y G Wang, T F Lei, L Lei, D P Fu, G L Yao, M Qiao,
Bo Zhang, IEEE member, Zhaoji Li
State Key Laboratory of Electronic Thin Films and Integrated Devices
University of Electronic Science and Technology of China
Chengdu, P.R.China
E-mail: xrluo@uestc.edu.cn

Abstract—A novel partial silicon-on-insulator (PSOI) high-voltage LDMOS is proposed and its breakdown mechanism is investigated numerically and experimentally. The PSOI LDMOS features double-sided charge trenches on the top and bottom interfaces of the buried oxide (BOX) (DTPSOI). In high-voltage blocking state, the charges located in the trenches enhance the electric field strength in the BOX, and a Si window makes the substrate share the vertical voltage drop and modulates the lateral field in the SOI layer. Both increase the blocking voltage (BV). A BV>700V DTPSOI LDMOS is realized on a 8μm-thick SOI layer over the 1.2μm BOX and 1.5μm-deep trench. Moreover, the Si window alleviates the self-heating effect.

I. INTRODUCTION

SOI technology offers many advantages, including low power loss, high speed and anti-radiation. However, it suffers a low vertical BV (BV_{ver}) and self-heating effect (SHE) because the BOX layer prevents the depletion region extending into substrate and cuts off the conduction heat path to heat sinker. The former is solved by enhancing the electric field strength in the BOX [1-4]. The ultra-thin SOI device improves the BV by using a lateral linear doping profile, while it is protected by Philips patents and there is a "hot spot" at the source side [1]. PSOI technology can increase BV and reduce SHE due to the Si window [5-6]. A BV>600V PSOI device was designed with a nonstandard doping on a thin SOI layer over 4μm BOX [6].

The PSOI LDMOS with double-sided charge trenches is proposed and fabricated. Its breakdown mechanism is investigated theoretically and experimentally. The trenches improve the BV_{ver}, and a silicon window releases the depletion region into substrate and modulates the lateral field distribution; the Si window also offers a conduction heat path. The structure therefore improves the BV and reduces the SHE.

II. STRUCTURE, MECHANISMS AND FABRICATION

Fig. 1 shows the cross sections of a DTPSOI LDMOS and a trench cell structure. t_S and t_I denote the thicknesses of the SOI layer and buried oxide, respectively. L_d and L_W are the lengths of the drift region and Si window. H, W, and D are the

height, width, and space of the trenches, respectively. N_d and N_{sub} are the doping concentrations of the SOI layer and Si substrate. The x- and y- direction are given in Fig. 1.

When a high positive voltage V_d is applied to the drain while the source, gate, and substrate are grounded, the substrate, buried oxide, and the SOI layer constitute a MIS (metal- insulator-semiconductor) -like structure. The inversion layer (hole layer) is thus formed on the bottom interface of the n-SOI layer, and oxide barriers (islands) partially prevent the extracting holes by the source. The holes are therefore located in the top trenches, with the charge density Q_I; accordingly, electrons are induced in the bottom trenches, as illustrated in Fig. 1(b). In the y-direction, Q_I enhances the field strength in the BOX, called E_I, from $\varepsilon_S E_S /\varepsilon_I$ of the normal PSOI to ($\varepsilon_S E_S + Q_I$) /ε_I. E_S and E_I are the field strengths of Si and BOX at their interface, ε_S and ε_I are their permittivity, respectively. In the x-direction, both the new electric field peak at the Si window and the hole with the density increasing from the source to drain modulate the field distribution. A Si window makes the substrate share the BV and offers heat conduction path.

The junction terminal technology is employed and the BV is thus determined by the vertical breakdown. BV is expressed by

$$BV = E_S t + \frac{t_I}{\varepsilon_I} Q_S + V_{sub} \qquad (1)$$

where $t = 0.5 t_S + \varepsilon_S t_I / \varepsilon_I$, V_{sub} is the voltage sustained by

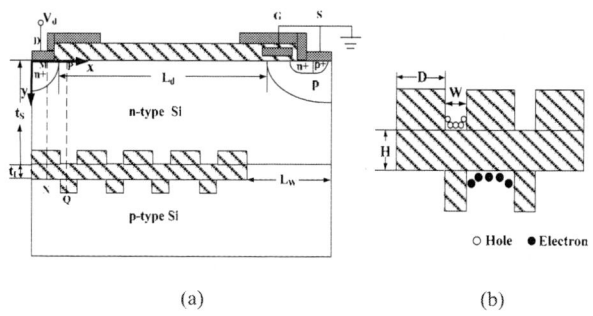

(a)   (b)

Fig.1 Schematic cross section of a DT PSOI LDMOS and a trench cell

978-1-4244-8425-6/11 $26.00 © 2011 IEEE

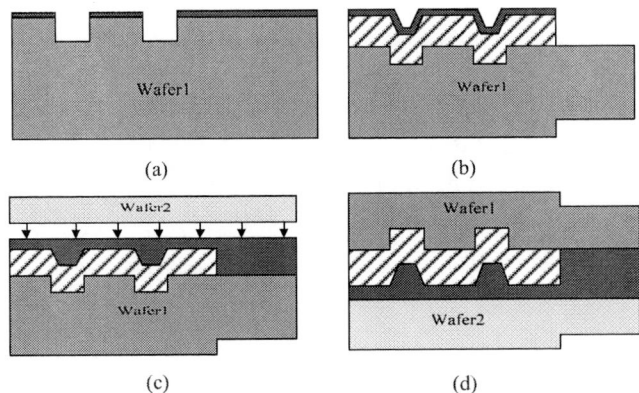

(a)   (b)

(c)   (d)

Fig. 2 Key fabrication processes of a DTPSOI wafer (a) etch Si trench; (b) thermally grow and deposit SiO_2, etch Si window, and dual-sided align; (c) deposit and planarize polysilicon, and bonding; (d) dual-sided alignment, thinning and dual-sided alignment

the substrate. The first term in (1) denotes the BV of a normal SOI LDMOS, and the second item is the enhanced BV caused by Q_I for a TPSOI LDMOS.

Fig. 2 shows the key fabrication processes for DTPSOI wafer. In order to align the surface device with the patterns of the BOX, Dual-sided alignment is used three times to mark the patterns of the BOX on the SOI layer surface. It is polysilicon instead of SiO_2 that it is planarized before bonding; consequently, the trenches at the back side are formed naturally.

III. RESULTS AND DISCUSSION

Fig. 3 shows the hole distribution in the top trenches and the electron distribution in the bottom trenches. It indicates that both the hole and electron concentrations increase from the source- to drain- side with the increasing interface potential. These carriers are mainly collected on the bottom of all trenches by the vertical field, and the charge concentration at the corners is higher due to the two-dimensional field crowding (as shown in Fig.4 (c)). The width of the bottom trenches, which is equal to the space D of the top trenches, is larger than that of the top trenches, resulting in a weakened effect on locating electrons, consequently, on the bottom of the trenches, the electron concentration is slightly lower than the hole concentration.

Figs. 4 (a)-(b) show the electric field distribution in the x-

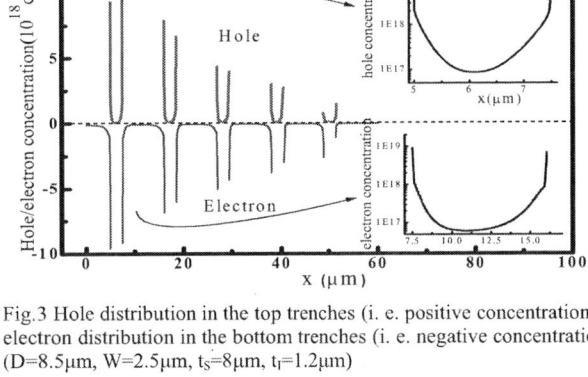

Fig.3 Hole distribution in the top trenches (i. e. positive concentration) & electron distribution in the bottom trenches (i. e. negative concentration). (D=8.5μm, W=2.5μm, t_S=8μm, t_I=1.2μm)

and y- direction. In the y-direction, compared with the PSOI LDMOS, the holes and electrons in the trenches greatly increase the field strength of the BOX for DTPSOI LDMOS. The field strengths decrease with the increasing distance from the hole layer in PQ line or from the electron layer in MN line in Fig. 4(a). In the x-direction, new electric field peaks are generated at the Si window and oxide barriers in Fig. 4(b). Both these new peaks and the increasing hole concentration from the source to drain improve the field distribution, as shown in Fig. 4(b). The vector field in one trench cell is given in Fig. 4(c), which indicates the 2D field distribution and the field crowding effect. The 738V DTPSOI LDMOS is obtained at D=8.5μm, W=2.5μm, H=1.5μm, t_I=1.2μm and t_S=8μm, while the BVs are 517V and 637V for PSOI LDMOS at t_I=1.2μm or t_I=1.2+H=2.7μm and t_S=8μm, respectively.

Fig. 5(a) shows the dependence of BV on N_d as the the functions of W and H. The BVs labeled circles are the maximum BVs (BV_{max}) for the DTPSOI LDMOSFETs with given device dimensional parameters, entitled a-, b-, c- and d-DTPSOI LDMOSFET. Fig. 5(b) gives the hole concentration in one trench under the drain for a-, b-, c- and d- LDMOSFET. A narrower trench (i. e. with a lower W value) has a stronger effect on locating carriers, the carrier concentration in each trench is thus higher (e. g. d-LDMOSFET); nevertheless the integral of charge density to W will decrease for a too narrow trench. The ratio of W/(W+D) denotes the effective region to locate charges. It will decrease and the continuity of the charge distribution degrades as W decreases or D increases. Therefore, the E_I and BV_{max} firstly increase and then decrease with the decreasing W, while monotonously decrease with the increase in the D value.

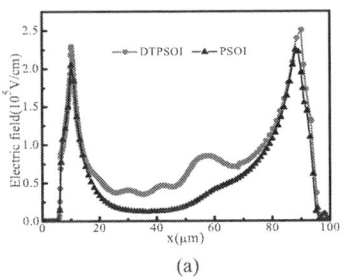

(a)

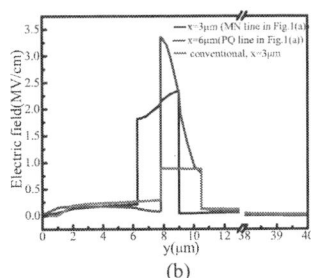

(b)

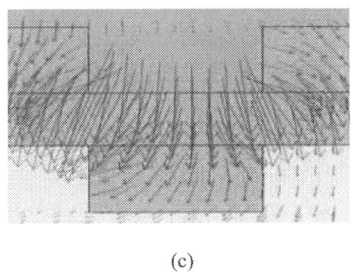

(c)

Fig.4 Electric field distribution (a) in the x-direction, (b) in the y-direction, and (c) vector field in a trench cell (the same dimensional parameters as those in Fig. 3, the Nd is optimized to obtain the maximum BV for each device)

978-1-4244-8425-6/11 $26.00 © 2011 IEEE

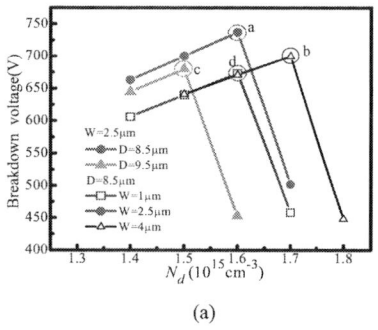
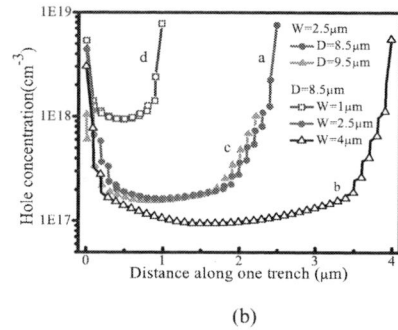

(a)   (b)

Fig. 5 (a) Dependence of BV on N_d as the functions of W and D; (b) hole concentration in one trench under the drain at BV_{max}

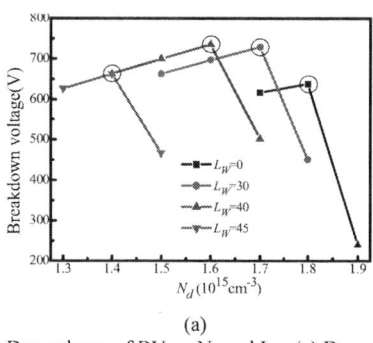
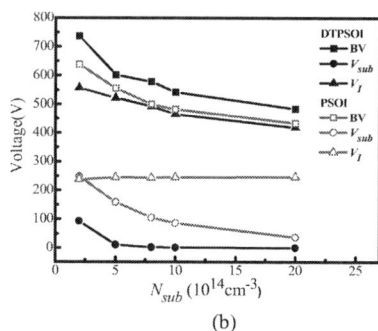

(a)   (b)

Fig. 6 Dependence of BV on N_d and L_w; (a) Dependence of BV on N_d as a function of L_w, and (b) influence of N_{sub} on BV at $L_W = 40 \mu m$

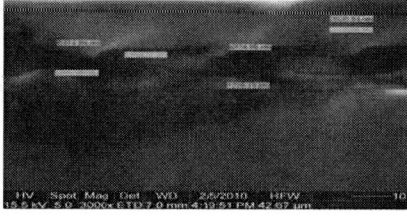
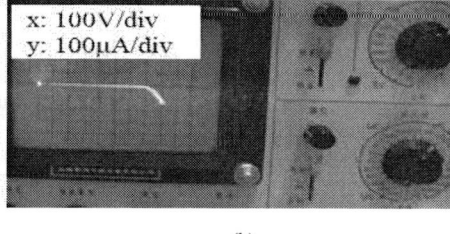
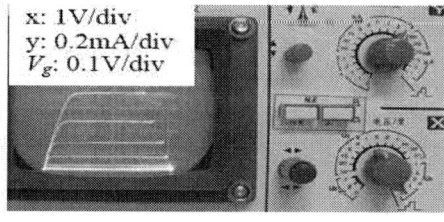

(a)   (b)   (c)

Fig. 7 (a) Scan electronic microscope graph, (b) off-state I-V curve and (c) on-state I-V curve

Fig. 6 shows the dependence of BV on L_W and N_{sub}. As L_W increases, both the number of the trench and Q_S in every trench reduce, resulting in the falls of E_I and V_I (the voltage sustained by the BOX), while the voltage shared by the substrate (V_{sub}) increases due to the extension of the depletion layer in the substrate. The BV_{max} (BV labeled circle) for a DTPSOI LDMOS firstly increases and then decreases with L_W in Fig. 6(a). A DTPSOI LDMOS with $L_W=0$ is an SOI LDMOS with double-sided trenches. The BV_{max} is a little lower than that of the DTPSOI device with an optimal L_W because of $V_{sub}=0$ for DTSOI LDMOS. For the DTPSOI LDMOS, the V_I accounts for 76%-87% of BV, and the V_{sub} can be ignored at $N_{sub}>5\times10^{14}cm^{-3}$ due to the premature breakdown at Si window, as shown Fig. 6(b). This breakdown makes potential contours located the region above the electron layer below the BOX under the drain (thus $V_{sub}\approx0$), and V_I decreases. The BV of the normal PSOI LDMOS decrease as N_{sub} increases because of a narrowed depletion width in the substrate. The V_I accounts for 37%-57% of BV and the V_{sub} for the PSOI LDMOS is always higher than that of the DTPSOI LDMOS with the same L_W.

We have fabricated the DTPSOI LDMOS by using the key process steps in Fig.2. The scan electronic microscope graph of the DTPSOI wafer is given in Fig. 7(a). The I-V curves in the on- and off-state are shown in Fig. 7(b) and Fig. 7(c). The 720V measured BV is obtained, which verifies the breakdown mechanism proposed.

IV. CONCLUSION

The PSOI LDMOS with double-sided charge trenches is proposed and fabricated. The charges located in the trenches enhance E_I, and a Si window makes the substrate share the vertical voltage drop and modulates the lateral field in the SOI layer, resulting in an enhanced BV. A 720V DTPSOI LDMOS is fabricated to verify the proposed mechanism, with the reduced SHE.

REFERENCES

[1] S. Merchant, E. Arnold, H. Baumgart, et al, Realization of high breakdown voltage (>700V) in thin SOI device, in Proc. ISPSD,1991, pp.31-35.

[2] Xiaorong Luo, Z. J. Li, B. Zhang, et al, "Realization of High Voltage (>700V) in New SOI Devices with a Compound Buried-Layer," Electron Device Lett., 29(12), pp.1395-1397, Dec. 2008,.

[3] Xiaorong Luo, T. F. Lei, Y. G. Wang, et al, "A high-voltage LDMOS compatible with high-voltage integrated circuits on p-type SOI layer," IEEE Electron Device Lett., 30(10), pp. 1093–1095, Oct. 2009.

[4] Xiaorong Luo, Y. G. Wang, H. Deng, et al, novel low-k dielectric buried-layer high-voltage LDMOS on partial SOI, IEEE Trans. Electron Devices, 57(2), pp. 535–538, Feb. 2010.

[5] F. Udrea, W. Milne, and A. Popescu, "Lateral insulated gate bipolar transistor (LIGBT) structure based on partial isolation SOI technology," Electron. Lett., vol. 33, no. 10, pp. 907–909, May 1997.

[6] R. Tadikonda, S. Hardikar, and E. M. S. Narayanan, "Realizing high breakdown voltages (> 600 V) in partial SOI technology," Solid State Electron., vol. 48, no. 9, pp. 1655–1660, Sep. 2004.

Proceedings of the 23rd International Symposium on Power Semiconductor Devices & IC's
May 23-26, 2011 San Diego, CA

Innovative Designs Enable 300-V TMBS® with Ultra-low On-state Voltage and Fast Switching Speed

Wesley Chih-Wei Hsu, Florin Udrea*, Pai-Li Lin, Yih-Yin Lin, Max Chen

R&D Department
Vishay General Semiconductor Taiwan
Taipei, Taiwan

*Engineering Department
University of Cambridge
Cambridge, UK

Abstract — **A 300-V TMBS® (Trench MOS Barrier Schottky) rectifier with novel active-cell and termination design is first proposed and demonstrated. The device features a combination of p- transparent anode and Schottky contact, considerably reducing on-state voltage and ensuring a high degree of performance uniformity. A floating p-layer is used in termination region under a trench area to significantly enhance the blocking capability to 95% of the ideal breakdown voltage. In addition, the new termination design is area-efficient, being only 30 to 40 μm in width, which is less than 33% of the conventional guard-ring terminations. Experimental results have shown that the new 300-V TMBS structure with the novel active cell and termination design exhibits ultra-low on-state voltage of less than 0.9 V at 250 A/cm², a fast turn-off time lower than 55 ns and a high breakdown voltage over 380 V. It is noteworthy that these new components in active and termination regions can be formed without any additional masks, and hence at no extra cost.**

I. INTRODUCTION

Since TMBS rectifier emerged in 1993 [1] and was first commercialised in 2005, it has been conceived as one of the fastest expanding types in the power diodes market. Its voltage range has been extended from 30V to 200V [2], and its applications are widespread in power supplies, portable devices and LED lighting. To drive TMBS applications into the higher voltage markets, the doping concentration needs to be decreased and the depth of the drift region needs to be enlarged. This in turn results in drawback of high on-state resistance. Note that the conventional TMBS is a unipolar device and the on-state resistance varies superlinearly with the breakdown voltage rating. In addition, the high electric field (e-field) present in the Schottky contact leads to the barrier lowering effect and causes high leakage currents at high blocking voltages. Although TMBS is capable of reducing surface e-field by the inherent trench pinch-off effect, the Schottky contact is still the weakest point of the active cell and needs to be protected at high voltage operation.

Another challenge of developing a high-voltage TMBS is the design of the termination structure, which should be able to distribute the surface potential smoothly towards the device edge so as to sustain high blocking voltages. Furthermore, the issue of hot carrier injection (HCI) [3] should be addressed because energised carriers (i.e. holes in this case) generated in the depletion region at high reverse bias are likely to be injected into oxide layers by the perpendicular e-field present at the oxide/silicon interface. This leads to oxide quality degradation and likely device failure. For that reason, a termination structure with low orthogonal e-field components is more preferable in terms of device reliability.

Conventional guard ring termination has been replaced by trench termination, which is effective, cost-saving and easy to make [4]. However, for high voltage applications, e.g. 300V and over, it is found that the efficiency of the trench termination is largely compromised and a new design is necessary to improve blocking capability.

II. DEVICE STRUCTURE AND PHYSICS

The design concept of the 300-V TMBS rectifier can be discussed in two parts:

(A) Active Cell

The device has a Schottky contact in conjunction with a transparent p-layer emitter, as shown in Figure 1. The injection of plasma is controlled by the "transparency" of the p-layer and further limited by the Schottky contact. By making the transparent layer more lightly doped, the transparency increases, allowing a higher fraction of electron current to penetrate through it and reach the anode contact. This results in less plasma formation in the on-state, and as a result, a faster reverse recovery response. By increasing the doping in the transparent layer, while still preserving the Schottky contact (i.e. suppressing significant tunneling

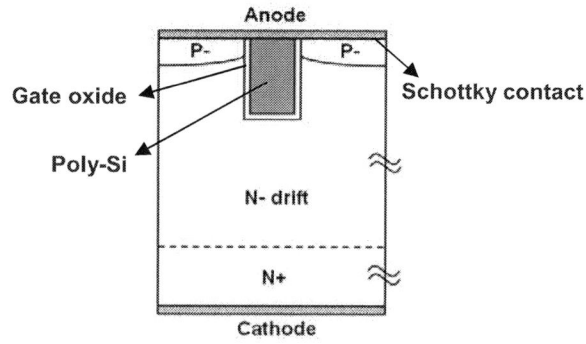

Fig. 1 Active cell structure of the 300-V TMBS

978-1-4244-8425-6/11 $26.00 © 2011 IEEE

(a)

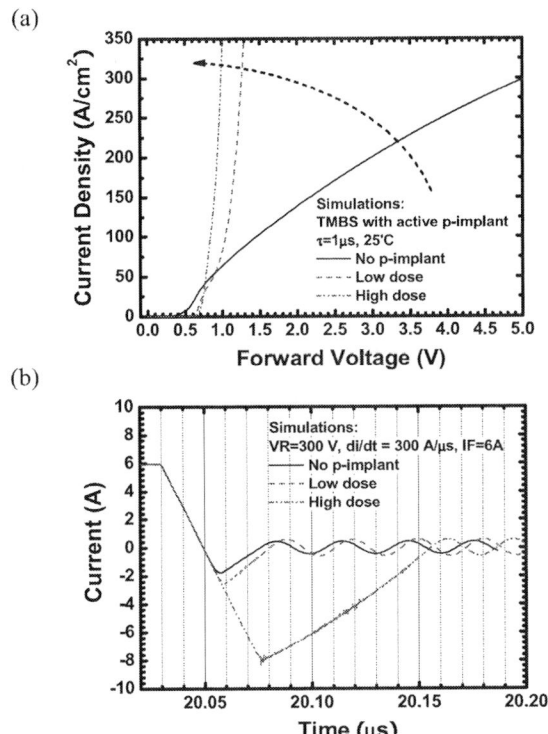

(b)

Fig. 2 TMBS with different p-implant doses in (a) forward conduction (b) reverse recovery characteristics

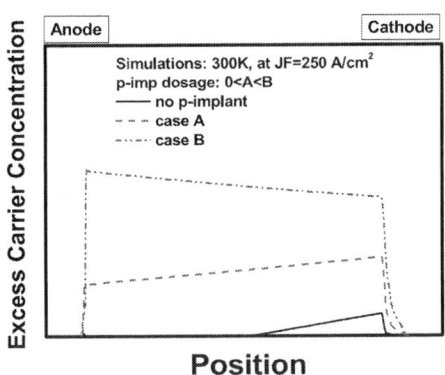

Fig. 3 Excess carrier concentration profiles of various p-implant doses

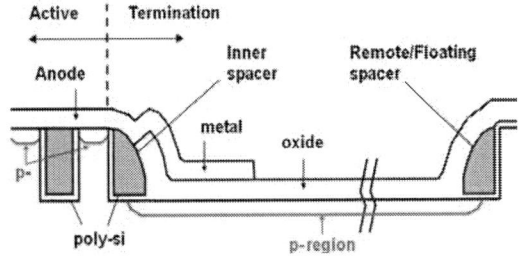

Fig. 4 TMBS with a new trench termination structure having a lightly-doped and floating p-region

specific to Ohmic contacts), the plasma level can be increased with a further enhancement in the on-state performance, though at the expense of increased switching losses. Generally speaking, the trade-off between on-state voltage and reverse recovery speed can be tailored by controlling the p-layer dose. In Figure 2 (a), it can be seen that at high current densities, e.g. 250 A/cm^2, the forward voltages decrease as the p- dose increases. This is due to stronger hole injection from p- regions. On the other hand, it should be noted that the junction voltages (at low current densities) go up as the p- dose increases. The reason is that the addition of p-type impurities raises the Schottky barrier height. Figure 2 (b) shows the simulated reverse recovery waveforms, indicating that the switching time (T_{RR}) and maximum reverse currents (I_{RM}) increase as the p- doses go up. It is noteworthy that a modest p- dose is enough to significantly reduce V_F and keep T_{RR} and I_{RM} almost unchanged.

An insight into the carrier distribution in on-state is shown in Figure 3. It can be seen the excess carrier concentration increases with the p-implant dose. In case A, the p- dose is low and the hole injection from the anode side is weak. However the carrier concentration is still larger than that without p- dose. In case B, the hole injection is much stronger due to its high p- implant dose. The excess charge concentration and its distribution inside the drift region determine the V_F and T_{RR} characteristics of the 300-V TMBS rectifier.

Similarly the transparent p-layer concept was applied in the so-called soft and fast recovery diode (SFD) [5] with shallow p- regions located between p+ area, instead of between trenches as shown in this paper, and therefore strong hole injection may take place at high current level. Nevertheless a transparent p- emitter, though with an Ohmic contact, has proved to be the major impetus in the development of Field-Stop IGBT (FS-IGBT) [6] and Soft-Punch-Through IGBTs (SPT-IGBTs) [7]. The introduction of the transparent emitter in IGBTs has led to ultra-low V_F and fast switching speed.

(B) Termination structure

In addition to the transparent p- Schottky contact in the active cell, the 300-V TMBS employs a new termination design which comprises a floating p-layer under the conventional trench termination, as shown in Figure 4. The inner poly-silicon spacer is connected to the anode electrode, while the remote spacer is floating. The floating p-region is formed by self-alignment technique without the need of mask definition. The design concept is similar to Junction Termination Extension (JTE) [8], and the difference is that the JTE's p- layer is connected to main junction, i.e. not floating. To some extent, the existence of the floating p-region is akin to the RESURF (Reduce Surface Field) [9] structure, helping to disperse the e-field at surface (in this case, in the whole termination area). Similarly to the design of the RESURF structure, the optimum blocking performance can be achieved when the p- layer dose is close to 2x10^{12} cm^{-2}.

978-1-4244-8425-6/11 $26.00 © 2011 IEEE   81

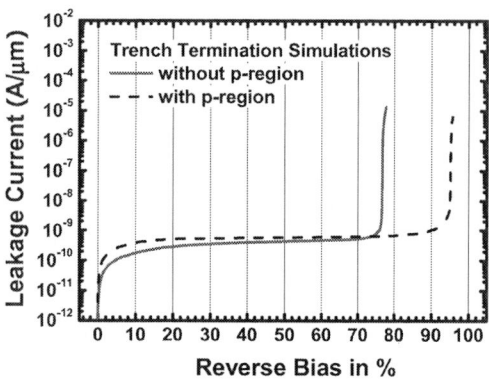

Fig. 5 Simulated reverse characteristics of the trench termination with and without p-region

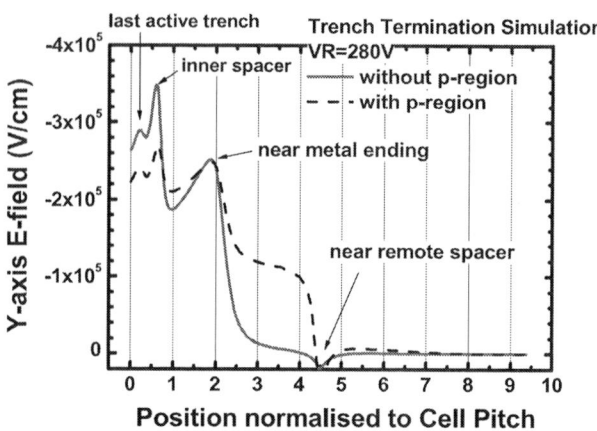

Fig. 7 Comparison of the simulated orthogonal e-field distributions

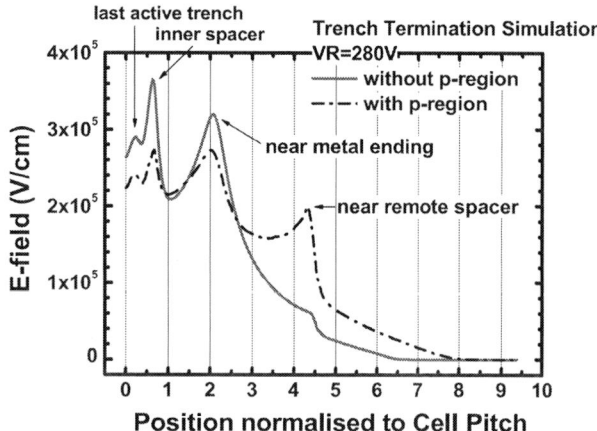

Fig. 6 Comparison of the simulated e-field distributions

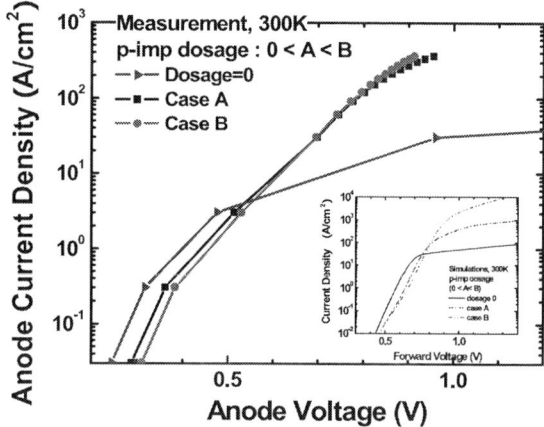

Fig. 8 Measured forward characteristics. The inset shows a similar trend in simulations.

Compared with the conventional trench termination, the new design can enhance the breakdown voltage (BV) from 75% to 95% of the ideal BV (i.e. the BV of active cell only) (Figure 5). The BV improvement comes from the fact that the e-fields increase along the floating p-layer and the peak e-fields in the active region are pulled down (Figure 6) to a level below 3×10^5 V/cm, that is approximately the critical e-field in silicon. The overall effect is that a smoother e-field distribution is achieved in comparison with that in the conventional trench termination, and therefore a higher BV is realised.

The new p-floating trench termination not only enhances the breakdown voltage but diminishes hot carrier injection (HCI) susceptibility. In Figure 7, it can be seen that the orthogonal e-fields decrease in the active region and increase in the p- floating layer. The reduction of peak e-fields means weaker HCI effect, and as a result better long-term oxide reliability is assured.

III. RESULTS AND DISCUSSION

The device is formed on an N-type epitaxy wafer with trench width 1μm. It is noteworthy that the transparent p-layers and termination p- regions are made without any additional masks. Figure 8 shows the measured forward characteristics of the 300-V TMBS rectifiers with different p-implant doses. As simulation results shown before, it can be seen that the forward voltages (V_F) increase at low current densities due to Schottky conduction dominance, while the V_F decrease at high current densities due to minority carrier injection, leading to a certain level of conductivity modulation. Figure 8 also includes an inset, which is obtained by simulation, that shows a similar trend to the measurement data.

From measurement results, it is found the addition of p-layer in the active region can also increase V_F uniformity: Figure 9 (a) show that the average V_F is as high as 6.1V at 250A/cm², 25°C and its distribution is relatively wide when there is no active p-implant. The V_F uniformity is much better when the p-layer is added with a modest dose and V_F decreases dramatically to 0.88V. In addition, the presence of

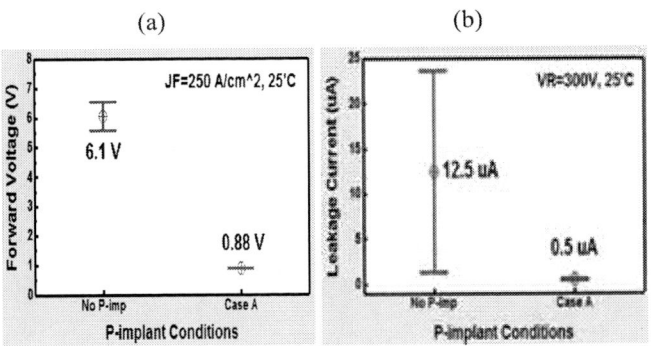

Fig. 9 Interval plots of the measured (a) forward voltages and (b) reverse leakage currents

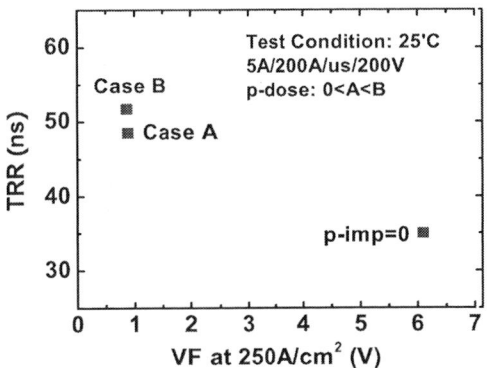

Fig. 10 Trade-off between V_F and T_{RR} among the devices with different active-region p- implant doses

the p-implant is able to reduce reverse leakage current (I_R) and tighten I_R distribution as shown in Figure 9 (b). The average I_R descends 25 times and its uniformity is greatly improved. The significant improvement in reverse leakage currents originates from the protection of the Schottky contact by the p- layer, helping reduce the surface e-field and therefore a relaxed barrier lowering effect and a lower leakage current.

The trench termination of the 300-V TMBS is 30-40 µm wide, less than 33% of the length of conventional guard-ring techniques. The anode metal, which functions as a field plate, should end at a certain position of the trench termination to gain the highest breakdown voltage. The measured BV of the 300-V TMBS is around 380V at 25°C while its leakage current is 1mA.

As mentioned above, the addition of active p- dose can adjust the performance between V_F and T_{RR}. The measurement of T_{RR} was done in an inductive-load tester, and the testing conditions are: V_R 200V, I_F 5A, and di/dt 200A/µs. In Figure 10, it can be seen that the V_F of TMBS decreases from 6.1V (no p- dose) to 0.88V (case A) and 0.86V (case B) while T_{RR} increases slightly from 35ns (no p- dose) to 49ns (case A) and 52ns (case B). Although T_{RR} is not the fastest among state-of-the-art products, the ultra-low V_F of 300-V TMBS offers considerable room for adjusting T_{RR} by implementing lifetime control techniques.

IV. CONCLUSIONS

A 300-V TMBS with transparent p-layer Schottky contact and p-implant trench termination is first proposed. By controlling the transparency of the p-layer, the forward voltage drop and reverse recovery characteristics can be tailored effectively to suit a particular application. With the addition of p-implant in the active region, it has been found that the V_F can be decreased to less than 0.9V at 250A/cm^2, 25°C. This is lower than other products on the market. The T_{RR} increases from 35ns to less than 55ns after adding p- dose. This can be reduced by applying lifetime control treatment.

The 300-V TMBS additionally features a floating p-implant trench termination, which helps to enhance its blocking capability to 95% of the ideal breakdown voltage. The floating p-layer makes the e-field distribution more uniform from the edge of active region to the cutting line, and hence lowers the peak e-fields in the active region, leading to a high BV. Moreover the p-implant trench termination reduces the orthogonal e-field, which is the root cause of hot carrier injection (HCI) for high-voltage electronic devices, and therefore assures long-term reliability for the device operations.

REFERENCE

[1] M. Mehrotra and B. J. Baliga, "The trench MOS barrier Schottky (TMBS) rectifier", IEDM Tech. Dig., p. 675-678, 1993

[2] M. Chen, H. Kuo and L. C. Kao, "The first commercial 200-V TMBS rectifier v.s. tranditional rectifier in telecom application", Proc. PCIM Europe, 2008

[3] L. Labate, S. Manzini, and R. Roggero, "Hot-hole-induced Dielectric Breakdown in LDMOS transistors", IEEE Tran. on Electron Devices, vol. 50, no. 2, p. 372-377, 2003

[4] W. C. W. Hsu, C. M. Liu, M. J. Kao, P. J. Kung, and M. J. Tsai, "A Novel Trench Termination Design for 100-V TMBS Diode Application", IEEE Electron Device Letter, vol. 22, no. 11, p. 551-552, 2001

[5] M. Mori, Y. Yasuda, N. Sakurai, and Y. Sugawara, "A Novel Soft and Fast Recovery Diode (SFD) with Thin P-layer Formed by Al-Si Electrode", Proc. of ISPSD, p. 113-117, 1991

[6] T. Laska, M. Munzer, F. Pfirsch, C, Schaeffer and T. Schmidt, "The Field-Stop IGBT (FS-IGBT). A new power device concept with a great improvement potential", Proc. of ISPSD, p. 355-358, 2000

[7] S. Dewar, S. Linder, C. V. Arx, A. Mukhitinov, and G. Debled, "Soft Punch Through (SPT) – Setting New Standards in 1200V IGBT", Proc. PCIM, Nuremmber, Germany, pp. 593, 2000

[8] B. J. Baliga, Fundamentals of Power Semiconductor Devices, Springer, 2008

[9] J. A. Appels, and H. M. J. Vaes, "HV Thin Layer Devices (Resurf Devices)", IEDM, p. 238-241, 1979

Proceedings of the 23rd International Symposium on Power Semiconductor Devices & IC's
May 23-26, 2011 San Diego, CA

Ultra Low Loss Trench Gate PCI-PiN Diode with $V_F < 350mV$

Motohiro Tsuda, Yasuaki Matsumoto, and Ichiro Omura

Department of Electrical Engineering and Electronics
Kyushu Institute of Technology
1-1 Sensui-cho, Tobata-ku, Kitakyushu-shi, Fukuoka, 804-8550, JAPAN
j349531m@tobata.isc.kyutech.ac.jp, omura@ele.kyutech.ac.jp

Abstract— **PiN diode forward voltage drop was reduced to as low as 325mV by the pulsed carrier injection (PCI) mechanism with trench MOS gate as the integrated injection control switch. The conventional PiN diodes have voltage drop of about 0.8V which is equivalent to 1%-2% energy loss in home appliances. The proposed PCI-PiN diode reduces the loss by more than 50% and the diode structure has process compatibility to conventional IGBTs and trench MOSFETs for easy implementation into mass production. The authors also confirmed PCI concept with the experiment with BSIT.**

I. INTRODUCTION

The conduction threshold voltage of 0.8V for conventional PiN diode (see Fig. 1 (a)) results in the substantial loss in rectifying circuit in power electronics systems. The estimated loss due to the PiN diode is 1%-2% for home appliances. The pulsed carrier injection (PCI) concept [1] has been proposed to reduce the conduction voltage drop to the level of 0.3V (see Fig. 1 (b)), which has potential to reduce the loss in rectifying circuit.

The PCI-PiN diode is different from conventional PiN diode in carrier injection control mechanism. In the forward condition for PCI-PiN diode, holes are injected into the i-layer in pulse wise with MHz range frequency, while the holes are continuously injected during forward bias condition for the conventional PiN diode. The PCI concept device has an extra N-layer next to P-emitter in PiN diode, low voltage switches to control the hole injection from P-emitter and electron conduction from the cathode to the extra N-layer. During electron conduction pulses, the stored carriers are reduced since holes are not injected. No PN junction appears along the current path during this pulse, the conduction threshold voltage of 0.8V is eliminated. However the original concept structure needs two external switches like MOSFETs for PCI control. The switch need other power supply to operate, therefore it is not practical.

This work newly proposes a practical structure for future production of the new low loss PCI-diode. The structure is integrated the PCI control switches. And, PCI concept is confirmed by the experiment with a feasible device.

	(a)Conventional PiN diode	(b)Original concept[1]	**(c)Proposed structure**
Diode structure	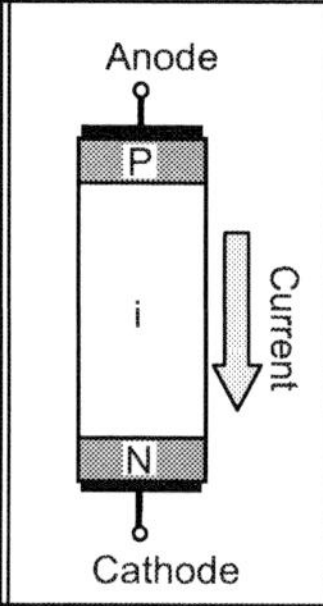	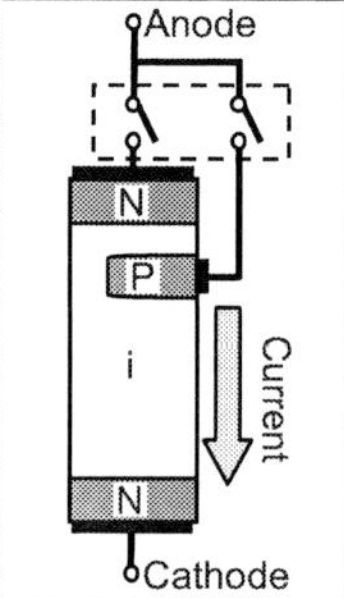	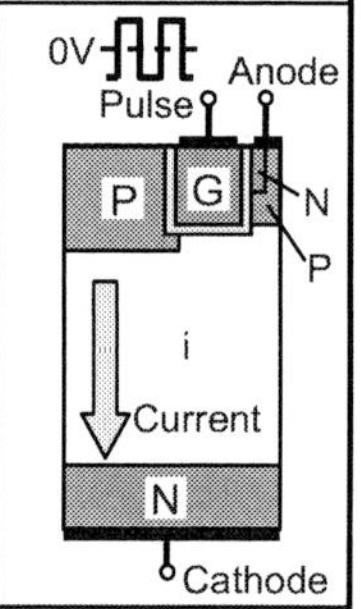
Carrier injection control	—	2 external switches	Switch integrated
V_F (J=50A/cm²)	0.8V	0.270V	0.325V
t_{RR}	1.10us	1.06us	1.29us

Figure 1. Schematic illustration of the conventional PiN diode, the original concept, and the practical structure. The practical structure has the trench instead of external switches.

II. SIMULATION

The new structure was demonstrated with 2-D TCAD simulations [2]. The structure has a trench MOS gate to switch the hole injection and the electron conduction in the anode of the diode during conduction state and the structure of the diode has the compatibility in fabrication process to state-of-the-art IGBTs and trench MOSFETs (see Fig. 1 (c)). Instead of external switches (see Fig. 1 (b)), the -10V/+10V square pulse voltage to form the P-channel/N-channel on the surface of the trench is applied to the gate. When the negative voltage -10V is applied, the holes are accumulated at the surface of

978-1-4244-8425-6/11 $26.00 © 2011 IEEE

gate oxide around the trench i.e. the P-channel is formed, and the holes are injected from anode side into the i-layer. This mode is similar to conventional PiN diode conduction operation. We call this mode the hole injection mode (see Fig. 2 (a)). Oppositely, when the positive voltage +10V is applied, the N-channel is formed, and only the electrons flow and it is expected that no PN junction appears along current path therefore the threshold voltage of 0.8V will not be appeared during this mode. We call this mode the electron conduction mode (see Fig. 2 (b)). The average forward voltage drop is decreased by repeating the hole injection mode and the electron conduction mode.

The optimum average forward voltage drop V_F is changed by injection control frequency. Figure 4 shows the average V_F for various frequencies as functions of hole injection time ratio. The conditions of simulations are as follows; the gate switch frequencies are fixed various values, the time of the hole injection mode and the electron conduction mode are changed respectively. The optimum average V_F was improved as increasing of the gate switch frequency and decreasing of the ratio of the hole injection mode time. The optimum V_F of 325mV was obtained at 1MHz under the hole injection time ratio of 0.2.

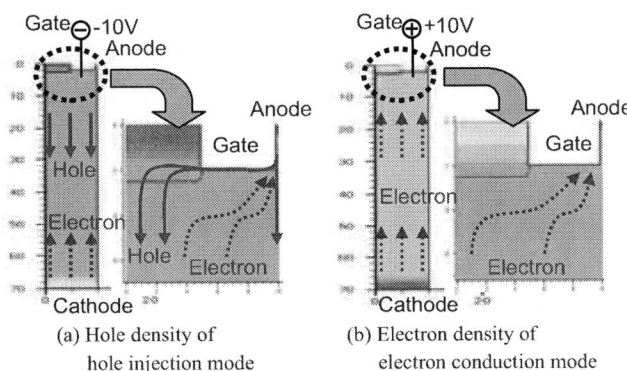

(a) Hole density of
hole injection mode

(b) Electron density of
electron conduction mode

Figure 2. Behavior of carriers at hole injection mode and electron conduction mode. Holes and electrons flow at hole injection mode, while only electrons flow at electron conduction mode.

The conduction threshold voltage of 0.8V is reduced by PCI concept in this structure. Figure 3 shows the transient forward voltage drop of the proposed diode at 1MHz pulse frequency. It was confirmed that the forward voltage drop was about 0.1V during the electron conduction mode, while the 0.85V voltage drop appears during the hole injection mode. The average forward voltage drop is as low as 362mV which is more than 50% lower than the conventional PiN diode.

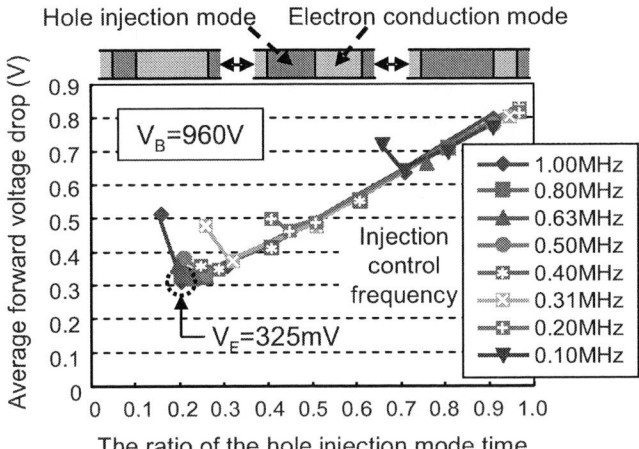

Figure 4. TCAD results of the average forward voltage drop for the proposed diode. The optimum value of 325mV was obtained with 1MHz pulse frequency at 50A/cm^2.

The comparison between the forward I-V_F characteristics of the proposed diode and that of conventional PiN diode, it seems that the conduction threshold voltage could be decreased to about 0.2V when hole injection mode time is 0.2us and electron conduction mode time is 0.8us. I-V_F characteristic is shown in Fig. 5. The simulated blocking voltage was 960V. This implies that the diode can dramatically reduce losses at diode bridge circuits in variety of appliance applications.

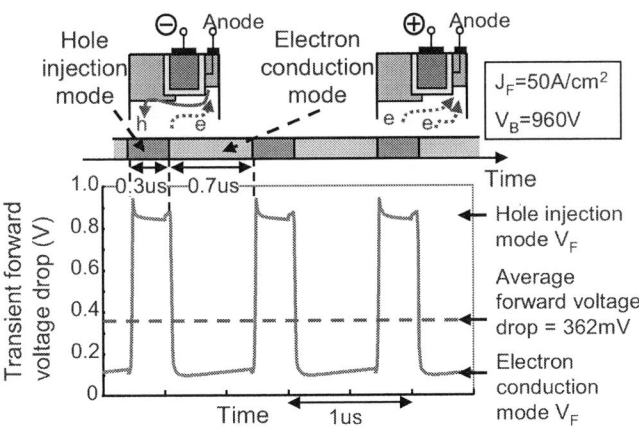

Figure 3. The transient forward voltage drop of the proposed diode at 1MHz pulse frequency at 50A/cm^2. The average forward voltage drop is as low as 362mV for this case.

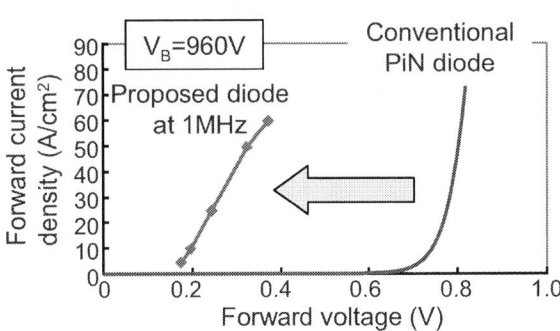

Figure 5. I-V_F characteristics. The conduction threshold voltage decreased to about 0.2V when hole injection mode time is 0.2us and electron conduction mode time is 0.8us.

The conduction losses at 50A/cm^2 including gate driving losses are shown in Fig. 6. As the injection control frequency is increased, the gate driving loss of the proposed diode is increased, while the conduction loss is gradually decreased. Thus, when the injection control frequency is extremely high, the average forward voltage drop is a little lower than the high frequency case, but the sum of these losses for the proposed diode is bigger. Considering these losses, the optimum injection control frequency is about 0.3MHz and the optimum V_F in that frequency is about 370mV. Then, the proposed PCI-PiN diode reduces the loss to about 50%.

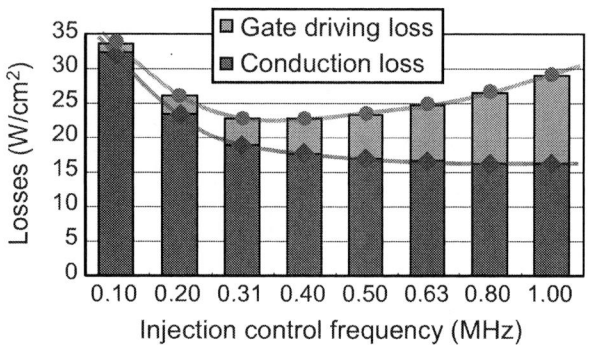

Figure 6. The conduction losses at 50A/cm^2 including gate driving losses. The optimum injection control frequency is about 0.3MHz.

The reverse recovery characteristic [3] of the proposed diode is shown in Fig. 7. After switching the hole injection mode and the electron conduction mode, reverse voltage is applied the proposed diode. Then, -10V is applied to the gate, the diode has the hole injection mode and is considered to be a usual PiN diode. The simulated reverse recovery time is about 1.29us which is sufficiently fast for rectifying 50Hz-60Hz AC current.

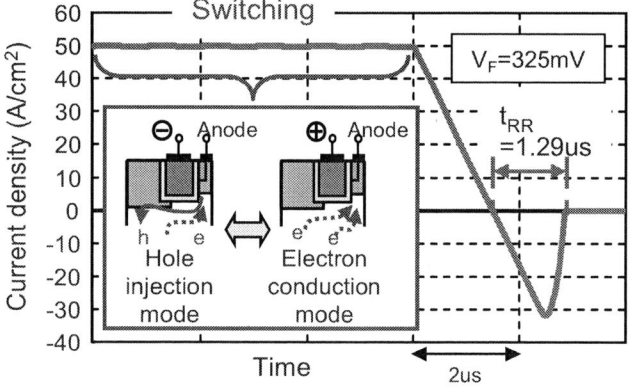

Figure 7. Reverse recoverry wave form for the proposed diode after switching. That time is about 1.29us which is sufficiently fast for rectifying 50Hz-60Hz AC current.

III. EXPERIMENT AND RESULT

The authors experimentally verified the PCI concept by using a BSIT (Bipolar Mode Static Induction Transistor) [4]. The structure of the BSIT and the experiment circuit with the BSIT are shown in Fig. 8. The BSIT has the source N-layer, the gate P-layer and the drain N-layer, which corresponds to the cathode N-layer, the anode P-layer and the anode N-layer for the PCI-diode structure so that the concept can be experimentally demonstrated with this particular device.

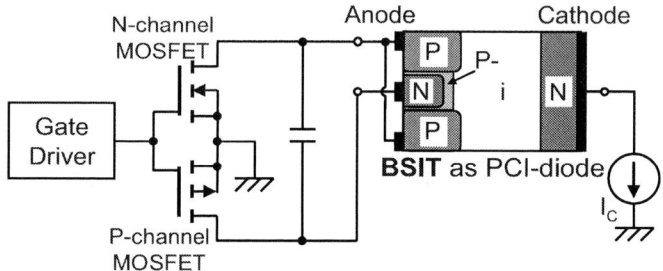

Figure 8. The structure of BSIT and the experiment circuit used BSIT. The circuit has two MOSFETs that switch the hole injection mode and the electron conduction mode. In addition, the circuit has a switch adjust the period passing a current BSIT.

The experiment circuit has N-channel MOSFET and P-channel MOSFET connected to the anode P-layer and the anode N-layer respectively, to switch the hole injection mode and the electron conduction mode. Two MOSFETs are alternately turned on and off.

When the N-channel MOSFET is turned on, the holes are injected from the anode P-layer into the i-layer in the BSIT. This mode is corresponding to the hole injection mode. Oppositely, when the P-channel MOSFET is turned on, only the electrons flow between cathode N-layer and the anode N-layer. This mode is corresponding to the electron conduction mode.

The PCI concept was experimentally demonstrated, as follows. Figure 9 shows the switching signal to two MOSFETs and the transient forward voltage drop of BSIT at 10A cathode current (I_C) at 0.1MHz pulse frequency. It was confirmed that the forward voltage drop was about 0.3V during the electron conduction mode, while the 1.1V voltage drop appears during the hole injection mode. It is verified the experiment that the PCI concept is feasible to reduce the losses with the forward voltage drop of a PiN diode. In this case, the average forward voltage drop is as low as 731mV.

The optimum average V_F is changed by switching signal frequency and I_C. Figure 10 shows the average V_F for various frequencies as functions of hole injection time ratio when I_C is 10A and the optimum average V_F when I_C is 5A and 18A. The optimum V_F of 620mV was obtained at 0.25MHz under the hole injection time ratio of 0.3. Moreover when I_C is 5A and 18A, optimum V_F of 346mV and 913mV was obtained at 0.25MHz and 0.5MHz respectively.

978-1-4244-8425-6/11 $26.00 © 2011 IEEE

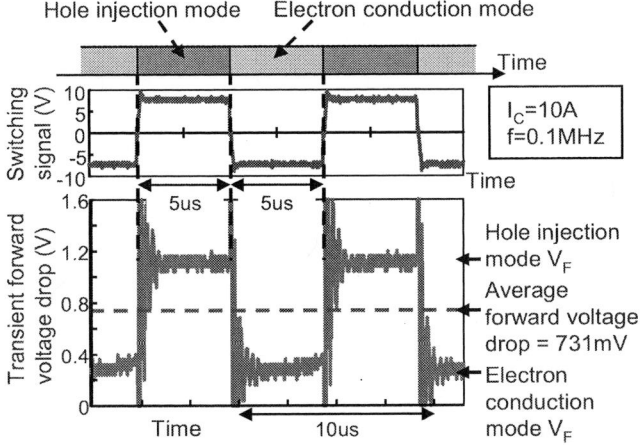

Figure 9. The transient forward voltage drop of BSIT at 0.1MHz pulse frequency when I_C is 10A. The average forward voltage drop is as low as 731mV for this case.

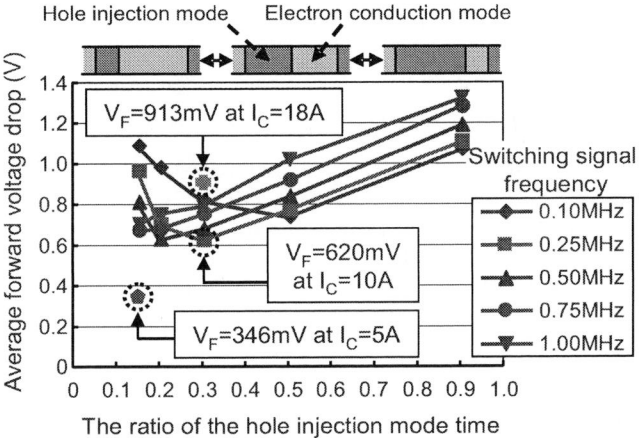

Figure 10. Experiment results of pulsed carrier injection concept with BSIT. The optimum average values of forward voltage drop are as low as 913mV at 18A and 620mV at 10A and 346mV at 5A.

It is different from the simulation that the average V_F is increased with switching signal frequency when the ratio of the hole injection mode time is high. From the transient V_F waveform, it is found the huge overshoot voltage due to the stray inductance in the circuit is appeared. The overshoot causes the increase in the average forward voltage drop at the switching (see Fig. 9).

As I_C is increased, the average V_F is increased because the voltage drops of the MOSFETs are increased by Ohm's law. It includes voltage drop due to N-channel MOSFET at the hole injection mode and P-channel MOSFET at the electron conduction mode. Therefore, the experimentally obtained average forward voltage drops in each mode are higher than the value from the simulation.

IV. CONCLUSION

Proposed trench gate PiN diode structure can achieve the PCI concept without external switches. The proposed PCI-PiN diode reduces the conduction loss to about 50%. This structure can be manufactured by present technologies. This concept potentially contributes to save energy because the waste of electricity is decreased and the electrical energy can be efficiently used. The PCI concept was also confirmed by the experiment.

REFERENCES

[1] Yasuaki Matsumoto, "Challenge to the Barrier of Conduction Loss in PiN Diode toward VF<300mV with Pulsed Carrier Injection Concept", Proc. of ISPSD 2010, pp.119-122, 2010.

[2] Sentaurus Device User Guide Ver.A-2007.12, 2007.

[3] Hansjochen Benda and Eberhard Spenke, "Reverse Recovery Processes in Silicon Power Rectifiers", Proc. of the IEEE, Vol. 55, No. 8, August 1967, pp.1331-1354, 1967.

[4] Masayasu Ishiko, Sachiko Kawaji, Hiroshi Tadano, Susumu Sugiyama, and Haruo Takagi, "A Normally-Off Bipolar Mode Static Induction Transistor (BSIT) with High Current Gains", Proc. of ISPSD 1992, pp. 92-97, 1992.

Proceedings of the 23rd International Symposium on Power Semiconductor Devices & IC's
May 23-26, 2011 San Diego, CA

Field Shielded Anode (FSA) Concept Enabling Higher Temperature Operation of Fast Recovery Diodes

S. Matthias, J. Vobecky, C. Corvasce, A. Kopta and M. Cammarata

ABB Switzerland Ltd Semiconductors
Lenzburg, Switzerland
Sven.Matthias@ch.abb.com

Abstract— **In this paper, we introduce the Field Shielded Anode (FSA) concept that enables higher temperature operation of fast recovery diodes with planar junction termination. Conventional diodes utilizing local lifetime control principles show excellent dynamic properties at the expense of a higher leakage current, which is generated during reverse blocking when the space charge region penetrates into the zone containing the radiation defects. In contrast to this, the FSA concept spatially separates the space charge region from the zone with the radiation defects. The ruggedness of conventional diodes can be exceeded with the new FSA concept, while the leakage current is reduced by a factor of ~4. This was achieved using a special junction extension introduced between the active area and the guard-ring termination. The design parameters and their influence on the softness and the safe-operating area are presented.**

I. INTRODUCTION

Fast recovery diodes are used in modern power converters. With continuously improved performances of switches such as the IGBT, the diode needs to follow in order to not become the limiting factor. The utilization of the chips in traction and drive applications requires high ruggedness, good surge-current capability, reverse recovery softness and low-loss operation. There are two main concepts of high power free-wheeling diodes established for applications, which require blocking voltages in the range of 1.2kV to 6.5kV and current ratings of 50A to 300A per chip. While the first one is based on the emitter-controlled principle [1], the second one utilizes local control of the axial carrier-lifetime to tailor the dynamic properties [2]. Both concepts have their inherent advantages and drawbacks. We will focus here on the principle of local carrier lifetime control. This concept is commonly implemented by an ion-irradiated anode to reduce the peak recovery current and the switching losses and provide soft reverse recovery. In addition, this is combined with an electron or ion irradiation to reduce the plasma in the n-base region in order to limit the reverse recovery losses under normal operation condition and dynamic avalanche under harsh switching transients to ensure a high safe operating area (SOA) performance [3]. Nevertheless, the localized point defects generated by the irradiations are a source of leakage current, when subjected to a space charge region of a reverse biased anode p-n junction. This effect is more pronounced for hydrogen or alpha-particle irradiated diodes compared to electron irradiated ones. The strong spatial localization of the defects from ion irradiation and the high defect introduction rate of the deep levels, which increases with the particle's size, are responsible for an increased leakage current. In this paper, we present the FSA concept, which eliminates the impact of these effects on the leakage current by locating the radiation defects outside the space charge region without sacrificing on any other parameters.

II. FIELD-SHIELDED-ANODE (FSA) CONCEPT

A conventional diode with locally incorporated deep levels to utilize local lifetime control is shown in Fig. 1. In this case the electric field evolving during reverse blocking is penetrating into the zone of radiation defects already at very low reverse voltages. This generates a high leakage current and limits the thermal stability of the chip.

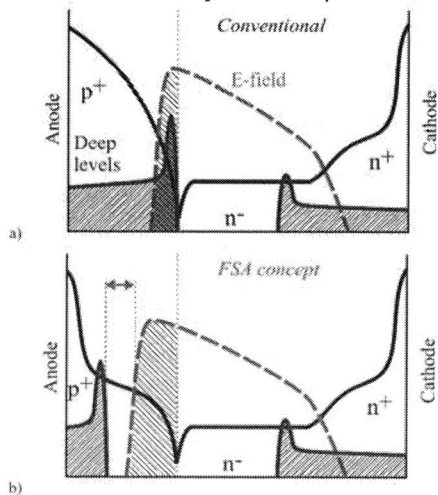

Figure 1. Schematic drawing (not to scale) of the cross-sections of two fast recovery diodes. a) The conventional diode has a single deep diffused anode-profile. b) Field shielded anode design separates the deep levels from the space charge region.

978-1-4244-8425-6/11 $26.00 © 2011 IEEE

Introducing a low-doped p-buffer region which acts as a Field-Shielding Anode (FSA) overcomes this issue inherently, because the space charge region is spatially separated from the zone of irradiation defects Fig. 1b. This kind of profile has been previously applied for large area discrete diodes [4]. However, their doping profile is much deeper and normally combined with a beveled angle junction termination, which can not be applied to the diode chips used here. Therefore, in the case of the chip-diodes, the doping profile has to comply with the requirements of a planar junction termination, consisting either of a guard-ring structure or a junction termination extension. Hence, the deeper part of the anode p-type layer is limited below 30 μm, while the doping concentration is increased to maintain the reverse blocking capability. Since the increase of the doping concentration can deteriorate the softness of reverse recovery, a careful optimization of the doping profile is of primary importance

III. EXPERIMENTAL

TABLE I.   TABLE OF DIODES USED IN THE PAPER

Table Head	FSA diodes			Convention-al diode
	A	*B*	*C*	*D*
FSA dose (a. u.)	150%	100%	50%	
$V_{breakdown}$ (V)	4100	4100	4000	4000
V_F (V) @ I_F = 125A	2.1	2.1	2.1	2.05
I_R (mA) V_R =3.3kV, T_j = 125°C	~0.6	~0.6	~0.7	~2

A conventional diode and three FSA-diodes variants have been used for this study. The FSA-diodes differ in the ion-implantation dose of the buffer anode implantation named FSA-dose according to Tab. I. The conventional diode "D" also differs in the position of the defect peak of the ion irradiation, which is situated within the p-n junction Fig. 1. The measured spreading resistance profiles of these diodes are shown in Fig. 2. Although the FSA-diode variants differ in the depth of the p-n junction (referred to as "buffer"-anode), the blocking voltage is not affected. It is more important to note the different levels of the p-buffer. In addition the surface concentration is of the same level as for the convention diode.

A. Static parameters – leakage current

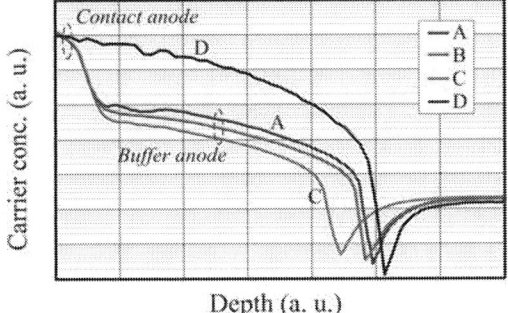

Figure 2.   The anode doping profiles of the diodes under study.

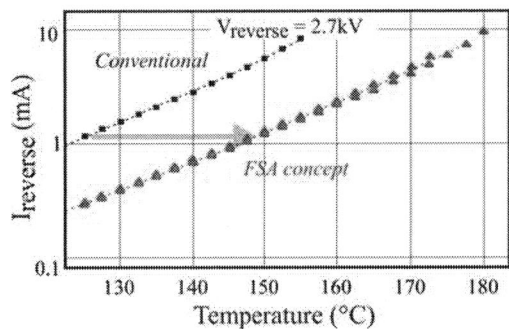

Figure 3.   Leakage current scaling with temperature for the conventional and the FSA diode (type "B").

It has been shown in Tab. I that both the forward voltage drop V_F and the blocking voltage $V_{breakdown}$ of the diode match the performance of the conventional diode. However, the superior static blocking behavior is fully maintained at elevated temperatures and reflects the advantage of the FSA-concept, as illustrated in Fig. 3. The arrow highlights the fact that the leakage current of the conventional diode at 125°C is equal to that of the FSA concept at 150°C. In addition, the FSA-diode offers stable reverse blocking even beyond a junction-temperature of 170°C.

The comparison of leakage currents between the devices A, B and C in Tab. I show that the leakage current can increase when the concentration of the anode buffer doping and junction depth decreases down to the level of diode "C". Consequently, diode "C" sets a lower limit on the buffer characteristic that should be designed to fully support the field before penetrating into the zone of irradiation defects.

B. Dynamic parameters – softness

An important performance parameter for a diode is the capability to switch-off low currents without oscillations over the whole temperature range, in particular at high DC-link voltages and under the influence of a high stray-inductance. This phenomenon is referred to as "softness". Fig. 4 shows the switching of the diodes "A" and "C" at a high DC-link voltage from an ON-state current which is one ninth of the nominal current and with a stray inductance which is double as high as for the rated IGBT module employing such chips. Diode "A" exhibits a reverse recovery maximum current I_{RM} in the range of 95A. During the subsequent tail-current phase the current snaps off at around I_F = -25A resulting in a high dI/dt causing a voltage overshot up to 4.5kV. In contrast to the behavior of diode "A", diode "C" shows a 20% lower I_{RM} as a result of a lower concentration of the ON-state plasma at the anode side of the n-base. During the following tail-current phase, the charge-carriers are extracted without a current snap-off and the maximum detected reverse bias overshoot voltage is limited to 3.1kV. In Fig. 4, the maximum amplitude of the generated voltage oscillations during reverse recovery is indicated by the grey bars for the four different diodes and considered as an indicator of the softness. It reflects the sensitive dependence of the softness and hence the maximum reverse recovery peak on the buffer-anode doping concentration and therefore on the associated emitter efficiency buffer-doping of the FSA-diodes. As a rule of thumb - the lower the buffer doping concentration the softer is the diode. This is a consequence of the purely emitter controlled mode of the low current operation. Diode "C" is in this setup matching the performance of the conventional (reference) diode "D". In contrast to the low current regime, the highly doped "contact" anode and the radiation defects of the local lifetime control determine the plasma levels at the anode side for higher current levels Fig. 1.

In summary it can be concluded that the required softness can be tuned by the doping-level of the deep buffer implant resulting in a tradeoff between the static reverse blocking behavior and the reverse recovery softness.

C. Dynamic parameters – safe operating area

The FSA-diode and in particular the buffer anode profile was discussed so far in the vertical direction only. However, a reliable diode operation under fast reverse recovery requires also a laterally optimized doping profile to ensure excellent dynamic properties like softness and high robustness. The latter requires the introduction of a junction extension in between the active area and the guard-ring termination Fig. 5a. This region resembles that of a resistive zone introduced previously [5]. However the junction extension utilized herein is formed by smaller guard rings having a laterally overlapping buffer doping while the high contact doping remains separated. A crucial parameter is the junction extension width.

Fig. 5b shows the dependence of the maximum (last pass) commutating dI/dt during diode turn-off. This graph confirms the necessity of the modified junction extension, because an early device failure is observed for a small width of the junction extension. For wider junction extension regions the device failure is located in the active area identifying that the full potential of the diode is utilized while exceeding the performance of the conventional diode.

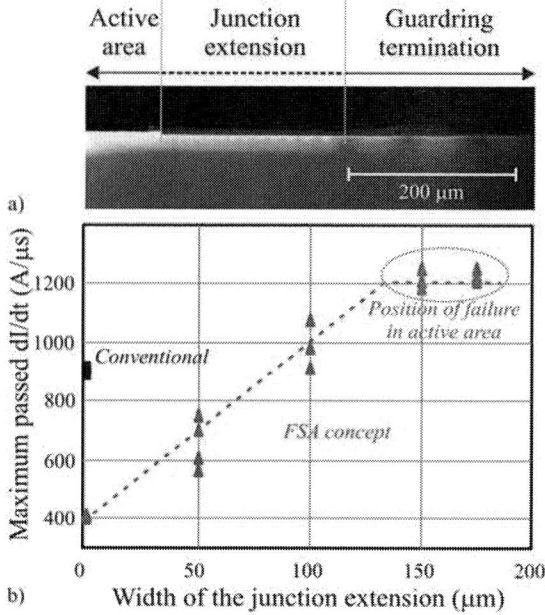

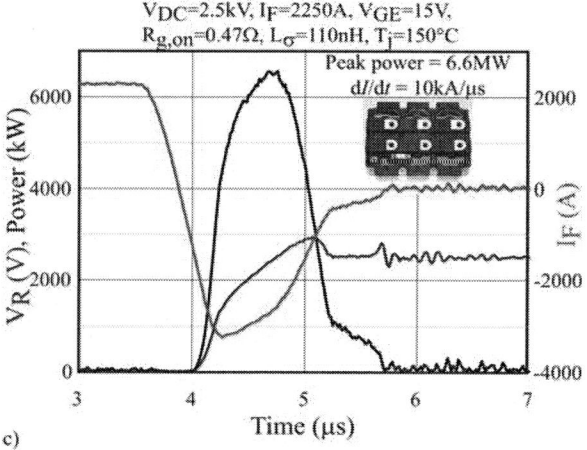

c)

Figure 5. a) Scanning electron image of a cross-section of the introduced main junction extension. b) Maximum passed dI/dt vs. width of the junction extension region (V_{DC}=2.5kV, L =1200nH, T_j=150°C) of a single chip. c) Measured diode SOA waveform on module-level under the above given conditions. The inset shows the 3.3kV / 1500A rated module.

The inset in Fig. 5c shows a high-temperature module rated at 3.3kV and 1500A with 12x diodes and 24x IGBTs inside. The high ruggedness on chip-level is well reproduced even under such heavy parallel conditions. The diodes are commutated with more than 10kA/µs. During the reverse current tail phase (Fig. 5c. between 4µs and 5µs) the diode is subjected to strong dynamic avalanche.

It has been shown experimentally that a junction extension ensures a rugged diode performance in terms of the reverse bias safe operation area. Contrary to the case of static reverse blocking, during reverse recovery and under an equivalent diode voltage the electric field is much higher due to the holes passing through the space charge region with saturation velocity. As a result, their positive space charge adds to that of the ionized donors, the gradient of the electric field grows and

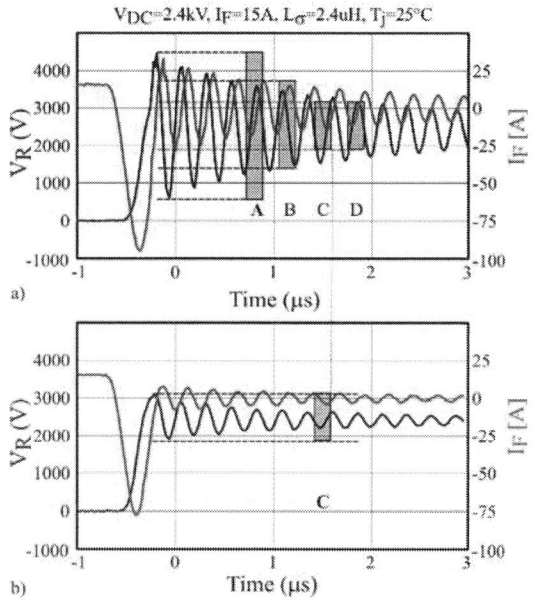

Figure 4.   Reverse recovery waveforms for the FSA-diodes. a) Diode "A". b) Diode "C".

978-1-4244-8425-6/11 $26.00 © 2011 IEEE

so does the peak electric field. As this field generates additional free carriers, this effect reflects in the waveform of current in the tail time. One can observe this in Fig. 5c after I_{RM} between 4µs and 5µs. As will be shown below, this effect is pronounced at the diode junction periphery.

A two-dimensional numerical simulation (Dessis$_{TCAD}$ device simulator) of two 1.7kV FSA-diodes is shown in Fig. 6 for the temperature of 400K during reverse recovery under identical operation conditions. The only difference between the simulated diodes is the junction extension region, which is introduced only in diode "2". As pointed out above, due to the dynamic avalanche the electric field at the p-n junction is enhanced in the positions with a high current density. Consequently, in diode "1" a high electric field and current crowding lead to an early device failure. In contrast to this, the junction extension suppresses the current crowding in this region effectively and hence lowers the electric field and yields a very robust device. Please note that the junction extension for these 1.7kV rated diodes in the simulation is about half the width required for the 3.3kV diode investigated experimentally.

D. Dynamic parameters – surge current

The last dynamic parameter addressed herein is the surge current capability which highlights the positive effect of the contact anode Fig. 2. The surface concentration on the anode side for the FSA variants is identical to the conventional diode resulting in a superior surge current protection as shown in Fig. 7. Even after 100x applied pulses of 10ms, each of them exceeding nine times the nominal current, no degradation was observed. This confirms the importance of the high contact anode doping that needs to be combined with the local axial lifetime control to ensure the contradicting parameters.

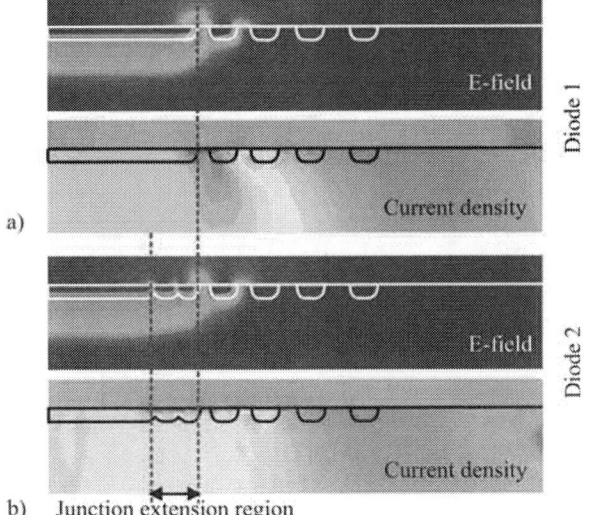

a)

b)   Junction extension region

Figure 6.   Cross-section of the electric field E and current density during dynamich avalanche of two 1.7kV FSA-diodes. All plots are with idenitcal color-scales. a) FSA without junction extension. b) FSA with junction extension.

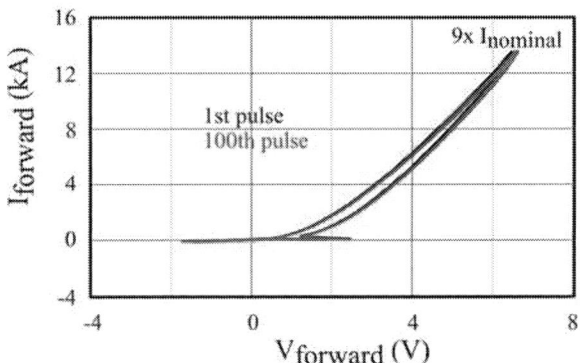

Figure 7.   Surge current capability of the FSA-diode on module-level. A 10ms surge current pulse is applied 100x times exceeding 9x the nominal current without degrading the device.

IV. CONCLUSION

In this paper we have introduced the field-shielded anode concept for chip-diodes. It was shown experimentally that the spatial separation of the zone with radiation defects and the space charge region forming during reverse blocking ensures superior static blocking capability compared to conventional diodes applying local axial carrier-lifetime principles. It was proven that a soft reverse recovery behavior is achievable by sensitive tuning of the buffer doping concentration matching the behavior of a conventional diode. The introduced junction extension region prevents current crowding efficiently. The confirmed diode SOA on chip-level exceeds the robustness of our conventional reference diode without the low-doped p-buffer region. In addition, the surge current is kept at the same level due to the high doping profile section of the new anode. In conclusion, the FSA concept combines the advantages of emitter-controlled principles in the low-current regime with the advantages of local axial carrier lifetime principles in high current regimes.

ACKNOWLEDGMENT

The authors thank Birgit Waishar and Yoichi Otani for the scanning electron micrographs and the spreading resistance profiles.

REFERENCES

[1] A. Porst, F. Auerbach, H. Brunner, F. Hille, G. Deboy and F. Hille, "Improvement of the diode characteristics using emitter-controlled principles (EMCON-diode)," Proc. ISPSD'97, pp. 213-216, 1997.

[2] J. Lutz and U. Scheuermann, "Advantages of the new controlled axial lifetime diode," Proc. PCIM'94, pp. 163, 1994.

[3] A. Kopta, M. Rahimo and U. Schlapbach ,"New Plasma Shaping Technology for Optimal High Voltage Diode Performance," Proc. EPE'07, pp 1-10, 2007.

[4] J. Vobecky, P. Hazdra and J. Homola, "Optimization of Power Diode Characteristics by Means of Ion Irradiation," IEEE Transactions on Electron Devices, vol. 43 No. 12, pp. 2283–2289, 1996.

[5] M. Nagasu, H. Kobayashi, T. Saiki, Y. Yasuda, R. Saitou and M. Mori; "3.3 kV IGBT Module having Soft Recovery Diodes with High Reverse Recovery dI/dt Capability (HiRC)," Proc. PCIM'98, pp. 175-178, 1998.

Proceedings of the 23rd International Symposium on Power Semiconductor Devices & IC's
May 23-26, 2011 San Diego, CA

Evaluation of 1.2kV Super Junction Trench-Gate Clustered Insulated Gate Bipolar Transistor (SJ-TCIGBT)

N. Luther-King, M. Sweet and E.M. Sankara Narayanan

Electrical Machines and Drives Research Group, The University of Sheffield, Mappin Street, S1 3JD, UK.

Fax: 0114 222 5196. Email: L.Ngwendson@sheffield.ac.uk

Abstract: **We report, for the first time, results of extensive 2D simulation evaluation of *the first* MOS controlled thyristor structure employing the Super Junction concept on a 1.2 kV field stop structure. In comparison to a standard device, simultaneous reduction in $V_{ce}(sat)$ and E_{off} can be achieved in a Super Junction Trench Clustered Insulated Gate Bipolar Transistor (SJ-TCIGBT). The simulation results show that up to 80% reduction in E_{off} is possible. Unlike the Super Junction Insulated Gate Bipolar Transistors, there is no significant increase in the saturation current with the anode voltage or the depth of the pillars. SJ-TCIGBT is a highly promising next generation device concept with record-breaking $V_{ce}(sat)$ - E_{off} trade-off enhancement to improve converter efficiency.**

I. INTRODUCTION

With increasing demand in energy efficiency of power electronic converters, there is a need to minimise the trade-off between the on-state Vce(sat) and the turn-off loss (E_{off}) in MOS-Bipolar devices. In devices such as the IGBTs and its variants, this is implemented by controlling anode injection efficiency through the anode thickness and doping and carrier life time. For example, The Field Stop (FS) is a classic concept with thin silicon, a transparent anode and buffer structure, which has greatly improved the $V_{ce}(sat)$ - E_{off} trade-off in IGBTs and it is widely used in industry [1].

The Super Junction (SJ) concept was first proposed to improve the specific on-state resistance (Ron(sp)) Versus breakdown voltage (BV) trade-off in high voltage power MOSFETs and commercial devices employing the concept are widely available [2]. SJ-IGBTs have also been proposed to improve the $V_{ce}(sat)$-E_{off} trade-off compared to the standard IGBTs [3,4]. However, in SJ-IGBTs there is a significant increase in the saturation current with anode voltages which bring about issues associated with turn-off and short-circuit performances in planar gate SJ-IGBTs with deep pillars. This can be attributed to the fact that the p-pillars are connected to the grounded p-base. The unavoidable existence of a strong current gain of the vertical PNP (P-base and P-pillar/N-drift/P-anode) transistor due to its narrow base (distance from pillars bottom to anode) causes poor current saturation behaviour. This is the case even though the P anode is designed to be a weak injector (i.e. a transparent anode). Attempts have also been to introduce an N layer (carrier enhancement layer) underneath the P pillars at the cathode end to supress the activation of the parasitic PNP transistor. To prevent the increase in saturation current with anode voltage, it is possible to locate the p pillars just below trench gates in a trench IGBT, which will require precise location of ultra-fine pitch p pillars underneath the trench gates [5].

The planar CIGBT and Trench gate CIGBT (TCIGBT) are a family of MOS-gate controlled thyristor structures experimentally proven at 1.2kV, 1.7kV and 3.3kV. They show current saturation characteristics and short circuit performance similar to IGBTs [6-9]. In a conventional CIGBT, a low 'self-clamping voltage' (V_{scl}) is required for wide FBSOA and necessitates careful control of the n-well implant dose. *Recently PMOS gates in addition to the NMOS gates have been proposed to relax this trade-off* [10].

The issues with SJ-IGBTs can be avoided in the CIGBT technology, because (i) the P-pillars are wide and truly floating which is necessary for safe turn-off and (ii) the unique 'self-clamping' of the cathode cell potential in the CIGBT technology can be used to control the saturation current density.

In this paper, we show for the first time, through 2D simulations that the SJ-TCIGBT concept can result in very low Vce(sat) and E_{off} simultaneously hence the possibility of a significantly improvement in the Vce(sat) – Eoff trade often necessary in MOS bipolar devices, while retaining all the other desired properties of a MOS-Bipolar power device. The Synopsys TCAD Package TMA Tsuprem4 and Medici have been used for process and device simulations [11].

II. DEVICE STRUCTURE AND OPERATION

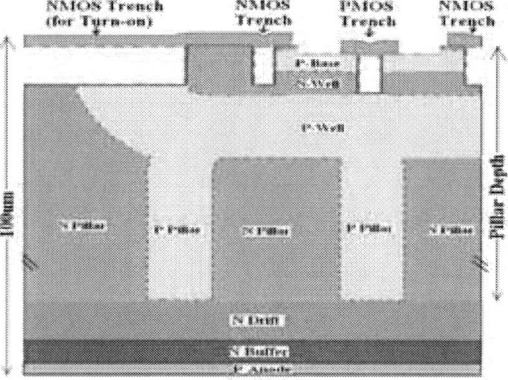

Fig. 1*: Schematic of the proposed SJ TCIGBT with deep PMOS trench channels. All the gates are connected together to form a three terminal device.*

Fig.1 shows the simulated half-cell of the FS SJ-TCIGBT. For a given N-pillar doping (typical values range from 1e15 to 8e15cm^{-3}); the corresponding P-pillar doping is chosen based on its width to achieve charge balance. It can be seen that the P-pillars are simply extensions of the floating P-well region to which they

978-1-4244-8425-6/11 $26.00 © 2011 IEEE

are connected while the N-pillars are similar to the N-drift region under the SJ region. PMOS gates connect the P well to the P base. The device thickness is 100μm, N-drift doping is 8e13cm^{-3} while the junction depth of the buffer is 10μm with 2e16cm^{-3} peak doping. The Vth of the NMOS (Vth$_{NMOS}$) and of PMOS (Vth$_{PMOS}$) are 5.5V and -1.0V respectively. The trench gates are all connected hence PMOS is activated only during turn-off, when $V_g <$ Vth$_{PMOS}$ [10].

The On-state performance of the SJ-CIGBT is the same as that of a conventional device. For on-state operation: a gate voltage $V_g >$Vth$_{NMOS}$, the cathode is grounded and $V_{anode} > 0.7V$ is applied. With increase in anode voltage, when the depletion from the reverse biased p-base/n-well junction reaches the p-well, the potential of the cathode cell does not increase further (or is 'clamped'). This 'self-clamping' of the cathode cell potential protects the cathode trenches from high anode voltages to realize wide FBSOA and fast switching. More details of CIGBT/TCIGBT operation can be found in published papers [5-10]. The presence of the SJ pillars only change the carrier dynamics within the device to further enhance turn-off speed.

III. BLOCKING VOLTAGE PERFORMANCE

To achieve the rated breakdown voltage of 1.2kV, the N-pillar and P-pillar doping and widths were adjusted to achieve charge balance using the charge-balance principle [1]. Simulation results show that the maximum BV achieved with the introduction of the SJ pillars into a conventional structure can be 17% higher. This means that wafer thickness can be further reduced to 85um. Fig.2 shows potential contours during the blocking mode. It can be seen the distance between

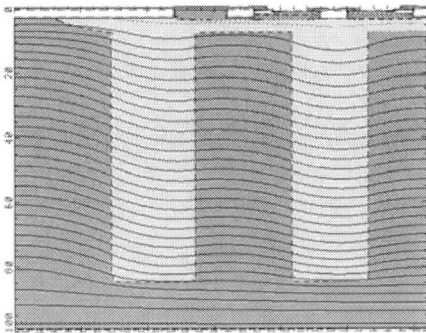

Fig. 2: *Equipotential lines across the device during the blocking state while supporting 1200V.*

two consecutive potential lines is constant throughout the depleted SJ region indicating that the electric field has a constant value. The lateral field component perpendicular to the vertical PN junctions is responsible for the depletion of the SJ pillars and generates the wave-like pattern of the equipotential contour lines. There are no indications that the blocking performance is limited by the second breakdown [12].

IV. ON-STATE PERFORMANCE

Fig.3 shows that for an identical anode and buffer, the SJ pillars do not influence the V$_{ce}$(sat). This is because the pillars doping in the range of 1 to 8e15cm^{-3} is at least an order of magnitude lower than the on-state excess carrier density due to thyristor mode of conduction. It is evident from fig.3 that a very low V$_{ce}$(sat) of 1.1V and 1.25V can be achieved at T$_j$=25^{0}C and T$_j$=125^{0}C respectively at

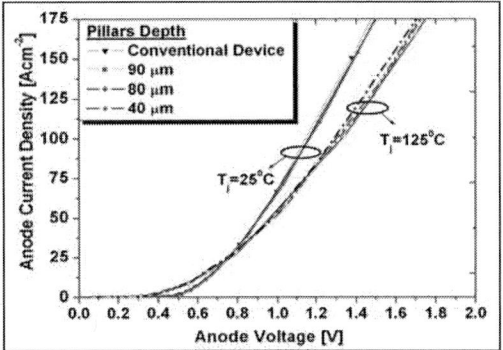

Fig. 3: *Typical on-state I(V) characteristics. Trench width 1μm, peak buffer doping=1e16cm^{-3} and τ=10μs.*

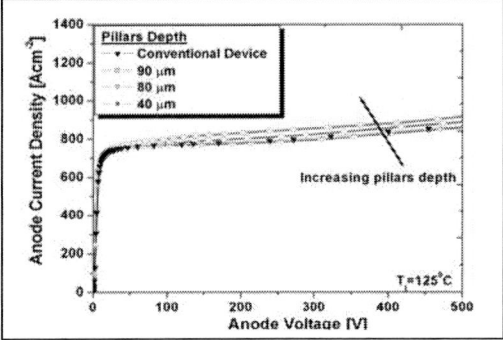

Fig. 4: *Typical I(V) characteristics in the on-state showing current saturation Current saturation is observed up to the rated BV.*

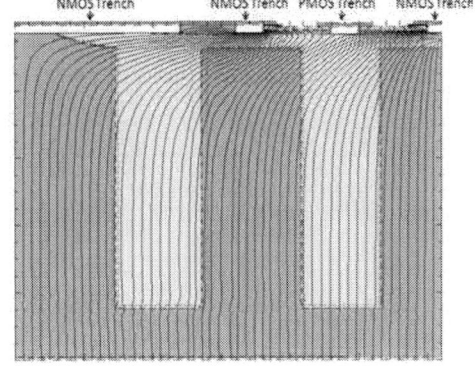

Fig. 5: *Current flow-lines in the on-state showing uniform distribution within the drift region. Vg=+15V.*

J=100Acm^{-2}. More importantly, the pillars depth has no influence on the V$_{ce}$(sat) and very slightly on J$_{sat}$. The

978-1-4244-8425-6/11 $26.00 © 2011 IEEE

magnitude of J_{sat} can be controlled with increased buffer doping. Fig.5 shows that in the on-state, the SJ regions are completely transparent to the carriers and current flow is uniform.

V. INDUCTIVE SWITCHING PERFORMANCE

Fig.6 shows that the V_{ce}(sat)-Eoff trade-off is significantly improved with deeper pillars. This can be understood by studying the influence of the P-pillars on the carrier dynamics during turn-off as shown in fig.7a-c. At low anode voltages such as 100V, fig7a, the spread of depletion layer from the P-well/N-pillar and P-pillars/N-pillar junctions constrain the holes to flow only within the P-pillars. At a high anode voltage of 600V, fig.7b, the pillars are fully depleted. The switching speed is increased because in the SJ section where the pillars are located there is vertical and lateral movement of the depletion layer from the P-well/N-pillar and P-pillar/N-pillar junctions respectively. Hence the N-pillar regions deplete faster as excess carriers are squeezed towards bottom of the pillars. With V_g now negative, the excess holes then flow only along the P-pillars into the cathode via the PMOS channel as shown in fig.7c. Therefore, the strength of the P-pillars in the SJ-TCIGBT device technology is that they act as extensions of the P-well region (i.e. like P-well fingers into the drift region) and increase the efficiency of excess holes collection from within the N-drift region. The PMOS channel short the P-well/P-pillars to the cathode and kills off the activity of the parasitic P-well/n-well/P-base transistor, during turn-off. Between the bottom of the pillars and the anode there is only a vertical movement of the depletion edge hence if the pillars are NOT sufficiently deep, they will have less influence on turn-off speed and E_{off} as shown in Fig.6.

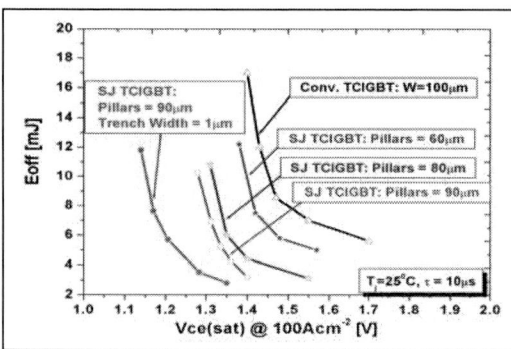

Fig.6: *Influence of pillars depth on turn-off energy loss Eoff). Each line is a variation of peak anode doping from 5e17cm^{-2} to 1e19cm^{-3} with x_j=1um V_{anode}=600V, τ=10μs, V_g = ±15V, L_{para}=150nH and R_g=22Ω.*

For very deep pillars and low anode doping (i.e. low excess carrier concentration) the switching speed can be very fast leading to oscillations. However, the switching speed can be slowed and oscillations in the turn-off waveforms can be avoided by increasing the

anode doping without significantly increasing the current tail as shown in fig.8. Increased anode-doping means

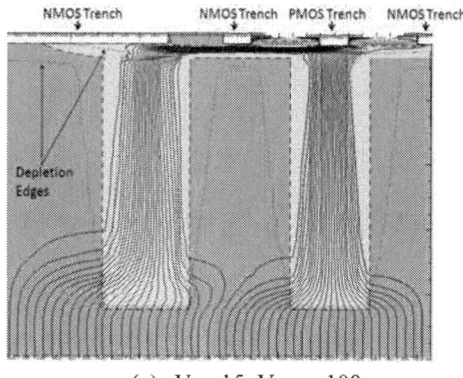

(a) V_g=-15, V_{anode}=100

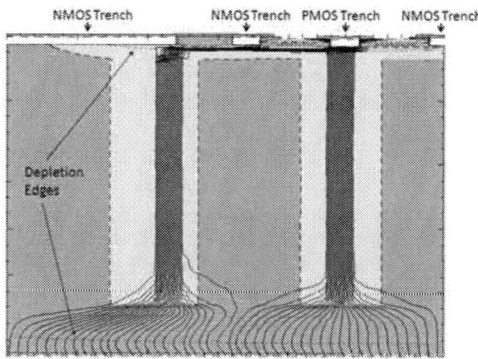

(b) V_g=-15, V_{anode}=600V

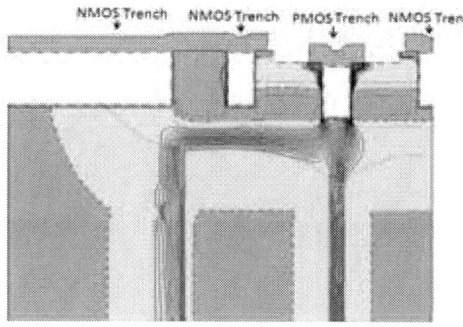

(c) V_g=-15, V_{anode}=600V

Fig.7: *Carrier flow and depletion edge movement during different phases of turn-off with V_g=-15V a) Voltage rise phase, V_{anode}=100V, b) Tail current phase, V_{anode}=600V and c) Holes flowing through the PMOS channels during the current tail phase*

reduction in V_{ce}(sat). The implication is that with deep pillars and increased anode injection, a more competitive V_{ce}(sat)-E_{off} trade-off with up to 80% lower E_{off} can be achieved with SJ-TCIGBT compared with the conventional device as shown in fig.9. Moreover, the combination of increased anode doping and increased effectiveness of P-pillars in removing excess holes from within the drift region mean smaller increase in current tail with increasing junction temperature, resulting in reduced temperature coefficient of V_{ce}(sat) and E_{off} as evident from fig. 9. Each line is a variation of anode doping from $5e17cm^{-3}$ to $1e19cm^{-3}$ with junction depth=1μm. Therefore, since excess holes are efficiently removed from the drift region through the p-pillars, the requirement of a highly transparent anode is not a necessity to achieve low E_{off}!

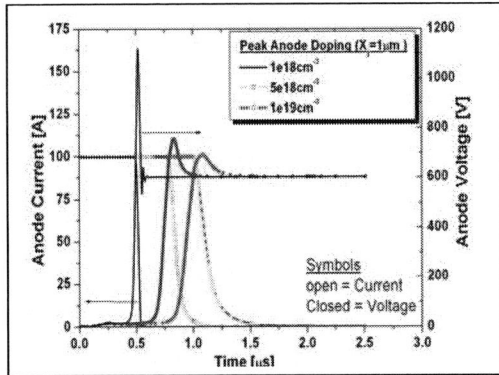

Fig.8: *Influence of peak anode doping on turn-off current and voltage waveforms.*

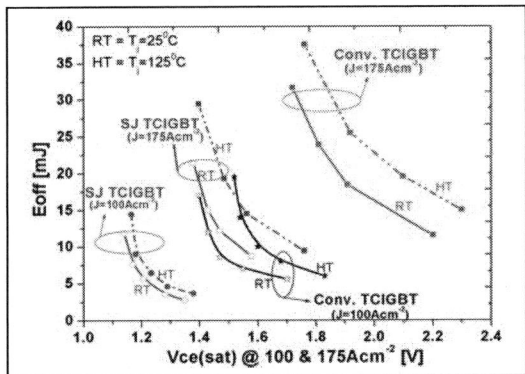

Fig.9: V_{ce}*(sat)-E_{off} trade-off at J=100 and 175Acm^{-2}. R_g=22Ω, pillars depth=90um, trench width=1um and τ=10μs.*

VI. SHORT CIRCUIT PERFORMANCE

A short circuit endurance time of 10μs at V_g=15V is the industry standard for MOS controlled devices. The short circuit endurance time before failure is directly related to the saturation current which is a function of the internal physics of the particular device. Fig.10 shows 20μs short circuit performance of SJ-TCIGBT with deep pillars of 80μm. It can be seen that the saturation current hence short circuit performance can be controlled by the buffer doping. The peak anode doping is $5e17cm^{-3}$. The room temperature Vce(sat) for increasing peak buffer doping of 1e16, 3e16 and $5e16cm^{-3}$ are 1.45V, 1.50V and 1.65V respectively, at J=100Acm^{-2}. More importantly, it can be seen that the SJ-TCIGBT can successfully turn-off up to 10 times the rated current at T_j=125^{0}C.

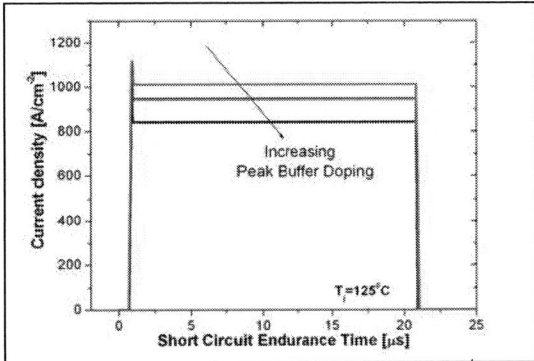

Fig.10: *Influence of buffer doping on the isothermal short circuit performance. The buffer doping are 1e16, 3e16 and 5e16 cm^{-3}, V_{anode} =600V and V_g=+15V.*

VII CONCLUSION

The results reported in this paper show that SJ-TCIGBT is a device concept that can provide very low V_{ce}(sat) and E_{off} simultaneously due to thyristor conduction and internal device physics. This is impossible in conventional transistor based technologies such as the IGBTs. Turn-off loss can be reduced by more than 80% and a significant reduction in the temperature coefficient of V_{ce}(sat) and E_{off} at T_j=125^{0}C is possible. While there is a very weak dependence of J_{sat} on the pillars depth, J_{sat} and hence short circuit withstand time can be controlled by the buffer doping.

VIII – REFERENCES

[1] T. Laska et. al., *The Field Stop IGBT*, ISPD'05, p.355-358

[2] G. Deboy, et. al., *A new generation of high voltage MOSFETs breaks the limit line of silicon*: Proc. IEDM, 1998. p. 683–5.

[3] F.D Bauer, *The SJBT*, ISPSD'02, pp. 197-200.

[4] M. Antonious, et. al., *The SJ- IGBT Optimization and Modelling*, IEEE Trans. on Elect Dev. Vol. 57, no. 3, March 2010,pp.594-600

[5] M. Antoniou1 et. al., *The 3.3kV Semi-SuperJunction IGBT for Increased Cosmic Ray Induced Breakdown Immunity*, ISPSD'09, pp.168-171.

[6] E. M. Sankara Narayanan, et al., *Clustered Insulated Gate Bipolar Transistor (CIGBT): A New Power Semiconductor Device*, Proceeding l0th IWPSD, 1999, pp. 1307-1312.

[7] K. Vershinin, et al., *Influence of the design parameters on the performance of 1.7 kV, NPT, planar Clustered Insulated Gate Bipolar Transistor (CIGBT)*, ISPSD'04, pp. 269-272

[8] K. Vershinin et al., *Experimental Demonstration of a 1.2kV Trench Clustered Insulated Gate Bipolar Transistor in Non-Punch-Through Technology*, ISPSD'06, pp.181-184

[9] M. Sweet et al., *Experimental Demonstration of 3.3kV Clustered Insulated Gate Bipolar Transistor in Non-Punch-Through Technology*, ISPSD'08, pp. 48-51

[10]. N. Luther-King, et. al., *Performance of a trench PMOS gated, planar, 1.2 kV CIGBT in NPT technology"*. ISPSD'08, pp.164-7

[11] Sysnopsys TCAD Sentaurus Device Package, 2007

[12] Zhang B, et. al., *Analysis of the forward biased safe operating area of the super junction MOSFET.*, ISPSD 2000, pp. 61– 64.

Proceedings of the 23rd International Symposium on Power Semiconductor Devices & IC's
May 23-26, 2011 San Diego, CA

Relaxation of Current Filament due to RFC Technology and Ballast Resistor for Robust FWD Operation

Akito Nishii, Katsumi Nakamura, Fumihito Masuoka and Tomohide Terashima
Power Device Works, Mitsubishi Electric Corporation
Fukuoka, 819-0192, Japan
Phone: +81-92-805-3417, Fax: +81-92-805-3881, e-mail (first author): Nishii.Akito@ea.MitsubishiElectric.co.jp

Abstract— **We have investigated the destruction mechanism of High Voltage (HV) Free Wheeling Diodes (FWD) during a reverse recovery operation. The most possible mode of the destruction phenomena originate in local heating due to current filament at the edge portion of the active area. To achieve a large reverse recovery Safe Operation Area (SOA), we focus on the boundary region between the active area and the termination area. To enforce our Relaxed Field of Cathode (RFC) concept [1, 2], it is more effective for the wider SOA to place a ballast resistance for avoiding the current from crowding around the anode region in the top surface of the diode.**

I. INTRODUCTION

For a fast switching device, such as the insulated gate bipolar transistor (IGBT) and the FWD, low switching power losses for new power applications are required. The diode technology requires having high turn off capability under high forward current density (J_A) and high dj_A/dt condition in terms of preventing the destruction of the power module. In the case of HV FWDs, the only possibility to solve the above problem is to increase N⁻ drift layer thickness (t_{N-}). Especially in the HV diode, the overall loss increases rapidly with t_{N-}. To obtain a diode with low overall loss, thin t_{N-} is required. However, in the case of thin HV diode, the snap-off and reverse recovery capability under the recovery operation decrease. To resolve this trade-off between overall loss, snap-off phenomena and reverse recovery capability, we have proposed the RFC diode [1, 2], which can eliminate the trade off limitation and has a good balance of these characteristics.

In the past few years, much research has been conducted into the failure mode during reverse recovery condition of the HV diode [3 ~ 6]. At ISPSD'10, we proposed an RFC diode with a new cathode structure of the edge termination region in terms of a large recovery SOA. However, these diode technologies have not been clarified regarding the destruction mechanism during reverse recovery of the HV diode.

Therefore, the focus of this paper is to verify the destruction phenomenon during the recovery operation of the HV diode. Further, we report an effective diode structure in terms of dynamic ruggedness and high HV FWD performance. The new RFC diode structure adopted a ballast resistance region at the edge of the active area, which can be achieved with low overall loss and increase recovery SOA. This new diode based on RFC technology will be the most promising structure as an Si FWD of the high-voltage class.

II. SIMULATED ANALYSIS FOR DESTRUCTION MECHANISM OF DIODE DURING REVERSE RECOVERY OPERATION.

A. Advanced RFC diode with ballast resistance region.

Fig. 1 shows simulated and experimental diode structure for this study. The RFC diode utilizes LPT(II) N buffer [7], which has a partial P layer on the cathode electrode of the active area, and a P layer at the cathode of the edge termination. The proposed new diode structure based on RFC technology is shown in Fig. 1-(c). The new diode has a ballast resistance region at the edge of the active area. The concept of the RFC diode is to enhance the effect of the suppression of the snap-off phenomenon during reverse recovery due to the relaxing electric field effect of the cathode, and to increase recovery SOA by controlling carrier concentration on the edge termination at the anode side during the on-state. Moreover, the advanced RFC diode adopting a ballast resistance region at the edge of the active area disperses the current concentration. As a result, the new RFC diode achieves high reverse recovery capability.

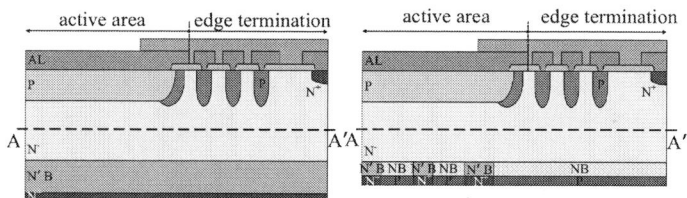

(a) conventional N buffer diode without ballast resistance.

(b) RFC planar anode diode without ballast resistance.

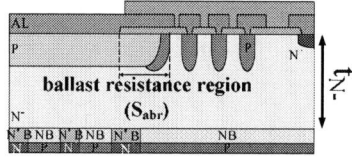

(c) advanced RFC planar anode diode with ballast resistance.

Fig. 1. Schematic cross-sectional view for various diodes of this study.

B. Destruction mechanism

Fig. 2 shows the reverse recovery waveforms of conventional and RFC diodes. The maximum temperature within the device during the reverse recovery is plotted. The maximum temperature during the reverse recovery operation of the conventional diode is over the critical temperature of Si (i.e. ≥800 K). On the other hand, the maximum temperature of the RFC diode is not over 800 K, and it is clear that it achieves high

978-1-4244-8425-6/11 $26.00 © 2011 IEEE

reverse recovery capability compared with the conventional diode.

Fig. 3 indicates simulated carrier concentration in the N⁻ drift region (A-A' line in Fig. 1). The carrier concentration of the RFC diode at edge termination is lower than that of the conventional one.

Fig. 4 presents simulated current distribution of various points during the reverse recovery. Fig. 5 shows simulated temperature distribution of various diodes at the maximum temperature point during the reverse recovery. From Figs. 3 ~ 5, the destruction phenomenon of the conventional diode occurs by the following mechanism.

1st step: the carrier of edge termination is flowed into the edge of the active area;

2nd step: the current filament appears at the boundary between the active area and edge termination;

3rd step: the temperature rises at the edge of the active area and the destruction occurs at the same point.

In the case of the RFC diode, the carrier of edge termination at the on-state is lower than that of the conventional diode, and reduces the current filament. The current filaments grow from the P cathode layer on the cathode inside the active area. To share local heating by multiple current filaments, the RFC diode achieves large recovery capability in comparison with the conventional diode. The destruction mechanism of the RFC diode is as follows.

1st step: current filament appears inside the active area;

2nd step: the current filament moves to the edge of the active area;

3rd step: the current filament gathers at the edge of the active area;

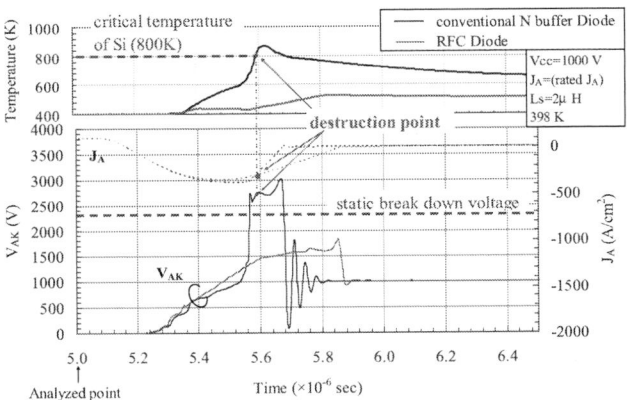

Fig. 2. Simulated reverse recovery waveforms and temperature of various diodes (at 398 K).

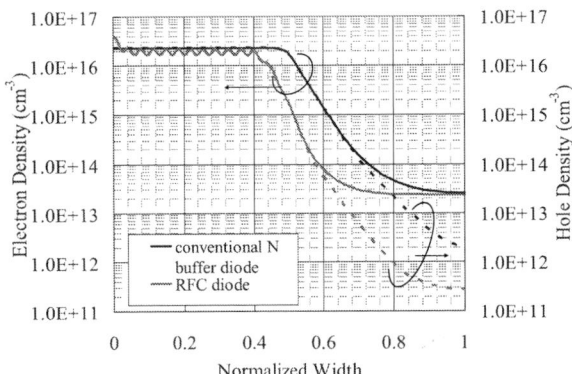

Fig. 3. Simulated carrier concentration in the N⁻ drift region (A-A' line in Fig. 1).

J_A (A/cm²)	con. N buffer diode without ballast resistance	RFC diode without ballast resistance	RFC diode without ballast resistance at destroyed condition.	advanced RFC diode with ballast resistance	advanced RFC diode with ballast resistance at destroyed condition.
	rated J_A	rated J_A	4x(rated J_A)	4x(rated J_A)	10x(rated J_A)
5.0μs (ON state)					
5.2μs					
5.3μs					
5.5μs					
5.6μs					
5.8μs	destruction				
5.9μs					
6.1μs					
6.4μs			destruction		
6.5μs					
6.7μs					
6.9μs					
7.1μs				destruction	

Fig. 4. Simulated current distribution of various diodes at various periods during reverse recovery (at 398K).

4th step: the temperature rises at the edge of the active area and the destruction occurs at the same point.

From those results, we focus on the structure of the boundary between the active area and edge termination. Fig. 6 shows the relationship of the ratio of the ballast resistance area and maximum temperature during the reverse recovery operation. Here, S_{abr} is the area of the ballast resistance region and $S_{active\ area}$ is the area of the active area. From Fig. 6, the optimized $S_{abr}/S_{active\ area}$ is 1.0 ~ 4.4.

Fig. 7 shows the reverse recovery waveforms of the RFC diode and advanced RFC diode adopting optimized ballast resistance region. The maximum temperature within device during the reverse recovery is plotted. The advanced RFC diode

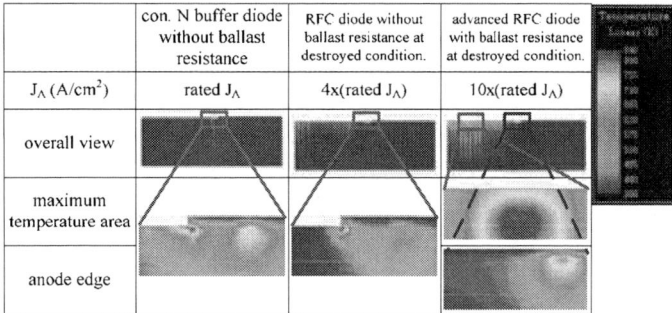

Fig. 5. Simulated temperature distributions of various diodes at point of maximum temperature.

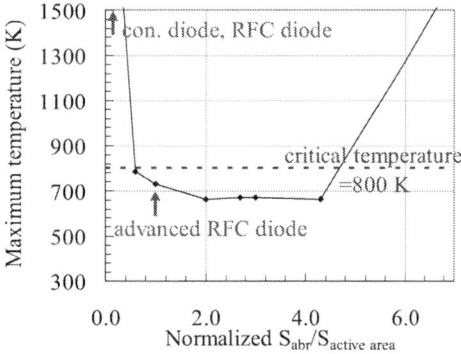

Fig. 6. Relationships of the ratio of ballast resistance area and maximum temperature during reverse recovery operation.

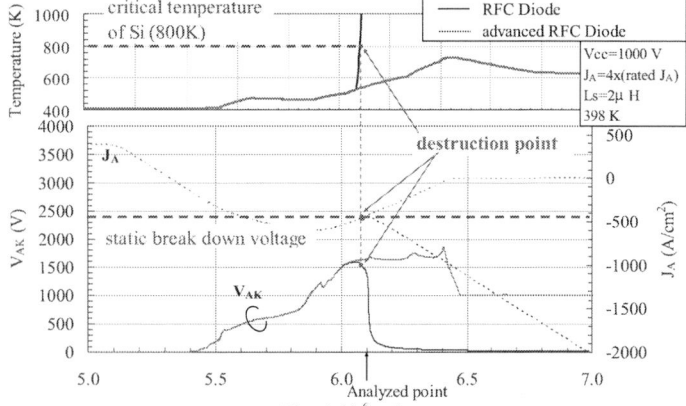

Fig. 7. Simulated reverse recovery waveforms and temperature (at 398K) of various RFC diodes.

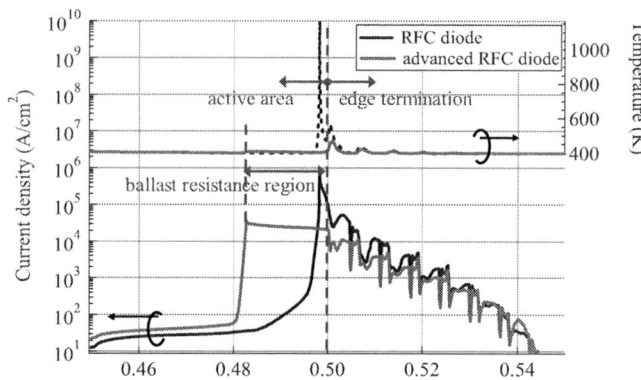

Fig. 8. Current and temperature distributions at destruction point in Fig. 7

decreases the maximum temperature within device.

Fig. 8 shows the current density and temperature distributions on the device surface at the destruction point shown in Fig. 7. The RFC diode without a ballast resistance region endures a local heating due to high current density at the edge of the active area. The advanced RFC diode with a ballast resistance region has much even current distribution in the ballast resistance region.

As shown in Fig. 4, the RFC diode appears to have strong current filament at the boundary between the active area and edge termination under higher J_A. On the other hand, the advanced RFC diode can relax the current filament at the edge of the active area. From Fig. 5, the destruction point of the advanced diode is located inside the active area instead of the edge of the active area as with the RFC diode at the point of maximum temperature.

From these results, we conclude that the destruction during the recovery operation of FWD originates in the current filament phenomena. The advanced RFC diode prevents the generation of current filament by reducing carrier concentration of edge termination at the on-state, and divides current density at the edge of the active area within ballast resistance region. As a result, the advanced RFC diode can have a high dynamic ruggedness.

III. EXPERIMENTAL REESULT FOR DYNAMIC BEHAVIOR OF THE ADVANCED RFC DIODE

In this section, we demonstrate the electrical characteristics of the advanced RFC diode for the 3300 V class. All diode data were 108A/cm² (at rated J_A) fabricated by FZ silicon wafer.

The trade-off characteristics between the forward voltage drop (V_F) and the reverse recovery loss (E_{REC}) of various 3300 V diodes are demonstrated in Fig. 9. The overall loss rapidly increases with t_N. The advanced RFC diode moves to high frequency region compared with the conventional one.

Fig. 10 indicates the reverse recovery waveforms under the worst-case condition in terms of snappy recovery. Due to the effect of RFC structure [1 ~ 2], the maximum voltage during reverse recovery operation under the snappy condition is suppressed in comparison with the conventional one.

Fig. 11 shows the reverse recovery waveforms under extreme conditions of various diodes with an optimized ballast resistance region. The maximum turn off peak power of the proposed RFC diode is 1.42 MW/cm² (at V_{CC}=2600 V, J_A=9x(rated J_A), Ls=4.6 μH). The reverse recovery capability of the advanced RFC diode is higher than that of the

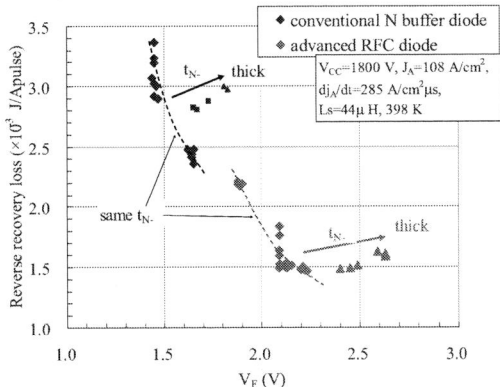

Fig. 9. Measured trade off characteristics between V_F and reverse recovery loss of various 3300V diodes

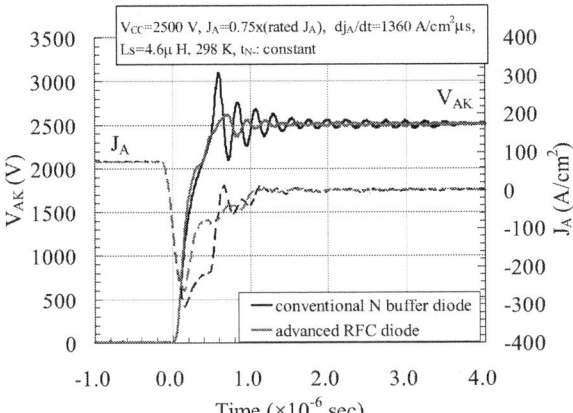

Fig. 10. Reverse recovery waveforms of various 3300V diodes with same t_{N-} under worst conditions in terms of snappy recovery (V_{CC}=2500V, J_A=0.75x(rated J_A), dj_A/dt=1360 A/cm²μs, dV_{AK}/dt=8.5 kV/μs, Ls= 4.6μ H, 298K).

conventional one.

Fig. 12 presents the measured recovery SOA of various 3300 V diodes. Here, J_A(break) and the maximum dj_A/dt are the maximum J_A and dj_A/dt without the diode's destruction, respectively. As shown in Fig.12, the advanced RFC diode can have larger reverse recovery SOA than the conventional N buffer diode. From this result, it is confirmed that the parameter of $S_{abr}/S_{active\ area}$ needs to be designed suitably.

IV. CONCLUSION

By using the numerical simulation, for both the conventional N buffer diode and our proposed RFC diode, we find out that the destruction mechanism during reverse recovery operation basically originate in the current filamentation at the edge of the anode area, which is the boundary region between the active anode region and the termination region. It is confirmed by both simulation and experiment our advanced RFC diode can be reinforced with the optimized ballast resistance laying between the anode and the edge termination region.

ACKNOWLEDGMENT

The authors wish to thank Mr. K. Sadamatsu and Mr. S. Kitajima for the development of the RFC diode utilized in this study and Dr. K. Satoh for his unrelenting encouragement.

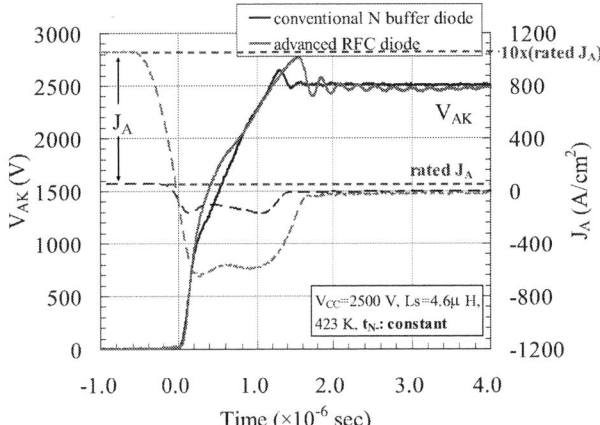

Fig. 11. Reverse recovery waveforms of various 3300V diodes with same t_{N-} under extreme conditions in terms of destruction (V_{CC}=2500 V, Ls=4.6μ H, 423 K).

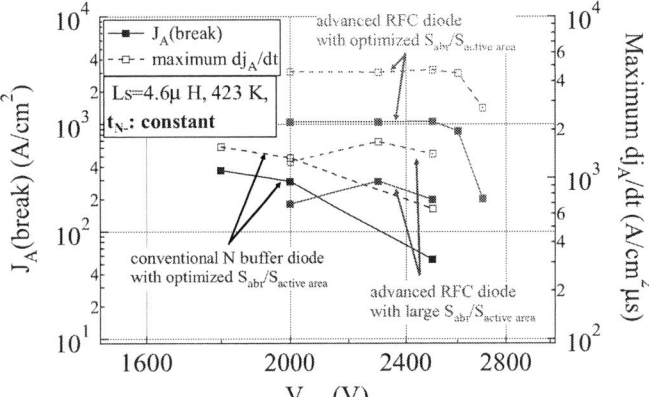

Fig. 12. Measured reverse recovery SOA of various 3300V diodes with same t_{N-}.

REFERENCES

[1] K. Nakamura, H. Iwanaga, H. Okabe, S. Saito and K. Hatade, "Evaluation of Oscillatory Phenomena in Reverse Operation for High Voltage Diodes," Proc. ISPSD'09, pp. 156-159, 2009

[2] K. Nakamura, F. Masuoka, A. Nishii, K. Sadamatsu, S. Kitajima and K. Hatade, "Advanced RFC Technology with New Cathode Structure of Field Limiting Rings for High Voltage Planar Diode," Proc. ISPSD'10, pp. 133-136, 2010

[3] R. Baburske, J. Luts, H.-J. Shulze and F.-J. Niedernstheide, "Analysis of the Destruction Mechanism during Reverse Recovery of Power Diodes," Proc. ISPS'10, 2010

[4] M. Domeij, J. Lutz and D. Silber, "On the Destruction Limit of Si Power Diodes During Reverse Recovery With Dynamic Avalanche," IEEE Trans. Electron Device, vol. 50, no. 2, 2003

[5] R. Baburske, B. Heinze, F.-J. Niedernostheide, J. Lutz and D. Silber, "On the Formation of Stationary Destructive Cathode-side Filaments in p+-n—n+ Diodes," Proc. ISPSD'09, pp. 41-44

[6] F.-J. Niedernosteide, E. Falck, H.-J. Shulze and U.Kellner-Werdehausen, "Influence of Joule Heating on Current Filaments Induced by Avalanche Injection," IEE Proc.-Circuits Device Syst., vol. 153, no. 1, pp. 3-10, 2006

[7] K. Nanamura, Y. Hisamoto, T. Matsumura, T. Minato, J. Moritani, "The Second Stage of a Thin Wafer IGBT Low Loss 1200V LPT-CSTBT™ with a Backside Doping Optimization Process," Proc. ISPSD'06, pp. 133-136, 2006

Proceedings of the 23rd International Symposium on Power Semiconductor Devices & IC's
May 23-26, 2011 San Diego, CA

Limits of Strongly Punch-Through Designed IGBTs

Thomas Raker, Hans-Peter Felsl, Franz-Josef Niedernostheide, Frank Pfirsch, and Hans-Joachim Schulze

Infineon Technologies AG
Neubiberg, Germany

Abstract— **We will focus on the turn-off behavior of strongly punch-through designed field-stop IGBTs. Our numerical simulations with a monolithic multi-cell structure show that the appearance of current filaments may limit the safe operating area (SOA) of very thin devices with a high resistivity of base material [1]. A detailed analysis of current densities and electric field distributions gives insight into the mechanisms resulting in the formation of current filaments. The limit for a filament-free turn-off behavior can be found in the thickness-*vs.*-resistivity phase diagram. It could be shown that also other device parameters, such as field-stop and p-emitter design, highly influence susceptibility for the appearance of current filaments during the turn-off phase.**

I. INTRODUCTION

Chip thickness reduction in combination with a higher resistivity of the starting material improves the fundamental trade-off relationship between static and dynamic losses. It is vital to avoid the appearance of current filaments within an IGBT during turn-off phase, because local heating may result in the destruction of the device. Therefore, it is of fundamental interest to investigate the silicon limits of the IGBT in order to provide a soft and filament free turn-off performance with an optimized trade-off for a given application.

We performed a detailed numerical analysis of the charge extraction process during turn-off. The simulations are carried out for a monolithic 1200V trench IGBT structure including up to eight identical IGBT cells. Thus, apart from inhomogeneities stemming from the cell structure, there is no further artificial inhomogeneity within the device [2].

II. RESULTS

Fig. 1 depicts turn-off waveforms of IGBTs for different thicknesses d and resistivities ρ. The IGBTs (a) and (b) show a turn-off behavior characterized by a uniform division of the turn-off current to the eight cells whereas during the turn-off phase of IGBT (c) a current filament arises (Fig. 2 and 3). The voltage trace V_{CE} for IGBT (c) shows a characteristic voltage dip which coincides with the formation of a filament in the base region of the IGBT. At the voltage maximum ($t=0.38\mu s$), charge carrier are extracted homogeneously out of the device. The total current in the space charge region near the cathode side is only supported by holes which steepen the electrical field at the cathode side and lead to dynamic avalanche (Fig. 2).

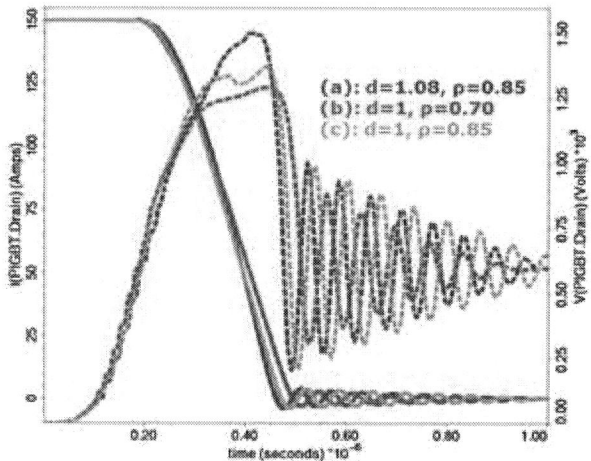

Figure 1. 3 IGBT turn-off waveforms @ thickness d and resistivity ρ. Switching conditions: 150A/cm² vs. 600V DC link voltage, 300K operating temperature, and suitable stray inductance for typical applications. The parameters d and ρ were normalized with respect to a reference IGBT.

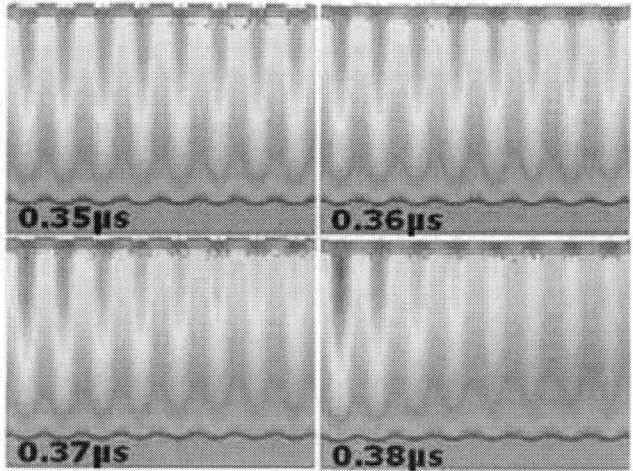

Figure 2. Impact ioniziation rate of IGBT (c) of Fig. 1. Timepoints are taken at the beginning of the voltage dip when filament formation begins.

Due to the high charge-carrier generation rate the collector emitter voltage V_{CE} drops and the current filament evolves at the left cell of the eight-cell structure. It extends over the whole

978-1-4244-8425-6/11 $26.00 © 2011 IEEE

space-charge region between the p-body at the cathode side and the anode emitter. The complete formation process is illustrated in Fig. 3. This process is autocatalytic because electrons moving to the IGBT backside cause in turn hole-injection from the anode emitter. The filament naturally disappears when the collector current I_C tends to zero.

The formation phase of filaments is always located below the gate regions of the monolithic structure and it is favored by numeric noise inherent to any numerical algorithm for solving differential equations.

To ensure that the formation of the current filament is not artificially caused by a coarse grid discretization in space, several simulations with higher spatial resolution of the device structure (i.e. smoother grid) were performed. Even in the case of highest possible resolution, the simulation shows still an evolving current filament. In this case, the filament is located on the left side (first cell) of the monolithic structure. In other simulations, the current filament arose at the right side (last cell). We also connected two monolithic structures in parallel and found the formation of multiple filaments in the middle of one monolithic structure.

The filament formation is favored by lateral electric fields focusing the carriers towards the center of the filament. Holes generated via impact ionization near the chip front side are lateral focused towards the filament center on their way to the cathode side. Even in the middle of the base region, slight lateral hole currents with direction into the center of the filament can be observed. Correspondingly, the electron current to the anode is spread near the anode emitter. Comparable charge carrier effects in current filaments are described in [2] and [3] for power diodes. This positive feedback loop between electron created in the avalanche region and holes injected from the anode emitter sustains the filament until the load current gets too low. Then, the charge carrier generation rate tends to zero with the consequence that the collector emitter voltage V_{CE} rises up again before converging to the DC link voltage of 600V.

Fig. 4 shows the thickness-*vs.*-resistivity phase diagram. Each point in this diagram represents a turn-off simulation. The solid line separates the area where filaments appear from the area with no filaments appearing during turn off. The limit, in terms of a filament free turn off, of strongly punch-through designed IGBTs can clearly be seen. Thinner silicon in combination with lower doping of the base material is not the right way for a performance improvement of IGBTs. Further measures are needed for a continuous reduction of on-state and dynamic losses of the IGBT without deteriorating the SOA of the IGBT. However, the softness must be considered. It gets worse in the case of thinner silicon and higher resistivity of the base material.

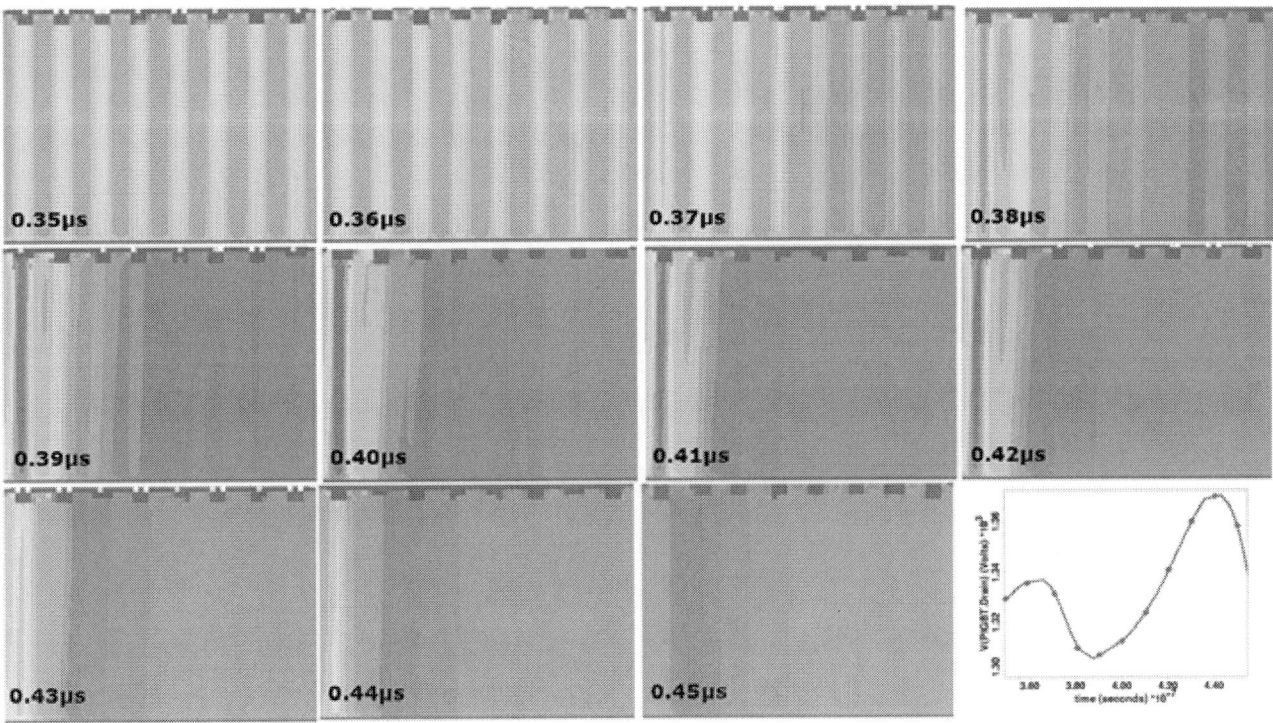

Figure 3.   Total current density distribution of IGBT (c) at selected time points (marked in the inset) during the turn-off period.

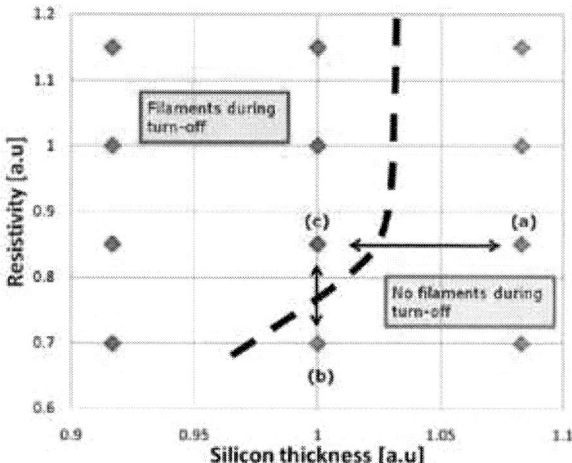

Figure 4. Thickness-vs.-resistivity phase diagram. Black arrows indicating the above three device structures discussed above. (a), (b), (c) mark the IGBT structure of Fig. 1. In the upper left area filaments appear during turn-off while in the lower right area no filament formation during turn-off was observed.

Further investigations show a significant influence of the field-stop and the p-emitter design. Filaments only appear in the case of low backside emitter efficiency, i.e. for a low doping concentration of the IGBT backside anode emitter or a high field-stop doping dose.

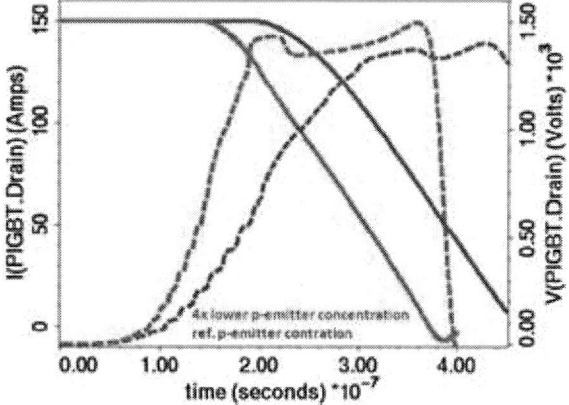

Figure 5. IGBT turn-off waveforms for two p-emitter designs @ $d=1$, $\rho=1.2$. Blue curve: reference emitter dose. Red curve: 4 times lower emitter dose

In the case of a weak p-emitter doping concentration, the collector emitter voltage V_{CE} rises steeper and a bigger voltage overshoot can be seen (Fig. 5). Again, local hole distribution enhances the electrical field at the front side and drives the device into dynamic avalanche. There are two reasons why the current filament for the lower p-emitter doping concentration is more pronounced (Fig. 6). Firstly, the absolute value of the electrical field strength near the cathode side is higher and

secondly the total current level at this stage is higher. Both result in a higher charge-carrier generation rate, a bigger voltage dip, and finally a higher filament current density. Even after the appearance of a current filament formation the impact generation rate is higher.

For all simulation results shown here room temperature (300K) was assumed. At higher temperatures (e.g. 423K), current filaments still occur with the difference that corresponding filament current densities are lower. This is due to the fact that impact ionization coefficients decrease with temperature.

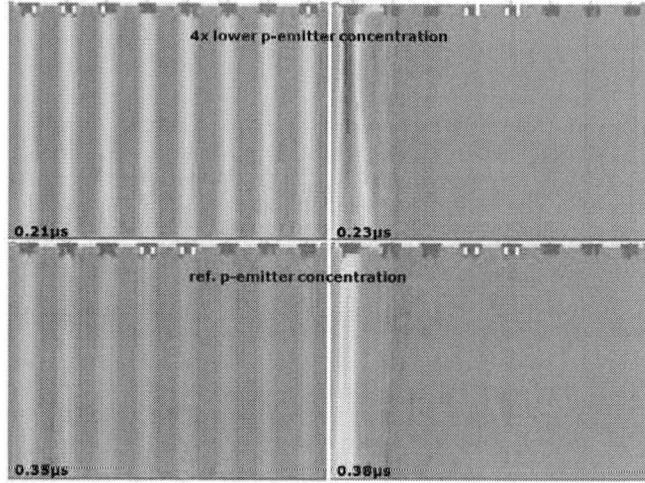

Figure 6. Current filament formation process for two IGBTs with different p-emitter doping concentrations. Shown times correspond to turn-off waveforms in Fig. 5.

From an application point of view, the important question is how such filaments can be avoided even for thin silicon. The gate resistor has a strong influence on filament formation. For all numerical results presented so far the gate resistor was so small that the n-channel was closed before filaments would evolve. Thus the Miller phase is terminated before load current begins to fall. In this case, the complete current is carried by holes. The situation changes if the gate resistor is increased and the Miller phase increases. If the n-channel is still open when the load current begins to drop, the electric field distribution is smoother (lower electric field crowding) so that dynamic avalanche is prevented. The appearance of current filaments and the characteristic voltage dip are effectively suppressed (Fig. 7). However, by using higher gate resistors turn-off losses are increased.

There is another way to prevent current filaments. Up to now, all simulations were done with the same cell pitch. If the cell pitch is bisected, the trend of current filament formation is lowered (Fig. 8). The reason is an increasing homogenization at the cathode side. Note, that this measure has consequences on the electrical performance of the IGBT. The short circuit current is if nothing else in the design is changed enhanced.

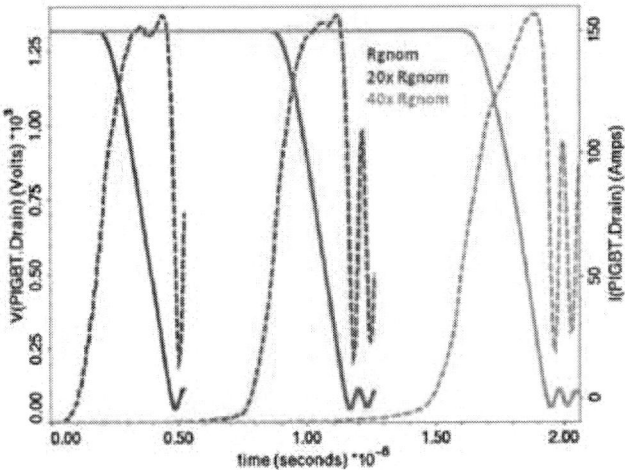

Figure 7. One IGBT design (d=1, ρ=1) turned-off with different gate resistors Rg.

III. CONCLUSION

As a result of these investigations, it is evident that great care has to be taken in changing the vertical structure of IGBTs. Thin silicon in combination with a high resistivity of the base material current limits the SOA of IGBTs. Device simulations show current filament formation in the case of turn-off for such aggressive chip designs. Especially for achieving lower static and dynamic losses, other chip design measures than the ones investigated have to be considered.

REFERENCES

[1] J.Oetjen, R.Jungblut, U.Kuhlmann, J.Arkenau, R.Sittig, "Current filamentation in bipolar power devices during dynamic avalanche breakdown", Solid State Electronics, vol. 44 (2000), pp.117-123.

[2] S.Milady, D.Silber, F.-J.Niedernostheide, H.P. Felsl, H.-J. Schulze, "Influences of Inhomogeneities and Cooling Mechanisms on the Movements of Avalanche Induced Current Filaments", S.Milady, D.Silber, F.-J.Niedernostheide, H.P. Felsl, H.-J. Schulze, Proceedings of the ISPS Prague, 20.August-1.September 2006, pp.63-70.

[3] H.P.Felsl, E.Falck, F.-J.Niedernostheide, S.Milady, D.Silber, J.Lutz, "Electro-thermal simulation of current ˉlamentation in 3.3-kV silicon p-i-n diodes with different edge terminations", Proceedings of ISPSD 2006, 4-8 June, Naples, Italy, pp. 13-15.

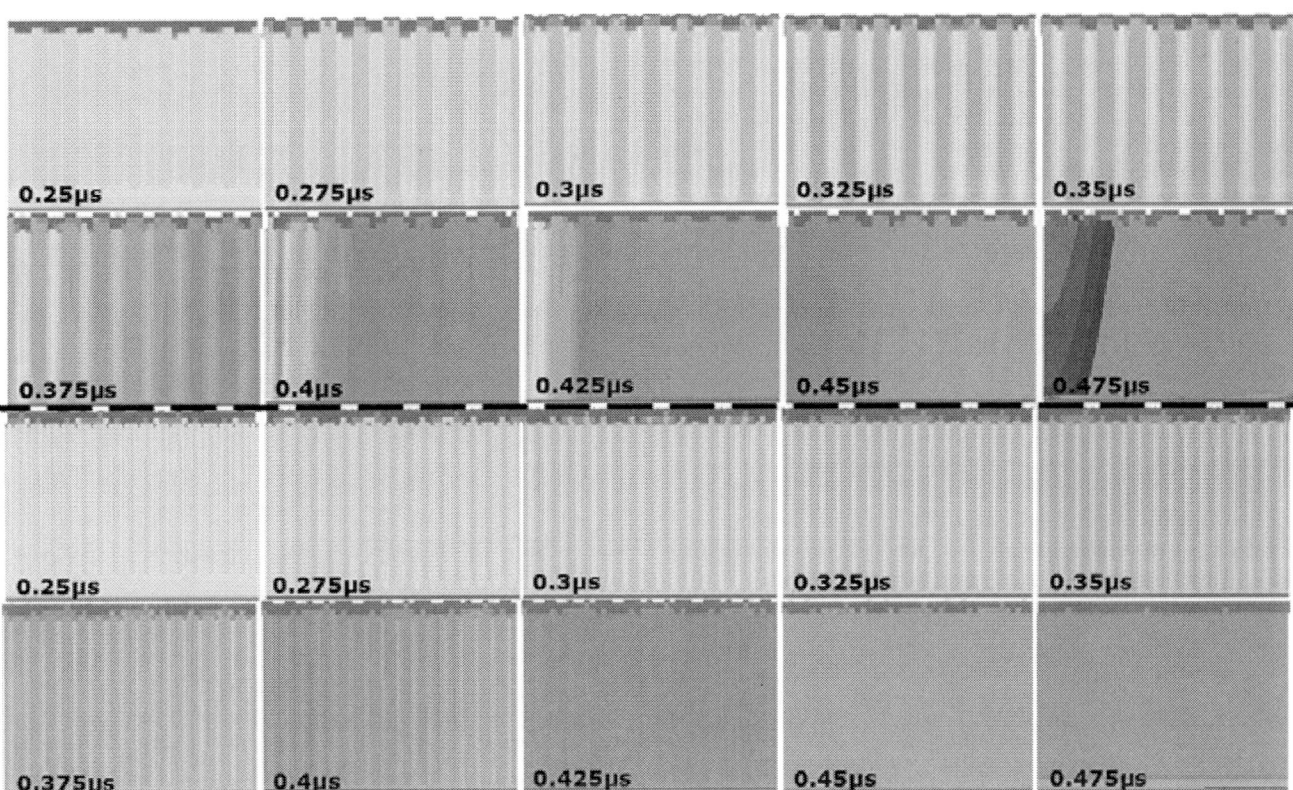

Figure 8. Two different IGBT (d=1, ρ=1) structures with different pitches at time points Above dotted black line: reference pitch. Below dotted black line: bisected reference pitch. Timepoints were taken around the maximum of V_{CE}. In the case of lower pitch the appearance of current filaments is drastically reduced.

Proceedings of the 23rd International Symposium on Power Semiconductor Devices & IC's
May 23-26, 2011 San Diego, CA

FILAMENT-INDUCED THERMOMIGRATION OF AN ALUMINUM DROP AT THE CATHODE-SIDE OF HIGH-VOLTAGE POWER DIODES

H.-J. Schulze, J.G. Bauer,
F.-J. Niedernostheide, H.P. Felsl

Infineon Technologies AG, Neubiberg,
Am Campeon 1-12
Munich, Germany
Hans-Joachim.Schulze@infineon.com

*J. Biermann, **J.Lutz,** R. Baburske
*Infineon Technologies AG, Warstein,·
*Warstein, Germany
**Chair for Power Electronics and EMC, TU Chemnitz
**Chemnitz, Germany

Abstract—**This paper shows how a buried aluminum eutectic drop at the cathode-side of a high-voltage power diode can affect the device behavior. The aluminum drop driven by thermo-migration, moves from the contact metallization some micrometers into the chip and forms a buried eutectic. Thermomigration [1] becomes stronger as the temperature gradient increases. High temperature gradients can be achieved at the cathode side if a single filament is triggered. Simulation results show that an early surface-punch-through at a spike may reduce the reverse-recovery ruggedness of the power diode. Such a detrimental filamentation can be avoided by a well-defined fabrication process of the device.**

INTRODUCTION

There are very fast switching IGBTs on the market. However, in a converter topology the inevitable freewheeling diode is often the limiting device. A clear understanding of the ruggedness limiting factors is essential. Current filaments can occur during the reverse-recovery process and can lead to destruction with a pin hole failure picture of the silicon chip. Filaments are considered to be relatively harmless with respect to device destruction, as long as they are moving along the contact area of the device. Such moving filaments are predicted by device simulations in case that the filaments appear at the anode side. However, a cathode-side filament tends to be fixed or to move only very slowly. Furthermore, it suppresses the transition into multiple cathode-side filaments [2]. Consequently, the appearance of a single cathode-side filament leads to a strong local heating at the cathode side which may lead to a melting of the metallization and finally to the formation of a buried aluminum eutectic. This eutectic moves towards the anode due to thermomigration.

SIMULATION CONDITIONS AND DIODE STRUCTURE

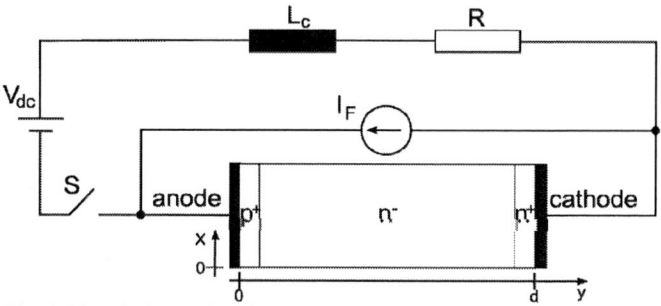

Fig.1 Simulation Circuit

To analyze the maximum temperature in filaments, electrothermal simulations of the reverse-recovery process of a p^+-n^--n^+ diode with a breakdown voltage of around 8 kV was performed with the device simulator Sentaurus$_{TCAD}$ [4]. A test circuit with an ideal switch S was used. At the beginning of the transient simulation, the switch S was open. The forward current I_F was flowing through the diode in the ON-state. After closing the switch S, the dc-link voltage V_{dc} enforced the reverse recovery of the diode. The test circuit also included the resistance R that reduced the diode voltage at high reverse currents and prevented an early snappy

978-1-4244-8425-6/11 $26.00 © 2011 IEEE

behavior enabling the investigation of ruggedness. The doping profile of the investigated diode had a base doping of $N_D = 6.7\times10^{12}$ cm^{-3}, a diode thickness of d = 670 µm, and identical Gaussian profiles of the p$^+$- and the n$^+$-regions ($N_{peak} = 2\times10^{19}$ cm^{-3}, FWHM = 0.59 µm). The lateral width was 800 µm. The values of the current were normalized to an effective contact area of about 3.5 cm^2. Edge termination was not taken into account. In addition to the silicon layer, a solder layer (150 µm thick) and a copper layer (300 µm thick) were added at the cathode side. The temperature at the bottom of the copper layer was fixed at the initial temperature of 400 K. Auger and Shockley-Read-Hall recombination, carrier-carrier scattering, doping-dependent, temperature-dependent and electric-field-dependent mobilities, and avalanche generation coefficients of Reggiani et al. [4] were considered. The charge-carrier lifetimes were homogeneously distributed over the middle region of the diode.

EXPERIMENTAL RESULTS AND SIMULATION RESULTS

To demonstrate the effect of thermomigration convincingly, a diode was stressed with an unrealistic high overstress condition. Fig. 1a shows the REM picture of an aluminum-rich layer grown into the silicon, Fig. 1b the EDX analysis of this area 10 µm below the surface revealing a significant aluminum concentration. Close to the surface, no aluminum was detected. The filament-induced thermomigration of the backside metal into the semiconductor chip can result in a significantly increased leakage current for applied voltages which are high enough so that the space charge region is in contact with the buried eutectic region formed by the migrating drop. To generate such a current filament, an extreme overload condition has been chosen.

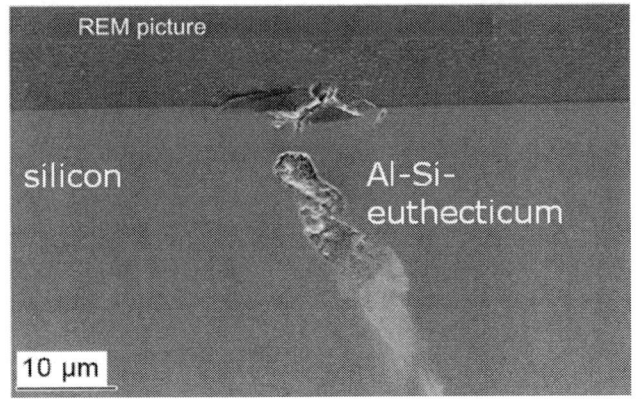

Fig. 1a REM picture of the molten chip area. Al-Si-euthecticum is found about 10µm below the wafer surface.

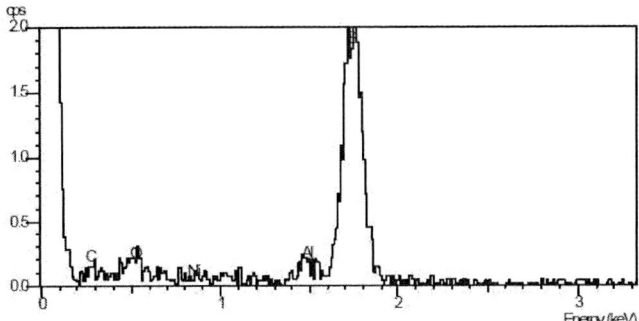

Fig. 1b EDX spectrum in the molten chip area 10 µm below the surface. The peak corresponding to the energy of 1.75keV results from the buried Al-Si-euthecticum.

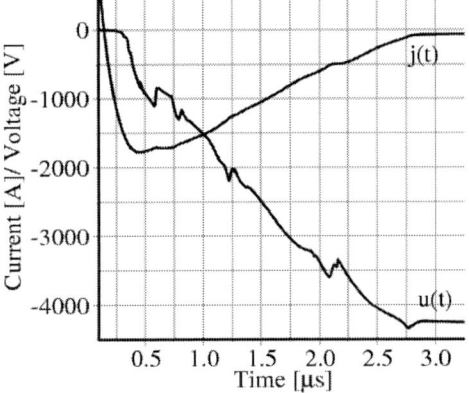

Fig. 2a : Reverse-recovery behavior of a diode. Diode voltage and current curves (L_c = 280 nH, V_{dc} = 4400 V, I_F = 800 A = 4 I_{rated}, $T_{initial}$ = 400 K, R = 2 Ω)

In Fig. 2a are shown the simulated reverse-recovery transients of a diode with a blocking capability of 8 kV.

During the reverse recovery period, two depletion layers arise, one near the anode side between the p$^+$-n$^-$ junction and the plasma layer and the other one near the cathode side between the plasma layer and n$^-$-n$^+$ junction, Fig.2b.

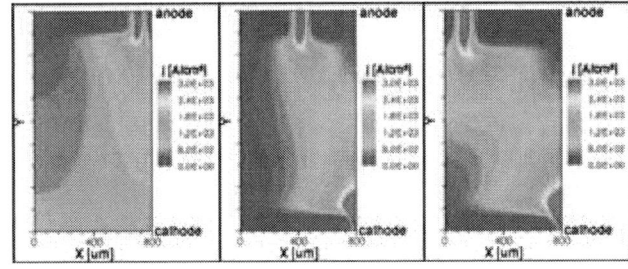

Fig. 2b: Current density distribution at 0.5 µs, 0.6 µs and 0.7 µs, showing a stationary cathode side filament

978-1-4244-8425-6/11 $26.00 © 2011 IEEE 105

Because of the high reverse current, filaments are triggered in both depletion layers. Due to the lateral movement of the anode-side filament, the maximum temperature at the anode-side reaches only the value of 460 K. However, the cathode-side filament stays fixed leading to local temperatures exceeding 1250 K at t=1.4µs. The maximum contact temperature reaches values greater than 800 K. This temperature is higher than the melting temperature of the eutectic region. That may explain the thermomigration process.

Due to the thermal gradient which occurs at the cathode side of the diode, metal out of the backside metallization can migrate into the base region of the diode. In Table 1, the temperatures occurring at the cathode-side filament are listed for different points in time for the reverse-recovery process shown in Fig. 2a.

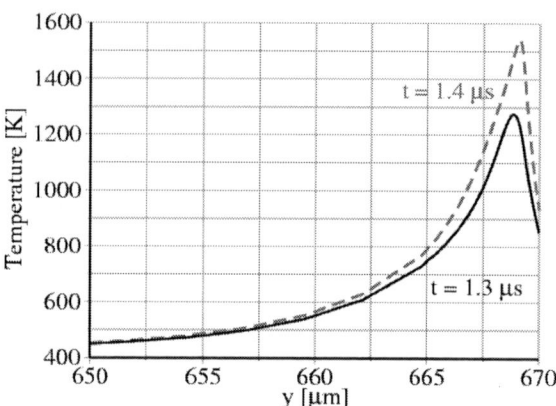

Fig. 3: Temperature in the area of the cathode-side filament during the reverse-recovery process shown in Fig. 2a, left.

t [µs]	$x_{Filament}$ [µm]	$T_{Kontakt}$ [T]	dT/dy [K/µm]	
0.5	800	400		
0.6	800	403		Temperature evaluation in the cross section through the filament at $x_{Filament}$
0.7	800	536		
0.8	800	644		
0.9	800	731	-420	
1.0	800	804	-444	
1.1	793.75	600	-151	
1.2	793.75	680	-194	Lateral movement of the filament begins between 1 µs and 1.1 µs, first movement in negative x direction; between 1.2 µs and 1.3 µs filament moves in positive x direction;
1.3	800	**855**	-367	
1.4	800	**934**	-729	
1.5	800	**997**	-650	
2.0	800	**1022**	-353	
3.0	800	819	5	
4.5	800	610	3	

Table 1. Temperature in the filament and at the cathode-side contact and temperature gradient in the filament

The temperature in the filament region at the cathode side drastically increases, which is coupled with a high increase in the temperature gradient. The maximum temperature gradient of 729 K/µm occurs at 1.4 µs. To illustrate in detail the temperature distribution and the temperature gradient, the temperature is plotted versus the y coordinate for a vertical cross section through the filament in Fig. 3. The maximum temperature is located nearly at the backside surface only in a depth of about 1 µm. The temperature decreases within a distance of about 20 µm from nearly 1400 K to 400 K.

What can be stated here is that a slowly moving or pinned cathode-side filament causes a hot spot directly in front of the cathode-side metallization layer. The metal out of the backside metallization can, driven by the temperature gradient, migrate in the base region of the diode, and forms a deep metal spike in the base region of the diode. This metal spike can have a depth of several µm up or more depending on overload conditions.

To analyze the effects of metal with respect to the behavior of current filaments and a possible destruction mechanism, a diode with spikes and a diode without spikes have been simulated under the same hard-switching conditions (Fig. 4a,b,c). The spike depth is 0.6 µm, the distance between the spikes is 20 µm, Fig. 4a.

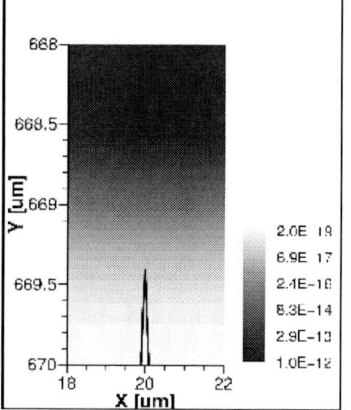

Fig. 4a Distribution of the doping concentration and spike shape in front of the cathode contact.

978-1-4244-8425-6/11 $26.00 © 2011 IEEE

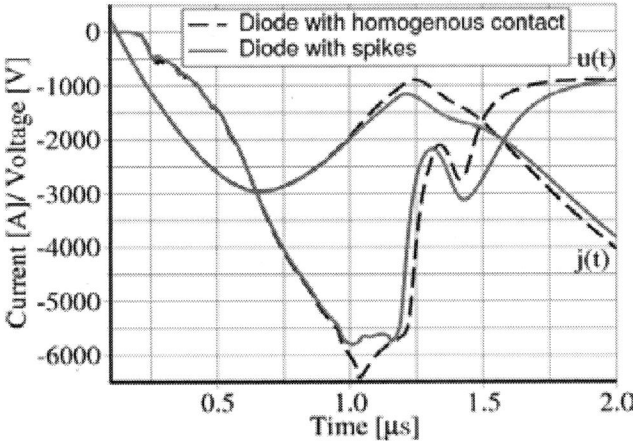

Fig. 4b Simulated reverse-recovery transients of a diode with spikes and a diode without spikes.

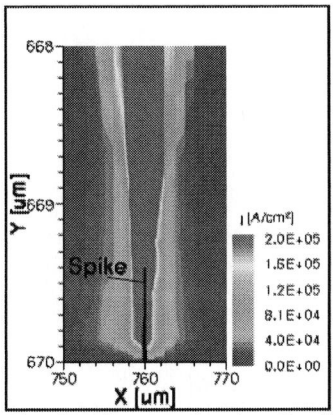

Fig. 4c Current-density distribution at the diode backside at 0.8 μs.

Thermally driven, the cathode-side filament moves very slowly. However, if the filament reaches the spike, it remains locally fixed at this position (Fig. 4c). A surface-punch-through of the cathode-side space-charge region appears, immediately leading to a high injection of charge carriers and

an earlier thermal runaway compared to the diode without any spikes (Fig. 4b). It is therefore supposed that spikes can contribute to the destruction mechanism suggested in [2]. In [2], unrealistic hard-switching conditions were necessary to create a destructive filament in simulation. If once a filament has driven a spike by thermomigration from the cathode side into the base region of the diode, the diode can be destroyed during the next pulse.

Conclusions

It has been demonstrated that in case of high electrical overload of power semiconductor devices, current filaments can appear and cause high temperature gradients in vertical direction of the device. Such extreme temperature gradients can effect thermomigration and, as a result, small metallic layers buried below the wafer surface may be formed. This indicates the necessity to avoid non-moving current filaments as far as possible and, if they occur to keep the power losses within these filaments as low as possible so that the maximum temperature within the filament does not exceed the eutectic temperature of the metallization.

REFERENCES

[1] B. Morillon, H.-M. Dilhac, G. Ganibal, A. Anceau: "Power Devices Insulation by Al Thermomigration", Proc. ISPS, 2002.

[2] R. Baburske, B. Heinze, F.-J. Niedernostheide, J. Lutz, D. Silber: "On the formation of stationary destructive cathode-side filaments in p^+-n^--n^+ diodes", Proc. ISPSD, 2009.

[3] TCAD (2007) Advanced Tcad Manual. Synopsys Inc. Mountain View, CA. [Online]. Available: http://www.synopsys.com

[4] S. Reggiani, E. Gnani, M. Rudan, G. Baccarani, C. Corvasce, D. Barlini, M. Ciappa, W. Fichtner, M. Denison, N. Jensen, G. Groos, M. Stecher, "Measurement and modeling of the electron impact-ionization coefficient in silicon up to very high temperatures", IEEE Trans. Electron Devices 55 (2005), pp. 2290-2299.

Proceedings of the 23rd International Symposium on Power Semiconductor Devices & IC's
May 23-26, 2011 San Diego, CA

Optimization of Diodes using the SPEED concept and CIBH

Manfred Pfaffenlehner, Hans-Peter Felsl,
Franz-Josef Niedernostheide, Frank Pfirsch,
Hans-Joachim Schulze

Infineon Technologies AG
Neubiberg, Germany
ManfredFranz.Pfaffenlehner@infineon.com

Roman Baburske, Josef Lutz

Chair of Power Electronics and EMC
Chemnitz University of Technology
Chemnitz, Germany

Abstract—**The surge current ruggedness of free-wheeling diodes can be improved by implementing the SPEED concept (S̲elf-adjusting P̲ E̲mitter Efficiency D̲iode). Experiments show that the switching ruggedness of such a diode is worse than that of a conventional diode. Simulations indicate that during diode turn-off filaments are pinned at the cathode side. These filaments can be avoided by implementing CIBH (C̲ontrolled I̲njection of B̲ackside H̲oles). It turns out that a necessary additional measure is to fully embed the p^+-areas of the SPEED anode in the low-doped p-type area to avoid high electrical field strengths and current crowding at the anode side. Combining these measures, the appearance of current filaments with an extremely high current density and their pinning to a certain area in the device during the turn-off period can be avoided.**

I. MOTIVATION

In high-power applications there is a need for free-wheeling diodes to withstand high current densities. As the output current of a given module is usually increased for a new chip generation, it becomes more and more difficult to reach the surge current targets.

There are several ways known to increase the surge current ruggedness of a free-wheeling diode. The most common methods are:

- Increase of the anode doping combined with a uniform reduction of the charge-carrier lifetime: This measure is usually easy to apply but leads to an unfavorable on-state carrier distribution in the diode. This may result in increased power losses and oscillations during turn-off of the diode.

- Increase of the anode doping combined with a local reduction of the charge-carrier lifetime in or below the p-emitter: A local lifetime killing is usually done by irradiation with light particles, like helium. This measure leads to a reduction of the hole injection of the anode respectively to a reduction of the excess carrier-charge density below the anode. Particle irradiation into the area of the space-charge region

leads to an increased leakage current at high operation temperatures. This counteracts the trend to higher operation temperatures.

Another way of improving the surge current ruggedness is the SPEED (S̲elf-adjusting P̲ E̲mitter Efficiency D̲iode) concept [1].

II. SPEED

A. Basics of the SPEED Concept

The SPEED uses an anode which consists of highly-doped p^+-areas which are located inside a low-doped p-emitter area (Fig. 1). At low current densities, hole injection comes mainly out of the low-doped p-emitter area. At high current densities the injection comes out of both p-doped areas but is mainly determined by the p^+-doped areas. Comparing a SPEED with a diode having a conventional anode (in the following labeled as conventional diode) but the same charge-carrier lifetime and the same forward-voltage drop at rated current as the SPEED, the SPEED has a much lower forward-voltage drop at very high current densities (Fig. 2). This leads to reduced power losses in the device during surge current and to a better surge current ruggedness of the SPEED.

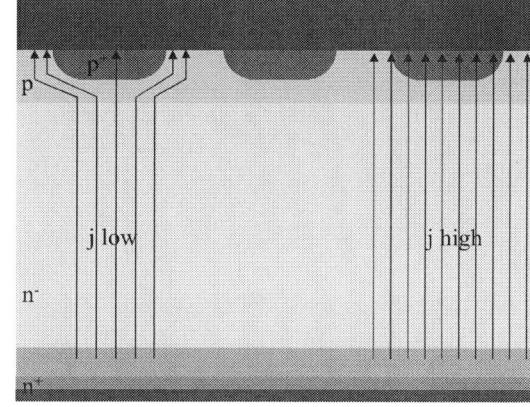

Figure 1. Cross section of the SPEED illustrating schematically the difference of current flow at low and high current densities

978-1-4244-8425-6/11 $26.00 © 2011 IEEE

B. Experimental Surge Current Results

Various SPEED diodes with a maximum blocking voltage of 3300 V have been manufactured and compared with a conventional diode. The SPEED variants differ in the lateral distance of the p^+-areas and the doping concentration of the p-area and the p^+-areas. To keep the process as simple as possible, the p-area and the p^+-areas were diffused with the same diffusion process. This means that in contrast to Fig. 1 the p^+-area is deeper than the p-area.

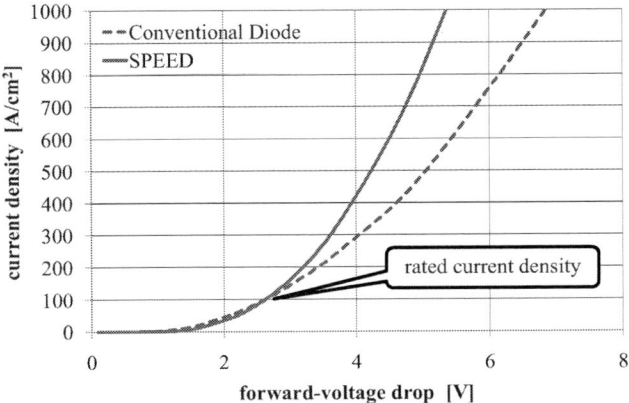

Figure 2.   Simulated forward-voltage drop of a SPEED and a conventional diode. The rated current density is 100 A/cm².

In Fig. 3, the surge current ruggedness of the SPEED variants and the conventional diode are plotted as function of the forward-voltage drop at rated current. By means of an extrapolation we found that a SPEED variant with the same forward-voltage drop at rated current as the conventional diode can withstand a 12 % higher surge current.

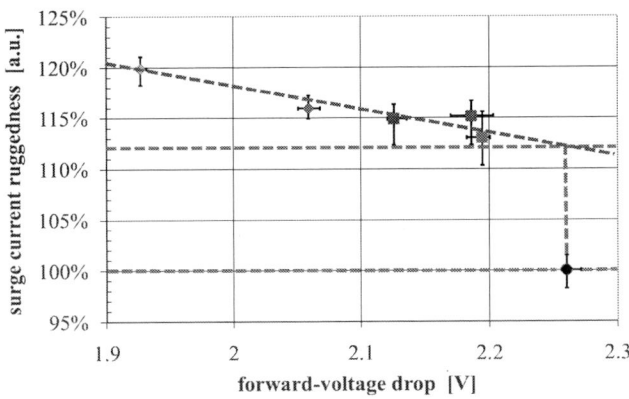

Figure 3.   Experimental results of the surge current ruggedness *vs.* the forward-voltage drop at rated current of various SPEED variants and a conventional diode (black dot on 100 % line).

C. Experimental SOA Results

We also evaluated the SOA (Save Operation Area) of the SPEED variants and the conventional diode. The switching ruggedness of all SPEED variants is much worse than that of the conventional diode: For the maximum of the power loss during switching (P_{max}) we found a deterioration of about 80 %. No major difference between the SPEED variants was observed. The point of destruction was randomly distributed in the anode area.

III.   IMPLEMENTATION OF THE CIBH CONCEPT

Simulations of the turn-off behavior of a SPEED were performed to gain an understanding of the poor switching ruggedness. As the point of destruction is located in the anode area and not in or near the junction termination structure, we performed device simulations of diode structures without any junction termination. The simulated switching conditions were based on the experimental switching conditions (Fig. 4).

The solutions of the 2-dimensional simulations showed the appearance of current filaments below the p^+-areas as soon as the diode takes over blocking voltage (Fig. 5). These filaments appeared, faded away and "jumped" between the p^+-areas. Shortly after the reverse current had reached its maximum, an additional filament on the cathode side appeared (Fig. 6). This filament did not fade away and did not move.

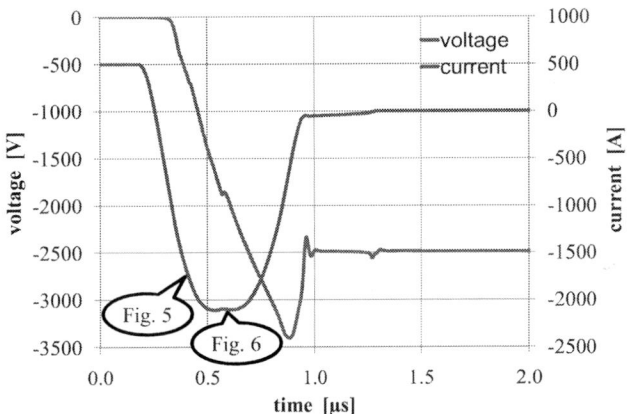

Figure 4.   Simulated turn-off behavior of a SPEED at a device temperature of 150 °C, two times rated current and a DC-link voltage of 2500 V.

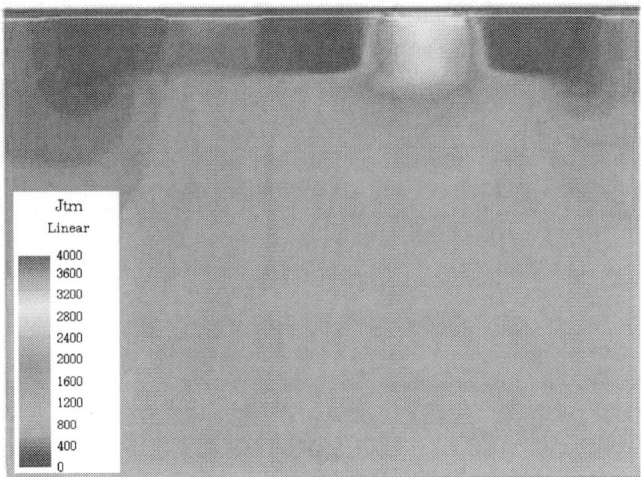

Figure 5.   Simulated total current density [A/cm²] in the SPEED during turn-off at the time marked in Fig. 4.

A. Simulation of a SPEED with CIBH

The occurrence of non-moving filaments is considered to be an important reason for the bad switching ruggedness of high-voltage devices. A potential measure to prevent the appearance of a cathode-side filament is the implementation of the CIBH (Controlled Injection of Backside Holes) concept [2]. Controlled injection of backside holes can be achieved by p-type regions located near the cathode of the diode. These p-type areas are not in contact with the cathode-side metallization of the diode. Under extreme switching conditions as, for example, a SOA test, avalanche breakdown occurs at the cathode-side p-n junction. Due to the charge carriers generated by this avalanche breakdown the appearance of a critical Egawa-like electric-field distribution is avoided and a better switching ruggedness of the diode is achieved [3].

Simulations of a SPEED in combination with the CIBH concept showed that the moving anode-side filaments still occur, but the pinned cathode-side filament can be prohibited as expected.

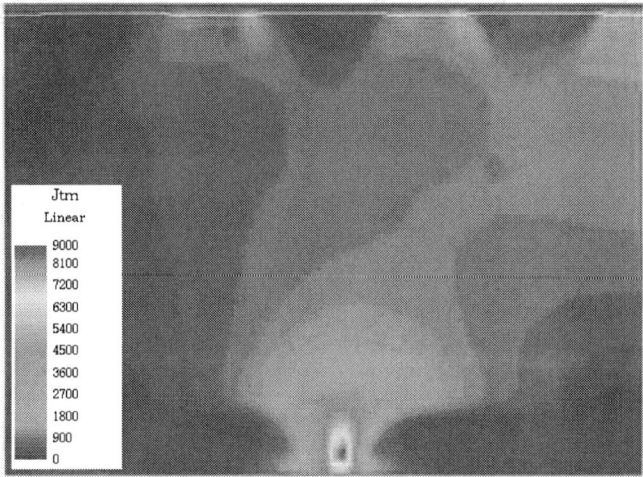

Figure 6. Simulated total current density [A/cm²] in the SPEED during turn-off at the time marked in Fig. 4.

B. Experimental SOA Results with CIBH

A diode implemented with both the SPEED and the CIBH concept was manufactured and the SOA was evaluated. However, no improvement of the switching ruggedness compared with the variants without CIBH was obtained. This indicates that the avoidance of the cathode-side filament is most likely a necessary requirement to gain a good switching ruggedness, but is not a sufficient one.

IV. OPTIMIZATION OF THE ANODE

Fig. 5 indicates a current crowding at the edges of the p^+-areas of the SPEED during switching. A simplified simulation setup for the speed which consists of only a single p^+-area was used to investigate the influence of the anode design of the SPEED. Apart from current crowding also a high electrical field strength can be found at the edge of the p^+-area (Fig. 7).

A. Variation of the Diffusion Depth of p^+ and p

For all SPEED variants considered so far, the p^+-areas and the p-areas of the SPEED were diffused with the same diffusion process. As a result, the p-n junction formed by the p^+-area and the n-base was deeper than that formed by the p-area and the n-base.

To optimize the anode of the SPEED, the process flow was changed so that the depths of the two p-n junctions could be varied to a certain extent independently from each other (Fig. 8).

The diffusion process used in the "OLD FLOW" was the process used in production for the conventional diodes. This process was implemented in the process simulation and was also used as the basis for the modifications performed in the "diffusion 1" and "diffusion 2" in the "NEW FLOW".

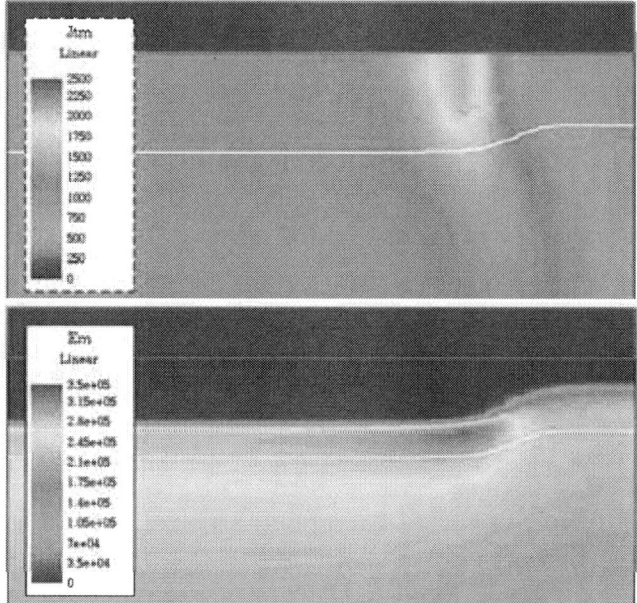

Figure 7. Current density and electrical field strength distribution near the edge of the p^+-area during diode turn-off at a device temperature of 150 °C, two times rated current and a DC-link voltage of 2500 V.

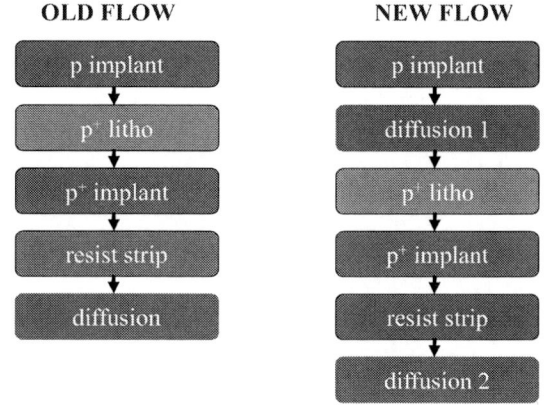

Figure 8. Changed flow of the process simulation for the anode of the SPEED.

For the "diffusion 2", the temperature budget of the base process was reduced step by step to gain a shallower p^+-profile. Two variants for "diffusion 1" were simulated: The standard process and a process with a strongly increased temperature budget which leads to a doubling of the depth of the p-n junction of the p-area. To compare the different anode variants the maximum of the electrical field strength during diode turn-off was extracted.

The results of the simulations show that the maximum of the electrical field strength can be reduced if the depth of the p^+-area is reduced with respect to the depth of the p-area (Fig. 9). The utmost left points on the two graphs still show the "OLD FLOW" as a reference. The other points are gained with the "NEW FLOW". The utmost right points are not simulated with a diffused p^+-profile but as a box profile. Such a doping profile can be realized e.g. by laser thermal annealing and is very shallow [4]. As can be seen by comparing the graphs for the two variants of the p-diffusion, increasing the p-depth has resulted in a reduced maximum electrical field strength compared with a conventional anode.

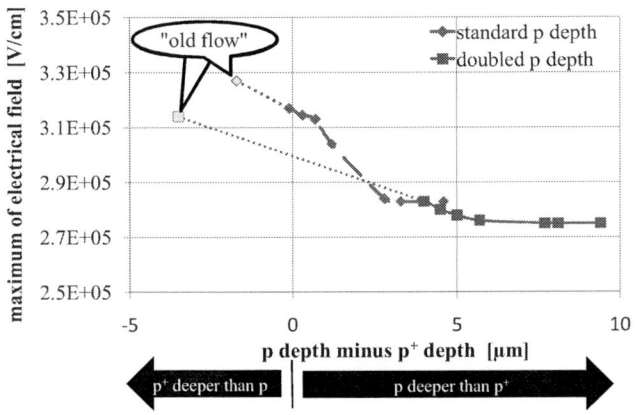

Figure 9. Maximum of the electrical field strength in the anode during diode turn-off for different anode variants of the SPEED. The maximum electric field strength in a diode with a conventional anode is 2.76E5 V/cm.

B. Variation of the Lateral Diffusion of the p^+-area

Another way to reduce the current crowding at the edge of the p^+-area is the implementation of a VLD (Variation of Lateral Doping) zone at the edge of the p^+-area.

The basic idea of implementing VLD in the SPEED is to gain a smoother lateral transition from the p^+-area to the p-area. The common way to manufacture VLD would be to use a fine patterned lithography structure to continuously reduce the p^+-doping concentration at the edge of the p^+-area.

However, in the simulation the lateral reduction of the p^+-area was reached by using an oxide edge with a slope with a very small angle as an implantation mask for the p^+-implantation.

The simulation was done with the "OLD FLOW". A significant reduction of the electrical field from 3.27E5 V/cm to about 3.08E5 V/cm was achieved. However, for a larger reduction of the electrical field strength very wide VLD-areas have to be implemented which lead to very wide p^+-areas. That is unfavorable for the forward characteristic of the SPEED.

V. CONCLUSION

Experiments show that the surge current ruggedness of free-wheeling diodes can be increased by implementing the SPEED concept. But the SPEED can lead to a degradation of the switching ruggedness of the anode area as confirmed by measurements. Three aspects were found in simulations that might be the cause of the poor switching ruggedness:

- A pinned cathode-side filament. This cathode side filament can be inhibited by implementing the CIBH concept.

- An increased current density, temperature and electrical field strength at the edges of the p^+-areas. This can be optimized by fully embedding the p^+-areas in the p-area.

- Moving filaments at the anode side below the $p+$-areas. There were no measures found to inhibit these filaments. But as they are not pinned, they are not considered to be the main factor limiting the switching ruggedness.

REFERENCES

[1] H. Schlangenotto, J. Serafin, F. Sawitzki, and H. Mauder, "Improved recovery of fast power diodes with self-adjusting p emitter efficiency", IEEE Electron Device Lett., vol. 10, pp. 322–324, 1989.

[2] M. Chen, J. Lutz, M. Domeij, H- P. Felsl, and H.-J. Schulze, "A novel diode structure with controlled injection of backside holes (CIBH)," Proc. ISPSD, pp. 9-12, 2006.

[3] J. Biermann, M. Pfaffenlehner, H.-P. Felsl, T. Gutt, H. Schulze, "CIBH Diode with Superior Soft Switching Bahavior in 3.3kV Modules for Fast Switching Applications", Proc. PCIM Europe, pp. 367-371, 2008.

[4] T. Gutt, H. Schulze, "Deep melt activation using laser thermal annealing for IGBT thin wafer technology", Proc. ISPSD, pp. 29-32, 2010

Proceedings of the 23rd International Symposium on Power Semiconductor Devices & IC's
May 23-26, 2011 San Diego, CA

Edge Termination Impact on Clamped Inductive Turn-off Failure in High-Voltage IGBTs under Overcurrent Conditions

X. Perpiñà, I. Cortés, J. Urresti-Ibañez, X. Jordà, J. Rebollo, J. Millán.

Instituto de Microelectrónica de Barcelona IMB-CNM (CSIC), Campus UAB, 08193 Bellaterra (Barcelona), Spain.

Abstract— **This work provides a physical insight into the failure of high-voltage IGBT modules for railway traction when an overload current event occurs during a clamped inductive turn-off. The inspection of failed IGBTs in power modules coming from the field reveals burnt-out points in the vicinity of the device edge termination. This physical signature has been also verified by experimental tests. To explore this result, physical TCAD simulations have been carried out considering, for the first time, the electro-thermal mismatch introduced by the edge termination. From simulation and experimental results, a destructive dynamic avalanche phenomenon at the last IGBT cell is identified as responsible for the observed failure.**

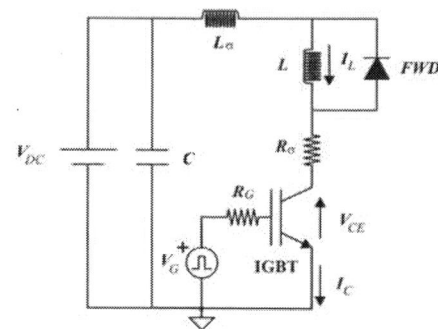

Figure 1. Schematic of the test-circuit used to reconstruct actual inverter switching conditions.

I. INTRODUCTION

When an overloading condition occurs, the reliability of railway traction inverters is extremely linked to the ruggedness of the IGBTs and diodes packaged together in multichip IGBT modules. One way to improve their ruggedness is by understanding the failure physics at chip level, as carried out with the edge termination of high-power diodes [1]. In such a case, the diodes' ruggedness during the reverse recovery is completely compromised by its design and several solutions have been addressed [2, 3]. However, as already observed in other working conditions under which IGBTs break down at their periphery [4], the edge termination could also have a negative influence on the IGBT inductive turn-off under overcurrent conditions. In this work, such an effect is studied at chip level on commercial IGBT modules (3.3kV/1200A, 24 IGBT devices) by following experimental tests [5] and electro-thermal simulations. First, tests up to failure have been performed to assess a physical signature when the Reverse Blocking Safe Operating Area (RBSOA) of IGBTs is overcome by an overcurrent event. Next, this failure is discussed on the basis of a simulation approach that contemplates the electro-thermal mismatch between IGBT cells due to the edge termination presence (device symmetry loss).

II. EXPERIMENTAL ANALYSIS

A. Robustness tests: Description and results

Tests up to failure have been performed on 5 IGBT power modules to determine and assess their electrical signature when their RBSOA limits are overcome by an overcurrent

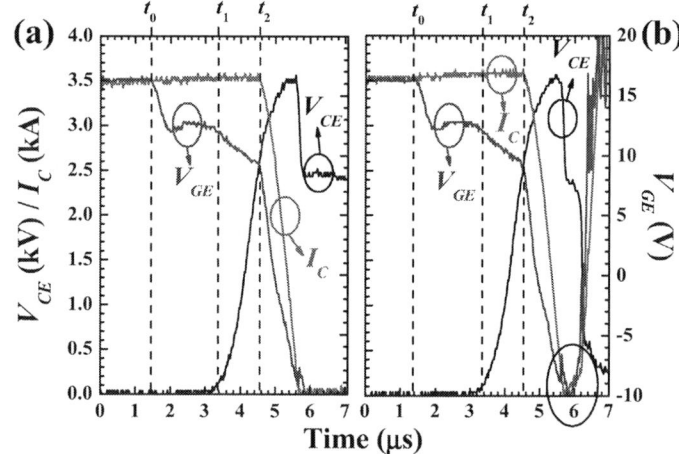

Figure 2. Experimental current and voltage waveforms for a safe, (a), and for a failed, (b), overcurrent turn-off.

event during their clamped inductive turn-off. Fig. 1 depicts the schematic of the used circuit (clamped inductive test-circuit). In our experiments, $V_{DC} = 2500$ V, $T = 398$ K, $L = 100$ µH, $L_\sigma = 285$ nH, and $R_G = 3.7$ Ω are considered. More details are available in [5].

The most interesting waveforms obtained from all tests are presented in Figs. 2 (a) and (b). On one hand, Fig. 2 (a) shows a safe overcurrent turn-off manifesting this behavior: When the turn-off signal is applied at time t_0, the gate-emitter voltage V_{GE} falls down to a plateau and remains constant until the end

978-1-4244-8425-6/11 $26.00 © 2011 IEEE

of the Miller effect [6]. Then, at time t_1, the collector-emitter voltage V_{CE} starts rising up while V_{GE} decreases. At time t_2, the diode of Fig. 1 (free-wheeling diode, FWD) starts conducting a current I_{FWD}, inducing the I_C decrease. Therefore, V_{GE} falls rapidly to zero. This coexistence of high-voltage and current ratings (dynamic avalanche process) finishes when I_L only passes through FWD and I_C becomes zero. At this point, V_{CE} falls abruptly to the nominal input value, V_{DC}. On the other hand, Fig. 2 (b) displays the experimental results corresponding to a failed turn-off process. The failure electrical signature is: V_{CE} is not sustained at V_{DC} after its overshoot and it falls towards zero (as observed in [5]), while I_C starts increasing again. Besides, the device is completely turned-off ($I_C = 0$) and I_C suddenly rises before the subsequent V_{CE} collapse and V_{GE} increase (see circle in Fig. 2 (b)).

B. Failure signatures observed within modules and devices

Figs. 3 (a) and (b) depict representative photographs related to the modules and devices destroyed during the tests. As Fig. 3 (a) illustrates, the visual inspection of a failed module reveals that its loss of functionality is due to a limited number of exploded IGBT devices. In all cases, the same failure pattern has been observed repeatedly, in which the package parasitic elements establish the position of the failed IGBTs (devices in the Leg 1 substrate (a)). This effect has been analyzed in previous works [7, 8], in which both the package parasitic elements and non-uniform thermal effects have been taken into account. Such studies put in evidence that an uneven current sharing between IGBTs determines the most electro-thermally stressed devices. Moreover, the destruction of an individual IGBT device normally gives rise to the destruction of adjacent IGBTs. Fig. 3 (b) shows the failure physical signature observed at chip level: A surface burnt-out point in the vicinity of the device edge termination [9], which reveals a hot spot formation at this location. This signature has also been reported in [10], in which other damages experienced by failed devices have been discussed.

III. FAILURE DISCUSSION

A. Simulated Structure Description

Physical simulations with Sentaurus TCAD tools [11] have been performed so as to reproduce the experimental failed turn-off in a single device with the purpose to provide a physical insight into the signature observed in Fig. 3 (b). The simulation approach has contemplated an electro-thermal mismatch by the structure itself; i.e.: Several IGBT basic cells together with the entire device edge termination have been considered (see Fig. 4), which contrasts to those approaches followed up to now [12, 13, 14]. The physical input parameters for the simulations have been adjusted in order for their results to match the IGBT experimental static and dynamic characteristics. In addition, the SPICE model corresponding to the FWD has been used in mixed-mode simulations to reduce the computation time.

B. Failure Interpretation based on Simulation Results

Fig. 5 demonstrates I_{FWD}, I_C, V_{CE}, V_{GE}, and the maximum temperature (T_{max}) waveforms for the simulated turn-off failure, highlighting several interesting time instants (steps).

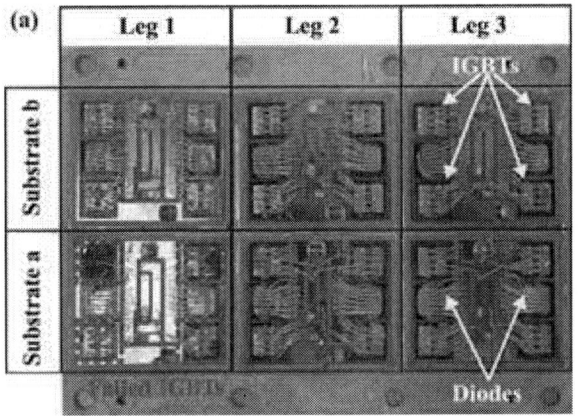

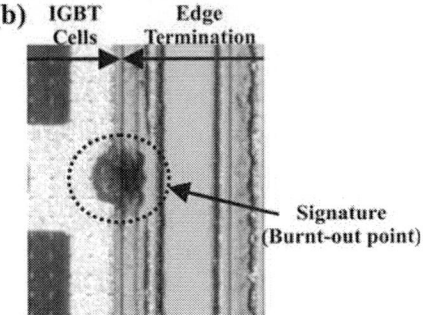

Figure 3. Photographs of a failed IGBT-module during an overcurrent turn-off (a) and zoom of the failure signature in an IGBT chip (b).

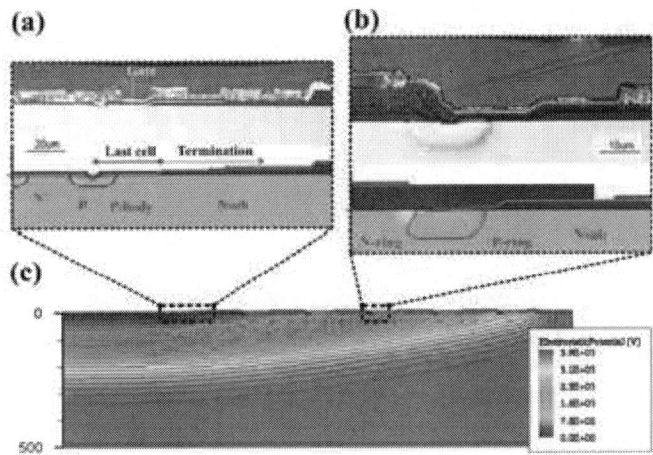

Figure 4. Simulated structure, showing details of: (a) Core, (b) edge termination, and c) the full structure in blocking state.

As Fig. 5 shows, V_{CE} first collapses and I_C suddenly increases when V_{GE} is still above the threshold voltage. Moreover, V_{CE} decreases as T_{max} increases, even when V_{GE} has reached its lower negative value. Figs. 6 and 7 depict the evolution of the current density mismatch during the failed turn-off event at the surface ($Y = 1$ nm) and deep inside the silicon bulk ($Y = 480$ µm) at several steps, respectively. Figs. 8 and 9 show the electric field, impact ionization, electron current flowlines, and temperature distribution just before the IGBT failure (step 8) and when I_C increases (step. 10). In fact, Fig. 8 shows three potential locations for the failure

978-1-4244-8425-6/11 $26.00 © 2011 IEEE

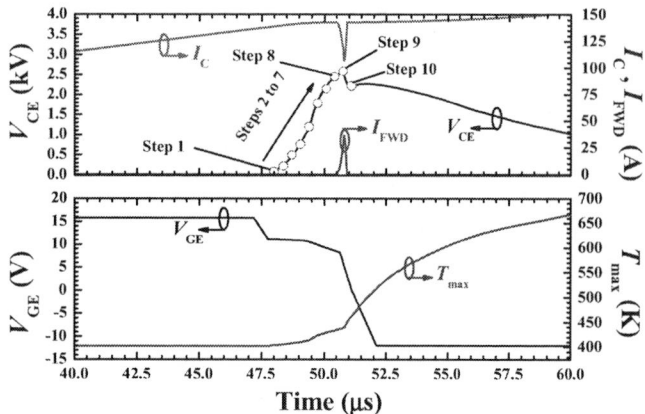

Figure 5. Simulated turn-off waveforms: Collector-emitter (V_{CE}) and gate-emitter (V_{GE}) voltages, IGBT collector (I_C) and FWD (I_{FWD}) currents, and maximum temperature (T_{max}), highlighting several time instants (see steps).

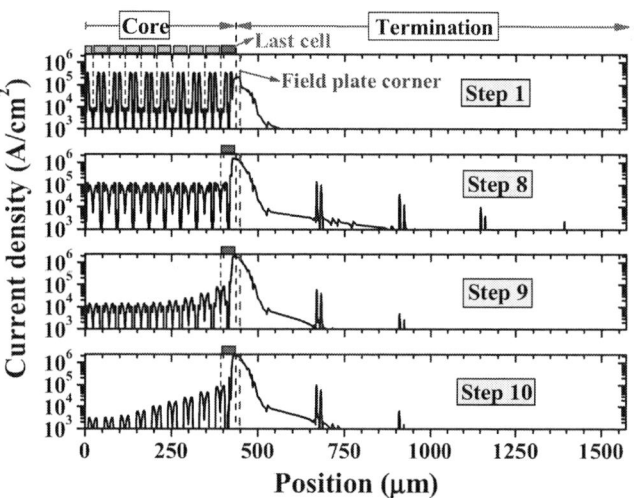

Figure 6. Total current density distribution on the simulated structure topside ($Y = 1$ nm) corresponding to the steps 1, 8,9 and 10 shown in Fig. 5.

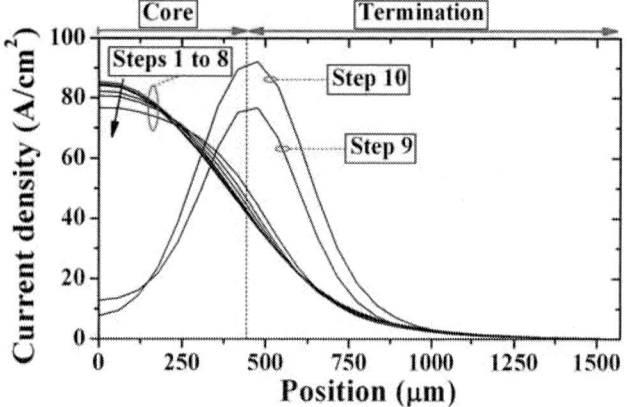

Figure 7. Total current density distribution in the simulated structure at $Y = 480$ μm corresponding to the steps highlighted in Fig. 5.

occurrence: Below the field plate corner (1), last cell's channel region closer to the edge termination (2), and the right corner of the P-body junction (3). Locations (1) and (3) exhibit

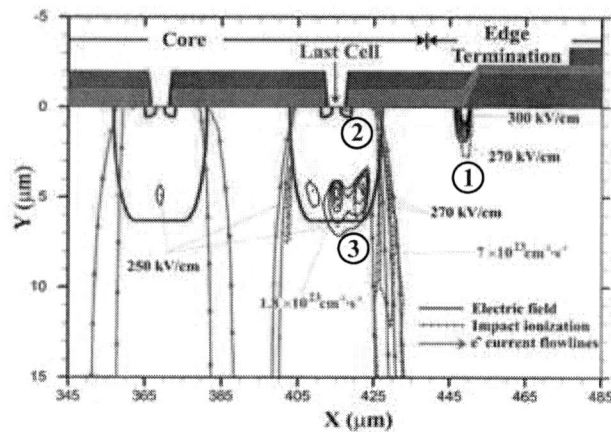

Figure 8. Electric field, impact ionization, and electron current flowlines in the last two cells just before failure (step 8).

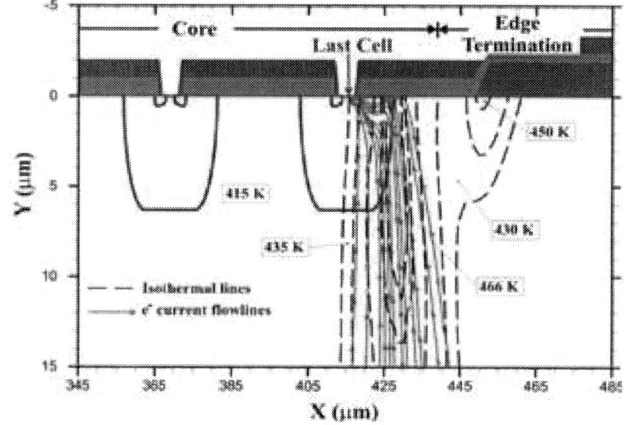

Figure 9. Temperature distribution and electron current flowlines at step 10.

high electric fields that may eventually induce the breakdown of the oxide or the P-body junction, respectively. Besides, location (2) may be affected by a thermal instability during the device turn-off due to the positive thermal coefficient of the channel mobility with temperature when V_{GE} is below a certain critical value [15]. This leads to the following behavior: The higher the temperature at (2), the lower the channel resistance. This situation produces the current mismatch shown in Figs. 6 and 7 (see steps 9 and 10). This corresponds to the situation depicted in Fig. 9: The channel is not completely switched off and originates a current crowding effect, as Figs. 6 and 7 indicate. In addition, the temperature distribution shown in Fig. 9 also denotes that the high electric field at (1) generates a hot spot which starts this electro-thermal effect (as also evidenced in Fig. 5). In order to verify whether this is the weakest point of the structure, simulations with the same working conditions have been conducted with only one IGBT cell (core situation), observing a safe turn-off.

From the comparison between experimental and simulation results, the differences between V_{CE} and I_C waveforms of Figs. 2 (b) and 5 come from the instant when the failure is produced (before or after the channel extinction) and the V_{CE} behavior (gradual or sudden collapse). The I_C increase observed in the case of experimental results can not be originated at (2), since the simulated V_{CE} shows a more gradual decrease than in the experiment. Therefore, as V_{GE} starts rising up after I_C (see Fig. 2 (b)), the failure should be originated at (3) (see Fig. 8), leading to the dynamic breakdown of the last IGBT cell. Such effect is promoted and mainly governed by the uneven current distribution among the IGBT cells due to the collected current coming from the area below the termination region (see step 1 in Fig. 6), which is more relevant when the dynamic avalanche process takes place. On a secondary level, carriers generated along the edge termination either by impact ionization or by a thermal process due to the eventual presence of hot spots (e.g., field plate corner) can also assist to the dynamic breakdown of the last cell (see Fig. 8). Both current contributions are collected by the last cell due to its proximity to the periphery, which lowers the resistance to the carriers' path. In such a situation, the critical electric field at (3) is reached at lower V_{CE} values, as the holes collected by the P-body junction modulate the electric field [5, 13, 16]. This explains the I_C rise and the V_{CE} abrupt collapse observed in the experimental waveforms (see Fig. 2 (b)): The dynamic breakdown of a single or few IGBT cells (local breakdown) close to the edge termination occurs and, subsequently, I_C rises up at the failed IGBT cells (first current increase highlighted in Fig. 2 (b)), leading to a hot spot formation. From this over-heated region, the heat-flux radially diffuses to the neighbouring cells, inducing the same electro-thermal effect as Fig. 3 (b) indicates. When a critical temperature value is locally reached [17], the second breakdown phenomenon is produced; i.e., the device losses the blocking voltage capability and V_{CE} suddenly collapses (see Fig. 2 (b)). Consequently, a single IGBT device not only collects all of the current passing through the IGBT module, but also takes the currents coming from the inductive load and FWD (see Fig. 1). In this later phase, a driver failure induced by a gate overcurrent peak and a likely latch-up phenomenon take place as Fig. 2 (b) displays: V_{GE} first rises and after, I_C still increases following a higher slope (second current increase). Eventually, the IGBT module explodes due to the high current values reached during the FWDs reverse recovery [10, 17].

IV. CONCLUSION

The operation of IGBT modules used in railway traction inverters are affected by abnormal events which impose severe stresses on the devices, particularly on the IGBTs. This is the case of the current overload conditions reached during IGBTs turn-off, for instance, when the train wheels glide. This work provides a physical interpretation for this failure by means of experimental (tests at limit) and simulation approaches. The failure is discussed on the basis of physical simulations considering several IGBT cells together with the edge termination to reproduce the electro-thermal mismatch during an inductive turn-off. Simulation and experimental results suggest that the failure mechanism is the dynamic breakdown of the last IGBT cell, which reinforced by a local temperature increase, gives rise to the IGBT secondary breakdown.

ACKNOWLEDGMENT

This work has been partially supported by the "Consejo Superior de Investigaciones Científicas" (CSIC) (under contract "Junta para la Ampliación de Estudios", JAE-Doc), the Spanish Ministry of Science and Innovation (Research Programs: THERMOS TEC2008-05577, Ramón y Cajal RYC-2010-07434, and RUE CSD2009-00046), and the European project PORTES (Power Reliability for Traction Electronics, MTKI-CT-2004-517224).

REFERENCES

[1] J. Lutz, and M. Domeij, "Dynamic avalanche and reliability of high voltage diodes", Microelectron. Reliab., vol. 48, no. 4, pp. 529-536, April 2003.

[2] S. Kameyama, T. Sugiyama, R. Tagami, and K. Nishiwaki, "Investigation of dynamic avalanche in the termination region for FWDs with high reverse capability", in Proc. ISPSD, Orlando (USA), pp. 137-140, May 2008.

[3] K. Nakamura, F. Masuoka, A. Nishii, K. Sadamatsu, S. Kitajima, and K. Hatade, "Advanced RFC technology with new cathode structure of field limiting rings for high voltage planar diode", in Proc. ISPSD, Hiroshima, (Japan), pp. 133-136, June 2010.

[4] U. Knipper, G. Wachutka, F. Pfirsch, T. Raker, and J. Niedemeyr, "Time-periodic avalanche breakdown at the edge termination of power devices", in Proc. ISPSD, Orlando (USA), pp. 307-310, May 2008.

[5] X. Perpiñà, J.F. Serviere, J. Urresti-Ibañez, I. Cortés, X. Jordà, S. Hidalgo, J. Rebollo and M. Mermet-Guyennet, "Analysis of Clamped Inductive Turn-off Failure in Railway Traction IGBT Power Modules under Overload Conditions", IEEE Trans. Ind. Electron., in press (published on-line, Digital identifier: 10.1109/TIE.2010.2077613).

[6] B. J. Baliga, Power Semiconductor Devices, PWS Publishing Company, Boston, MA, 1996.

[7] R. De Maglie, G. Lourdel, P. Austin, J.-M. Dienot, J.-L. Schanen, and J.-L. Sanchez, "Use of Accurate Chip Level Modeling and Analysis of a Power Module to establish Reliability Rules", in Proc. ISIE, pp. 1571-1576, Montreal (Canada), July 2006.

[8] A. Castellazzi, M. Ciappa, W. Fichtner, G. Lourdel, and M. Mermet-Guyennet, "Compact modelling and analysis of power-sharing unbalances in IGBT-modules used in traction applications", Microelectron. Reliab., vol. 46, no. 9-11, pp. 1754–1759, September/November 2006.

[9] X. Perpiñà, J.F. Serviere, X. Jordà, A. Fauquet, S. Hidalgo, J. Urresti-Ibañez, J. Rebollo, and M. Mermet-Guyennet, "IGBT module failure analysis in railway applications", Microelectron. Reliab., vol. 48, no. 8-9, pp. 1427-1431, August/September 2008.

[10] M. Ciappa, "Selected failure mechanisms of modern power modules", Microelectron. Reliab., vol. 42, no. 4-5, pp. 653-667, April/May 2002.

[11] TCAD TOOL Suite. http://www.synopsys.com/products/tcad/tcad.html. Synopsys, 2006.

[12] A. Kopta, M. Rahimo, U. Schlapbach, N. Kaminski, and D. Silber, "Limitation of the short-circuit ruggedness of high-voltage IGBTs", in Proc. ISPSD, Barcelona (Spain), pp. 33-36, June 2009.

[13] P. Rose, D. Silber, A. Porst, and F. Pfirsch, "Investigations on the stability of dynamic avalanche in IGBTs", in Proc. ISPSD, Santa Fe (USA), pp. 165-168, June 2002.

[14] Y. Mizuno, R. Tagami, and K. Nishiwaki, "Investigations of inhomogeneous operation of IGBTs under unclamped inductive switching condition", in Proc. ISPSD, Hiroshima (Japan), pp. 137-140, June 2010.

[15] P.Spirito, G. Breglio, V. d'Alessandro, and N. Rinaldi, "Analytical model for thermal instability of low voltage power MOS and SOA in pulse operation", in Proc. ISPSD, Santa Fe (USA), pp. 269-272, June 2002.

[16] M. T. Rahimo, and N. Y. A. Shammas "Freewheeling diode reverse-recovery failure modes in IGBT applications", IEEE Trans. Ind. Appl., vol.37, no. 2, pp. 661-670, March/April 2001.

[17] T. Puritis, "Problems related to the avalanche and secondary breakdown of silicon P-N Junctions", Microelectron. Reliab., vol. 35, no. 5, pp. 713-719, May 1997.

Proceedings of the 23rd International Symposium on Power Semiconductor Devices & IC's
May 23-26, 2011 San Diego, CA

Hybrid Isolation Process with Deep Diffusion and V-Groove for Reverse Blocking IGBTs

Haruo Nakazawa,Masaaki Ogino,Hiroki Wakimoto,
Tsunehiro Nakajima and Yoshikazu Takahashi
Fuji Electric Holdings Co.,Ltd.
Matsumoto,Nagano,Japan
nakazawa-haruo@fujielctric.co.jp

David Hongfei Lu

Fuji Electric Systems Co.,Ltd.
Matsumoto,Nagano,Japan

Abstract— **We newly developed a 1200V Reverse Blocking (RB)-IGBT used to form bi-directional switches in advanced Neutral-Point-Clamped (A-NPC) 3-Level modules. It featured a hybrid through-silicon isolation structure combining wafer front-side boron deep diffusion with back-side V-groove etching. Collector layer was implanted into the back-side and the surface of the V-grooves, and electrically connected to the front-side boron diffusion after activation to achieve reverse-blocking capability. Thermal budget for the surface deep boron diffusion was thereby shortened more than a half of that in full diffusion case to improve both throughput and yield. Sufficient reverse blocking capability was experimentally verified.**

I. INTRODUCTION

Recently, a variety of efforts have been made worldwide to reduce carbon dioxide emission. The ever more important emphasis in the field of power electronics is on the reduction of power loss and the improvement of efficiency in power conversion systems. In this end, multi-level power converters have been previously proposed as one of the most effective approaches. Several types of neutral-point-clamped (NPC) 3-Level power converters have been proposed [1]. Among them, one type utilizing diodes for the clamping of the output to the neutral point has been commercialized to reduce both the switching loss and the size of the filters [2]. However, these systems have disadvantages in terms of using too many semiconductor devices.

To solve above problems, advanced NPC (A-NPC) 3-Level converters, in which bidirectional switches are used for clamping the output to the neutral point, has been proposed. Reverse-Blocking IGBTs(RB-IGBTs)[3-5], first developed for matrix converter applications, suit well in A-NPC converters to reduce the total number of the devices, therefore the total loss, by eliminating the diode in the conventional serial connection between an IGBT and a diode.

In this paper, we describe a new hybrid through-silicon isolation process and the electrical properties for 1200V RB-IGBTs, and demonstrate the advantages of an A-NPC 3-Level converter using the RB-IGBTs, and our 6th generation 1700V IGBTs and FWDs (Free Wheeling Diodes).

II. OVERVIEW OF RB-IGBT

There are, most typically, two methods to form bi-directional switch as shown in Figure 1. The conventional bi-directional switch consists of two IGBTs and two diodes as shown in Figure 1(a). The series diodes are necessary to block reverse bias due to the lack of reverse blocking capability in conventional IGBTs, and they cause additional on-state voltage drop (conduction loss). On the other hand, the RB-IGBTs can achieve the same bi-directionality while eliminating the discrete diodes in the circuit and reducing the total loss of the converter, Figure 1(b).

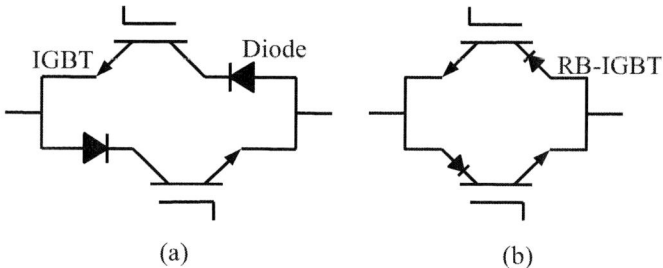

Figure 1: Bi-directional switches composed with (a) IGBTs and diodes, (b) RB-IGBTs.

As shown in Figure 2(a), our previous 1200V RB-IGBT realizes its reverse blocking capability by extending the thermal budget used for 600V devices to form a deeper (~200um) p+ isolation layer, which penetrates the n- drift layer and reaches the backside p+ collector by thermal diffusion of boron from the front surface. The thermal diffusion requires tremendously lengthy time at high temperature, and suffers limitation to higher breakdown voltage RB-IGBTs with deeper isolation layer, since the depth increases only with the square root of the diffusion time. Furthermore, lateral diffusion of the isolation layer also increases the die size. For higher voltage devices, it is inappropriate to simply extend the deep diffusion isolation process.

RB-IGBTs with isolation layers formed by V-groove etching through the whole silicon layer from wafer back side

978-1-4244-8425-6/11 $26.00 © 2011 IEEE

and subsequent ion implantation and activation on the surface of V-grooves have been demonstrated in [8]. However, this method lacks in both yield stability from large depth of the V-groove, and the compatibility of supporting substrate to conventional backside metallization units.

In this work, a robust hybrid isolation process was newly developed for 1200V RB-IGBT.

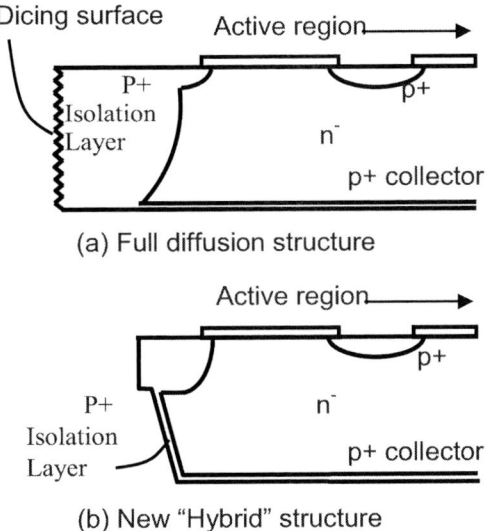

(a) Full diffusion structure

(b) New "Hybrid" structure

Figure 2: Cross-sectional structure of RB-IGBTs (a) full diffusion method, (b) new "Hybrid" method.

III. HYBRID ISOLATION PROCESS FOR RB-IGBT

Figure 2(b) shows the cross-sectional view of a RB-IGBT with the new isolation layers. The new hybrid isolation layers are fabricated as shown in Figure 3. First, boron diffusion is conducted from the front-side of wafer before the formation of surface cell structures. Second, after backside wafer thinning, V-groove etching is performed form the backside. Third, Boron was implanted into the back-side and the surface of the V-grooves. The collector layer is electrically connected to the front-side boron diffusion after boron activation. Finally, backside metallization is performed.

Figure 4 shows a SEM image of the anisotropic etched V-groove. An excellent V-groove shape has been obtained. Compared with that in the case of full diffusion for 1200V devices, the thermal budget of the front-side surface deep boron diffusion has been significantly shortened to less than a half in this hybrid scheme.

For higher blocking voltage devices, it is necessary to apply hybrid isolation process. Besides the merit of shorter diffusion time, yield loss associated with crystal defects in extreme thermal budget has also been reasonably suppressed.

IV. ELECTRICAL CHARACTERISTICS OF 1200V RB-IGBT

Bi-directional blocking characteristics of 1200V RB-IGBT with our new hybrid method are shown in Figure 5. Forward

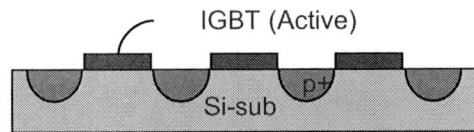

(a) IGBT with diffused p+ Isolation

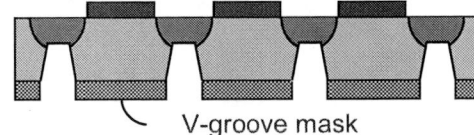

(b) Anisotropic Si-Etching from backside

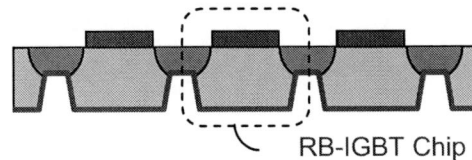

(c) Collector electrically connected with diffused p+ Isolation.

Figure 3: Schematic process flow of the new "Hybrid" RB-IGBT.

Figure 4: SEM image of a V-groove formed by means of anisotropic silicon wet etching

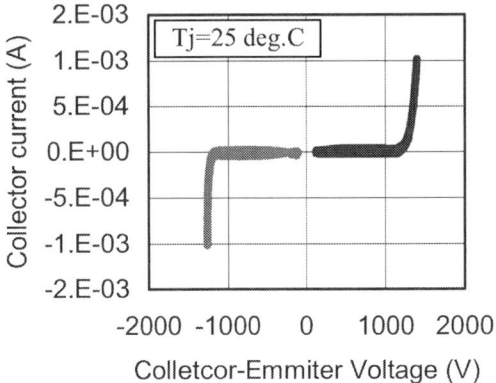

Figure 5: Bi-directional blocking characteristic of the new "Hybrid" RB-IGBT.

and reverse blocking abilities more than 1200V have been achieved

Figure 6 shows the I-V characteristics of our new 1200V RB-IGBT at V_{GE}=15V and Tj=125°C. As a reference, the corresponding characteristics of serially connected IGBT and diode of our 6th generation is also shown in comparison. At a rated current of 50A, the on-state voltage drop (Von) of RB-IGBT is 3.0V, while that of the serially connected pair is 3.9V.

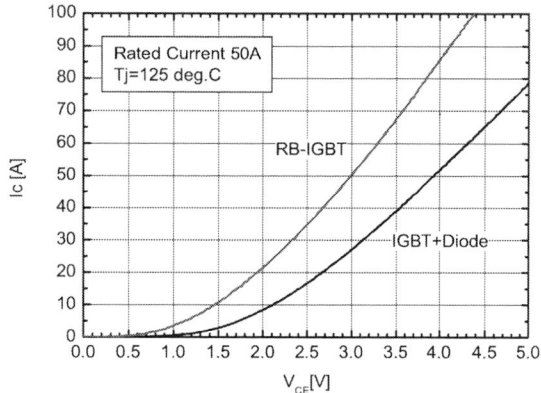

Figure 6: I-V characteristics.

Figure 7 shows the trade-off characteristics between on-state voltage drop (Vce(sat)) and turn-off energy(Eoff). E_{off} was measured at a bus voltage of 600V. It has been achieved the superior trade-off characteristics of RB-IGBTs.

Figure 8 shows the reverse leakage characteristics at V_{GE}=0V and Tj=125°C for our new RB-IGBT. Sufficient reverse blocking voltage is obtained for 1200V device at a low leakage current level.

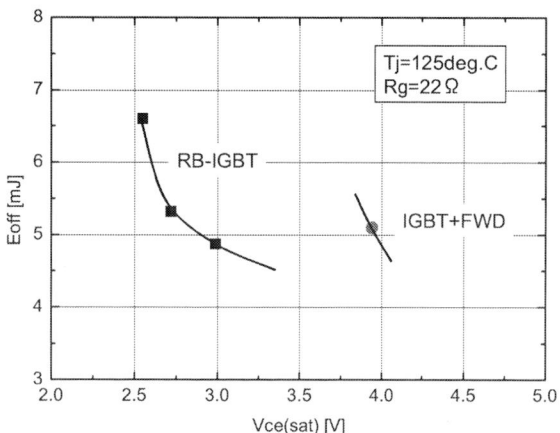

Figure 7: Trade-off relationships between Vce(sat) and Eoff.

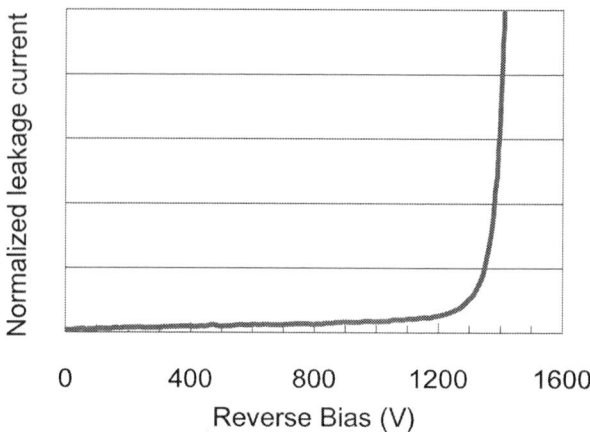

Figure 8: Leakage characteristics of the new "Hybrid" RB-IGBT at 125DegC.

V. NEWLY DESIGNED 3-LEVEL MODULE

Exterior view and the equivalent circuit of newly developed 3-Level module (one–arm) for A-NPC are shown in Figure 9 (a) and (b), respectively. Package size is 110mm ×80mm×30mm. The module incorporates all devices necessary for one-phase operation. It has two main switches, T1 and T2 of our 6th generation IGBT and FWD rated at 1700V/200A, and two anti-parallel connected clamping switches, T3 and T4 of RB-IGBTs rated at 1200V/200A.

Large stray inductance causes excessive voltage surge at the turn-off of IGBTs. The sum of stray inductance in packages and bus bar becomes more than 100nH, should the single-phase configuration of A-NPC 3-Level converter be formed with a standard 2-in-1 IGBT module, integrating T1 and T2, and a 1-in-1 bi-directional switch module, integrating T3 and T4. By integrating all devices for one-phase operation in single module, the stray inductance of each pair of main terminals, P, N, M, and U is made less than 40nH, which is similar to that for conventional 2-in-1 IGBT modules. It is realized by the optimization of the shape of terminals and by shortening the current passes between P-N, P-M and M-N.

Figure 10 shows the calculated power conversion efficiencies and power losses for four types of inverters, i.e. a 2-Level inverter, a diode-clamped NPC inverter, an A-NPC inverter with bidirectional switches shown in Figure 1(a), and an A-NPC inverter with RB-IGBTs. The operating conditions for these inverters are: a carrier frequency of 3 kHz, a DC-link voltage of 900V, and an output current of 100A. Electrical characteristics of our 6th generation V-series 1700V and 1200V IGBT are used in 2-Level inverter and diode-clamped NPC inverter. Stray inductance is assumed the same in both the NPC inverter and the A-NPC inverter.

As shown in Figure 10, power conversion efficiency of A-NPC 3-Level inverter using RB-IGBTs is 98.57%. Compared with 2-Level inverter, power loss reduction of A-NPC and NPC inverters are 31% and 15%, respectively.

978-1-4244-8425-6/11 $26.00 © 2011 IEEE

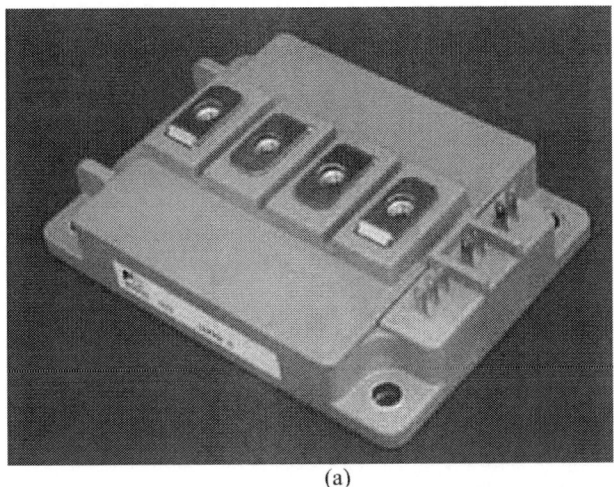

(a)

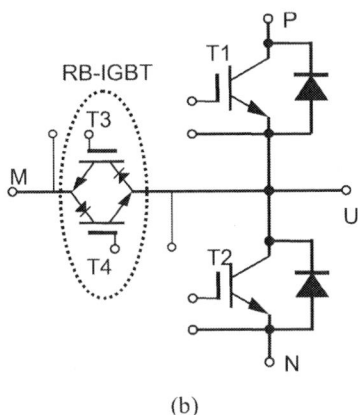

(b)

Figure 9: (a) The external view and (b) the equivalent circuit of A-NPC 3-Level modules (one–arm).

VI. CONCLUSIONS

We developed a new hybrid isolation process to overcome the problems associated with extra-long time thermal diffusion process, and applied it in the fabrication of a 1200V RB-IGBT. An advanced-NPC 3-Level inverter using our new 1200V RB-IGBTs as bidirectional switch, and 1700V IGBTs and FWDs as main switches, achieves its power conversion efficiency as high as 98.57%. Compared with those in 2-Level inverter and conventional NPC 3-Level inverter, power loss in A-NPC inverter has been reduced by 31% and 15%, respectively.

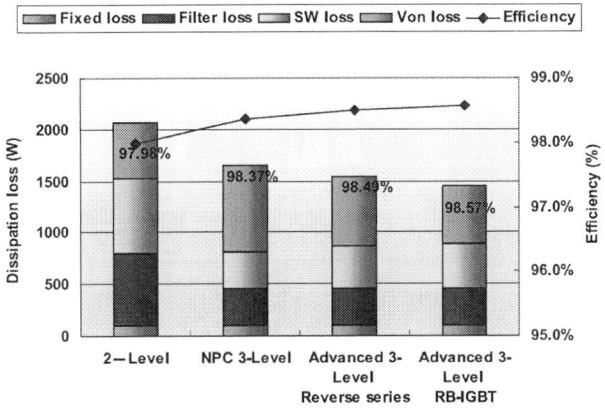

V_{DC}=900V,Io=100A,Fc=3kHz,Tj=125DegC

Figure 10: Power losses calculated for normal 2-Level, NPC 3-Level, A-NPC using reverse series connected IGBT and FWD, and A-NPC using RB-IGBT.

ACKNOWLEDGEMENT

The authors wish to thank the process development, and manufacturing staffs in Fuji Electric Systems Co., Ltd. for their invaluable support and helpful advices on device and module fabrication.

REFERENCES

[1] A.Nabae,I.Takahashi,H.Akagi,"A New Neutral-Point-Clamped PWM Inverter",IEEE Trans.on industrial applications,vol.1A-17,no.5,pp518-523,1981.

[2] "IGBT Power Modules for 3-level UPS Inverters", http://www.semikron.com/,July 2008.

[3] M. Takei, T. Naito and K. Ueno, "The Reverse Blocking IGBT for Matrix Converter with Ultra Thin Wafer Technology", ISPSD '03 proceedings, pp. 156-159,2003.

[4] T. Naito, M. Takei, M. Nemoto, T. Hayashi and K. Ueno, "1200V Reverse Blocking IGBT with Low Loss for Matrix Converter", ISPSD '04 proceedings, pp. 125-128,2004.

[5] N.Tokuda,M.Kaneda and T.Minato,"An ultra-small isolation area for 600V class Reverse Blocking IGBT with Deep Trench Isolation process(TI-RB-IGBT)",ISPSD'04 proceedings, pp. 129-132,2004.

[6] M.Yatsu,K.Fujii,S.Takizawa,Y.Yamakata,K.Komatsu,H.Nakazawa, Y.Okuma, "A Study of High Efficiency UPS Using Advanced Three-level Topolopy", PCIM'10 Europe, Proceedings, pp.550-555,2010.

[7] K.Komatsu,M.Yatsu,S.Miyashita,S.Okita,H.Nakazawa,S.Igarashi, Y.Takahashi,Y.Okuma,Y.Seki and T.Fujihira,"New IGBT Modules for Advanced Neutral-Point-Clamped 3-Level Power Converters",IPEC'10 proceedings, pp. 523-527,2010.

[8] K.Shimoyama, M.Takei, Y.Souma, A.Yajima,S.Kajiwara and H.Nakazawa,"A New Isolation Technique for Reverse Blocking IGBT with Ion Implantation and Laser Annealing to Tapered Chip Edge Sidewalls", ISPSD'06 proceedings, pp. 124-127,2006.

Proceedings of the 23rd International Symposium on Power Semiconductor Devices & IC's
May 23-26, 2011 San Diego, CA

Reduction of the temperature dependence of leakage current of IGBTs by field-stop design

H.-J. Schulze, S. Voss, H. Huesken, F.-J.
Niedernostheide

Infineon Technologies AG
D-81726 München; Germany

Abstract— **In this paper we propose a new method to reduce the temperature dependence of the leakage current of IGBTs by reducing the temperature dependence of the anode-side current gain α_{pnp}. The temperature dependence of α_{pnp} can be reduced by using field-stop zones that contain doping atoms with deep levels in the band gap of silicon. We demonstrate how the temperature dependence of the leakage current is influenced when using deep-level donors instead of shallow-level donors in the field-stop zone.**

I. Introduction

The temperature dependence of the anode-side current gain α_{pnp} of IGBTs determines the temperature dependence of the leakage current. The leakage current at high temperature is a limiting factor in cases of overload like a short circuit which may lead to device destruction. The temperature dependence of α_{pnp} is influenced by the emitter efficiency γ_E of the backside emitter and the transport factor α_T (without avalanche generation)

$$\alpha_{pnp} = \gamma_E \alpha_T . \qquad (1)$$

The backside emitter efficiency γ_E can be expressed as [1, 2]

$$\gamma_E = 1 - \left(\frac{N_B}{N_E} \right) \left(\frac{D_E}{D_p} \right) \left(\frac{W}{L_E} \right) \qquad (2)$$

N_B, N_E denote the dopant densities of the base and of the backside emitter, D_E the minority-carrier (electron) diffusion coefficient, D_p the hole diffusion coefficient, the W base width and L_E the minority-carrier diffusion length in the emitter.

The usual temperature dependence of α_{pnp} is positive due to the temperature dependence of the transport factor. When an impurity with a deep donor level – typically with an energy level > 100 meV below the conduction band – is used for doping the field-stop zone, the electrically active doping dose

of the field-stop zone increases with temperature. In conjunction with a doping of the backside emitter with a shallow level dopant (< 100 meV above the valence band) the temperature dependence of the ratio N_B/N_E is dominated by the temperature dependence of N_B as

$$\frac{N_B}{N_E}(T) = \left(\frac{N_B(T)}{N_E(T) \approx const.} \right) \propto N_B(T) \qquad (3)$$

The temperature dependence of the activated doping dose in the novel field stop can be used to reduce this positive temperature coefficient in a well-controlled way.

The field-stop doping profile is a major part of the base doping profile N_B. Thus, the temperature dependence of $N_B(T)$ is mainly determined by the temperature dependence of the concentration of the ionized deep donors $C_{ion}(T)$ of the field-stop doping, according to

$$N_B(T) \propto C_{ion}(T) . \qquad (4)$$

The temperature dependence of the concentration of ionized deep donors $C_{ion}(T)$ can be calculated for a deep level using the charge neutrality equation and Fermi-statistics as shown in earlier work [3, 4]. Figure 1 shows the activation ratio of ionized to neutral donors C_{ion}/C_o as a function of temperature for a deep donor with an energy level of 0.2eV below the conduction band. The temperature dependence of this ratio is shown for various doping concentrations C_{tot} which is the sum of C_{ion} and C_o. Figure 1 shows that the activation state of a deep donor changes significantly with temperature if the dopant concentration is high enough.

978-1-4244-8425-6/11 $26.00 © 2011 IEEE

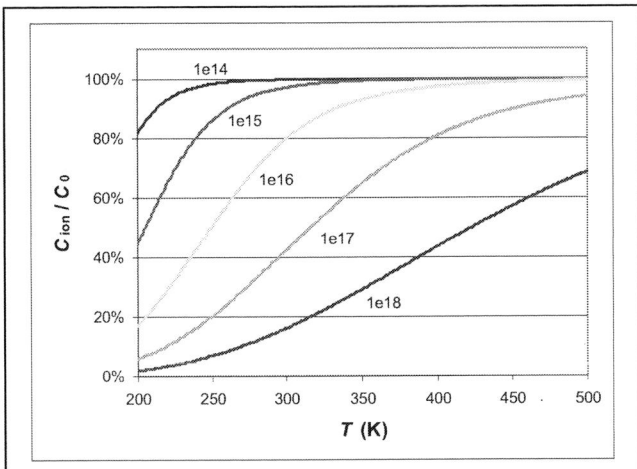

Figure 1. Activation ratio of ionized to neutral donors C_{ion}/C_0 as a function of temperature for a deep donor with an energy level of 0.2 eV below the conduction band. The parameter is the total doping concentration C_{tot} of the deep level which is the sum of C_{ion} and C_0.

II. RESULTS

We carried out leakage-current measurements at field-stop IGBTs that are rated for a breakdown voltage of 600 V. The leakage currents were measured at voltages between 100 V and 600 V in 100-V steps as a function of temperature. Three groups having field-stop zones with different donor types and different backside-emitter efficiencies were measured. The field-stop zone of IGBT1 was doped with a deep-level donor, whereas IGBT2 and IGBT3 were doped with a shallow-level donor. Typical donors for a shallow-level doping are phosphorous or arsenic with trap levels about 0.1 eV below the conduction band edge; typical deep donors are selenium or sulfur each with two donor levels of about 0.2 eV and 0.4 eV below the conduction band edge [5, 6]. Furthermore, the groups differ in the backside-emitter efficiency tuned by the respective doping dose of the p-emitter. IGBT1 and IGBT3 have an identical dose, whereas IGBT2 has a reduced dose and hence reduced emitter efficiency. Apart from that, the IGBTs of all groups have the same chip size, thickness and cell structure.

TABLE I. DESIGN PARAMETERS OF IGBTS

IGBT	backside-emitter dose	energy level of field-stop donor
IGBT 1	medium	deep
IGBT 2	low	shallow
IGBT 3	medium	shallow

The trade-off between on-state losses and turn-off losses are shown in Figure 2. IGBT1 and IGBT3 show identical switching losses at room temperature, the reduced emitter efficiency of IGBT2 is reflected in higher saturation voltage and lower switching losses. Hence, IGBT1 and IGBT3 have the same current gain at room temperature. A first indication of the effect of the energy level of the field-stop donor is given

by the different temperature behavior of IGBT1 and IGBT3: The current gain of IGBT3 is increasing more strongly with temperature, resulting in lower saturation voltage and higher switching losses at 175 °C compared to IGBT1. This was already demonstrated in an earlier work from our group [4].

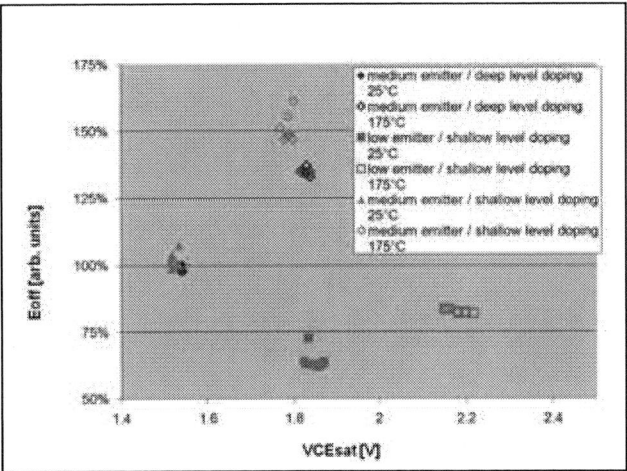

Figure 2. Trade-off curve of IGBTs under investigation. The devices were measured at a current density of 250 A/cm² for junction temperatures of 25 °C and 175 °C, respectively.

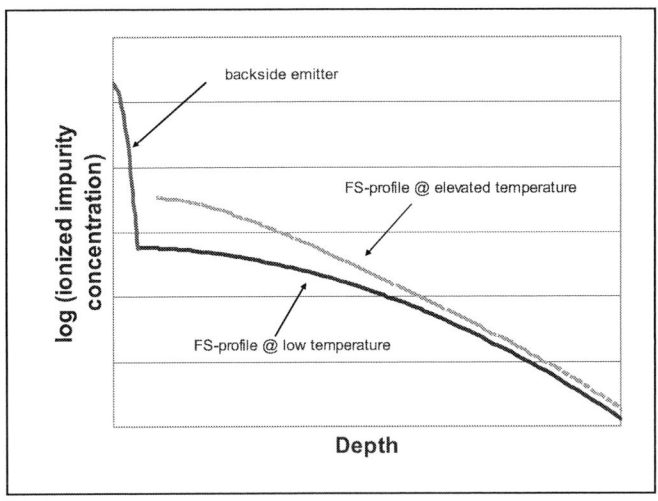

Figure 3. Schematic view of doping profiles for the backside emitter and the field-stop zone with deep-level donors. Whereas the concentration of ionized emitter doping atoms is constant with temperature, the concentration of the ionized impurities of the field-stop region increases with temperature as a function of concentration.

Typical doping profiles close to the collector side of the IGBTs with different field-stop doping levels and p-emitter concentration are shown schematically in Fig. 3.

Figure 4 shows the leakage-current distribution of the three IGBTs as a function of temperature. For each single device and voltage, the leakage current at a certain temperature is set in relation to the leakage current at 125 °C. The plot shows a probability plot of the natural logarithm of the leakage current. In this plot style, every data point is assumed to reflect a quantile of the distribution and is then assigned a sigma value, which relates this quantile value to a Gaussian distribution. A perfect Gaussian distribution would translate into a straight line under this transformation. As is easily seen, the leakage-current distributions follow a Gaussian trend quite well. The difference in temperature coefficient of the leakage current as a function of field-stop dopant is quite obvious. IGBT1 has significantly lower temperature dependency compared to IGBT2 and IGBT3, which have the same temperature coefficient despite strong variation in emitter efficiency.

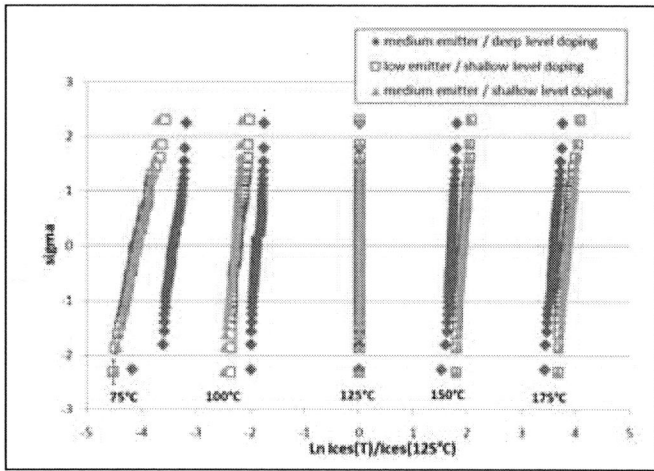

Figure 4. Distribution of leakage current levels (normalized to 125 °C) for a shallow-level doping and a deep-level doping of the field-stop zone as a function of temperature. Data points are transformed to a probability format.

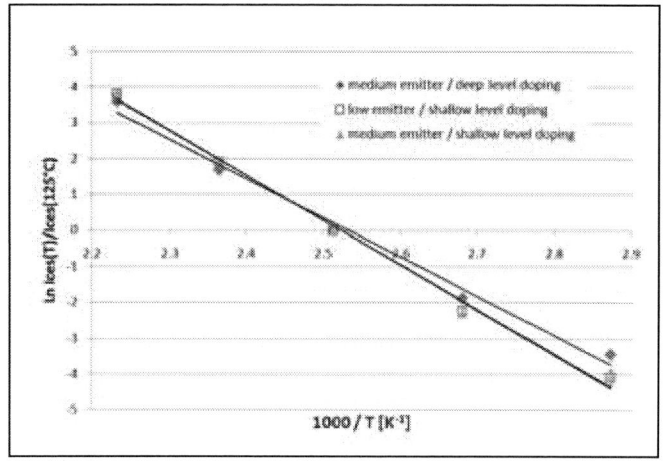

Figure 5. Leakage current (normalized to 125 °C) for a shallow-level doping and a deep-level doping of the field-stop zone as a function of temperature. Shown are mean values, comprising up to 48 data points (8 devices times 6 voltages) for all device groups. Typical variations are indicated as error bars. The temperature dependence of the leakage current is independent of the emitter doping concentration, confirming that the increase of leakage current is determined by the field-stop doping level.

In Figure 5 the logarithm of the normalized leakage current is shown as a function of reciprocal temperature. Plotted are the mean values of the distributions shown in Figure 4. The normalized leakage current follows an Arrhenius law. This leads us to the conclusion that the generation of the leakage current as a function of temperature follows the same mechanism. A more detailed inspection of the Arrhenius plot reveals that the slopes of IGBT2 and IGBT3 (deep donors) are reduced, i.e. the activation enthalpy determining the temperature dependence of the leakage current is reduced but independent from emitter doping concentration.

This effect is still more pronounced on condition of higher emitter doping concentration and higher field-stop doping concentration because the temperature dependence of the field-stop doping activation increases with increasing deep level donor concentration.

The effect of the temperature dependent field-stop doping concentration can also be used for an improved softness during turn-off of the IGBTs and for a reduced temperature dependence of the turn-off power losses. In Figure 6 it is demonstrated that for a field-stop zone doped with a deep donor the softness at room temperature can be significantly improved. In this case the donor doping concentration was chosen such that the softness during turn-off is much more pronounced at room temperature compared to higher temperatures due to the special temperature dependence of the transistor current gain for the case of field-stop zones with deep donors. The temperature dependence of the softness during turn-off can be minimized by a proper selection of the field-stop dose.

Furthermore, the temperature dependence of other relevant electrical parameter as e.g. the turn-on behavior and the short-circuit behavior can be significantly reduced.

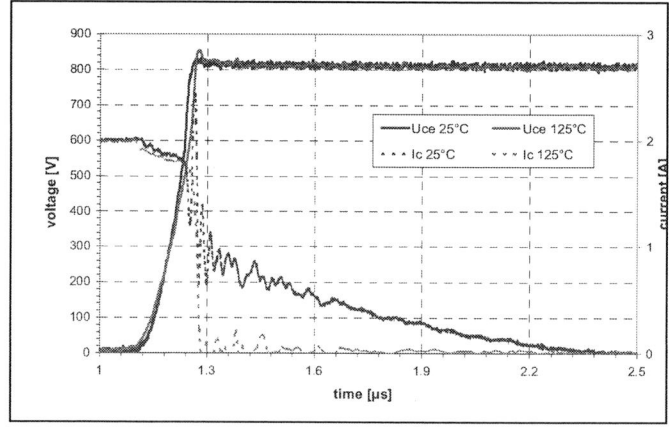

Figure 6. Switching curves of an IGBT with a field-stop zone doped with a deep level donor.

III. Summary

IGBTs with field-stop layers created by deep or shallow donor levels have been processed and their electrical behavior has been characterized. Our measurement results clearly reveal that deep-level field-stop IGBTs have a superior temperature dependence of the leakage current to shallow-level field-stop IGBTs. Furthermore, the analysis of the temperature dependence of the E_{off}-V_{cesat}-trade-off relationship and the switching behavior shows an excellent performance of the deep-level field-stop IGBTs.

References

[1] J.Lutz, Halbleiter-Leistungsbauelemente, ISBN: 978-3-540-34206-9, 2006 Springer

[2] Vinod Kumar Khanna, The Insulated Gate Bipolar Transistor (IGBT) Theory and Design, ISBN: 0-471-23845-7, © 2003 Institue of Electrical and Electronics Engineers, John Wiley & Sons, Inc., Publication

[3] S.Voss, N.A.Stolwijk, and H.Bracht, J.Appl.Phys., 92(8), 4809, (2002)

[4] S.Voss, H.-J. Schulze, F.-J. Niedernostheide, Proceedings of The 22nd International Symposium on Power Semiconductor Devices   ICs, Hiroshima 2010, pp.141-144

[5] P. Wagner, C. Holm, E. Sirtl, R. Oeder, and W. Zulehner, Adv. Solid State Phys. XXIV, 191, (1984)

[6] S.D.Brotherton, M.J.King, and G.J.Parker, J.Appl.Phys. 52, 4649 (1981)

Proceedings of the 23rd International Symposium on Power Semiconductor Devices & IC's
May 23-26, 2011 San Diego, CA

Electro-thermal instability in multi-cellular Trench-IGBTs in avalanche condition: experiments and simulations

M. Riccio, A. Irace, G. Breglio, P. Spirito, E. Napoli
Dept. of Biomedical, Electronic and Telecommunication Engineering
University of Naples Federico II
Naples, Italy
Email: michele.riccio@unina.it

Y. Mizuno
Toyota Motor Corporation
Kirigahora 543, Nishihirose,
Toyota 470-0309, Japan
Email: mizuno@yoshihito.tec.toyota.co.jp

Abstract—This paper reports on the results of a study on electro-thermal instability induced in multi-cellular Trench-IGBTs in avalanche condition. Experimental measurements, made on T-IGBTs, show possible inhomogeneous current distribution under Unclamped Inductive Switching (UIS) confirmed by transient infrared thermography measurements. Together with this, an analytical modeling of avalanche behavior has been included in a compact electro-thermal simulator to study the interaction between a large numbers of elementary cells of T-IGBTs forced in avalanche condition. Electro-thermal simulations qualitatively replicate the possible inhomogeneous operation observed experimentally. Finally a possible theoretical interpretation of the instability in avalanche condition for T-IGBT is given.

I. INTRODUCTION

In recent years power devices performances and reliability have become an extremely important target in industry and research sector. This is especially true for automotive applications as electrical (EV) or hybrid vehicles (HEV) or railway traction where the operating environment of power devices is harsh; so development has concentrated on securing ruggedness during high voltage and high current operations. Recent works have been presented where two-dimensional simulation of some tens of cells has been performed in UIS condition [1]. Anyway, where three-dimensional device simulations are needed, the limit imposed by convergence complexity limits the system to few elementary cells. In this case, to take into account electro-thermal feedback taking place in a large number of IGBT elementary cells, a new simulation approach is needed together with an extensive experimental analysis to verify possible instability in avalanche condition.

Previously, it has been reported that, during UIS transient, IGBTs can be prone to inhomogeneous operation [2]. Therefore, this study examined IGBT behavior under UIS conditions by means of infrared thermal analysis and compact electro-thermal simulations.

II. EXPERIMENTAL IR ANALYSIS

To get a deeper insight about possible electro-thermal instability in avalanche condition, that causes inhomogeneous

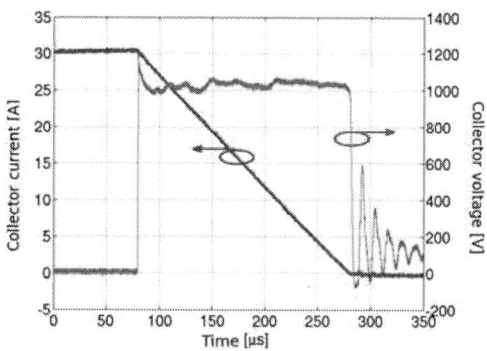

Fig. 1. Electrical waveform for an UIS test with $I_{MAX} = 30A$ and $L = 4.7mH$.

current operation, a detailed analysis of the behavior in UIS test for a given device family has been done by using an ultra-fast IR thermography system [3] and detecting in the same time the electrical current and voltage waveforms. The tested devices are Trench-IGBTs with $900\ V - 1000\ A$ peak current capability. A 4.7 mH load inductor has been used as a reasonable compromise between electrical and thermal stress during the UIS test discharging time.

In what follow thermal maps obtained on two sample at the maximum rated UIS currents will be reported. These results can be considered representative for the used device family because has been found that electro-thermal behavior during UIS transient are qualitatively similar for all the tested devices.

A. Measurement results

In Fig. 1 the electrical waveform for an UIS test are reported. The V_{CE} voltage waveform during the UIS phase presents some oscillations of the sawtooth type superposed to the V_{BR} level, that increase both in amplitude and in frequency as the UIS current increases.

Fig. 2 shows the normalized temperature increase maps $\Delta T/\Delta T_{MAX}$ taken on the first sample at different times from the turn-off of the T-IGBT (i.e. *from the starting of UIS phase*)

978-1-4244-8425-6/11 $26.00 © 2011 IEEE

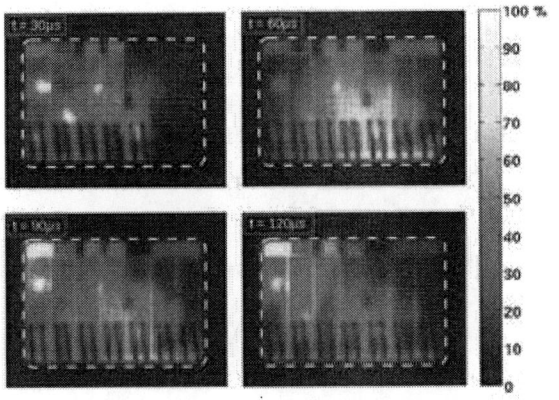

Fig. 2. Normalized increase temperature maps taken on the first device for 30 Amps UIS test at different times (indicated on the maps) from the turn-off of the IGBT.

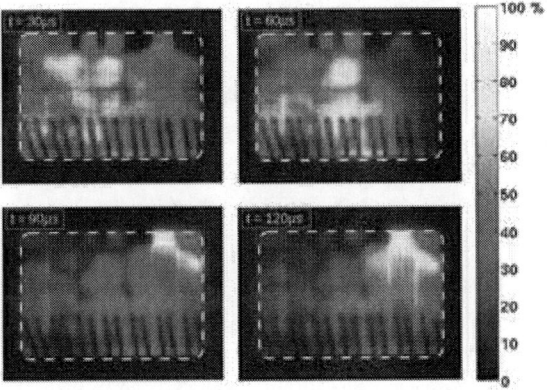

Fig. 3. Normalized increase temperature maps taken on the second device for 30 Amps UIS test at different times (indicated on the maps) from the turn-off of the IGBT.

for peak UIS current $I_{CMAX} = 30\ A$.

In these IR maps the dynamics of the heating along the sample can be noted: after $30\mu s$ is mainly located on three small areas, and afterwards extends over the center part of the chip area ($t = 60\mu s$). At the time instant $t = 90\mu s$ and $t = 120\mu s$ the current moves again in restricted areas proving an inhomogeneous avalanche operation.

The thermal maps in Fig. 3 refer to the second device tested at the same current and load inductor. It can be noted an initial increase in temperature with quite uniform distribution along the chip, followed by a creation of an hot spot in a restricted area, that shrinks furthermore in subsequent times ($t = 90\mu s$ and $t = 120\mu s$).

The demonstration of an increasing temperature (and current) focalization during UIS, strongly suggest the effect of a regenerative mechanism, by a negative resistance region, taking place during the UIS phase, that could lead to a failure. This point will be investigated using electro-thermal simulations of the UIS transient and the relevant mechanism will be more clearly assessed, as discussed in the following.

III. ELECTRO-THERMAL SIMULATION

As said before, where three-dimensional simulation of multi-cellular structured device is needed, the limit imposed by convergence complexity limits the system to few elementary cells.

A valid alternative, used in this work, is to describe the single cell with an accurate compact one-dimensional electrical model and a three-dimensional thermal model [4]. Then the multicellular T-IGBT can be treated as a number of equivalent single cells all of them sharing the same collector and gate voltages and with the emitters connected through a metal layer with a finite resistivity.

A. IGBT compact model

The electrical description of the single cell is a one-dimensional improvement of the Hefner's model [5], [6] with the addition of the parasitic bipolar *npn*-transistor and a better description for the carrier mobility and lifetime in epi-region. The obtained on-state model has been calibrated to fit the experimental characteristics of the same device used previously for thermal analysis [7].

1) Avalanche modeling: Analytical modeling of avalanche behavior has also been included in the compact model.

Considering that the breakdown voltage for an IGBT is determined by the open-base, collector-emitter breakdown voltage of the bipolar transistor, the BV_{CEO} is reached when the product of the carrier multiplication factor and the common base current gain approach unity. Some expressions, apart the simple Miller relation, have been proposed in literature for multiplication factor. In this work we have used an analytical expression for the multiplication factor M proposed by Spirito in [8]. The complete expression is:

$$M = \frac{1}{1 - \frac{V_{CE}}{BV_{CBO}}\exp\left[\frac{A}{\sqrt{N_B'BV_{CBO}}}\left(1 - \sqrt{\frac{BV_{CBO}}{V_{CE}}}\right)\right]} \quad (1)$$

where A is a constant depending only from the used semiconductor and N_B' is the sum of N_B (doping in the epi-layer) and N_{sat} that accounts for the velocity saturation effect responsible of the space-charge concentration in the base-collector depletion region.

This avalanche model has been calibrated using a single cell TCAD simulation of the same device used to calibrate the on-state model. This step has been very critical for the presence of a negative resistance region in the blocking I-V curve ($V_{GE} = 0V$) with a subsequent "S-shaped" trend.

Considering the influence of the space-charge concentration in the base-collector depletion region, a little modification has been proposed to the avalanche multiplication model described before. Since the device to model has a trench structure it is reasonable that the particular shape of the electrical field near the collector-base junction leads to a non uniform distribution for the generated electron-hole pairs. To take into account this phenomenon has been proposed to introduce a fitting parameter α to reduce the area used to calculate the amount

978-1-4244-8425-6/11 $26.00 © 2011 IEEE

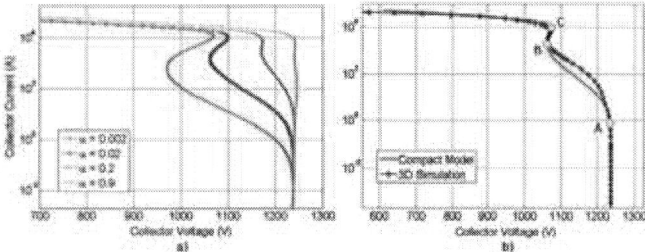

Fig. 4. I-V output curve for different α values and comparison between I-V curve simulated with 3D software package and proposed compact model.

of charge-carrier caused by carrier velocity saturation. Then an "effective area" is used instead of the junction area:

$$N_{sat} = \frac{I_C}{qA_{eff}v_{psat}} - \frac{I_{mos}}{qA_{eff}v_{nsat}} \qquad (2)$$

where

$$\begin{cases} A_{eff} = \alpha A \\ 0 < \alpha \leq 1 \end{cases} \qquad (3)$$

In Fig. 4-a) the impact of α on the I-V curve for different values is reported . As expected larger α implies more extended negative resistance region. A key point is that α, in some way, take into account the device geometry and change drastically the peak and valley points of the reverse I-V curve.

Using the simulated I-V curve obtained with a TCAD, the model parameters has been extracted to have the best fit. Fig. 4-b) shows the calibration result. The good result is clearly visible. The two curves have the same shape and the model predict with high accuracy the most important points: peak, valley and latch-up (named A, B and C respectively on picture). The negative resistance is also present with a good approximation on the slope.

B. Avalanche simulation results

During UIS transient the DUT is current forced by the load inductor and the electrical solution for each cell is coupled with that of other $n - 1$ cells. At each time step, the implemented simulator solves the following electrical problem:

$$\sum_{i,j=1}^{n_x,n_y} I_{i,j}\left(V_{CE}\left(i,j,t_k\right),T\left(i,j,t_k\right)\right) = I_L\left(V_{CE}\left(i,j,t_k\right)\right) \qquad (4)$$

In what follows two dynamic avalanche simulations on T-IGBT will be presented. The scope of this section is then to investigate the involved phenomena and to analyse the correlation between electro-thermal instability and interplay between a large number of elementary cells composing the real device.

The presented simulations have been performed at high current to take into account also the limit imposed by the latch-up of the parasitic thyristor. To this purpose, $I_{MAX} = 100\ A$ (first and second case), $L = 4.7\ mH$ and $T_0 = 300\ K$. The electro-thermal simulation starts when the T-IGBT is turned-off during UIS tests. The initial conditions are exactly the same

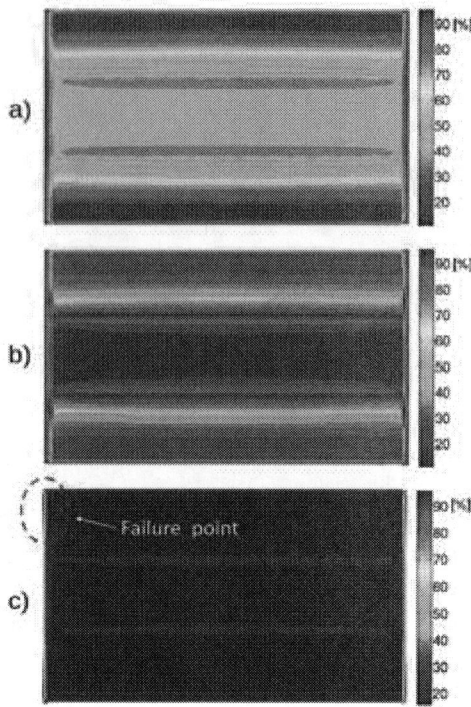

Fig. 5. Normalized collector current density at different time instant: a) $11\mu s$; b) $13\mu s$; c) $15\mu s$.

for all cells (equal current and temperature). The simulated structure is composed by a reduced layout divided into 50×80 all identical macro-cells. In this way it is simple to identify instable mechanisms to unbalance the initial condition. Fig. 5 reports, for the first case, the normalized collector current density at different time instants.

From the current density maps we can see that increasing the time, the current initially distributed on the device area, begin to be constricted in the periphery away from the two rows of the emitters PAD ($t = 11\mu s$). The current density in that reduced area grows and the relative temperature increases, so that the current moves to the device center ($t = 13\mu s$). Considering the high current constriction and the temperature dynamic, after $15\mu s$ the device fails with a single cell goes into latch-up status. This simulation result clearly shows inhomogeneous device operation during avalanche phase and a current filament development. Obviously the ideal structure and layout symmetry make the electro-thermal dynamics quite different compared to the experimental case. To better see what happens when the discharging transient starts; Fig. 6 shows the collector current density in three cells. The results are useful to explain how the current, initially equally distributed over all the device area, is constricted in a reduced area as consequence of the electro-thermal instability induced by negative resistance region in the I-V curve. Moreover in Fig. 6 it is reported the superimposition of the T-IGBT I-V curve and the trajectory of the same three points during the transient. This diagram clearly shows the bifurcation in the I-

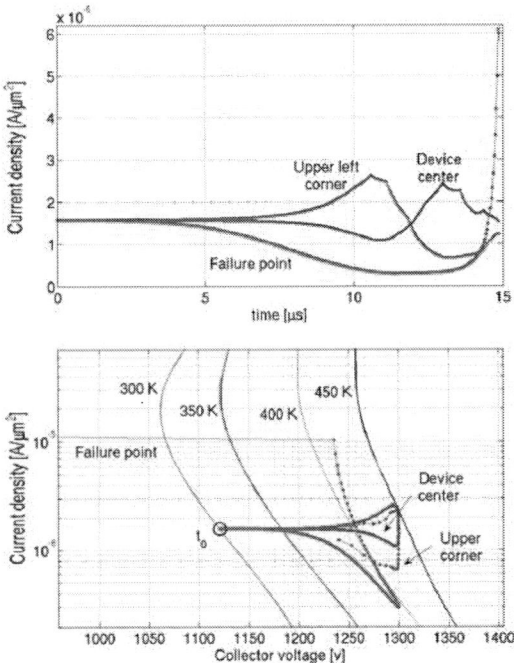

Fig. 6. Collector current density during time in three cells and relative trajectory into the I-V plane.

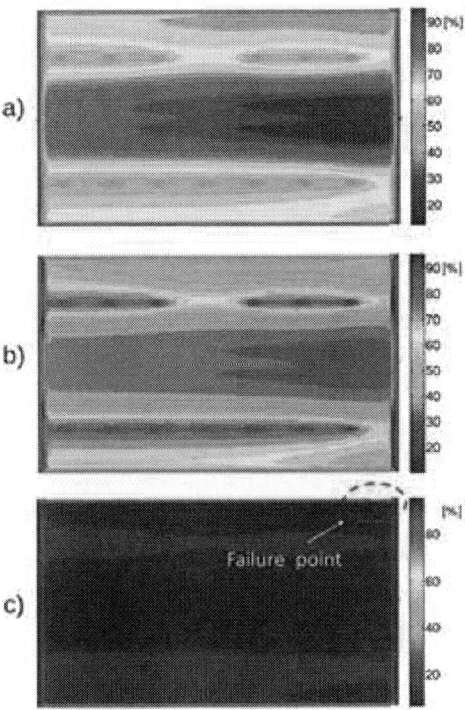

Fig. 7. Normalized collector current density at different time instant: a) $13\mu s$; b) $15\mu s$; c) $17\mu s$.

V plane confirming the instability mechanism. For the second simulation the emitter pads has been reduced in number and asymmetrically arranged to introduce a little unbalance in the device layout. Using the above structure, we were able to see the movement of the current from an area to another in a time scale of tens of μs (Fig. 7), quite similar to the previous simulation, but with a more realistic shape compared to the experimental thermal maps.

For both simulations the dynamic that governs current movement is affected by the device layout and parasitic parameters. Anyway this phenomenon can exist only if an electro-thermal instability holds a positive feedback between thermal and electrical dynamics.

IV. CONCLUSION

This paper has presented IR experimental analysis and compact electro-thermal simulations of T-IGBTs operation under UIS conditions. This experimental analysis proofs that the oscillations seen on the voltage waveform can then be attributed to the temperature increase of the reduced device area conducting the whole current. The lateral movement of the current form the heated area to a new one (that is at lower temperature) changes dynamically the current distribution within the device. The period of these oscillation is linked to the thermal time constant that controls the temperature increase in the filament: this latter could change depending on the conducting area.

The simulation performed with a multi-cellular reduced structure confirm the hypothesis on the possible electro-thermal instability induced by a negative resistance branch on the output I-V curve of the single cell. The used analytical avalanche model is accurate to take into account some relevant phenomena involved into avalanche operation. The simulations results also suggest that the variation in breakdown voltage can not be the main mechanism that will lead to a inhomogeneous current distribution.

REFERENCES

[1] Y. Mizuno, R. Tagami, and K. Nishiwaki, "Investigations of inhomogeneous operation of IGBTs under Unclamped Inductive Switching Condition," in *Proc. International Symposium on Power Semiconductor Devices & ICs (ISPSD)*, 2010, pp. 137–140.

[2] C.-C. Shen *et al.*, "Failure Dynamics of the IGBT During Turn-off for Unclamped Inductive Loading Conditions," *IEEE Transactions on Industry Applications*, vol. 36, pp. 614–624, 2000.

[3] M. Riccio *et al.*, "An equivalent-time temperature mapping system with a 320×256 pixels full-frame 100 khz sampling rate," *Review of Scientific Instruments*, vol. 78, pp. 106 106–3 pages, 2007.

[4] A. Irace, G. Breglio, and P. Spirito, "New developments of THERMOS³, a tool for 3D electro-thermal simulation of smart power MOSFETs," *Microelectronics Reliability*, vol. 47, pp. 1696–1700, 2007.

[5] A. R. Hefner, "Analytical modeling of device-circuit interactions for the power Insulated Gate Bipolar Transistor (IGBT)," *IEEE Transaction on Industry Applications*, vol. 26, pp. 995–1005, 1990.

[6] ——, "A dynamic electro-thermal model for the IGBT," *IEEE Transactions on Industry Applications*, vol. 30, pp. 394–405, 1994.

[7] M. Riccio, M. Carli, L. Rossi, A. Irace, G. Breglio, and P. Spirito, "Compact electro-thermal modeling and simulation of large area multicellular Trench-IGBT," in *27th International Conference on Microelectronics*, 2010, pp. 379–382.

[8] P. Spirito, "An Analytical Expression for the Multiplication Factor M in Semiconductors with Equal Ionization Coefficients," *IEEE Transactions on Electron Devices*, vol. 8, pp. 951–953, 1972.

Proceedings of the 23rd International Symposium on Power Semiconductor Devices & IC's
May 23-26, 2011 San Diego, CA

On Chip ESD Protection of 600V Voltage Node

Vladislav A. Vashchenko[1], Antonio Gallerano, Andrei Shibkov[2]

National Semiconductor Corporation
2900 Semiconductor Drive, M/S E-155, Santa Clara, CA 95052

[1] *With National Semiconductor at the time of first abstract submission*

[2] *Angstrom Design Automation*

Abstract— **This study presents for the first time ESD protection solutions in integrated silicon process technologies for the voltage range up to 600V. The ESD protection clamp is implemented using a NLDMOS-SCR type ESD device architecture. The study presents both reversible triggering I-V characteristics suitable for package level ESD protection as well as dependence of the ESD device characteristics upon the structure parameters and the state of a control electrode.**

I. Introduction

High voltage integrated process technologies have been developed and become commercially available recently. These integrated processes are expected to facilitate progress in LED lighting, motor drivers, solar energy, galvanic isolation and factory automation due to a higher level of circuit design integration. An example of a photovoltaic circuit is presented in Fig 1. In this case, the dc voltage level generated by the solar panels needs to be converted to the standard ac current of the grid with the key components being 4 discrete devices used in the T-bridge with a voltage tolerance over 600V. Solutions for on-chip ESD protection in integrated silicon process technologies are typically designed for voltage tolerances below 250V. Thus ESD protection for voltage above the 600V range is unknown. In spite of possible expectations that Ultra High Voltage (UHV) pins may not require ESD protection the package level spec for such ICs still needs to be met.

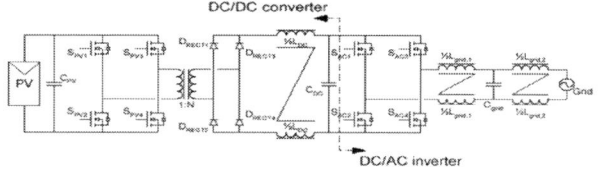

Fig. 1 Power circuit of the photovoltaic inverter with Ultra high Voltage components

II. Device and Experiments

A. Ultra High Voltage Foundry Process

The output characteristics of a 500V UHV NLDMOS obtained with Transmission Line Pulse (TLP) are shown in Fig. 2. The measurements at a gate bias $V_{GS}= 0V$ show the pulsed triggering voltage (VT1) over 600V. At higher gate bias the triggering voltage decrease is observed due to avalanche-injection instability in the parasitic bipolar structure. In the ESD time domain this phenomenon is a major limiting factor for the device pulsed Safe Operating Area (SOA) (Fig. 2). The failure mechanism is local overheating due to current filament formation [1, 2]. The architecture explored for the UHV ESD protection device was based upon the conventional NLDMOS "race track" layout with UHV pad in the middle. The additional p-emitter region has been added. Various options for the layout have been studied to implement different level of p-emitter isolation. In particular four separate p+ regions have been inserted in the central drain n+ diffusion to serve as emitters for the parasitic PNP formed with the body of the UHV NLDMOS, using the drain area as base (Fig.3, a).

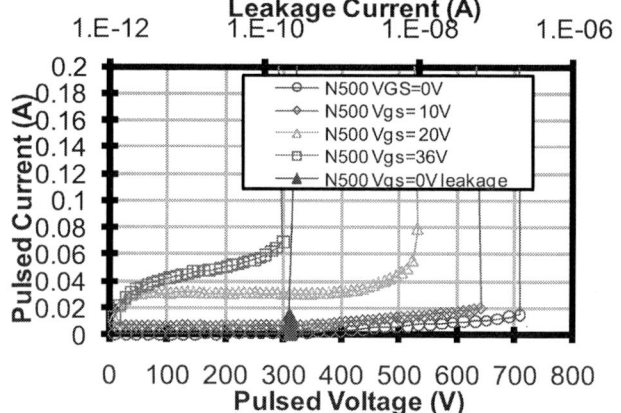

Fig. 2 TLP characteristics of a typical 500V device at different gate bias

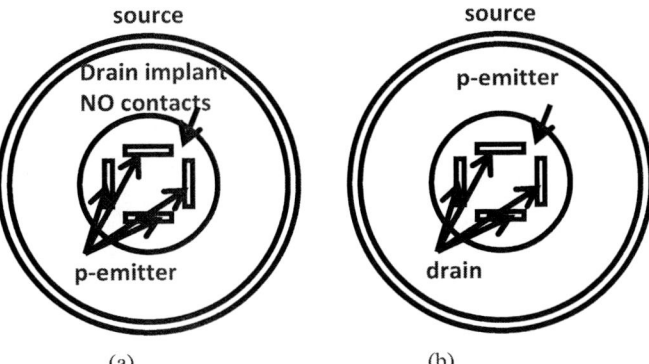

Fig. 3 Simplified layout view of a UHV NLDMOS-SCR for UHV voltage tolerance

The ESD device so constructed combines the parasitic NPN with an additional parasitic PNP structure formed by p-emitter, n-drift and p-body region, acting as collector. An internal silicon controlled rectifier structure is thus formed transforming the NLDMOS into a NLDMOS-SCR ESD device. In transient operation with channel and avalanche current the forward bias conditions of the p+ emitter junction are achieved and the structure turns on into a double injection conductivity modulation mode [2]. The device triggering characteristics are dependent upon the level of p-emitter isolation and the level of the transient gate coupling. The last effect can easily be observed in the measurements of the NLDMOS-SCR triggering characteristics at different constant gate bias (Fig. 4). According to the data in Fig.4, for the high level of p-emitter isolation no reversible snapback operation is observed. It is worth noting that while the modified device has the same width as the original UHV NLDMOS, carrier carrying capability is greatly enhanced by the conductivity modulation action introduced by the p+ emitter even at relatively low voltages. The current level at which the device fails (IT2) increases monotonically with the gate voltage and thence the amount of channel current.

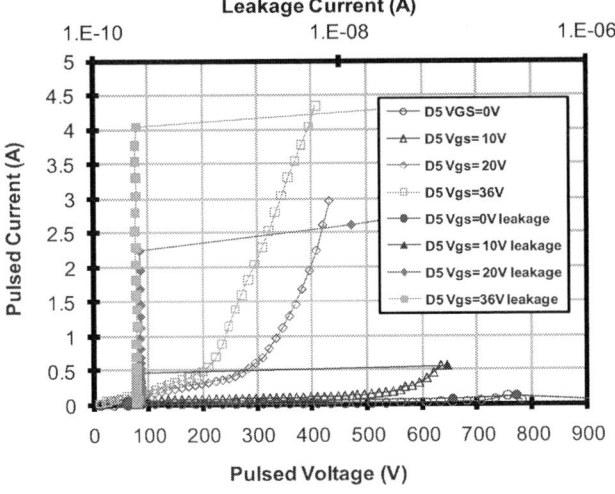

Fig. 4 TLP Output characteristics of the UHV NLDMOS-SCR with the layout design shown in Fig. 3, a

The irreversible triggering characteristics of the device indicate that the maximum current for a 50 Ohm TLP load is above the physical maximum level of the 700um wide device. Alternative design techniques may involve a high side reference, an isolated avalanche diode for example, in order to increase the transient gate voltage during the triggering according to Fig. 4 [3].

An attempt to reduce p-emitter isolation by elimination of the floating drain region (Fig. 3, b) resulted in significant reduction of the voltage tolerance due to p-substrate to p-emitter punch through effect at the level of ~400V.

In summary: a first attempt for straightforward implementation of a NLDMOS-SCR device architecture in a foundry process was essentially unsuccessful, demonstrating the complexity of the ESD device design.

Nevertheless, proof of concept of a UHV tolerant ESD device has been explored using a different UHV process with lateral superjunction and mixed-mode numerical simulations results are presented in the following sections.

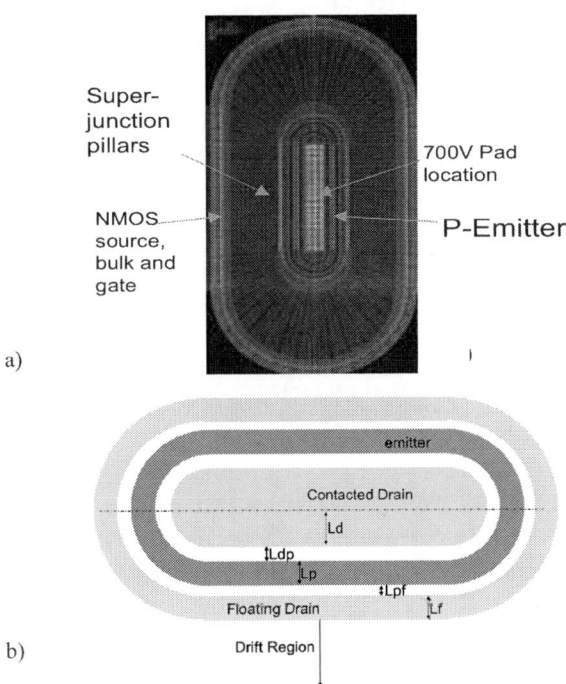

Fig. 5 Layout view (race track style) of a UHV SJ nLDMOS-SCR (a) and device parameter definition (b)

B. NLDMOS-SCR in Research Stage Super Junction Process

High Voltage processes using multi-RESURF or super junction (SJ) techniques have become increasingly more popular in recent years as alternative to conventional RESURF techniques [4]. In this process the lateral super-junction drift region is formed by lightly doped n- and p-pillar regions to achieve optimal breakdown voltage level and on-state resistance [5]. The layout of a super junction UHV NLDMOS modified to become an NLDMOS-SCR is shown in Fig. 5, a. Similarly to the conventional UHV NLDMOS-SCR devices, described in the previous section, the central drain region has been partially transformed into a p+ emitter.

The TLP characteristics for different layout options with corresponding parameter variations (Fig.5, b) show reversible snapback operation indicating full turn on of the parasitic SCR structure (Fig. 6). The high maximum on-state current IT2 is above 12A. Unlike the conventional UHV NLDMOS SCR, high IT2 is achieved in the grounded gate configuration.

It has been experimentally shown that the triggering voltage can be changed by varying the layout parameters of the drain/p-emitter region (Fig. 6) or the length of the drain drift region (Fig. 7). A wide VT1 range, from around 300V to 520V is obtained with the layout options described in Table 1. The large VT1 variation is the result of different degrees of electrical isolation achieved around the p+ emitter. The device with the lowest VT1, V07 has the least amount of isolation since the p+ emitter directly faces the depletion region developed in the off state and will need the least amount of current and therefore voltage to become forward biased with respect to the surrounding n drain region. In device V06 an n+ donut doped region is added around the p+ emitter with the effect of retarding the establishment of p+ emitter forward bias. In V02 the p+ emitter region width is reduced compared to V06 allowing for a smaller voltage drop from the point of highest potential at the center of the structure. Finally in V03 the longest n+ doped floating drain region length *Lf* provides the highest VT1 due to higher emitter isolation.

With further p+ emitter isolation, introduced by increasing the floating region length *Lf* to 15 and 20μm, a destructive snapback operation for *Lf* ≥ 15um is observed, similarly to the reference UHV NLDMOS devices.

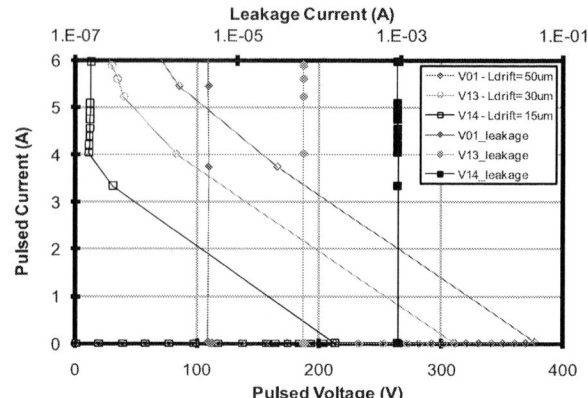

Fig. 7 TLP characteristic for devices with different drift region length *Ldrift*

III. Mixed-Mode Simulation Analysis of the ESD Device Operation

Mixed mode analysis of the device operation has been completed using a new mixed-mode simulation tool [6, 7]. Instead of process simulation the parameterized device cross-sections have been created by defining the structure regions, mesh and analytical doping profiles (Fig. 8). The simulations were performed for different doping profiles, geometrical and circuit element parameters. The device characteristics have been analyzed using both simple transient voltage ramp (Fig. 9, a) as well as HBM pulse circuit (Fig. 9, b).

TABLE 1 SUMMARY OF THE DRAIN/P+ EMITTER GEOMETRIES AND THE CORRESPONDING VT1 AND HOLDING VOLTAGE VALUES. THE MEANING OF THE PARAMETERS IS EXPLAINED IN FIG. 5

Device Name	Lf (um)	Lpf (um)	Lp (um)	Vt1 (V)	Vsp (V)
V07	0	0	10	301	27.1
V06	5	0	10	360	33
V02	5	0	5	438	33
V03	10	0	5	535	39

Fig. 6 TLP Characteristics of the UHV SJ NLDMOS-SCR for various drain/p+ emitter layout options

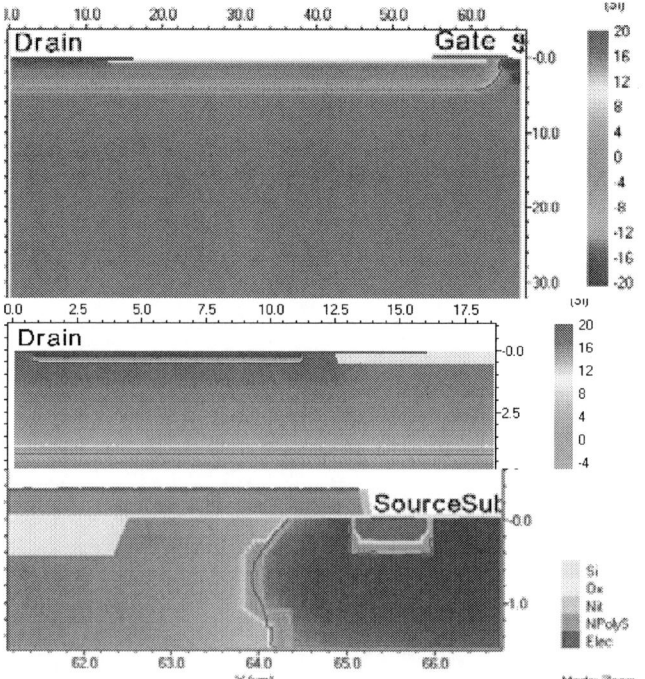

Fig. 8 Full cross-section of the parameterized NLDMOS-SCR device and zoomed view of the drain and source regions

978-1-4244-8425-6/11 $26.00 © 2011 IEEE

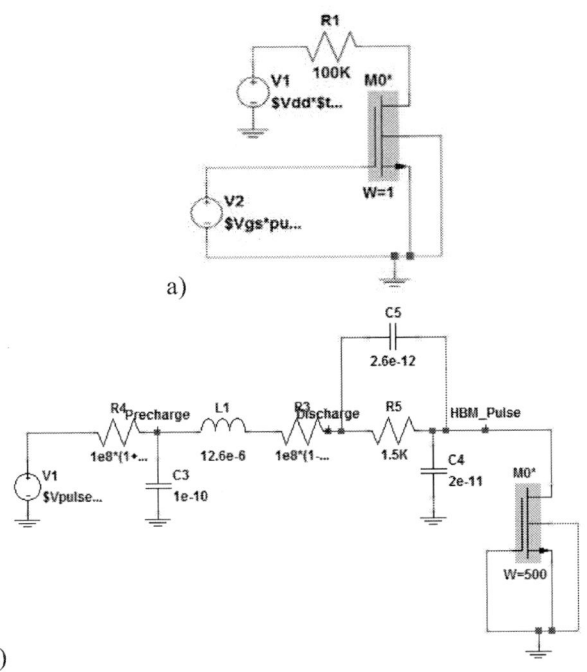

a)

b)

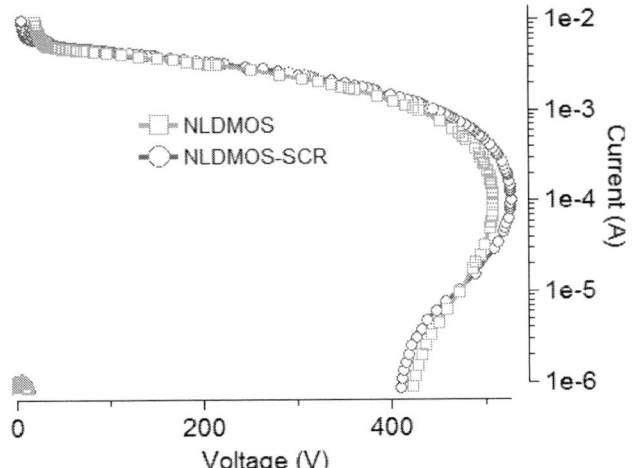

Fig. 9 Circuit editor views for interactive mixed-mode circuits used for the transient curve trace of the breakdown characteristics and HBM pulse simulations

Comparison of the breakdown characteristics of simplified 400V versions of NLDMOS and NLDMOS-SCR with reduced drift region are presented in Fig.10. In this structure for simplicity metal field electrodes where not taken into account. It is shown that the most critical part of the NLDMOS-SCR is P-emitter isolation resulting in reduction of the breakdown voltage at low current level. At the same time due to the floating drain region the critical voltage for negative differential resistance is higher in case of NLDMOS-SCR (Fig.10). Comparison of voltage waveforms for a 2kV HBM pulse shows that the NLDMOS-SCR provides adequate voltage waveforms for the ESD protection of the UHV pins (Fig.11).

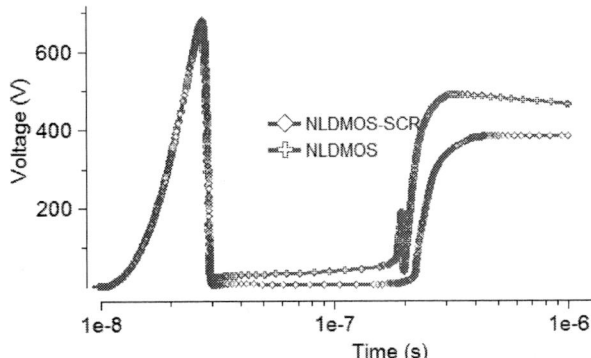

Fig. 11 NLDMOS and NLDMOS-SCR transient voltage waveforms for a 2kV HBM zap

IV. Conclusions

In this study the viability of the NLDMOS-SCR as ESD clamp solution for UHV applications has been demonstrated. Conventional drain extension techniques can be employed to achieve the desired ESD robustness level. Reversible snapback characteristics with high IT2 have been obtained for NLDMOS-SCR manufactured in Super Junction process technology for the transient triggered clamps with grounded gate configuration. The experimental results have been further validated by mixed-mode numerical simulation.

References

[1] M. Mergens et al, "Analysis of lateral DMOS power devices under ESD stress conditions," IEEE Trans. On Elec. Dev., vol. 47, no. 11, 2000, pp. 2128–2137.

[2] V. Vashchenko and V. Sinkevitch,"Physical Limitations of Semiconductor Devices," Springer, 2010.

[3] N. Olson, V. Vashchenko, E. Rosenbaum and P Hopper,"Small footprint trigger voltage control circuit for Mixed-Voltage applications," EOS/ESD Symp., 2008, pp. 196-203

[4] J. Appels and H. Vaes," High Voltage thin layer devices (RESURF devices)," IEDM, 1979, pp. 238-241

[5] S. Nassif-Khalil, L. Hou and C. Salama,"SJ/RESURF LDMOST," IEEE Trans. On Elec. Dev., vol 51, no 7, 2004, pp. 1185-1191

[6] DECIMMTM User Manual, Angstrom Design Automation, 2010. www.angstromda.com

[7] V.A. Vashchenko, A.A. Shibkov, ESD Design for Analog Circuits, Springer, (2010), www.analogesd.com

Fig. 10 Mixed-mode circuit for the curve trace of the breakdown characteristics and HBM pulse mixed-mode simulation

Proceedings of the 23rd International Symposium on Power Semiconductor Devices & IC's
May 23-26, 2011 San Diego, CA

CSTBT[TM](III) having wide SOA under high temperature condition

Yusuke Fukada, Kenji Suzuki, Tetsuo Takahashi, Tatsuo Harada, Hidenori Fujii
Shinichi Ishizawa, Junichi Yamashita, John F Donlon* and Tomohide Terashima
Power Device Works, Mitsubishi Electric Corporation,
* Powerex, Inc USA
1-1-1 Imajukuhigashi Nishi-Ku Fukuoka-City 819-0192 JAPAN
E-mail: Fukada.Yusuke@ap.MitsubishiElectric.co.jp

Abstract— **This paper presents high temperature performance of CSTBT[TM] (III) and its main parameters. The key for high temperature operation is suppressing the parasitic NPN transistor action. N+ emitter width, P+ diffusion layer depth and gate oxide thickness are main parameters for suppressing the parasitic action. The optimized 1200V CSTBT[TM](III) succeeded in 200°C operation without any thermal runaway or turn-off failure.**

I. INTRODUCTION

Recently, continuous improvement of power devices has become not only issues for customer requirements but responsibilities for all of us in the world because of the conservation of fossil fuels and the sustainable economic development points of view.

IGBT, as the key device of power chip, has more room for current density and operation temperature range.

As one of the solution for these requirements, we have already proposed the concept of CSTBT[TM](III) which has finer cell pitch and retrograde doping profile on carrier stored layer(CS-layer) [1-3]

In this report, we evaluate the effects of dominant structural parameters for high temperature operation of CSTBT(III) in 1200V class, and the optimized CSTBT(III) demonstrate superior performance.

II. DEVICE STRUCTURE

Fig.1 shows the schematic structure of CSTBT(III). CSTBT(III) is fabricated by using LSI fine-pattern technology for improving the tradeoff relationship between on-state voltage (V_{CE}(sat)) and turn-off loss (E_{SW}(off)).

CS-layer formed just below p-base region enhances hole concentration, and the effect increases with narrowing the trench pitch. In addition, CS-layer has retro grade profile by using high energy implantation, which suppresses the affection on MOS channel doping profile.

In order to further improve the performance especially for high temperature condition, suppressing the parasitic NPN action which causes latch-up phenomenon is most effective measure. We found three main parameters throughout

evaluation of device structures and characteristics. The parameters are as follows.

(A) N+ emitter layer width.
This parameter effects resistance just below N+ emitter layer. The finer N+ emitter pattern can be used to reduce the gain of parasitic NPN-transistor.

(B) Depth of P+ diffusion layer.
P+ diffusion layer also influences resistance of P-base layer near the n+ emitter. We examine the depth of P+ diffusion to optimize the structure.

(C) Gate oxide thickness
Gate oxide thickness and P-base concentration are parameter of threshold voltage. Thinner gate oxide structure reduces P-base resistance.

These parameters influence electrical characteristics such as. V_{CE}(sat) - E_{SW}(off) tradeoff relationship. We evaluate and optimize these parameters without increasing power losses.

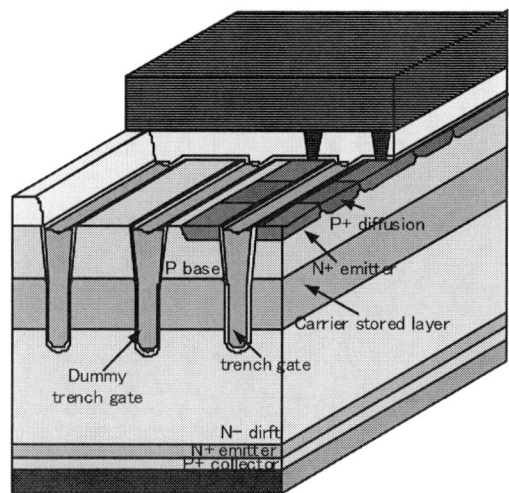

Figure.1 Schematic structure of CSTBT[TM](III)

III. SIMULATION AND EXPETIMENTAL RESULTS

A. Effect of finer patterning of N+emitter layer

Fig.2 shows schematic waveforms of inductive load switching. The circuit used in simulation is shown in Fig.3. Analysis point A means the period when corrector current reduces at 33percent(1/3). Fig.4 indicates current density along gate channel surface at the period of A. The fine-pattern technology realizes narrow N+ emitter region with keeping total channel width. Thus, as the N+ emitter width becomes fine, resistance between P-base layer just below each N+ emitter and emitter electrode reduces, and current density at the portion also reduces. These effects drastically suppress parasitic NPN transistor action because of high potential barrier between N+ emitter and P-base region. 0.637V and 0.641V in Fig.4 means Fermi potential height from P- base layer just below N+ emitter region to N+ emitter region. Structure (b) keeps higher potential height by 0.004V.

To evaluate this effect, we fabricated test CSTBT(III) chips with various N+ emitter width (normalized = 0.6, 1.0, 2.0) and chip size. Table 1 summarizes test chip variation (see table 1).

Fig.5 shows measured maximum controllable current at the condition of V_{cc}=600V, V_{GE}=20.0/-15.0[V], Ta=150°C. The maximum controllable current is greatly improved, and it reaches sufficient value by using finer N+ emitter pattern, regardless of chip size. However, N+ emitter width less than 1.0 (normalized value) are poor dimensional control.

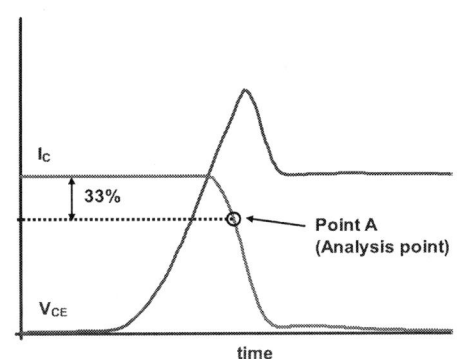

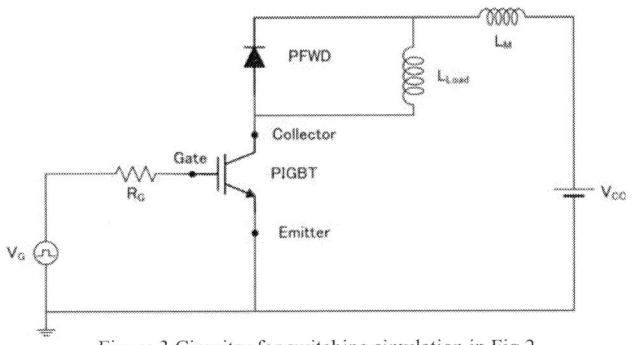

Figure.2 Schematic waveformes of inductive load switchiing. Two dimensional data at analysis point is shown in Fig. 4 and Fig. 6.

Figure.3 Circuitry for switching simulation in Fig.2

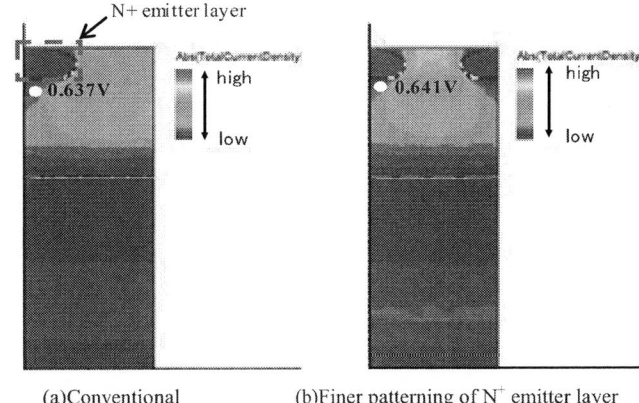

(a)Conventional   (b)Finer patterning of N⁺ emitter layer

Figure.4 Current density distribution and potential benith N⁺ emitter under turn-off phase.

Table1. Test chip variation

	Chip size(normalized)
Type A	1.0 (Small)
Type B	8.3 (Medium)
Type C	12.0 (Large)

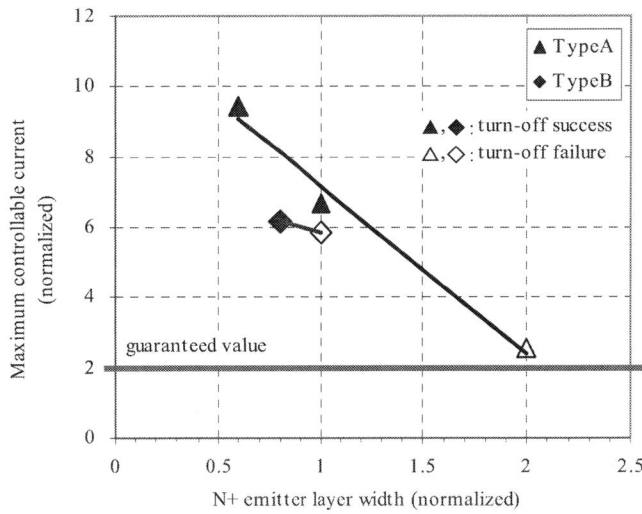

Figure.5 Relationship between maximum controllable current and normalized N+ emitter layer width (experiment). Maximum controllable current is normalized by rated current.

B. Effect of deeper P+ diffusion layer

P+ diffusion layers formed between N+ emitter layers reduce resistance P-base layer which act as base region of the parasitic NPN transistor. Fig.6 also shows the potential height from P- base layer to N+ emitter region at period of A(in Fig.2) with regard to depth of P+ diffusion layer.

Doping of P+ diffusion layer and N+ emitter layer are compensated each other so that final doping profile is sensitive to depth of P+ diffusion layer. In the case of shallow layer, the parasitic NPN transistor is comparatively turned on easily. In opposite situation, threshold voltage is altered because of side diffusion of deep P+ diffusion layer. We confirmed these effects on the CSTBT(III) with different implantation energy of P+ diffusion layer (normalized = 1.0, 1.1, 1.2, 1.3). Fig.7 shows relationship between maximum controllable current and implantation energy of P+ diffusion layer. Maximum controllable current at 1.0 (normalized value) steeply drops by means of latch-up phenomenon. It keeps sufficient values at 1.1 or higher value. 1.1 or 1.2 is best value for fabrication, because the deep P+ diffusion layer affects threshold voltage over 1.3.

the capacitance of gate oxide. If the thinner gate oxide is fabricated, P-base dosage should be concentrated to obtain the same threshold voltage. As the P-base dosage is increased, resistance of P-base region reduces. Therefore the thinner gate oxide can be used to reduce the gain of parasitic NPN-transistor. However, thinner gate oxide affects gate capacitance and breakdown characteristics of gate oxide.

In this experiment, the lowest P-base dosage sample only shows turn-off failure, and thin gate thickness samples show high turn-off capabilities. Sufficient controllable current is achieved by selecting relevant gate oxide thickness and threshold voltage to reduce P-base resistance under N+ emitter region.

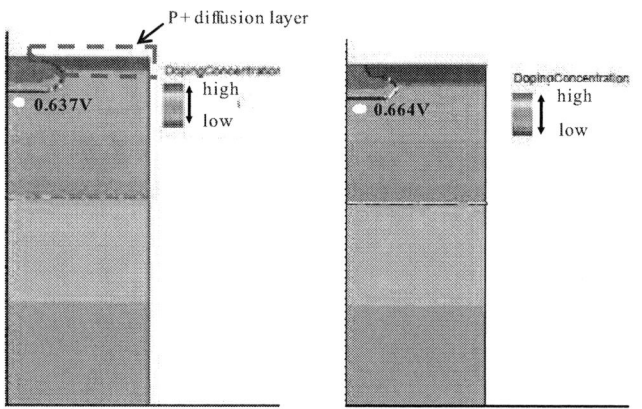

Figure.6 Doping concentration and potential of P base beneath N+ emitter.

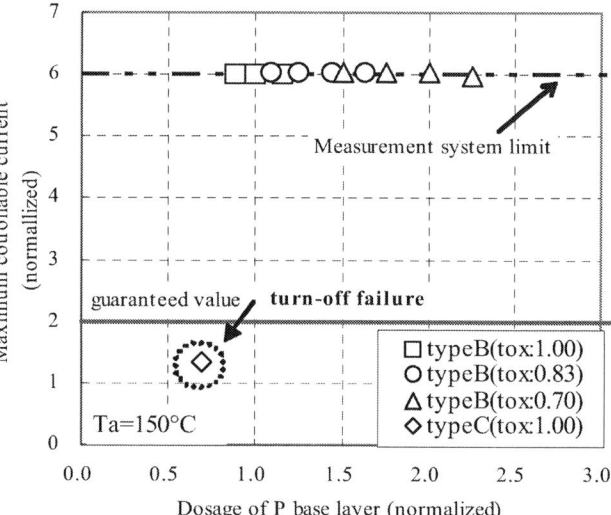

Figure.8 Relationship between maximum controllable current and dosage of P base layer under various gate oxide thickness(tox : normalized value). (experiment)

D. Temperature dependence of maximum contorollable current and lekage current

We fabricated CSTBT(III) optimized for high temperature operation. Table.2 summarizes parameters for the CSTBT(III). We reach the conclusion that ten percent deeper P+ diffusion layer is only needed for the device, because ten percent deeper P+ diffusion overlaps on N+ emitter layer, and it reduces P-base resistance drastically. In Fig.9, the maximum controllable current is six times higher (measurement system limit) than rated current from 25°C to 200°C temperature range.

Fig.10 shows temperature dependency of leakage current at 1200V. Type C has large chip size, which has disadvantage in self heating, thermal runaway is not observed even at 210°C.

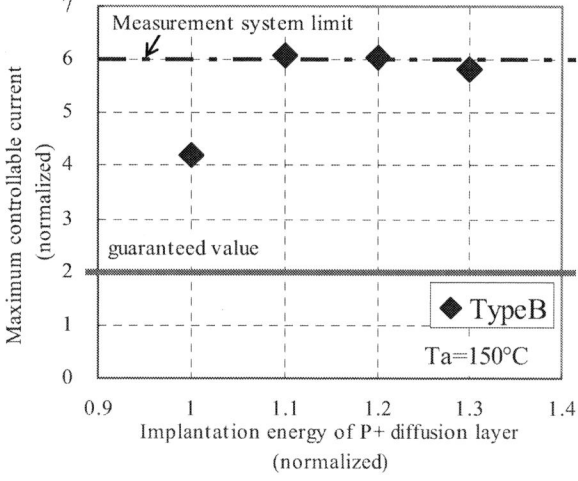

Figure.7 Relationship between maximum controllable current and P⁺ diffusion layer depth (experiment).
Maximum controllable current is normalized by rated current.
Type B is defined in Table. 1.

Table.2 Optimized structural parameters for CSTBT(III)

	conventional	Proposed structure of CSTBT^{TM}(III)
N⁺ emitter layer width	1.0	1.0
Depth of P+ diffusion layer	1.0	1.1
gate oxide thickness	1.0	1.0

C. Effect of thinner gate oxide thickness

Fig.8 shows experimental relationship between maximum controllable current and P-base dosage of various gate oxide thicknesses. The threshold voltage is inversely proportional to

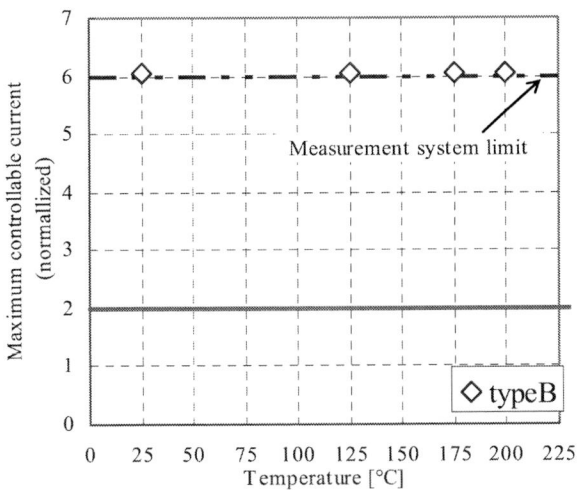

Figure.9 Temperature dependence of maximum contorollable current of proposed structure of CSTBT(III) (experiment)
Type B is defined in Table 1.

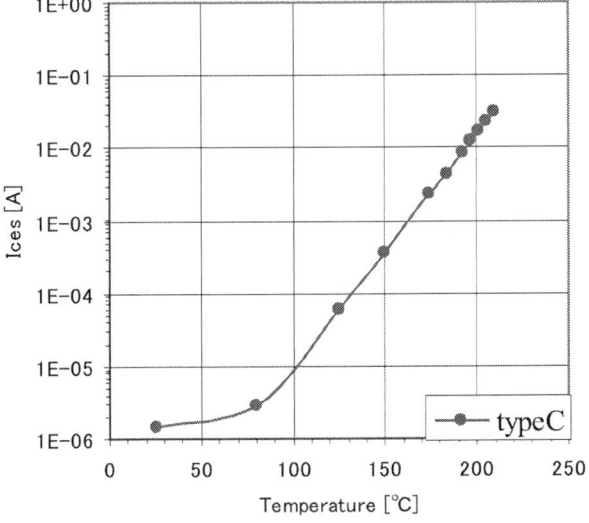

Figure.10 Temperature dependence of leakage current of proposed structure of CSTBT(III)(experiment). Type C is defined in Table.1 .

IV. CONCLUSION

We examined the high temperature performance of 1200V class CSTBT[TM](III) by simulation and experiments. We demonstrated that N+ emitter width, P+ diffusion layer depth and gate oxide thickness are closely related to high temperature characteristics, and optimized 1200V CSTBT[TM](III) succeeded in 200°C operation without any thermal runaway and turn-off failure.

REFERENCES

[1] T. Takahashi, Y.Tomomatsu, K.Sato, "CSTBT[TM](III) as the next generation IGBT," ISPSD'08, pp.72-75 (2008)

[2] K.Suzuki, T.Takahashi, R.Fujii, K.Tsurusako, Y.Tomomatsu, "A new 1.7kV CSTBT(III)[TM] for the next generation power module," PCIM'10, pp.227-331 (2010)

[3] H. Takahashi, H. Haruguchi, H. Hagino, T. Yamada, "Carrier Stored Trench-Gate Bipolar Transistor (CSTBT) - A Novel Power Device for High Voltage Application -", ISPSD'96, pp.349-352 (1996)

Proceedings of the 23rd International Symposium on Power Semiconductor Devices & IC's
May 23-26, 2011 San Diego, CA

Physical Analysis of Carrier Lifetime Controlled IGBT [II]

Chihiro Tadokoro*, M. Kaneda, K. Takano*, S. Kusunoki, T. Minato, J. Yahiro* and K. Hatade
Power Device Works, Mitsubishi Electric Corporation
*Power Chip Design Sec., Design & Technology Dept., Fukuryo Semicon Engineering Corporation
1-1-1 Imajukuhigashi Nishi-Ku Fukuoka-City 819-0192 JAPAN
Phone: +81-92-805-3827,   Fax: +81-92-805-3881.
E-Mail:Tadokoro.Chihiro@zd.MitsubishiElectric.co.jp

Abstract

For IGBTs, there are strong requirements for high current density usage from the cost reduction point of view and high speed operation from the system efficiency point of view. Strong carrier lifetime control is needed to reduce a turn-off loss (Eoff) of IGBT for high frequency usage. It seems to be insufficient for our previous report [1-2] to deeply understand about correlation between electric characteristics and Cathode Luminescence (CL), which stands for free conduction carrier trap levels inside Si band structure. Clearer physical model is necessary to improve an agreement for both high current density and high speed operation. Therefore, we applied another analysis method of PL (Photo Luminescence) to ensure the physical model for carrier lifetime controlling method to combine relatively heavy dose of Electron Beam (EB) irradiation and high temperature thermal annealing.

Introduction

Recently, the operation frequency area of IGBT has extended from 5-20kHz to around 50kHz. Corresponding to high speed operation, the strong carrier lifetime control is needed to decrease the turn-off switching loss (Eoff). On the other hand, the IGBTs' chip size shrink is very important from the cost reduction point of view this is the reason why we should consider the device characteristics' stability in the high current density operation. Against this strong carrier lifetime control, the extremely high current density conduction, which seems to be some kind of a "stress" to recover a carrier lifetime as a thermal annealing process, would cause electrical characteristic shifts after a long term use especially under high temperature condition more than 150degC.

Then we studied by CL analysis to understand the physics of carrier lifetimes' behavior [1-2]. CSTBT[TM] [3], which was our special but standard structure of IGBT in mass production, was chosen as a sample, and the correlation between the electrical characteristics' shift and the CL spectrum for both before and after DC conduction current stress was analyzed.

■ Described as our previous paper, the changing ratio of ON state forward voltage drop V_{CE}(sat) shift after DC current stress was several percent small. And the milder the carrier lifetime control was, the fewer the shift was. So DC conduction current has the same kind of effect as a thermal annealing.

■ The electron trap revels, that increased by EB irradiation, decreased by thermal annealing process, but a few trap revel was remained and grown by DC current stress. Consequently it was considered that the turnoff tail current increased because the number of electrons decreased, and the number of holes increased relatively.

■ From the result of the CL analysis, we supposed that a minus V_{CE}(sat) shift after DC current stress was caused by the defect at energy level of Valence band side, whom it was not possible to observe with CL.

■ It was proved that the electrical characteristic shift of IGBT after DC current stress was possible to control by a combination of high dose of EB irradiation and high temperature thermal annealing that we performed to adjust an initial characteristic of IGBT.

PL analysis

CSTBT[TM] [3] is chosen to be evaluation same as the previous works. Fig.2 shows the experimental procedure, in which IGBT applied EB irradiation and the thermal annealing   after passing a conventional process. At a point of approximately a half distance of N-drift layer thickness from the top side of device shown in Fig. 1, PL spectrums were obtain for both "before" and "after" DC current stress in the order of several ampere per square-centimeters as the acceleration condition to simulate the long term stability of electrical characteristics of device

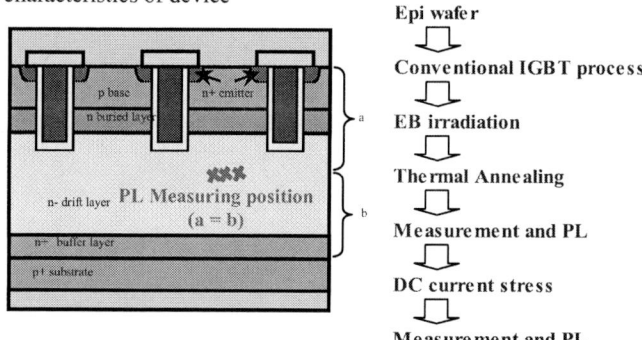

Fig.1 Schematic cross-sectional view of IGBT applied CSTBT[TM] structure

Fig. 2 Experimental Procedure

The four samples with a different combination of EB irradiation, thermal annealing process, and the annealing temperature were prepared shown in Table 1. The strength relationship of carrier lifetime control is HCL > CCC > HHS > woLC.

Three levels, 0.87, 0.71 and 0.27 eV shown in Fig. 3, are well known as the carrier trap levels induced by EB irradiation for the standard carrier lifetime control process, but two another levels, 0.79eV C-line for Ci-Oi and 0.97eV G-line for Cs-Sii-Cs, are mainly discussed in this analysis, here Ci stands for Carbon interstitial, Oi for Oxygen interstitial, Cs for Carbon substantial and Sii for Silicon interstitial.

978-1-4244-8425-6/11 $26.00 © 2011 IEEE

Table1: The summary of carrier lifetime control process, electrical characteristics and the characteristic peak intensity in the PL spectrum of the IGBT chips before and after the DC current stress

Sample	EB irradiation dose*	Annealing Temperature**	Annealing Time***	VCE(sat)[V] (at rated current) Before stress	Eoff[mJ/cm2] (at rated current) Before stress	Qtail[uC] (at rated current) Before stress	Analyzed position	C-line			G-line			Band Edge		
								Before stress	After stress	Before-After	Before stress	After stress	Before-After	Before stress	After stress	Before-After
woLC	Nothing	Nothing	Nothing	1.289	16.361	40.8		0	0	0	0	0	0	2442	699	1743
HHS	Heavy	Heavy	Short	1.385	5.28	7.8	middle of n-drift layer	202	88	114	0	833	-833	1776	1367	409
CCC	Conventional	Conventional	Conventional	1.811	1.509	0.6		286	127	159	0	0	0	949	1307	-358
HCL	Heavy	Conventional	Long	2.04	1.179	0.5		-	-	-	-	-	-	-	-	-

"Heavy" dose is about two times larger than 'Conventional'. "High" temperature is 25% higher than "Conventional". "Short" time is a quarter length of "conventional".

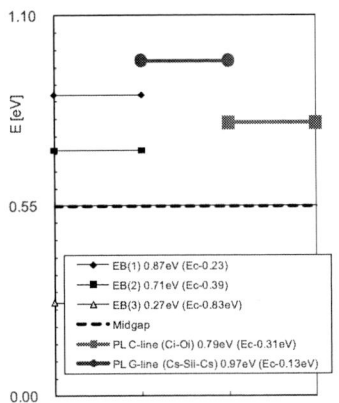

Fig. 3 C-line and G- line carrier trap energy levels inside an energy band gap of Si.

In our previous studies [1-2], we found out some difficulties for constructing the physical model for carrier lifetime behavior in case of taking account of the widely spreading CL signal, which is standing almost full range of above Fermi level from 0.8eV to 1.1eV band edge, shown in Fig. 4. In Fig. 4, CL spectrum was several times magnified to directly compare with PL one, and only the conventional condition sample of CL spectrum was plotted.

This widely spread CL signal including the main three peaks of C-line, G-line, and Band Edge was assigned as the luminescence by photon coupling and/or scattering, but this broad signal made it difficult to understand. This is the reason why, we renewed a physical analysis method from CL to PL that is easy to measure and to obtain very accurate result in this study. To ensure PL spectrum has enough accuracy, PL spectrum of the sample chip without any carrier lifetime control is shown in Fig.5. For both "before" and "after" DC current stress case, there was no energy level around both C and G center, which are in the carrier lifetime controlled one. The difference between "before" and "after" appeared at the Band-Edge (BE), and decreasing of BE-luminescence after current stress might turn into the in-luminous center like a dangling bond. Only the very shallow levels just underneath the bad-edge coming from sample preparation damage caused by getting a cross-section of chip. The PL peaks observed in two levels, 0.81eV and 0.98eV were caused by photon coupling and/or scattering.

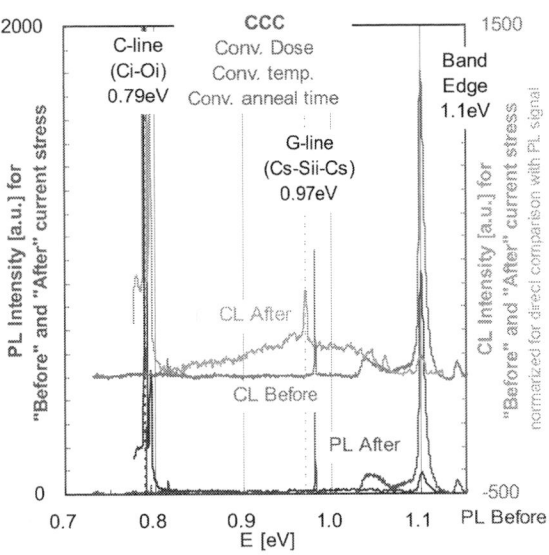

Fig. 4 PL vs. CL for the conventional carrier lifetime control condition

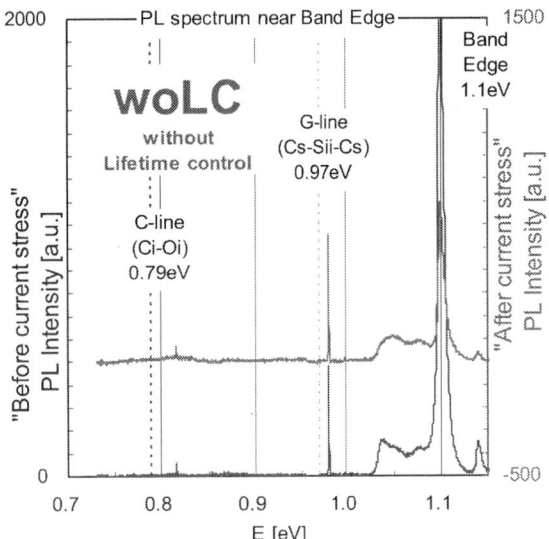

Fig. 5 PL spectrum for "without" carrier lifetime control
There is no special peak around G-line or C-line.

978-1-4244-8425-6/11 $26.00 © 2011 IEEE

Correlation of electrical characteristic and PL

To easily understand phenomena during DC current conduction stress [1-2], output I-V curves and turn-off collector current, collector voltage and gate voltage waveforms are shown in Fig. 6 for both "before" and "after" current stress. Continuous DC current conduction of very long term, several hundred hours, at the extremely high current density, several hundred amperes per square-centimeter, has same kind of effect as a thermal annealing for carrier lifetime recovering process, in which V_{CE}(sat) of "after" current stress is lower than "before" one. And the tail current of "after" current stress is rising up from "before" one even in collector voltage and gate voltage waveforms being almost the same. But details are different from each other, i.e. thermal curing effect and current healing effect, like a discussion below. As the first order approximation for the fundamental tradeoff relationship between an V_{CE}(sat) and Eoff shown in Fig. 7, an Eoff replaces into a turn-off tail charge (Qtail), shown in Fig. 8, as a representative of ambipolar carrier lifetime of bipolar action like a conductivity modulation in the high injection level operation mode, where Qtail is a time integration of turn-off collector current from a time of Vg restart point to decrease from Vth to the end of current going to zero.

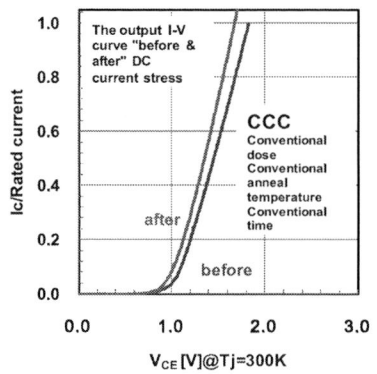

Fig.6-(a) Output I-V curve before and after DC current stress

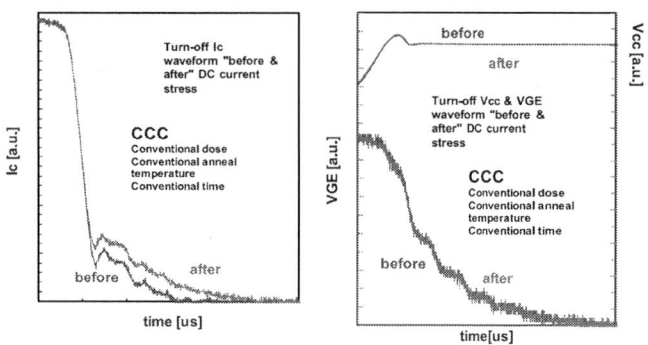

Fig.6-(b) Turn-off tail current Ic, collector voltage and gate voltage before and after DC current stress

From this point of view, three typical PL peak's, especially Band-Edge luminescence (BE-signal), behaviors are plotted on Fig. 9 as one of the most important figures in this study. BE-signal is not linearly but clearly proportional to Qtail. As we discussed in CL result [1-2], this main BE-signal reduction can be interpreted as the transmutation from the "luminous" to "in-luminous" recombination center such as dangling bond. Unfortunately this kind of luminous physics cannot allow to directly show the asymmetry of recombination rate between the electron - hole pair, the above hypothesis is also supported by our previous CL analysis. And another peak of C-line and G-line grow up as Qtail decreasing or the carrier lifetime shortening, shown in Fig. 9.

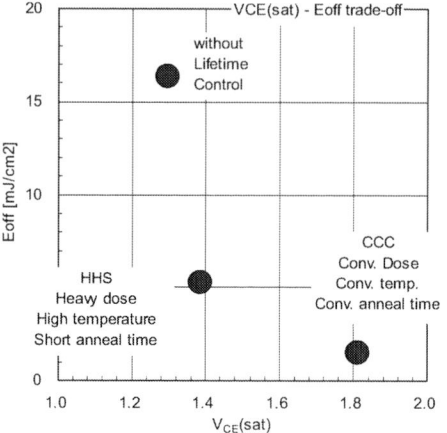

Fig.7 V_{CE}(sat)-Eoff trade-off relationship

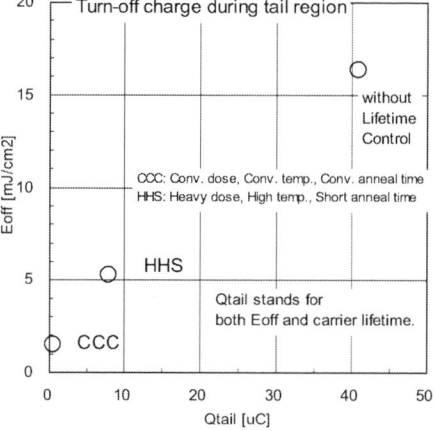

Fig.8 Qtail(turn-off tail charge) vs. Eoff

DC current stress causes the bi-directional effects for the carrier lifetime shifts (Fig. 10). The positive shift of Eoff shows the carrier lifetime being cured, as if the additional thermal annealing process were applied. This is characterized by the raising of C-line (Fig. 11-(a)) at the very deep level for conduction electrons. On the other hand, the negative shift is against the conventional characteristics of bipolar device, and it means the carrier lifetime reduction. This is also characterized by G-line arising (Fig. 11-(b)). So, from the macroscopic point of view, carrier lifetime control consisted of EB irradiation and thermal annealing process was explained by the Band-Edge luminescence and the "in-luminescence" center like a dangling bond. And, from microscopic point of view, DC current stress was described by the deep-trap levels such as C-line or G-line, and had bi-directional effects for electrical characteristics of devices.

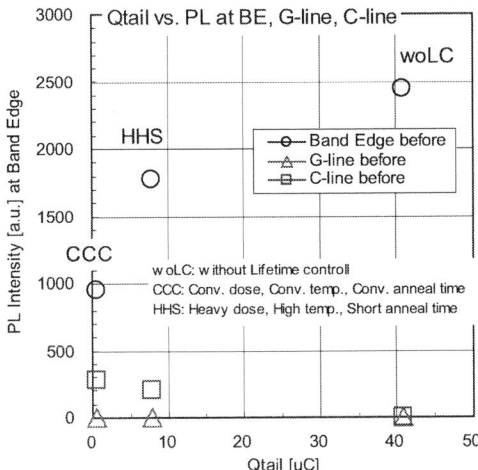

Fig.9 Qtail(turnoff charge) vs. PL spectrum

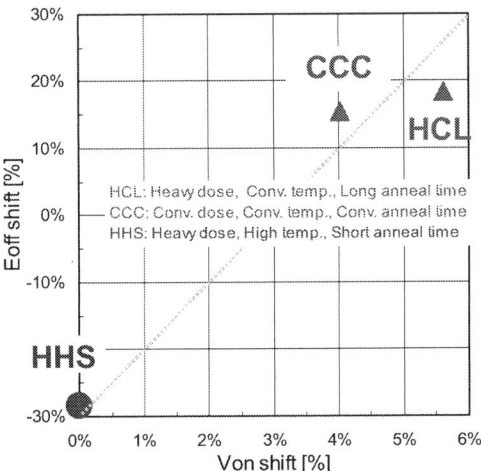

Fig. 10 Von shift vs. Eoff shift

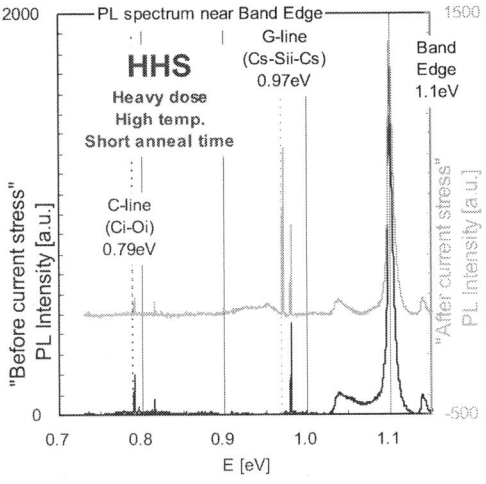

Fig. 11-(a) PL spectrum of the sample HHS before current stress
The G-line grew up after the current stress.

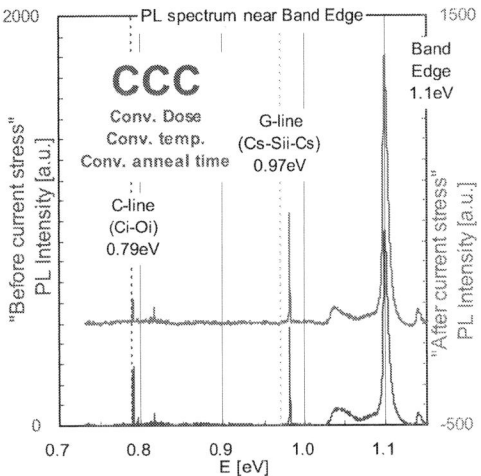

Fig. 11-(b) PL spectrum of the sample CCC before current stress
The C-line was cured by the current stress.

Conclusion

PL analysis is more effective than the previous CL analysis [1-2] to understand the physics of carrier lifetimes' behavior during the high injection level described by the correlation between the turn-off tail charge (Qtail) and energy levels, which includes three characteristic luminescence of the conduction Band-Edge, C-line and G-line, inside of band-gap. The shifts of Qtail, which stands for both turnoff loss Eoff and carrier lifetime, corresponded to the changing patterns of G-line and/or C-line before and after DC current stress. From the PL analysis, the gross lifetime could be described by the sum of the Band-Edge luminescence and/or the recombination center "in-luminescence". We also confirmed that our previous conclusion is still effective after this deep insight, i.e. one of the best solution for carrier lifetime control is the combination of heavy EB irradiation and high temperature thermal annealing.

Acknowledgement

We thank Dr. K. Sato for affording an opportunity to do this study and their encouragement.

References

[1] M. Kaneda et. al., EPE 2009, paper No.471
[2] C.Tadokoro et. al., ISPSD2010, pp. 145-148
[3] H.Takahashi et al., ISPSD1996, pp. 349-352
[4] R. Sauer, Appl. Phys. A36, 1 (1985)
[5] G. Davies, Phys. Rev. B73, pp. 165202-165210 (2006)
[6] F. Niwa, EDD-08-46, (2008)

Proceedings of the 23rd International Symposium on Power Semiconductor Devices & IC's
May 23-26, 2011 San Diego, CA

High temperature wafer bonding technique for the realization of a voltage and current bidirectional IGBT

A. Bourennane[1,2], H. Tahir[1], J-L. Sanchez[1], L. Pont[1], G. Sarrabayrouse[1], E. Imbernon[1]

[1]CNRS; LAAS, 7 avenue du colonel Roche, F-31077 Toulouse, France
[2]Université de Toulouse ; UPS, INSA, INP, ISAE; LAAS ; F-31077 Toulouse, France

Abstract— **For applications requiring voltage and current bidirectional switches like the matrix converter application, the bidirectional switch can be obtained by associating unidirectional switches. However, this arrangement leads to an increase of the on-state voltage drop and increases the circuit complexity. In order to overcome these drawbacks, it is necessary to design a monolithically integrated MOS controlled current and voltage bidirectional switch. This paper focuses on the monolithically integrated bidirectional IGBT. This well known structure consists of MOS sections on each side of the wafer. The difficulty with this structure mainly resides in its realization and control. This paper is an attempt to demonstrate the feasibility of a bidirectional IGBT using a double side photolithography technique on one hand and proposes a high temperature bonding technique that could be used for the realization of a bidirectional IGBT using the Si/Si wafer bonding technique on the other hand.**

I. INTRODUCTION

Silicon to silicon direct wafer bonding was reported and detailed for the first time by Lasky in 1986 for the purpose of SOI technology and by Shimbo *et al* for the realization of bonded PIN diode [1][1]. This technique was also used in the microelectronics domain, to enhance the performances of integrated circuits [2]. Moreover, it offers to the designer an extra degree of flexibility to trade off the single wafer concentrated complexity for multiwafer solutions [3].

For vertical power device applications, the direct wafer bonding was used to realize a PIN diode [4] and it could be used to develop an intelligent vertical power device by isolating the main power device from the control circuit [5]. It was also reported that it can be used to realize a monolithically integrated current and voltage bidirectional power device [6][7][8][9].

The direct wafer bonding can be classified into three kinds: UHV (Ultra High Vacuum) wafer bonding, hydrophilic and hydrophobic wafer bonding [2]. The wafer bonding technique used for the realization of the power device is the hydrophobic direct wafer bonding technique. This choice is due to the absence of an interfacial oxide layer after bonding which can prevent the current to flow through the device drift region. The

hydrophobic wafer bonding involves typically the following steps, which are briefly detailed below:
1. Chemical Mechanical Polishing (CMP) of the silicon surfaces intended for bonding to obtain flat and polished surfaces.
2. Surface preparation to obtain clean and hydrophobic surface.
3. Prebonding at room temperature in air to initialize bonding.
4. Bond annealing to reinforce the bond.

Regarding the fourth step, one could carry-out the annealing at low or high temperature. The annealing at low temperature (T< 400 °C) overcomes some problems such as doping profile broadening, thermal stress introduction, defect generation, and possible contamination [10]. However, a large number of voids can appear at the bonding interface [11]. Figure 1 (a) shows an example of hydrophobic wafer bonding realized at LAAS-CNRS clean room at a temperature of 600°C. The characterization of the interface recombination properties by analyzing the μPCD technique shows that it is non-uniform as shown in figure 1a. The regions of high carrier lifetime could be non-bonded regions in which the silicon surface is passivated by Hydrogen and/or Fluor present after the last chemical processing. Therefore, the mechanical solidity of the structure is degraded. For the purpose of comparison, two silicon wafers were bonded at high temperature (1100°C) and it can easily be seen by analyzing μPCD that the voids have completely disappeared (figure 1b).

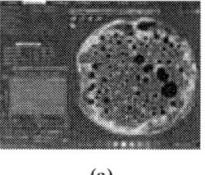 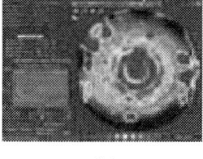

(a)　　　　　　　　　　(b)

Figure 1. μPCD characterization of wafers bonded:
(a) at low temperature (b) at 1100 °C

II. REALIZED STRUCTURES

In order to demonstrate the feasibility of the bidirectional IGBT structure using the double side photolithography technique and using the high temperature bonding technique,

978-1-4244-8425-6/11 $26.00 © 2011 IEEE　　140

the structures (a) and (c) were realized and characterized. However, the process for the realization of the bidirectional IGBT using the bonding technique (b) is still in progress.

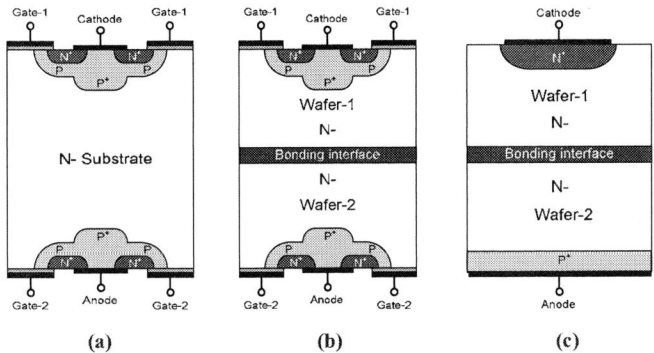

(a)   **(b)**   **(c)**

Figure 2. Bidirectional IGBT realized by: (a) the double side photolithography technique (b) direct wafer bonding (c) PIN diode realized by direct wafer bonding technique.

A. Requirements on the position of the bonding interface

We have carried-out 2D simulations using Sentaurus on the bidirectional IGBT structure that uses the bonding technique. The bonding interface is represented by a region of a low carrier lifetime and it is placed in the N- drift region at different distances *"d"* from the cathode. We can easily notice that as soon as the depletion region reaches the bonding interface, the leakage current increases rapidly. This is due to the numerous recombination centers at the bonded interface which promote the transition of carriers from an energy band to another and consequently increasing the leakage current [12]. The bonded interface must be located so far so that the depletion region in the blocking state cannot reach this interface for the nominal voltage.

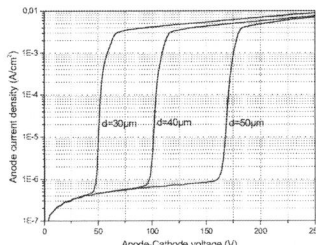

Figure 3. I-V characteristics in the blocking state for different values of the distance d separating the bonding interface from the cathode

B. Integration of the bonding step in the flexible IGBT process

As mentioned previously, the realization of the bonding at high temperature allows to achieve a voids free bonded interface. Moreover, if one takes advantage of an already existing IGBT process step, then one would overcome the drawback of an additional high temperature step as reported in the literature. Indeed, the analysis of the flexible IGBT process developed at LAAS-CNRS shows that one can make use of the necessary N+ annealing step (at 1100°C) in order to reinforce the bond. Therefore, this step has no consequence on the thermal budget of the IGBT process (figure 4).

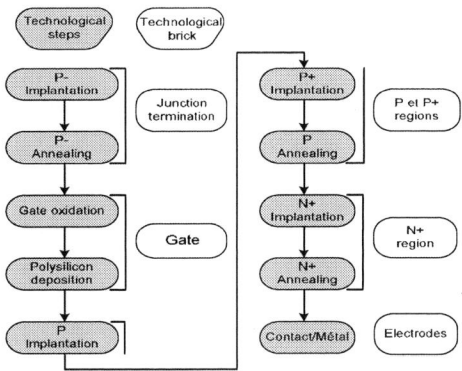

Figure 4. The flexible IGBT process steps (LAAS-CNRS).

III. HIGH TEMPERATURE DIRECT WAFER BONDING TECHNIQUE

The bonding process is carried-out in four steps as detailed hereafter:

A. Chemical Mechanical Polishing (CMP) of the silicon surface intended for bonding

Hydrophobic wafer bonding requires some strict conditions on parameters such as flatness and roughness of the two silicon surfaces intended for bonding [13][14][15]. The CMP is one solution to obtain surfaces having flatness and roughness that meet the necessary requirements for a successful bonding [16][17][18]. For our application this CMP step is necessary not only for flatness and roughness of silicon surfaces but also to remove mechanically the impurities which can be accumulated during the IGBT process. Indeed, the backside of the wafer intended for bonding cannot be protected from the oxidation and polysilicon deposition before the bonding step. Consequently, it is very difficult to obtain the polished initial silicon surface quality after a necessary RIE etching step to remove the undesirable deposited polysilicon layer and a necessary chemical etching step to remove the oxide layer. It is to note that it is reported that a bonding of good quality requires surface roughness lower than 0.3 nm [19].

B. Surface preparation

The cleanliness of silicon surfaces is very important for bonding. It allows to improve the bonding energy and prevent the formation of voids by the organic or metallic impurities. Generally a standard RCA wet cleaning is used to remove organic and metallic contaminants [2]. The wafers are then immersed into nitride acid (HNO_3) to grow a few monolayer of chemical oxide which is finally etched in buffered HF to obtain the desired hydrophobic surfaces and dried with nitrogen gas without rinse.

C. Prebonding

Prebonding step has to follow immediately after drying the wafers step. It is carried-out in the AML-AWB-04 at room temperature under vacuum (10^{-5} mbar).

978-1-4244-8425-6/11 $26.00 © 2011 IEEE   141

D. Annealing step

Annealing at high temperature is necessary to form Si-Si bond as already mentioned [15]. It is done at a temperature of 1100°C.

IV. REALIZATION OF THE BIDIRECTIONAL IGBT

A. The double side photolithography technique

This technique consists on the realization of the MOS sections of the bidirectional IGBT (figure 2a) on both sides of the same wafer. This technique, compared to the bonding technique, offers the advantages that one wafer is needed and that the electrical properties of the N- substrate are not affected by the process. However, the different regions of the top side and the backside of the wafer cannot be realized at the same time, so the electrical characteristics in the direct and reverse conducting states can present slight differences. Moreover, with this technique, the wafer is more exposed to contamination.

- *Experimental results*

We have realized, in the LAAS-CNRS clean room, using the flexible IGBT process (figure 4) a bidirectional IGBT structure shown in figure 2a by the double side photolithography technique. Figure 5 shows the front side and backside microphotographs of the realized bidirectional IGBT. The forward and reverse on-state I_V characteristics were obtained by carrying on-wafer characterization. The wafer is characterized in the forward conducting state and then was turned over to carry-out characterizations in the reverse conducting state.

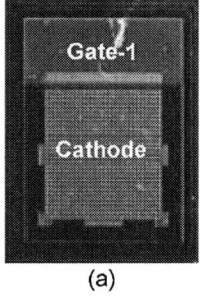 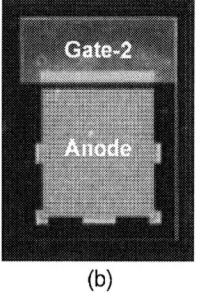

(a)   (b)

Figure 5. Front side (a) and backside (b) microphotographs of the realized bidirectional IGBT by the double side photolithography technique

Figure 6 shows the forward and reverse I_V characteristics in the conducting state of the realized bidirectional IGBT as well as the anode current as function of gate-cathode voltage(forward conducting mode) and as function of gate-anode voltage (reverse conducting mode).

We can easily remark that the set of curves for the forward and reverse conducting I_V characteristics for different gate cathode voltages are not symmetrical (figure 6a). This could be attributed to the technique used for this realization. Indeed, this technique could result in gate oxide qualities for MOS sections on the front side and those on the backside of the wafer that are slightly different. The I_V characteristics in forward and reverse conducting states for the same gate voltage are unfortunately not identical in this realization. The threshold voltage of the front side and backside MOS sections of the realized bidirectional IGBT (figure 6b) are different.

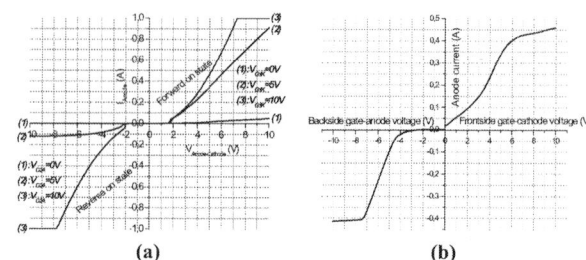

(a)   (b)

Figure 6. (a) Experimental forward and reverse conducting I-V characteristics of the realized bidirectional IGBT. (b) $I_A(V_{GK})$ and $I_A(V_{GA})$ characteristics of the front side and backside MOS sections

This difference could be due to the fact that the gate oxide qualities on both sides were affected during the wafer processing.

B. Si/Si direct wafer bonding technique

The direct wafer bonding technique described previously was used for the realization of a bidirectional IGBT (figure 2b). Two wafers are processed separately in order to realize the front side and backside MOS sections. Afterwards, the two wafers are bonded to realize the bidirectional IGBT structure (figure 2b). A microphotograph of the bonded wafers is given in figure 7.

The main advantages of this technique are:

- It allows to realize the different regions of the front side and backside MOS sections in the same conditions which permit to obtain symmetrical characteristics in forward and reverse state.
- It uses the technological process exactly as in the case of conventional IGBT (unidirectional IGBT). This reduces the time of realization and allows the protection from eventual contaminations.

The drawback of this technique resides in the fact that it needs bonding which can degrade the device performance both in the conducting and blocking states [20].

 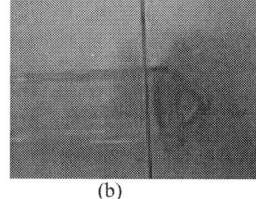

(a)   (b)

Figure 7. Realized bidirectional IGBT using the wafer bonding technique: (a) one face of the bonded wafer (b) edge of the bonded wafer

- *Validation of the direct wafer bonding technique by the realization of the PIN diode*

We carried-out bonding experiments on PIN diodes because of the simplicity and rapidity of its process. Indeed, for this realization we needed one implantation on the front side of each wafer for realizing P+ and N+ regions. Moreover, for bonding, we didn't have to carry out the polishing step, because the backsides of the two wafers were not processed. A microphotograph of the realized PIN diode is given in figure 8a.

The characterization results for two realized PIN diodes are given in figure 8b. The first diode has an N- total substrate thickness of 600 um. The second diode has an N- total substrate thickness of 500 um.

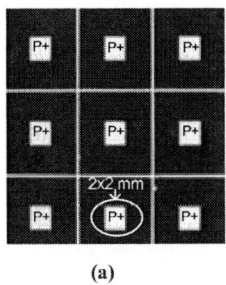
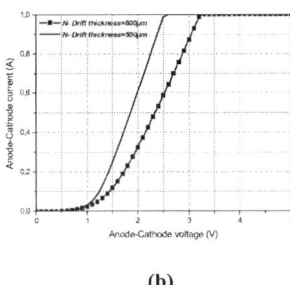

(a) **(b)**

Figure 8. (a) Microphotograph of the realized PIN diode by the direct wafer bonding technique. (b) I_V characteristics for N- drift total substrate thickness of 600μm and 500μm

As can be expected, the on-state voltage drop depends on the substrate thickness. The higher the thickness of substrate the higher the on-state voltage drop. This result on the PIN diode allows to validate the wafer bonding technique.

V. CONCLUSION

In this paper, the advantage of the high temperature direct wafer bonding was highlighted through an experimental realization. The main steps of the wafer direct bonding were detailed briefly in order to highlight the conditions necessary for carrying-out a successful bonding. The bonding technique was validated on a PIN diode. It is also used for the realization of a bidirectional IGBT. The technological process is still in progress. The advantage of using a high temperature bonding technique is related to the fact that one can make use of the already existing N^+ annealing step (1100°C) in the flexible IGBT technological process for reinforcing Si-Si bond.

The realization of the bidirectional IGBT was also carried-out using the double side photolithography technique but unfortunately the experimental results obtained by this realization showed that the I-V characteristics of the realized bidirectional IGBT in the forward and reverse conducting states were note symmetrical. This could be attributed to a difference in the quality of gate oxides. Indeed, with the double side photolithography technique, only one face of the wafer can be processed at a time for each process step.

REFERENCES

[1] Drago Resnik, Danilo Vrtacnik, Uros Aljancic, Slavko Amon, "Study of low-temperature direct bonding of (111) and (100) silicon wafers under various ambient and surface conditions," Sensors and Actuators A: Physical. vol. 80, pp. 68–76, March 2000.

[2] Silke H. Christiansen, Rajendra Singh, Ulrich Gosele, "Wafer Direct Bonding: From advanced Substrate Engineering to future applications in micro/nanoélectronics", Proc IEEE., vol. 94., 2006, pp.2060–2106.

[3] JIwei Jiao, Deren Lu, Bin Xiong, Welyuan Wang "Low temperature silicon direct bonding and interface behaviour," Sensors and Actuators A: Physical, vol 50., 1995, pp. 117–120.

[4] E.P Burte, R. Wiget, "A comparaison between fast power diodes fabricated on substrates made by silicon to silicon direct bonding and on epitaxial substrates", Proc. Power Electronics and Drive Systems, 1997, pp. 6-11.

[5] C. Harendt, W. Wondark, U. Apel, H.G. Graf, B.Hofflinger, J.Korec and E. Penteker," Wafer bonding for intelligent power ICs: integration of vertical structures", Proc. IEEE International SOI Conference, 1995, pp. 152-153.

[6] K.D Hobert, F.J KUB, G. Dolny, M. Zafrani, J.M. Neilson, J. Gladish, C. McLachlan, "Fabrication of a double side IGBT by very low temperature wafer bonding". ISPSD'99, 1999. Pp. 45-48.

[7] K.D Hobert, F.J KUB, M. Ancona, J.M. Neilson, P.R. Waind, "Trantient analysis of 3.3Kv double side double gate IGBTs". ISPSD'04, 2004. pp. 273-276.

[8] J.M. Neilson, F.J KUB, K.D Hobert, K. Brandmier. M. Ancona, "Double side IGBT phase leg architecture for reduced recovery current and turn-on loss". ISPSD'02, 2002. pp. 141-144.

[9] K.D Hobert, F.J KUB. M. Ancona, J.M. Neilson, , K. Brandmier. P.R. Waind, "Characterisation of Bi-directional double side double gate IGBT fabricated by wafer bonding". ISPSD'01, 2001. pp. 125-128.

[10] Qin-Yi Tong, Giho Cha, Roman Gafiteanu, Ulrich Gosele, "Low temperature wafer direct bonding,", Journal of Microelectromecanical System, vol 3., 1994, pp. 29-35.

[11] Xuan Xiong Zhang, Jean-Pierre Raskin ," Low-Temperature Wafer Bonding: A Study of Void Formation and Influence on Bonding Strength," Journal of Microelectromecanical System, vol 14., 2005, pp. 368-382.

[12] L. Phung, D.Valente, W. Vervisch, N.Batut, F. Alkayal, L. Ventura. "Soudure directe Silicium sur Silicium : application au contrôle de la durée de vie de commutation des diodes bipolaires de puissance". EPF 2006 Grenoble – France.

[13] H. H. YU, Z. SUO, "A model of wafer bonding by elastic accomodation," J. Mech. Phys. Solids, vol. 46, pp. 829–844, 1998.

[14] S. S. Iyer and A. J. Auberton-Herve, " Silicon wafer bonding technology for VLSI and MEMS applications", Processing series, London, EMIS, vol. 1, 2002.

[15] U. Ggosele, Q.Y. Tong, "semiconductor wafer bonding," Annu. Rev. Mater. Sci. pp. 215-241, 1989.

[16] K.H. Park, H.J Kim, O.M. Chang, H.D. Jeong,"Effect of pad properties on material removal in chemical mechanical polishing," Journal of Materials Processing Technology, pp. 73-76, 2006.

[17] Markus Forsberg, " Effect of process parameters on material removal rate in chemical mechanical polishing of Si (100)," Microelectronic Engineering, pp. 319-326, 2004.

[18] Yuling Liu, Kailiang Zhang, Fang Wang, Weiguo Di, " investigation on the final polishing slurry and technique of silicon substrat in ULSI," Microelectronic Engineering, pp. 438-444, 2002.

[19] H. Takagi, R.Maeda, T.R. Chung, N.Hosodaand T. Suga," Surface Roughness on Room-Température Wafer Bonding by Ar Beam Surface Activation", J. Appl. Phys. Vol. 37, pp. 4197-4203, Part 1, No 7, 1998.

[20] A. Laporte, M. Bagneres, D. Strutzenberger, J.M. Reynes, G. Sarrabayrouse, " Influence of bonded interface on the DC characteristics of a high voltage bipolar transistor ," ISPSD'95, pp. 279-282, 1995.

Proceedings of the 23rd International Symposium on Power Semiconductor Devices & IC's
May 23-26, 2011 San Diego, CA

Effects of back-side He irradiation on MOS-GTO performances

C. Ronsisvalle, V.Enea
STMicrolectronics, Catania, Italy

C.Abbate, G.Busatto, F.Iannuzzo,
A.Sanseverino
DAEIMI University of Cassino, Italy

G.A.P.Cirrone
LNS-INFN, Catania, Italy

Abstract— **A new version of a 1200V-20A MOS-GTO with 0.4 cm² die area is presented. The improvements of the switching performances have been achieved thanks to a new He irradiation technique performed from the back side of the device in order to avoid the degradation of the surface gate oxide. The high energy He irradiation allowed us to kill the lifetime in proximity of the N⁻/N⁺ interface in such a way to significantly improve the device switching performances as suggested by the simulations. The irradiation did not affect the on-state characteristics. Instead, a reduction by a factor ~4 in the storage time and more than 30% decrease in the turn-off energy losses have been measured on irradiated samples with respect to not irradiated ones.**

I. INTRODUCTION

Insulated Gate Bipolar Transistor (IGBT) devices are widely used in power conversion, but their performances are limited in high voltage applications, because the conductivity modulation of the N⁻base present in their vertical structure is sustained only by the collector junction. As a consequence, for devices with blocking voltages larger than 3 kV, the required wide base causes the ON state voltage drop to become relatively high. To overcome this limitation, in the last years, many thyristor-like structures have been proposed [1-3]. Among them, the MOS-GTO (Metal Oxide Semiconductor Gate Turn-Off thyristor) seems to be very promising especially in high voltage applications, thanks to a very low voltage drop and good switching performances. The MOS-GTO is the monolithic functional integration of an emitter switched GTO [4] and ensures better performances than those of IGBT, even in its clustered version [5], thanks to the fact that the conductivity modulation of the N base region of the MOS-GTO is sustained by both anode and cathode junctions. In addition, this new device is not affected by the presence of parasitic thyristors which was the main limitation of other MOS controlled thyristors [2, 3] because of their negative impact on the safe operations of those devices. As for the other bipolar devices, the switching performances of a MOS-GTO are affected by the carriers lifetime in the N base and a trade off between on-state and switching performances can be achieved thanks to the

implementation of lifetime killing techniques [6, 7].

The objective of this paper is to present the results of a study executed on 1200V-20A MOS-GTO samples irradiated with He in order to improve their switching performances. It is shown that irradiation can be executed from the back side of the die in order to avoid any degradation of the gate oxide and of the active region of the structure. Simulation and experimental results show much better performance of the irradiated devices in terms of storage time, switching losses

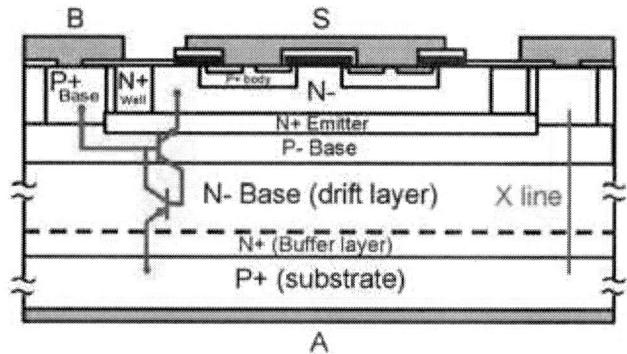

Figure 1.   Elementary cell of the MOS-GTO.

and voltage drops, compared to not irradiated ones.

II. MOS-GTO STRUCTURE

Fig. 1 shows the monolithic structure of the MOS-GTO elementary cell. It is obtained by integrating low-voltage diffused MOS (DMOS) cells inside the emitter region of the high voltage thyristor. Each DMOS cell is surrounded by N⁺ wells that have the function of inhibiting the action of the lateral PNP parasitic transistor. The presence of an N⁺ buffer layer makes this device a PT (punchthrough) version, however it is possible to fabricate non-PT devices without the buffer layer. In spite of MOS-GTO apparent complexity the perspective cost of the device is expected to be comparable with that of similar ratings IGBTs [1]. In fact the fabrication process of the MOS-GTO requires only slight modifications

978-1-4244-8425-6/11 $26.00 © 2011 IEEE   144

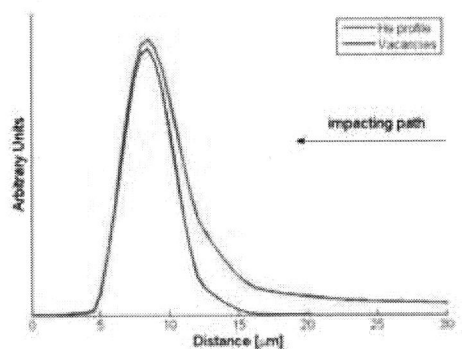

Figure 2. Typical He distribution and vacancies profile in a silicon target

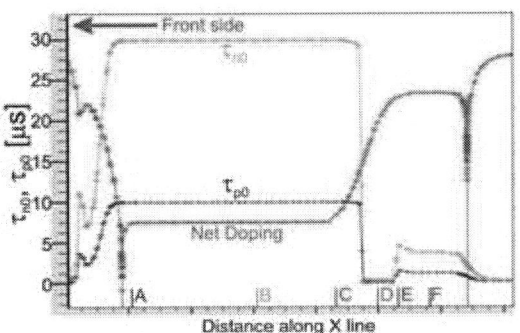

Figure 3. Lifetime and doping profile inside the elementary cell along the X line marked in Fig. 1.

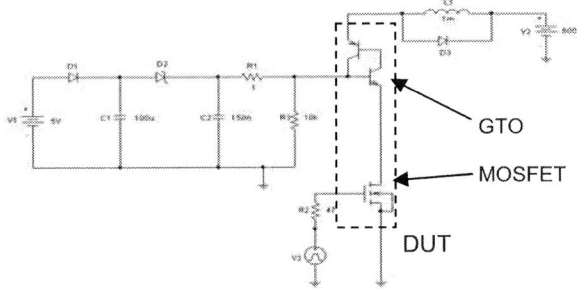

Figure 4. Circuit used for the switching study. The DUT (Device Under Test) is the dashed rectangle and includes a GTO section and a MOSFET section.

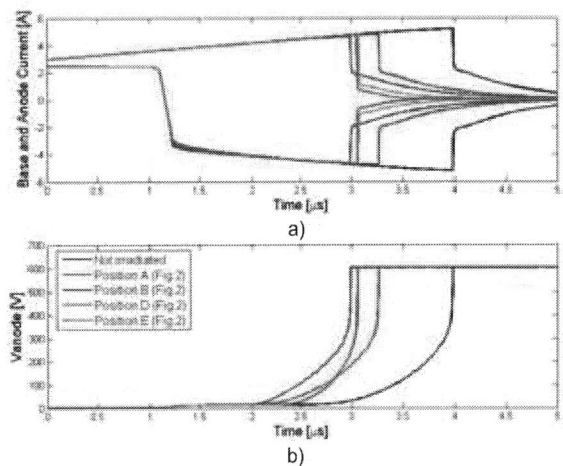

Figure 5. Simulated (a) anode and base currents and (b) anode voltage waveforms, for not irradiated device and for A, B, D and E positions of the He peak concentration.

III. OPTIMIZATION OF THE MOS-GTO PERFORMANCES BY HE IRRADIATION

A preliminary study has been performed by means of the 2D SILVACO ATLAS software [9] in order to study the effects of He irradiation on the performances of a 1200V-20A MOS-GTO. As well known, this lifetime killing technique can be used to modify the carrier lifetime in a very localized region whose depth can be chosen by using a proper energy of the impacting particles [10]. Typical He profile and primary vacancies distribution in the silicon target, obtained by SRIM [11] simulations, are reported in Fig. 2.

The doping profile of the simulated device, reported in Fig. 3, was obtained by ATHENA [9] process simulator applied to the fabricated device. The profile refers to the X line marked on the cell of Fig. 1, starting from the P^+Base region. In Fig. 3, the lifetime profiles of holes and electrons for one of the simulated cases are reported too. The width of the modified lifetime region has been chosen of about 6 μm according to SRIM simulations and experimental results reported in [10]. A reduction of two orders of magnitude in the lifetime profile of majority and minority carriers was imposed according to the results shown in [12] for common dose values used in lifetime killing techniques. Six different positions of the He peak concentration, indicated from A to F in Fig. 3, were considered in the simulations.

The schematic of the circuit used for the switching simulation is reported in Fig. 4. The driving circuit includes a standard MOSFET driver and a passive network: V_1, D_1, D_2, C_1, C_2, R_1 and R_2. This network is able to supply a forward base current to trigger the GTO at the turn-on and to keep the voltage across the MOSFET below the breakdown limit during the extraction of the base current at the turn-off [1].

The anode and base currents (a) and the anode voltage (b) waveforms are reported in Fig. 5 for positions A, B, D and E of He peak concentration. For comparison, the waveforms of a not irradiated device is reported too. The test conditions are: V_2=600V, $I_{B,ON}$=2A, I_A=6A @ turn-off.

When the He peak concentration is located in the N base, moving the peak from the surface toward the buffer layer (positions A and B) causes a significant reduction of storage time and voltage rise time. When the peak is located across the N/N$^+$ transition (positions D and E), current tailing and voltage rise time are significantly reduced, whereas the storage time has a weak increase compared to results obtained for the position B.

The overall effects of the variation of the position of the He peak concentration is summarized in Fig. 6, where the

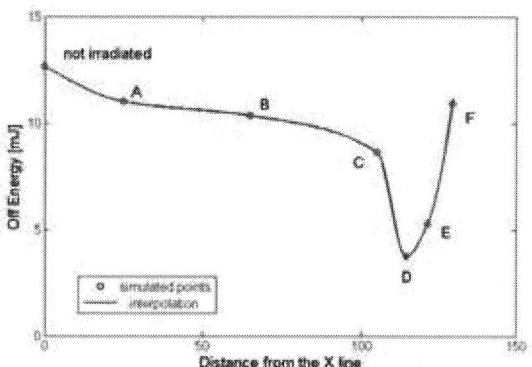

Figure 6. Turn-off energy losses as a function of the position of the He peak concentration.

simulated turn-off energy losses are reported as a function of the peak position. The minimum corresponds to the case when the lifetime is reduced at the N/N$^+$ interface (position D). In this case, a reduction of about 75% of the turn off energy is observed with respect to the not irradiated device.

Further simulations, not reported here for brevity, have shown that He irradiation has almost no effects on the on-state characteristics and on the turn-on switching performances of the device.

IV. EXPERIMENTAL RESULTS

The samples used in the experiment are rated at 1200V-20 A with a die area of ∼ 0.4 cm^2. He irradiation was executed at high energy, ∼ 40MeV, at the Cyclotron facility at the Laboratory Nazionali del Sud, Catania, Italy. On the basis of SRIM simulations, the energy of the beam was chosen in order to obtain an expected localization of the He peak concentration at the transition between N-drift and N$^+$ buffer layers (position D of Fig. 3).

The irradiation was executed from the back side of the wafer in order to avoid damages at the gate oxide and inside the active region of the device. The wafer were irradiated at the end of all technological process, metallization included, before cut and packaging.

The duration of the irradiation was changed in order to achieve tree different doses (namely, 2.5·10^9, 5·10^9 and 1·10^{10} atoms/cm^2) on the samples placed at the centre of the beam. Moreover, the radial variation of the ion flux was used to obtain other intermediate doses. A thermal annealing was performed after the irradiation in order to stabilize the defects induced.

The first interesting result of the experimental characterization is the confirmation that the on-state characteristics of the MOS-GTO are practically independent of the He irradiation dose, as shown in Fig. 7 which reports the I-V characteristics measured at I$_B$=100mA and V$_{GS}$=10V on not irradiated (black curve) and 1·10^{10} atoms/cm^2 He irradiated (red curve) devices.

The turn-on waveforms are reported in Fig. 8 for not irradiated device (black curve), and devices irradiated with 0.75·10^9 Atoms/cm^2 (blue curve) and 3.1·10^{10} atoms/cm^2 (red

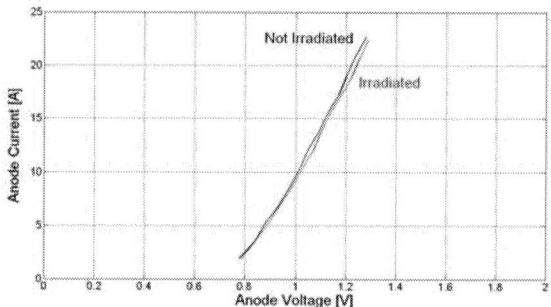

Figure 7. Experimental IV characteristics of not irradiated (black) and 1·10^{10} atoms/cm^2 He irradiated (red) devices

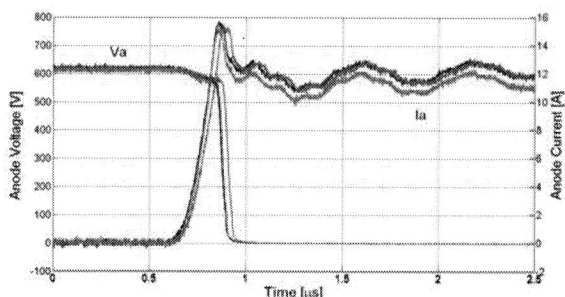

Figure 8. Anode current and voltage waveforms during turn-on of not irradiated (black), 0.75·10^9 (blue) and 3.1·10^{10} (red) Atoms/cm^2 He irradiated devices.

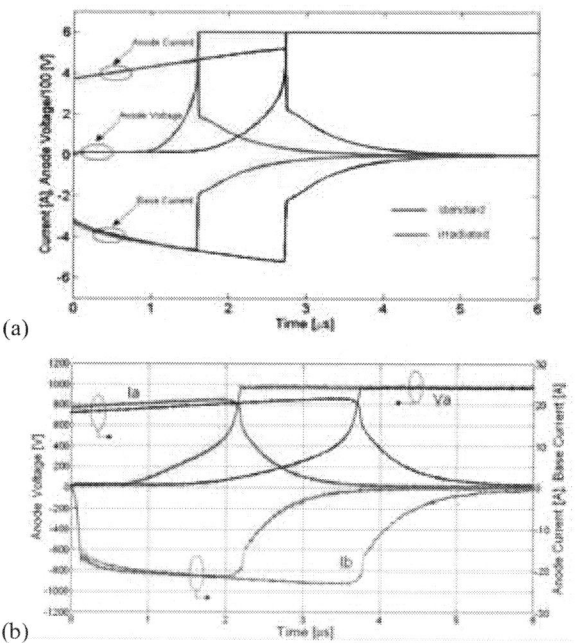

(a)

(b)

Figure 9. Simulated (a) and experimental (b) anode voltage, anode and base current waveforms of not irradiated (black) and He irradiated (red) devices.

curve). The test conditions are: V$_2$=1000V, I$_{B,ON}$=100mA, I$_A$=12A @ turn-on. Also in this case only weak variations of the current rise time are observed.

The simulated and experimental turn-off waveforms are

reported in Fig. 9 a) and b), respectively. The test conditions are: V_2=1000V, $I_{B,ON}$=100mA, I_A=22A @ turn-off. The waveforms refer to not irradiated (black curves) and $1 \cdot 10^{10}$ atoms/cm^2 He irradiated (red curves) devices. They demonstrate that, in these conditions, He irradiation is very effective in reducing the storage time and in speeding up the voltage rise, but it produces a lower effect in the tailing current. The speed up of the anode voltage causes a reduction of the turn-off energy of about 30%.

The improvement of MOS-GTO switching performances is quantified in Fig. 10 and Fig. 11, where storage time and turn-off energy, respectively, are reported as a function of the irradiation dose for three values of the anode current. The storage time is reduced by a factor ~4 and the turn-off energy losses are reduced by about 30% for all collector currents.

The experimental results show that the improvement of the switching performances obtained after the irradiation are lower than what we expected from the simulations. We attribute this difference to the fact that the He peak concentration obtained after the irradiation was located in the base region instead of being placed at the N/N$^+$ transition, due to an energy of the ions impacting the target higher than that obtained by SRIM. This behavior is consistent with the measurement of the spreading resistance performed on fresh and irradiated samples, not reported here for brevity. In fact the resistivity profile of the irradiated devices shows a significant reduction in the base region having the peak localized in vicinity of the position B of Fig. 3. This consideration is in agreement with the simulation results of Fig. 4 which shows in that case a 30% reduction of the turn-off energy. Much better performances can be expected with a localization of the He peak concentration at the N/N$^+$ transition.

V. CONCLUSION

The paper demonstrates that it is possible to perform He irradiation from the back side of the wafer thus avoiding any damages at the gate oxide of MOSFET placed on the device surface. Experimental results show that He irradiation can significantly improve the turn-off switching performances of the MOS-GTO without affecting its on-state and turn-on switching characteristics. Simulation results indicate that a proper localization of He peak concentration, obtained by changing the parameters of the irradiation (energy and dose), can be used to optimize the device characteristics and to improve significantly its switching performances. A reduction by more than 70% of the turn off energy losses compared to not irradiated devices can be expected thanks to He irradiations.

ACKNOWLEDGEMENTS

The authors wish to acknowledge Laboratori Nazionali del Sud, INFN, Catania, Italy, for having permitted the experiment. Special thanks goes to the laboratory technicians for the valuable assistance supplied during the irradiations and to Dr. Markus Italia from IMM-CNR, Catania, Italy, for the spreading resistance measurements.

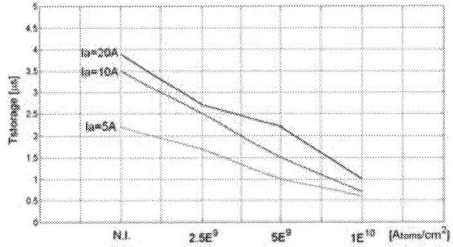

Figure 10.   Storage time at different irradiation doses.

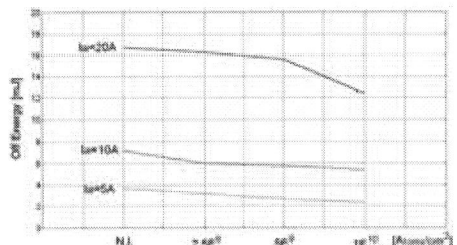

Figure 11.   Turn off energy at different irradiation doses.

REFERENCES

[1]   C. Ronsisvalle, V. Enea, C. Abbate, G. Busatto, A. Sanseverino, "Perspective Performances of MOS-Gated GTO in High-Power Applications", IEEE Trans. Electron Devices, Vol. ED- 57, pp. 2339-2343, September 2010.

[2]   V.A.K. Temple," MOS-Controlled Thyristor – A new class of power devices", IEEE Transaction on Electron Devices, vol.33, 1986, pp. 1609-1618.

[3]   M. Nandakumar, B.J. Baliga, M.S. Shekar, S. Tandon, A. Reisman, "A New MOS-Gated Power Thyristor Structure with Turn-off Achieved by Controlling the Base Resistance", IEEE Electron Devices Letters, vol.2, no.5, May 1991.

[4]   M.S. Shekar, B.J. Baliga, M. Nandakumar, S. Tandon and A. Reisman, "Characteristics of the Emitter-Switched Thyristor", IEEE Transaction on Electron Devices, vol.38, no.7, July 1991.

[5]   E.M.Sankara Narayanan, M.R. Sweet, N. Luther-King, K. Vershinin, O. Spulber, M. De Souza and J.V. Subhas Chandra Bose , "A Novel Clustered Insulated Gate Bipolar Transistor for high power applications", Semiconductor Conference, CAS 2000.

[6]   K. Nishiwaki, T. Kushida, A. Kawahashi, "A fast and soft recovery diode with ultra small Qrr (USQ-Diode) using local lifetime control by He ion irradiation", Proc. 13th International Symposium on Power Semiconductor Devices and ICs, 2001. ISPSD '01, pp. 235 – 238.

[7]   J. Vobecky, P. Hazdra, "Fast Recovery Diode With Novel Local Lifetime Control", Proc. 18th International Symposium on Power Semiconductor Devices and ICs, 2006. ISPSD '06, pp. 1 – 4.

[8]   C.Ronsisvalle, V.Enea, "The ESBT: A new monolithic power actuator technology devoted to high voltage and high frequency applications", Proc. CIPS, Nuremberg, Germany, 2008.

[9]   TCAD-OMNI User's Manuals: Device Simulation Software, SILVACO Int., Santa Clara, CA.

[10]   P. Hazdra, V. Komarnitskyy, "Local lifetime control in silicon power diode by ion irradiation: introduction and stability of shallow donors", Circuits, Devices & Systems, IET Volume: 1, Issue: 5, 2007, pp. 321 – 326.

[11]   SRIM "The Stopping and Range of Ions in Matter" http://www.srim.org

[12]   P. Spirito, S. Daliento, A. Sanseverino, L. Gialanella, M. Romano, B.N. Limata, R. Carta, L. Bellemo, "Characterization of recombination centers in Si epilayers after He implantation by direct measurement of local lifetime distribution with the AC lifetime profiling technique ", IEEE Electron Device Letters, Volume: 25 , Issue: 9, 2004, pp. 602 – 604.

Proceedings of the 23rd International Symposium on Power Semiconductor Devices & IC's
May 23-26, 2011 San Diego, CA

Temperature Dependence of Switching Performance in IGBT Circuits and Its Compact Modeling

Masataka Miyake, Masaya Ueno,
Junichi Nakashima, Hiroki Masuoka, Uwe Feldmann,
Hans Juergen Mattausch and Mitiko Miura-Mattausch
Graduate School of Advanced Sciences of Matter
Hiroshima University
Higashi-Hiroshima, Hiroshima, 739–8530 Japan
Email: http://www.hisim.hiroshima-u.ac.jp

Takaoki Ogawa and Takashi Ueta
Toyota Motor Corporation
Toyota, Aichi, 471–8571 Japan
Email: ueta@fq.tec.toyota.co.jp

Abstract—We have developed the compact IGBT model HiSIM-IGBT, based on a complete solution for the potential distribution, which connects the surface-potential of the MOS-FET part to the bipolar part by an iterative procedure in a self-consistent way. Here we report the self-heating extension of HiSIM-IGBT, a compact model for power diode including the reverse recovery effect and the model application to accurate prediction of experimental switching characteristics.

I. INTRODUCTION

The IGBT (Insulated Gate Bipolar Transistor) has been developed for high voltage applications which require low switching power and fast switching speed at several hundreds to thousands volts [1]. Since power-electronic circuits are a key target for current energy-saving efforts, accurate prediction of power dissipation in the circuit-design phase is urgently needed. Fig. 1a shows measured switching characteristics for the circuit depicted in Fig. 1b. It is seen that the collector current (I_c) peak determines maximum power dissipation. Fig. 2 shows the temperature dependence of this switching characteristic, which verifies significant changes in the switching behavior including peak position and height. To predict power dissipation of circuits accurately, these measurements must be reproduced.

II. HiSIM-IGBT

We have developed a trench-gate IGBT model for circuit simulation [2], [3], named HiSIM-IGBT, which is based on the potential distribution along the device as shown in Fig. 3, where all node potentials are solved iteratively to obtain self-consistent solutions [3]. Fig. 4 shows good agreement of the IGBT current-voltage characteristics between experiments and HiSIM-IGBT. A consistent solution is very important to predict the behavior of the dominating bipolar part during switching, which may vary according to the node potentials. Fig. 5 demonstrates this effect with the calculated potential node of V_{cb} during switching. Even though the absolute changes are quite small, they contribute exponentially to the current flow.

Furthermore, an accurate potential distribution during switching is important for the accuracy of the dynamic internal

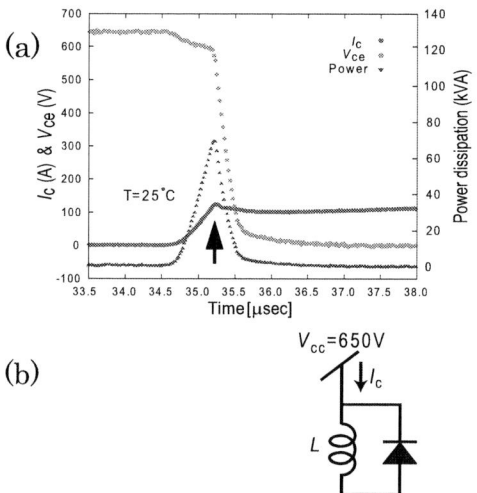

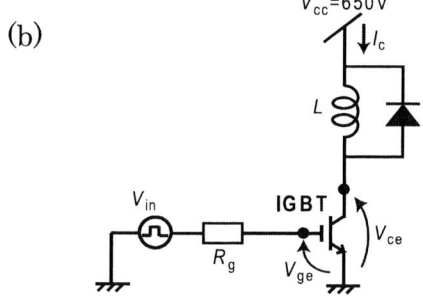

Fig. 1. (a) Measured collector current I_c, collector voltage V_{ce} and power dissipation P, of the IGBT during switch-on; (b) test inverter circuit with an inductive load and a free-wheeling diode.

capacitances of the IGBT, which are important to correctly predict switching performance. Moreover, for the accurate gate capacitance, the surface-potential-based MOSFET-model concept [4], [5], [6], [7], [8] is applied to the MOSFET part of the IGBT.

III. SELF-HEATING EFFECT AND TEMPERATURE DEPENDENCE

Fig. 6 shows 2D-device simulated distribution of the power density and the lattice temperature within the IGBT along the cut line shown in the figure. As shown in the figure, the lattice temperature distribution is almost even from the MOSFET channel to the BJT forward junction through the base region. Therefore, the RC network with just one temperature node

978-1-4244-8425-6/11 $26.00 © 2011 IEEE

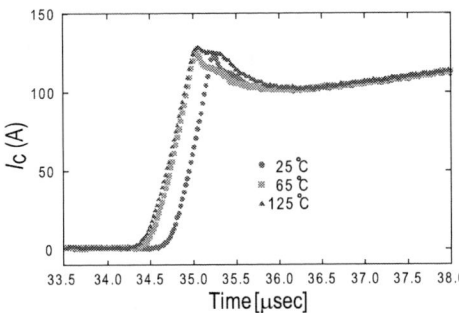

Fig. 2. Measured temperature dependence of collector current I_c during IGBT switch-on for the circuit shown in Fig. 1b.

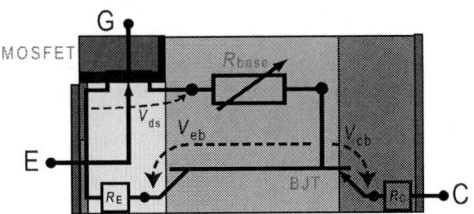

Fig. 3. Equivalent circuit of HiSIM-IGBT.

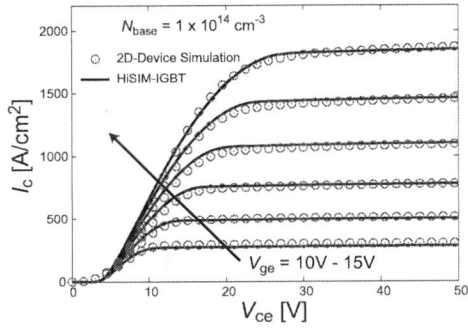

Fig. 4. Reproduction of IGBT current-voltage characteristics.

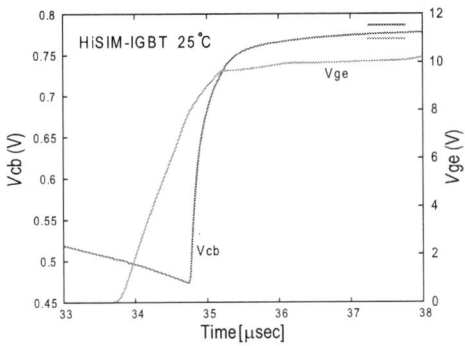

Fig. 5. Response of node-potential V_{cb} during V_{ge} switching with node potential V_{ge} (see Fig. 1b).

shown in Fig. 7a can be applied to model the self-heating effect within the IGBT. Figs. 8a and 8b show the collector current and the temperature increase of the studied IGBT [10] as a function of the collector voltage including the self-heating

effect with the equivalent circuit shown in Fig. 7a [9]. Actual temperature rise is less drastic as predicted here due to the heat-radiation effect which can also be considered in the reported HiSIM-IGBT version.

In MOSFETs, there exists an operating drain-voltage point with temperature-independent current, and this drain-voltage point is a function of the gate voltage as can be seen in Fig. 9. Above and below this voltage, the current reduces and increases with rising temperature, respectively. On the contrary, the bipolar temperature dependence is different and current normally increases with rising temperature, but becomes complicated in an IGBT due to the strong temperature dependence of the base resistance. For the p-i-n reverse-recovery-protection diode, the current-decreasing influence of a low-doped resistive region is demonstrated in Fig. 10. Measured data are in good agreement with simulation results from the compact equivalent circuit of Fig. 7b, which is implemented in HiSIM-Diode. The nonlinear series resistance R_s within the equivalent circuit is responsible for the current reduction with increased temperature.

Convergence and CPU time of circuit-level simulation is very important for the practical use of compact models. Figs. 11a and 11b show simulation results of a test inverter circuit similar to Fig. 1b with a 2D-device simulator and HiSIM-IGBT, respectively. Power dissipation and increased temperature due to the self-heating effect are verified in the figures. Here it should be noted that it took more than 10 hours to calculate Fig. 11a with a commercial 2D-device simulator, while it took much less than 0.1 second (much more than 360,000 times faster than the former) for Fig. 11b with HiSIM-IGBT implemented in the original noncommercial circuit simulator SPICE3f5 [11]. The same Linux workstation is used for the both simulations. In addition, much more switching cycles are tried to simulate, however the 2D-device simulation cannot converge. On the other hand, HiSIM-IGBT implemented in SPICE3f5 very easily converges as shown in Fig. 11c.

IV. MODEL VERIFICATION

The switching performance is complicated by the opposite temperature dependences of the IGBT's MOSFET and bipolar parts and also the interaction with the diode's reverse recovery. Fig. 12 compares the calculated switching performance of the circuit of Fig. 1b at room temperature with measurements and Fig. 13 shows magnified results at different temperatures. The largest temperature dependence is observed during switch-on of the IGBT. The temperature dependence in the initial stage is determined by the sub-threshold characteristics of MOSFET, whereas the temperature dependence of the peak is determined by the diode entering into the reverse recovery. The temperature dependent current increase is drastic up to 75°C. Beyond 75°C the switching waveform becomes nearly independent of temperature. This characteristic is well reproduced with the developed HiSIM-IGBT.

978-1-4244-8425-6/11 $26.00 © 2011 IEEE

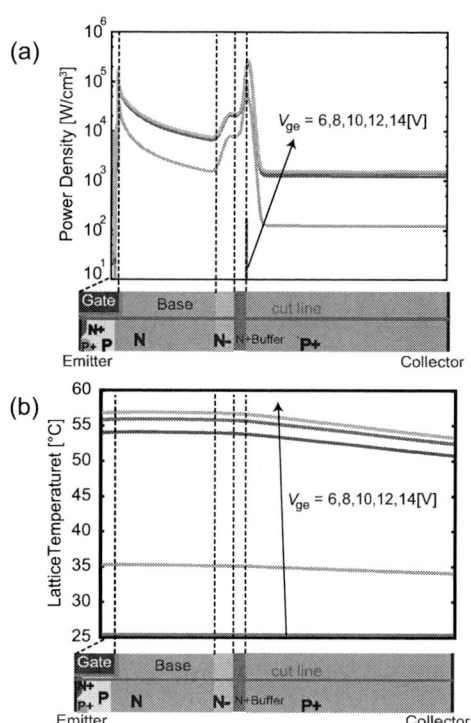

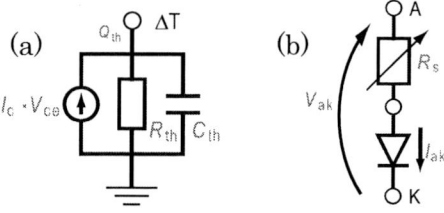

Fig. 6. IGBT temperature distribution.

Fig. 7. (a) *RC* network for the self-heating model (b) equivalent circuit of the HiSIM-Diode compact model.

V. CONCLUSION

We have introduced the self-heating-effect model to the compact trench-gate IGBT model HiSIM-IGBT, with the thermal network approach. Complete solution for the potential distribution, connects the surface potential of the MOSFET part to the bipolar part by an iterative procedure in a self-consistent way. Together with a compact model for power diode including the reverse recovery effect, the developed IGBT model is applied to the circuit simulation of a test inverter circuit. The switching performance is complicated by the opposite temperature dependences of the IGBT's MOSFET and bipolar parts and also the interaction with the diode's reverse recovery. In this work, accurate reproduction of the measured switching performance is achieved due to correct temperature-dependence modeling of each component and most importantly by the self-consistent solution for the potential distribution within each device, which determines the relative strength of each contribution.

ACKNOWLEDGMENT

The authors would like to thank Toyota Central Laboratory and Toyota Motor Corporation for their financial and technical supports.

REFERENCES

[1] D. Ueda, K. Kitamura, H. Takagi, and G. Kano, "A New Injection Suppression Structure for Conductivity Modulated Power MOSFETs", in Proc. 18th Int. Conf. Solid State Devices and Materials, pp. 97-100, 1986.

[2] A. Oohashi, M. Miyake, M. Yokomichi, H. Masuoka, T. Kajiwara, T. Kojima, N. Sadachika, U. Feldmann, H. J. Mattausch, and M. Miura-Mattausch, "Toward Predictable IGBT Model for Optimization of Device Parameters", The 5th International Workshop on Compact Modeling, pp. 53-55, South Korea, January 2008.

[3] M. Miyake, A. Ohashi, M. Yokomichi, H. Masuoka, T. Kajiwara, N. Sadachika, U. Feldmann, H. J. Mattausch, M. Miura-Mattausch, T. Kojima, T. Shoji, and Y. Nishibe, "A Consistently Potential Distribution Oriented Compact IGBT Model", in Proc. 39th IEEE Annual Power Electronics Specialist Conf., pp. 998-1003, Rhodes, Greece, June 2008.

[4] M. Miura-Mattausch, H. J. Mattausch, and T. Ezaki, "THE PHYSICS AND MODELING OF MOSFETS (Surface-Potential Model HiSIM)", World Scientific Publishing, ISBN 978-981-256-864-9 (981-256-864-6), 2008.

[5] M. Miura-Mattausch, N. Sadachika, D. Navarro, G. Suzuki, Y. Takeda, M. Miyake, T. Warabino, Y. Mizukane, R. Inagaki, T. Ezaki, H. J. Mattausch, T. Ohguro, T. Iizuka, M. Taguchi, S. Kumashiro, and S. Miyamoto, "HiSIM2: Advanced MOSFET Model Valid for RF Circuit Simulation", IEEE Trans. Electron Devices, Vol. 53, no. 9, pp. 1994-2007, September 2006.

[6] N. Sadachika, M. Miyake, M. Yokomichi, T. Kajiwara, A. Oohashi, T. Minami, Y. Oritsuki, T. Sakuda, T. Yoshida, T. Murakami, H. Kikuchihara, U. Feldmann, H. J. Mattausch, M. Miura-Mattausch, T. Ohguro, T. Iizuka, M. Taguchi, S. Miyamoto, R. Inagaki, and Y. Furui, *HiSIM HV 1.1.1 User's Manual*, (http://home.hiroshima-u.ac.jp/usdl/HiSIM HV/), Hiroshima University & STARC, 2009.

[7] M. Miura-Mattausch and H. Jacobs, "Analytical model for circuit simulation with quarter micron metal oxide semiconductor field effect transistors: Subthreshold characteristics," Jpn. J. Appl. Phys., vol. 29, pp. L2279-L2282, Dec. 1990.

[8] M. Miura-Mattausch, U. Feldmann, A. Rahm, M. Bollu, and D. Savignac, "Unified complete MOSFET model for analysis of digital and analog circuits," IEEE Trans. CAD/ICAS, vol. 15, pp. 1-7, Jan. 1996.

[9] H.A. Mantooth, et al., IEEE Trans. on Power Electron., vol. 12, no. 3, pp. 474-484, May 1997.

[10] V.K. Khanna, *IGBT Theory and Design*, Wiley Interscience, ISBN 0-471-23845-7, August 2003.

[11] Spice webpage within the Department of Electrical Engineering and Computer Science (EECS) at the University of California, Berkeley, "http://embedded.eecs.berkeley.edu/pubs/downloads/spice/".

[12] P.R. Palmer, E. Santi, J.L. Hudgins, Xiaosong Kang, J.C. Joyce and Poh Yoon Eng, "Circuit simulator models for the diode and IGBT with full temperature dependent features," IEEE Trans. Power Electronics, Vol. 18, Issue 5, pp. 1220-1229, Sept. 2003.

[13] R. Chibante, A. Araujo and A. Carvalho, "Finite-Element Modeling and Optimization-Based Parameter Extraction Algorithm for NPT-IGBTs," IEEE Trans. Power Electronics, Vol. 24, Issue 5, pp. 1417-1427, May 2009.

[14] S. Machida, T. Sugiyama, M. Ishiko, S. Yasuda, J. Saito, K. Hamada, "Investigation of correlation between device structures and switching losses of IGBTs," 21st Int'l Symposium on Power Semicond. Dev. & IC's (ISPSD), pp. 136-139, Barcelona, June 2009.

[15] A. R. Hefner and D. M. Diebolt, "An experimentally verified IGBT model implemented in the Saber circuit simulator", IEEE Trans. Power Electron., vol. 9, p. 532, 1994.

[16] B. J. Baliga, M. S. Adler, P. V. Gray, R. P. Love, and N. Zommer, "Insulated gate rectifier (IGR): A new power switching device," in Proc. Tech. Dig. IEDM, 1982, pp. 264-267.

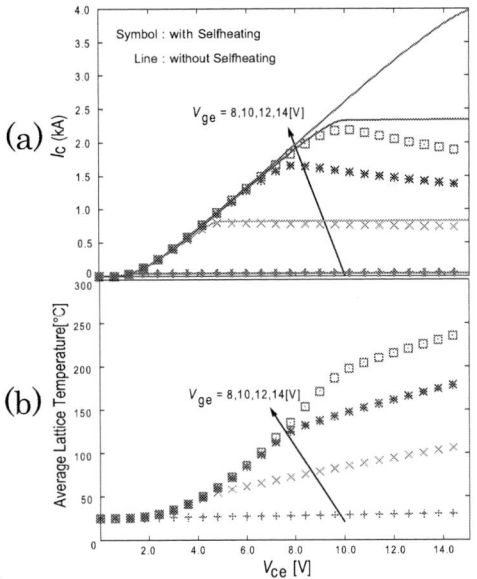

Fig. 8. (a) Characteristics of collector current I_c as a function of collector voltage including the self-heating effect as calculated with HiSIM-IGBT. (b) Averaged IGBT temperature for the calculation in (a).

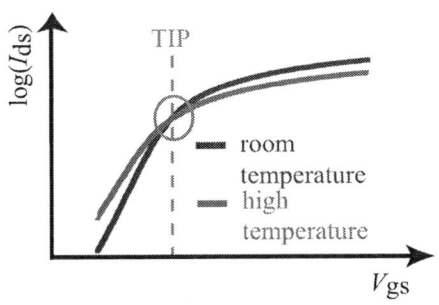

Fig. 9. Temperature-independent point (TIP) in the MOSFET-part operation.

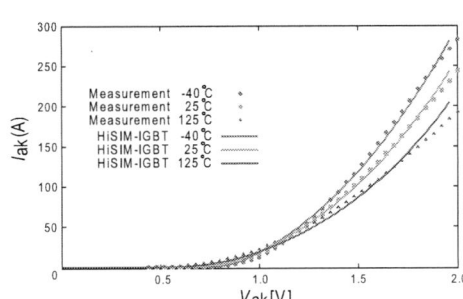

Fig. 10. Temperature dependence of p-i-n power diode current.

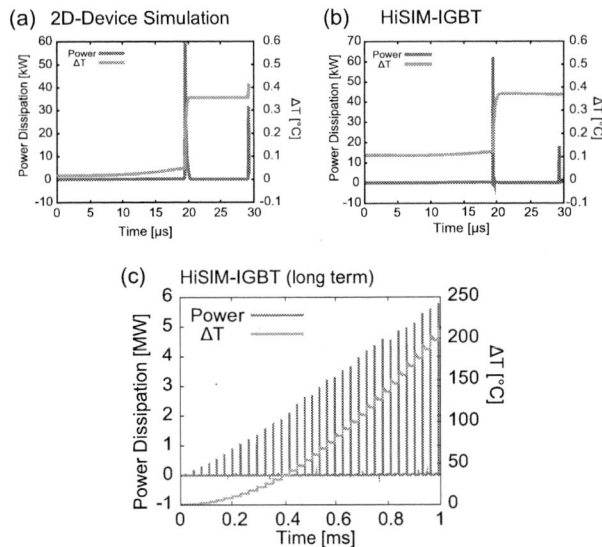

Fig. 11. (a) 2D-device simulation took more than 10 hours. (b) HiSIM-IGBT implemented in SPICE3 took only less than 0.1 second. (c) Only HiSIM-IGBT can stably converge the simulation with many cycles.

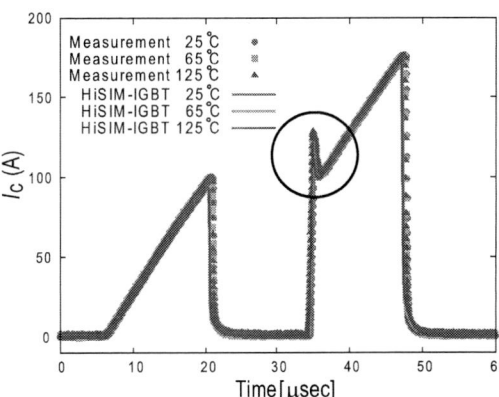

Fig. 12. Switching characteristics of the circuit shown in Fig. 1b as calculated with HiSIM-IGBT in comparison to the measured data at three different temperatures.

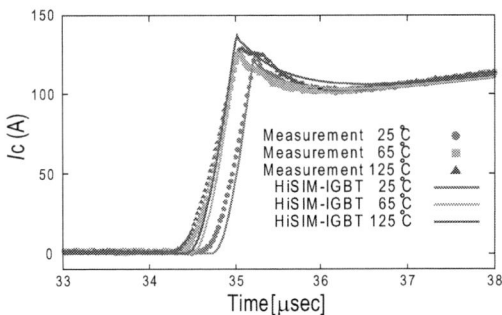

Fig. 13. Magnified switch-on calculation results for different temperatures from $25°$ to $125°$.

978-1-4244-8425-6/11 $26.00 © 2011 IEEE

Proceedings of the 23rd International Symposium on Power Semiconductor Devices & IC's
May 23-26, 2011 San Diego, CA

Full Understanding of Hot-Carrier-Induced Degradation in STI-based LDMOS transistors in the Impact-Ionization Operating Regime

S. Poli, S. Reggiani, G. Baccarani, E. Gnani, A. Gnudi

ARCES and DEIS, University of Bologna
Bologna, Italy, email: spoli@arces.unibo.it

M. Denison, S. Pendharkar, R. Wise

Texas Instruments, Inc.
Dallas, Texas

Abstract— Hot-carrier-injection (HCI) effects are studied in n-channel rugged LDMOS transistors in high current-voltage biases, by monitoring the linear and saturation regimes. Experimental data reveal that the degradation effects responsible for the HCI parameter drifts are mainly localized in the channel and in the drift region close to the drain. The temperature dependence of the HCI degradation is analyzed to gain understanding in the underlying physics. TCAD simulations aimed at investigating the sensitivity of the current shift to different local distributions of trapped charges have been carried out, and a compact model for the linear current has been developed for the purpose of extracting the effective-mobility degradation in the channel and the charge trapped in the drift region. The overall methodology represents a new approach to the HCI analysis suitable for device structures with STI in the drain extension region.

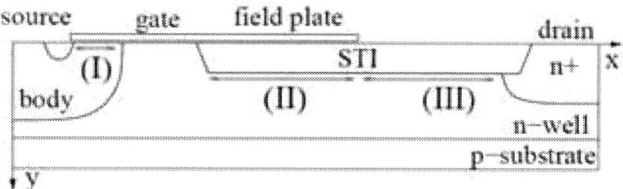

Figure 1. Schematic view of the STI-based LDMOS with buried body implant. Three different regions are monitored in the high stress conditions, corresponding to high field peaks at the Si/SiO2 interface: (I) channel, (II) STI under field plate and, (III) STI at the drain side.

I. INTRODUCTION

Lateral DMOS transistors are widely used as integrated high-voltage switches and drivers in mixed-signal integrated circuits. The extended drain structure with shallow trench isolation (STI) and the "RESURF" technology [1] are widely used to obtain high-voltage capability while keeping low on-resistance so to optimize the specific resistance versus breakdown voltage trade-off.

In addition, the use of a buried body implant [2] is capable of suppressing the snapback effect in the *I-V* characteristics: this important feature enables an extension of the electrical safe-operating-area to higher drain voltages, when impact ionization at the drain side of the drift region becomes dominant [3, 4].

High operational drain and gate biases make the LDMOS device vulnerable to the damage caused by hot-carrier-injection (HCI), and the reliability characterization in STI-based LDMOS devices have recently drawn much attention [5-7]. However, a thorough investigation of the degradation mechanisms in the impact-ionization regime is still missing. In a previous work [8], the HCI-induced degradation of the rugged LDMOS has been investigated for different stress biases and ambient temperatures, focusing on the

enhancement of the HCI effect in the channel region. In this work, the analysis has been extended to the degradation effects in both linear and saturation regimes. The use of TCAD simulations in combination with a compact model for the linear current is proposed as a new methodology for the qualitative and quantitative investigation of HCI drift in STI-based devices.

II. DEVICE MODEL AND SIMULATION SETUP

Fig. 1 shows the schematic cross-section of the transistor investigated in this paper: an n-channel lateral DMOS transistor with shallow trench isolation and a buried body implant under the source [2]. The device sketch shows the three regions that are mainly contributing to performance degradation under HCI: the channel region (I), the STI to silicon interface under the gate field-plate (II), and the drain side of the STI (III). Each region experiences large electric field peaks at the Si/SiO2 interface under stress conditions. A 2D simulation deck has been used for the TCAD numerical investigation using Synopsys' tools [9]. A first definition of the concentration profiles within the device cross-section has been derived from SIMS measurements and process simulations [4]. Comparisons with experimental results in

---

Work supported by the SRC Research Contract No. 2007-VJ-1667

978-1-4244-8425-6/11 $26.00 © 2011 IEEE

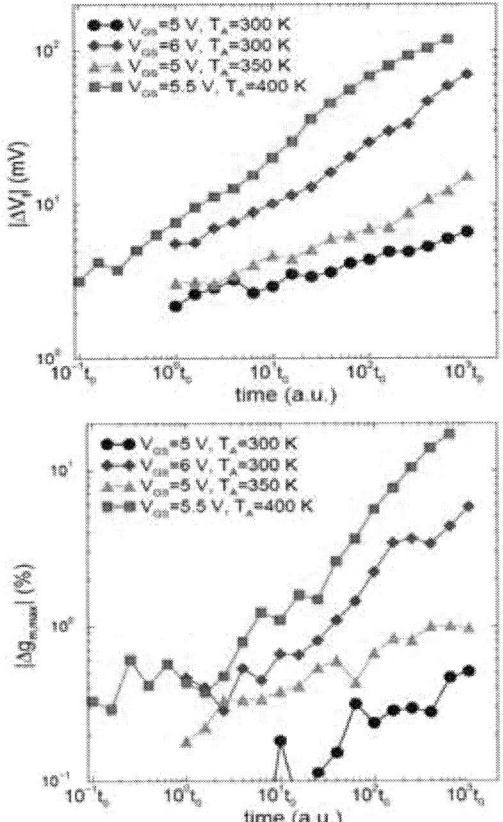

Figure 2. Experimental (top) threshold voltage shift and (bottom) maximum transconductance shift as a function of stress time at $V_{DS} = 70$ V and variable V_{GS} and T_A.

different operating regimes and with different characterization techniques (i.e., DC and TLP measurements) have been carried out to validate the overall structure [8]. Simulations have been performed using the drift-diffusion transport model and the heat transfer equation available within the electro-thermal tool. Special care has been devoted to the tuning of the self-heating effects through a calibration of the thermal boundary conditions, obtaining a good agreement between the experimental and numerical characteristics up to large drain and gate biases and over a wide range of ambient temperatures (T_A).

III. EXPERIMENTAL DATA AND TCAD ANALYSIS

Hot-carrier induced degradation measurements have been performed under DC stress conditions for a fixed drain bias ($V_{DS} = 70$ V) and different gate voltages ($V_{GS} \geq 5$ V) and ambient temperatures ($T_A = 25\text{-}125$ °C). In all biases, the device operates in its impact-ionization regime. The drift of the main electrical parameters has been monitored at regular intervals during the stress, evaluating the threshold shift (ΔV_t), the maximum transconductance shift in linear regime ($\Delta g_{m,max}$), and the linear and saturation current shift ($\Delta I_{D,lin}$ and $\Delta I_{D,sat}$, respectively). The linear current is measured at $V_{DS} = 0.1$ V, whereas the saturation current is extracted at $V_{DS} = 35$ V, close to $V_{DD}/2$, with V_{DD} the nominal voltage of the device. As a first approach to the analysis of the

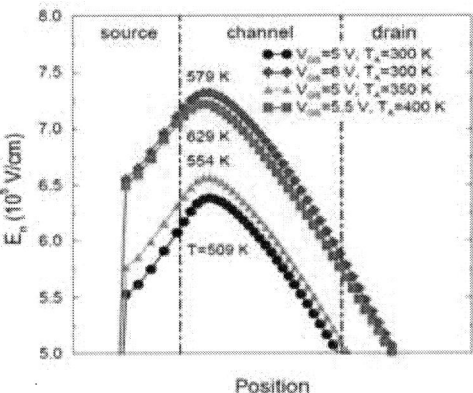

Figure 3. Normal electric field vs. position at the oxide interface along region I for two gate biases and ambient temperatures. Labels: local lattice temperature, slowly varying along the channel. Low values are observed for the parallel component of the electric field in the same region.

experimental data, TCAD simulations at the stress bias condition have been performed, monitoring the physical quantities impacting on the HCI degradation in each critical region. By this way, the impact of degradation in each region on the device parameters has been separately investigated as well.

A. Degradation effects along the channel

Fig. 2 shows the experimental ΔV_t and $\Delta g_{m,max}$ shifts as a function of stress time for different stress biases and ambient temperatures. As previously observed in [8], an increase of the degradation level for short stress times is reported by increasing the stress gate voltage or T_A. In addition, an increase of the curve slope is observed, with similar bias dependencies. With particular reference to the V_t shifts, stress and recovery measurements have been performed to verify the negligible impact of the recovery effects.

The large ΔV_t and $\Delta g_{m,max}$ is ascribed to an enhancement of the interface trap formation along the channel, and TCAD analysis has shown a clear correlation between the observed degradation features and the local lattice temperature and electric field in this region. In Fig. 3, the normal component of the electric field is reported along the channel and the adjacent source and drain regions at the considered biases. High electric field peaks close to the source junction explain the strong interface and oxide trap formation in this area [10]. Further, for short stress times, similar values of the V_t shift are found to be associated to close values of the electric field. On the other side, the increase in the slope is found to be thermally driven, with the steepest slope for the stress condition with higher channel lattice temperature (shown in the legend). It is worth noticing that large self-heating effects are reported due to both the large saturation current and the additional hole flux generated by impact-ionization and collected at the body region.

B. Degradation along the drift region

Fig. 4 shows $\Delta I_{D,lin}$ and $\Delta I_{D,sat}$ shifts for the same stress conditions. With reference to the $\Delta I_{D,lin}$ case, large degradation levels are found when increasing V_{GS} and T_A. As reported in

978-1-4244-8425-6/11 $26.00 © 2011 IEEE

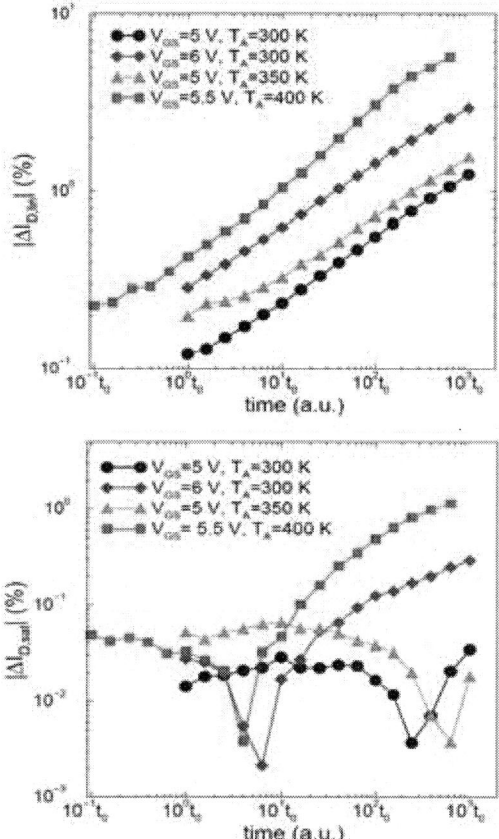

Figure 4. Experimental (top) linear current shift and (bottom) saturation current shift as a function of stress time at $V_{DS} = 70$ V and variable V_{GS} and T_A. The saturation current is evaluated at $V_{DS} = 35$ V.

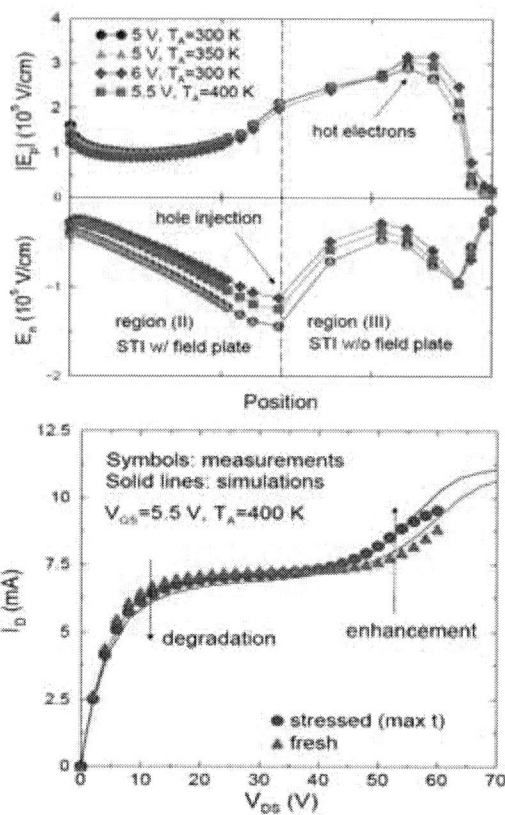

Figure 5. (Top) Normal and parallel electric field components along the drift region at the considered stress biases. (Bottom) Output characteristics at the maximum V_{GS} showing the different effect of interface traps at medium-low and high voltages. Symbols: experiments. Lines: TCAD simulations with fixed N_{it} along region III.

[8], the linear current degradation shifts under high V_{GS} stresses can be larger than those reported in the case of low V_{GS} stresses, when degradation is mainly occurring at the source side of the STI (region II), and with an opposite temperature dependence. Thus, the $\Delta I_{D,lin}$ degradation in Fig. 4 can be ascribed mainly to an enhancement of the degradation occurring along the STI interface close to the drain (region III). This is further confirmed by the analysis of the electric field local distributions. As shown in Fig. 5, high parallel electric field components are observed along region III, causing interface-trap formation by hot electrons. It is important to note that, as further analyzed in the following, the maximum contribution given by the channel degradation to the total linear current drift is less than 30%.

For the saturation current degradation, two different trends are observed for short and long stress times. At short stress times, a slight enhancement (negative $\Delta I_{D,sat}$) of the current is reported. This effect can be ascribed to holes generated by impact-ionization in the drift region and injected by a large normal field at the STI close to the field-plate edge (Fig. 5, top). At longer times, the degradation effect due to hot electrons at the drain side of the STI becomes dominant, recovering a trend identical to that observed for $\Delta I_{D,lin}$. Furthermore, considering the output characteristics at a fixed gate voltage ($V_{GS} = 5.5$ V), a different impact of the trap formation along the drift region is observed in the quasi-

saturation and in the impact-ionization regime, as shown in Fig. 5, bottom. In the medium-low voltage limit, the trap formation along the drift causes a resistance increase degrading the saturation current. At larger drain voltages, the modulation of the local electron density induced by the acceptor-type interface traps causes an enhancement of the impact-ionization current [11], which explains the observed overall current increase. This effect is well captured by TCAD simulations including interface traps with a fixed concentration (N_{it}) along region III. On the contrary, if large trap formation is assumed along the interface in region II, the positive feedback on the impact-ionization current would be lost, and the degradation effect would be dominant over the entire V_{DS} range.

IV. ANALYTICAL MODELING

In order to have an additional insight on the degradation mechanisms, and to separately extract the contributions to degradation in the different regions of the device, an analytical model of the current in the linear regime has been developed. The total source-to-drain resistance has been divided in local resistive contributions and each of them has been separately modeled and calibrated against numerical results extracted by TCAD simulations. In each region, the resistance model accounts for the local doping, the mobility dependencies (e.g., by modeling the effective mobility as a function of normal

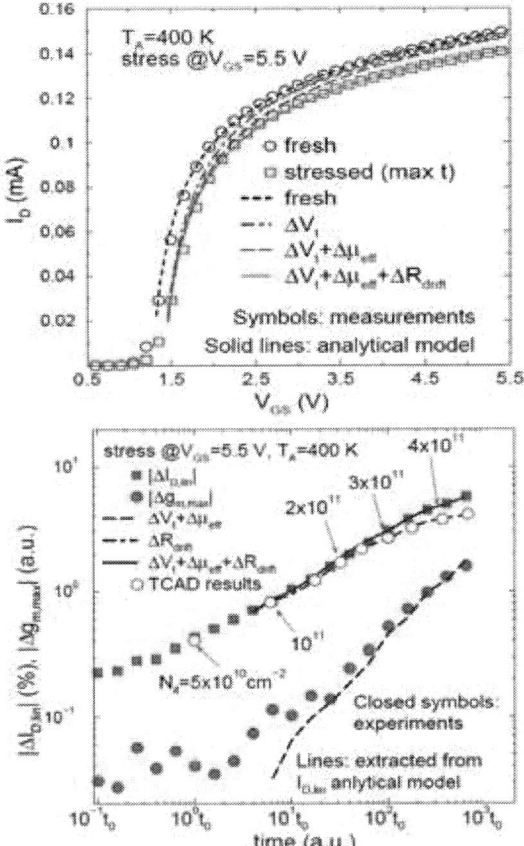

Figure 6. (Top) Turn-on characteristics of a fresh and stressed device (symbols) compared with the analytical model (solid lines). The analytical model is reported by separately accounting for the effects of different parameter shifts, namely, ΔV_t, $\Delta \mu_{eff}$, and ΔR_{drift}. (Bottom) $\Delta I_{d,lin}$ extracted from the turn-on characteristics at different stress times (lines) compared with experiments (closed symbols): the contributions of channel degradation only ($\Delta V_t + \Delta \mu_{eff}$) and of drift region only (ΔR_{drift}) are separately reported. Open symbols: TCAD simulations with fixed average N_{it} in region III. The extracted average N_{it} values are given in legend.

effective field, doping and T_A) and the geometrical effects. The dependencies on the applied gate bias have been checked separately for each contribution by means of the extracted TCAD results. In Fig. 6, top, an example of the use of the model is reported. First, a calibration is performed, perfectly matching the turn-on characteristic of a fresh device. Then, the degradation of the different parameters is considered in order to obtain the stressed device *I-V* curve. In Fig. 6, the effects of the shift of the threshold voltage, of the channel mobility, and of the drift region at long stress times are separately reported. A quantitative extraction of the channel mobility degradation is also obtained, with a mobility reduction of about 11%. A separation between the channel and drift contributions to the total on-resistance increase is shown in Fig. 6, bottom: the extraction is performed for different stress times, showing the time evolution of the different degradation mechanisms. A dominant contribution of the drift degradation over the channel degradation is reported. The channel resistance increase, mainly associated to the mobility degradation effects, is found to well correlate with the measured $g_{m,max}$ shifts. Finally, TCAD simulations with fixed interface traps along

region III have been performed for extracting the average N_{it} along the drift region (open symbols), obtaining a validation of the analytical extraction. It is worth noting that the presented methodology, allowing for a separated evaluation of the N_{it}-induced mobility degradation and N_{it} localization along the drain side of the STI, represents an approach to the HCI analysis which is especially suited for STI-based devices, where other conventional measurement techniques like, e.g., the charge-pumping approach, could fail due to the presence of a thick oxide covering the degraded interface, which limits the gate-induced voltage sweep.

V. CONCLUSIONS

In this work a study of the hot-carrier effects on the device parameters of a STI-based rugged LDMOS is presented, which is based on the extensive use of TCAD results and analytical modeling. The role played by the different regions within the device to the linear- and saturation-current degradation curves has been separately extracted, as well as the information on the effective mobility degradation and the average charge trapped along the STI interface. All the experimental observations have been clearly explained by using the proposed approach.

REFERENCES

[1] A. W. Ludikhuize, "A Review of RESURF Technology," in *Proc. of ISPSD'00*, Toulouse (France), May 22-25, 2000, pp. 11-18.

[2] P. Hower, J. Lin, S. Pendharkar, B. Hu, J. Arch, J. Smith, and T. Efland, "A Rugged LDMOS for LBC5 Technology," in *Proc. of ISPSD'05*, Santa Barbara (CA), May 23-26, 2005, pp. 327-330.

[3] J J. Lin and P. L. Hower, "Two-carrier current saturation in a lateral DMOS," in *Proc. of ISPSD'06*, Napoli (Italy), June 4-8, 2006, pp. 1–4.

[4] S. Reggiani, G. Baccarani, E. Gnani, A. Gnudi, M. Denison, S. Pendharkar, R. Wise, and S. Seetharaman, "Explanation of the Rugged LDMOS Behavior by Means of Numerical Analysis", in *IEEE Trans. on Electron Devices*, 56(11), pp. 2811-2818, November 2009.

[5] J.F. Chen, K.-S. Tian, S.-Y. Chen, K.-M. Wu, C.M. Liu, "On-Resistance Degradation Induced by Hot-Carrier Injection in LDMOS Transistors With STI in the Drift Region", in *IEEE Electron Dev. Lett.*, 29(9), pp. 1071-1073, 2008.

[6] J.F. Chen, K.-S. Tian, S.-Y. Chen, K.-M. Wu, J.R. Shih, K. Wu, "An Investigation on Anomalous Hot-Carrier-Induced On-Resistance Reduction in n-type LDMOS Transistor", in *IEEE Trans. On Dev. And Mat. Rel.*, 9(3), pp. 459-464, 2009.

[7] J.F. Chen, S.-Y. Chen, K.-M. Wu, J.R. Shih, and K. Wu, "Convergence of Hot-Carrier-Induced Saturation Region Drain Current and On-Resistance Degradation in Drain Extended MOS Transistors," *IEEE Trans. On Electron Devices*, 56(11), pp. 2843–2847, 2009.

[8] S. Poli, S. Reggiani, G. Baccarani, E. Gnani, A. Gnudi, M. Denison, S. Pendharkar, R. Wise, S. Seetharaman, "Investigation on the temperature dependence of the HCI effects in the rugged STI-based LDMOS transistor," in *Proc. of ISPSD'10*, Hiroshima (Japan), June 6-10, 2010, pp. 311-314.

[9] Synopsys Inc., "Sentaurus device simulator (release C-2009.06)," 2009.

[10] S. Aresu, W. De Ceuninck, G. Van De Bosch, G. Groeseneken, P. Moens, J. Manca, D. Wojciechowski, and P. Gassot, "Evidence for source side injection hot carrier effects on lateral DMOS transistors," in *Microelectronics Reliability*, 44(9-11), pp. 1621-1624, 2004.

[11] S.-Y. Chen, J.F. Chen, J.R. Lee, K.-M. Wu, C.M. Lie, and S.L. Hsu, "Anomalous Hot-Carrier-Induced Increase In Saturation-Region Drain Current in n-Type Lateral Diffused Metal-Oxide-Semiconductor Transistors", *IEEE Trans. On Electron Devices*, 55(5), pp. 1137–1142, 2008.

Proceedings of the 23rd International Symposium on Power Semiconductor Devices & IC's
May 23-26, 2011 San Diego, CA

Avalanche Instability in Oxide Charge Balanced Power MOSFETS

J. Yedinak, R. Stokes, D. Probst*, S. Kim*, A. Challa*, S. Sapp**

Fairchild Semiconductor Corporation

Wilkes Barre, PA, USA; *Salt Lake City, Utah, USA; ** San Jose, California, USA

Ph: (570) 706-4023, Fax: (570) 706-4030, Email: joe.yedinak@fairchildsemi.com

Abstract

Power MOSFET designs have been moving to higher performance particularly in the medium voltage area. (60V to 300V) New designs require lower specific on-resistance (R_{SP}) thus forcing designers to push the envelope of increasing the electric field stress on the shielding oxide, reducing the cell pitch, and increasing the epitaxial (epi) drift doping to reduce on resistance. In doing so, time dependant avalanche instabilities have become a concern for oxide charge balanced power MOSFETs. Avalanche instabilities can initiate in the active cell and/or the termination structures. These instabilities cause the avalanche breakdown to increase and/or decrease with increasing time in avalanche. They become a reliability risk when the drain to source breakdown voltage (BV_{dss}) degrades below the operating voltage of the application circuit. This paper will explain a mechanism for these avalanche instabilities and propose an optimum design for the charge balance region. TCAD simulation was employed to give insight to the mechanism. Finally, measured data will be presented to substantiate the theory.

Introduction

Medium voltage rated Oxide Charge Balanced (Shielded Gate) MOSFETs similar to the one shown in Fig 1 have become the MOSFET of choice in DC to DC converters, Fly Forward, and Power Over Ethernet (POE) applications because of their lower R_{SP} and Q_{gd}.

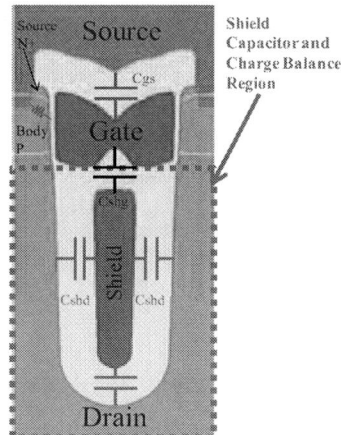

Fig 1 Typical Shielded Gate Charge Balance Structure

For oxide charge balance devices the breakdown voltage and a portion of the R_{SP} is controlled by the charge balance set up by the shielding oxide thickness and drift region doping charge in the mesa. The higher the voltage rating for the MOSFET, the bigger the impact this charge balance has on reducing R_{SP}. For a given cell pitch and

shielding oxide thickness, the designer makes a choice of the silicon mesa charge and considers factors like process thermal budget, doping segregation, and fixed oxide charge of the shielding oxide. The designer develops trade-off curves for R_{SP}, BV_{dss}, and gate to drain charge (Q_{gd}) as displayed in Fig 2. Other characteristics like Unclamped Inductive Switching (UIS) [1], Gate input capacitance (C_{iss}), Miller capacitance (C_{rss}), and gate charge (Q_g) may also be considered.

Simulation Analysis

Shielded Gate MOSFETs similar to the one shown in Fig 1were fabricated and the device structures were analyzed using SYNOPSYS® TSUPREM-4™ and Medici Mixed Mode TCAD simulation tools. Fig 2 reveals BV_{dss} to be relatively flat over a possible design range of epi doping. If the designer's only interest is the best trade-off for BV_{dss} and R_{SP} (Case1), an epi doping concentration greater than 1.0 times the normalized optimum value would most likely be chosen because increasing Q_{gd} is of secondary importance to the final design performance. Designing in this area causes the avalanche location within the cell to move toward the P_{body} junction and trench sidewall. This can result in charge injection into the gate oxide or a reduction in UIS capability. [1], [2] The characteristics for BV_{dss} instability designing in this area of charge balance is for BV_{dss} to decrease with increasing time in avalanche. This is because avalanche injected holes into the shield oxide shift the device to a more N-type rich charge balance state and the electric field in the mesa region increases with increased hole injection in the shield oxide.

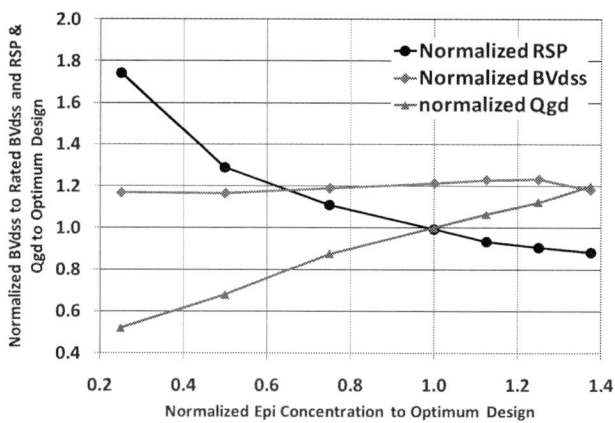

Fig 2 Normalized Plot of 10V R_{SP} @ 200A/cm2, BV_{dss}, & Q_{gd} @ 200A/cm2, Vd=50% rated for shield oxide fixed charge of 1e11

978-1-4244-8425-6/11 $26.00 © 2011 IEEE

By designing for better switching performance and dv/dt immunity by lowering Q_{gd} (Case2), an epi concentration to the left of 1.0 normalized value would be desirable in Fig 2. Increasing Rsp is of secondary importance to improving the BV_{dss} versus Q_{gd} trade-off. Previous work showed that this area of charge balance has the most stable avalanche location pinning it to the trench bottom during UIS [1]. The characteristic for BV_{dss} instability designing in this area of charge balance is for the BV_{dss} to increase with increasing time in avalanche. This is because injected charge moves the mesa to a higher N-type charge balance state. Provided the shield oxide thickness at the trench bottom can support the increase electric field, the voltage in the mesa between the trenches increases.

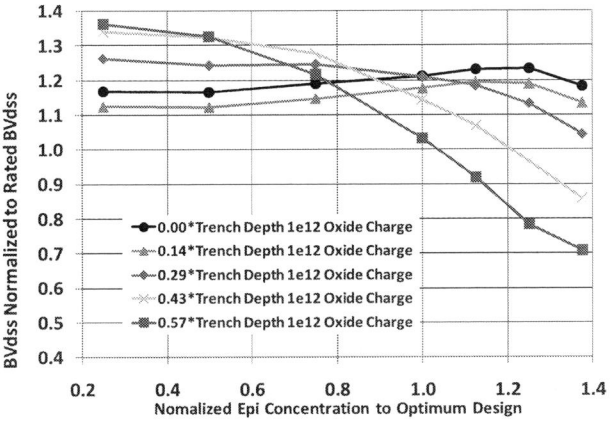

Fig 3 Simulated normalized BV_{dss} vs normalized epi concentration for oxide charge increased to 1e12 from 5e10 a fixed distance from the trench bottom

Fig 3 shows a BV_{dss} trade-off for epi doping and fixed shield oxide charge increasing from 5e10 to 1e12 a fixed distance from the trench bottom. Utilizing Fig. 2 & 3 the optimum charge balance condition for a fixed shield oxide thickness for balancing the trade-offs between R_{SP}, BV_{dss}, Q_{gd}, & BV_{dss} stability is the normalized epi concentration of 1.0.

As shown in Fig 3 for Case 1, BV_{dss} decreases as the holes in the oxide, simulated as fixed oxide charge, near the trench bottom increases for epi concentrations to the right of 1.0 normalized epi concentration. The electric potential first compresses at the trench bottom increasing the electric field in the silicon and shield oxide. BV_{dss} decreases as shown by the 0.14*Trench Depth curve. Reduced voltage supported below the trench results as shown in Fig 4(a) & 4(b). This injected charge continues to build in the oxide as time in avalanche increases causing the avalanche point to move up the trench side wall towards the top of the shield electrode. Thus BV_{dss} degrades even further as shown by the 0.29, 0.43 and 0.57 * Trench Depth curves. Figs 4(b) through 4(e) show that

the reason for this is the mesa supports less voltage towards the trench bottom and shifts to a more N-type rich charge balance condition.

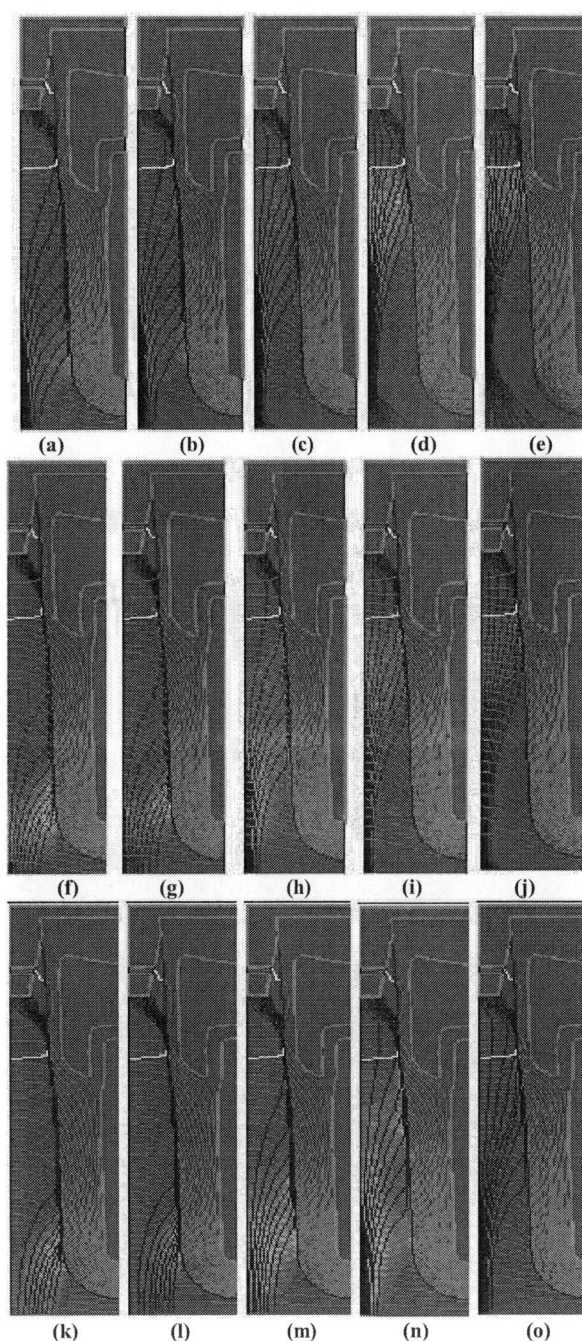

Fig 4 Simulated 1e12 Oxide Charge Impact Ionization contours, Current flow lines, and Potential contours: (a), (b), (c), (d), & (e) are normalized epi concentration 1.25, (f), (g), (h), (i), & (j) are normalized epi concentration 1.0, & (k), (l), (m), (n), & (o) are normalized epi concentration 0.75. Left to right the shielding oxide charge is increased to 1e12 from 5e10 for each epi concentration a distance normalized to the total trench depth of 0, 0.14, 0.29, 0.43, 0.57

For Case 2, BV_{dss} again first decreases for the same reason described above for Case 1 as shown in Fig 4(k) and 4(l). This injected charge continues to build in the oxide as time in avalanche increases causing the avalanche point to move up the trench sidewall. The result is BV_{dss} continues to increase as shown by the 0.29, 0.43, and 0.57 * Trench Depth curves. When the avalanche point and the injected charge level moves up the trench sidewall, BV_{dss} increases resulting from an improved charge balance condition for epi concentrations less than 0.8. Thus, the electric field in the mesa increases and voltage increases within the mesa as shown in Fig 4(l) through 4(o). The designer must choose the best trade-off for BV_{dss}, R_{SP}, Q_{gd}, & avalanche stability. This is most likely a concentration between 0.9 and 1.0 normalized epi concentration and shown in Fig 4(f) through 4(j).

Figs 5(a) through 5(c) plot the sensitivity to fixed oxide charge as the distance the fixed charge increases from the trench bottom to the top of the shield electrode for three charge balance conditions. For simulation, the oxide charge has a uniform distribution along the length of the shield oxide. (This is most likely not the reality in experimental devices.) The point at which the change in the distance of the fixed charge reaches the top of the shield electrode is around a normalized distance of 0.6. At the top of the shield electrode the fixed oxide charge stops having a significant impact on BV_{dss} due to the modification of the charge balance. Fig 5(a) shows a 1.25X optimum drift epi doping concentration (Case 1), the BV_{dss} starts to fall as soon as the positive oxide charge is increased to 5e11 and continues to fall as the distance the charge increases from the trench bottom toward the top of the shield electrode is increased. For Case 2 Fig 5(c) shows that when the drift epi doping concentration is 0.5X the optimum, BV_{dss} becomes less sensitive to injected charge close to the trench bottom. Only at an extreme oxide charge value of 3e12 or when the entire shield oxide along the trench sidewall has a 2e12 charge shown by the 0.57 normalized distance does BV_{dss} fall below the rated value. In order to obtain optimal trade-offs between R_{SP} and Q_{gd} and good oxide charge insensitivity, the optimal design in Fig 5(b) is chosen. This design has good charge immunity to 1e12 charge across the entire length of the shield oxide and a good compromise for R_{SP} and Q_{gd} shown in Fig 6.

Experimental Results
The maximum measured degradation in BV_{dss} for charge injection resulting from a one minute time in avalanche with 33mA/cm^2 on the experimental devices was -12% of rated BV_{dss} for devices fabricated similar to Case 1 and +0.7% of rated BV_{dss} for devices fabricated similar to Case 2. From Fig 5(a) and 5(c) for these shifts in BV_{dss} the charge injected uniformly along the length of the shield oxide would be 5e11 or less. Also, from Fig 5(a)

and 5(c) for an injected charge of 1e12 the distance from the bottom of the trench that the charge injection could extend would be 0.2 to 0.3 the normalized depth. Fig 6 shows how the oxide charge and the distance the oxide charge is changed from the trench bottom can have an adverse effect on Q_{gd} while reducing R_{SP}. Measurement of such parameters after the BV_{dss} has shifted is difficult because the injected oxide charge can decay, shift position, or sweep out due to the time delay and/or the applied fields from the subsequent measurements.

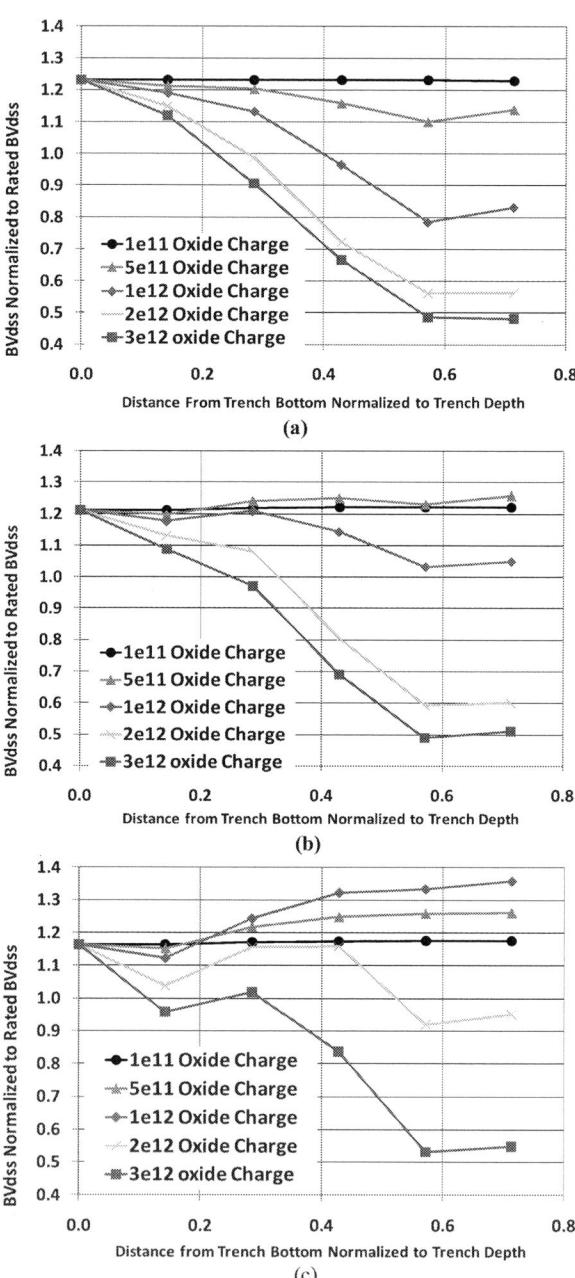

Fig 5 Simulated normalized BV_{dss} vs normalized distance from trench bottom that charge is increase from 5e10: (a) normalized epi concentration 1.25, (b) normalized epi concentration 1.0, & (c) normalized epi concentration 0.5.

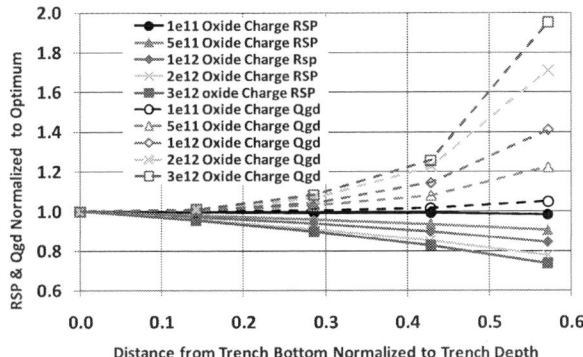

Fig 6 Simulated normalized R_{SP} and Q_{gd} at 50% rated voltage for various oxide charge for the optimum epi doping concentration vs normalized distance the oxide charge is increased to from 5e10 a fixed distance from the trench bottom

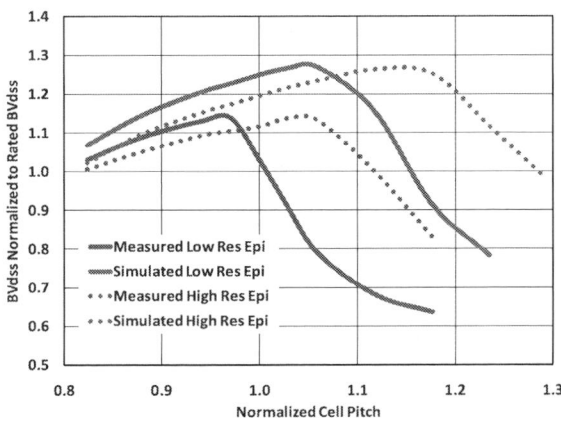

Fig 7 Measured cell plus termination BV_{dss} and simulated cell BV_{dss} normalized to rated voltage vs normalized cell pitch for two epi resistivities : High 1.11 normalized and Low 1.35 normalized epi concentration

Measured data in Fig 7 was on devices fabricated with the same shield oxide thickness, trench width, and two different concentrations of epi. To obtain different mesa charges without running several different epi doping concentrations, wafers were processed with a matrix mask having different silicon mesa widths. The results show that measured BV_{dss} falls off faster with increasing cell pitch, mesa charge, than the simulated values. Hot spot analysis of the measured low mesa charge example (high resistivity epi) shown in Fig 8(a) confirms that when the cell pitch is less than the normalized value 0.9, BV_{dss} occurs within the active cell and increases with time in avalanche. In Fig 8(b), when the normalized cell pitch is 0.9 -1.0 the BV_{dss} occurs in the active area and is stable. If the normalized cell pitch is greater than 1.0, a high mesa charge (low resistivity epi), the onset of avalanche occurs in the perimeter termination region shown in Fig 8(c) and BV_{dss} decreases with increased time in avalanche. The discrepancy between the BV_{dss} fall off at larger cell pitch (higher mesa charge) for the simulated BV_{dss} shown in

Fig 6 can now be explained by the termination region having higher BV_{dss} sensitivity to increasing charge in the shield oxide.

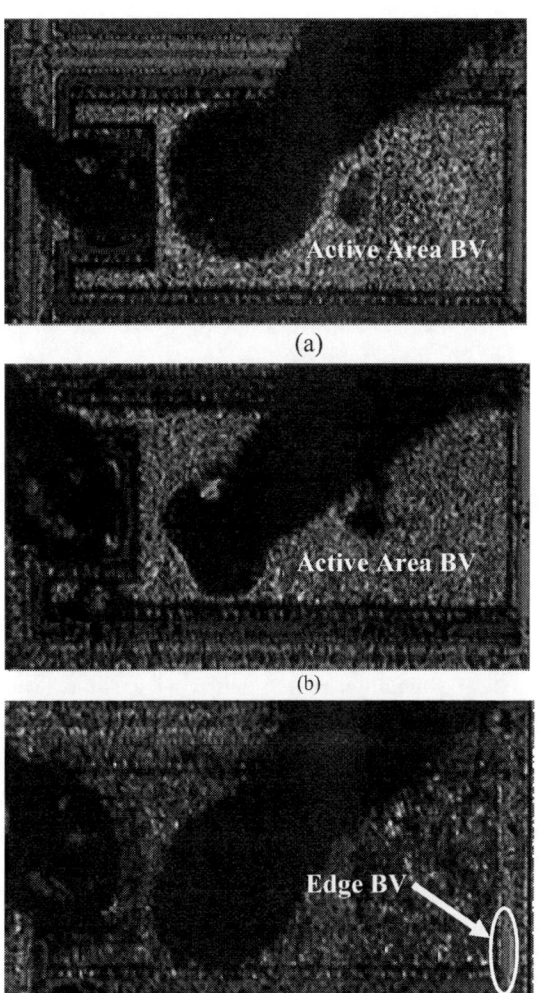

Fig 8 Hot Spot Analysis of fabricated die (a) 0.8, (b) 1.0, & (c) 1.1 normalized cell pitch.

Conclusions

A possible theory to explain the instability that causes BV_{dss} to shift downward or upward with increased time in avalanche for shielded gate charge balanced devices has been described. The combination of avalanche hole injection and the position the charge is injected into the shield oxide effects the charge balance within the device resulting in the BV_{dss} instabilities. The optimum stable BV_{dss} design point may not be at the charge balance condition that yields the peak BV_{dss} but at a lower charge condition.

References

[1] J. Yedinak et. al., ISPSD'10 Proceedings, pp. 333-336, 2010
[2] B. J. Baliga, Silicon RF Power MOSFETS, World Scientific Publishing 2005, pp 80-81

Proceedings of the 23rd International Symposium on Power Semiconductor Devices & IC's
May 23-26, 2011 San Diego, CA

Prognostics of Power MOSFET

José R. Celaya, Abhinav Saxena
SGT Inc. at NASA Ames Research Center
Prognostics Center of Excellence
Moffett Field, CA, 94035
Email: {jose.r.celaya, abhinav.saxena}@nasa.gov

Vladislav Vashchenko
SGT Inc.
Moffett Field, CA, 94035

Sankalita Saha
MCT at NASA Ames Research Center
Prognostics Center of Excellence
Moffett Field, CA, 94035

Kai Goebel
NASA Ames Research Center
Prognostics Center of Excellence
Moffett Field, CA, 94035

Abstract— **This paper demonstrates how to apply prognostics to power MOSFETs (metal oxide field effect transistor). The methodology uses thermal cycling to age devices and Gaussian process regression to perform prognostics. The approach is validated with experiments on 100V power MOSFETs. The failure mechanism for the stress conditions is determined to be die-attachment degradation. Change in ON-state resistance is used as a precursor of failure due to its dependence on junction temperature. The experimental data is augmented with a finite element analysis simulation that is based on a two-transistor model. The simulation assists in the interpretation of the degradation phenomena and SOA (safe operation area) change.**

I. INTRODUCTION

Prognostics is an engineering discipline focused on predicting the time at which an in-service component will fail or no longer perform its intended function. Predictions are made *in-situ* on individual in-service components. This is in contrast to statistical reliability methods that produce mostly a priori life estimates. The science of prognostics is based on the analysis of failure modes, detection of early signs of wear and aging, and fault conditions. These signs are then correlated with a damage propagation model and suitable prediction algorithms to arrive at a *remaining useful life* (RUL) estimate. The discipline that links studies of failure mechanisms to system lifecycle management is often referred to as prognostics and health management (PHM). PHM techniques have recently enjoyed considerable attention, for example, in the aerospace domain where the assessment of *in-situ* health of components and subsystem enables safe operations. Although the emphasis in PHM has so far been on mechanical components, the ability to perform health assessment of electronic components becomes essential as more safety-critical functionality is assumed by electronics. To that end, an in-depth understanding of aging mechanism and their manifestation is vital. The work reported here contributes to this undertaking.

In this paper a prognostics technique is presented for a power MOSFET based on an accelerated aging methodology. The methodology utilizes thermal and power cycling and was validated with tests using 100V power MOSFET devices. The major failure mechanism for the stress conditions is die-attachment degradation, typical for discrete devices with lead-free solder die attachment. It has been identified that ON-state resistance changes due to its dependence on junction temperature and can be used as a precursor of failure for the die-attach failure mechanism in the stress conditions. It has been shown that this particular degradation process provides characteristics to which data-driven prognostics algorithm can be applied. The experimental data is supported by a finite element analysis simulation. The numerical simulation assumes a two-transistor model. Results are used to interpret the phenomena of device degradation and SOA change. A Gaussian process regression framework is used for prediction of time to failure. The features used in the algorithm are based on normalized ON-resistance computed from *in-situ* measurements of the electro-thermal response. Results are presented from experiments on power MOSFET IRF520Npbf in a TO-220 package. The choice of the particular component is mainly due to its common use in switched mode power supplies in aerospace systems like radars and navigation equipment.

A. Related work

In [1] a model-based prognostics approach for discrete IGBTs was presented. RUL prediction was accomplished using a particle filter algorithm where the collector-emitter current leakage has been used as the primary precursor of failure. A prognostics approach for power MOSFETs was presented in [2]. There, the threshold voltage was used as a precursor of failure; a particle filter was used in conjunction with an empirical degradation model. The latter was based on accelerated life test data.

This work was sponsored by NASA Aviation Safety Program, IVHM Project.

978-1-4244-8425-6/11 $26.00 © 2011 IEEE

Identification of parameters that indicate precursors to failure for discrete power MOSFETs and IGBTs have received considerable attention in the recent years. Several studies have focused on precursor of failure parameters for discrete IGBTs under thermal degradation due to power cycling overstress. In [3], collector-emitter voltage was identified as a health indicator; in [4], the maximum peak of the collector-emitter ringing at the turn of the transient was identified as the degradation variable; in [5] the switching turn-off time was recognized as failure precursor; and switching ringing was used in [6] to characterize degradation. For discrete power MOSFETs, on-resistance was identified as a precursor of failure for the die-solder degradation failure mechanism [7][8]. A shift in threshold voltage was named as failure precursor due to gate structure degradation fault mode [2][9].

There have been some efforts in the development of degradation models that are a function of the usage/aging time based on accelerated life test. For example, empirical degradation models for model-based prognostics are presented in [1] and [2] for discrete IGBTs and power MOSFET respectively. Gate structure degradation modeling discrete power MOSFETs under ion impurities was presented in [10].

II. ACCELERATED AGING EXPERIMENTS

Accelerated aging approaches provide a number of opportunities for the development of physics-based prognostics models for electronics components and systems. In particular, it allows for the assessment of reliability in a considerably shorter amount of time than running long-term reliability tests. The development of prognostics algorithms face some of the same constrains as reliability engineering in that both need information about failure events of critical electronics systems. These data are rarely ever available. In addition, prognostics requires information about the degradation process leading to an irreversible failure; therefore, it is necessary to record *in-situ* measurements of key output variables and observable parameters in the accelerated aging process in order to develop and learn failure progression models.

Thermal cycling overstress leads to thermo-mechanical stresses in electronics due to mismatch of the coefficient of thermal expansion between different elements in the component's packaged structure. The accelerated aging applied to the devices presented in this work consists of thermal overstress. Latch-up, thermal runaway, or failure to turn ON due to loss of gate control are considered as the failure conditions. Thermal cycles were induced by power cycling the devices without the use of an external heat sink. The device case temperature was measured and directly used as control variable for the thermal cycling application. For power cycling, the applied gate voltage was a square wave signal with an amplitude of ~15V, a frequency of 1KHz and a duty cycle of 40%. The drain-source was biased at 4Vdc and a resistive load of 0.2Ω was used on the collector side output of the device. The aging system used for these experiments is described in detail in [4]. The accelerated aging methodology used for these experiments is presented in detail in [8].

Figure 1 shows an X-ray image of the device after degradation. It can be observed that the die solder has

migrated and that voids have formed. This confirms that the thermal resistance from junction to case has increased during the stress time resulting in increase of the junction temperature and ON-resistance. Figure 2 presents a plot of the measured $R_{DS(ON)}$ as a function of case temperature for several consecutive aging tests on the same device. For each test run, the temperature of the device is increased from room temperature to a high temperature setting thus providing the opportunity to characterize $R_{DS(ON)}$ as a function of time at different degradation stages. It can be observed how this curve shifts as a function of aging time, which is indicative of an increased junction temperature due to poor heat dissipation and hence degraded die-attach.

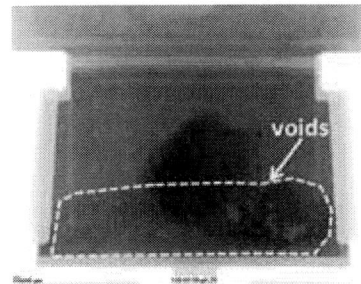

Figure 1. X-ray of degraded device.

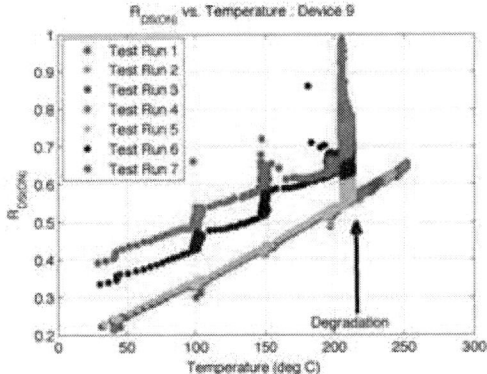

Figure 2. R_{DSON} degradation process due to die-attach damage.

III. MODELING

A. Mixed-mode simulation

Numerical analysis was performed using a finite element model (FEM) representation of the device under consideration (figure 3). This numerical analysis provided I-V characteristics at different values of gate bias *Vgs* for a device with generic simulation parameters roughly close to tested devices.

The electrical response was obtained with a mixed-mode circuit-device simulation using software DECIMM™ from Angstrom Designs Automation [11][12]. The mixed-mode circuit presented in figure 4 is simulated in conjunction with the FEM of the MOSFETs. This was implemented both with a single transistor and with two transistors as shown in Figure 4. Results for both models are discussed below. A voltage-controlled voltage source circuit was used to auto bias the gate voltage. This prevents the device from running outside the SOA.

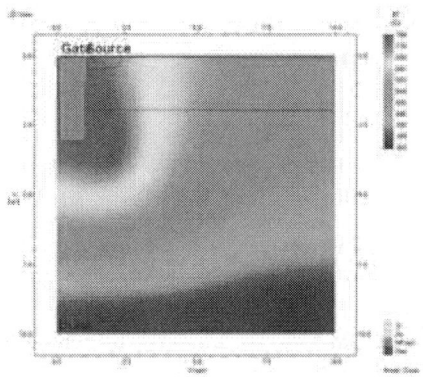

Figure 3. Finite element model vertical DMOS device cross-section.

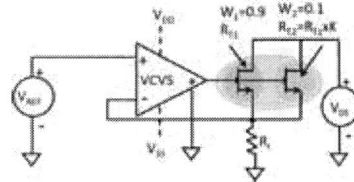

Figure 4. Two-transistor mixed-mode simulation circuit.

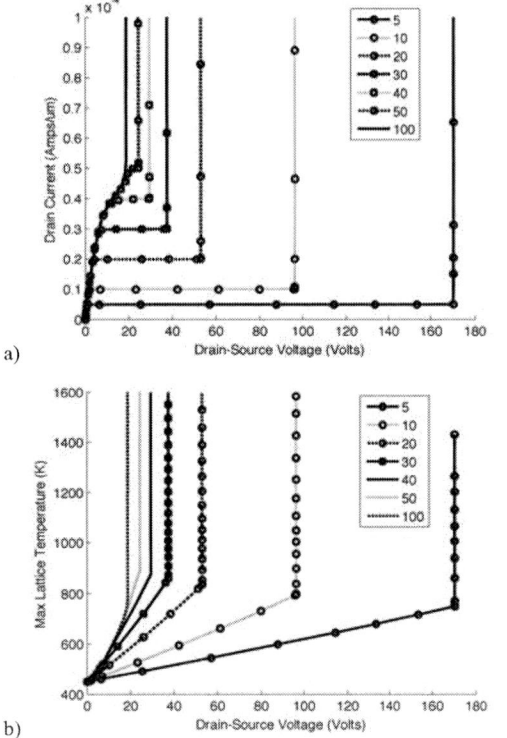

Figure 5. Single transistor with auto bias reference (uV): a) SOA, b) extracted maximum temperature.

The electro-thermal SOA for a single transistor mixed-mode simulation with auto bias control of the gate voltage is presented in figure 5 for two conditions a) slow transient pulse with constant gate bias; b) slow transient pulse with auto bias circuit. The observed instability points represent the critical voltages and currents limiting the safe operation area of the electrical regime.

B. Two transistor degradation model

The two-transistor model physically represents the device with partial area die-attachment degradation. The first transistor has original default parameters including the thermal resistance R_{T1} and area factor 90% while the second transistor depicts degradation due to electro-thermal stress represented by 10% of area with deviation of the thermal resistance coefficient K (figure 4). As can be seen from the simulation results in figure 6, even a small deviation in the thermal resistance of the second transistor ($R_{T2}=K \times R_{T1}$) results in significant reduction of the critical voltage in auto bias conditions (figure 6).

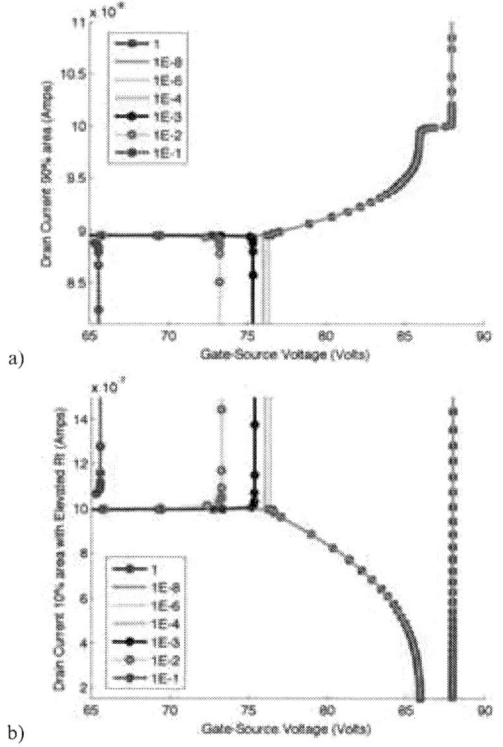

Figure 6. Results of numerical analysis for different thermal resistance parameters of the 10% second transistor model region at 450K heat sink $R_{T2}=K \times R_{T1}$.

IV. PREDICTION OF REMAINING USEFUL LIFE

Gaussian Process Regression (GPR) is a data-driven technique that can be used to estimate future fault degradation based on training data collected from measurement data. First, a prior distribution is assumed for the underlying process function that may be derived from domain knowledge [13]. Then this prior is tuned to fit available measurements which is used with the probabilistic function for regression over the training points [14]. The output is a mean function to describe the behavior and a covariance function to describe the uncertainty. These functions can then be used to predict a mean value and corresponding variance for a given future point of interest. The behavior of a dynamic process is captured in the covariance function chosen for the Gaussian process. The covariance structure also incorporates prior beliefs of the underlying system noise. A covariance function consist of various hyper-parameters that define its properties. Proper tuning of these hyper-parameters is key in the performance. While a user typically needs to specify the type of covariance function, the corresponding hyper-parameters can be learned from training data using a gradient based optimization (or other optimization) such as maximizing the marginal likelihood of the observed data with respect to hyper-parameters [14].

In the application here, the ON-resistance was computed as the ratio of voltage and currents between the drain and source terminals of the device. By estimating the relationship between operational temperature and ON-resistance of the device, the computed ON-resistance was normalized to eliminate temperature effects. The signal was filtered by computing the mean of every one minute long window. Since the complexity of GPR is $O(n^3)$, computational effort increases with number of data points and hence it is important to keep the number of training points low. Therefore a uniform sampling of the curve was carried out to select the desired number of training points to train the GPR and make predictions. This process was repeated 35 times and the results were aggregated to produce final prediction. As shown in the figure, predictions were made at four (somewhat arbitrarily chosen) time instances: 160, 180, 200, and 220 minutes into aging. Subtracting the time when the prediction was made from the time when the dashed lines crosses the failure threshold gives the estimated remaining component life. As more data becomes available, the predictions become more accurate (as indicated by the proximity of the predicted dashed lines to the crossing of the failure threshold by the ON resistance) and the prediction spread becomes more precise (uncertainty cones are more narrow for later predictions).

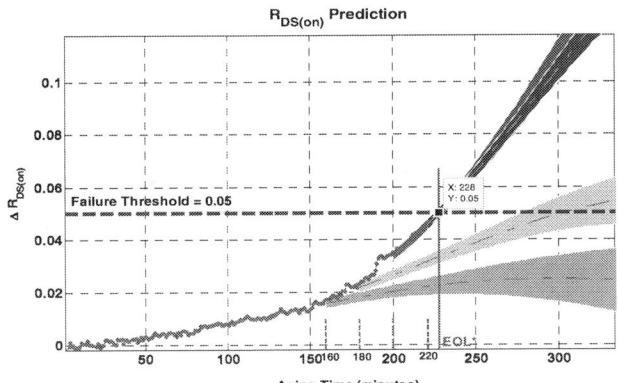

Figure 7. Prediction of RUL for aged device using Gaussian process regression technique.

V. DISCUSSION

The proposed prognostics technique reports on preliminary work that serves as a case study on the prediction of remaining life of power MOSFETs. There are several strong assumptions that need to be challenged in order to make the proposed process practical. For instance, the future operational conditions and loading of the device are considered constant at the same magnitudes as the loads and conditions used during accelerated aging. In addition, the algorithm development is conducted using accelerated life test data. In real world

implementation, the degradation process of the device would occur in a considerably larger time scale. This is a topic of future work.

The proposed two-transistor model is shown to be a good candidate for a degradation model for model-based prognostics. The model parameters K, and W1 could be varied as the device degrades as a function of usage time, loading and environmental conditions. Parameter W1 defines the area of the healthy transistors, the lower this area, the larger the degradation in the two-transistor model. In addition, parameter K serves as a scaling factor for the thermal resistance of the degraded transistors, the larger this factor, the larger the degradation in the model.

REFERENCES

[1] B. Saha, J. R. Celaya, P. F. Wysocki, and K. F. Goebel, "Towards prognostics for electronics components," in *Aerospace conference, 2009 IEEE*, pp. 1–7, 2009.

[2] S. Saha, J. R. Celaya, V. Vashchenko, S. Mahiuddin, and K. F. Goebel, "Accelerated aging with electrical overstress and prognostics for power mosfets," in *IEEE EnergyTech 2011*, (submitted), 2011.

[3] N. Patil, J. Celaya, D. Das, K. Goebel, and M. Pecht, "Precursor parameter identification for insulated gate bipolar transistor (igbt) prognostics," *IEEE Transactions on Reliability*, vol. 58, no. 2, pp. 271–276, 2009.

[4] G. Sonnenfeld, K. Goebel, and J. R. Celaya, "An agile accelerated aging, characterization and scenario simulation system for gate controlled power transistors," in *IEEE AUTOTESTCON 2008*, pp. 208–215, 2008.

[5] D. Brown, M. Abbas, A. Ginart, I. Ali, P. Kalgren, and G. Vachtsevanos, "Turn-off time as a precursor for gate bipolar transistor latch-up faults in electric motor drives," in *Annual Conference of the Prognostics and Health Management Society 2010*, (Portland, OR.), 2010.

[6] A. Ginart, M. Roemer, P. Kalgren, and K. Goebel, "Modeling aging effects of igbts in power drives by ringing characterization," in *IEEE International Conference on Prognostics and Health Management*, 2008.

[7] J. R. Celaya, N. Patil, S. Saha, P. Wysocki, and K. Goebel, "Towards accelerated aging methodologies and health management of power mosfets (technical brief)," in *Annual Conference of the Prognostics and Health Management Society 2009*, (San Diego, CA.), 2009.

[8] J. Celaya, A. Saxena, P. Wysocki, S. Saha, and K. Goebel, "Towards prognostics of power mosfets: Accelerated aging and precursors of failure," in *Annual Conference of the Prognostics and Health Management Society 2010*, (Portland, OR.), 2010.

[9] J. Celaya, P. Wysocki, V. Vashchenko, S. Saha, and K. Goebel, "Accelerated aging system for prognostics of power semiconductor devices," in *IEEE AUTOTESTCON 2010*, pp. 1–6, 2010.

[10] A. E. Ginart, I. N. Ali, J. R. Celaya, P. W. Kalgren, S. D. Poll, and M. J. Roemer, "Modeling SiO2 ion impurities aging in insulated gate power devices under temperature and voltage stress," in *Annual Conference of the Prognostics and Health Management Society 2010*, (Portland, OR.), 2010.

[11] DECIMMTM User Manual, Angstrom Design Automation, 2010. www.angstromda.com

[12] V.A. Vashchenko, A.A. Shibkov, *ESD Design for Analog Circuits*, Springer, (2010), www.analogesd.com.

[13] K. Goebel, B. Saha, and A. Saxena, "A comparison of three data-driven techniques for prognostics," in *Proceedings of the 62nd Meeting of the Society For Machinery Failure Prevention Technology (MFPT)*, vol. 2008., (Virginia Beach VA), pp. 119–131, May 2008.

[14] C. E. Rasmussen and C. K. I. Williams, *Gaussian Processes for Machine Learning*. Cambridge MA: MIT Press, 2006.

Proceedings of the 23rd International Symposium on Power Semiconductor Devices & IC's
May 23-26, 2011 San Diego, CA

Modeling the 3D Self Ballasting Behavior and Filamentation Under High Current Stressing in DeNMOS

Amitabh Chatterjee (1), Sameer Pendharkar(2), Charvaka Duvvury (2), Forrest Brewer (1)

[1]Department of Electrical and Computer Engineering, University of California, Santa Barbara, CA - 93106, USA

[2]Silicon Technology Development, Texas Instruments Inc Dallas, TX -75243, USA

Abstract—**A critical understanding of self ballasting behavior due to current crowding of avalanche generated carriers in a DeNMOS is developed. Then we study its performance under the gate biased conditions. The impact of flow of holes and electron in the bulk and across the surface - on the snapback-back features has been critically evaluated through variations in the device structure (associated with process parameter) which has also been extensively studied through 2D & 3D TCAD simulations. We demonstrate that after an initial homogeneous triggering (due to bipolar snapback), self heating preferentially activates the 2D array of bipolars in the bulk and subsequently current instability under negative resistance regime (as the bipolar turns on) leads to inhomogeneous triggering in the 3D.**

I. INTRODUCTION

Poor high current performance in a De-NMOS (a critical nano-meter high voltage I/O device) in advanced low voltage technologies (fig. 1), which is characterized by low It2 and large variations amongst the dies (fig. 2)[1]. The triggering of the parasitic bipolar has been explained through sequential turn-on of an *array of 2D bipolars* first across the *surface*, which is followed by 2D turn-on in the *bulk* and eventually the 3D localization, as it leads to electro-thermal runaway due to *filamentation* (fig. 2).

and the bulk) triggering in-homogeneity and causing 3D localization has been inadequately addressed. Moreover, the physics behind current crowding phenomenon which leads to potential buildup (which we call *intrinsic ballasting*) is relatively less understood, and has lead to un-optimized design of De-NMOS particularly under the gate biasing. Through detailed 2D &3D TCAD simulations, we present a clear under-standing of the role of surface and bulk parsitic bipolar under gate biased conditions as we study the impact of source ballasting and also deactivate the emitter (i.e. source open). In this abstract, we critically evaluate the impact of intrinsic ballasting due to current crowding mechanisms as we optimize the high current performance in the De-NMOS.

II. PHYSICS OF INTRINSIC BALLASTING & ESD PERFORMANCE UNDER BIASING TEMPLATE

Intrinsic ballasting phenomenon under an ESD event can be explained due to two reasons: (a) initially due to current crowding of holes (near the pinch-off region), which flow into the substrate (i.e behaves as non-linear resistor which leads to buildup of potential) and (b) subsequently as the bulk bipolar turns on, triggers crowding of current due to electrons (at the edge of drain contact PZ) (fig. 3 & fig. 4)). The jumps (i.e. snapback) observed in the TLP characteristics can be explained (as observed through simulations) due to the potential buildup, as the accumulated holes near the surface (across the pinch-off region) and subsequent modulation of the space charge region (constituted by mobile carriers) triggers snapback as the bulk bipolar (path 2) gets activated.

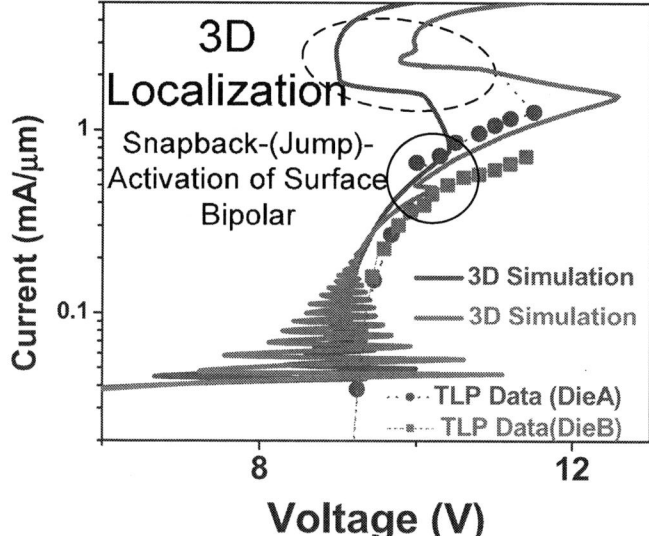

Fig. 2 Strong snapback is charecterized by activation of bulk bipolar and a 3D localization which can cause permanent damage. Sample A early activation as device survives snapback

Fig. 1 De NMOS is charecterized sudden switching of electron injection from path1 (surface) to path2 (bulk)

Previously, the subtle role of source ballasting which limits the 2D localization was explained on an unbiased structure (i.e. gate or substrate), where we had shown an improvement both in die-die variations and the voltage buildup across the oxide at the drain end. Also, previously it was shown the *strong snapback* activates the 2D bipolar, as it also triggers 3D localization, and leads to permanent damage in the devices with N-Well drain extension. However, so far the subtle role of 2D bipolars (i.e. both the surface

978-1-4244-8425-6/11 $26.00 © 2011 IEEE

164

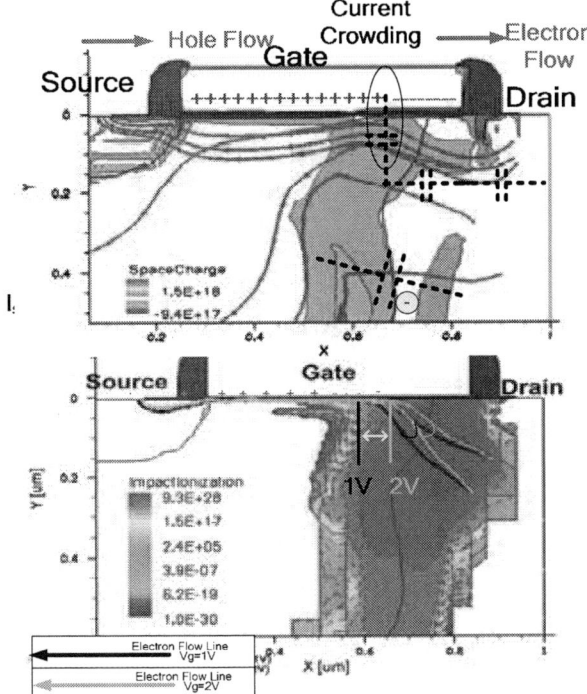

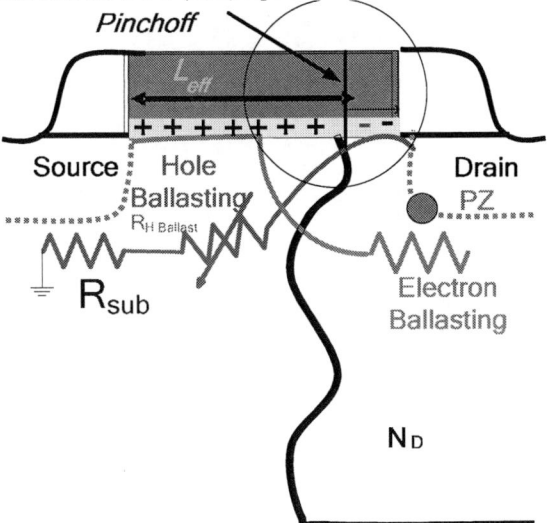

Fig 3. Current crowding due to flowlines of electrons and holes near pinchoff leads to buildup of potential and prevents snapback till surface currents $J_{surface}=qV_sN_D$ lead to space charge modulatio

Now at a critical value of electron current at the surface (It_1), the space charge modulation under injection of minority carriers (i.e. electrons) compensates the buildup of potential due to the space charge of majority carriers (i.e. holes). Modulation of the space charge near the *pinch-off* triggers snapback, in the process it activates the lower (bulk) bipolar.

Fig. 4 Intrinsic ballasting due crowding of holes below the surface and electrons near the edge

Under these conditions each additional hole which are injected into the substrate is compensated by an injected electrons. The activation process is highly sensitive to both (i) the electrostatics near the pinch-off region and also the thermal generation. Interestingly, the I-V slope beyond the jump (i.e snapback) is primarily determined by current crowding of electrons near the drain contact, where-in positive biasing on the gate

aggravates the current density and triggers regenerative avalanche injection.

To summarize, the physics of localization in the 3D can be explained through a simplified phenomenological model, i.e where the onset is driven by dynamic formation of a localized hot spot, and once the surface bipolar is turned on (initially uniformly in the 2D array), it can leads to non-uniform conduction in the array of 2D bipolars across the 3D (as it trigger filamentation).

More-over, at these preferential sites (i.e *dynamic* hot spot) (fig. 5 & fig. 6) the surface bipolar turns-on early and goes into low voltage high current state, while the remainder of the device is forced into relatively lower conduction state. The global coupling, is achieved through a common potential V_D, across the drain, now forces the remainder of the device into lower conducting branch (i.e. also due to hysteresis in the IV curve). Moreover, it is a very strong function of temperature and triggers localization in 3D.

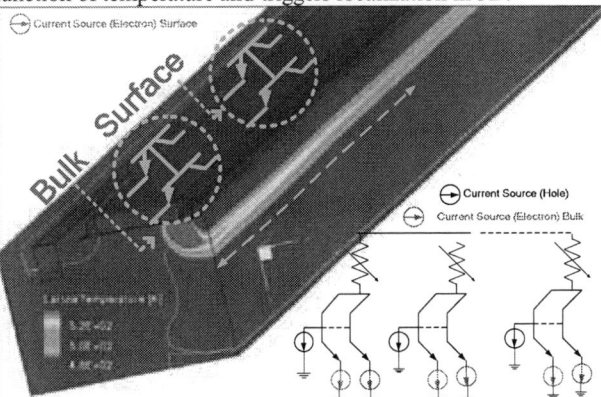

Fig 5 Physics of 3D localization and improvement under biasing. Bipolar can be explained due to activation of surface bipolar

Fig. 6 Electro-thermal instability triggered by activation of surface bipolars prolonged negative resistance as the surface bipolar is activated strongly at a preferential site

a) Dual Role of Gate Biasing

Primarily the gate biasing improves the flow of holes, which helps to alleviate the dynamical process involving formation of localized hot spot (as the TLP current builds up) as it delays the onset of *3D localization* (fig. 7). Therefore, the onset of *3D localization* (filamentation) is delayed as the array of 2D surface bipolars is activated uniformly under gate biasing. However, the subtle electrostatics due to mobile charges around the drain and its image across the gate is determined by ambipolar current crowding mechanism (i.e flow of electrons and holes) (fig. 4). This can be extremely severe in these ultra-thin gate oxide devices due to self heating below the thin-oxide as energetic holes accumulate near the surface. Therefore, the gate bias has a deleterious impact on the flow of electrons near the drain region whereby the devices exhibit duality in the performance.

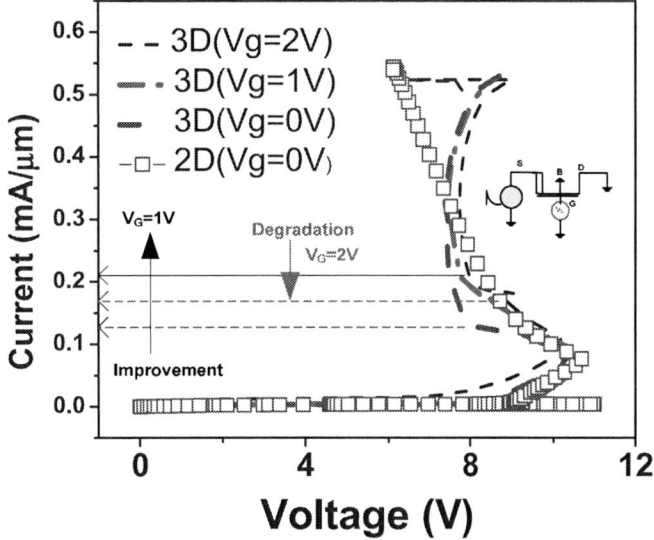

Fig. 7 2D vs 3D showing improvement in performance due to efficient activation of surface bipolar leading to alleviation of filamentation under gae bias

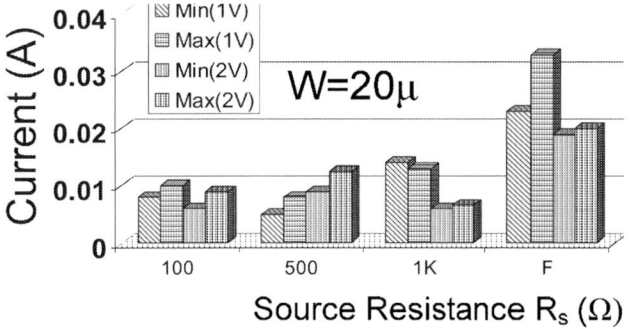

Fig. 8 Variaitions due to gate bias under 2V & 1V and improved performance at $R_L=100\Omega$

b) Optimized Performance using Gate biasing and Source ballasting

More critically, the gate biasing also impacts die to die variations, due to the above duality. In contrast to the improvement shown earlier, (i.e without gate biasing and higher R_s); these variations are exacerbated for higher values of source resistor R_s under the gate bias (Fig. 8). This can be explained due to the fact that when R_s is large, the electrostatic coupling between gate and drain, is exacerbated under high current stressing (hence higher self-heating). However, irrespective of an improvement in It2, when the source is open, it leads to buildup of voltage across the oxide, which is outside the ESD design window (fig. 10-11). Therefore one needs an added substrate bias to improve variations within die in addition to source ballasting.

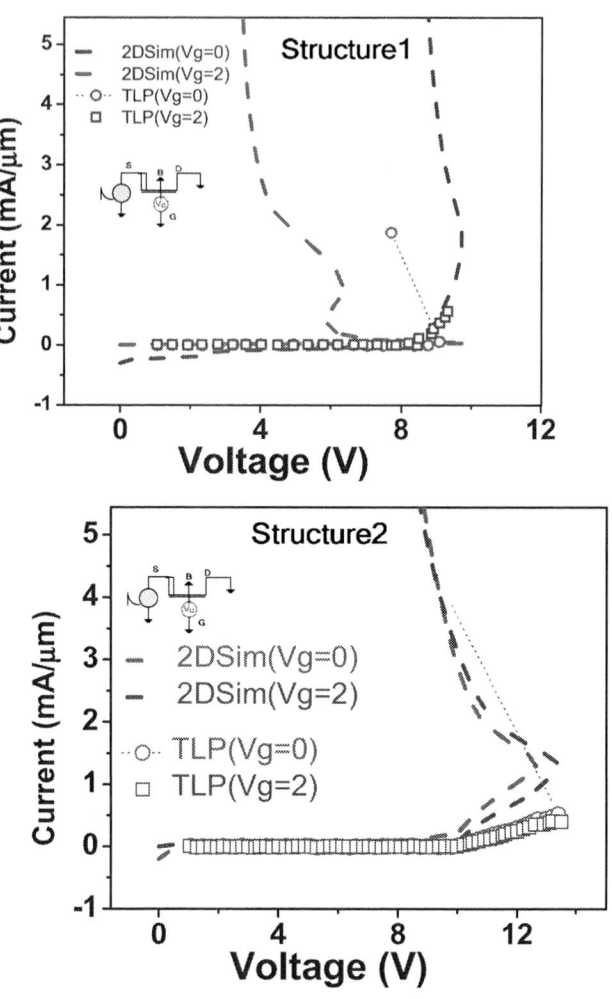

Fig. 9 Poor snapback current It1 (i.e *failure current for DeNMOS*) for *Structure 1* (w.r.t fig 12a&b) and improved performance with gate biasing *Structure 1*. However, *Structure 2*, improved snapback current, *lesser impact of gate biasing*

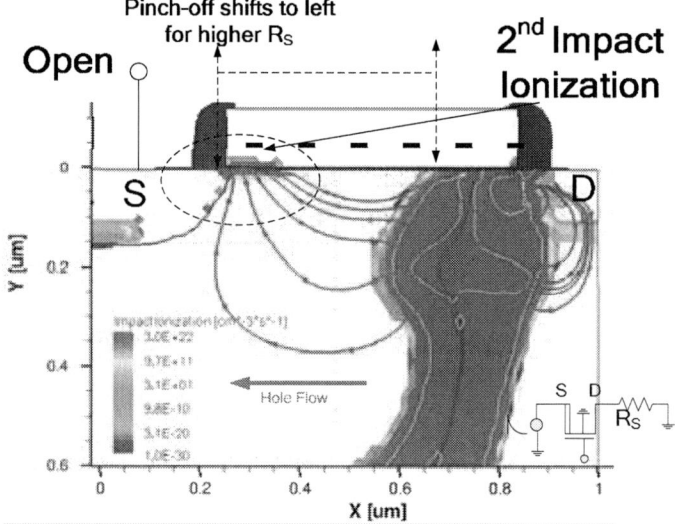

Fig. 10 Voltage buildup under source open conditions lead to 2nd pocket of impact ionization

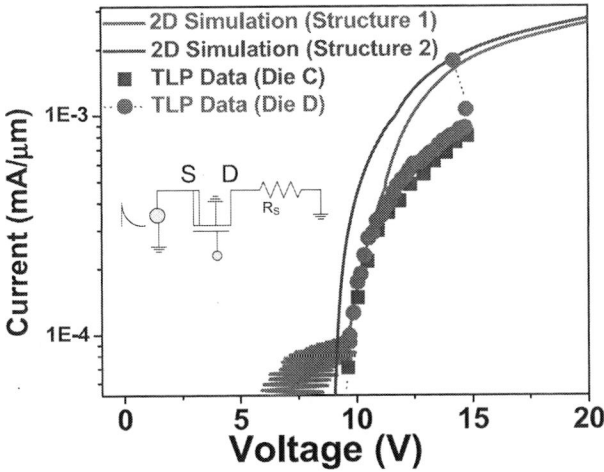

Fig. 11 Improved performance when source is open i.e R_s large. Higher voltage across the device prevents it as protection device

III. ESD PERFORMANCE IN DeNMOS: IMPACT OF DEVICE VARIATION (PROCESS PARAMETERS) ON BALLASTING ACTION

Another interesting feature of these variations in failure current (i.e It2) is that while some of the devices show *remarkable improvement(10X)* under gate biasing, the others remain largely *unaffected* by the biasing (Fig.9 and Fig 12a&b). To understand this *anamolous trend*, 2D simulations were performed by varying both the Halo implant and the P-well implant and co-relating it with observed TLP data under biasing (Fig. 9). Now, a slight variation in junction profile not only regulate (fig 12a & 12b) (i) how the hole flow lines are established in the bulk (in the process they also determine the established potential in the substrate) but also (ii) the amount of depleted charge imaged by the gate (i.e δq) near the pinch-off across the extended drain region.

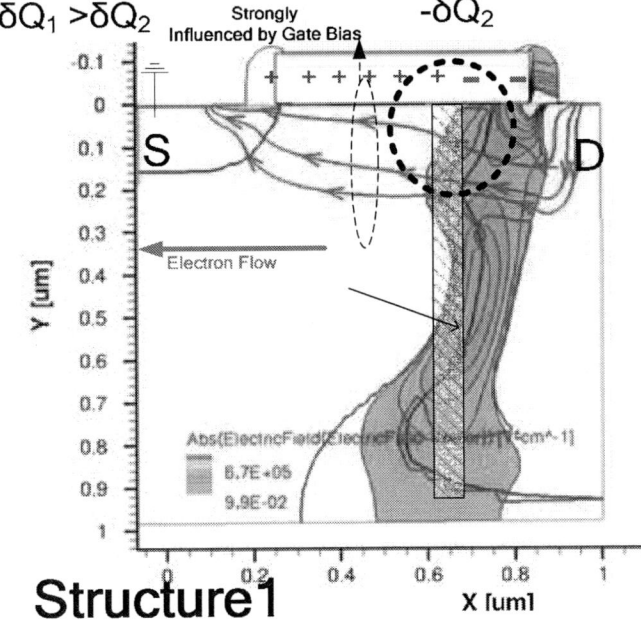

Fig 12a *Structure 1*: Greater bottom curvature Impacts of flow lines with gate bias as the lower path is not properly activated

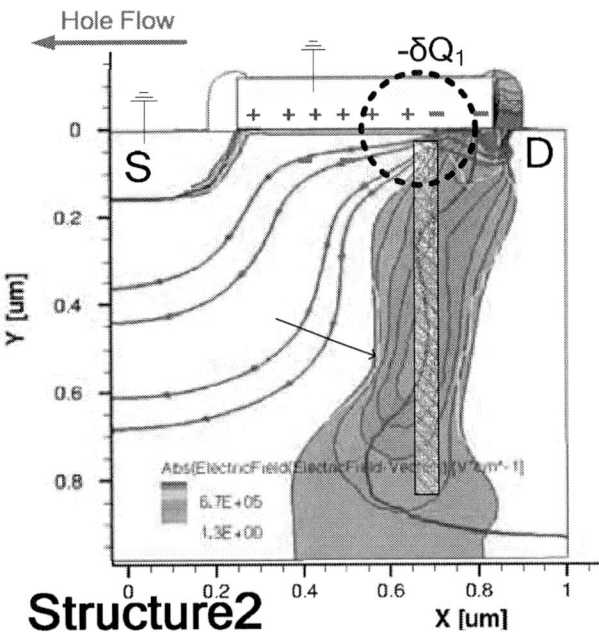

Fig 12b *Structure 2*: Lower bottom curvature: Lessens the impact of gate biasing on flow

This electrostatic coupling results in minority carrier injection across two paths (fig.1), which determines how they modulate the established space charge due to the flow of holes. Now, the gate bias prevents the *activation of the bulk bipolar* (in *structure 2*) as higher number of charges are imaged on the drain side(i.e $\delta q_1 > \delta q_2$) and in the process prevents early modulation due to injected minority carriers (the phenomenon is interpreted as the *kirk effect* – due to the E-field modulation across the surface as the injected electron conc. equals ionized background dopants N_D (fig. 6)).

IV. CONCLUSION

The sequential process which involves localization- first in the symmetric 2D plane due to activation of surface bipolar and subsequent localization in the 3D, as the activity shifts into the bulk has been analyzed for the first time under biased conditions. Intrinsic ballasting behavior in the DeNMOS has been modeled and correlated with crowding of hole current (across the surface) and electron current (in the bulk). We study the performance of the parasitic bipolar under gate biasing, which can optimized to enhance the failure performance by adding a suitable ballast resistor and an appropriate gate bias. In this work we present, first ever experimental evidence and the co-relation of gate bias dependence on the structural variations in the DeNMOS, which leads to variations within the dies.

REFERENCES

[1] G. Boseli et. al IRPS, 2007

[2] R. Steinhoff et. al,*EOS/ESD* 2003

[3] A. Chatterjee et. al., IRPS 2010

[4] A. Chatterjee et. al., ISPSD 2010

[5] V. Vashchenko and V.F Sinkevitch *Springer Verlag, 2008*

[6] P. Hower et. al.*ISPSD 2001*

[7] C. Duvvury et. al. ESD/EOS, 2001

[8] P. Hower et. al. *TED*, 1970

[9] V. Vassilev et. al Micro Elect Rel 2005

[10] K. Esmark, *PhD Thesis*, ETH, 2001

[11] S. Pendharkar et. al. *ISPSD* 2002

Proceedings of the 23rd International Symposium on Power Semiconductor Devices & IC's
May 23-26, 2011 San Diego, CA

Low-On-Resistance Strain-Controlled LDMOS Transistors

for 0.25-μm Power ICs

Masafumi Miyamoto, Nobuyuki Sugii[1], Yukihiro Kumagai[2], and Yoshinobu Kimura[1]

Mixed Signal LSI Development Department, Micro Device Division, Hitachi, Ltd.,
Ome, Tokyo 198-8512, Japan
e-mail: masafumi.miyamoto.ef@hitachi.com
[1]Central Research Laboratory, Hitachi, Ltd.,
Kokubunji, Tokyo 185-8601, Japan
[2]Mechanical Engineering Research Laboratory, Hitachi, Ltd.,
Hitachinaka, Ibaraki 312-0034, Japan

Abstract—We have developed a new 12 V LDMOS transistor for 0.25 μm power ICs, which is designed from the viewpoint of mechanical stress to reduce on-resistance. A critically low resistance substrate has been developed to reduce the resistance from the surface source to the backside of the transistor, avoiding compressive stress due to high boron doping in the substrate. A buried-polysilicon sinker is utilized to apply tensile stress to the channel and the offset-drain region. The existing mechanical stress distribution is confirmed by two-dimensional UV-Raman spectroscopy. The transconductance of the LDMOS transistor is increased by 12% owing to the tensile stress and the total on-resistance is reduced by 16% owing to the channel and source resistance reduction, which directly leads to a higher efficiency of analog power circuits.

I. INTRODUCTION

Silicon laterally diffused metal-oxide-semiconductor (LDMOS) transistors have been integrated to many types of analog power IC such as pulse width modulated (PWM) amplifiers, DC-DC converters and RF amplifiers, because the device characteristics such as high-frequency switching and low feedback charge enable us to design circuits easily and the fabrication process has excellent compatibility with the analog/digital complementary metal-oxide-semiconductor (CMOS) process [1, 2]. As a characteristic of power circuits using LDMOS transistors, the power loss is determined by the total on-state resistance (R_{on}) from drain to source. Therefore, reducing R_{on} is the key to reducing the total power loss in analog power circuits, and the low R_{on} has been continuously required as a figure of performance.

Recently, power transistors using other semiconductor materials such as SiC and GaN have been investigated to achieve higher performance using their advantage of material physicality. However, they currently have difficulty in integrating digital circuits and less dependability and rather a higher cost than Si. On the other hand, the locally strained Si technique is commonly introduced to nanoscale CMOS transistors for achieving higher drive current [3, 4], which means a lower on-state resistance. A globally strained LDMOS transistor using a strained-Si/SiGe substrate [5] and

a locally strained power MOS transistor [6] have been proposed. The globally strained LDMOS transistor and the locally strained power MOS transistor reduced their on-state resistance by up to 20% from those of reference Si power MOS transistors. Especially, the locally strained power MOS transistor has almost the same dependability and cost as reference Si technologies.

II. DEVICE DESIGN

A schematic cross section of the strain-controlled LDMOS transistor and its three on-state resistance components, offset-drain resistance (R_O), channel resistance (R_C), and the resistance from the surface source to the backside of the transistor (R_S) are shown in Fig. 1. The LDMOS transistor is fabricated at the surface of a p⁻ epitaxial layer on a high-boron-concentration p⁺⁺ substrate. The gate length is 0.23 μm and the offset-drain length of 0.65 μm is used to achieve a 12 V drain breakdown voltage. To reduce the effects of source parasitic resistance and inductance, a highly boron-doped buried-polysilicon (BP) sinker is utilized to connect the source at the surface and backside of the transistor. The space from the source edge to the BP is reduced to 1 μm because the boron diffusion length is much smaller than that of a conventional sinker, which is fabricated by diffusion from the wafer surface. Obviously, a smaller space is more efficient for reducing the chip area and for achieving a strong stress effect in the channel region, but boron diffusion from the BP will affect the channel characteristics in the smaller space.

The entire LDMOS transistor consists of mirror repetitions of this schematic cross section to achieve a desired gate width. The total on-state resistance (R_{on}) is determined not only by the channel region but also by the offset-drain region to maintain the drain breakdown voltage and the source resistance from the surface to the backside. The largest part in the total R_{on} is R_S, which occupies 30%. The second largest part is R_O of 27%, and the third is R_C of 16%. The residual resistance component is an unknown part caused by features such as contact and interconnect resistance.

978-1-4244-8425-6/11 $26.00 © 2011 IEEE

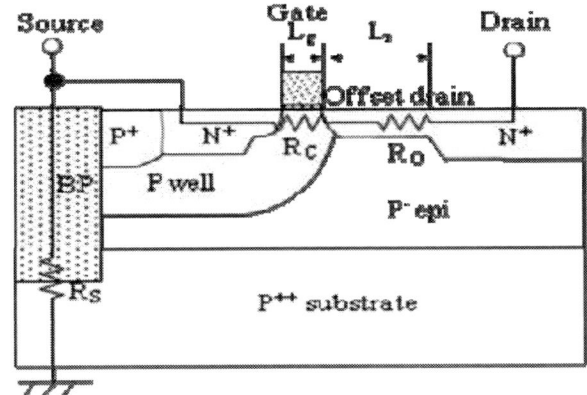

Fig. 1. Schematic cross section of the fabricated LDMOS transistor.

First, we considered reducing the resistance from the surface source to the backside of the transistor (R_S) because it is the largest. To reduce R_S, doping a higher boron concentration in the substrate is a simple method. However, the compressive stress at the surface p⁻ epitaxial layer increases in the high-boron-concentration substrate, as shown in Fig. 2. The substrate-resistivity dependence of surface stress is measured on wafers of various boron concentrations by UV-Raman spectroscopy [7, 8]. The stress values in Fig. 2 are the average of those measured at the center and edge of the wafers. Because boron atoms are smaller than silicon atoms, substituting boron in silicon reduces the lattice constant and produces compressive stress at the surface p-epitaxial layer. Compressive stress reduces electron mobility in the channel region and increases R_{on}. Therefore, the boron concentration in the substrate is limited not to produce compressive stress. Eventually, Rs is reduced to 1/2 from that of the conventional LDMOS transistor.

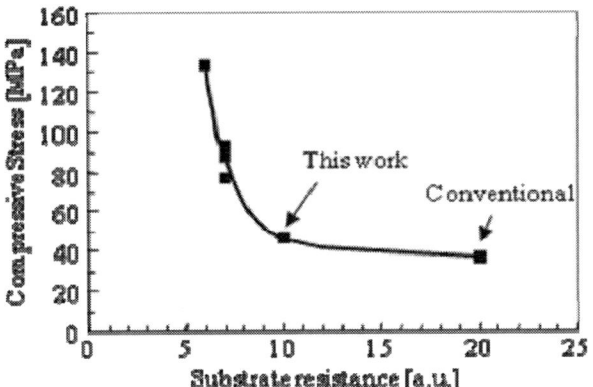

Fig. 2. Substrate stress measured by UV-Raman spectroscopy as a function of substrate resistance.

Next, we considered the second largest offset-drain resistance (R_O) component. This region is considered as a lightly-doped resistor. The impurity concentration of the offset-drain region is determined to maximize BV_{ds} / R_{on},

where BV_{ds} is the drain breakdown voltage. We measured the stress dependence of the sheet resistance of this offset-drain region, as shown in Fig. 3. Tensile stress is applied by bending the chip mechanically. The dependence is about -1.4%/100 MPa, which is much lower than that of the channel region. The electron mobility in the offset-drain region is mainly determined by impurity scattering and is less dependent on the effective mass of an electron.

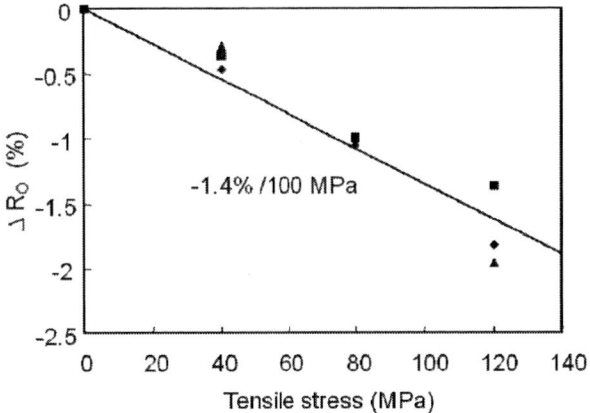

Fig. 3. Resistance change of offset-drain region caused by tensile stress.

Finally, the channel resistance (R_C) is considered. This resistance component is the same as the channel resistance in a usual NMOS transistor. Applying tensile stress to the channel region using an etch-stop SiN layer is often utilized for nanoscale NMOS transistors [4]. However, this technique is not suitable for LDMOS transistors because it will produce compressive stress in the offset-drain region and increase R_{on}. Therefore, we utilized a buried-polysilicon sinker to apply tensile stress to the entire surface of the LDMOS transistor including the offset-drain region.

A deep trench is formed at first on the previously described p⁻ epitaxial layer on a p⁺⁺ substrate by dry etching. Then, highly boron-doped amorphous silicon is deposited on the wafer and fills the deep trench. Excess amorphous silicon is etched back, following the conventional LDMOS transistor fabrication process flow. In the thermal process steps of LDMOS transistor fabrication such as annealing and oxidation, the amorphous silicon is crystallized to polysilicon and shrinks in volume. Thus, a tensile stress of about 600 MPa is produced around the BP area. The simulated surface stress distribution from the source edge to the drain is shown in Fig. 4. An uniform tensile stress of 600 MPa is assumed in the BP area, which is 1 μm outside the source edge. The mirror boundary condition at both sides of the BP and drain is used in the stress simulation because the actual LDMOS pattern is repeated by mirror inversion. As a simulation result, a tensile stress of about 250 MPa is uniformly applied from the channel to the offset-drain region. Stress fluctuations at both sides of the channel are caused by sidewall spacers.

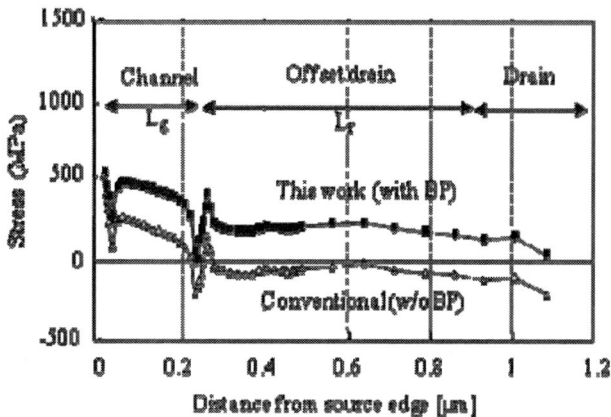

Fig. 4. Simulated surface stress distribution from source edge to drain in this work..

III. EXPERIMENTAL RESULTS

The existing tensile stress is confirmed by two-dimensional UV-Raman measurement on transistor cross sections [7,8]. Figure 5 shows the stress distribution around the BP and the Raman shift along the cross section at 0.5 μm depth. The laser wavelength and power used are 532 nm and 0.3 mW, respectively. The spatial resolution is about 0.25 μm in this measurement.

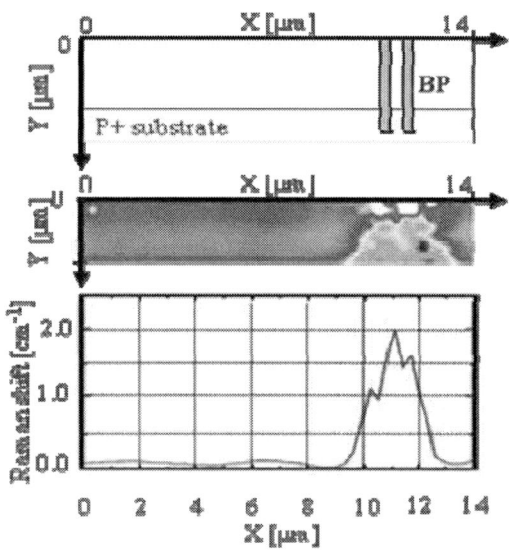

Fig. 5. Stress distribution around BP area using two-dimensional UV-Raman spectroscopy.

The stress around the BP area is observed clearly, and the Raman peak shift is about 2 cm^{-1}. Therefore, a tensile stress of about 500 MPa is estimated in the center of the BP area. As the distance from the BP edge to the source edge is 1 μm, the tensile stress in the channel region is reduced by 1/2 from the peak amount (250 MPa).

A comparison of 75 measuring points in a wafer of this work and those of the conventional transistor using the R_{on}-V_{th} plot is shown in Fig. 6. The average R_{on} is decreased

from 3.05 to 2.9 Ω mm by 0.15 Ω mm and the average V_{th} is lowered by 28 mV. The V_{th} shift corresponds to a Ge concentration of 3-5% in a strained-Si/SiGe substrate [9-11]. The shape of this R_{on}-V_{th} plot is not markedly changed, thus, almost the same V_{th} and R_{on} shifts occur at each LDMOS transistor. As a matter of fact, 3 sigma of V_{th} in a wafer is slightly increased from 30.8 to 31.2 mV, and that of R_{on} is also slightly increased from 60.8 to 61.1 Ω mm.

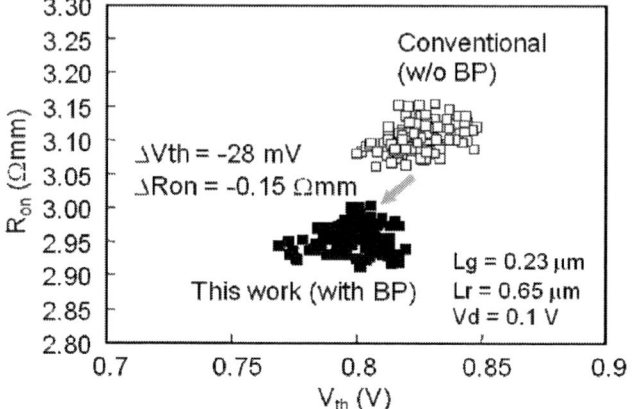

Fig. 6. Comparison of the LDMOS transistors fabricated with BP and without BP using R_{on}-V_{th} plot.

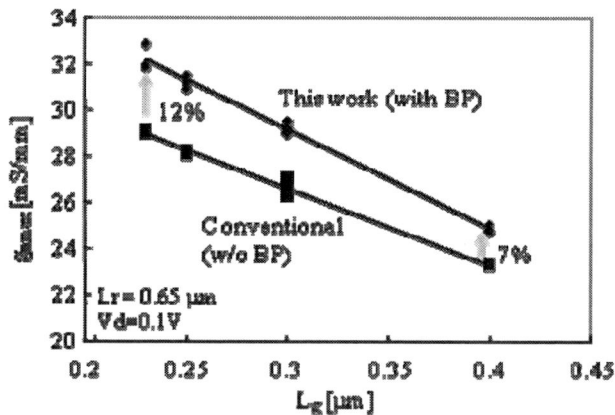

Fig. 7. Gate length dependence of g_{mmax} in comparison of this work and the conventional LDMOS transistor.

A comparison between the L_g dependence of the maximum transconductance (g_{mmax}) in this work and that in the conventional LDMOS transistor is shown in Fig. 7. The g_{mmax} is increased by 12% at Lg = 0.23 μm, but it is reduced to 7% at L_g = 0.4 μm. This phenomenon indicates the gate length dependence of the tensile stress in the channel region.

From the extraction results of R_C and R_O, and the direct measurement of R_S, the total resistance components are summarized in Fig. 8 and compared between this work with BP and the conventional LDMOS transistor. The resistance from the surface source to the backside (R_S) is reduced from 1.3 to 0.8 Ω mm using a critically high concentration substrate. The substrate resistance itself is reduced by 1/2 from that of the conventional substrate, but R_S is not reduced to 1/2 because the BP resistance component is included in R_S. The total on-resistance from the drain to the backside is reduced from 4.4 to 3.7 Ω mm by reducing the substrate resistance critically and applying tensile stress to the channel and the offset-drain region, utilizing the buried polysilicon sinker as a stressor.

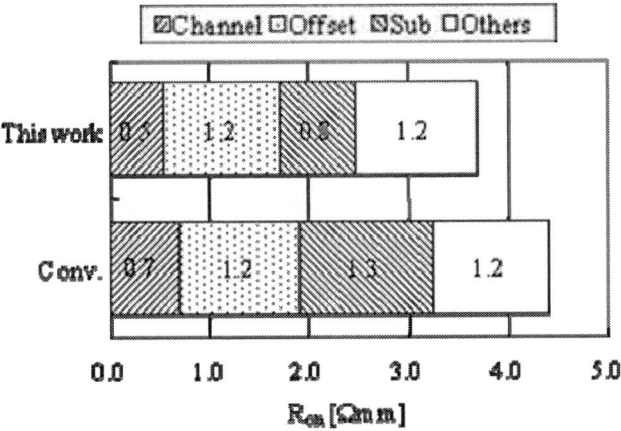

Fig. 8. Comparison of resistance components in this work and the conventional LDMOS transistor.

Finally, we mention the result of reliability measurement. We measured the drain breakdown voltage and the time-dependent dielectric breakdown (TDDB) of the gate oxide and hot carrier lifetime. There are no distinct changes from those of the conventional LDMOS transistor.

IV. CONCLUSIONS

We have demonstrated a new low-on-state-resistance LDMOS transistor with a critically low resistance substrate to reduce the resistance from the surface source to the backside of the transistor and with a buried-polysilicon sinker to apply tensile stress to the channel and the offset-drain region. The boron concentration in the substrate is determined so as not to produce compressive stress at the surface of the wafer. The buried-polysilicon sinker is selected to apply tensile stress to both the channel and the offset-drain region. The existing mechanical stress is confirmed by two-dimensional UV-Raman spectroscopy. The transconductance of the LDMOS transistor is increased by 12% owing to the tensile stress and the resistance from the surface source to the backside is reduced by 38% owing to the critically low resistance substrate. The total on-state resistance from the drain to the backside of the transistor is reduced by 16% owing to channel and source resistance reduction, which directly leads to a higher efficiency of analog power amplifier or converter.

ACKNOWLEDGMENT

The authors wish to thank Tokyo Instruments, Inc. for measuring the stress distribution using 2D UV-Raman spectroscopy. They also thank Taro Higashide and Masaya Iida for their cooperation in device fabrication and Yoshinori Yoshida for fixing amorphous-silicon deposition conditions.

REFERENCES

[1] K. Sakamoto, M. Shiraishi, and T. Iwasaki, "Low on-resistance and low feedback-charge lateral power MOSFETs with multi-drain regions for high-efficient DC/DC converters," Proc. Int. Symp. Power Semiconductor Devices and ICs, 2002, p. 25.

[2] I. Yoshida, M. Katsueda, Y. Maruyama, and I. Kohjiro, "A highly efficient 1.9-GHz Si high-power MOS amplifier," IEEE Trans. Electron Devices **45**, 1998, p. 953.

[3] J. Welser, J. L. Hoyt, and J. F. Gibbons, "Electron mobility enhancement in strained-Si n-type metal-oxide-semiconductor field-effect transistors," IEEE Electron Device Lett. **15**, 1994, p. 100.

[4] F. Ootsuka, S. Wakahara, K. Ichinose, A. Honzawa, S. Wada, H. Sato, T. Ando, H. Ohta, K. Watanabe, and T. Onai, "A high dense, high-performance 130-nm node CMOS technology for large scale system-on-a-chip applications," IEDM Tech. Dig., 2000, p. 575.

[5] M. Kondo, N. Sugii, M. Miyamoto, Y. Hoshino, M. Hatori, W. Hirasawa, Y. Kimura, Y. Kondo, and I. Yoshida, "Strained-silicon MOSFETs for analog applications: utilizing a supercritical-thickness strained layer for low leakage current and high breakdown voltage," IEEE Trans. Electron Devices **53**, 2006, p. 1226.

[6] P. Moens, J. Roig, F. Clemente, I. De Wolf, B. Dosete, F. Bauwens, and M. Tack, "Stress-induced mobility enhancement for integrated power transistors," IEDM Tech. Dig., 2007, p. 877.

[7] J. Chen and I. De Wolf "Theoretical experimental Raman spectroscopy study of mechanical stress induced by electronic packaging," IEEE Trans. Components Packag. Technol. **28**, 2005, p. 484.

[8] A. Ogura, D. Kosemura, K. Yamasaki, S. Tanaka, A. Kitano, and I. Hirosawa, "Measurement of in-plane and depth strain profiles in strained-Si substrates," Solid-State Electron. **51**, 2007, p. 219.

[9] J. S. Goo, Q. Xiang, Y. Takamura, F. Arasnia, E. N. Paton, P. Besser, J. Pan, and M. R. Lin, "Band offset induced threshold variation in strained-Si n-MOSFETs," IEEE Electron Device Lett. **24**, 2003, p. 568.

[10] W. Zhang and J. G. Fossum "On the threshold voltage of strained-Si-Si$_{1-x}$Ge$_x$ MOSFETs," IEEE Trans. Electron Devices **52**, 2005, p. 263.

[11] H. M. Nayfeh, J. L. Hoyt, and D. A. Antoniadis, "A physically based analytical model for the threshold voltage of strained-Si n-MOSFETs," IEEE Trans. Electron Devices **51**, 2004), p. 2069.

Proceedings of the 23rd International Symposium on Power Semiconductor Devices & IC's
May 23-26, 2011 San Diego, CA

0.25μm, 20V High Performance Complementary Bipolar Transistor with Dual EPI and Oxide-Filled Deep Trench Isolation for High Frequency DC-DC Converters

T. Kwon, S. Haynie*, A. Sadovnikov*, P. Allard, J. Strout and A. Strachan*

Advanced Process Technology Development Group, National Semiconductor, South Portland, ME, USA

*Santa Clara, CA, USA

Taehun.Kwon@nsc.com

Abstract— **Power supply designers must increase the switching frequency of converters to meet industry demands for small sizes. In order to handle high switching frequency, a closed-loop DC-DC converter needs a high-speed error amplifier with low Rdson.Qg LDMOS power switches. In this paper, 0.25um, 20V high performance complementary bipolar transistors were developed for the high-speed error amplifier design. Dual epi was used to suppress parasitic bipolar behavior that leads to a latch-up. Also, an oxide-filled deep trench isolation was used to minimize parasitic capacitance. As a result, robust 5GHz NPN and 3GHz PNP transistors were integrated with a low Rdson.Qg LDMOS.**

I. INTRODUCTION

One of the major trends in recent DC-DC converters is increasing the switching frequencies as described in figure 1. The high frequencies result in the passive components of the converters – capacitors and inductors – becoming smaller thereby reducing the total size and weight of the equipment.

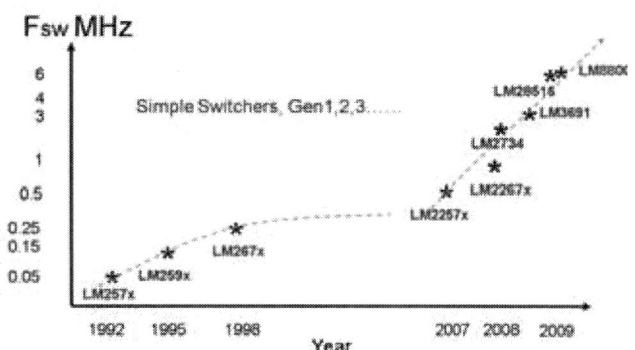

Figure 1.   Trends in switching frequencies. It flattened out, then took off in recent years due to the demands for small sizes.

Figure 2 shows a closed-loop step down DC-DC converter. In order to increase the switching frequency of the converter, low switching loss of the power switches and high frequency handling capability of the compensation circuitry are necessary. A low Rdson.Qg LDMOS transistor [1] was

developed for the power switches and high performance complementary bipolar transistors in this paper were developed for the high speed error amplifier in the compensation circuitry.

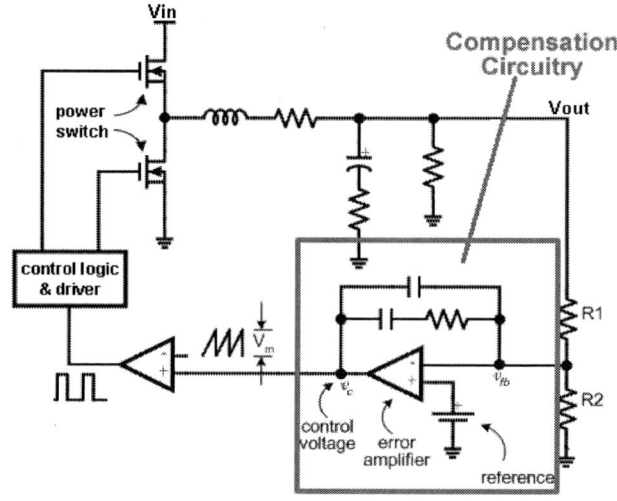

Figure 2.   A closed loop DC-DC converter that requires low Rdson.Qg LDMOS and high-speed error amplifier for high switching frequency.

II. DEVICE AND ISOLATION STRUCTURES

A. PNP Transistor

A conventional vertical PNP transistor has parasitic bipolar transistors as seen in figure 3 and the parasitic transistors can cause a substrate induced latch-up [2]. Figure 4 shows the N-isolation doping profile of the conventional vertical PNP transistor. In order to suppress the parasitic PNP behavior, heavily doped N-isolation is necessary. However, high N-isolation doping increases the sheet resistance of the P-buried layer and decreases the breakdown voltage between the P-buried layer and the N-isolation. In order to increase the N-isolation doping concentration without these shifts, dual epi process was used. Figure 5 shows a cross section of the improved high performance vertical PNP transistor. The

978-1-4244-8425-6/11 $26.00 © 2011 IEEE   172

doping profile of the N-isolation is shown in figure 6. The N-isolation was implanted prior to the 1st epi, and the P-buried layer was implanted after the 1st epi. The epi thickness was optimized by the breakdown voltage between the P-buried layer and the N-isolation. The parasitic PNP beta of the conventional PNP transistor was about 10 ~ 20 due to the low N-isolation doping concentration, but heavily doped N-isolation was achieved with dual epi in the new vertical PNP transistor. The parasitic PNP beta went down below 0.1.

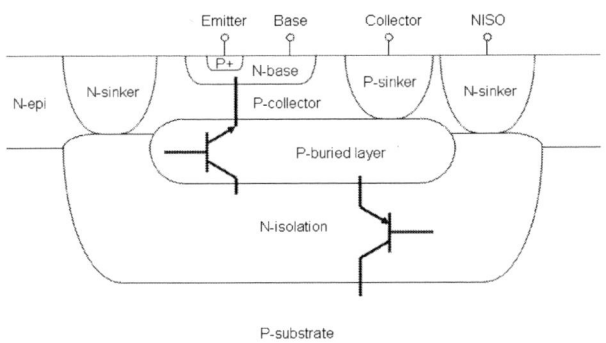

Figure 3.   A conventional vertical PNP transistor includes high gain parasitic PNP to p-substrate.

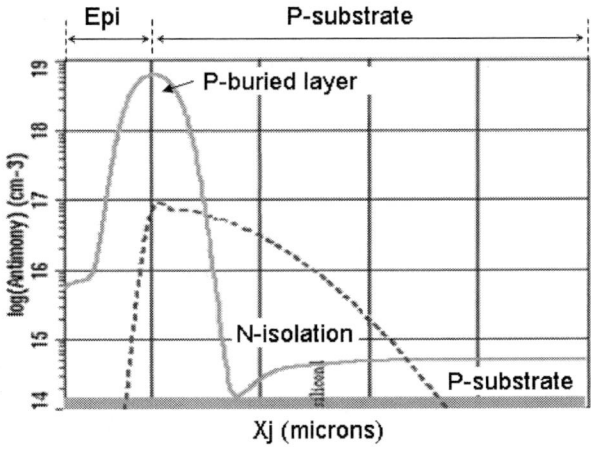

Figure 4.   A conventional vertical PNP transistor includes high gain parasitic PNP to p-substrate.

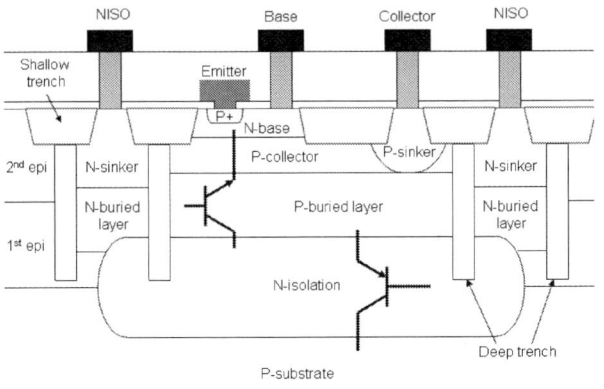

Figure 5.   The improved high performance vertical PNP includes a heavily doped N-isolation region to suppress latch-up.

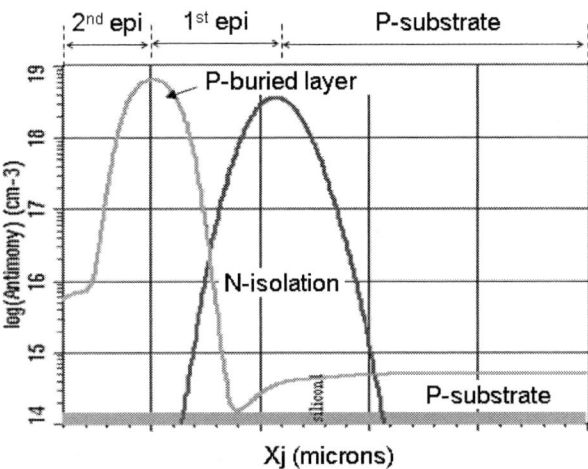

Figure 6.   The N-isolation doping profile of the improved vertical PNP transistor. The 1st epi thickness is carefully chosen to balance breakdown voltage with parasitic N-isolation resistance.

B.   NPN Transistor

Figure 7 shows a SEM cross section of the NPN transistor. Oxide-filled deep trench isolation with air gap was used to reduce parasitic capacitance [3].

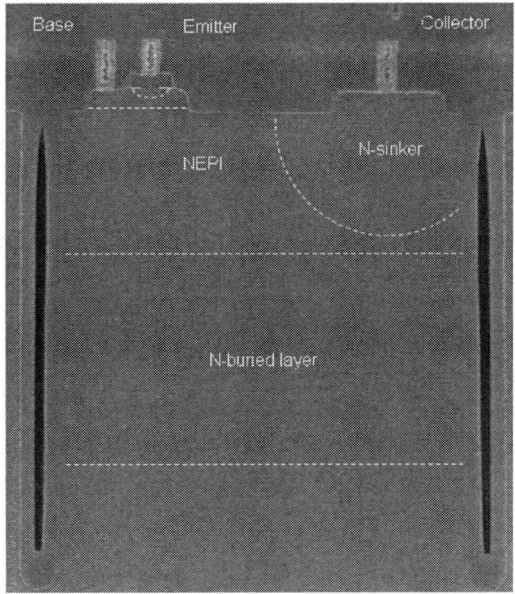

Figure 7.   A SEM cross section of the NPN transistor that shows the oxide filled deep trench with air gap.

C.   Deep Trench Isolation

The oxide filled deep trench isolation in this paper reduces parasitic capacitance effectively, but the sealed trench can be opened by subsequence wet etch processes such as resist strip and pad oxide etch. Once the trench is opened, wet chemicals go into the deep trench and pop up during diffusion processes. This can cause a yield issue. The oxide fill process was carefully optimized to prevent the trench from the opening.

Figure 8 shows the results of the optimized and not optimized oxide fill processes.

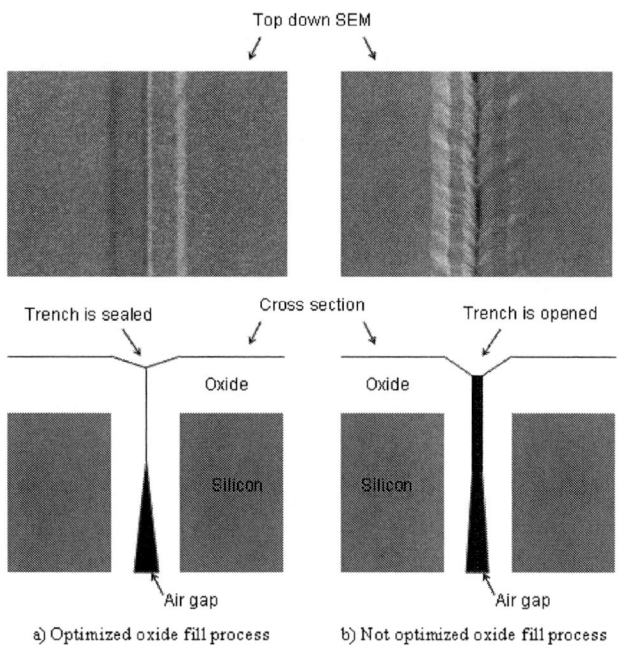

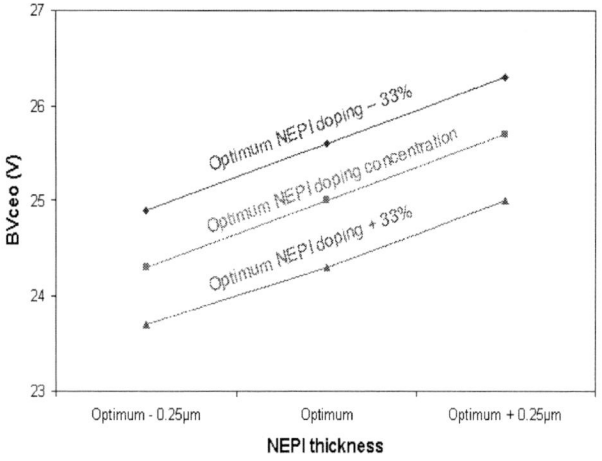

Figure 8. Comparison of the oxide fill processes that received subsequent wet etch processes.

III. ELECTRICAL CHARACTERISTICS

The NEPI thickness and doping concentration were optimized for the NPN transistor. Figure 9 shows the BVceo for a different NEPI.

Figure 9. NPN BVceo for a different NEPI thickness and doping concentration.

The base widths of the transistors were optimized by the base implant energy. Figure 10 and figure 11 show the beta and early voltage for a different base implant energy.

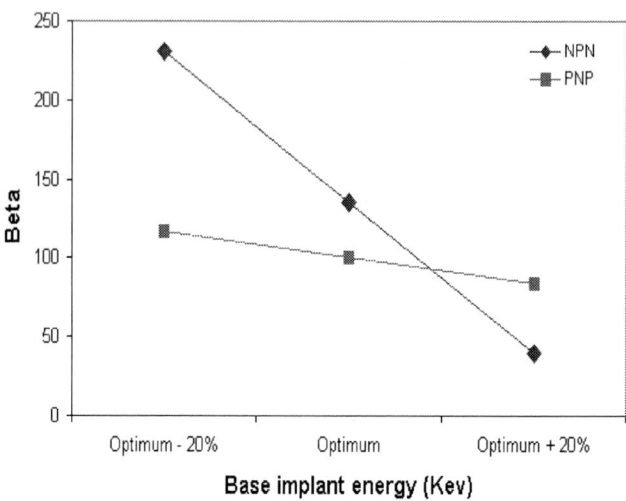

Figure 10. Beta for a different base implant energy.

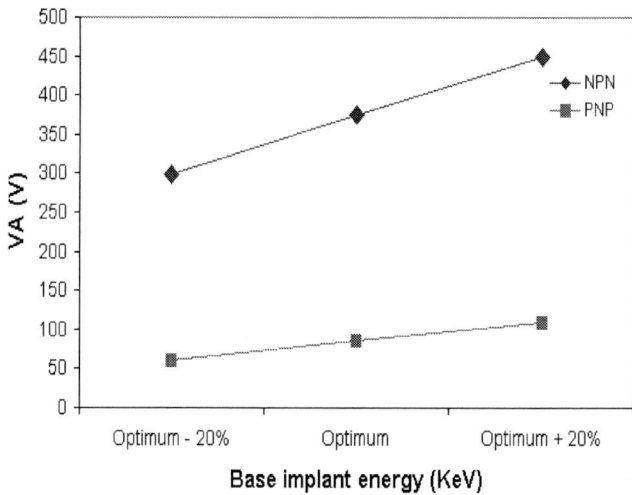

Figure 11. VA (Early voltage) for a different base implant energy.

The electrical characteristics of the complementary bipolar transistors are described in table 1. A cut off frequency (fT) of 5GHz of NPN and 3GHz of PNP were achieved with 25V and 38V BVceo. The P-collector of the PNP transistor was optimized not only for the PNP transistor but also for other devices such as PLDMOS so its BVceo is higher than that of the NPN transistor. The Beta * VA (early voltage) of the NPN transistor is higher than 50,000V. This is comparable to higher performance complementary BiCMOS technology [4]. Forward output characteristics of the complementary bipolar transistors are shown in figure 12 and figure 13.

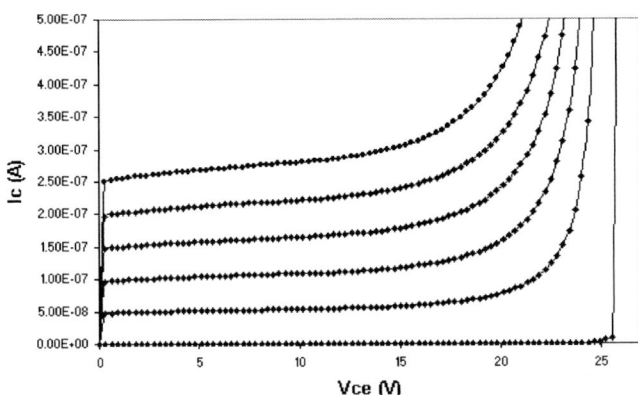

Figure 12. Forward output characteristics of the NPN transistor.

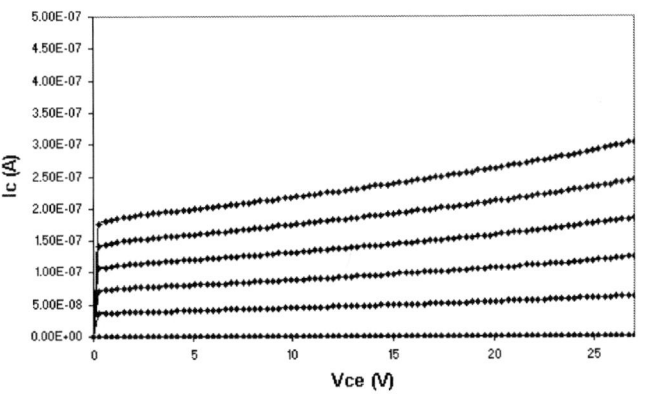

Figure 13. Forward output characteristics of the PNP transistor.

TABLE I. ELECTRICAL CHARACTERISTICS MEASURED.

Parameter	NPN Tr.	PNP Tr.
Beta (Vce = 5V, peak fT)	135	100
Early Voltage (Vbe =0.65V)	375V	85V
BVceo (Vbe = 0.7V, Ib = 0A)	25V	38V
fT peak (Vce = 5V)	5GHz	3GHz
Emitter Size	0.25µm X 1µm	0.25µm X 1µm

IV. CONCLUSION

High performance complementary bipolar transistors were integrated with a low Rdson.Qg LDMOS transistor to increase switching frequencies of DC-DC converters. Dual EPI prevents latch up caused by substrate injection current and deep trench isolation with air gap minimizes parasitic capacitance. These high performance complementary bipolar transistors enable the design of a 500MHz GBWP error amplifier. The switching frequency of corresponding DC-DC converter is 25MHz.

REFERENCES

[1] S. Haynie, A. Gabrys, T. Kwon, P. Allard, J. Strout, A. Strachan "Power LDMOS with Novel STI Profile for Improved Rsp, BVdss and Reliability" ISPSD2010, pp.241 ~ 244.

[2] A. Strachan "Substrate Current Injection and Latchup in Complementary Vertical Bipolar Processes", BCTM2006, pp.61 ~ 66.

[3] L.J. Choi, E.Kunnen, S. Van Huylenbroeck, A. Piontek, A. Sibaja-Hemandez, F. Vleugels, T. Dupont, P. Leray, K. Devriendt, X.P. Shi, R. Loo, S. Vanhaelemeersch and S. Decoutere "A Novel Deep Trench Isolation Featuring Airgaps for a High-Speed 0.13um SiGe:C BiCMOS Technology" VLSI-TSA 2006, pp. 88 ~ 89

[4] S. J. Harrington, A. Bousquet, S. Nigrin1, S. Suder and B.M.Armstrong "A High Performance 36V Complementary Bipolar Technology on Low Thermal Resistance Compound Buried Layer SOI Substrates" BCTM2010, pp.37 ~ 40.

Proceedings of the 23rd International Symposium on Power Semiconductor Devices & IC's
May 23-26, 2011 San Diego, CA

Integration of 100V LDMOS Devices in 0.35µm CMOS Technology

Soon Tat Kong, Paul Stribley, Chris Lee, * Michaelina Ong

X-FAB Semiconductor Foundries AG, Tamerton Road, Roborough, Plymouth, UK, PL6 7BQ

*X-FAB Sarawak Sdn. Bhd. Kuching, Sarawak, Malaysia

E-mail: Chris.Kong@xfab.com Ph: +44 1752693235, Fax: +44 1752693200

Abstract—**Successful integration of 100V LDMOS devices in 0.35µm CMOS technology is presented in this paper. These integrated devices are enhanced N-type and P-type LDMOS which are compatible with thin (14nm) and thick (40nm) layers of gate oxide. A breakdown voltage of more than 100V with R_{DS} (ON) =200/180mΩ.mm² for N-type LDMOS and R_{DS} (ON) =690/640mΩ.mm² for P-type LDMOS with 14nm/40nm gate oxide thickness.**

1. INTRODUCTION

Today, integration of High-Voltage (HV) Lateral Diffused MOS transistors (LDMOS) and low-voltage CMOS devices are the key solution for the smart-power, mixed signals ASICs (Application-Specific Integrated Circuits) and power IC applications [1-3]. Also, there are ever increasing demands for routinely integrating low voltage and high voltage devices on the same wafer driven by the trend of System on Chip (SOC) technology. Such integration provides the benefits of miniaturisation, low energy consumption, high-level of integration, performance and cost effectiveness that enable SOC manufactures to create lower-power, high-density, higher-performing systems.

Basically, an integrated circuit may contain either an HV N-type LDMOS or HV P-type LDMOS structure or even both, commonly used for output power stages of an integrated circuit that comprise a signal processing and control circuitry [4-5]. In this work, we propose 100V HV N-type and P-type LDMOS devices in standard 0.35µm technology [6].

From the design of HV devices point of view, the following hot topics need to be mentioned: - firstly, the device breakdown and R_{DS}(ON) resistance; secondly, the maximum voltage or current boundaries, and finally temperature dependent characteristics during operation. Basically, reduced R_{DS}(ON) resistance gives a smaller temperature rise and therefore reduces the risk of device destruction during switching. Also, it can reduce the cost of packaging and avoids the need for a cooling system. In addition, maximum voltage or current boundaries of HV devices are defined as the voltage and current conditions within which the device can be expected to operate reliably for long time periods. Beyond maximum voltage or current boundaries, the transistors suffer electrical wear-out and possibly device destruction through charge capture, self-heating or breakdown mechanisms. These effects are present only at very large drain voltages and

can be easily observed and identified by electrical characteristics, e.g. a breakdown on the output characteristics. Device power dissipation has also been shown and cause "self-heating" effects which reduce the efficiency of the device and modify its electrical properties.

2. DEVICE STRUCTURES AND OPERATIONS

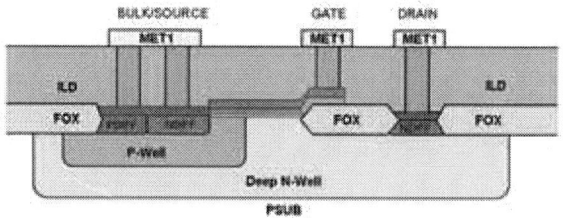

Fig.1: The simplified cross-section of 100V N-type LDMOS

Fig.1 and 2 show the simplified cross-sections of 100V N- type and P-type transistors respectively. These devices can be easily fabricated with low-voltage CMOS circuitry on the same chip. The cross-section of the N-type transistor with linearly graded doping (Deep N-well) in the whole drift region is used, in order to achieve very high breakdown voltage. Also, the deep N-well region is used under the P-Well region to give a high side breakdown voltage. In general, this device structure is appropriate for high and low side circuits.

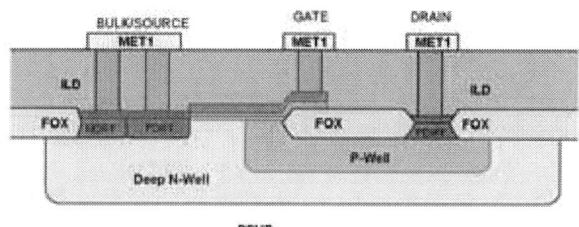

Fig.2 : The simplified cross-section of 100V P-type LDMOS

The 100V P-type LDMOS structure is designed for high side only. The deep N-well doping has been optimized to achieve low R_{DS}(ON) resistance for 100V breakdown. The P-well diffuses both vertically and laterally to optimize the drift region of the P-type LDMOS.

978-1-4244-8425-6/11 $26.00 © 2011 IEEE   176

3. EXPERIMENTAL RESULTS

The enhanced 100V N-type and P-type transistors were tested on a Signatone probe station with a Keithley Model-4200SCS semiconductor characterization system. The maximum applicable gate voltage is 5V for the devices with gate oxide thickness of 14nm. The device dimensions (W x L) of the N-type and P-type LDMOS devices are 20 μm x 1.1μm and 20 μm x 2μm respectively.

A. Breakdown Voltage

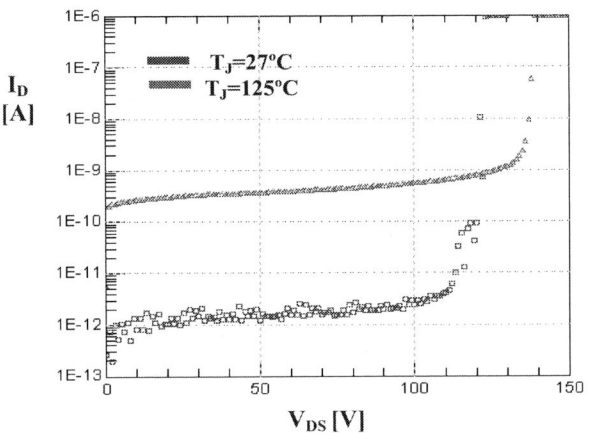

(a) N-type LDMOS with 14nm gate oxide thickness

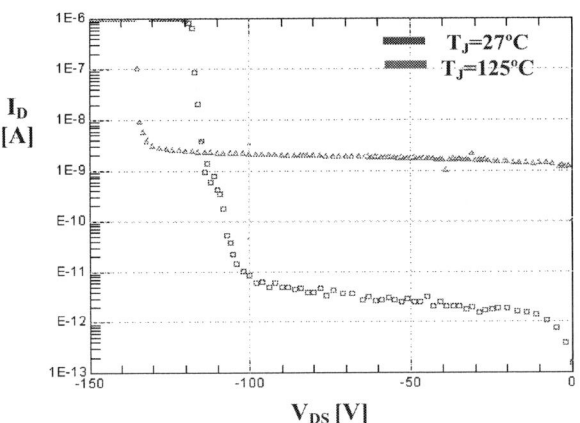

(b) P-type LDMOS with 14nm gate oxide thickness

Fig. 3: Typical Drain-Source breakdown voltage of 100V LDMOS devices at $T_J=27^oC$ and $T_J=125^oC$

The breakdown voltage characteristic of N-type and P-type LDMOS are measured as shown in Fig.3 (a) and (b). In this case, the leakage currents of N-type LDMOS at $T_J=27^oC$ (blue trace) and 125^oC (red trace) are approximate equal to 2pA and 0.5nA respectively. The breakdown voltage of N-type & P-type LDMOS are approximately 120V & 135V at $T_J=27^oC$ and 125^oC.

B. Transfer characteristic and Transconductance(G_M)

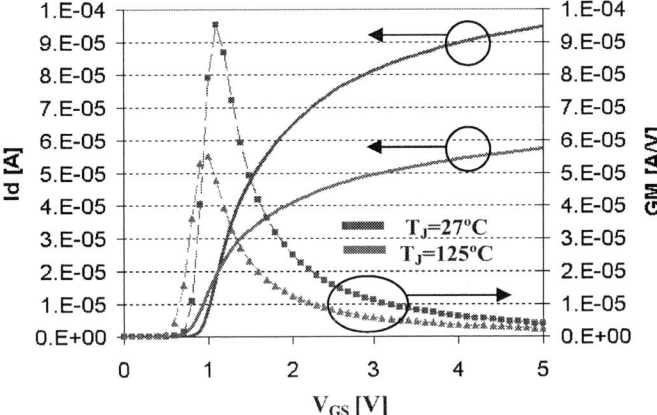

(a) N-type LDMOS with 14nm gate oxide thickness

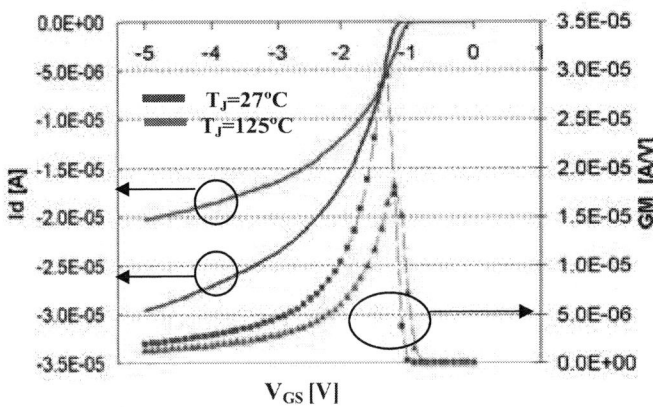

(b) P-type LDMOS with 14nm gate oxide thickness

Fig. 4 Typical transfer characteristics of 100V N-type and P-type LDMOS transistors at $T_J=27^oC$ and $T_J=125^oC$

Fig. 4(a) and (b) show the transfers characteristic curves of N-type and P-type devices at room temperature (blue trace) and high temperature (red trace). Threshold voltage was extracted from these characteristics at a current level of 1μA. It can be seen that the threshold voltage of 100V devices changes with temperature and that the drive current reduces.

At $T_J=27^oC$, it was shown that P-type LDMOS had a higher threshold voltage of about -1.12V compared to 0.88V for the N-type LDMOS. This variation comes from the difference in doping density and channel length in the P-type LDMOS channel compared with that in the N-type LDMOS channel. The measured peak transconductance values of the N-type and P-type transistors at $T_J=27^oC$ are 9.6E-5 A/V and 3.0E-5 A/V respectively.

978-1-4244-8425-6/11 $26.00 © 2011 IEEE

C. Output Characteristics

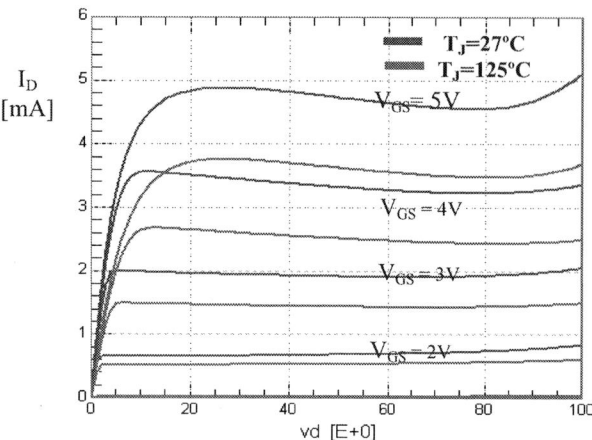

(a) N-type LDMOS with 14nm gate oxide thickness

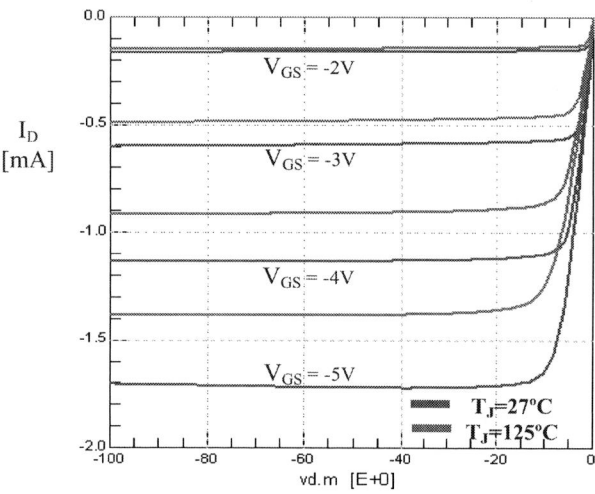

(b) P-type LDMOS with 14nm gate oxide thickness

Fig. 5: Typical output characteristics of 100V N-type and P-type LDMOS transistors at T_J=27°C and T_J=125°C

Fig. 5 (a) and (b) show the measured output characteristics of the 100V N-type and P-type transistors at T_J=27°C and 125°C. These curves were obtained by sweeping the drain voltage from 0 to 100V for each value of gate voltage sweeping from 0V to 5V in steps of 1V. Self heating is observed in the negative slope of the curves at high V_{GS}. It was shown that both devices can operate up to the maximum V_{GS} rating of 5 V with V_{DS} = 100V. Table 1 shows the summary of the R_{DS}(ON) resistance of N-type and P-type transistors for the thin gate oxide (14nm) and thick gate oxide (40nm) structures. The maximum applicable gate voltages of the thin and thick gate-oxide devices are 5V and 12V respectively.

Table 1: Typical R_{DS}(ON) resistance of N-type and P-type transistors for thin gate oxide (14nm) and thick gate oxide (40nm) structures

Device type (BV=100V)	Specific R_{DS}(ON) resistance [mΩ.mm²]	
	Thin gate oxide thickness (14nm)	Thick gate oxide thickness (40nm)
N-type	200	180
P-type	690	640

D. 1/f Flicker Noise

It is well known that flicker noise is strongly dependent on the interface quality of the gate oxide [7]. Thus, the flicker noise of the 100V N-type and P-type devices with gate oxide thickness of 14nm was measured with a BTA 9812A system. It consists of a controller unit and an amplifier/filter unit used for detection and amplification of the noise current generated by the 100V devices.

Flicker noise model, which is always associated with a power spectral density (S_{id}). It has been described in SPICE noise models [8] as:-

$$S_{id}(f) = \frac{KF \cdot I_{DS}{}^{AF}}{C_{OX} \cdot L_{eff}{}^{2} \cdot f^{EF}} \qquad (1)$$

Where KF is a flicker noise coefficient, AF is a flicker noise exponent (typically between 0.5-2) and f is the analysis frequency. The frequency exponent (EF) is bias-dependent. C_{OX} is the gate oxide capacitance and L_{eff} is the effective channel length of the 100V LDMOS devices. It is important to note that reduced LDMOS dimensions are accompanied by an increased level of noise.

Fig.6 (a) and (b) show a typical drain current noise characteristic measured in the saturation region (V_{DS}= 20V) for the 100V N-type and P-type devices. The measured frequency range is between 25 Hz to 100K Hz. It was shown that the flicker noise spectrum for both devices clearly exhibits good low-frequency noise spectra.

In this work, the flicker noise of the 100V N-type and P-type devices have been modeled based on equation 1. A pragmatic approach has been taken to include the effect of this form of noise in 100V transistors such that the modelled flicker noise can be included in circuit simulations. Thus, the LDMOS devices were simulated using HSPICE simulator with BSIM3V3 models for a fixed combination of the drain (+/- 20V) and the gate voltages at T_J=27°C.

The measured noise characteristics are compared with the simulations. The KF and AF values of N-type LDMOS used in the simulations are 2.82E-28 and 1.1 respectively. For P-type LDMOS, parameter extraction values of KF and AF are 1.06E-28 and 1.42. The EF parameter values of N-type and P-type LDMOS devices are 0.96 and 1.11 respectively.

978-1-4244-8425-6/11 $26.00 © 2011 IEEE

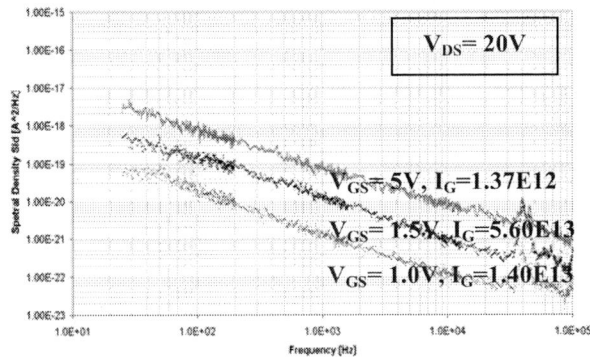

(a) N-type LDMOS

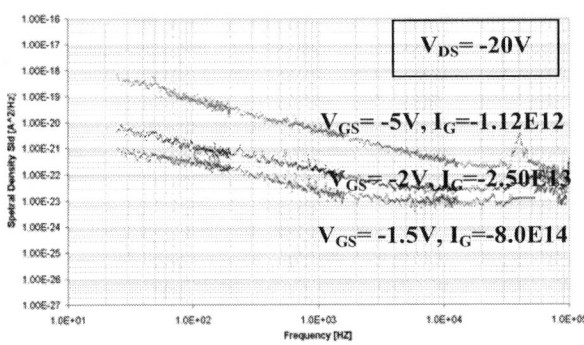

(b) P-type LDMOS

Fig. 6: Typical drain current flicker noise characteristics of 100V N-type and P-type LDMOS transistors at T_J=27°C

4. CONCLUSION

The 100V high-voltage LDMOS transistor structures integrated within 0.35µm technology have been successfully implemented and tested. This process development is compatible with CMOS standard processing. Experimental results show that the LDMOS structure has a breakdown voltage of more than 100 volts for both N-type LDMOS and P-type LDMOS transistor structures. Moreover, these 100V devices characteristics show excellent performances in term of breakdown voltage, R_{DS}(ON) resistance, high temperature and normal flicker noise.

ACKNOWLEDGMENT

The authors would like to thank Suba Subramania, Ian Macpherson and Roberto Gaertner for their support of this work.

4. REFERENCES

[1] J.S. Writter, "A modular BICMOS technology includilng 85V DMOS devices for analogue/digital ASIC applications", Solid State Device Research Conference, 1992. ESSDERC '92. 22nd European, PP.555-558.

[2] M.Elwin et al, "Optimization of 100V high side LDMOS using multiple simulation techniques" ISPSD 2009, PP. 104-107.

[3] J. Van der Pol et al., "A-BCD, An economic l00V RESURF silicon-on-insulator BCD technology for consumer and automotive applications, ISPSD 2000, PP. 327-330.

[4] P.Igic et al, "Perspective on Power IC technology: From design lab to wafer fab", MIEL 2010, PP. 73-77.

[5] Xiaorong Luo et al, "A High-Voltage LDMOS Compatible With High-Voltage Integrated Circuits on p-Type SOI Layer", IEEE Electron Device Letters, 2009, PP.1093-1095.

[6] XFAB Process & Device Specification XH035

[7] K.K.Hung et al, "Flicker noise characteristics of advanced MOS technologies", IEDM 1988, PP.34-37.

[8] N.H.Hamid et al, "Time-domain modeling of low-frequency noise in deep-sub micrometer MOSFET", IEEE Transactions on Circuits and Systems, PP. 245-257.

Proceedings of the 23rd International Symposium on Power Semiconductor Devices & IC's
May 23-26, 2011 San Diego, CA

High-Voltage Thick Layer SOI Technology for PDP Scan Driver IC

Ming Qiao[1], Lingli Jiang[1], Meng Wang[1], Yong Huang[1,2], Hong Liao[2] , Tao Liang[1,2], Zhen Sun[2], Bo Zhang[1], Zhaoji Li[1], Guangzuo Huang[2], Yuanyuan Zhao[1], Li Lai[1], Xi Hu[1], Xiang Zhuang[1], Xiaorong Luo[1], and Zhuo Wang[1]

[1] State Key Laboratory of Electronic Thin Films and
Integrated Devices
University of Electronic Science and Technology of China
Chengdu, P.R.China
qiaoming@uestc.edu.cn

[2] Changhong Electric Co., Ltd.
Mianyang, P.R.China

Abstract—**Based on 11-µm-thick silicon layer and 1-µm-thick buried oxide layer, a novel high-voltage thick layer SOI technology has been developed for driving plasma display panels (PDP). HV pLDMOS, nLDMOS, nLIGBT and LV CMOS are compatible with deep trench isolation. The length T, Y for HV pLDMOS and TD for HV nLDMOS are optimized to reduce the device size and satisfy the off-state breakdown voltage simultaneously. Interdigitated N+&P+ and a deep P+ are adopted in the source region of HV nLDMOS and cathode region of HV nLIGBT to suppress parasitic NPN action and gain better on-state characteristics. A PDP scan driver IC using the developed high-voltage thick layer SOI technology shows that the rise and fall times of the output stages are about 17.6 ns and 16.6 ns respectively.**

I. INTRODUCTION

In recent years, color plasma display panel has attracted a great deal of attention with its superior contrast ratio and wide viewing angles. It has become an excellent choice for high-definition video displays and computer monitors. As the next generation of large-screen flat-panel display, the color PDP has been developed towards larger screen. The number of the high-voltage driver circuits in one color PDP has been increasing. Therefore, PDP driver ICs are required to have multiple output stages [1-6]. The cost reduction is extremely important and the merits of the high-voltage integration technology play a significant role in it. Silicon-on-Insulator (SOI) technology has been preferred in such multi-output applications for its compact dielectric isolation and suppression of parasitic bipolar components among device/circuit blocks.

In this paper, we introduce a novel high-voltage thick layer SOI technology which is developed to realize 96-bit output PDP scan driver IC. HV pLDMOS, nLDMOS, nLIGBT with the breakdown voltage above 200 V and LV CMOS are compatible with deep trench isolation. Some key parameters of the high-voltage device have been optimized and experimentally investigated not only to shrink the device area but also to improve the device performance.

II. HIGH-VOLTAGE THICK LAYER SOI TECHNOLOGY

Figure 1 shows the schematic cross-sectional view of HV pLDMOS, HV nLDMOS, and HV nLIGBT using the proposed high-voltage thick layer SOI technology, which is based on a standard 1 µm rule low-voltage CMOS process.

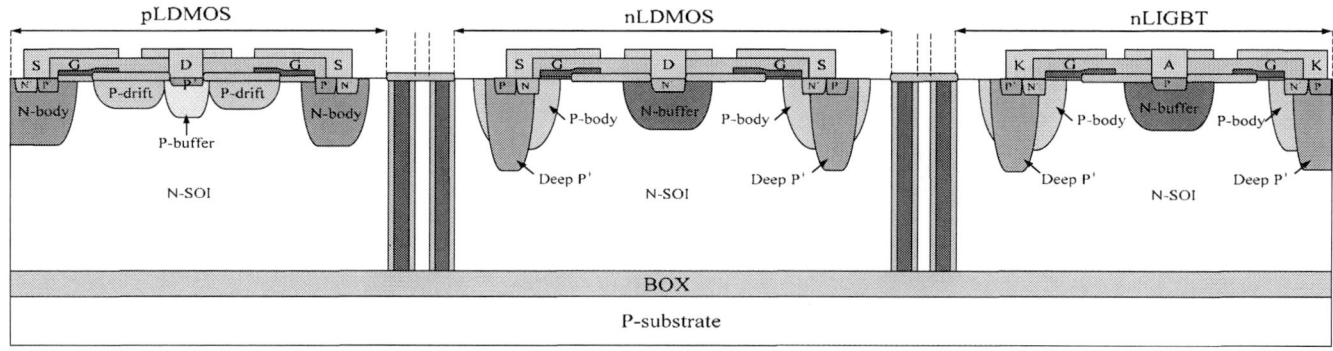

Fig.1. Schematic cross-sectional view of HV pLDMOS, nLDMOS and nLIGBT using thick layer SOI.

978-1-4244-8425-6/11 $26.00 © 2011 IEEE

The thickness of the SOI layer and buried oxide layer are 11 μm and 1 μm respectively. The high-voltage devices are isolated by deep trenches. The trenches are filled with poly-silicon.

HV pLDMOS

Drain-centered structure is adopted for HV pLDMOS, while the bias of the source electrode of the pLDMOS is higher than that of the SOI layer outside the adjacent isolation trench, thus the trench termination must be designed to support sufficient applied voltage. Two types of HV pLDMOS have been developed as shown in Fig.2, one with single trench and the other with double trenches. For the pLDMOS with double trenches, the SOI layer between the two trenches is shorted to the source electrode. Figure 3 shows the potential contours of HV pLDMOS in the off-state with single trench and double trenches respectively which were obtained from 2D device simulations using SILVACO. For the pLDMOS with single trench, premature avalanche breakdown would happen because of the crowded surface electric field in the vicinity of the trench isolation. Higher off-state breakdown voltage will result in a longer distance T_1 ($T_1=T-Y$) between the trench and N-body at the expense of larger chip area. For the pLDMOS with double trenches, the potential of the silicon area between the isolation trenches has been made the same as the source electrode through the metal interconnection to alleviate the crowded surface electric field, thus a shorter length T_1 could be obtained. The outer trench of this structure is used for sustaining the applied high voltage, but it increases extra device area. For these two types of structures with different design rules, the one with smaller device area is preferred.

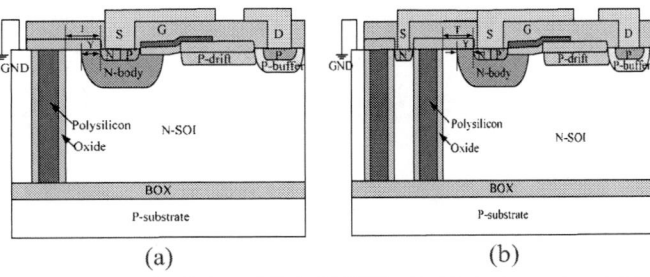

(a) (b)

Fig.2. Cross-sectional view of HV pLDMOS with (a) single trench and (b) double trenches.

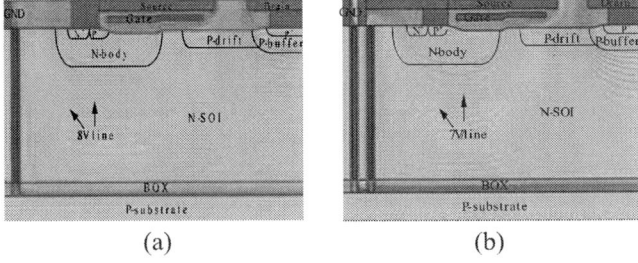

(a) (b)

Fig.3. Simulated potential contours of HV pLDMOS with (a) single trench and (b) double trenches under typical biasing condition.

The length T (the distance from trench to N^+) and Y (the distance from the boundary of N-body to N^+) have a significant impact on the off-state breakdown voltage of the pLDMOS. Figure 4 (a) and Figure 4 (b) show the

experimental results of off-state breakdown voltage for HV pLDMOS as a function of length T and Y_1 ($Y=Y_1+\triangle L_d$ ($\triangle L_d$: lateral diffusion length of N-body)) with different implant dose of the drift region respectively. With the length T above 6.5 μm and Y_1 below 0 μm, the off-state breakdown voltage of over 200 V can be achieved. The output characteristics of the HV pLDMOS are shown in Fig.4 (c) (Vgs from -20 V to -140 V with applied voltage step of -30 V).

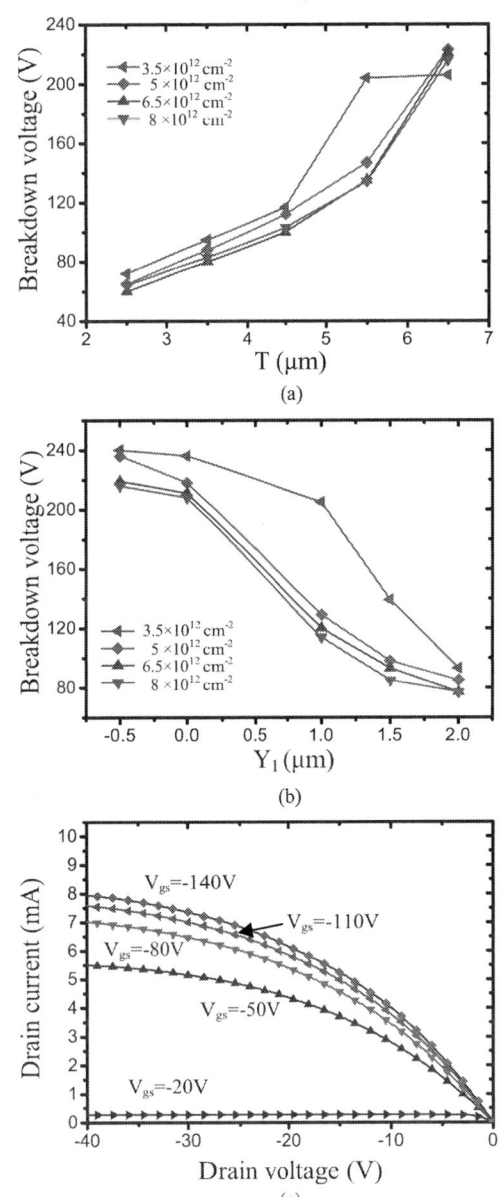

(a)

(b)

(c)

Fig.4. Measured off-state breakdown voltage of HV pLDMOS with drift region implant dose of 3.5×10^{12} cm^{-2}, 5×10^{12} cm^{-2}, 6.5×10^{12} cm^{-2}, and 8×10^{12}cm^{-2} as a function of (a) T and (b) Y_1. (c) Measured output characteristics for HV pLDMOS.

HV nLDMOS

To decrease the device size, we have developed drain-open type HV nLDMOS, in which partial device areas are cut out to lower the specific on-state resistance as shown in Fig.5. In the source region, there are interdigitated N^+&P^+ which could help

978-1-4244-8425-6/11 $26.00 © 2011 IEEE

to suppress parasitic NPN action. A deep P^+ is also adopted in the source region, which could provide better on-state characteristics. The device breakdown depends strongly on the length TD (the distance from trench to the n-buffer region).

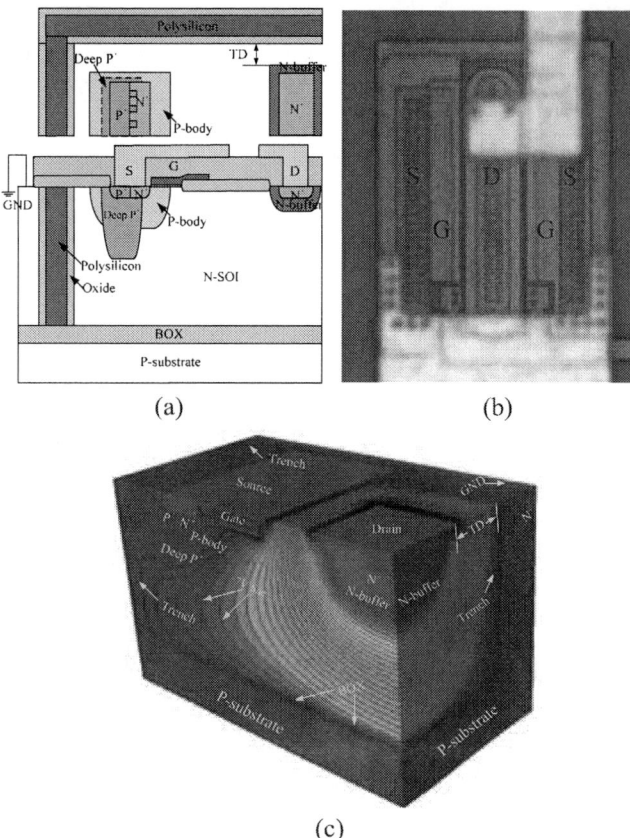

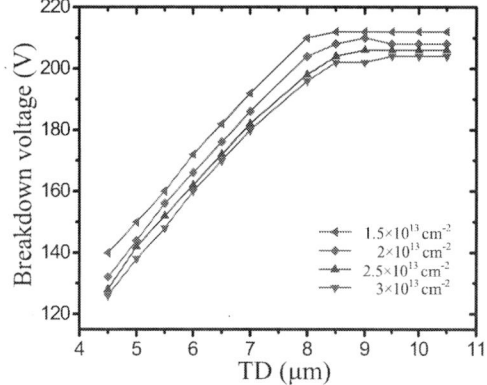

(c)

Fig.5. (a) Cross-sectional and top view of HV nLDMOS. (b) Microphotograph of HV nLDMOS. (c) Simulated 3D potential contours of HV NLDMOS under typical biasing condition.

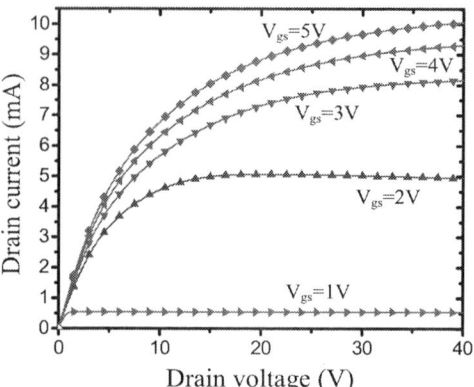

Fig.7. Measured output characteristics for HV nLDMOS.

HV nLIGBT

Anode-centered type HV nLIGBT has been adopted in the proposed high-voltage thick layer SOI technology. Similar to HV nLDMOS, in the cathode region of HV nLIGBT, there are interdigitated $N^+\&P^+$ and a deep P^+ as shown in Fig.8. Figure 9 shows the measured output characteristics for the HV nLIGBT. Figure 10 shows the normalized current of HV pLDMOS, nLDMOS and nLIGBT. It indicates that the current abilities of the HV nLIGBT are superior to that of the HV LDMOS, which could meet the requirements of the high voltage and large current output stages.

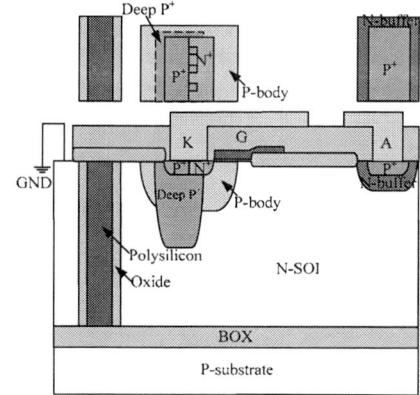

Fig.8. Cross sectional and top view of HV nLIGBT.

Figure 6 shows the measured off-state breakdown voltage as a function of the length TD with different implant dose of the N-buffer region. With the length TD over 8.5 μm, the off-state breakdown voltage of above 200 V can be realized. Figure 7 shows the measured output characteristics for HV nLDMOS.

Fig.6. Measured breakdown voltage of HV nLDMOS with N-buffer implant dose of 1.5×10^{13}cm^{-2}, 2×10^{13}cm^{-2}, 2.5×10^{13}cm^{-2}, and 3×10^{13} cm^{-2} as a function of TD.

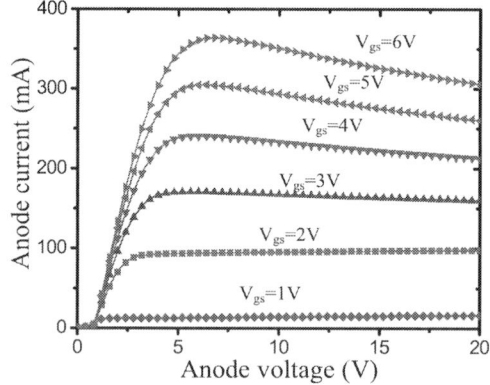

Fig.9. Measured output characteristics for HV nLIGBT.

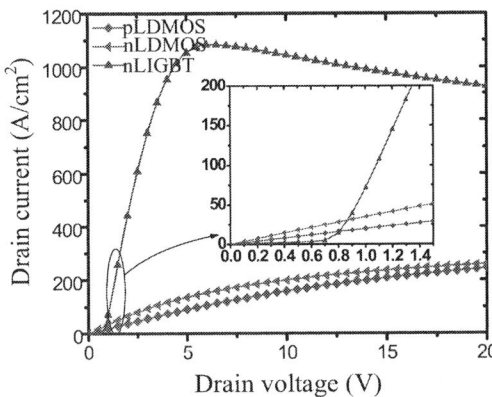

Fig.10. Normalized current of HV pLDMOS, nLDMOS and nLIGBT.

III. APPLICATION TO PDP SCAN DRIVER IC

A 96-bit out PDP scan driver IC has been fabricated using the developed high-voltage thick layer SOI technology. The microphotograph of the PDP scan driver IC is shown in Fig.11 (a). It includes 96-bit shift register, 96-bit latch, selector, and level shifter. Figure 11 (b) shows a 50-inch color PDP TV using the developed 96-bit output PDP scan driver IC. The switching waveforms of the scan driver IC, which is tested in the 50-inch color PDP, are shown in Fig.11 (c). The rise and fall times of the output stages are about 17.6 ns and 16.6 ns respectively.

IV. CONCLUSION

A novel high-voltage thick layer SOI technology has been developed based on 11-μm-thick silicon layer and 1-μm-thick buried oxide layer. Deep trench isolation technology is adopted to avoid cross talk between HV devices. Key parameters of the HV devices are optimized to realize high off-state breakdown voltage and reduce the device size. Interdigitated N^+&P^+ and a deep P^+ are adopted in the source region for HV nLDMOS and cathode region for HV nLIGBT to suppress parasitic NPN action and gain better on-state characteristics. High blocking capability for both off-state and on-state breakdown voltages have been obtained with the developed high-voltage thick layer SOI technology. A 96-bit output PDP scan driver IC has been developed using the novel high-voltage thick layer SOI technology. The rise and fall times of the output stages for the scan driver IC used in a 50-inch color PDP TV are about 17.6 ns and 16.6 ns respectively. Low cost, low power consumption, high operating frequency and sufficient margin to ensure noise immunity can be realized for the requirements of the PDP scan driver IC based on the proposed high-voltage thick layer SOI technology.

ACKNOWLEDGMENT

Project supported by the National Natural Science Foundation of China (No. 60906038) and the Science-Technology Foundation for Young Scientist of University of Electronic Science and Technology of China (No. L08010301JX0830).

(a)

(b)

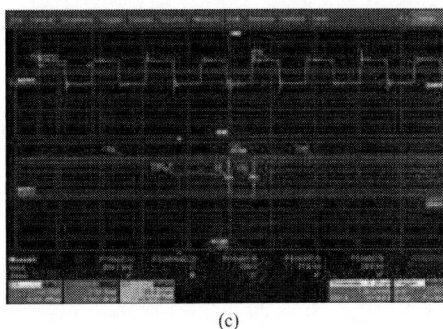

(c)

Fig.11. (a) Microphotograph of the 96-bit output PDP scan driver IC using high-voltage thick layer SOI technology. (b) A 50-inch color PDP TV using the developed scan diver IC. (c) Switching waveforms of the developed scan driver IC.

REFERENCES

[1] K. Kobayashi, H. Yanagigawa, K. Mori, S. Yamanaka, and A. Fujiwara, "High Voltage SOI CMOS IC Technology for Driving Plasma Display Panels," in Proc. of ISPSD, Kyoto, Japan, June 1998, pp.141-144.

[2] J. Kim, T.M. Roh, S.G. Kim, Q.S. Song, D.W. Lee, J.G. Koo, K.I. Cho, and D.S. Ma, "High-Voltage Power Integrated Circuit Technology Using SOI for Driving Plasma Display Panels", IEEE Trans. Electron Devices, vol. 48, pp. 1256-1263, 2001.

[3] T. Nitta, S.Yanagi, T. Miyajima, K. Furuya, Y. Otsu, H. Onoda, and K.Hatasako, "Wide Voltage Power Device Implementation in 0.25μm SOI BiC-DMOS," in Proc. of ISPSD, Naples, Italy, June 2006, pp. 341-344.

[4] H. Sumida, K. Maiguma, N. Shimizu, and H. Kobayashi, "250V-Class Lateral SOI Devices for Driving HDTV PDPs," in Proc. of ISPSD, Jeju, Korea, May 2007, pp. 229-232.

[5] M. Qiao, B. Zhang, Z. Q. Xiao, J. Fang, and Z. J. Li, "High-Voltage Technology Based on Thin Layer SOI for Driving Plasma Display Panels," in Proc. of ISPSD, Orlando, Florida, USA, May 2008, pp. 52-55.

[6] D.H. Lu, T. Mizushima, H, Sumida, M. Saito, and H. Nakazawa, "High Voltage SOI P-channel Field MOSFET Structures", in Proc. of ISPSD, Barcelona, Spain, June 2009, pp. 17-20.

Proceedings of the 23rd International Symposium on Power Semiconductor Devices & IC's
May 23-26, 2011 San Diego, CA

Considerations on the optimal power stage segmentation algorithm for MHz integrated synchronous Buck DC-DC converters

Xiaopeng Wang, Alex. Q. Huang

NSF FREEDM Systems Center, Department of Electrical and Computer Engineering
North Carolina State University
Raleigh, NC. USA
thomas_wang_ncsu@yahoo.com, aqhuang@ncsu.edu

Abstract— **For those MHz integrated synchronous Buck DC-DC converters (ISBC), a power stage segmentation technique might be applied for the sake of improving light load efficiency. The paper discusses the difference about efficiency in the case that losses contributions from Cds and Cgd in inactive power FET subcells are considered or not and indicates the existence of efficiency optimization's saturation effect in respect to the number of active power FET cells. After that, the paper presents the variation characteristics of power FET rdson using On-Semi SCN05 technology's eight manufacturing runs and temperature shift as two example cases. The variation of rdson implies that practical efficiency might deviate from an expected one, provided that the number of active power FET subcells is selected to be linearly proportional to the load current as that implemented in nowadays power FET width segmentation algorithms. Finally, the paper suggests a novel segmentation algorithm with automatic rdson compensation ability.**

I. INTRODUCTION

Power FET width of integrated synchronous Buck DC-DC converter (ISBC) is normally determined via efficiency optimization at a specific high load current condition. Consequently, for the sake of improving light load efficiency, power stage width segmentation technique [1,2,3,4,5,6] might be implemented so that the number of active power FET subcells can be dynamically adjusted in relevant to load current and switching loss of the ISBC can be significantly decreased at light load condition. However, many nowadays segmentation efficiency analysis and implementation algorithms [1,2,3,4] ignored the impact of Cds and Cgd in inactive power subcells and did not notice an associated saturation effect about the optimal number of active power FET subcells. Also, the impact of rdson variation due to temperature and technology on segmentation algorithm are not addressed.

In section II, the paper firstly presents load efficiency curves of an ISBC with seven different numbers of active power FET subcells, in which the data come respectively from Cadence simulation, chip test and losses prediction model on the basis of an event based switching losses analysis method [7, 8, 9]. The curves illustrates that the Cds and Cgd in inactive power FET subcells significantly change the trend of power efficiency in relation to load current, provided that we

compare them with those curves which is based on the losses model [1,2,3,4] ignoring Cds and Cgd in the inactive power FET subcells. Section II also indicates that the Cds and Cgd in inactive power FET subcells are responsible for a saturation effect which happens during the derivation of the optimal number of active power FET subcells in theoretical. The variation characteristics of rdson of power FET against manufacturing run and temperature are investigated in section III, with the aid of which it is concluded that load current derived segmentation algorithms [1, 3, 5] might not yield an optimal efficiency in practical. A Vsw pinning segmentation algorithm is proposed in section IV. Sction V is the conclusion.

II. SEGMENTATION, EFFICIENCY AND SATURATION EFFECT

A. Segmentation and load efficiency curves

An ISBC chip with a specification listed in Table-I is fabricated in On Semi SCN05 technology by MOSIS program and the die micrograph is illustrated in Fig. 1.

TABLE- I
SPECIFICATION OF AN EXAMPLE ISBC CHIP

IC Technology	AMI06
	SCN3ME_SUBM
Switching frequency	4MHz
Input voltage	3.0V
Output voltage	1.5V
Peak efficiency load current	0.4A
Power capacitor	9.4uF
Power inductor	1.2uH
Total Power FET subcells (M_T)	32
Selectable number of active Power FET subcells (M)	8, 10, 12, 14, 18, 24, 32

Four groups of load efficiency curves for the ISBC chip are respectively given in Fig. 2, Fig. 3, Fig. 4 and Fig. 5, which are depicted on the basis of the data coming from Cadence simulation, chip experimental, events based switching losses model [7,8,9] and regular losses model ignoring Cds and Cgd in inactive power FET subcells [1,2,3,4].

978-1-4244-8425-6/11 $26.00 © 2011 IEEE
184

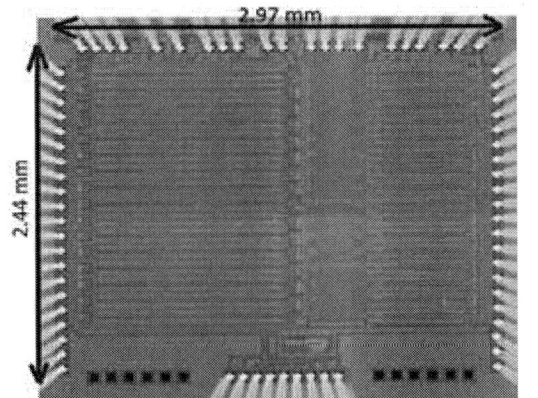

Fig. 1. Die micrograph of the ISBC chip fabricated by MOSIS in On Semi SCN05 technology.

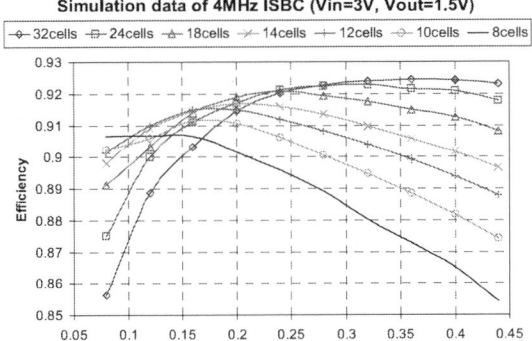

Fig. 2. Load efficiency curves based on Cadence simulation.

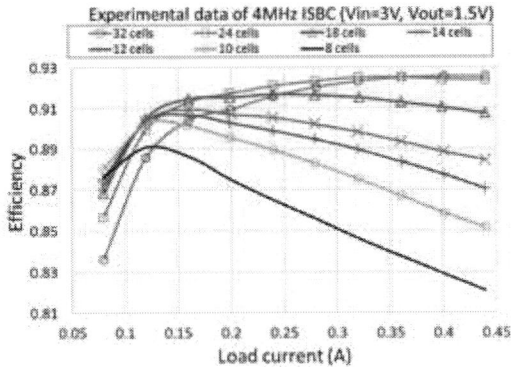

Fig. 3. Load efficiency curves based on chip test.

Comparing Fig. 5 with Fig. 2, Fig.3 and Fig. 4, we notice that Cds and Cgd in inactive power FET subcells degrades the efficiency at light load condition to a significant extent so that the efficiency of the ISBC will keep the trend of dropping despite of an effort to inactivate numerous power FET subcells.

B. Power FET optimal sizng

The part of power loss in an ISBC dependent on power FET sizing is given in (1) [7,8,9].

$$P_{Loss}^{S} = \frac{I_{nom}^{2} R_{dson}}{M W_{cell}} + M W_{cell} X + M_{T} W_{cell} Y_{Cds} + (M_{T} - M) W_{cell} Z_{Cgd} \quad (1)$$

in which I_{nom} is the nominal load current with an expectation of peak efficiency; W_{cell} is the width of an unit power FET subcell; M_{T} is the number of the total subcells and M accounts for the active ones; R_{dson} is the unit width rdson of power FET; X refers to the unit width switching losses of an active subcell excluding the charging/discharging losses relevant to Cds; Y_{Cds} is the unit width charging/discharging losses relevant to Cds; Z_{Cgd} is the unit width charging/discharging losses determined by Cgd of an inactive power FET subcells.

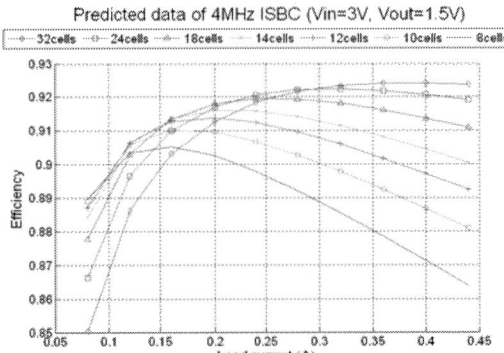

Fig. 4. Load efficiency curves based on losses model where losses from Cds and Cgd of inactive subcells are included [7, 8, 9].

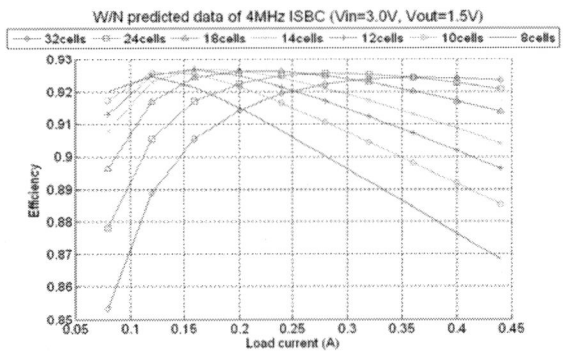

Fig. 5. Load efficiency curves based on losses model where losses from Cds and Cgd of inactive subcells are not included [1,2,3,4].

When M_{T} is set by design, power FET sizing is to take $M = M_{T}$ and derive the minimal value of P_{Loss}^{S} taking W_{cell} as a variant. Consequently, the optimal value of W_{cell} is given in (2).

$$W_{cell} = \frac{I_{nom}}{M_{T}} \sqrt{\frac{R_{dson}}{(X + Y_{Cds})}} \quad (2)$$

C. Saturation effect in segmentation

With the segmentation technique, part of subcells will be disabled when the magnitude of operating load current I_{o} is smaller than that of I_{nom}. From the consideration of efficiency optimization, the number of active subcells M is given in (3) which is derived from (1).

$$M = \frac{I_o}{W_{cell}} \sqrt{\frac{R_{dson}}{(X - Z_{Cgd})}} \qquad (3)$$

Fig. 6 illustrates the optimal value of M in different operating load current I_o for the ISBC in Table-I. The dot and rectangle symbols respectively depict the calculated value in (3) and its round version. Fig. 6 shows that when the load current approaching its nominal value (I_{nom}=0.4A), the optimal number of active subcells following (3) will be larger than M_T which is 32 in the example ISBC, that is, the optimization of M will become saturated because of the Cds and Cgd in inactive subcells. The triangle symbols in Fig. 6 depict the practical solutions of number optimization.

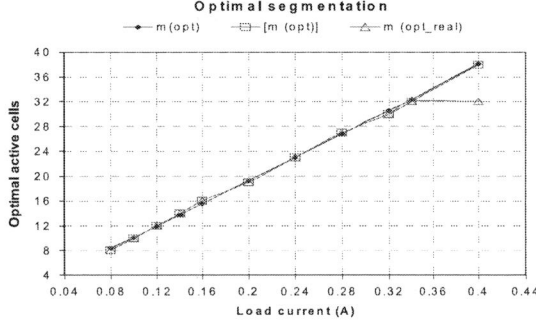

Fig. 6. Saturation effect of the optimal active subcells because of the Cds and Cgd in inactive subcells.

III. RDSON VARIATION AND SEGMENTATION ALGORITHM

A. Rdson variation

Due to manufacturing error, the MOSFET's Rdson will be somewhat different from the specified values. Fig. 7 shows the data statistics of rdson relevant parameters in ON-SEMI SCN05 standard CMOS technology [10].

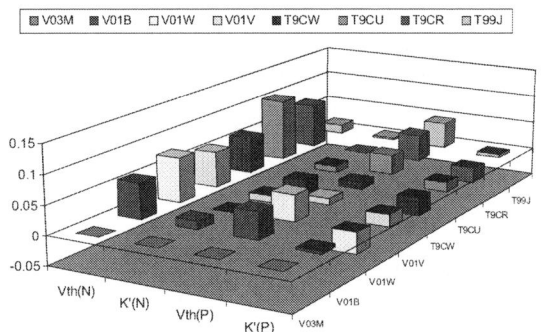

Fig. 7. Rdson variation in eight manufacturing runs [10]

Fig. 8 shows the measured Rdson of the ISBC in Cadence simulation environment with MOSIS BSIM3V3.1 level 49 Spice models for different Ids, Vds, corner models and

temperatures. It is observed that Rdson has more than 50% of change in the simulation when the temperature shifts from 0°C to 100°C; Rdson deviation is in the range of 4% when the Ids changes from 0.1A to 0.4A; The maximal Rdson deviation is in the range of 10% when the process corner changes. Thus, temperature presents dominant impact on the Rdson in comparison with other factors.

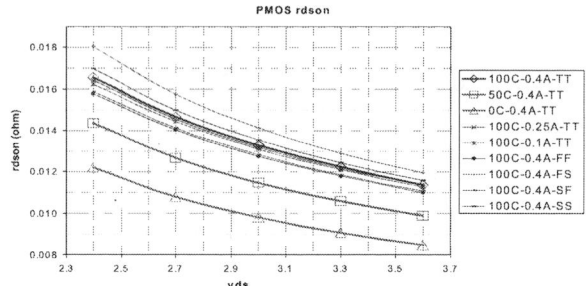

Fig. 8. PMOS rdson variations in corner simulation

B. Segmentation algorithm

The equation (3) indicates a linear relationship between M and I_o. The relationship was actually implemented in the nowadays power FET segmentation algorithms [2, 4], that is, load current information segmentation algorithms (CIS) as illustrated in Fig. 9 and given in (4).

$$M_{CIS} = K_{CIS} I_o \qquad (4)$$

in which K_{CIS} is a constant coefficient to be specified.

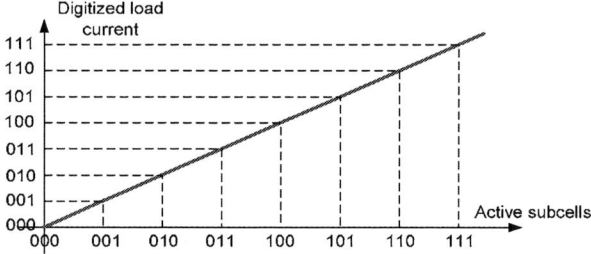

Fig. 9. Nowadays segmentation algorithms[2,4]

The equation (3) also indicates that the value of M should depend on R_{dson}. Thus, if we follow Fig. 9 and only utilize the current information to determine M, the obtained efficiency might not be optimal because of the variation of R_{dson} due to manufacturing error or temperature.

Applying small perturbation signal ΔR_{dson} into (3), the necessary adjustment ΔM about the number of active subcells is in (5) in order to maintain the optimal efficiency.

$$\Delta M = \frac{\Delta R_{dson}}{R_{dson}} M \qquad (5)$$

This kind of rdson related adjustment is not available with the CIS algorithms in (4).

IV. VSW PINNING SEGMENTATION ALGORITHM

A Vsw pinning segmentation (VswPS) technique [9] is proposed to overcome the variation of rdson. Vsw is the voltage level of the phase node in an ISBC. The concept of the VswPS is illustrated in Fig. 10, where the sampled value of V_{SW} is regulated to stay in a specific zone characterized by two threshold voltages V_{high} and V_{low} in all load conditions via adjusting the value of R_{dson} by turning on or off part of power FET subcells.

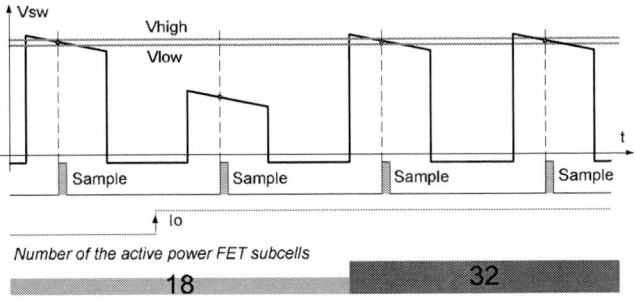

Fig. 10. Conceptual Vsw pinning automatic segmentation [9]

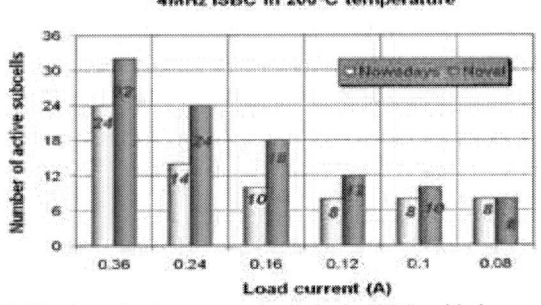

Fig. 11. Number of active power subcells in the ISBCs with the nowadays [2,4] and Vsw pinning segmentation algorithms [9].

The number of active subcells M_{VswPS} on the basis of the VswPS algorithm in Fig. 10 is in (6) and its sensitivity to PMOS rdson is in (7).

$$M_{VswPS} = \frac{R_{dson}^P I_o}{W_{cell} K_2} \qquad (6)$$

in which K_2 is a constant coefficient to be specified and .

$$\Delta M_{VswPS} = \frac{\Delta R_{dson}^P}{R_{dson}^P} M_{VswPS} \qquad (7)$$

After comparing (7) with (5), we conclude that VswPS algorithm has an expected automatic rdson compensation ability. Fig. 11 shows what happens to the number of active subcells of the ISBC with CIS or VswPS algorithm when the operating temperature in simulation is increased from 90 ° to 200°. With the CIS, the number of active subcells will not change; while with the VswPS, the number of active sub cells will be increased when the temperature increases.

V. CONCLUSION

The paper discusses several efficiency related characteristics of the ISBC with segmentation techniques. The impact of the Cds and Cgd in inactive power FET subcells on the efficiency of ISBC and the associated saturation effect are firstly introduced. After that, the rdson variation characteristics due to manufacturing run and temperature are presented. It is also discussed that present CIS segmentation algorithm cannot compensate rdson variation and might not yield the optimal efficiency. Finally, a VswPS segmentation algorithm is introduced as one solution with the capability of rdson compensation for the benefit of efficiency.

ACKNOWLEDGMENT

The authors want to express their sincere thanks to the support of MOSIS educational program and NCSU power management consortium (PMC) members.

REFERENCES

[1] S. Musunuri, and P. L. Chapman, "Improvement of light-load efficiency using width-switching scheme for CMOS transistors," IEEE Power Electro. Letters, Vol.3, No.3, pp.105-110, Sep. 2005.

[2] H. Lee, K. Chang, K. Chen, and W. T. Chen, "Power saving of a dynamic width controller for a monolithic current-mode CMOS DC-DC converter," Proceedings of the fifth international workshop on system-on-chip for real-time applications, 2005, pp. 352-357.

[3] H. Huang, K. Chen, and S. Kuo, "Dithering skip modulation, width and dead time controllers in highly efficiency DC-DC converters for system-on-chip applications," IEEE Journal of Solid-State Circuits, Vol. 42, No. 11, pp.2451-2465, Nov. 2007.

[4] T. Y. Man, P. K. T. Mok, and M. Chan, "Analysis of Switching-Loss-Reduction Methods for MHz-Switching Buck Converters," IEEE Electron devices and solid-state circuits, EDSSC'07, pp. 1035-1038, 2007.

[5] O. Trescases, G. Wei, A. Prodic, and W. T. Ng, "Predictive efficiency optimization for DC-DC converters with highly dynamic digital loads," IEEE Trans. Power Electron. Vol.23, No.4, pp.1859-1869, July 2008.

[6] V. R. H. Lorentz, S. E. Berberich, M. Marz, A. J. Bauer, H. Ryssel, P. Poure and F. Braun, "Light-load efficiency increase in high-frequency integrated DC-DC converters by parallel dynamic width controlling," Analog Integr. Circ. Sig. Process, Vol. 62, No. 1, pp. 1-8, Jan. 2010.

[7] X. Wang, J. Park, E. R. B. Van, and A. Q. Huang, "Switching losses analysis in MHz integrated synchronous Buck converter to support optimal power stage width segmentation," IEEE Energy Conversion Congress and Exposition, ECCE'10, pp. 2718-2724, Sep. 2010.

[8] X. Wang, and A. Q. Huang, "Capacitor Energy Variation Based Designer-Side Switching Losses Analysis for Integrated Synchronous Buck Converters in CMOS Technology," IEEE Applied Power Electronics Conference, APEC'11, pp. , Mar. 2011.

[9] X. Wang, "Power Efficiency Conscious Design and Implementation of High Frequency Integrated Synchronous Buck DC-DC Converters for Portable Electronics Applications," Ph.D. dissertation, North Carolina State Univ., Elec. Eng. Dept., Raleigh, 2010 (public available after Sep. 2011).

[10] The MOSIS service, http://www.mosis.com/Technical/Testdata/ami-c5-prm.html

Proceedings of the 23rd International Symposium on Power Semiconductor Devices & IC's
May 23-26, 2011 San Diego, CA

The ESD Failure Mechanism Of Ultra-HV 700V LDMOS

Jian-Hsing Lee[1], Tzu-Cheng Kao[2], Chien-Liang Chan[2]

[1]Independent ESD/EOS/Latch-up consultant

[1]Vancouver, Canada
anolee49@yahoo.com.tw

Jin-Lian Su[2], Hung-Der Su[2], and Kuo-Cheng Chang[2]

[2]Technology Development Division
[2]Richteck Technology Corporation
[2]Cubei City, Taiwan
peter_kao@richtek.com

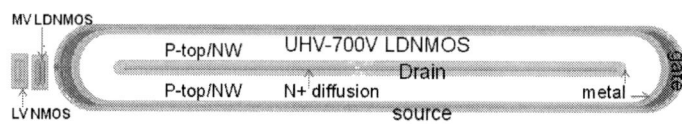

Fig. 1 Top views for UHV-700Vdevice, MV device and LV device

Abstract—**A new kind of ESD failure mechanism is found in the UHV 700V LDNMOS during the HBM ESD zapping event. The device is damaged by its own charges and board stored charges, not damaged by the HBM stress current. The device junction capacitor and test-board capacitor store the charges from the ESD tester before the avalanche breakdown occurring. After the avalanche breakdown, the two capacitors discharge the stored charges to give the additional currents to stress the device. This phenomenon is called the charged-capacitor model (CCM) [1].**

I. INTRODUCTION

Now, the ultra-high voltage (UHV) process is commonly used for switch power supply products since it often provides various devices for designing all functions in one chip [2]. Except the UHV device, the UHV technology often incorporates the low voltage (LV) device and median voltage (MV) device. However, the UHV device is susceptible to the electrostatic-discharge (ESD) damage during the assembling and the reliability test [2]. In this paper, that the UHV device is very vulnerable to the human-body model (HBM) ESD stress is found. Does the UHV device during these tests suffer the same kind of issue? Although the UHV device is nearly one hundred times larger than the MV device and the LV device (Fig. 1), it still cannot pass the HBM specification (2KV). On the contrary, the MV device and LV device all can pass the HBM specification. Why this is not an issue for LV device and MV device and becomes a big problem in the UHV device. Can the ultra-high breakdown voltage (UHV$_{BK}$) and huge dimension or any weak point of the structure make the UHV device vulnerable to the ESD stress? From this study, an analytical model is derived to find out the issue caused by the UHV$_{BK}$, huge dimension and weak point of the structure. Regardless of the DC stress or ESD test, the test-board and the UHV device will accumulate a lot of charges as it is biased at the UHV. The higher the breakdown voltage, the more the charges are stored. This makes the test no longer a pure test and transfer to another kind of stress when the device is triggered on accidentally.

II. EXPERIMENTS

In order to investigate the more detailed insight interaction between the UHV device and HBM, the real-time IV characteristics of the UHV device during the HBM zapping event is measured. The apparatus used to capture the real-time discharge behavior of the UHV device during the HBM zapping event is a 500 MHz digitizing oscilloscope with 4G/sec sampling rate.

A. Device Discharge Behavior For Zapping Voltage Below The Turn-on Threshold Volatge

Fig. 2 shows the voltage and current waveforms of the UHV device under the 1.0KV HBM zapping event and the zero-loading current waveform of a 1.0KV HBM. The zero-loading current I_{ZL} starts from zero and rises toward a peak value and then decreases to zero from its peak value with a time constant RC. For UHV device under the 1.0KV HBM zapping event, the device behaves as a capacitor since the zapping voltage is still too small to turn on the device. Thus, the device current I_{DEV} also starts from zero and rises toward a peak value and then decreases to a finite value (~20mA) from its peak value with a time constant RC. This charges up the capacitors to cause the device voltage V_{DEV} rising toward a peak value. Compared to the I_{ZL}, the RC time constant and the peak current of the I_{DEV} are smaller. Moreover, the I_{DEV} does not decrease to zero and is kept as a constant. This provides the leakage path to sink the stored charges of the capacitors to result in the V_{DEV} slightly titling to its peak value.

B. Device Discharge Behavior For Zapping Voltage Beyond The Turn-on Threshold Volatge

Fig. 3 shows the voltage and current waveforms of the UHV device under the 1.5KV HBM zapping event and the zero-loading current waveform of a 1.5KV HBM. Similar to

978-1-4244-8425-6/11 $26.00 © 2011 IEEE

the device under the 1.0KV HBM zapping event, the I_{DEV} still starts from zero and rises toward a peak value and then decreases from its peak value until the occurrence of the avalanche breakdown. Subsequently, the V_{DEV} is clamped at 920V, and the I_{DEV} increases more and more. As the I_{DEV} is increased to the value that can forward the P-substrate to source junction ($V_{SUB} = I_{DEV} \times R_{SUB} \geq 0.9V$) [3], the source begins to inject the electrons to the channel. Then, these electrons diffuse through the channel and are collected by the high field drain to modulate the channel region as a heavy conductivity region gradually [4]. The corresponding transit time is $T_B = L^2/2D_B$ [5]. where L is the channel length, D_B is an electron diffusion constant. Then, a positive feedback phenomenon [6] is formed if the substrate potential ($V_{SUB} = M \times I_{DEV} \times R_{SUB} \geq 0.9V$) can be kept. At last the device becomes a low impedance device to cause the I_{DEV} increasing sharply and goes into a stable snapback region instantaneously. It can find that the I_{DEV} is even increased much higher than the peak value of the I_{ZL}. Furthermore, the V_{DEV} drops down to the low voltage instantaneously. After this transient, the I_{DEV} follows the trend of the I_{ZL} to decrease with a time constant RC.

C. Device Discharge Behavior Versu Sequent HBM's

Fig. 4 shows the voltage and current waveforms of the UHV device under three sequent 1.5KV HBM zapping events. The response waveforms of the UHV device are varied after each HBM zapping event. For the 2nd HBM zapping event, the V_{DEV} still follows the same trajectory of the UHV device under the 1st HBM zapping event. This implies that the 1st HBM zapping event does not induce any impact on most regions of the UHV device. However, the occurrences of the peak current and the snapback phenomenon are postponed. This implies that the weakest region of the UHV device might be degraded by the 1st HBM zapping event. So, it cannot create the same amount of the electron-hole pairs to sustain the positive feedback phenomenon in the same time. For the UHV device during the 2nd HBM zapping event, the critical current to drive it into the snapback region is apparently larger compared to the UHV device during the 1st HBM zapping event ($I_{crit2} > I_{crit1}$).

For the UHV device during the 3th HBM zapping event, the peak current I_{p3} appears earlier compared to the UHV device during the 1st HBM zapping event (I_{p1}). Apparently, the weakest region of the UHV device was damaged by the 2nd HBM zapping event. Thus, the breakdown voltage of the device is decreased to the lower voltage compared to that for the 1st HBM zapping event.

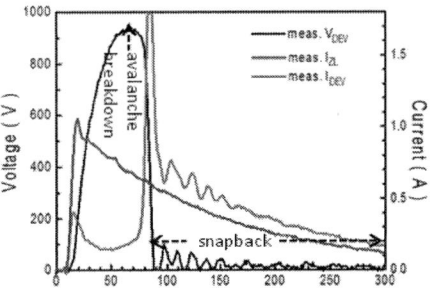

Fig. 3 Voltage and current waveforms of a UHV device during a 1.5KV HBM zapping event and zero-loading current waveform of a 1.5KV HBM.

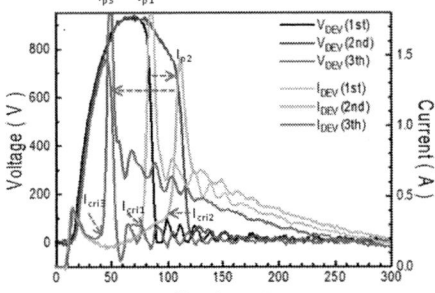

Fig. 4 Voltage and current waveforms of a UHV device during three sequent 1.5KV HBM zapping events.

III. DISCHARGE MODELS FOR UHV 700V LDNMOS

A. Zero-Loading Discharge Current

The equivalent circuit for a device under the ESD test (DUT) proposed by Leo [7] can be simplified to Fig. 5a. The C_O and C_B are the ESD discharge-head capacitor and test-board capacitor respectively. From Fig. 5a, the discharge current $I_O(t)$ for zero-loading is governed by the equation

$$L\frac{d^2 I_O(t)}{dt^2} + R_0\frac{dI_O(t)}{dt} + \frac{I_O(t)}{C} = 0 \tag{1}$$

The solution of Eq. (1) is given by

$$I_O(t) = \frac{V_O}{2\beta L_O}[\exp((-\alpha + \beta)t) - \exp(-(\alpha + \beta)t)] \tag{2}$$

where $\alpha = \dfrac{R_o}{2L_o}$, $\beta = \dfrac{\sqrt{(R_o C_o)^2 - 4L_o C_o}}{2L_o C_o}$

Fig. 5b shows that the simulated discharge currents (circles) based on Eq. (2) all can fit the measured discharge current (lines) very well. This verifies that the simplified equivalent-circuit is valid for simulating the discharge behavior of the device under the ESD zapping event.

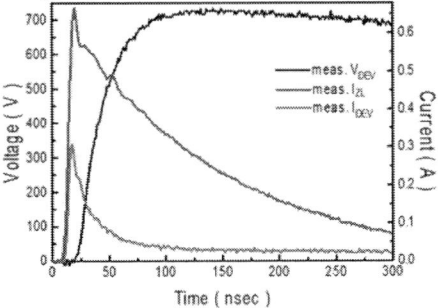

Fig. 2 Voltage and current waveforms of the UHV device under a 1.0KV HBM zapping event and zero-loading current waveform of a 1.0KV HBM.

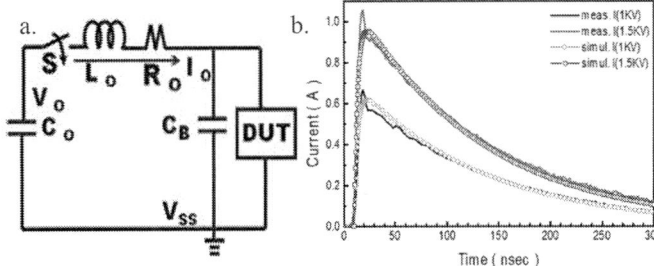

Fig. 5 a. Simplified equivalent circuit of a device under the ESD test, b. simulated and measured zero-loading HBM discharge currents

B. Device Discharge Behavior Before Breakdown

Based on Fig. 5a, the equivalent circuit of the UHV device before the occurrence of the avalanche breakdown can be simplified to Fig. 6. Before the snapback, the substrate potential $R_{sub} \times I_{DEV}$ (<0.9V) is too small and can be neglected. Thus, the $I_O(t)$ from the C_O is governed by the equation

$$L_o \frac{d^2 I_O(t)}{dt^2} + R_o \frac{dI_O(t)}{dt} + (\frac{1}{C_O} + \frac{1}{C_J} - \frac{C_B}{C_J(C_J + C_B)})I_O(t) = 0 \quad (3)$$

Substitute the initial condition, the $I_O(t)$ can be obtained

$$I_O(t) = \frac{V_o}{2\beta L_o}\{\exp[(-\alpha + \beta)t] - \exp[-(\alpha + \beta)t]\} \quad (4)$$

where $\alpha = \frac{R_o}{2L_o}$, $\beta = \frac{\sqrt{R_o^2 - 4L_o(\frac{1}{C_O} + \frac{1}{C_J} - \frac{C_B}{C_J(C_J + C_B)})}}{2L_o}$

The currents flowing through the device I_{DEV} and test-board I_B are

$$I_{DEV}(t) = I_O(t)C_J/(C_J + C_B) \quad (5)$$

$$I_B(t) = I_O(t)C_B/(C_J + C_B) \quad (6)$$

The drain voltage V_{DEV} of the device can be expressed

$$V_{DEV}(t) = \frac{-V_o}{2\beta L_o(C_B + C_J)}\{\frac{1 - \exp[(-\alpha + \beta)t]}{-\alpha + \beta} + \frac{1 - \exp[-(\alpha + \beta)t]}{\alpha + \beta}\}$$
(7)

Based on Eq. (5)-Eq. (7), the IV curves of the UHV device under the +1.0KV HBM zapping event can be evaluated as shown in Fig. 7. The measured IV curves are plotted as the solid lines and the simulated IV curves are plotted as the circles. Except the region after the constant I_{DEV} appearing, the simulated curves all can match the measured curves very well. This verifies that the board capacitor C_B and device junction capacitor C_J can discharge and store the charges coming from the HBM capacitor C_O before the avalanche breakdown occurring on the UHV device.

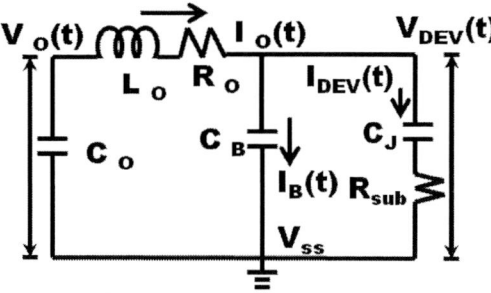

Fig. 6 Equivalent circuit of a UHV device under HBM test before the occurrence of the avalanche breakdown.

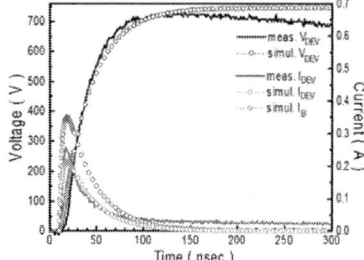

Fig. 7 Simulated and measured voltages and currents of the UHV device zero-under 1.0KV HBM zapping event.

C. Device Discharge Behavior At The Snapback

From Fig. 3, it can find that the snapback voltage is nearly a constant, though the discharge current decreases with the time. Due to the same potential, the two capacitors C_B and C_J no longer sink the charges coming from the HBM capacitor C_O or discharge any charge to the device. Thus, the device operated at the stable snapback region can be modeled as an on-resistance R_{on} in series with a voltage source V_{sp} as shown in Fig. 8a. Based on Fig. 8a, the discharge current in the snapback region can be obtained

$$I_O(t) = I_{DEV}(t) = A\exp[(-\alpha + \beta)t] + B\exp[-(\alpha + \beta)t] \quad (8)$$

where $A = \frac{1}{2\beta}[\frac{(V_f - V_{sp} - R_o I_f)}{L_o} + (\alpha + \beta)I_f]$,

$$B = -\frac{1}{2\beta}[\frac{(V_f - V_{sp} - R_o I_f)}{L_o} + (\alpha - \beta)I_f]$$

where V_{sp} is the device snapback voltage, V_f is the voltage before the snapback, and I_f is the current from the C_o before the snapback.

Based on Eq. (5)-Eq. (8), the voltage and currents for the UHV device under the +1.5KV HBM zapping event can be evaluated. The measured IV curves are plotted as the solid lines and the simulated IV curves are plotted as circles as shown in Fig. 9a. Except the regions at the avalanche breakdown and at the beginning of the snapback, the simulated results all can match the measured results well. For the transient at the beginning of the snapback, only the weakest region of the device enters the low-voltage snapback region, but most regions of the device are still not turned on yet [8]. With the high voltages, the two capacitors C_J and C_B start to discharge their stored charges to the weakest region of the UHV device as shown in Fig. 8b. During this transient, the bonding wire of the package and interconnection metal of the UHV device (Fig. 1) behave as the inductors L_W and L_M. Apparently, the loops for board-discharge current I_B and device-discharge current I_J are also composed of RLC's, which are similar to the equivalent circuit of the device at the stable snapback region in Fig. 8a. The only differences for the three loops are the RLC values. Thus, the board-discharge current I_B and the device-discharge current I_J all can follow Eq. (8). If the I_B is taken account into the I_{DEV}, it becomes excellent fit between the simulated current I_{DEV} and the measured current I_{DEV} at the beginning of the snapback as shown in Fig. 9b. This verifies that Eq. (8) also can represent the discharge behaviors of the two capacitors C_J and C_B at the beginning of the snapback. Fig. 10 shows the simulated I_J flowing from the device, which is higher and shorter than the I_{DEV} flowing from outside the device.

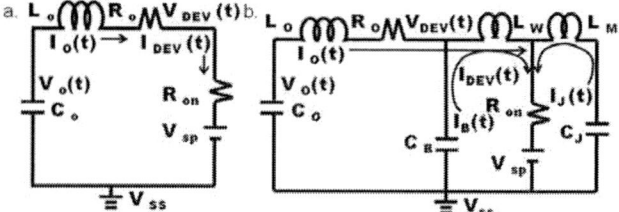

Fig. 8 Equivalent circuit of a UHV device biased at a. the stable snapback region, b. the beginning of the snapback.

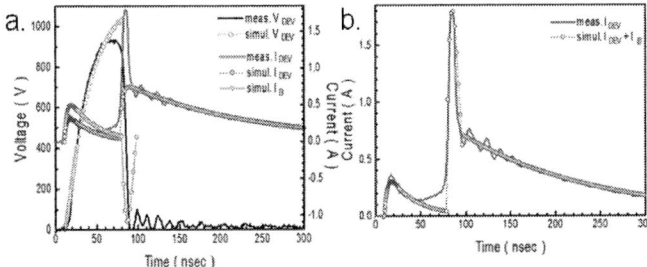

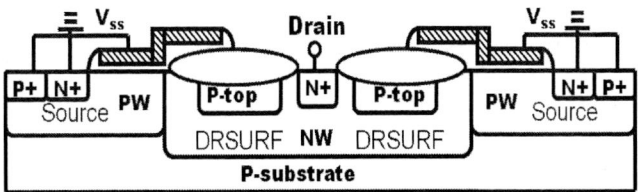

Fig. 12 Cross-section of the UHV 700V LDNMOS along the line A to B in Fig. 11.

Fig. 9 a. Simulated and measured voltages and currents of the UHV device zero-under 1.5KV HBM zapping event, b. measured current I_{DEV} and simulated current I_{DEV} taken account into I_B at the beginning transient of the snapback.

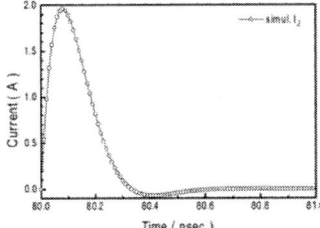

Fig. 10 Simulated device-discharge current I_J.

IV. FAILURE ANALYSIS AND HBM TEST RESULT

A. Failure Analysis

Fig. 11a shows the back-side emission-microscope (EMMI) photograph for the UHV device biased at the avalanche-breakdown region. The hot spot is found at the edge of the N+ diffusion of the drain (ENDD) since this region suffers the two-dimension double RESURF (DRESURF) effect. Unlike other N+ diffusion regions of the drain between two DRESURF regions (Fig. 12), the ENDD is surrounded by three DRESURF regions (A, B, C in Fig. 11). This will make the electrical field of the ENDD higher than that of any other region of the UHV device. Thus, the avalanche breakdown will occur at the ENDD as the V_{DEV} is pulled higher than the avalanche-breakdown voltage of the UHV device. However, both the unexpected-discharge currents I_B and I_J are the ultra-short current pulses from Fig. 9 and Fig. 10. The duration times of the I_B and I_J are nearly 15nsec and 0.3nsec respectively, which are too short to turn on other regions of the UHV device [1]. Thus, the only region that can discharge the I_B and I_J is the ENDD of the UHV device. Subsequently, all currents are crowded at this small region to result in the thermal-run away occurring in there as shown in Fig. 11b.

B. HBM Test Result

From the failure analysis, the UHV device during the HBM zapping event won't be damaged by the unexpected currents if the drain voltage V_{DEV} can be kept below the avalanche-breakdown voltage. Based on Eq. (7), the V_{DEV} is proportional to the HBM zapping voltage and is the reciprocal of the capacitance of the UHV device. Thus, the increase in the breakdown voltage V_{BK} or the number of the UHV device all can increase the device HBM threshold voltage as shown in Table I. Fig. 13 shows the modified UHV device, which moves the P-top implant away from the ENDD (P2) instead of the nearby ENDD (P1) to eliminate the two-dimension DRESURF effect. After this modification, the HBM threshold voltage of the UHV device can be increased to 3.5KV as shown in Table I.

Table I: HBM Test Result

Stru.	No. of Dev.	V_{BK}	HBM
Fig. 1	1	920V	1.0KV
Fig. 1	2	920V	1.3KV
Fig. 1	4	920V	1.5KV
Fig. 1	1	1150V	1.2KV
Fig. 1	2	1150V	1.6KV
Fig. 1	4	1150V	2.1KV
Fig. 11	1		3.5KV

V_{BK} is defined by the clamped voltage of the device during HBM zapping event, not the value measured by DC meter.

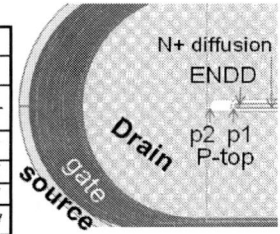

Fig. 13 Modified UHV 700V LDNMOS.

REFERENCES

[1] Jian-Hsing Lee, J.R. Shih, H.P. Kuan, and Kenneth Wu, "The Influence of Decoupling Capacitor on The Discharge Behavior of Fully Silcided Power-Clamped Device Under HBM ESD Event," in Proc. 17th IPFA, 2010.

[2] Tsung-Yi Huang, et al., "Mobile charge induced breakdown instabiliy in 700V LDMOSFET," in VLSI-TSA Symposium, pp. 105–108, 2008.

[3] Dao-Hong Yang, Jone F. Chen, Jian-Hsing Lee, and Kuo-Ming Wu, "Dynamic Turn-On Mechanism of the n-MOSFET," IEEE Electron Dev. Lett., pp. 895–897, 2008.

[4] Jian-Hsing Lee, Wu-Te Weng, Jiaw-Ren Shih, Kuo-Fen& Yu, and Tong-Chern Ong, "The Positive Trigger Voltage Lowering Effect for Latch-Up," in Proc. 11th IPFA, pp. 85-89, 2004 .

[5] B. Van Zeghbroeck, Principles of semiconductor devices, 2007, p. 5.5.5, URL: http//ece-www.colorado.edu/~bart/book/book/index.html.

[6] Jim-Hsing Lee, Jiaw-Ren Shih*, Yi-Hsun Wu, Boon-Khim Liew, and Huey-Liang Hwang, "An analytical model of positive H.B.M ESD current distribution and the modified multi-finger protection structure," in Proc.7th IPFA, pp. 162-167, 1999.

[7] Leo van Roozendaal, Ajith Amerasekera, Peter Bos, and WillemBaelde, "Standard ESD testing of Integrated Circuits," in Proc. 12th EOS/ESD Symp., pp.119-130, 1990.

[8] Jian-Hsing Lee, Kuo-Ming Wu, Shao-Chang Huang, Chin-Hsin Tang, "The dynamic current distribution of a multi-fingerd GGNMOS under high current stress and HBM ESD events," in Proc. 44th IRPS, pp. 629-630, 2006.

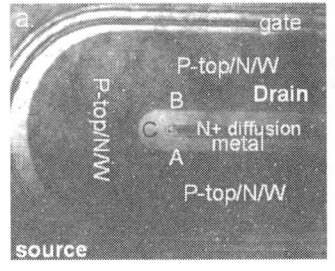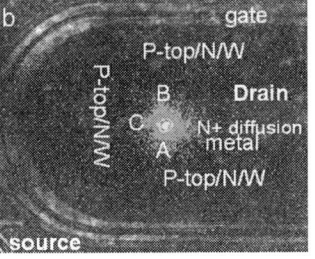

Fig. 11 Hot spots are all on the ENDD for the UHV device a. biased at the avalanche-breakdown region, b. after the 1.5KV HBM zapping.

Proceedings of the 23rd International Symposium on Power Semiconductor Devices & IC's
May 23-26, 2011 San Diego, CA

Techniques to Prevent Substrate Injection Induced Failure During ESD Events in Automotive Applications

Amaury Gendron, Chai Gill, Craig Aykroyd and Carol Zhan

Freescale Semiconductor, 2100 East Elliot Road, Tempe, AZ 85284 USA

Abstract—**This paper presents several techniques to improve ESD robustness for high voltage IO designs in automotive applications. SCR-based ESD clamps designed on isolated wells can generate high level of substrate injection during ESD events, causing false triggering and irreversible failures of internal components. To mitigate substrate injection effects, we have defined design strategies leading to a set of designs rules for proper integration with ESD clamps.**

I. INTRODUCTION

In the arena of automotive applications, the reliability requirements are increasingly rigorous to meet a wide array of industry standards [1]. Most of us are aware of the danger and high cost of failing electrical components in a vehicle even as more ICs are integrated into complex automotive ECUs to automate our driving experience. In this context, developing efficient and compact ESD protections has become increasingly challenging. And providing ESD robust designs has become a major competitive advantage for semiconductor manufacturers. To protect analog IOs against ESD, SCR based local clamps are extremely attractive as they provide high robustness and low on-state resistances [2]. However, there is a parasitic PNP inherent to SCR structures, with the grounded P-type substrate forming the collector [3]. As a result, a high hole current can be injected in the substrate from the ESD clamp. Without proper integration, the hole current can lead to disturbances and potential damages in the surrounding circuit.

Smart power technologies offering high density logic and power management capabilities allow designing complex analog systems at a very competitive cost. The present study was conducted on a Freescale SmartMOS© technology based on a 0.25 µm CMOS platform with high voltage extension up to 80 Volts [4]. The process has very low ppm failures with high temperature capabilities as required by automotive OEMs. High voltage isolation leverages deep trenches with n-type buried layers (NBL) on a low-doped p-type epitaxy grown from a high-doped p-type substrate. The trench depth is designed to force any parasitic carriers' injection to flow in the highly doped substrate. Thus, the process guarantees a high recombination rate for electrons, hence reduces the risks associated with external latch-up. On the counterpart, a hole current can disturb the circuit far from the injection source as it spreads in the substrate without generating a high voltage drop.

The purpose of this study is to address substrate injection issues during ESD events. We will investigate an automotive chip with low ESD compliance reported on several IO pins. By leveraging failure analysis and TCAD simulations, we will be able to identify the damage on a high voltage (HV) diode, and to trace the root cause to the hole substrate current injected by the ESD clamp. To resolve this reliability issue, three different strategies for mitigating the substrate injection will be proposed. Each of them will be characterized from test structures containing ESD clamps connected with HV diodes in different configurations. The characterization results will allow a comparison on ESD robustness of each strategy.

II. CASE STUDY

A particular challenge of ESD design was to protect 40 V IO pins of a typical current-voltage monitor using Hall effects for ignition system. The selected 40 V clamp has been successfully proven in different configurations on numerous chips. However, the integration with a specific HV diode consistently generated weak HBM performance.

A. IO cell protection methodology

The ESD protection circuit consists of a 40 V local ESD clamp connected between the IO and ground pins, in order to provide a low impedance current path during an ESD event. The HV diode in reverse biased mode is connected in parallel with the 40 V clamp (Figure 1).

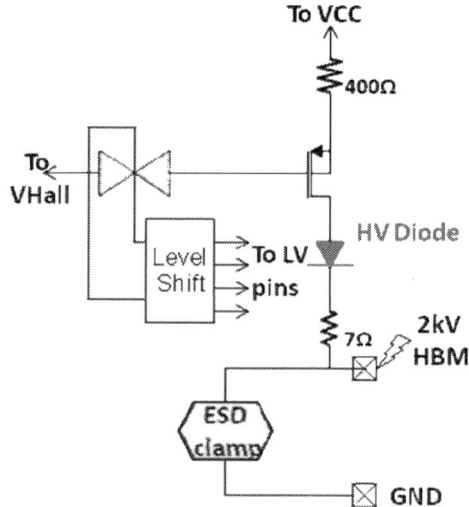

Figure 1. Schematic of the VHall IO block including a HV diode to be protected by a local 40 V ESD clamp.

978-1-4244-8425-6/11 $26.00 © 2011 IEEE

The transient responses of the ESD clamp and the HV diode during an ESD were characterized with a transmission-line pulse (TLP) tester. The standalone 40 V clamp triggers at 49 V (V_{t1}), snaps back to 28.5 V (V_H) and is capable of sustaining up to 5.5 A (I_{t2}). The standalone HV diode with V_{t1} at 54.5 V constrains the protection margin. Based on direct interpolation between the V_{t1}'s of the HV diode and the 40 V clamp, there is a significant margin for ESD protection up to approximately 8 kV HBM. However, the integration of the HV diode with the 40 V clamp on product resulted in I_{t2} of 1.2 A corresponding to 1.8 kV HBM compliance (Figure 2).

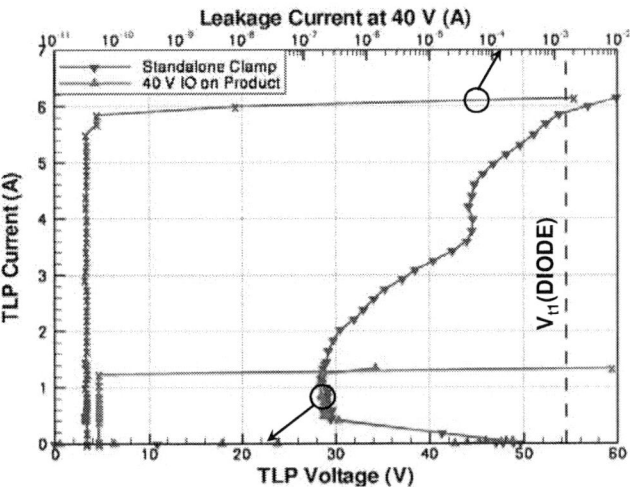

Figure 2. TLP of HV diode with clamp versus standalone clamp as constrained by V_{t1} of HV diode.

B. Devices cross-sections

The 40 V ESD clamp is based on a SCR structure (Figure 3) [5]. The cross-section includes two P-wells isolated from each other by N-sinker rings which are connected to the NBL. Shallow but heavily-doped N and P-types diffusions are implanted in each P-well and connected to external IO and ground electrodes. The SCR triggering is controlled by the avalanche breakdown at the lateral junction between the P-well tied to the ground and an N-well implanted over the middle N-sinker. The breakdown voltage can be easily adjusted by tweaking the distance, Sp, between the N and P wells. A parasitic PNP is inherent to this cross-section. The emitter is formed by the P-well tied to the IO, the base by the NBL and the collector by the substrate. When the clamp turns on, the SCR triggering induces the parasitic PNP triggering, resulting to hole injection in the substrate.

The HV diode has been developed from an isolated N-type LDMOS [6], in which the source implantation has been replaced by an extension of the body implantation (Figure 4). In this configuration, the body acts as the anode and the drain as the cathode. The NBL is shorted to the anode in order to prevent hole injection in the substrate when the HV diode operates in forward mode. The drawback of this isolation technique is the formation a parasitic NPN with the NBL as the emitter, the anode as the base and the cathode as the collector. This NPN could be at risk of turning on if the anode

(base) to NBL (emitter) junction becomes forward-biased while the HV diode is reverse-biased.

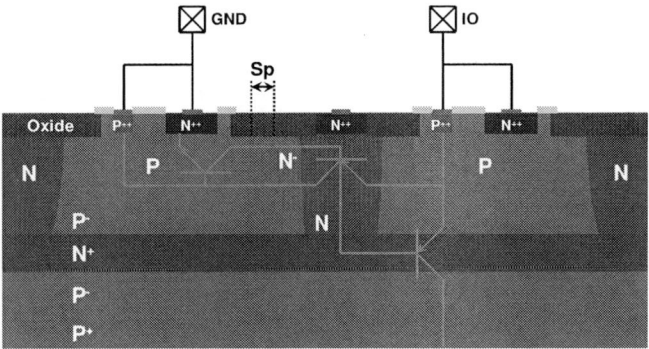

Figure 3. SCR-based ESD clamp cross-section with the parasitic PNP substrate.

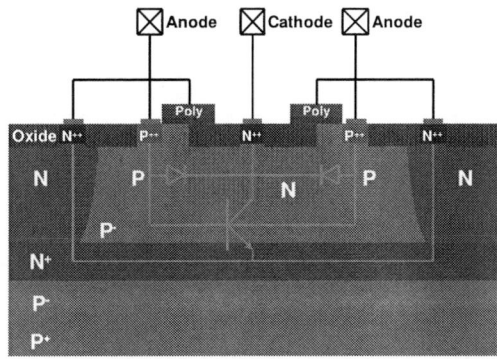

Figure 4. HV diode cross-section with the parasitic NPN.

C. Failure analysis

After 2kV HBM tests, the failed pin was deprocessed for optical inspection of the silicon surface. Damages were clearly identified on the cathodes of the HV diodes while the ESD clamps were intact (Figure 5).

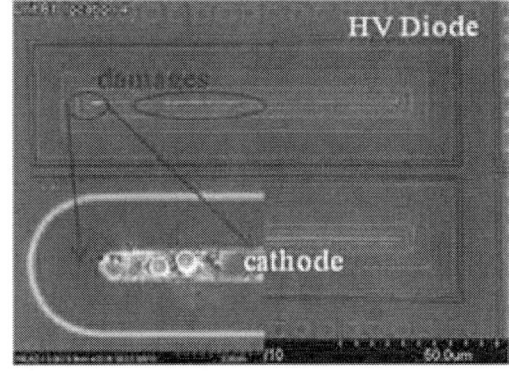

Figure 5. FA image showing the damage on the HV diode, after a 2kV HBM stress.

D. Failure mechanism

To analyze the failure mechanism, ESD clamps with reduced distances Sp and hence lower triggering voltages were designed and connected in parallel with HV diodes on test vehicles. The purpose was to investigate the impact on the robustness of the transient voltage overshoot of the clamp. During an ESD event, high voltage ESD clamps typically trigger at a higher voltage than the quasi-static TLP V_{t1} [7]. This voltage overshoot increases with the current of the discharge and can be harmful for IO devices to be protected. However, TLP results did not show any failure current improvement with higher V_{t1} margin. Instead, the failure occurs always at very low current, between 0.4 and 0.5A, irrelevant to the V_{t1} margin (Figure 6).

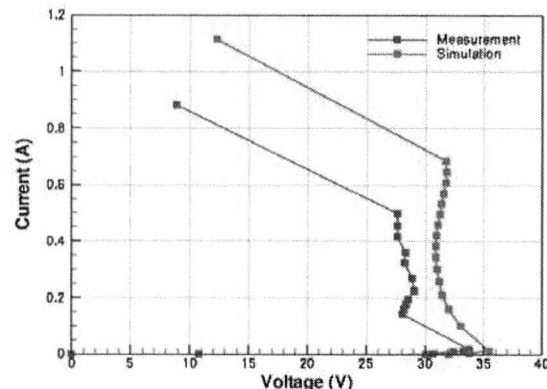

Figure 7. Comparison of simulated TLP characteristic with measurement results for an ESD clamp (Sp=0.80 μm) and a HV diode in parallel.

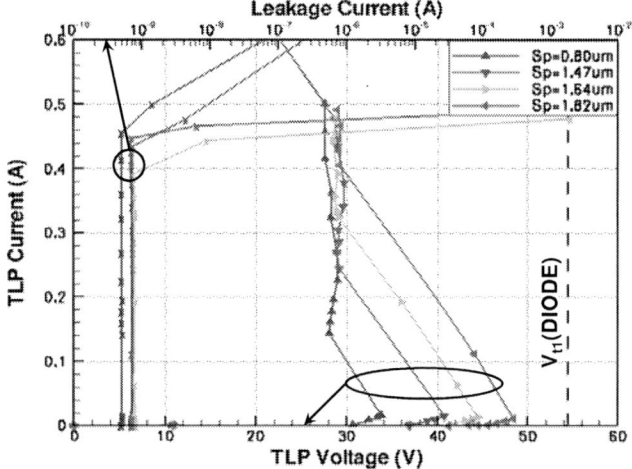

Figure 6. TLP characteristics of ESD clamps hooked up to the HV diodes for different clamps V_{t1}.

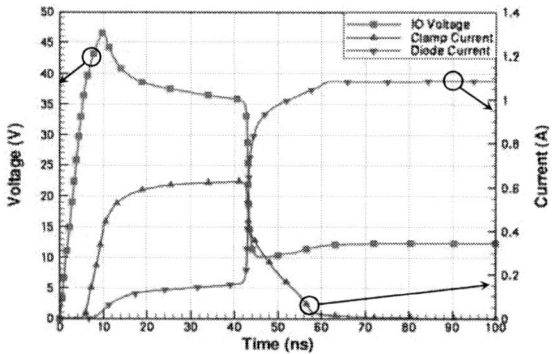

Figure 8. TLP voltage and current waveforms after the 2nd snapback.

After having excluded the possibility of a HV diode failure induced by the clamp overshoot, we have carried out a TCAD simulation study in order to gain insight in the physical mechanisms during an ESD event. Process simulations were leveraged to generate a 2D cross-section of the ESD clamp (Sp=0.80 μm) and the HV diode in the same silicon bulk. Next electro-thermal simulations were performed to replicate the TLP characteristic. The results are in good agreement with the measurement data. In particular, the second snapback leading to the failure is reproduced with 38% precision on the current (Figure 7). This precision level is in the range of expectation for a 2D simulation, which gives only an approximation of the current distribution in the substrate bulk.

The simulated voltage and current waveforms during the TLP pulse after the second snapback (Figure 8) show the ESD clamp turns on first and snaps back above 35 V, then, between 40 and 50 ns, the clamp turns off while the HV diode shunts all the current and the voltage drops to 10 V. When the clamp turns on, the current in the HV diode increases up to 100mA range. The simulated hole current distribution demonstrates that the HV diode operates as the collector for substrate current injected by the ESD clamp (Figure 9). Eventually, the voltage drop in the HV diode biases in forward the anode to NBL junction, leading to the triggering and the snapback of the parasitic NPN (Figure 10). As this parasitic NPN is not designed to operate during an ESD, it gets irreversibly damaged by typical current crowding near small contact areas.

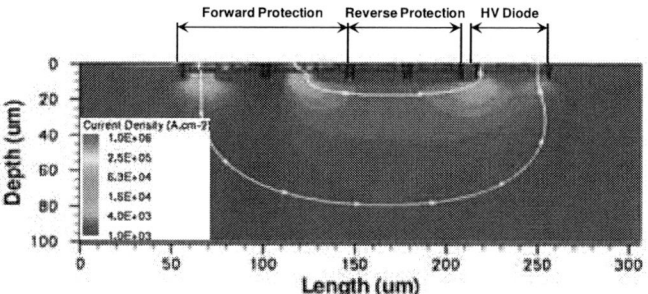

Figure 9. Hole current distribution during ESD clamp operation at t=20 ns.

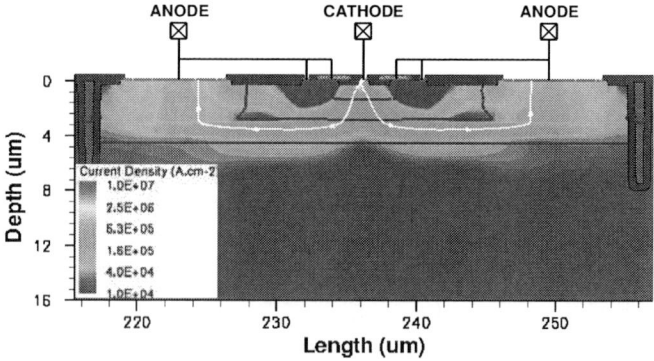

Figure 10. Electrons current distribution in the HV diode at t=75ns.

978-1-4244-8425-6/11 $26.00 © 2011 IEEE

III. STATEGIES TO MITIGATE SUBSTRATE INJECTION RISKS

To meet high ESD robustness specifications, appropriate strategies had to be implemented to deal with the clamp hole injection. Due to time constraint, developing a new isolated ESD clamp was not realistic. Instead, we proposed to mitigate the failure risks by either preventing the substrate current to reach the HV diode or by avoiding the snapback of HV diode parasitic NPN. Based on these approaches, three sets of designs guidelines were defined.

A. Substrate current collection

To prevent the hole from spreading in the substrate, grounded substrate ties should be drawn in the surrounding of the ESD clamp [8]. This design allows collecting the hole before they reach the HV diode. However, the substrate ties have poor collection efficiency due to the low doped epitaxy layer on top of the substrate. In particular, substrate ties are less efficient in collecting holes than the HV diode as they are situated in heavily-doped P-wells much shallower than NBL.

Test structures were designed for varied distances between the ESD clamp and the HV diode. The distance between the two devices was populated with 2 μm wide substrate ties placed every 15 μm. For the control structure, the HV diode was placed at minimum distance to clamp with a 4 μm wide substrate tie ring. The minimum distance configuration showed failure current ~0.16A while configurations with higher substrate tie densities are recorded in TABLE I. As expected, the robustness is enhanced with the number of substrate ties. However, this strategy requires a large increase of footprint due to increasing substrate tie density.

TABLE I. FAILURE CURRENT IMPROVEMENT FOR SUBSTRATE TIES TEST STRUCTURES.

Clamp to Diode Spacing	70 μm	140 μm	210 μm
Failure Current Improvement	x 1.55	x 1.75	x 2.55

B. HV Diode isolation biasing

One technique to avoid the snapback of the HV diode parasitic NPN is to limit the hole collection through the NBL. It can be done by inserting a resistor between the anode of HV diode and ground. Any current flowing through the resistor tends to cause reverse bias of the NBL to substrate junction thus keeping away the hole from NBL.

The results for varied resistance values are reported in TABLE II. This strategy is very efficient in improving the failure current, especially for 50Ω and higher resistances. With 200 Ω on anode of HV diode, the configuration achieved I_{t2}~5.25X higher than without resistor on anode of HV diode.

TABLE II. FAILURE CURRENT IMPROVEMENT FOR TEST STRUCTURES WITH ISOLATION RESISTANCE.

Isolation Resiatance	15 Ω	25 Ω	50 Ω	100 Ω	200 Ω
Failure Current Improvement	x 2.70	x 3.50	x 4.60	x 4.75	x 5.25

C. HV Diode size

Another strategy consists of increasing the size of the HV diode hence a higher hole current is required to trigger the NPN into snapback. In terms of footprint, this design can be an attractive alternative to a substrate tie density increase for better hole current collection. For a HV diode ten times larger than the control one, characterization results showed I_{t2}~4.5X improvement.

IV. CONCLUSION

In summary, SCR based ESD clamps can inject high levels of hole current into the substrate during an ESD event. Based on the above test case, we have demonstrated that failures can be induced in the core device of the VHall IO circuit. A particular concern is isolated HV diode with its NBL tied to ground which can act as the collector for hole substrate current. Three methods were presented to mitigate the risks: increasing the substrate tie density for better hole collection, inserting a resistor along the HV diode ground path in order to reverse bias the NBL to substrate junction and enlarging the size of the HV diode to increase the triggering current of any parasitic devices.

ACKNOWLEDGMENT

We would like to thank our colleagues Olin Hartin and Dan Blomberg for their support on TCAD simulations. We also gratefully acknowledge fruitful discussions with our colleague Mike Baird.

REFERENCES

[1] M. Mergens, M. Mayerhofer, J. Willemen, and M. Stecher, "ESD protection considerations in advanced high-voltage technologies for automotive," in Proc. EOS/ESD Symposium 2006, pp. 54-63.

[2] V. Vashchenko, A. Concannon, M. Ter Beek, and P. Hopper, "Comparison of ESD protection capability of lateral BJT, SCR and bi-directional SCR for hi-voltage BiCMOS circuits," in Proc. BCTM 2002, pp. 181-184.

[3] V. Vashchenko, V. Kuznetsov, and P. Hopper, "Implementation of dual-direction SCR devices in analog CMOS process," in Proc. EOS/ESD Symposium 2007, pp. 75-79.

[4] V. Parthasarathy, R. Zhu, V. Khemka, T. Roggenbauer, A. Bose, and P. Hui, "A 0.25μm CMOS based 70V smart power technology with deep trench for high-voltage isolation," in Proc. IEDM 2002, pp. 459-462.

[5] A. Gendron, C. Gill, C. Zhan, M. Kaneshiro, B. Cowden, and C. Hong, "New high voltage ESD protection devices based on bipolar transistors for automotive applications," in Proc. EOS/ESD Symposium 2011, in press.

[6] V. Khemka, R. Zhu, T. Roggenbauer, and A. Bose, "LDMOSFETs with current diverter for smart power technologies," in Proc. ISPSD 2006, pp. 345-348.

[7] A. Delmas, A. Gendron, M. Bafleur, N. Nolhier, and C. Gill, "Transient voltage overshoots of high voltage ESD protections based on bipolar transistors in smart power technology," in Proc. BCTM 2010, pp.253-256.

[8] F. Farbiz, and E. Rosenbaum, "Modeling of majority and minority carrier triggered external latch-up," in Proc. IRPS 2008, pp. 270-277.

Proceedings of the 23rd International Symposium on Power Semiconductor Devices & IC's
May 23-26, 2011 San Diego, CA

IGBT Driver Chip Set With Advanced Digital Signal Processing

J. Lehmann, G. Katzenberger, G. Königsmann, M. Roßberg, R. Herzer

SEMIKRON Elektronik GmbH & Co. KG
Nuremberg, Germany
jan.lehmann@semikron.com

Abstract—**This article describes a new approach to control and drive electronic power switches (such as IGBTs). It uses two ASICs – one on the primary (control) side of the driver and one on the secondary (power) side of it. Both chips exchange control and status signals with each other across the insulation barrier between both sides. The system on the power side digitizes all analog sensor inputs (such as the temperature, DC-link voltage, V_{CEsat} voltage) and transmits this data and other status information to the control side. The other way round, the primary side sends repetitive and differential control signals to the power side. To achieve the required data rates across the insulation barrier, the characteristics of the physical layer are improved. As a result, this new topology greatly improves the protection of the control (primary and client) side in case of a fatal failure of the power side of the system.**

I. Introduction

IGBT drivers guarantee optimum control, monitoring and potential separation between the control side and the power side of the driver. Because of the required potential separation, two essential tasks have to be solved: on the one hand the signal transmission across the potential separation, on the other hand the power supply for the secondary side circuits. State-of-the-art IGBT drivers transmit a minimum of control and status signals via the necessary potential separation [1], [2]. Generally, one pulse transformer is used to transmit control signals from the primary side to the secondary side as well as error signals reversely. This bidirectional use of the pulse transformer leads to limitations, which are pictured in Fig. 1. On the one hand, a detected error can only be transmitted at the ON state. The polarity of the error signal must be the same as the control signal for the ON state. So an OFF-state pulse is clearly identifiable. Furthermore, the ability of a feedback channel is bought dearly by a single pulse transfer mode. A control signal must be sent once, a repetitive signal transmission is not possible. The pulse transformer has not to be controlled by low-resistance switches like an H-bridge. In that case, the feedback channel would be disabled. Because of this, the pulse transformer is a part of a serial resonator circuit. The first pulse edge has a high contour accuracy, but the second edge is affected by a long-lasting

back-swing behavior. During the back swing time an opposite control signal pulse is not allowed (see Fig. 1). In that concept is it impossible to send repetitive control signals or even transmit data streams.

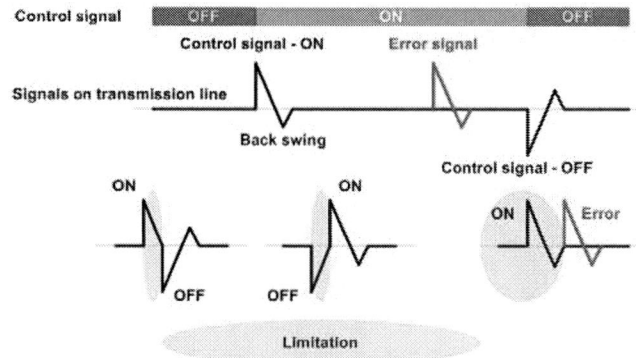

Figure 1. Signal transmission via pulse transformer of a state-of-the-art driver.

II. New Driver Concept

The main focus of the new driver concept is the global digitalization of sensor signals and their processing inside the IGBT driver. The main aim is the identification of the actual operating conditions and an optimized controlling of the power switches, e.g. IGBTs.

A. Signal transmission

The requirement for the new concept is the opportunity to transmit a lot of information (e.g. control signals, digitized sensor signals) via the insulation barrier. This is realized by two separate and bidirectional transmission channels via pulse transformer. Thus, the former discussed limitations in Fig. 1 are eliminated. Fig. 2 shows the principles of information transmission via pulse transformers, like Frequency Modulation (FM), Pulse Width Modulation (PWM) and bipolar binary Amplitude Shift Keying (ASK; uses the polarity of the signal). Using the ASK, control signals are transmitted with short delay times. On the other hand, digitized sensor signals can be transmitted as a serial data

978-1-4244-8425-6/11 $26.00 © 2011 IEEE

stream by using PWM or FM. Furthermore, the repetitive transmission of the control signal enhances the immunity to interference.

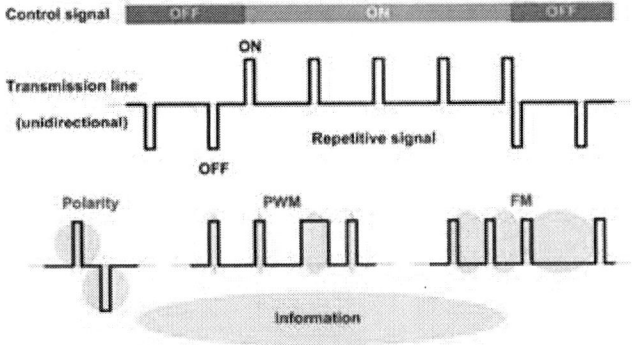

Figure 2. Signal transmission via pulse transformer: New driver concept.

In Fig. 3 is shown an example of an analog sensor signal transmission from the secondary side to the primary side. The DBC-temperature sensor signal is processed by an analog-to-digital converter into a 10bit value. This digitized value can be transmitted potential-free as a serial data stream by the pulse transformer. In former driver concepts, the DBC-temperature sensor signal had to be processed on the primary side [3]. The necessary potential separation must be guaranteed over the whole signal path. Due to the new concept, the potential separation between the power and control level is always guaranteed also in cases of power module destruction. The illustrated example of temperature signal transmission can also be also adapted to other sensor signals, e.g. DC link voltage, the current sensor signal, collector-emitter saturation voltage V_{CEsat} of the switch.

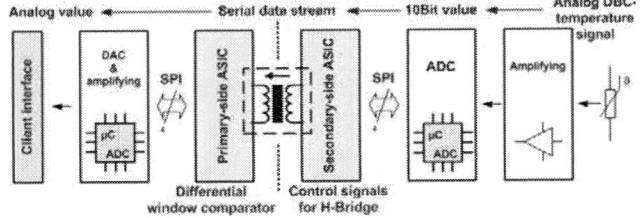

Figure 3. Analog temperature sensor signal transmission.

B. Driver concept

A block diagram of the new driver concept is given in Fig. 4. For the pictured topology of an inverter circuit, one primary-side and two secondary-side Application Specific Integrated Circuits (ASIC) are needed. Each ASIC is controlled by a microcontroller via Serial Peripheral Interface (SPI). Both ICs contain digital and peripheral functionality, such as power stages for driving signals to the pulse transformer (part of the modulator), differential window comparators (part of the demodulator), General Purpose Input/Output pins (GPIO), miscellaneous analog circuits, high-voltage power stages and an ASIC-internal Wishbone bus. The new concept aims at a high flexibility of driver functions and driver behavior.

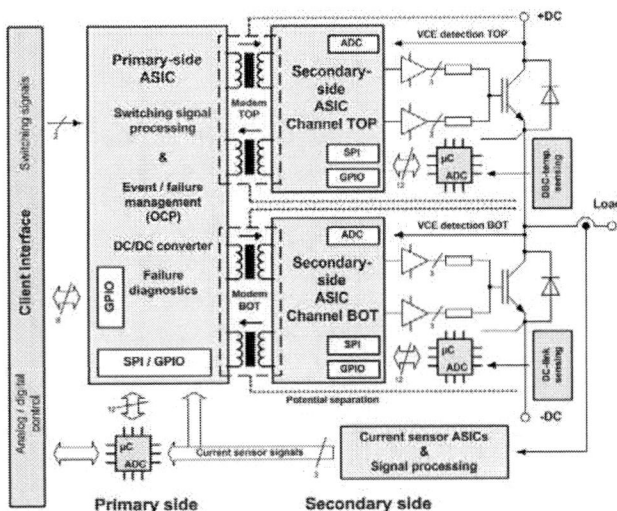

Figure 4. Block diagram of the new driver concept with advanced digital signal transmission.

These targets are a result of IC-internal wishbone bus architecture and a coupled SPI. The SPI is used to configure the driver functions and to transfer digitalized signals into the ASICs. The well-known driver functions, like short-pulse suppression, dead time of interlock between two switches of an inverter, interlock function, the monitoring of forbidden oscillations on the control input pins, event or error management, etc. are programmable. The configuration of functions via SPI and the wishbone bus allows a wider range than former concepts with a few configuration pins. Furthermore, due to the possibility having a lot of sensor information inside the driver, there is a chance to analyze the operation conditions of the switches or the power module. This allows manipulating the switching behavior depending on the operating state. This fact is used by the new "IntelliOff" function. "IntelliOff" uses the DBC-temperature, DC-link voltage and current sensor information for an optimum switching control in the power module's critical or worst cases. If defined threshold values for the temperature, the collector current and the DC-link voltage are exceeded and the turn-off behavior of the switch will be changed and the maximum emitter-collector overvoltage V_{CEmax} and the dv_{CE}/dt will be reduced. Furthermore, a "soft-turn-off" function is implemented, in case an over-current is measured by the current sensor or a short circuit is detected by the V_{CEsat}-monitoring.

III. REALIZATION

The basis for the new driver concept is a dedicated modem (modulator/demodulator) between the primary and secondary side. The back swing and the resulting dead time, the biggest disadvantages of the former solutions (see Fig. 1), can be eliminated by an H-bridge controlled pulse transformer. Fig. 5 shows the improved transmission line as part of the modem. The polarity of the transmitted signal can be detected by the comparator outputs. If there is a repetitive signal on the transmission line, either the "modem_rx_up" (see Fig. 5) or the "modem_rx_down" output signal will toggle. Additionally, a transmitted FM or PWM signal will be

978-1-4244-8425-6/11 $26.00 © 2011 IEEE

reproduced at one of the outputs, too. In order to gain high flexibility the digital core has been designed in the Very High Speed Integrated Circuit Hardware Description Language (VHDL) [4]. In difference to previous solutions, the ASICs contain additional registers for recording a lot of events and failure states. That can be useful for diagnostic purposes. The state of the registers can be read out by the SPI, too. The client-side system can write into the registers of the ASICs to configure its functions.

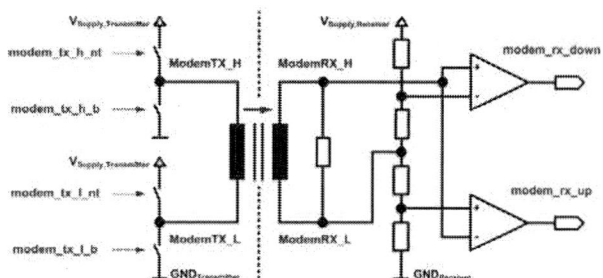

Figure 5.   Transmission line: Part of the modulator and demodulator.

A.   Primary-side ASIC

The primary-side ASIC has been designed in CMOS technology with high-voltage capability to control two IGBT switches. For applications, which use the both switches in an inverter topology (half bridge), well-known safety-relevant functions are integrated, e.g. the realization of an interlock function and a dead time between both IGBTs. Furthermore, an input signal's short pulse suppression and oscillation detection are realized in the primary side's digital core.

Figure 6.   Chip photograph of the primary-side IC.

The digital core contains the following functions, programmable by VHDL:

- error and event management
- signal generation: modem signals (see. Fig. 5 – left and right side), control signals of the DC/DC converter and GPIOs
- adjusting of the on-chip oscillator
- analyzing the signals of the ADC
- bridge between SPI and wishbone bus: configuration of the driver's properties and reading out register information

Fig. 6 shows a chip photograph of the primary-side IC, with the well visible digital core in the middle of the IC. Around the digital core there are the analog circuits (ADC's, differential window comparators as part of the demodulator, reference voltage regulator), the power stages (e.g. DC/DC converter), the interface circuits (e.g. schmitt-trigger inputs) and the IC-pins.

B.   Secondary-side ASIC

The secondary-side ASIC (see Fig. 7) has to control one switch. Main core functions are the same as the described primary-side IC functions. A new part of the secondary-side IC is the control of the switching behavior of IGBT and freewheeling diode. The IC contains 6 power-stages for an adaptive turn-on and turn-off, e.g. "IntelliOff" or "soft-turn-off". These stages are shown in the chip photo (see Fig. 7 on the left side). For the reliability of the system it is necessary that the IGBT works in its safe operating area (SOA). An important safety-relevant function is the monitoring of the collector-emitter saturation voltage of the IGBT. The identification of a desaturation protects the IGBT against overload and short-circuit conditions. The desaturation monitoring is realized by a programmable ADC.

Figure 7.   Chip photograph of the secondary-side IC.

C.   IGBT driver PCB with ASIC

An example of a primary-side ASIC's application is shown in Fig. 8. The photograph shows a part of an IGBT driver PCB with the ASIC and the coupled microcontroller (µC). Also visible are the pulse transformer package for one channel as part of two modems. Furthermore, the power transformer package as part of the DC/DC converter is placed on the right side of the photograph.

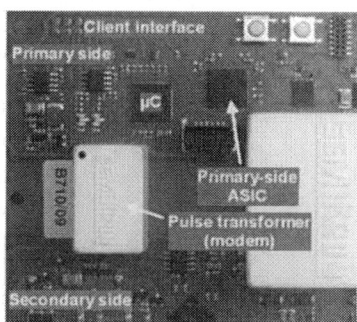

Figure 8.   Photograph of an IGBT driver PCB with primary-side ASIC in a QFN-64 package.

IV. RESULTS

Measured signals of the transmission line are presented in Fig. 9. The switching signal (see Fig. 9 - Ch1) of the client-side microcontroller passes the primary-side ASIC, is transferred by the modem and is reconstructed on the secondary side (see Fig. 9 - Ch3, Ch4).

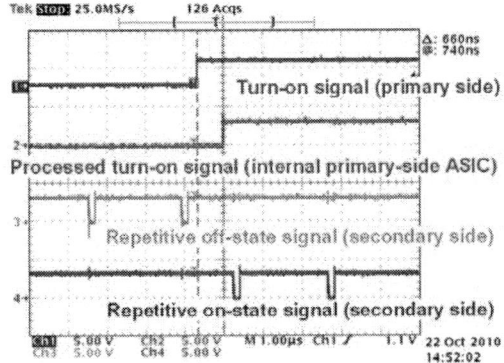

Figure 9. Switching signal transmission from the primary to the secondary side.

An example of the configuration via SPI is shown in Fig. 10. The screenshot of the oscilloscope shows a reset of the event manager control registers. In case of configuration, the ASIC is a SPI slave and must be selected by the master (e.g. microcontroller) via the low-active Chip Select (CS) signal (see Fig. 10 – Ch1). The MOSI-signal (see Fig. 10 – Ch3: Master Out Slave In) is the information, which is written into the ASIC by the microcontroller. First, a write order is sent, followed by the target wishbone address, and after that the register's values. If the write order is done correctly, the ASIC as slave will send a high-active acknowledgement signal via the MISO output (see Fig. 10 – Ch4). After the data transfer, the CS signal goes to the high level again and the MISO output returns to the tri-state level.

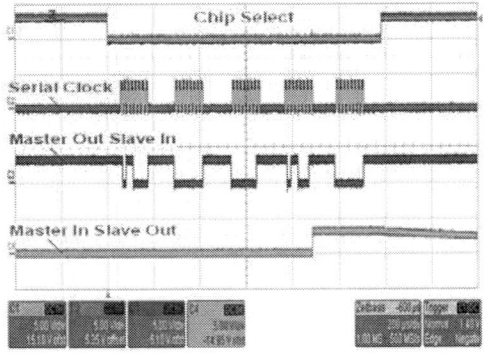

Figure 10. Configuration of an ASIC via SPI (event manager control registers' reset)

The operation of a SKiiP4 system in a normal, "IntelliOff" and "soft-turn-off" mode is presented in Fig. 11. In the normal mode the IGBT is controlled by a lower gate turn-off resistor ($R_{g(off)}$) to minimize turn-off losses (E_{off}). If I_C exceeds a value of 90% of the rated current and $V_{DC\text{-}link}$ is higher than 1200V there will be a risk of a collector-emitter over-voltage. Consequently, the IGBT should be turned-off with a higher

$R_{g(off)}$, that reduces the over-voltage with the drawback of higher turn-off losses. If I_C exceeds a value of 1.6 times of the rated current the over-current protection threshold will be reached. In that case the IGBT would be turned-off in the "soft-turn-off" mode by the highest $R_{g(off)}$.

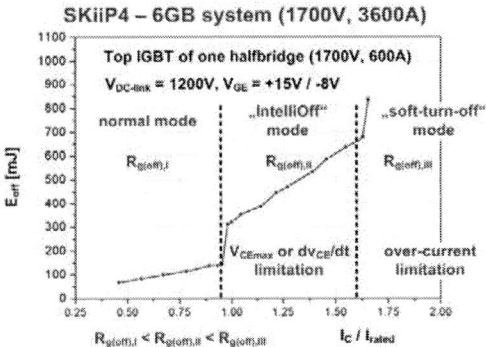

Figure 11. Turn-off losses of an IGBT switch in normal, "IntelliOff" and "soft-turn-off" mode.

V. CONCLUSION

An ASIC chip set has been developed for the implementation of the new driver concept with digital signal processing and transmission of data streams across the isolation barrier. Transmission of data streams is essential for advanced data processing. Processing of various sensor signal data inside the IGBT driver and an adapting of the switching behaviour of the switch can be useful for the characteristics of the switch (e.g. E_{on}, E_{off}, V_{CEmax}, dv_{CE}/dt) and the handling of worst or critical cases close to the limits of the SOA. Both ASICs contain digital and peripheral functionality, such as power stages for driving signals to the pulse transformer (part of the modulator), ADC's, differential window comparators (part of the demodulator), a reference voltage regulator, supply voltage monitoring circuits and high-voltage power stages (e.g. DC/DC converter). Based on a hardware description language (VHDL), the digital core is more flexible and can be adapted on different applications. The ASICs contain internal bus architectures with a coupled serial interface, additional registers for recording many events and failure states.

ACKNOWLEDGMENT

The authors would like to thank Mr. H. Hahn, X-FAB Semiconductor Foundries AG, for the useful support.

REFERENCES

[1] J. Lehmann, S. Pawel, R. Herzer, R. Bittner, S. Boigk, "Compact High Power System Design with Smart Power ICs", Proc. PCIM Europe 2002, pp. 347–352.

[2] J. Thalheim, "Chipset for Flexible and Scalable High-Performance Gate Drivers for 1200V – 6500V IGBTs", Proc. ISPSD 2008, pp. 197–200.

[3] A. Wintrich, U. Nicolai, W. Tursky, T. Reimann, "Application Manual Power Semiconductor", pp. 111-113, Verlag ISLE, Ilmenau, 2010.

[4] H. Kaeslin, "Digital Integrated Circuit Design", p. 175ff., Cambridge University Press, New York, 2008

Proceedings of the 23rd International Symposium on Power Semiconductor Devices & IC's
May 23-26, 2011 San Diego, CA

Solutions to Improve Flatness of Id-Vd Curves of Rugged nLDMOS

S. Mouhoubi, F. Bauwens, J. Roig, P. Gassot, P. Moens, M. Tack

Power Technology Centre, ON Semiconductor, Oudenaarde, Belgium

samir.mouhoubi@onsemi.com

Abstract—**This work summarizes results of TCAD simulations aiming to reduce/suppress the bump in the output characteristics of rugged nLDMOS devices. It is shown that the origin of the bump is not due to bipolar activation. Thus, by simple variations of the geometrical parameters and/or process variations, the intrinsic MOS of the nLDMOS could be driven in a regime allowing a drastic improvement of its Id-Vd flatness with limited impact on the sRon-Vbd trade-off.**

I. INTRODUCTION

Over the last years, the need of robust Lateral DMOS transistors (LDMOS) has led to several investigations in order to improve their Safe Operating Area (SOA). The key to achieve this is to suppress the parasitic bipolar formed by the drain drift region, the source and the body by increasing the doping level of the base (body). In that way, the activation of the parasitic bipolar is delayed and thus, better SOA's are obtained [1]. However, the extension of the SOA limits was accompanied with an unusual degradation of the flatness of the Id-Vd curves (a sudden increase of the current followed by a saturation). Several studies were carried out to explain the physical mechanism behind the degradation of the Id-Vd flatness [2][3] but there is a lack of solutions proposed to get rid of it or diminish it. For example, it is claimed that suppressing the bump is possible by shunting the base-emitter path of the bipolar by inserting hole (h+) collection sites in the body by means of dotted openings through the poly-gate [4]. This paper demonstrates that the bump is not linked to the activation of the parasitic bipolar and proposes ways to improve the flatness of the Id-Vd curves by means of either process variations or variations of the device geometrical parameters.

II. DEVICE DESCRIPTION & MODELING

The device used in this study is an n-type Lateral DMOS (nLDMOS) integrated in a 0.18μm smart power trench-based isolation technology [5] for a voltage tier of 45V. Fig. 1 depicts a 2D cross section of the nLDMOS built with the Synopsys TCAD tools. The device has separate Source and Bulk connections to allow collecting Bulk current. Drain and BLN are shorted to avoid reach-through of the vertical structure Nwell-Pfield-Nepi. The most relevant parameters for the current study are the lengths of the channel, accumulation

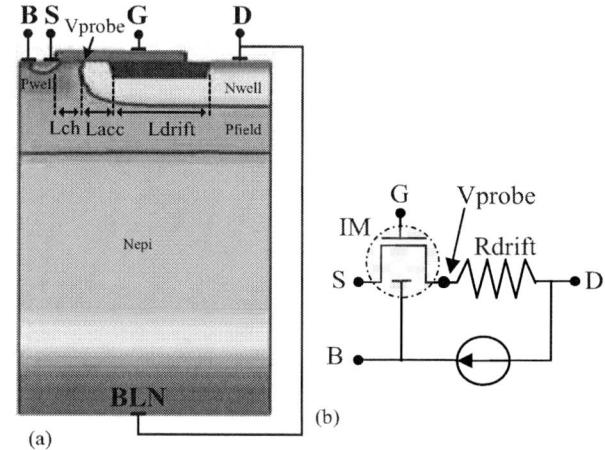

Figure 1: (a) Cross section of the simulated nLDMOS used in the current study (BLN connected to Drain, Source and Bulk grounded). (b) Equivalent circuit model of the nLDMOS. IM=intrinsic MOS.

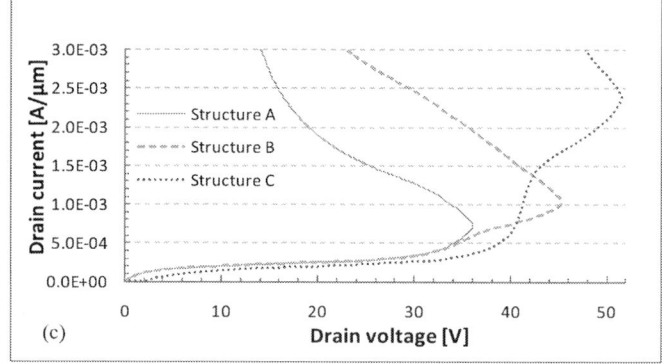

Figure 2: Simulated Snap-Back curves at Vgs=3V for the nLDMOS with weak and robust Pwell (Structure A and B resp.). Structure C represents the nLDMOS with the Source floating (Emitter of the parasitic bipolar disconnected).

and drift region (represented in Fig. 1 by Lch, Lacc and Ldrift respectively). Fig. 2 shows an example of output characteristics obtained with the Pwell of the CMOS technology. At Vd=36V (Structure A), the device snaps back due to the activation of the parasitic NPN bipolar formed by

978-1-4244-8425-6/11 $26.00 © 2011 IEEE

the Nwell (collector), Pwell (base) and Source-N$^+$ (emitter). This bipolar triggers due to the holes generated in the drift region that flow towards the Bulk contact leading to the forward biasing of the emitter-base junction. A way to address this issue is is to reduce the base-emitter resistance by implanting a highly doped body as proposed in [2]. In that way, it is possible to enlarge the SOA allowing the device to achieve both higher voltage capability and higher current density. However, a new phenomenon occurs: an unusual enhancement of the current followed by a saturation [1-3]. This is reproduced by simulation as depicted in Fig. 2: With the robust Pwell, it is possible to postpone the snapback by approximately 10V at the expense of a less flat Id-Vd curve. To study this phenomenon, a simplified equivalent circuit was used [1][3]. Fig. 1b shows an example of such equivalent circuit in which a voltage probe (Vprobe) is inserted in order to be used as the drain voltage of the Intrinsic MOS (IM).

III. ORIGIN OF THE BUMP

In the past, trials were carried out to explain the mechanism behind the alteration of the Id-Vd curves. The analysis focused mostly on the charge compensation occurring in the drift region which drive the device into a compression regime (Kirk-effect [6]) or an expansion regime in case of a robust Pwell [1]. The saturation of the curves after the sudden increase was attributed to the poly-gate above the drift region that acts as a control-gate which stabilizes the current provided by the impact ionization. In [4], another way of enlarging the SOA was presented: instead of reducing the forward current gain of the parasitic bipolar, hole collectors are placed in the channel region. It is claimed that this method suppresses the bipolar without showing the unusual bump.

This section aims to demonstrate that the suppression of the bump is not linked to the suppression of the parasitic bipolar. Indeed, a simple experiment could confirm it: if the emitter (Source) is disconnected, the Id-Vd curves should become flat since there is no more bipolar involved. However the simulation shows it is not the case; the bump is still present as depicted in Fig. 2 (Structure C).

The flatness of the output characteristics reported in [4] are not due to the bipolar but are a consequence of inserting the hole collectors in the channel region. Indeed, in order not to alter the breakdown of the device, the collectors must be spaced from the drift region with the minimal allowable channel length. Which means, by construction of the device, the channel length is increased; and this is the direct cause of the bump suppression as will be shown in the next sections.

IV. SOLUTIONS TO IMPROVE THE FLATNESS OF THE CURVES

As explained in [3] the bump is directly linked to the operating regime of the IM (which is mainly determined by Vprobe). Thus, a better control of this MOS characteristics is key to solve the issue. One of the possibilities to reduce the bump would be to force the IM to rapidly operate in saturation. This could be obtained by allowing higher potential at the drain channel edge or by increasing the channel resistance. The former situation could be reproduced by enlarging Lacc to allow more potential lines to enter the accumulation region and increase the potential of the IM drain

faster. The latter situation could be mimicked simply by increasing the channel length and/or doping concentration. Another way to suppress the bump is to split the gate at the edge of the STI (as will be shown). This option has the advantage to keep Vprobe constant up to very high Drain voltages (which stabilizes the IM). All these solutions are developed in the following sections.

A. Enlargement of accumulation region

By enlarging Lacc, one can permit to offer more space to the potential lines to spread over the accumulation region (as depicted in Fig. 3). This results in a faster increase of the potential Vprobe at the drain of the IM (refer to Fig. 4); which induces a faster saturation of the IM (see Fig. 5b). And thus, the Id-Vd curves of the nLDMOS are flattened (as shown in Fig. 5a). From Fig. 5 one can see that the bump appears at approximately an nLDMOS Drain voltage of 30V (for the shortest Lacc). This is seen in the IM output curves: The onset of the bump occurs exactly when the IM changes its regime from the linear part to the saturation. Contrarily, for the largest Lacc, the IM is already in saturation before the nLDMOS Drain voltage reaches 10V (which guarantees flat curves for the nLDMOS).

As shown in Fig. 5a, the larger Lacc the flatter the output characteristics of the nLDMOS. However, one has to keep in mind that an excessive increase of Lacc will inevitably lead to reliability issues. Indeed, the potential lines entering the accumulation area will induce a higher electric field across the gate oxide. This could cause a poor oxide lifetime, or even

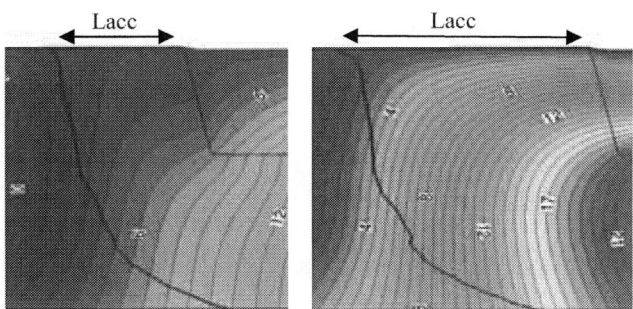

Figure 3: Potential lines distribution for two different Lacc (The inserted numbers represent the values of the electrostatic potential in Volts).

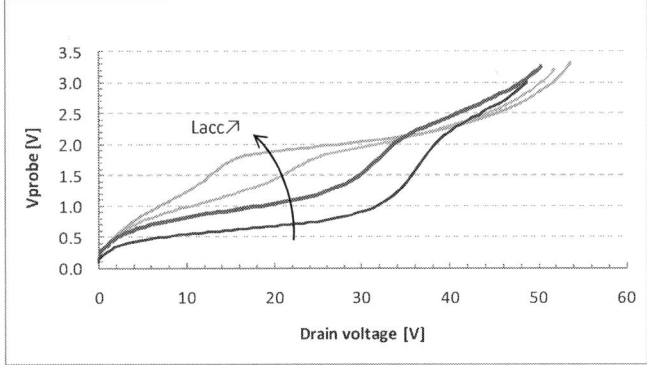

Figure 4: Evolution of Vprobe with the voltage of the nLDMOS Drain for increasing Lacc (with a step of 0.1 μm).

worse, an early oxide breakdown. An optimal Lacc value should be a compromise between Id-Vd curves flatness, BVdss and oxide integrity (that is imposed by the reliability constraints depending on the application aimed).

Note that this solution presents the advantage of reducing the on state resistance sRon. A larger Lacc results in less current crowding.

B. Increase of channel length

A longer channel has two benefits on the one hand, it provides a faster saturation and on the other hand, it limits the current increase when the IM changes its regime from linear to saturation. The improvements are clearly visible on Fig. 6. A channel length increase of 0.3µm compresses the IM characteristics and flattens considerably the curves of the nLDMOS. Unfortunately, this simple and efficient solution has the drawback of increasing the specific on state resistance due to channel length increase and device dimension increase. However, it remains interesting for the applications where robustness is the prime parameter and sRon is less critical. It is also very useful for the high voltage devices where the drift region determines most of the device resistance. Furthermore, comparing this solution with the one proposed in [4] shows a substantial advantage: a pure channel increase will have less impact on sRon compared to the dotted channel structure where not only the channel has increased but also the effective gate width is reduced (due to the inserted hole collectors).

C. Increase of channel doping concentration

Instead of enlarging the channel to increase its resistance, it is possible to get the same effect by increasing the channel implanted dose. Of course, this leads to a higher threshold voltage Vth. Fig. 7a displays the Id-Vd curves of the nLDMOS for two different channel doping profiles and Fig. 7b depicts the output curves of the IM. For Dose2 (which has the highest surface doping concentration), the Id-Vd curves are much flatter and the IM characteristics saturate faster. The effect is more pronounced than what is obtained by channel length increase. However, a body with higher channel dose will diffuse more towards the Nwell leading to a slightly longer effective channel. To have a fair estimate of the impact of a pure channel dose increase, it would be necessary to compare structures with slightly different "drawn channel". In the case of Fig. 7b, it is possible to do so by comparing the light continuous curve of Dose1 with the dark discontinuous line of Dose2: The improvement is still considerable. As an example, the Vth of the structures simulated in Fig. 7 are 0.7V and 1.3V for Dose1 and Dose2 respectively.

D. Split gate

So far, the solutions proposed were based on process variation or on changes in the device geometrical parameters. In this section a structural modification is performed on the device. The split gate concept (or discontinuous gate). Such a structure is shown in Fig. 8. It consists of splitting the poly-gate on top of the accumulation region. One of the resulting parts of the poly acts still as a gate, while the other one keeps

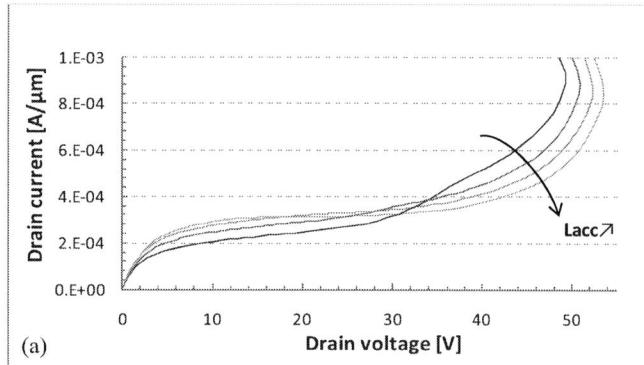

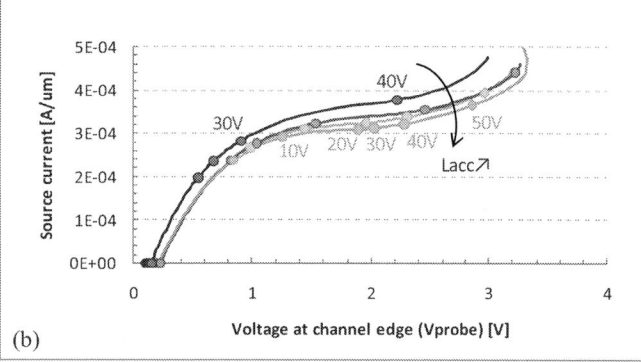

Figure 5: (a) Impact of Lacc on the nLDMOS characteristics (Id-Vd). (b) IM behavior. Simulation details: Vgs=3V, Vbs=0V, Lacc step of 0.1µm.

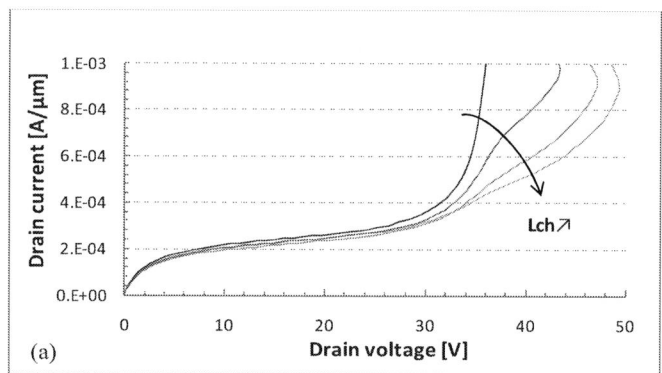

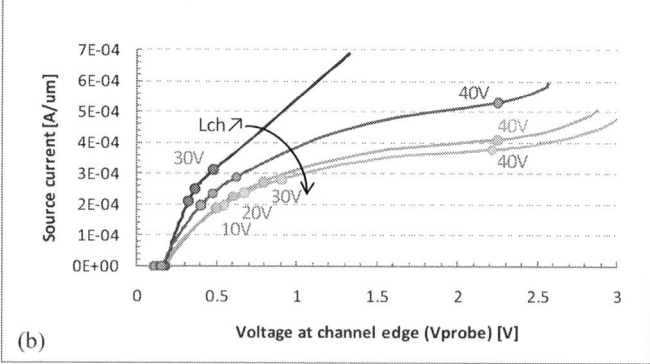

Figure 6: (a) Impact of Lch on the nLDMOS characteristics (Id-Vd). (b) IM behavior. Simulation details: Vgs=3V, Vbs=0V, Lch step of 0.1µm.

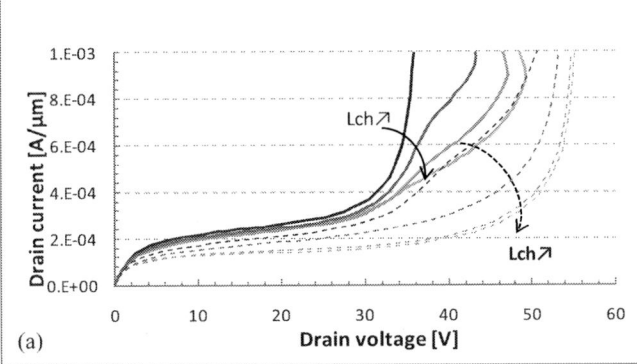

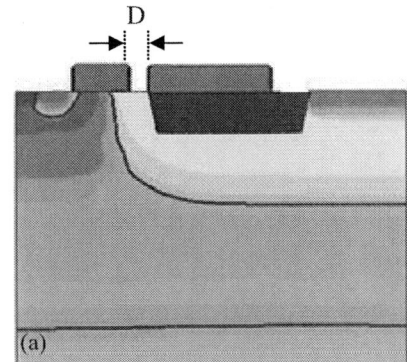

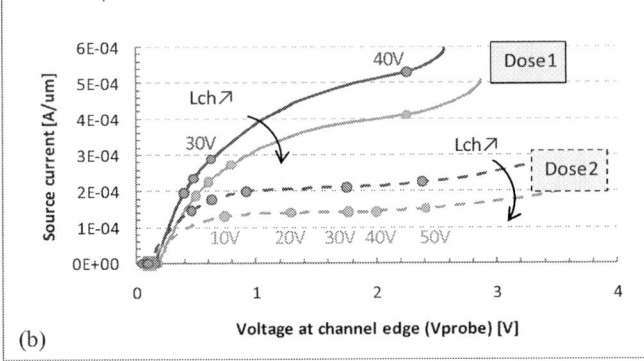

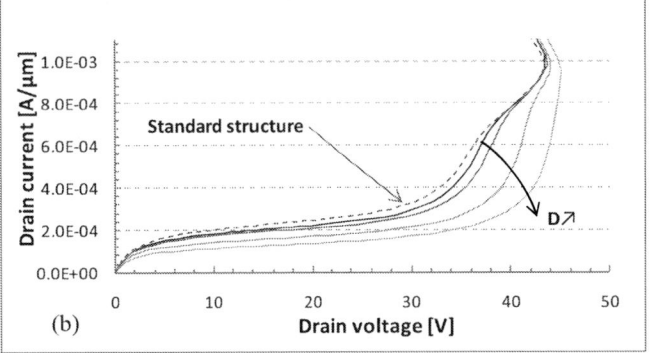

Figure 7: (a) Simulated Id-Vd curves for two different channel doping profiles and varying Lch values. (b) IM behavior. Thin dashed lines correspond to Dose2 and thick continuous lines correspond to Dose1 (Dose2 > Dose1).

Figure 8: (a) Cross section of the new nLDMOS structure with a "Split Gate" concept. (b) Comparison of the output characteristics of the standard nLDMOS and the Split Gate structure for different spacing values "D".

the field plate effect active. This option has the advantage (thanks to the poly-gate) to keep Vprobe constant up to very high Drain voltages. This helps to stabilize the IM. Increasing the distance "D" diminishes the bump progressively as shown in Fig. 8b: the snap back voltage is not really increased, but the flatness of the curves is substantially improved.

This is accompanied by an increase of sRon and BVdss. The former is due to the fact that the accumulation layer is reduced. Indeed, increasing "D" reduces the poly-gate overlap on the Nwell; which means that the accumulation layer formed is less. This could be seen as a reduction of the cross section through which the current has to flow. The latter is due to lower impact ionization levels (due to less current). Once again, a compromise between SRon, BVdss and flatness of the Id-Vd curves has to be made.

By its topology, this structure could be sensitive to misalignments (poly to poly and poly to STI). However in advanced technologies (and especially for specific layers like Poly), the photolithography control is quite precise.

V. CONCLUSION

This work presents solutions to improve the flatness of the output characteristics of rugged nLDMOS structures. The origin of the bump is proven not be related to the parasitic bipolar, but rather to the regime at which the intrinsic MOS evolves. By means of TCAD simulations, different solutions

were explored and their advantages/drawbacks were assessed. It is shown that by simple process and/or geometrical variation, it is possible to improve drastically the flatness of the Id-Vd curves. The Split-Gate concept which is a simple structural modification of the poly looks promising and allows a very good control of the intrinsic MOS.

The solutions presented in this work have the advantage of adding no process complexity and thus no extra cost: the choice of one of them can be based only on their added value from an electrical point of view.

REFERENCES

[1] P. Hower, J. Lin, S. Pendharkar, B. Hu, J. Arch, J. Smith, and T. Efland,, "A Rugged LDMOS for LBC5 Technology", ISPSD, May 23-26 2005, pp. 327–330.

[2] John Lin and Philip L Hower, "Two-Carrier Current Saturation in a Lateral Dmos", ISPSD, June 4-8 2006, pp. 89-91.

[3] S. Reggiani et al, "Explanation of the Rugged LDMOS Behavior by Means of Numerical Analysis", IEEE TED, vol. 56, no. 11, Nov. 2009, pp. 2811-2818.

[4] T. Khan, V. Khemka, R. Zhu and A. Bose, "Rugged Dotted-channel LDMOS structure", IEDM, December 15-17 2008, pp. 1-4.

[5] R. Charavel et al, "Next generation of Deep Trench Isolation for Smart Power technologies with 120 V high-voltage devices", Microelectronics Reliability, vol. 50, 2010, pp. 1758–1762.

[6] A. W. Ludikhuize, "Kirk effect limitations in high voltage IC's," in Proc. ISPSD, 1994, pp. 249–252.

Proceedings of the 23rd International Symposium on Power Semiconductor Devices & IC's
May 23-26, 2011 San Diego, CA

The vertical voltage termination technique – characterizations of single die multiple 600V power devices

Kremena Vladimirova*, Jean-Christophe Crebier*, Christian Schaeffer*, Delphine Constantin**

* Grenoble Electrical Engineering Lab (G2Elab)
961 Houille Blanche
38402 St Martin d'Hères, France

**3 parvis Louis Néel
BP 257
38016 Grenoble, France

e-mails: vladimirova@g2elab.grenoble-inp.fr, crebier@g2elab.grenoble-inp.fr

Abstract— **Deep trench terminations are commonly known as a technique to achieve ideal breakdown voltages for high voltage devices. This paper presents the use of deep trench terminations as an original concept to integrate multiple vertical power devices on a common die. The concept is based on the creation of vertical deep trench terminations on the periphery of the devices, thus allowing to separate the drift regions and to completely insulate the multiple power devices sharing the same backside contact electrode. Power diodes in the range of 600V are fabricated and experimentally tested to validate the concept. The prototypes demonstrated excellent forward and reverse biased static characteristics.**

I. INTRODUCTION

Deep trench terminations have been widely investigated for high voltage device edge termination and are commonly known as a solution providing ideal voltage handling capability whereas consuming small junction termination area [1, 2]. Recently, a new technology for power integrated circuits using a combination of top and back trenches was presented. The demonstration was made for vertical power devices in the range of 5V [3]. Lately, [4] presented the evolution of this approach for high voltage power devices with breakdown levels in the range of 600V and higher. Several vertical PIN diodes sharing the same backside cathode were integrated in the same power die. Each power diode was individually separated, from its "neighbor" with a peripheral vertical voltage termination. This concept consisted of creating a deep trench vertical termination on the periphery of the devices, allowing to reduce surface electric field peaks but also to separate the different active regions. Thus, effective insulation between neighbor devices is guaranteed although they share a common backside contact electrode. Multidiode devices in the range of 600V are fabricated, packaged and passivated with silicone dielectric gel. The experimental validation of the concept showed near ideal breakdown voltage with small leakage current ratings but also a highly resistive behavior of the forward biased diodes.

This paper presents deeper investigation of this original concept for wafer-level integration of single die multiple vertical power diodes.

The paper focuses on the special care taken to improve the characterization of the manufactured power dies while adding new data and characteristics.

More specifically, the paper presents the new results obtained for the static characteristics of the fabricated devices and an evaluation of the reverse blocking characteristics as function of the temperature. In order to improve the static characterization of single and multiple diodes power die, special packages have been developed in order to use Kelvin probe measurement technique. The resulting characteristics matched completely with the theoretical ones, allowing to validate the approach. Power devices with deep trench edge terminations are known to exhibit larger leakage current than planar devices [5]. Therefore, the reverse biased characterization is carried out in a thermal chamber to investigate more specifically the evolution of the leakage current and the breakdown voltage levels with respect to the temperature.

II. DEVICE STRUCTURE AND FABRICATION

This section shortly recalls the main concept and the realization of the single die multiple 600V vertical power diodes.

A. Single die mulitple power vertical diodes

The concept of the vertical voltage termination relies on the creation of a deep trench termination on the periphery of the devices. The same trench termination serves not only to ensure the voltage handling capability of the device but also to separate the active regions of the different devices sharing the same backside electrode terminal. The trench termination is then filled with a dielectric material and thus effective insulation between the different devices is guaranteed. Fig. 1

This work is supported in part by the French Research National Agency (ANR) with projects MOBIDIC ANR-06-BLAN-0204-03 and ECLIPSE ANR-09-BLAN-0036-01.

978-1-4244-8425-6/11 $26.00 © 2011 IEEE

shows a schematic view of the principle using vertical power diodes as example.

Figure 1. Schematic view of multiple power diodes integrated in a common die using the vertical voltage termination technique

As fig. 1 demonstrates with this technique multiple vertical power devices can be "islanded" on a common power die. This approach could be of great interest for the building of multiphase 3D power module assemblies [6]. Significant reductions in interconnect number as well as self aligned device chip are among the advantage of such approach. It appears as a key element of the package of power devices at wafer levels where several components are interconnected throughout the technological process, leading to reproducible and highly reliable the assembly.

B. Main manufacturing steps

In order to demonstrate the feasibility and the effectiveness of the proposed technique, vertical power diodes with vertical voltage terminations were fabricated. In our case 500μm N+ type (100) substrate plus 55μm lightly doped (20 Ω.cm) epitaxial layer was used for the realization of the prototypes. These parameters were chosen so that the breakdown voltage of the optimum plane junction should be 800V. All process steps are performed full wafer. After the deposition of 3μm thick layer of aluminum on both sides of the silicon substrate the different patterns are distinguished by a lithography step followed by wet etching of the metal on the surface and the trenches realization. The trenches were realized with Deep Reactive Ion Etching. After the DRIE process step the depth and the width of the trenches are 100μm and 150μm respectively. We used the aluminum metal layer as masking material. No post processes for surface or wall treatment were performed after the silicon etching step. Fig. 2 and 3 show the realized prototypes.

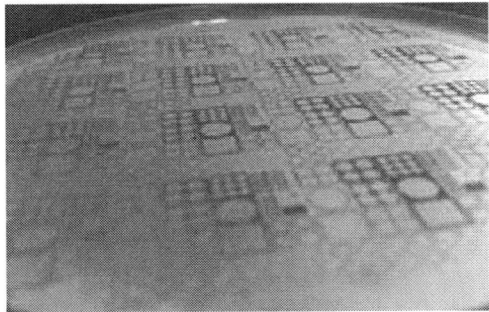

Figure 2. Photograph of the manufacuterd wafer : various patterns can be seen containtg multiple "islanded" diodes

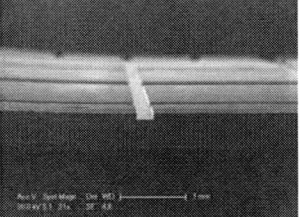

Figure 3. Top surface and cross sectionnal view showing multiple power diodes integrated on a common silicon die and separated by vertical voltage terminations

After the wafer sawing the chips containing several devices are individually packaged and passivated with silicone oil or gel.

III. RESULTS AND DISCUSSION

A. Forward characterizations

The forward biased characterization requires the use of Kelvin probes technique in order to optimize the quality of the measurements. A specific packaging architecture must be designed in order to optimize the power interconnections and to add two sensor terminals. A schematic of the designed packaging is given in fig. 4 below. The important issue here is related to the wire bond interconnects. Since our equipment is only able to solder wires with 33μm diameter, it is important to solder several wires but also to spread them at the surface of the chip in order to improve the homogenous distribution of the current over the cross section of the power devices.

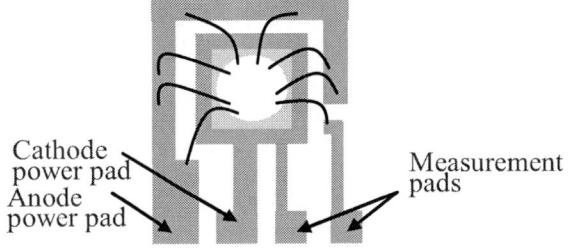

Figure 4. Schematic of the package for effective forward characterization.

Two packages have been developed. One is specifically used to maximize the current distribution and the quality of the measures of a single diode component. With this aim, wire bonds are connected to a metal track which is almost completely surrounding the power device. In such manner, the distribution of the power current is assumed to be ideal even if the wires are not fully adapted to the current ratings used in this experiment. A picture of the package is given in fig. 5 were it can be seen that the power device is soldered on a PCB track and surrounded by another track and two Kelvin probes on one side of the die.

978-1-4244-8425-6/11 $26.00 © 2011 IEEE

Figure 5. Pictures of the packages dedicated for forward bias characterization of single power diodes with vertical voltage terminations.

A second package has been designed for the characterization of a single die, four 600V vertical power diodes, each of them being connected on the anode side with several wire bonds spread at the surface of the metal layer plus a Kelvin probe on each of them. Fig. 6 presents a picture of the package in this case.

Figure 6. Picture of the package dedicated for forward bias characterization of single die 4 diodes component including multiple Kelvin probes.

The forward characterization of the single diode component is given in fig. 7 below. It is a 1.4x1.4cm² large diode (prototype fig. 5) with blocking capability in the range of 600 to 800V.

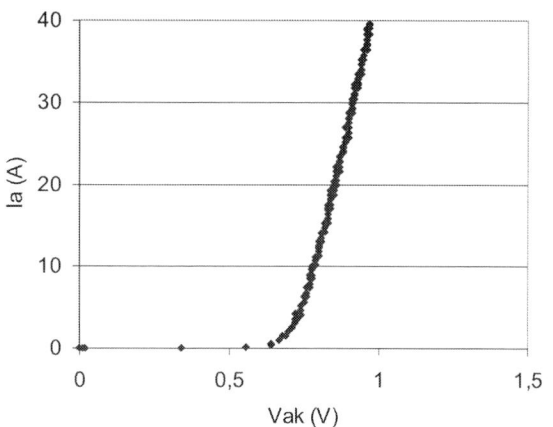

Figure 7. Forward biased characteristic of a single diode with vertical voltage termiantions.

Testing the device under forward biased polarization with a power curve tracer HP 371A, 40A forward current was

measured under 1V forward voltage drop. This result is fully in accordance with our expectations, allowing to demonstrate the quality of the manufactured devices and the measurement technique that has been used.

Nonetheless, it must be mentioned that above a certain current level the devices passivated with silicon oil demonstrated a non expected behavior. Deeper investigations are required to identify the reason of such operation. Besides, this outstanding forward characteristic hides probably poor dynamic behavior. This will have to be investigated in the future.

B. Reverse biased characterization

Considering the reverse characterization of the power devices terminated with vertical edges manufactured in DRIE, the observations of three main data were carried out:

- -the ability of single die multiple diodes to withstand voltage while terminated and separated by the vertical edge termination technique

- -the ability of single or multiple diodes dies to withstand nominal voltage breakdown levels under minimized leakage current level

- -the ability to minimize leakage current level at nominal reverse voltage levels and under high temperatures (up to 100°C)

1) Single die multiple diode reverse characteristics

Fig. 8 shows the reverse characteristics in the case of testing the monolithic integration of prototype fig. 6. The four curves correspond to the four power diodes integrated in the same die. All four devices fulfill the initial design specifications considering the breakdown voltage rating of 600V.

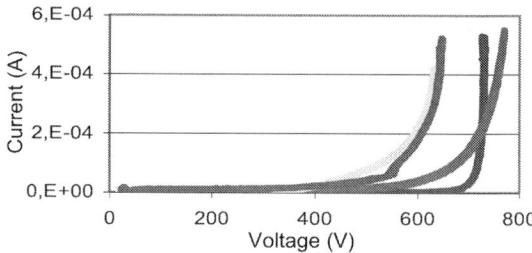

Figure 8. Static characteristics of reverse biased four power diodes sharing the same power die

The results show clearly that all four diodes have different voltage handling capabilities. It can be noticed in fig. 8 that one of the diodes has a leakage current of 10µA at 600V while two others have leakage currents in the range of 200 µA.

2) Reverse biased characterization at room temperature

Fig.9 represents one of the most significant achievements of blocking capability of a device with vertical voltage terminations. The device demonstrated a breakdown voltage of 720V with a leakage current of 1.7μA at 600V.

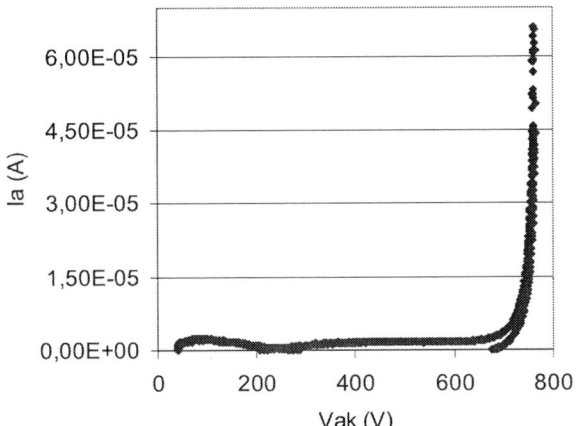

Figure 9. Reverse biased characteristic of one of the best single die multiple diodes

3) Reverse blocking characteristics as function of the temperature

Fig.10 shows the leakage current measurements at high temperature. The current level is in complete accordance with our expectations with leakage current dependence of temperature similar to that for planar devices. These results consolidate the choice of the technology and the measurement approach that we have presented in this paper.

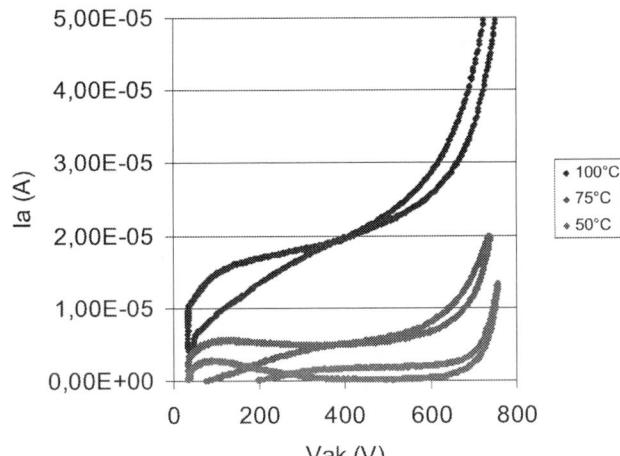

Figure 10. Reverse biased characteristics as function of temperature

IV. CONCLUSION

This paper presented the practical results of the continuation of the static characterization of single die multiple vertical diodes terminated and islanded with vertical edge terminations. Precise static forward and reverse biased characterizations have proven the ability of the termination technique to offer outstanding voltage breakdown with minimum leakage current at nominal voltage levels of 600V. Islanded multiple diodes were identified functional and exhibiting satisfactory performance levels. The robustness of the implementation still needs further investigations as well as the dynamic responses of the power devices. Nevertheless, this edge termination technique provides the possibility of wafer level packaging with collective assemblies and interconnections. Besides, with such surface treatment and edge terminations power devices could benefit from 3D power chip on chip packaging configuration, advantageous for effective high frequency double side cooling implementations.

ACKNOWLEDGMENT

The authors would like to thank CIME-nanotech and RTB centers PTA and FEMTO-ST for their technical support.

REFERENCES

[1] C. Park, J. Kim, T. Kim, and D.J Kim," Deep trench terminations using ICP RIE for ideal breakdown voltages", in *Proc*. IEEE Int. Symp. Power Semiconductor Devices and IC's, ISPSD 2003, pp.199 – 202

[2] L. Theolier, H. Mahfoz-Kotb, K.Isoird,and F. Morancho, "A new junction termination technique: The Deep Trench Termination (DT2)",in Proc. IEEE Int. Symp. Power Semiconductor Devices and IC's, ISPSD 2009, pp.176 – 179

[3] P. Igic, P. Holland, S. Batcup, R. Lerner, and A. Menz, "Technology for power integrated circuits with multiple vertical power devices", in Proc. IEEE Int. Symp. Power Semiconductor Devices and IC's, ISPSD 2006, pp.1 – 4

[4] K. Vladimirova, J.C. Crebier, Y. Avenas, and C. Schaeffer, "Single die multiple 600V power diodes with vertical voltage terminations and isolation", in Proc. IEEE ECCE 2010, 12-16 Sept. 2010, pp. 2200-2205

[5] R. Kuhne, and E. Kasper, "Reduction of leackage currents in silicon mesa devices," in IEEE Trans. On Microwave theory and techniques, vol.46, no.5, 1998

[6] E. Vagnon, P.O. Jeannin, J.-C. Crebier, and Y. Avenas, "A Bus-Bar-Like Power Module Based on Three-Dimensional Power-Chip-on-Chip Hybrid Integration," in IEEE Trans. Industry Applications, vol.46, no.5, 2010

Proceedings of the 23rd International Symposium on Power Semiconductor Devices & IC's
May 23-26, 2011 San Diego, CA

Investigation of Parasitic BJT Turn-on Enhanced Two-Stage Drain Saturation Current in High-Voltage NLDMOS

Chih-Chang Cheng, H.L. Chou, F.Y. Chu, R.S. Liou, Y.C. Lin, K.M. Wu, Y.C. Jong, C. L. Tsai, Jun Cai, H. C. Tuan
Analog/RF & Specialty Technology Division, Taiwan Semiconductor Manufacturing Company (TSMC)
Hsin-Chu, Taiwan, Email: chengcc@tsmc.com

Abstract—**A two-stage drain current phenomenon in saturation region, named as I_d-V_d hump, has been investigated in high-voltage NMOS transistor. A parasitic BJT turn-on enhanced I_d-V_d hump model is proposed and characterized by using a two-dimensional device simulation. By optimizing channel/drift-region process conditions, both parasitic BJT and impact-ionization generation can be suppressed. Both measured result and simulated result of the optimized device are presented.**

I. INTRODUCTION

Device I-V curve behavior is a fundamental index in the understanding of the physical limitation of device safe operation area (SOA). Due to high power operation, Kirk-effect and parasitic BJT will become more and more important in high voltage devices. In this work, an abnormal two-stage drain saturation current (I_d) phenomenon, called I_d-V_d hump, is described and analyzed by using a 2D device simulation. A combination of both Kirk-effect and parasitic BJT turn-on model is proposed to explain the enhanced I_d-V_d hump effect. New device with a modified dosage/energy in both channel and drift region is fabricated to reduce the I_d-V_d hump effect without significant impact to breakdown-voltage (BV) and on-state resistance (R_{dson}).

II. EXPERIMENTAL RESULTS AND DISCUSSION

A. Conventional device:

Fig. 1 shows a conventional device cross section of high-voltage NMOS transistor. The device was processed in a 0.15um BCD technology with a 5V gate oxide thickness. A butted P^+/N^+ contact at source-side is used to minimize bulk resistance (R_b) and device pitch size. Drain operating voltage is 40V with BV=63V and R_{dson}=43mΩ-mm². Fig.2 shows a comparison of I_d-V_d curve at V_g=5V. A 2D numerical device simulation with calibration work is performed to demonstrate accurate I_d-V_d and I_s-V_d in fig. 2. The I_d-V_d hump current starts to increase at V_d=25V and exhibits a saturation at V_d=35V. Source current shows the same tendency of drain current but a different current level at a higher V_d bias (fig.2). Based-on the simulation result, two bias conditions are chosen for the purpose of I_d-V_d hump analysis; one is V_d=20V (no I_d-V_d hump) and the other is V_d=40V (with I_d-V_d hump).

Fig. 3 shows the spatial distribution of (a) electron current, (b) hole current, and (c) electric-field at V_d=20V (no I_d-V_d

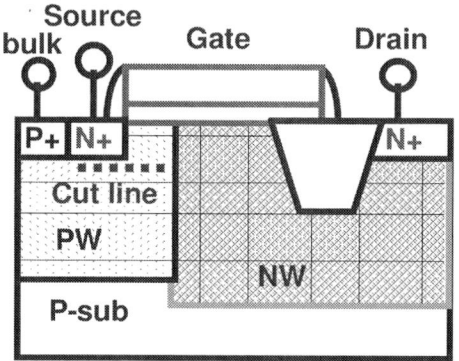

Figure 1. Cross-section of 40V high voltage NMOS. A lateral cut line is indicated for the following analysis.

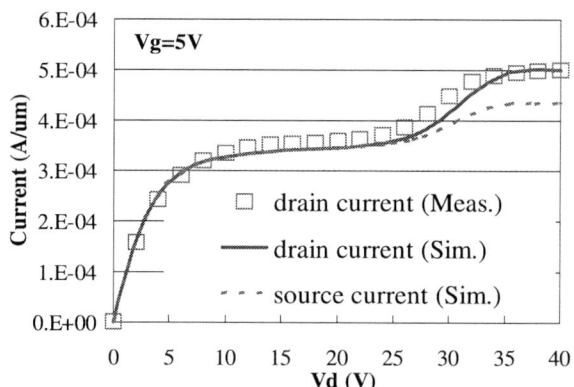

Figure 2. Comparison of drain current and source current versus drain voltage at Vg=5V in the measurement and in the simulation. Calibration work is performed to get accurate result between simulation and measurement.

hump) and at V_d=40V (with I_d-V_d hump). Hole carrier is generated by impact-ionization at drain-side (fig.3 (c)), which is referred to the well-know phenomenon "Kirk-effect"[1, 2]. As device is turned-on at a higher V_g, the increasing amount of electron current alters the space charge distribution in a reverse biased p-n junction[2]. At high currents thus the base widens and the highest electrical field shifts from PW/NW junction to the n^-/n^+ junction at drain-side[2]. This electrical-field relocation creates stronger impact ionization at drain-side and thus resulting in an extra electron/hole current to drain and bulk respectively.

978-1-4244-8425-6/11 $26.00 © 2011 IEEE   208

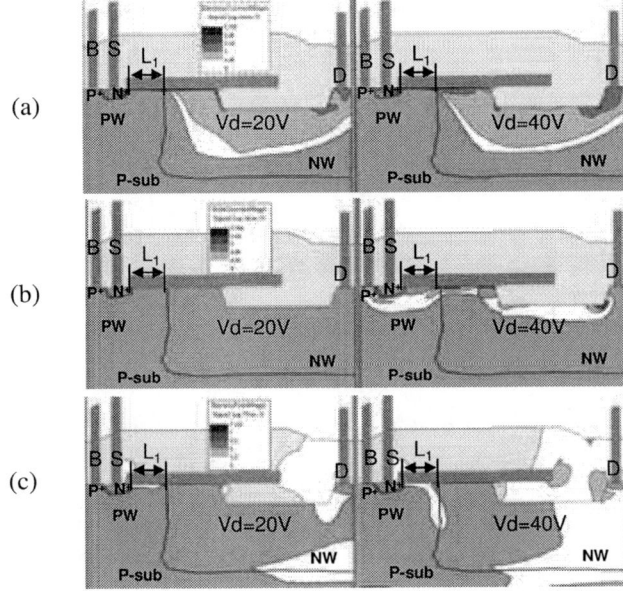

Figure 3. Spatial distribution of (a) electron current, (b) hole current, and (c) electrical-field at Vg/Vd=5V/20V (left side) and at Vg/Vd=5V/ 40V (right side). Electron current flow (fig. 3(a)) is confined in the channel region and spreads out in the drift region. Hole current flows from drain to source and finally pickup by P+ bulk terminal (fig. 3(b)).

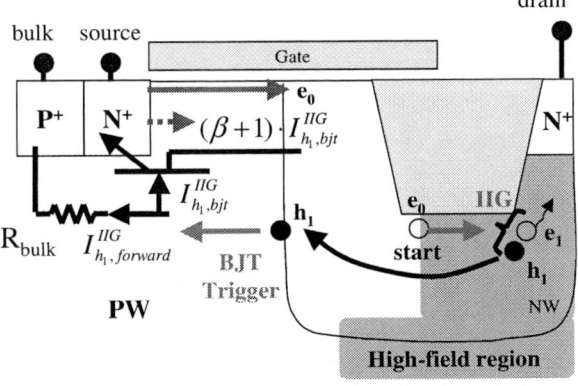

Figure 4. Illustration of parasitic BJT turn-on enhanced I_d-V_d hump process. High field region is drawing according to right-side figure of fig. 3(c) and filled with gray color. R_{bulk} is bulk resistance. Ie_0 is device intrinsic electron current. Electron/hole pairs (e_1/h_1) are generated by impact-ionization generation (IIG) at drain-side. h_1 current in PW region can be divided into two parts; one is a hole current forming a forward-voltage drop and the other is a bjt base current. β is the beta gain of parasitic BJT.

B. I_d-V_d / I_s-V_d hump model:

To provide a better understanding of abnormal I_d-V_d and I_s-V_d, a parasitic BJT model combined with Kirk-effect is proposed to explain both I_d-V_d and I_s-V_d hump effect. Fig. 4 illustrates the process of parasitic BJT turn-on enhanced I_d-V_d hump. As an intrinsic electron carrier (e_0, current is represented by I_{e0}) flows in the drift region, Kirk-effect created high electrical-field region provides an environment to generate electron (e_1) /hole (h_1) pairs at drain. The generated electron current ($I_{e_1}^{IIG}$) and hole current ($I_{h_1}^{IIG}$) will flow toward drain terminal and bulk terminal, respectively. As bulk

resistance (R_{bulk}) is high, the $I_{h_1}^{IIG}$ can be divided into two parts; one is $I_{h_1,forward}^{IIG}$ and the other is $I_{h_1,bjt}^{IIG}$. The $I_{h_1,forward}^{IIG}$ current creates a forward voltage drop ($I_{h_1,forward}^{IIG}$ x R_{bulk}) between N$^+$/PW junction. When the forward-potential is built, a parasitic NPN BJT (N$^+$/PW/NW) will be turned-on with a base current ($I_{h_1,bjt}^{IIG}$). If the parasitic BJT beta gain is β, device current can be expressed by the following equation (1) ~ (4).

$$I_D = I_{e_0} + I_{e_1}^{IIG} + \beta \cdot I_{h_1,bjt}^{IIG} \qquad (1)$$

$$I_S = I_{e_0} + (\beta + 1) \cdot I_{h_1,bjt}^{IIG} \qquad (2)$$

$$I_{bulk} = I_{h_1,forward}^{IIG} \qquad (3)$$

$$I_{h_1}^{IIG} = I_{h_1,forward}^{IIG} + I_{h_1,bjt}^{IIG} \qquad (4)$$

According to simulation result, $I_{h_1,forward}^{IIG} \gg I_{h_1,bjt}^{IIG}$ and $I_{e_1}^{IIG} \cong \beta \cdot I_{h_1,bjt}^{IIG}$. It implies that the parasitic BJT effect is as important as Kirk-effect in our case.

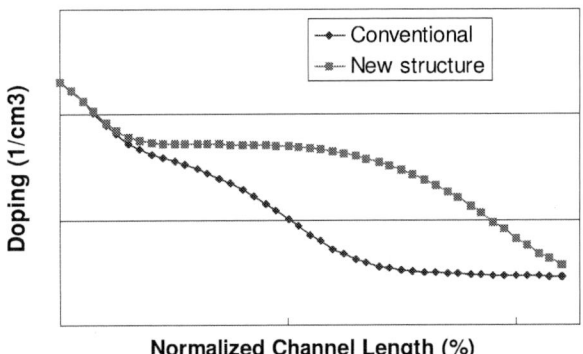

Figure 5. Substrate doping versus normalized channel length in conventional device and in new device. The cut line is indicated in fig.1.

C. New device:

A new device with a similar cross-section of Fig. 1 but modified dosage/energy at both channel/drift regions is fabricated. Since I_d-V_d hump effect combines both Kirk-effect and parasitic BJT, a separated dosage/energy optimization at channel/drift regions is required. Fig. 5 shows a lateral substrate doping distribution before and after the optimization of channel-region. With an optimized process in channel region, bulk resistance is reduced to suppress the beta gain of parasitic BJT. A similar work is also performed at drift-region. By choosing a proper dosage/energy in drift region, impact-

978-1-4244-8425-6/11 $26.00 © 2011 IEEE   209

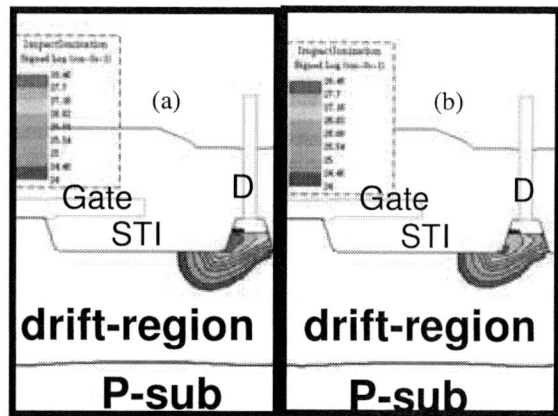

Figure 6. Impact-ionization distribution at Vg/Vd =5V/40V, (a) before process optimization, (b) after process optimization. A smaller impact-ionization generation is occurred after process optimization.

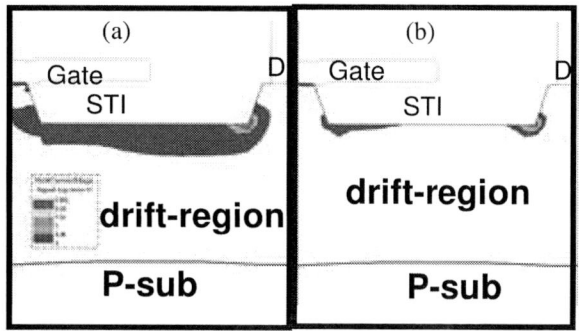

Figure 7. Hole current distribution (a) before process optimization, and (b) after process optimization. A smaller hole current flows in the drift region.

ionization generation (Fig. 6) can be reduced; indicating that fewer hole carriers flow to source (Fig. 7), and thus resulting a smaller $I_{h_1}^{IIG}$ in eq.(4).

The potential contour of Fig. 8 provides more information of how to optimize the device. Three potential lines (0V, 4V, 8V) are shown for the conventional device (Fig. 8(a)) and for the new device (Fig. 8(b)). A comparison of each potential line of before/after process optimization is shown in Fig. 8(c). As potential is small (0V and 4V), the potential line of the optimized device will shift toward drain, which reduces the bulk resistance and hence the forward bias of N+/PW of the parasitic BJT. However, at a higher potential (8V), potential line will shift toward source, which implies a smaller electrical field at drain side and hence IIG in Fig.6. When the optimized process conditions can achieve the opposite shift of potential line, the I_d-V_d hump effect can be completely eliminated (Fig. 9) with <10% R_{dson} sacrifice and no impact on BV.

III. CONCLUSION

A parasitic BJT turn-on enhanced I_d-V_d hump model is characterized by using a 2D simulation. Several simple

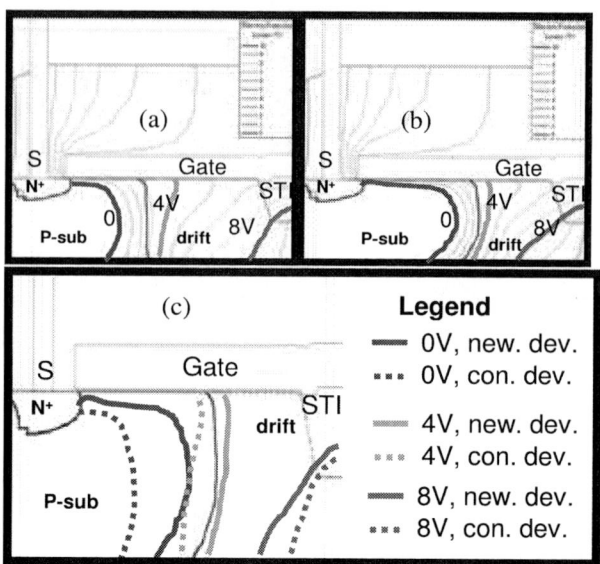

Figure 8. Electron potential contour at V_g/V_d= 5V/40V, (a) conventional device, (b) new device, and (c) a comparison of both conventional device and new device. Potential lines of 0V, 4V, 8V are emphasized. The 0V and 4V contour line of new device are shift toward drain but the 8V contour line is shift toward source.

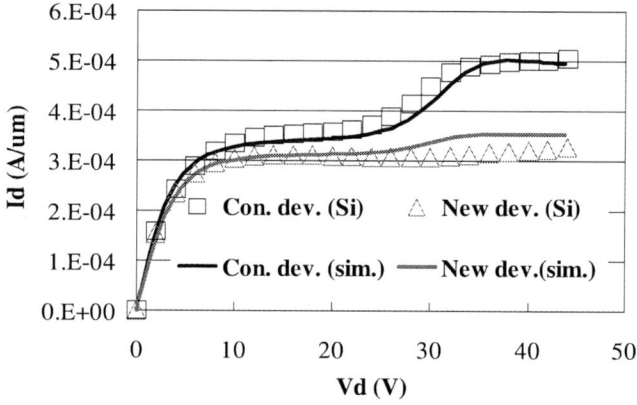

Figure 9. Comparison of drain current versus drain voltage at V_g=5V in conventional device and in new device. Measurement result is represented by symbol. Simulation result is represented by solid line. I_d-V_d hump effect is completely eliminated as the process optimization is performed.

equations are proposed to get the better understanding of parasitic BJT. By optimizing the process conditions of channel/drift region, I_d-V_d hump effect can be completely eliminated without significant BV/R_{dson} sacrifice.

IV. REFERENCE

[1] John Lin et. al, "Two-Carrier Current Saturation in a Lateral Dmos", ISPSD, pp. 1-4, 2006.

[2] A. W. Ludikhuize, "Kirk effect limitations in high voltage IC's", ISPSD, pp. 249-252, 1994.

Proceedings of the 23rd International Symposium on Power Semiconductor Devices & IC's
May 23-26, 2011 San Diego, CA

High Vgs MOSFET Characteristics with thin gate oxide for PMIC Application

Jaehan Cha, Kyungho Lee, Sungoo Kim, Juho Kim, Namkyu Park, Taejong Lee

CE & SMS division, MagnaChip Semiconductor Corporation

1 Hyangjeong-dong Heungdeok-gu, Cheongju, Chungcheongbuk-do, 361-728 Korea

Email:jaehan.cha@magnachip.com

Abstract—**we present the characteristics of a newly developed MOSFET which has thin gate oxide but sustains high gate voltage. It consists of MIM capacitor coupled floating gate poly and source/drain junction formed by low voltage well. The characteristics of the MOSFET depend on the choice of capacitance coupling ratio which is determined by the ratio of MIM (Metal-Insulator-Metal) capacitance and gate oxide capacitance. This means that the proper choice of layout can control the device performances; threshold voltage, I_D-V_D, and I_D-V_G characteristics. Some of smart power management ICs requires of a low density non-volatile memory for the purpose of PMIC trimming. To meet this purpose, NVM embedding on PMIC solution should be cost effective. This newly proposed high gate voltage MOSFET with thin gate oxide can be an effective solution for the periphery circuits of single poly EEPROM operation.**

I. INTRODUCTION

BCD technologies have been widely used in a variety of areas such as mobile applications, battery chargers, various portable power management products, display driver ICs on advanced logic platforms, automotive applications, and storage controller chips, etc. In recent years, significant research effort for HV power ICs has been directed to integrate power management, logic, and audio & communication functions in a single chip (so called system-on-a-chip). Also, the strong demand for PMIC (Power management IC) is focused on the BCD (Bipolar-CMOS-DMOS) technology development which includes cost effective solution, high DMOS performance, option devices availability. Basically BCD technology is formed on low voltage CMOS process platform, and high voltage devices are embedded, especially high voltage BJT and LDMOS (Lateral Double-Diffused MOSFET), EDMOS (Extended Drain MOSFET) [1-6]. To reduce process steps and remove additional thermal budget which is an original difficulty in SOC (Silicon On a Chip) technology, thin CMOS platform gate oxide is used as DMOS gate oxide. But, some of DDI (Display Drive IC) applications need the high gate voltage for gate driver and also the periphery circuits to operate single poly EEPROM which needs high gate voltage drivability. We propose the embedding solution of high gate voltage MOSFET (HGV MOSFET) which has logic thin gate oxide on aBCD18-2 (2nd generation of advanced 0.18um technology node BCD platform in Maganchip, which has single poly EEPROM IP and deep trench isolation for 12V to 40V DMOS operation) technology without additional thermal budget and

process [7-8]. To achieve this purpose, MIM capacitor is directly connected to poly silicon gate of MOSFET, and gate voltage is controlled by this MIM capacitor top plate voltage. In this paper we present the MOSFET operation characteristics in theoretical view point. From these analyses, we extracted the device design parameters to enhance device performance and suppress the off state leakage current.

II. DEVICE STRUCTURE AND PRINCIPLES OF MOSFET OPERATION

Fig. 1 illustrates the device structure of HGV N-type MOSFET with thin gate oxide.

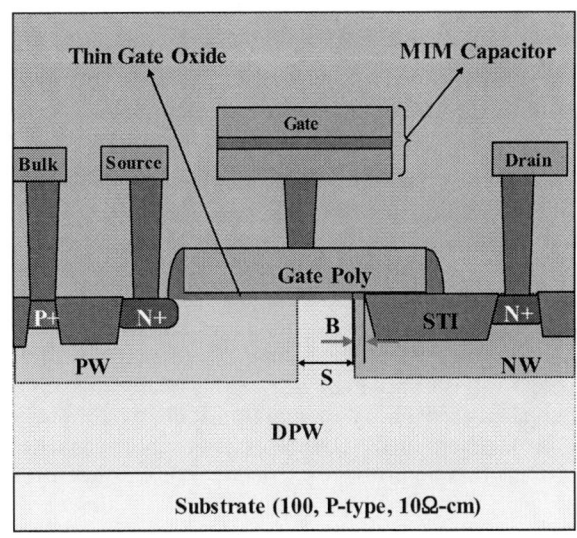

Fig. 1 Cross section view of HGV MOSFET with thin gate oxide.

Drain junction is formed with CMOS platform N-well over high voltage deep P-well which is formed to make DMOS structures in aBCD18-2. By using offset STI under gate poly silicon edge at drain region, high drain breakdown voltage can be achieved. "S" in Fig. 1 is the spacing between CMOS platform N-well and P-well, which is a key factor to determine drain junction breakdown voltage. The design rule "B" in Fig. 1 is gate poly silicon overlap distance to CMOS platform N-well beneath gate oxide region. The design rule "B" is a key factor to determine the device performance of

978-1-4244-8425-6/11 $26.00 © 2011 IEEE   211

HGV MOSFET with thin gate oxide. The thickness of gate oxide is 125Å, which is formed by the gate oxidation process of CMOS platform. Poly silicon gate is directly connected with MIM capacitor and the capacitance of MIM is 1fF/um². High gate voltage on top plate of MIM capacitor is reduced at poly silicon gate through capacitance coupling effect which is originated from charge conservation law. Therefore this new proposed device structure can sustain high gate voltage and drain voltage without additional thick gate oxide generation process. That is to say, there is no additional thermal budget for thick gate oxide formation.

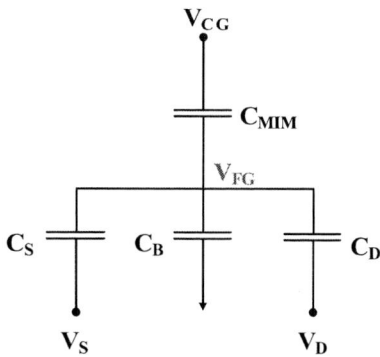

$$V_{FG} = \frac{C_{MIM} \cdot V_{CG} + C_S \cdot V_S + C_D \cdot V_D}{C_S + C_D + C_B + C_{MIM}}$$

Fig. 2 equivalent capacitance model and the voltage of poly silicon gate by capacitance coupling effect.

Fig. 2 shows the equivalent capacitance model of HGV MOSFET with thin gate oxide. V_{CG}, V_S, V_D, V_{FG} are the gate control voltages on the top plate of MIM capacitor, source voltage, drain voltage and floated poly silicon gate voltage respectively. C_S, C_B, C_D, C_{MIM} are each the gate-source overlap capacitance, the gate-bulk overlap capacitance, the gate-drain overlap capacitance and the capacitance of MIM capacitor. V_{FG} is easily induced by using charge conservation of equivalent capacitors. From this equation, we simply know that V_{FG} is determined by not only the ratio of capacitances but also the bias voltage of each node of MOSFET. Generally V_S is zero, therefore V_{FG} is determined by the ratios of capacitances, V_{CG} and V_D. Threshold voltage of HGV MOSFET with thin gate oxide is increased by the inverse coupling ratio;

$$V_{Th1} \cong \frac{C_S + C_D + C_B + C_{MIM}}{C_{MIM}} \cdot V_{Th0} \qquad (1)$$

Here, V_{Th1} is the threshold voltage of HGV MOSFET with thin gate oxide and V_{Th0} is the threshold voltage of normal MOSFET which has the same junction structure of HGV MOSFET with thin gate oxide but does not have the MIM capacitor coupled gate. This equation (1) is induced by ignoring V_S and V_D in V_{FG}.

III. DEVICE PERFORMANCES

Fig. 3 (a) shows the threshold voltage dependency on the channel length. "S" is 0.7um and "B" is 0.1um; constant for each channel length. C_{MIM} and C_B increased increasing channel length. Therefore coupling ratio is different for each channel length. Fig. 3(b) shows the threshold voltage dependency on the inverse coupling ratio for each channel length. Inverse coupling ratio is calculated by using layout design size and 125Å gate oxide thickness and 1fF/um² MIM capacitance. From this figure, we know that the equivalent capacitance model and equation (1) are excellent solution for the threshold voltage estimation of this newly proposed HGV MOSFET with thin gate oxide. If inverse coupling ratio is 1, threshold voltage is changed from V_{Th1} to V_{Th0}. Measured V_{Th0} is 0.843V, and estimated V_{Th0} is 0.834V by using equation (1).

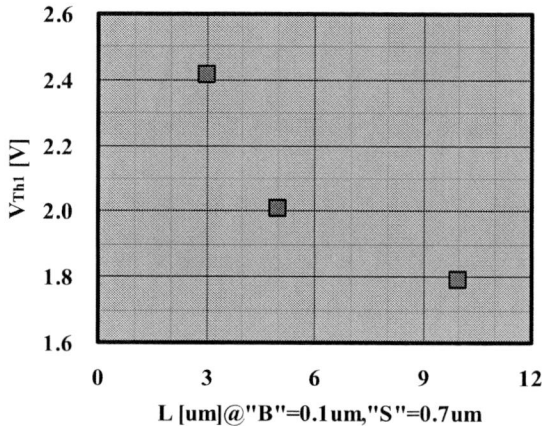

Fig. 3(a) Channel length dependency of V_{Th1}

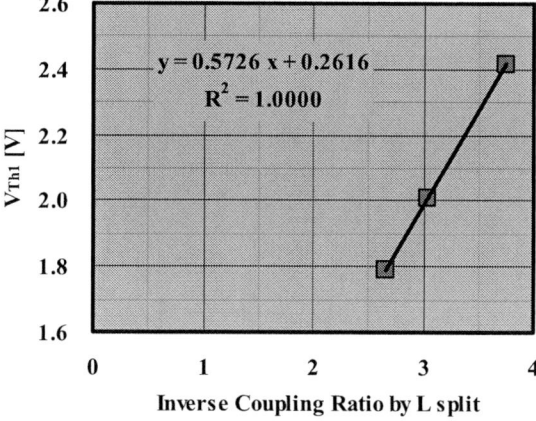

Fig. 3(b) Inverse coupling ratio dependency of V_{Th1}

Fig. 4 (a) shows the threshold voltage dependency on the key design rule, "B". Device width and length are 20um and 5um respectively and "S" is 0.7um constant for all "B" split. Fig. 4(b) shows the threshold voltage dependency of inverse coupling ratio. It also shows that the equivalent capacitor model and threshold voltage equation (1) are proper to explain the threshold voltage of HGV MOSFET.

Fig. 5 shows the I_D-V_G characteristics of HGV MOSFET with thin gate oxide according to the design rule "B". In case of low V_D, V_D=0.1V, there is no off-current issue. At high drain voltage, there is very large off-current over 100pA/um for the case of "B" > 0.4um. This can be interpreted as abnormal DIBL (Drain Induced Barrier Lowering) effect [9]. But this case is not the DIBL effect. It is originated from Non-Zero gate voltage by capacitive coupling of drain voltage. For the transistor off-state, V_{CG} = 0V and V_S = 0V.

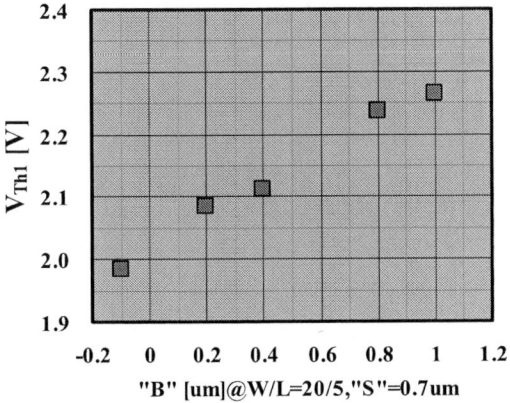

Fig. 4(a) key design rule "B" dependency of V_{Th1}

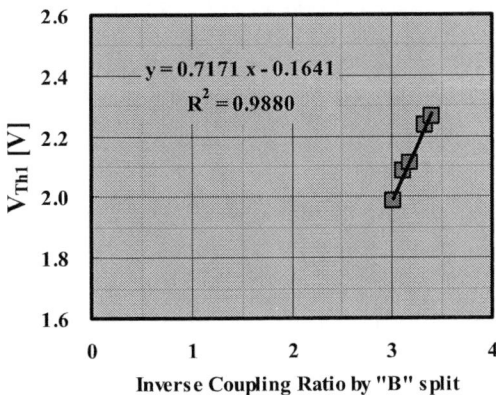

Fig. 4(b) Inverse Coupling Ratio dependency of V_{Th1}

Therefore the voltage of poly silicon gate (V_{FG}) is only determined by the ratio of capacitances and drain voltage.

$$ V_{FG} = \frac{C_D}{C_S + C_D + C_B + C_{MIM}} \cdot V_D \qquad (2) $$

Equation (2) clearly explains this large off-current by increasing design rule "B". As design rule "B" increases, C_D of HGV MOSFET with thin gate oxide increases. The poly silicon gate voltage, V_{FG}, increases with the production of drain voltage and off-state coupling ratio. From this result, we know that the suppression of off-current can be achieved by selecting proper design rule for the HGV MOSFET with thin gate oxide. Fig. 6 shows the DIBL dependency on off-state coupling ratio for 12.1V drain voltage.

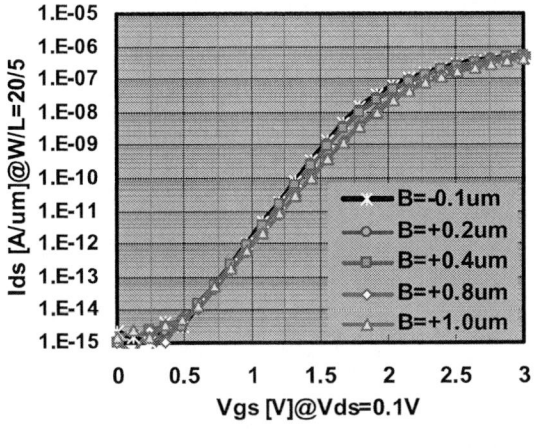

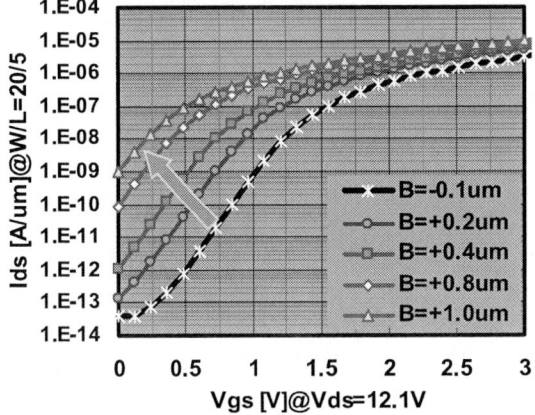

Fig. 5 Id-Vg Characteristics of HGV MOSFET by design rule "B" dependency

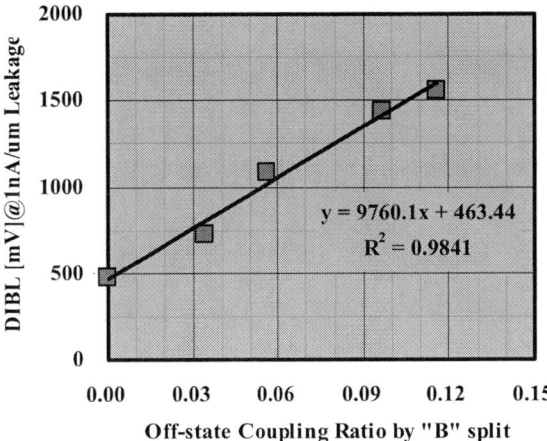

Fig. 6 DIBL dependency on off-state coupling ratio

Fig. 6 shows the reason why "B" is called the key design rule of HGV MOSFET with thin gate oxide. By optimizing channel length and key design rule "B" we can use this MOSFET without off current issue.

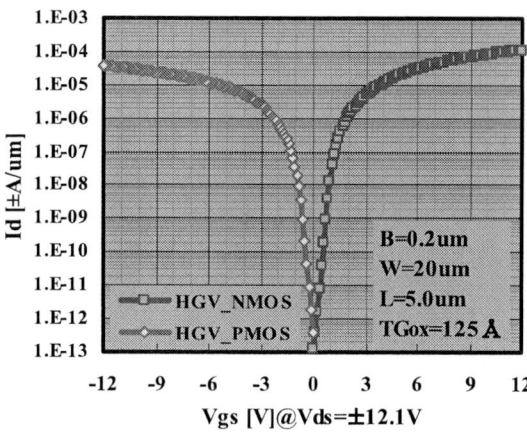

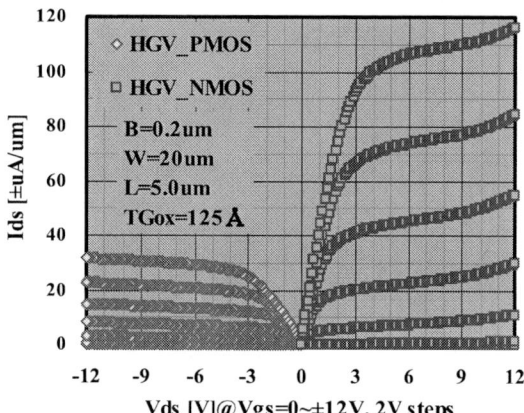

Fig. 7 I-V Characteristics of HGV C-MOSFET with thin gate oxide

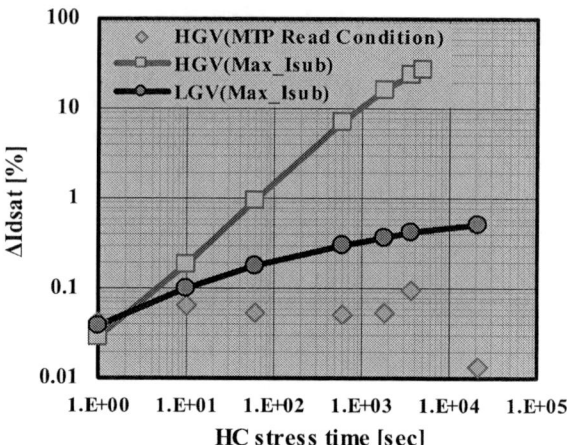

Fig. 8 Hot Carrier Immunity Characteristics of HGV NMOSFET with thin gate oxide

Fig. 7 shows the current-voltage characteristics for high gate voltage complementary MOSFET with thin gate oxide with the key design rule "B = +0.2um". Although the device structure of PMOSFET is same as HGV NMOSFET with thin gate oxide it has the opposite doping type. But it shows good current-voltage characteristics.

Fig. 8 shows the hot carrier immunity characteristics of HGV MOSFET and low gate voltage MOSFET which has the same device structure as HGV MOSFET without MIM capacitor connected to the gate. There are 3 hot carrier stress test conditions. Black circled sample has the maximum substrate current generation condition (V_G = 3.0V, V_D = 12V) in low gate voltage MOSFET Blue squared sample has the maximum substrate current generation condition (V_G = 6.0V, V_D = 12V) in HGV MOSFET, and Green diamonded sample has a MTP (Multi-Time Programmable single poly EEPROM) read condition (V_G = 3.0V, V_D = 1.0V) in HGV MOSFET. This figure clearly shows that this newly proposed HGV MOSFET with thin gate oxide can be a good candidate for cost effective MTP-embedded technology to a BCD without additional thermal budget.

IV. SUMMARY

Newly proposed high gate voltage complementary MOSFET with thin gate oxide has a good current-voltage performance with proper choice of the key design rule "B" which determines the off state leakage current. These devices can be used for the periphery circuit for the operation of single poly EEPORM which is used for trimming of PMIC product, and have sufficient hot carrier immunity. Therefore these devices can be a good solution for embedding NVM to BCD for PMIC application.

REFERENCES

[1] K.Y. Ko et al., "BD180LV – a 0.18um BCD with Best-in-Class LDMOS from 7V to 30V," *Proc. of ISPSD*, pp. 71-74, 2010

[2] Damiano Riccardi et al., "BCD8 from 7V to 70V : a new 0.18um technology platform to address the evoluation of applications towards smart power Ics with high logic contents," *Proc. of ISPSD*, pp. 73-76, 2007

[3] T. Uhlig et al., "A18 – a novel 0.18um smart power SOC IC technology for automotive appllication," *Proc. of ISPSD*, pp. 237-240, 2007

[4] B. Desoete, A. De Smet, and P. Moens, "A multiple deep trench isolation structure with voltage divider biasing," *Proc. of ISPSD*, pp. 237-240, 2007

[5] P. Moens et al., "Reliability assessment of deep trench isolation structures," *Proceeding of the International Reliability Physics Symposium(IRPS)*, pp. 573-577, 2005

[6] B. Elattari et al., "Impact of charging on breakdown in deep trench isolation structures," *Proceeding of the European Solid-State Devices Research Conference (ESSDERC)*, pp. 513-517, 2003

[7] N.K. Park et al., " aBCD18-an advanced 0.18um BCD Technology for PMIC Application" *Proc.of ISPSD*, pp. 231-234, 2009

[8] E. Carman et al., "Single Poly EEPROM for Smart Power IC's" *Proc.of ISPSD*, pp.177-179, 2000

[9] S. Wolf, "Silicon Processing for the VLSI Era Volume 3 – The Submicorn MOSFET", p213, Lattice Press, 1995

Proceedings of the 23rd International Symposium on Power Semiconductor Devices & IC's
May 23-26, 2011 San Diego, CA

Drift Design Impact on Quasi-Saturation & HCI for Scalable N-LDMOS

Yun Shi, Natalie Feilchenfeld, Rick Phelps, Max Levy, Martin Knaipp*, Rainer Minixhofer*

Analog and Mixed Signal Technology Development, IBM Microelectronics Division,
Essex Junction, Vermont, 05452, USA
*R&D Department, Austriamicrosystem, AG, Unterpremstaetten, Austria

Abstract—In this paper, we discuss the scalable NLDMOS design in a 0.18μm HV-CMOS technology. The design impacts in quasi-saturation are compared between the 25V and 50V NLDMOS to demonstrate the implications in output and f_T characteristics. The STI depth sensitivity in DC, *ac* and HCI characteristics is investigated. The results prove a very robust design, featuring <10% I_{dlin} shift over 10 year lifetime for +/-10% STI depth variations.

I. INTRODUCTION

The integrated high voltage CMOS (HV-CMOS) has proven a promising technology with low process complexity and matching performance to BCD [1]. As HV-CMOS continues to find use in diverse applications and at increasing operation voltages, it is important to have a comprehensive understanding of the drift region optimization with respect to device parametrics, process variation and hot carrier stability [2]-[5]. Based on the 0.18μm HV-CMOS technology presented in [1], we investigate the drift design impact on scalable NLDMOS, as well as its sensitivity on the shallow trench isolation (STI). The 25V-rated and 50V-rated LDMOS are used as examples, which represent partially depleted and fully depleted drift designs respectively. The output characteristics, g_m, f_T and hot carrier injection (HCI) are studied in details to show the design trade-off. In order to address quasi-saturation and its impact on DC and *ac* performance, FIELDAY [6] is utilized to simulate internal JFET voltage versus V_{gs} and V_{ds}. The STI process sensitivity is explored with varying etching depth by +/-10%. The R_{on}, BV_{dss}, f_T, f_{max} matrix and I_{dlin} degradation are analyzed carefully. The results demonstrate a robust NLDMOS design that is not sensitive to variations in STI depth.

II. DEVICE AND PROCESS

A schematic cross section of the isolated N-LDMOS is shown in Fig. 1. The device is built in standard 0.18μm HV-CMOS technology [1]. The drift region comprises an implanted high voltage N-well and shallow trench isolation (STI). The HV N-well also provides body isolation. The STI process is unchanged from 0.18μm base CMOS technology. With the scaling of STI length and the charge in drift region, we are able to achieve 25V~50V rated N-LDMOS in the same

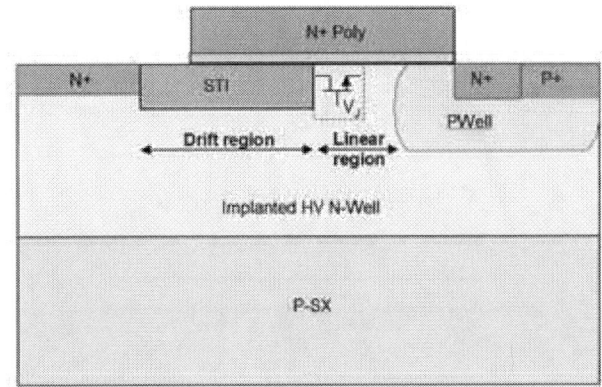

Fig. 1. A schematic cross-section of the isolated N-LDMOS. The JFET gate voltage is defined as V_J, and $V_J \approx 0$ at the onset of quasi-saturation.

III. RESULTS AND DISCUSSION

A. Quasi-Saturation & Output Characteristics

First, the output curves for 25V- and 50V-rated NLDMOS are shown in Fig. 2 (a) and (b). The knee voltage of 25V design is noticeably shifted to higher value at higher V_{gs}. Eventually, no strong pinch off is observed at V_{gs}=5V up to device breakdown. In contrast, the 50V design shows pinch off at low V_{ds} over the full V_{gs} sweep. The internal device characteristics, including potential, depletion boundary, current flowlines, are simulated at V_{gs}=5V and V_{ds}=25V for both designs, and compared in Fig. 3. The 25V NLDMOS drift region is fully un-depleted, the current tends to flow closer to STI interface. In contrast, the 50V NLDMOS shows a well depleted drift region under the field plate, where the current is pinched at the STI bottom corner when I_d saturates [2], and spreads more into the bulk. Thus the 50V NLDMOS demonstrates an improved output impedance. The fully-depleted design is inevitably associated with the Kirk effect [7][8], which can be observed in body current (I_B). Fig. 4 compares the measured I_B versus V_{gs} at V_{ds}=25V for the 25V-NLDMOS and V_{ds}=50V for the 50V-NLDMOS. The onset of the Kirk effect leads to a lowered field at STI corner and the high field is shifted towards drain. Consequently, I_B reduces distinctively in 50V-NLDMOS for the intermediate gate voltage and eventually increases again when the peak

978-1-4244-8425-6/11 $26.00 © 2011 IEEE

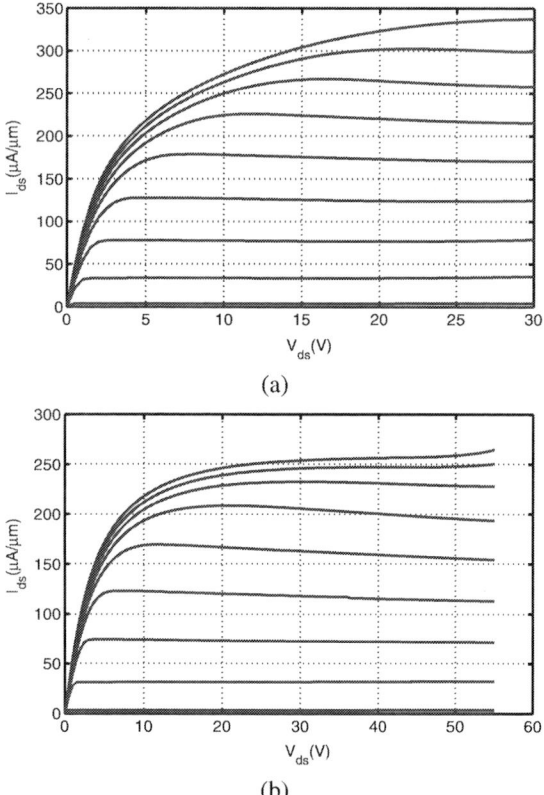

(a)

(b)

Fig. 2. I_d vs. V_{gs} curves for (a) 25V and (b) 50V NLDMOS. V_{gs} sweeps from 0.5V to 5V, in the step of 0.5V. Quasi-saturation is observed in both devices and a degraded output impedance at higher V_{gs} is observed in 25V NLDMOS.

field at the drain side increases further. This I_B characteristics for 50V N-LDMOS is well known to favor robust reliability performance.

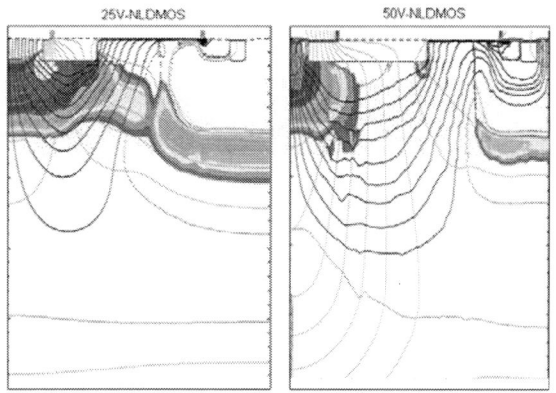

Fig. 3. Comparison of TCAD simulation results at V_{gs}=5V and V_{ds}=25V. Blue lines: potential, red lines: depletion boundary, black lines: flowlines, and color contours: impact ionization rate. The drift region under field plate is well depleted for 50V design. I_{ds} pinches off at STI bottom corner.

Besides, the quasi-saturation is observed in both designs at

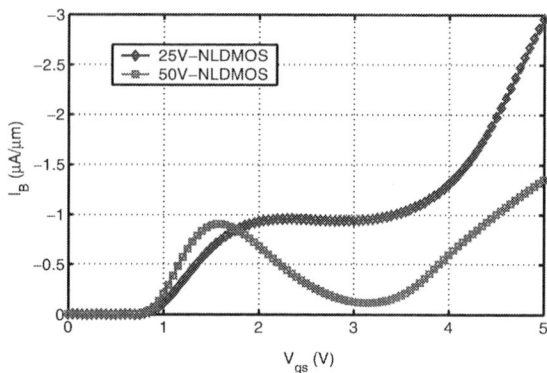

Fig. 4. Comparison of body current I_B between two scalable NLDMOS. V_{ds}=25V for 25V NLDMOS and V_{ds}=50V for 50V NLDMOS. 50V design shows I_B reduction due to the Kirk effect and favors improved HCI.

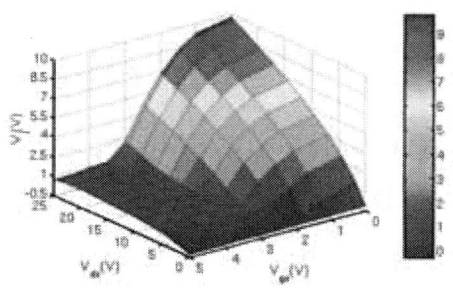

Fig. 5. TCAD simulated 25V NLDMOS internal V_J as function of V_{gs} and V_{ds}. V_{gs} sweeps from 0 to 5V and V_{ds} sweeps from 0 to 25V. $V_{J,max}$=9.8V at V_{gs}=0V and V_{ds}=25V. At higher V_{gs}, no V_{ds} dependence is observed.

the bias dependence, we define the JFET gate voltage, V_J, as shown in Fig. 1, By running simulations, we are able to measure this internal V_J. The V_{gs} and V_{ds} dependence are shown in Fig. 5 and 6. At low V_{gs}, increasing V_{ds} can pinch off the JFET, shown as an increase in V_J. However, the charge in the linear region cannot support the high electron current when $V_{gs} \geq 3V$. therefore current crowds along STI side-wall, $V_J \approx 0V$ and is eventually no longer a function of V_{ds}. As discussed, because of the Kirk effect associated with 50V drift design, the high E-field shifts towards drain side, effectively lowers the V_{ds} impact. As a result, the 50V NLDMOS shows weaker V_{ds} dependence, lower V_J, and is prone to quasi-saturation.

B. Quasi-Saturation & ac Characteristics

The transconductance (g_m) and cut-off frequency (f_T) are compared between two designs. The open-short de-embedding method is used to extract GSG pad parasitics. f_T is calculated using "-20dB/decade" slope in ac current gain $|h_{21}|$. Fig. 7 shows the measured g_m curves at V_{ds}=5V and 25V, respectively. At low V_{gs}, two designs show similar g_m curves due

978-1-4244-8425-6/11 $26.00 © 2011 IEEE

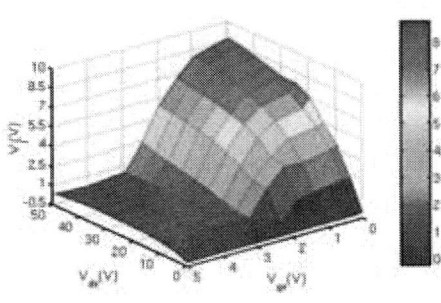

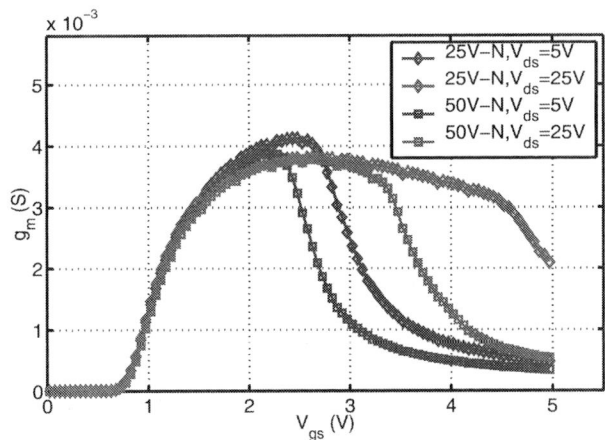

Fig. 6. TCAD simulated 50V NLDMOS internal V_J as function of V_{gs} and V_{ds}. V_{gs} sweeps from 0 to 5V, and V_{ds} sweeps from 0 to 50V. Data show lower $V_{J,max}$ of 8.5V and weaker V_{ds} dependence than 25V design.

Fig. 7. g_m vs V_{gs} curves for two scalable NLDMOS at V_{ds}=5V and 25V. 50V NLDMOS shows g_m roll-off at lower V_{gs}, due to a stronger quasi-saturation effect.

to the common body designs. A g_m roll-off caused by quasi-saturation is observed at high V_{gs}. At V_{ds}=5V, both designs have similar V_{gs} dependence on quasi-saturation as shown in Fig. 8 (a), therefore we observe the similar g_m roll-off. However, at V_{ds}=25V, the 50V design shows stronger quasi-saturation, which can be explained by the Kirk effect. The g_m roll-off in the 50V NLDMOS is 1V earlier than in the 25V NLDMOS, as demonstrated in Fig. 7 and Fig. 8 (b).

In the first order, the f_T of a FET can be expressed by (1), showing a strong dependence on g_m. Therefore, the f_T curves in Fig. 9 show a roll-off consistent to the g_m curves. Increasing V_{ds} delays the onset of high injection impact. The 50V design shows a stronger quasi-saturation effect, with f_T roll-off at lower V_{gs}. Due to scaling of STI, field plate and linear region, the C_{gd} is increased for 50V design. Therefore the 50V NLDMOS has a 1 GHz lower f_T than the 25V NLDMOS.

$$f_T = \frac{g_m}{2\pi \times (C_{gs} + C_{gd})} \qquad (1)$$

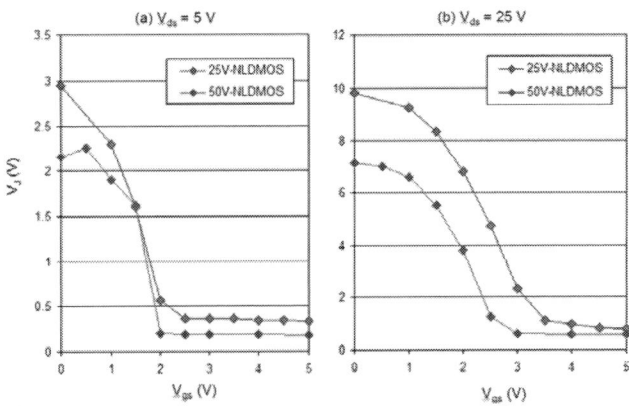

Fig. 8. The comparison of V_J vs. V_{gs} between 25V and 50V NLDMOS at (a) V_{ds}=5V, and (b) V_{ds}=25V. 50V NLDMOS shows stronger quasi-saturation under higher V_{ds}.

C. STI Process Sensitivity and Hot-Carrier Stability

Given the inclusion of the STI in the drift design, a change in its depth impacts the charge in drift region, hence the E-field distribution. Therefore the STI depth becomes a key process factor. In order to investigate STI depth sensitivity, two experiments are implemented with a 10% deeper STI and a 10% shallower STI. Fig. 10 compares R_{on}, BV_{dss}, f_T and f_{max} between control and STI depth splits. With STI varying between -10% to +10%, the impact on R_{on} is less than 5% and the impact on BV_{dss} is less than 3V. There is no impact on f_T and f_{max}, because the C_{gd} is determined by the overlap in the linear region and less sensitive to STI depth.

A 30,000-second HCI stress is conducted under peak I_B condition for each device. The STI depth modulation in I_{dlin} degradation is compared in Fig. 11 and Fig. 12 for two de-

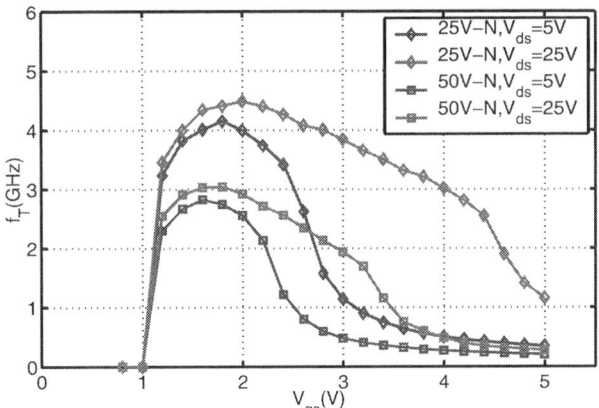

Fig. 9. f_T vs V_{gs} curves for two scalable NLDMOS at V_{ds}=5V and 25V. Scaling causes 1 GHz f_T reduction for 50V NLDMOS.

978-1-4244-8425-6/11 $26.00 © 2011 IEEE   217

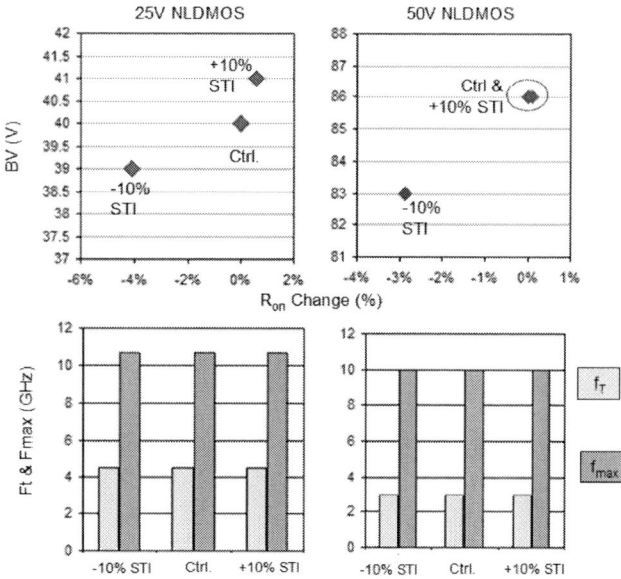

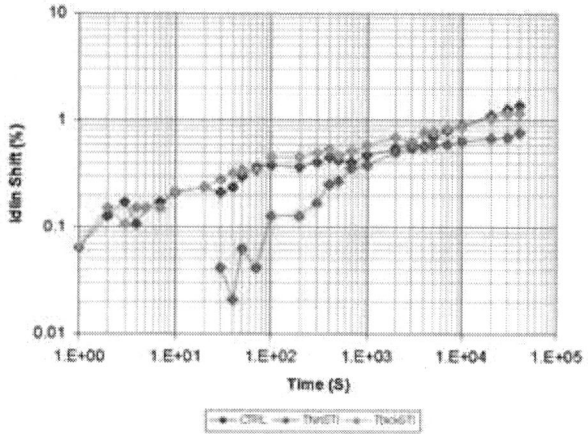

Fig. 12. I_{dlin} degradation for 50V NLDMOS. Shift is well below 1% at 10^4 sec, without STI depth sensitivity.

Fig. 10. STI depth impact on R_{on}, BV_{dss}, f_T and f_{max}. Data show negligible impact on DC performance, and no impact on ac performance.

IV. CONCLUSION

We have discussed DC, ac, and reliability performance for scalable N-LDMOS in a HVCMOS technology. The drift region trade-off in device characteristics is compared between 25V and 50V rated NLDMOS. The STI depth sensitivity in NLDMOS scaling is studied. The fully depleted 50V design shows very robust HCI performance.

ACKNOWLEDGEMENTS

The authors gratefully thank Ted Letavic for the helpful discussions and acknowledge Jui-chu Lee and Jay Rasco for S-parameter measurement and RF data extraction.

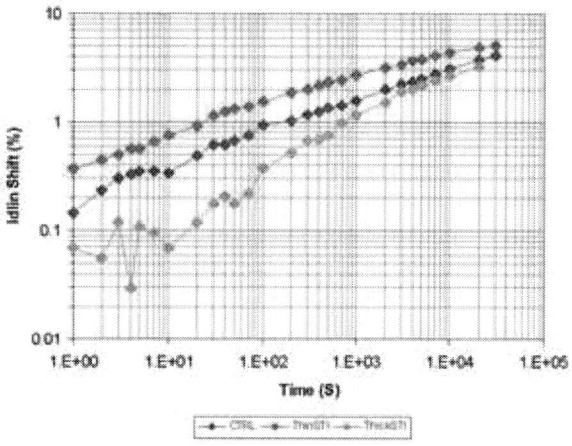

Fig. 11. I_{dlin} degradation for 25V NLDMOS. Ctrl. shows 3% degradation at 10^4 sec. STI depth variations show slight impact on degradation.

ΔI_{dlin}=3% at 10^4 second. The 10% STI depth modulation only slightly impact ΔI_{dlin} at 10^4 second by 1%.

The 50V LDMOS results are very insensitive to the changes in STI depth. As discussed and shown in Fig. 3, the 50V NLDMOS design features a fully depleted drift region under field plate. The current flow is pulled away from STI bottom by E_y, favoring less electron injections into STI interface. As demonstrated in Fig. 12, ΔI_{dlin} is well below 1% at 10^4 second, and no STI depth sensitivity is observed. With voltage scaling to 50V and above, the balance between quasi-saturation, reliability and process stability can be achieved

REFERENCES

[1] R. Minixhofer, N. Feilchenfeld, M. Knaipp, G. Rohrer, J.M. Park, M. Zierak, H. Enichlmair, M. Levy, B. Loeffler, D. Hershberger, F. Unterleitner, M. Gautsch, K. Chatty, Y. Shi, W. Posch, E. Seebacher, M. Schrems, J. Dunn, D. Harame, "A 120V 180nm High Voltage CMOS smart power technology for System-on-Chip integration", *Proceedings of the 2010 ISPSD*, pp. 75-78, 2010.

[2] T. Letavic, S. Sharma, R. Cook, R. Brock, A. Gondal, C .Mandhare, W. van Noort, "A Field-Plated Drift-Length Scalable EDPMOS Device Structure", *Proceedings of the 2009 ISPSD*, pp. 108-111, 2009.

[3] J. Roig, P. Moens, F. Bauwens, D. Medjahed, S. Mouhoubi, P. Gassot, "Accumulation Region Length Impact on 0.18μm CMOS Fully-Compatible Lateral Power MOSFETs with Shallow Trench Isolation *Proceedings of the ISPSD 2009*, pp. 88-91, 2009.

[4] S. Haynie, A.Gabrys, T. Kwon, P. Allard, J. Strout, and A.Strachan, "Power LDMOS with Novel STI Profile for Improved R_{sp}, BV_{dss}, and Reliability" *Proceedings of the ISPSD 2010*, pp. 241-244, 2010,

[5] A. Mai, H. Rcker, R. Sorge, "Impact of the Drift Region Profile on Performance and Reliability of RF-LDMOS Transistors" *Proceedings of the 2009 ISPSD*, pp. 100-103, 2009.

[6] FIELDAY USER Manual. IBM Technology Simulation Group.

[7] A. W. Ludikhuize, "A Review of RESURF Technology," *Proceedings of the 2000 ISPSD*, pp. 11 - 18, 2000.

[8] F. Udrea, State-of-the-Art Technologies and Devices for High-Voltage Integrated Circuits, *IET Circuit, Devices and Systems*, pp. 357 365, Issue 5, Oct. 2007

Proceedings of the 23rd International Symposium on Power Semiconductor Devices & IC's
May 23-26, 2011 San Diego, CA

A Versatile 30V Analog CMOS Process in a 0.18μm Technology for Power Management Application

Yong-Keon Choi, Il-Yong Park, Hyun-Chol Lim, Mi-Young Kim, Chul-Jin Yoon

Nam-Joo Kim, Kwang-Dong Yoo, and Lou N. Hutter

Analog Foundry Business Unit
Dongbu HiTek, Bucheon, Kyeonggi-Do, South Korea
Email:yongkeon.choi@dongbu.com

Abstract—a versatile 30V analog CMOS process in a 0.18 μm technology node has been developed by using cost-effective and modular fashion. To reduce the thermal budget deep NWELL isolation is formed after CMOS well formation. The drain-extended (DE) CMOS from 7V to 30V shows very competitive trade-off performance between the breakdown voltage and the specific on-resistance. In addition, low 1/f noise of 5V CMOS can be obtained by pure gate oxide process.

I. INTRODUCTION

Power management is becoming highly growing market in the semiconductor industry. Especially, switching regulator and power management interface product are expected to be the highest growing area. There are huge numbers of products with relatively small volume in the area. And each product needs to be optimized in terms of process, device, and mask layers as well as circuit design. To save the development time for the process and optimize the mask layer for various kinds of products, they need to use one process platform which can be used for several process options such like 5V CMOS, 1.8V/5V CMOS, 5V/HV, and 1.8V/5V/HV process options. Therefore, the modularity is the key feature of the process

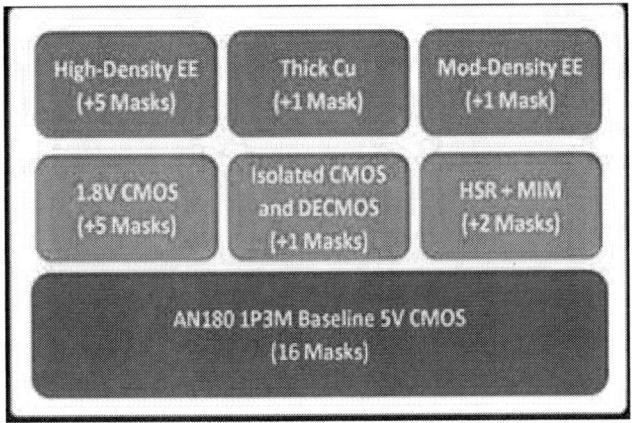

Figure 1. Process modularity for AN180.

development for the purpose. In addition, up-to-date process technology for power management has to have good analog characteristics in view of matching and noise.

In this paper, we present newly developed AN180 process which has designed to satisfy the cost-effectiveness, good analog performance, and high modularity. The modularity and process flow, electrical characteristics for key components, reliability, and single-poly EEPROM are discussed in the following chapter.

II. MODULARITY OF THE PROCESS

The key feature of the AN180 process is the modularity and compatibility. Fig. 1 shows the process modularity for the AN180 technology. The baseline process is 5V CMOS and there are several optional processes to add 1.8V CMOS module, DECMOS and isolation module, HSR (High Sheet-Rho Resistor) and MIM (Metal-Insulator-Metal) capacitor module. A moderate density EEPROM uses one additional layer to make bit cell size with 94 μm^2. Thick copper process for high current application is added with one additional layer. High density EEROM with bit cell size of 2.1 μm^2 is under development. Due to the effort of the modularity, we achieved various kinds of process options, such as 5V CMOS, 1.8V/5V CMOS, 5V/HV, and 1.8V/5V/HV, supported by one SPICE model. This modular fashion of the process options can help designers to choose the cost-effective process for their product. For example, designer can take 1P3M baseline 5V CMOS, DECMOS module, and HSR and MIM module with 19 layers. The technology can provide minimum two levels of metal process.

III. ELECTRICAL CHARACTERISTICS

This chapter describes the electrical characteristics for the key components.

A. 5V CMOS

As discussed in the previous chapter, the baseline process of AN180 technology is 5V CMOS which is widely used for analog circuit in power management, especially battery

978-1-4244-8425-6/11 $26.00 © 2011 IEEE   219

powered applications. Noise is very important parameter in designing analog circuits not only for low frequency linear analog circuits such like bias circuit and audio amplifiers but also wide range of applications such as RF circuits and CMOS image sensors [2]. Even though many design techniques have been developed to overcome flicker noise, lower noise performance of the MOSFET will play a critical role in overall analog circuits. In MOSFET, the gate oxide and its oxide quality with channel affects the 1/f noise. Because AN180 process has dual gate oxide for 1.8V and 5V, even if only 5V CMOS process options must have 1.8V oxide thermal process in the baseline process to keep the compatibility. To improve the 1/f noise of 5V CMOS, AN180 process employed pure gate oxide process for 1.8V oxide while the 0.18 μm BCD process used NO oxide in the previous technology [3]. Fig. 2 shows the 1/f noise comparison data between pure oxide and NO oxide for 5V NMOS and 5V PMOS. The pure oxide process achieved the lower noise performance than NO oxide process.

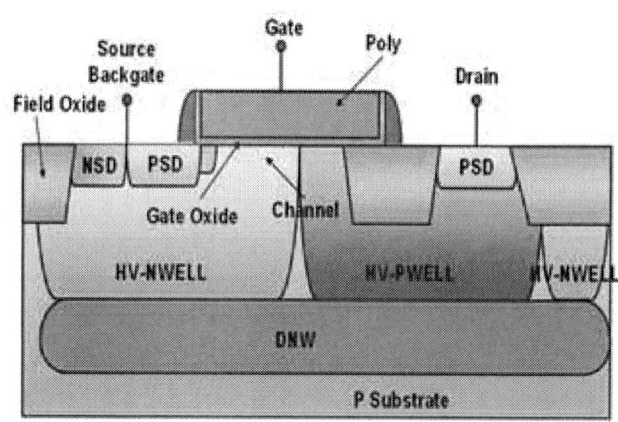

Figure 3. Cross-sectional view of DEPMOS.

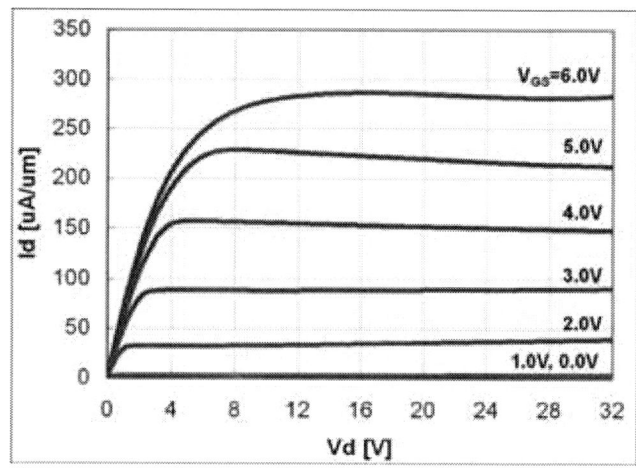

Figure 4. Ids-Vds characteristics for the 30V DENMOS transistor.

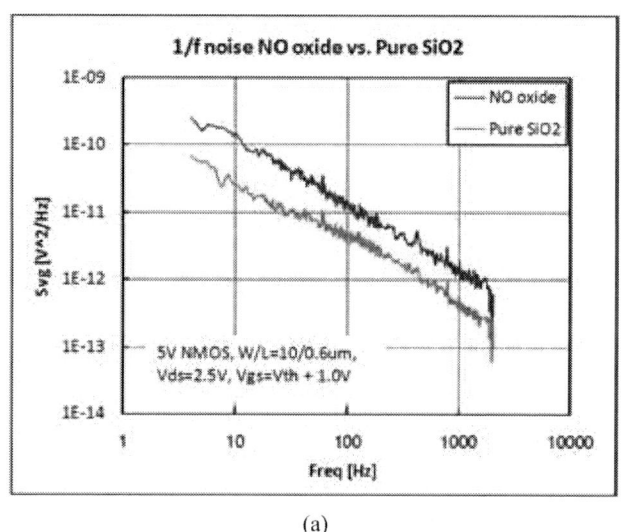

(a)

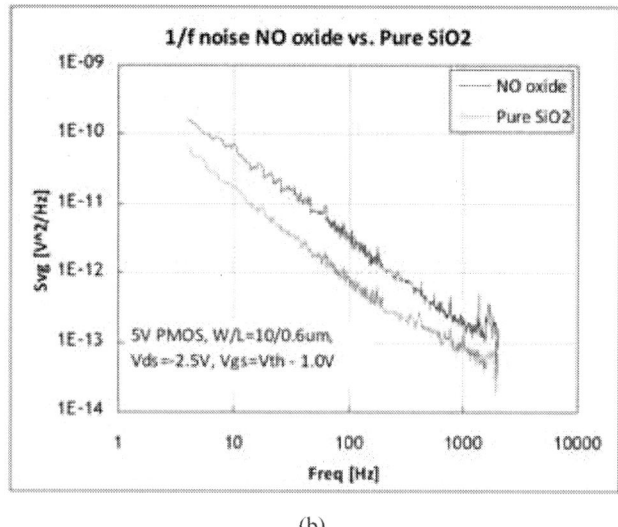

(b)

Figure 2. 1/f noise characteristics between NO oxide and pure oxide for

B. DE(Drain-Extended) CMOS

Based on this 5V CMOS baseline, DECMOS module can be added by using only one additional mask layer which is deep NWELL (DNW) for isolation. DNW formation is carried out by high energy implants of phosphorus just after CMOS well formation to reduce the thermal budget. 5V NWELL and 5V PWELL are used as a body and drift of the DEMOS. The doping concentrations of 5V wells are optimized to satisfy 5V CMOS characteristics and DECMOS as well. Fig. 3 shows the cross sectional view of the DEPMOS which shows all the high-voltage layers. DNW implants with 2 MeV can isolate 5V PWELL from the p-substrate. The same isolation scheme can be used for 1.8V and 5V CMOS isolation. The maximum breakdown voltage from the DNW to p-substrate is 46V which is high enough to float the isolation guard ring up to 30V rated voltage from the substrate. Fig. 4 shows the Ids-Vds characteristics of 30V DENMOS from 0V to 6V of gate voltage with 1V spacing. The threshold voltage is 0.65V and the saturation current is 220 μA/μm. I-V curve shows stable characteristics up to Vds of 32V at all gate voltages as shown in Fig. 4. The trade-off characteristics between the breakdown voltage and the specific on-resistance are shown in Fig. 5. DECMOS transistors in AN180 show very competitive performance compared to 0.13 um BiCMOS technology [1]. For example, 20V DENMOS and DEPMOS have a breakdown voltage of 40V and -30V and the specific on-resistance of 44 and 55 mΩ·mm^2, respectively. The specific on-resistance of the DEPMOS shows fairly linear to the breakdown voltage from 7V to 24V transistors and then 30V transistor shows a bit higher on-resistance from the trend. In case of DENMOS, linear trend only fits from 7V to 20V. 24V and 30V DENMOS has higher on-resistance than the linear trend. This results show that the optimum device in this technology has the breakdown voltage between 35V and 40V. Above that voltage range, the device sacrifices the on-resistance to obtain high breakdown voltage. The important electrical specifications for DECMOS are summarized in Table 1.

TABLE I. KEY ELECTRICAL SPECIFICATIONS FOR DECMOS.

Device	Parameters			
	BV [V]		Rsp [mΩ·mm2]	
	N-ch	P-ch	N-ch	P-ch
7V DEMOS	15	-13	8.8	20
12V DEMOS	18	-17	12	28
20V DEMOS	40	-30	44	55
24V DEMOS	43	-35	56	68
30V DEMOS	45	-40	70	90

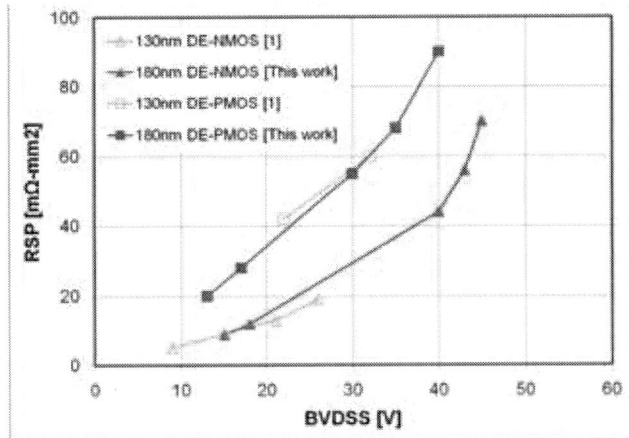

Figure 5. Trade-off characteristics between the breakdown voltage and the specific on-resistance of DECMOS transistors.

For the long-term reliability of DENMOS, HE (Hot Electron) SOA (Safe Operating Area) is used to provide a reference data showing maximum allowable voltage which the degradation of worst case parameter is less than 10% up to 10 years. To get the HE-SOA we measured HCI (hot-carrier injection) degradation at several gate voltages for each component. One example of HCI degradation at a certain gate voltage is shown in Fig. 6. During the HCI stress test, we measured threshold voltage (V_T), transconductance (g_m), saturation current (Id,sat), and linear current (Id,lin) at some

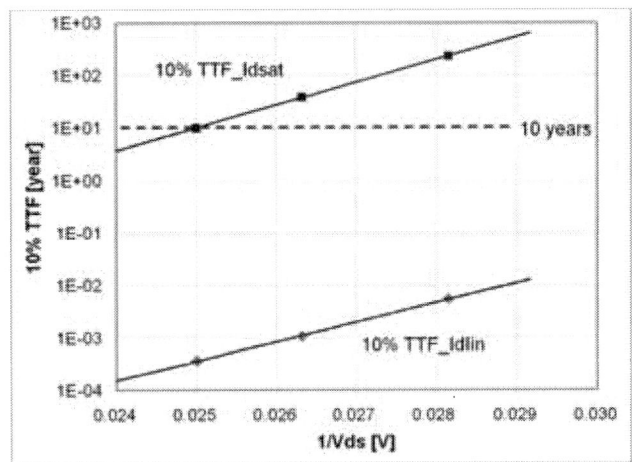

Figure 6. Lifetime estimation for 24V DENMOS within 10% drift of the worst case parameter (Id,lin).

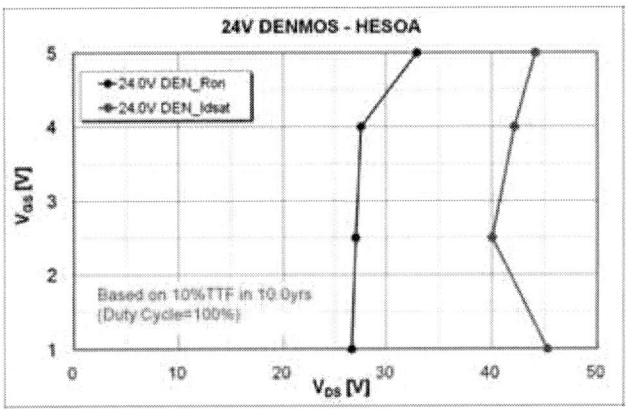

Figure 7. HE-SOA for 24V DENMOS based on 10% TTF within 10 years. The device was stressed at Vgs=1V, 2.5V, 4.0V, and 5.0V to figure out the safe operating area.

HCI degradation. For the 10 years lifetime at DC, the 24V DENMOS is able to be used up to Vds of 27V. Fig. 7 shows the HE-SOA graph for 24V DENMOS based on 10% TTF of worst-case parameter satisfying 10 years life time. The device was stressed at Vgs=1V, 2.5V, 4.0V and 5.0V to obtain the safe operating area. Finally, the designer can use 24V DENMOS at all gate and drain voltages within the rated voltage range up to 10 years within less than 10 % shift of worst case parameter.

C. Single-Poly EEPROM

Fig. 7 shows the endurance test results for the single-poly EEPROM with one additional mask for tunneling oxide of 80 Å. Because AN180 has dual gate oxides, which are 30Å for 1.8V and 125Å for 5V, conventional single poly EEPROM had to use 125 Å for the tunneling oxide. The operating voltage and time of this device is 17V and 10 msec, respectively. To drive this bit cell, we have to use field transistor to sustain 17V and this makes the bit cell size very big due to the size of field transistor. To reduce the IP size, we

978-1-4244-8425-6/11 $26.00 © 2011 IEEE

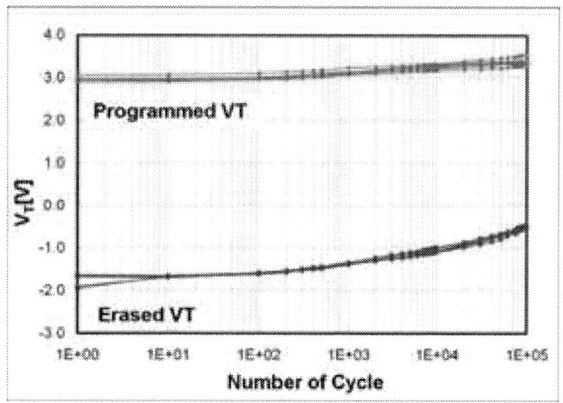

Figure 8. Endurance test results for the single-poly EEPROM with bit cell size of 94 μm² using one additional mask layer.

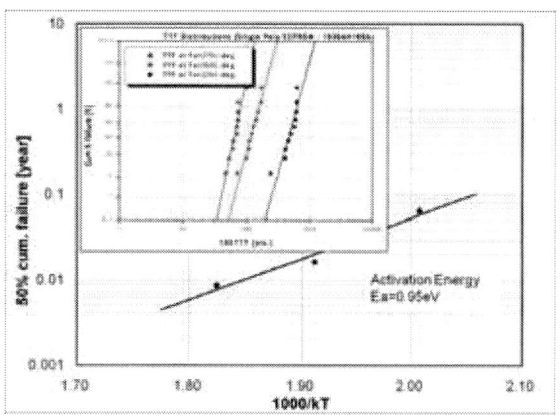

Figure 9. Data retention test results for the single-poly EEPROM.

used one additional mask and thermal process to obtain 80Å tunneling oxide of single-poly EEPROM. Due to the use of thinner oxide and the isolation scheme, we could use smaller size transistor for the driving circuitry and thus reduce the IP size. The operating voltage and time are +/- 5.5V and 10 msec for program and +/- 5.5V and 100 msec for erase, respectively. Fig. 8 shows the endurance test results for the single-poly EEPROM with 80 Å tunneling oxide. The specification for program and erase Vt are 3.0V and -1.6V, respectively. After 100K cycles the gap between the Vtp and Vte maintains higher than 3.5V as shown in Fig. 8. In addition, data retention test was carried out by using fresh sample as shown in Fig. 9. The inset in Fig. 9 shows the cumulative % failure at Ta=225 °C, 250 °C, and 275 °C. The activation energy for the EEPROM is 0.95 eV which shows the bit cell normally operates at high temperature acceleration. The estimated lifetime based on 10% TTF (time to failure) at 125 °C is more than 15 years from the Fig. 9.

IV. SUMMARY

A versatile 30V analog CMOS technology in a 0.18 um technology node has been developed by using cost-effective and modular fashion i.e., 5V CMOS, 1.8V/5V CMOS, 5V/HV, and 1.8V/5V/HV process options. To reduce the thermal budget, deep NWELL for isolation is formed after CMOS well formation. And all the high voltage devices use existing CMOS wells for their body and drift region to achieve low number of mask layers. The drain-extended (DE) CMOS from 7V to 30V shows very competitive trade-off performance between the breakdown voltage and the specific on-resistance. Including bipolar, diodes, MIM capacitor, high sheet resistor, and single-poly EEPROM, the process provides best and cost-effective solution for applications with high-density logic, good analog, and reasonable power capability. The new process uses only one additional mask layer, which is DNW, on the 1.8V/5V logic process while offering a good DECMOS transistors and various kinds of process options.

ACKNOWLEDGMENT

This work was supported by the R&D innovation program of the Korea Ministry of the Knowledge. The authors would like to thank J.O. Lee and J.H. Kim for a great help to get the useful data for this work.

REFERENCES

[1] R. Pan et al., "High voltage (up to 20V) devices implementation in 0.13 um BiCMOS process technology for System-On-Chip (SOC) design," Proc. of ISPSD, pp. 349-352, 2006.

[2] H Tian and A. E. Gamal, "Analysis of 1/f noise in switched MOSFET circuits," IEEE Trans. On Circuits and Systems, Vo. 48, No. 2, pp. 151-157, Feb. 2001

[3] I.Y. Park et al., "BD180 – a new 0.18 um BCD (Bipolar-CMOS-DMOS) technology from 7V to 60V," Proc. of ISPSD, pp. 64-67, 2008.

Proceedings of the 23rd International Symposium on Power Semiconductor Devices & IC's
May 23-26, 2011 San Diego, CA

1kV AlGaN/GaN Power SBDs with Reduced On Resistances

Kiyeol Park, Younghwan Park, Shinwhan Hwang, and Woochul Jeon

Samsung Electro-Mechanics
Suwon, South Korea
E-mail: woochul.jeon@samsung.com

Junghee Lee

School of IT Engineering
Kyungpook National University
Daegu, South Korea

Abstract—**Lateral Schottky Barrier Diodes (SBDs) with reduced on resistances (Ron) and breakdown voltages (Vbd) of higher than 1kV were fabricated on AlGaN/GaN/Si HEMT epi wafers. To improve the forward characteristics of the SBDs, an additional ohmic metal deposition process (2nd ohmic process) has been inserted between the first ohmic metallization process and a Schottky metal deposition process. To minimize the increase of the reverse leakage current, various 2nd ohmic metal patterns under the Schottky metal or a patterned Schottky electrode structure have been used to figure out optimum SBD designs. The proposed SBDs have achieved 25~75% lower on resistances at the operating voltages of 1.5~1.8V, and the leakage current increase due to the additional 2nd ohmic structures reduced to 1~3 orders of magnitude increase from 5 to 6 order of magnitude increase.**

I. INTRODUCTION

Gallium Nitride (GaN) power devices are drawing more attentions in high power switching applications such as switch mode power supplies (SMPSs) and inverters due to their superior power densities, efficiencies, and switching speeds to current silicon power devices. As current silicon power devices are almost touching their material limits, to further improve the efficiency and the power density of a power conversion system, power switches with new materials such as wideband gap semiconductors are required. The switching speed of a GaN power device is high enough to reduce the size and the complexity of a power conversion system, and its high power density and low leakage current enhances the power conversion efficiency at the same time [1]. Because of these promising properties, GaN power devices would be one of the best candidates as power switches in the future power conversion systems.

A typical SMPS includes a power factor correction circuit, which consists of at least one FET and one diode. To improve the efficiency of the system, a lot of power conversion topologies have been suggested. However, due to the limited switching properties of the silicon power devices in the system, the efficiency improvements by using various power conversion topologies are limited. To further improve the efficiency and to reduce the size of the current systems, the silicon power switches in the system need be replaced by normally-off AlGaN/GaN FETs and (AlGaN/)GaN SBDs, if silicon power devices do not break their material limits. A lot of research groups have been developing high power AlGaN/GaN FETs. However, only a few groups have been developing high power AlGaN/GaN SBDs [2][3], as controlling the turn-on voltage and the reverse leakage current is not a trivial process. To increasing the current density of a SBD, vertical GaN SBDs fabricated on a bulk GaN wafer or a GaN epitaxial wafer have been suggested. [4][5] However, due to the high fabrication costs of GaN bulk substrates and the high vertical defect densities of GaN epitaxial wafers, vertical GaN SBDs may not be useful in a typical power conversion system, and lateral AlGaN/GaN SBDs are more suitable to be used in commercial applications.

An AlGaN/GaN SBD normally has high Schottky barrier height in the contact between a Schottky metal and an AlGaN layer. Higher Schottky barrier height would be superior property in reducing the reverse leakage current and increasing the breakdown voltage. However, due to the high Schottky barrier height, the turn-on voltage becomes above 1.5V, which is not suitable in a typical power conversion system. As the required operating voltages in commercial switching systems are 1.5V~1.8V, the turn-on voltage of the AlGaN/GaN SBD should be lower. To lower the turn on voltage of the SBDs, different Schottky metals, such as titanium, can be used. However, as titanium could be considered as an ohmic contact metal, the reverse leakage current significantly increases, which makes the devices useless.

In this paper, to reduce the turn-on voltage and to prevent the leakage current increase at the same time, SBDs with various 2nd ohmic metal patterns deposited under the Schottky metal electrode are presented. (Fig. 1) To block the lateral

978-1-4244-8425-6/11 $26.00 © 2011 IEEE
223

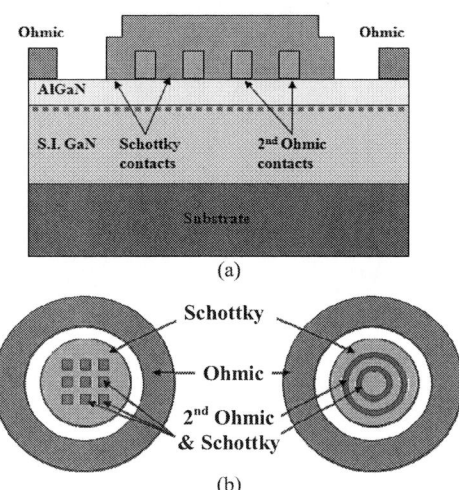

(a)

(b)

Figure 1. Cross section view (a) and top view (b) of ohmic-grid-Schottky barrier diode

reverse leakage current, SBDs with 2nd ohmic metal under the saw-tooth edge patterned Schottky electrodes are also presented. (Fig. 2)

II. DEVICE DESIGN AND FABRICATIONS

A. Device Design I - Ohmic-Grid-Schottky Barrier Diodes

Ohmic-Grid-Schottky Barrier Diodes (OGSBDs) were fabricated on a Sapphire wafer with an AlGaN/GaN HEMT structure, which was grown by metal-organic chemical vapor deposition (MOCVD). To fabricate the devices, at first, the devices were isolated by an ICP-RIE MESA etch process. Ohmic contact electrode, which consists of Ti/Al/Ni/Au, were deposited by an electron beam evaporation process and defined by a liftoff process. Then the wafer has been annealed by a rapid thermal annealing (RTA) process. A titanium evaporation process has been used to define the 2nd ohmic metal structure. Finally, Schottky contact electrode, which consists of Ni/Au, were deposited by an electron beam evaporation process and patterned by a liftoff process. As a reference device, a normal SBD without a 2nd ohmic pattern has been fabricated on the same wafer. The shape of the Schottky electrode of the fabricated ring SBD was a 100μm diameter circle, and the distances between the Schottky contact and the Ohmic contact were 15μm ~ 20μm. The 2nd ohmic metal structures under the Schottky metal were 2x2

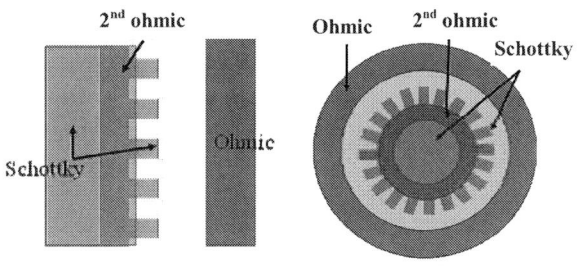

Figure 2. Top view of Edge patterned SBD

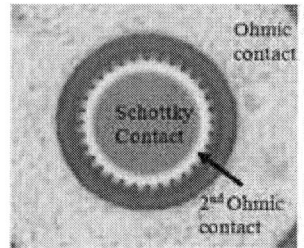

Figure 3. Fabricated Edge patterned SBD

μm², 3x3μm², 4x4μm² square arrays with similar 2nd ohmic contact areas, multiple rings with 2μm line width and 50μm ~ 80μm diameters, or 40~50μm diameter circles.

B. Device Design II - Edge patterned SBDs

Edge patterned SBDs were fabricated on AlGaN/GaN HEMT wafers grown by MOCVD on a 150mm conductive (111) silicon substrates, which were supplied by AZZURRO Semiconductors AG. The devices were fabricated using the same fabrication processes used to fabricate the OGSBDs. The pattern of the 2nd ohmic structure under the Schottky electrode is a donut shaped electrode with 10um line width and 85~95um diameters. The edge of the Schottky electrode has saw tooth shape patterns to suppress the reverse leakage current flow through the 2nd ohmic contacts by expanding the depletion region between two Schottky patterns. Fig.3 shows an actual device fabricated.

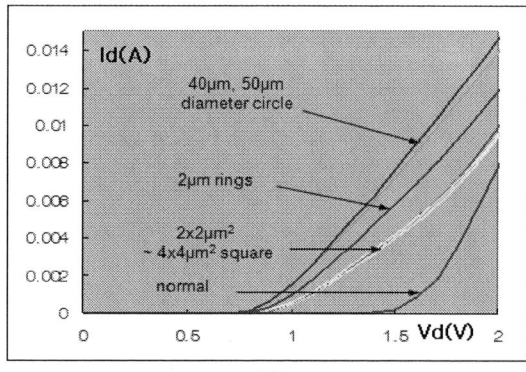

(a)

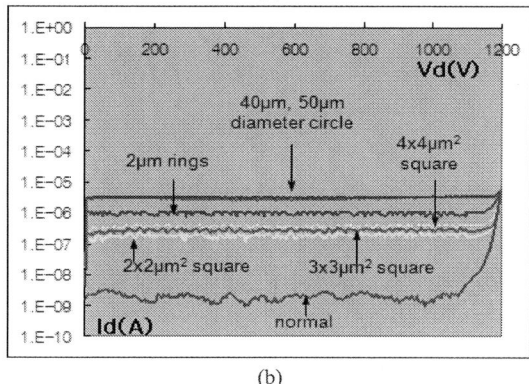

(b)

Figure 4. Measured forward (a) and reverse (b) I-V curves of the fabricated SBDs with 2nd ohmic structures (Fig. 1) and a normal SBD (fabricated with GaN on Sapphire wafer)

III. RESULTS AND DISCUSSION

Fig.4 shows the measured I-V curves of the fabricated SBDs with patterned 2nd ohmic metal structures. The turn-on voltages of the proposed OGSBDs decrease to 0.9V from 1.5V, and, as a result, the on resistances at the applied voltage of 1.8V become 3/4 to 1/2 lower than that of the normal SBD. The reverse leakage currents are 2 to 3 orders of magnitude higher than the normal SBD. The smaller the grid pattern size, the lower reverse leakage current has been achieved. For the comparison, if most of the Schottky contact area is covered by an un-annealed ohmic metal (a leaky SBD), the reverse leakage current is higher than 1mA, which is about 6 orders of magnitude higher than that of the normal SBD.

The Edge patterned SBD shows the turn on voltage shift from 1.5V to 1V, and the on resistance becomes 1/2 of that of the normal SBD. The reverse leakage current increases only an order of magnitude, which is acceptable in high power switching applications. (Fig.5)

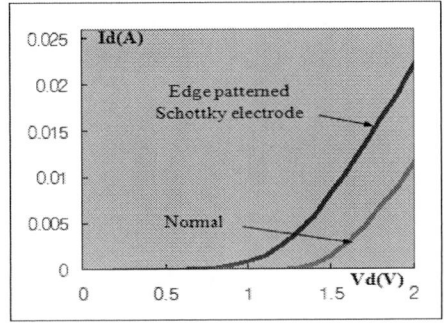

(a)

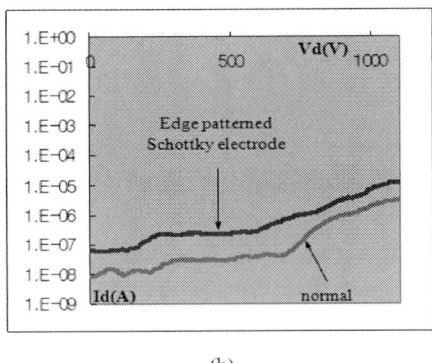

(b)

Figure 5. Measured forward (a) and reverse (b) I-V curves of the fabricated SBDs with a Edge patterned electrode (Fig. 2) and 2nd ohmic structures and a normal SBD (fabricated with GaN on Silicon wafer)

Fig. 6 explains the forward and reverse operations of the Edge patterned SBD. When the applied forward voltage is lower than the turn-on voltage of the Ni/Au Schottky contact, the current flows through the 2nd ohmic contacts only (Fig.6 (a)), and if the applied voltage is higher than the turn-on voltage, the forward current flows through both of the Schottky contact and the 2nd ohmic contact (Fig.6 (b)). When a reverse bias is applied to the diode, the depletion regions under the edge of the Ni/Au Schottky contacts expand both lateral and vertical directions and cover the leakage paths of

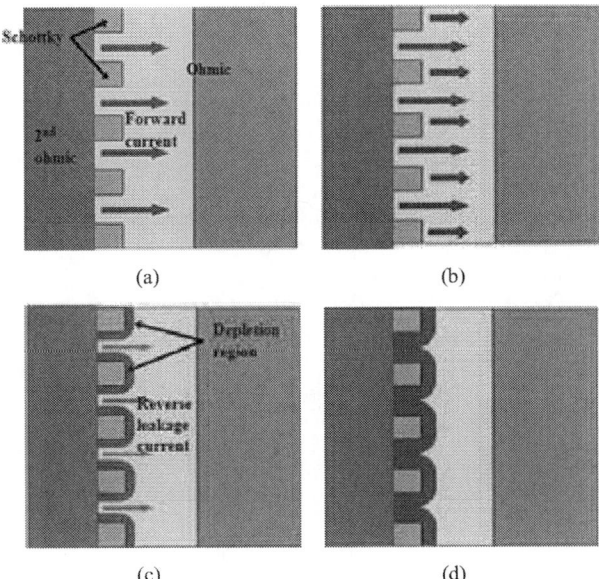

Figure 6. Forward (a),(b) and reverse (c)(d) operations of Edge patterned SBD

the 2nd ohmic contacts, which suppress the leakage current flow through the 2nd ohmic contacts (Fig.6 (c)). Fig. 7 shows the forward and reverse operations of the OGSBDs, which are similar to those of the Edge patterned SBDs. In OGSBDs, the reverse leakage currents through the 2nd ohmic patterns are mainly blocked by vertical expansions of the depletion regions.

If the 2nd ohmic metal contact under the Schottky contact occupies more than a half of the Schottky contact area, the on resistance could be much lower than the proposed SBDs. However, the reverse leakage current increases more than 4 orders of magnitude, which may not be used as power switching devices.

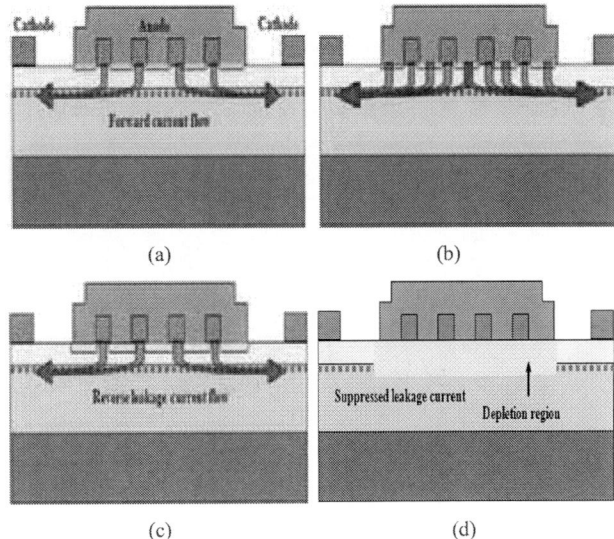

Figure 7. Forward (a),(b) and reverse (c)(d) operations of OGSBD

978-1-4244-8425-6/11 $26.00 © 2011 IEEE   225

IV. CONCLUSION

Two types of SBDs with reduced on resistances have been fabricated on AlGaN/GaN HEMT Epitaxial wafers. The proposed SBDs have achieved 25~75% lower on resistances at the operating voltages of 1.5~1.8V, and the leakage current increase due to the additional 2nd ohmic structures reduced to 1~3 orders of magnitude increase from 5 to 6 orders of magnitude increase. The breakdown voltages of both devices were higher than 1kV. The Edge patterned SBDs showed better forward and reverse I-V characteristics than those of OGSBDs. To offer an additional control over the trade-off between a low on resistance and a low reverse leakage current, an optimal 2nd ohmic pattern sizes and locations need to be decided in the design of OGSBDs. To further improve the I-V properties of the Edge patterned SBDs, the size and the shape of the Schottky edge pattern and the location and the contact area of the 2nd ohmic ring need to be optimized.

ACKNOWLEDGMENT

The authors would like to thank Sangki Yoon and the technical staffs at Samsung Electro-Mechanics for the support on device fabrications. The authors further acknowledge Alexander Loesing and AZZURRO Semiconductors PG for the technical support on GaN on Si epi wafers.

REFERENCES

[1] S. Iwakami, O. Machia, I. Yoshimichi, B. Ryohei, Y. Masataka, E. Toshihiro, K. Nobuo, G. Hirokazu, and I. Akio, "Evaluation of AlGaN/GaN Heterostructure Field-Effect Transistors on Si Substrate in Power Factor Correction Circuit," Japanese Journal of Applied Physics, Vol. 46, No. 29, pp. L721-L723, 2007.

[2] H. Ishida, D. Shibata, M. Yanagihara, Y. Uemoto, H. Matsuo, T. Ueda, T. Tanaka, D.Ueda, "Unlimited High Breakdown Voltage by Natural Super Junction of Polarized Semiconductor," IEEE Electron Device Letters, vol. 29, no. 10, pp. 1087-1089, 2008.

[3] W. Chen, K. Wong, W. Huang, and K. J. Chen, "High-performance AlGaN/GaN lateral field-effect rectifiers compatible with high electron mobility transistors," Applied Physics Letters, Vol. 92, pp. 253501_1 - 3, 2008.

[4] T. Horii, T. Miyazaki, Y. Saito, S. Hashimoto, T. Tanabe, M. Kiyama, "High-Breakdown-Voltage GaN Vertical Schottky Barrier Diodes with Field Plate," Materials Science Forum, vol. 615-617, pp. 963–966, 2009.

[5] A. Kamada, K. Matsubayashi, A. Nakagawa, Y. Terada, and T. Egawa, "High-Voltage AlGaN/GaN Schottky Barrier Diodes on Si Substrate with Low-Temperature GaN Cap Layer for Edge Termination," Proc. ISPSD'08, pp. 225-228, 2008

Proceedings of the 23rd International Symposium on Power Semiconductor Devices & IC's
May 23-26, 2011 San Diego, CA

3.7 mΩ-cm^2, 1500 V 4H-SiC DMOSFETs for Advanced High Power, High Frequency Applications

Sei-Hyung Ryu, Lin Cheng, Sarit Dhar, Craig Capell,
Charlotte Jonas, Robert Callanan, Anant Agarwal, and
John Palmour

Cree Inc.
4600 Silicon Drive, Durham, NC, USA
sei-hyung_ryu@cree.com

Aivars Lelis, Charles Scozzie, and Bruce Geil

U.S. Army Research Laboratory
2800 Powder Mill Rd, Adelphi, MD, USA

Abstract— **We present our most recent developments in 4H-SiC DMOSFETs. A 4H-SiC DMOSFET with an active area of 0.1 cm^2 showed a specific on-resistance of 3.7 mΩ-cm^2 with a gate bias of 20 V, and an avalanche voltage of 1500 V with gate shorted to source at 25°C. A threshold voltage of 3.5 V was extracted from the DMOSFET, and a subthreshold swing of 200 mV/dec was measured. The device was successfully scaled to an active area of 0.5 cm^2, and the resulting device showed a drain current of 377 A at a forward voltage drop of 3.8 V at 25°C.**

I. INTRODUCTION

Recent developments in wide bandgap materials and device technologies have resulted in superior power switching devices that can revolutionize the power electronics industry. It has been proven that SiC JBS diodes and Power MOSFETs offer substantial advantages over commercially available silicon devices [1], which can easily justify the costs of SiC devices. Very high reliability (> 100 billion device-hours) was demonstrated in the field for the SiC JBS diodes. We are confident that SiC power device technology will provide components with excellent reliability and performance, which will replace silicon devices in many applications. Recently, the first generation of 1200 V rated SiC MOSFETs have been released to the market [2]. Further improvement in the performance is desired, which can result in a reduction in manufacturing costs and make these devices even more attractive for a wider range of applications. In this paper, we present our most recent developments in 4H-SiC DMOSFETs – 3.7 mΩ-cm^2 4H-SiC DMOSFETs capable of blocking 1500 V, which provide sufficient voltage margin for 1200 V rated power switches.

II. DEVICE STRUCTURE AND FABRICATION

Fig. 1 shows a simplified cross-section of the 4H-SiC DMOSFET. The MOS channel length, defined by the p-well and n$^+$ source regions, is approximately 0.5 μm. When a positive gate bias, in excess of the threshold voltage, is

applied, the MOS channel turns on and electrons flow laterally from the n$^+$ source, through the MOS channel on the implanted p-well. They then flow through the JFET region formed by two adjacent p-well regions and, finally, through the lightly doped n$^-$ drift region into the backside drain. The MOS channel disappears when the gate electrode is shorted to the source or when a negative gate bias is applied. With the MOS channel off, the 4H-SiC DMOSFET behaves as a 4H-SiC PiN diode with implanted anodes, which is referred to as the body diode. The device supports the high voltage by reverse biased pn junction formed by the implanted p-wells and the thick n$^-$ drift layer.

The devices were fabricated on a 10 μm thick 4H-SiC n-type drift layer, with a doping concentration of 6 x 10^{15} cm^{-3}. A self-aligned implantation technique [3] was used to achieve a channel length of 0.5 μm. After a high-temperature implant activation, a gate oxide layer was formed, followed by NO post oxidation anneal. A degenerately doped polysilicon layer was then deposited and patterned to form gate electrodes. Then, an Inter-Metal-Dielectric (IMD) layer was deposited, and Ni based ohmic contacts were formed on the source and drain regions. Finally, a thick layer of aluminum was deposited and patterned to form gate and source wire bonding pads.

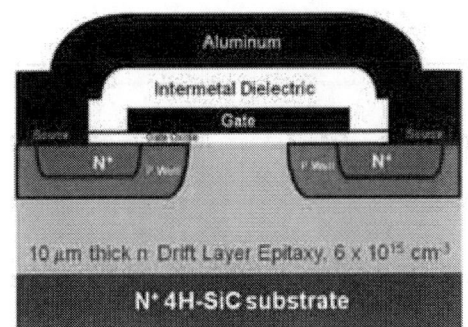

Figure 1. A simplified cross-section of the 4H-SiC DMOSFET.

This work was supported by internal research funds from Cree, Inc., and Army Research Laboratory/Honeywell Subcontract no.4202621207.

978-1-4244-8425-6/11 $26.00 © 2011 IEEE

III. EXPERIMENTAL RESULTS

A. Static characteristics

Fig. 2 shows representative on-state IV characteristics of the 1500 V SiC DMOSFETs, measured at room temperature using a Tektronix 371 curve tracer. The device area, including the dicing streets and pad areas, was approximately 0.16 cm^2, and the active area was approximately 0.1 cm^2. The specific on-resistance ($R_{on,sp}$), with respect to the active area, was 3.7 mΩ-cm^2 at room temperature with a V_{GS} of 20 V, which increased to 4.3 mΩ-cm^2 when V_{GS} was reduced to 15 V. The forward voltage drop (V_F) at an I_D of 30 A, which corresponds to a current density of 300 A/cm^2, was 1.1 V with a V_{GS} of 20 V, and 1.3 V with a V_{GS} of 15 V, respectively. In the off-state, the device showed an avalanche voltage of 1500 V, which is sufficient for 1200 V class devices, at room temperature with a V_{GS} of 0 V, as shown in Fig. 3. The theoretical parallel plate E-field for the given drift layer doping concentration and thickness at this voltage is 2.05 MV/cm, which is slightly lower than the previously reported values [4]. The blocking characteristics did not change with application of negative gate biases, indicating the leakage current was mostly originating from the junction area, instead of the MOS channel. This suggests further optimization of edge termination structure is needed for maximum blocking performance.

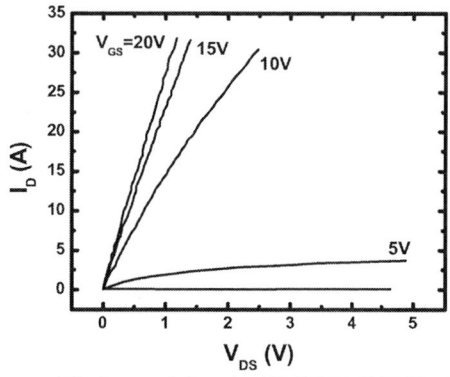

Figure 2. On-state IV characteristics of a 4H-SiC DMOSFET with an active area of 0.1 cm^2 at room temperature.

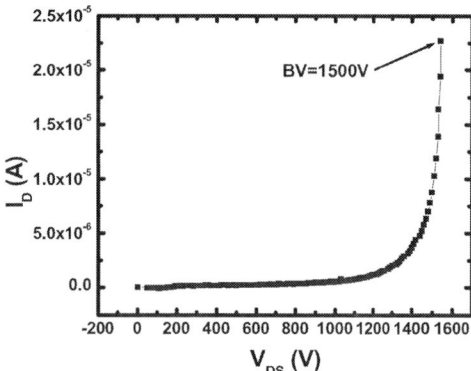

Figure 3. Blocking characteristics of a 4H-SiC DMOSFET with an active area of 0.1 cm^2 at room temperature. A V_{GS} of 0 V was used.

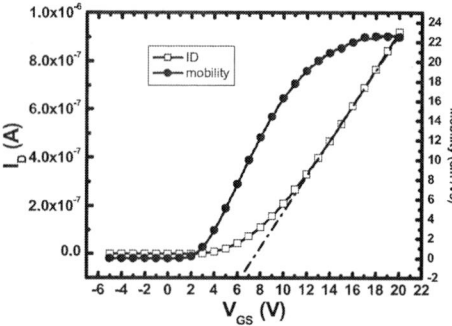

Figure 4. I_D-V_{GS} characteristics measured from a 200 μm/200 μm test MOSFET (V_{DS}=50 mV).

Fig. 4 shows the I_D-V_{GS} characteristics measured from a test lateral MOSFET with a (W/L) of 200 μm / 200 μm, fabricated on the same wafer as the SiC DMOSFETs. V_{DS} was set at 50 mV to keep the device in the linear region. A peak MOS channel mobility of 23 cm^2/Vs was observed at a V_{GS} of 18 V, and the threshold voltage extracted from the linear portion of the I_D-V_{GS} curve was approximately 7 V.

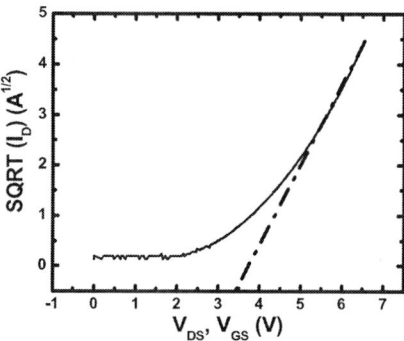

Figure 5. Threshold voltage measurement from the 0.1 cm^2 4H-SiC DMOSFET.

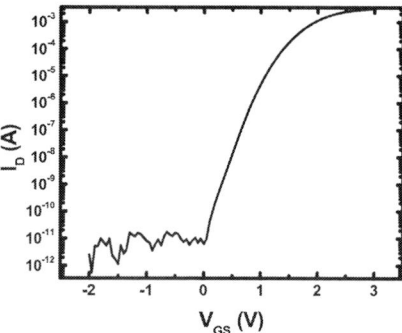

Figure 6. Subthreshold I-V characteristics measured from the 0.1 cm^2 4H-SiC DMOSFET. A V_{DS} of 50 mV was used.

Threshold voltage measurements were also performed on the power DMOSFETs, as shown in Fig. 5. In this measurement, V_{GS} was swept from 0 V to 6.5 V, while the gate electrode was shorted to the drain electrode ($V_{GS} = V_{DS}$). The threshold voltage was extracted from the linear extrapolation of the SQRT (I_D) vs. V_{GS} characteristics. The SiC DMOSFET threshold voltage was approximately 3.5 V, which is significantly lower than that of the test MOSFET. The difference can be attributed to the short channel structure of the SiC DMOSFET. Fig. 6 shows the subthreshold I-V characteristics of the SiC DMOSFET, with a V_{DS} of 50 mV. I_D was reduced to 1 µA at a V_{GS} of 0.9 V. This value is fairly close to zero. A positive shift of the I-V characteristics is strongly desired to ensure normally-off behavior at all operating temperatures. A subthreshold swing of 200 mV/dec was observed at this bias point.

The SiC DMOSFETs were scaled up to 7 mm x 8 mm, which is the same size reported in [4]. The active area of the resulting device was approximately 0.5 cm². An on-wafer characterization showed that the blocking capability was the same as the 0.1 cm² device, as shown in Fig. 7. However, a meaningful on-state characterization could not be done on wafer because the chuck was adding parasitic resistances, which is significant as compared to the on-resistance of the SiC DMOSFET. Hence, the on-state characterization had to be performed on a packaged device. A multi-lead package, as shown in Fig. 8 was used. The package enabled chip-level Kelvin measurements, which eliminate the effects of wire bonds, that can be significant for the current levels used for device characterization.

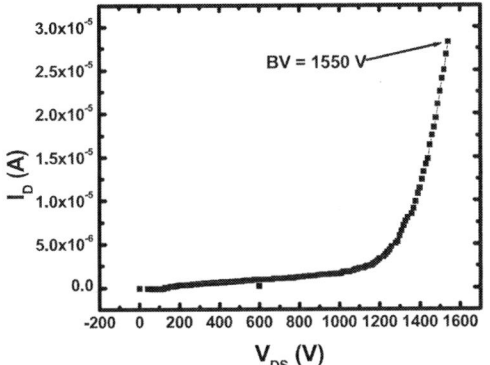

Figure 7. Blocking characteristics of a 7 mm x 8 mm 4H-SiC DMOSFET (active area = 0.5 cm²) at room temperature. A V_{GS} of 0 V was used.

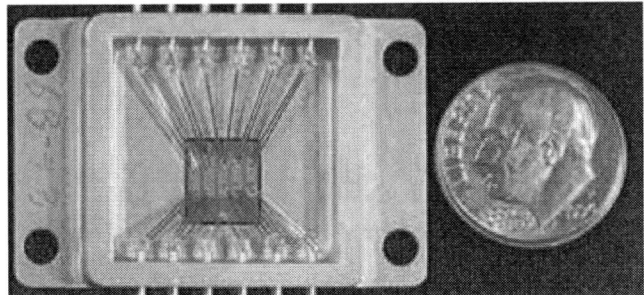

Figure 8. Packaged 7 mm x 8 mm (0.5 cm² active area) SiC DMOSFET used for on-state characterization.

Fig. 9 shows the on-state characteristics of the 7 mm x 8 mm (0.5 cm² active area) SiC DMOSFET at room temperature, measured using a Tektronix 371 curve tracer. The specific on-resistance, measured at a V_{DS} of 0.5 V, was

3.6 mΩ-cm² with a V_{GS} of 20 V (I_D = 56 A), and 4.2 mΩ-cm² with a V_{GS} of 15 V (I_D = 47.5 A). These values are very close to those measured from the 0.1 cm² devices, indicating that the vertical DMOSFET structure in 4H-SiC can be easily scaled to any current level desired. At room temperature, a drain current of 377 A (= 750 A/cm²) was measured at a drain voltage of 3.8 V, with a V_{GS} of 20 V. The power dissipation at this bias point, normalized to the chip size, is approximately 2.6 kW/cm². This value exceeds the dissipation capabilities of most available packaging methods. Developments of high performance packaging techniques, such as presented in [5], are absolutely necessary to fully utilize the potential of these devices.

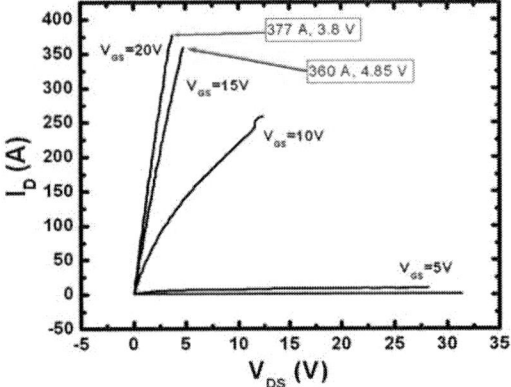

Figure 9. On-state IV characteristics of a 4H-SiC DMOSFET with an active area of 0.5 cm², at room temperature.

B. Switching Characteristics

The 7 mm x 8 mm SiC DMOSFET was subjected to a room temperature inductive load double pulse switching test. A rectifier built with two 50 A, 1200 V SiC JBS diodes [6] in parallel configuration, was used as the freewheeling diode for this test. A supply voltage of 600 V, and a 4.7 Ω gate resistor were used. A V_{GS} of 20 V was used to turn on the transistor, and a V_{GS} of -2 V was used to turn-off the device. Figs. 10 and 11 show the turn-on and turn-off transients, respectively.

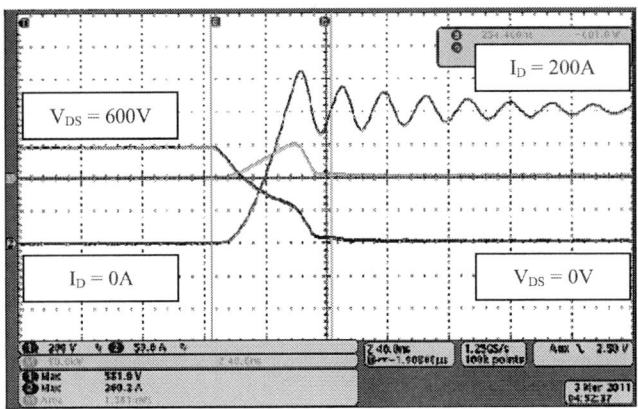

Figure 10. Turn-on transients for the 7 mm x 8 mm 4H-SiC DMOSFET in an inductive load switching configuration. Time scale: 40 ns/div.

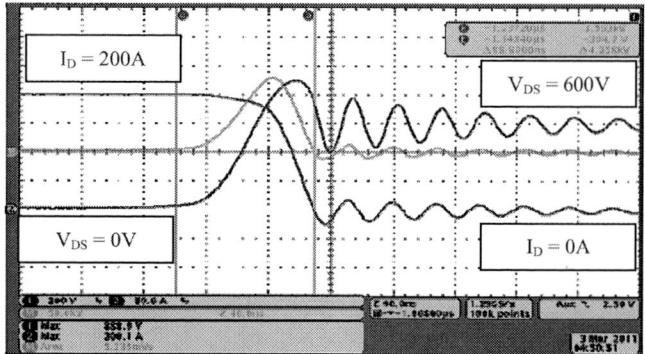

Figure 11. Turn-off transients for the 7 mm x 8 mm 4H-SiC DMOSFET in an inductive load switching configuration. Time scale: 40 ns/div.

A voltage fall time (90% to 10%) of 52 ns was observed from the turn-on transients, and a voltage rise time (10% to 90%) of 40 ns was observed from the turn-on transients, respectively. A turn-on energy of 1.58 mJ, and a turn-off energy of 5.24 mJ were extracted from the waveforms. This result translates into a switching loss of 136 W at a switching frequency of 20 kHz, which is approximately 0.11 % of the power (600 V x 200 A = 120 kW) the device is processing. The results suggest that the device is capable of much higher switching frequencies, provided that the ringing caused by parasitic inductances can be minimized.

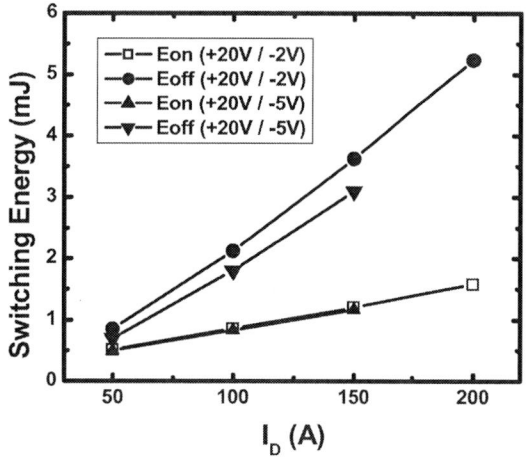

Figure 12. Energy losses during switching events for the 7 mm x 8 mm SiC DMOSFET, extracted from the double pulse measurement waveforms. (+20V/-2V) means V_{GS}=20 V to turn on, V_{GS}=-2V to turn off. (+20V/-5V) means V_{GS}=20 V to turn on, V_{GS}=-5V to turn off.

Switching energy is plotted as a function of drain current in Fig. 12. Switching losses increased linearly with increasing drain current, as predicted in [7]. Fig. 12 also shows that the switching losses are dominated by the turn-off losses. An attempt was made to reduce the turn-off losses by using a V_{GS} of -5 V, instead of -2 V. Significant reduction of turn-off losses was observed. However, the drain current was limited to 150 A due to excessive drain voltage overshoot, caused by ringing due to parasitic inductances. This indicates that

switching frequency and efficiency can be significantly improved by using more negative gate biases for the turn-off, in conjunction with optimization of switching circuit by minimizing the parasitic components.

IV. FUTURE WORK

The MOS interface needs further development. In current devices, the V_{GS} values required to completely turn-off the MOSFETs are closer to zero than would be ideal. To ensure normally-off operation, a significant positive shift of the IV characteristics is desired. Further improvements in MOS channel mobilities in power MOSFET structures are also desired. Although very low specific on-resistances values have been achieved, those devices require relatively high values of V_{GS} or oxide field (E_{ox}). For more affordable devices, a higher channel mobility is desired to achieve even lower values of specific on-resistance.

It has become obvious that one of the major performance limiting factors for the high performance SiC devices is the packaging technology. Conventional packaging technologies developed for silicon devices do not provide adequate performances required by advanced SiC devices. Significant resources should be focused on high performance packaging technology in the near future.

V. SUMMARY

1500 V 4H-SiC DMOSFETs are presented. A 4H-SiC DMOSFET with an active area of 0.1 cm^2 showed a specific on-resistance of 3.7 mΩ-cm^2 with a gate bias of 20 V, and an avalanche voltage of 1500 V. A threshold voltage of 3.5 V was extracted from the DMOSFET, and a subthreshold swing of 200 mV/dec was measured. The device was successfully scaled to 7 mm x 8 mm chip size (active area = 0.5 cm^2), and the resulting device showed a drain current of 377 A at a forward voltage drop of 3.8 V. In an inductive load switching configuration, the 7 mm x 8 mm SiC DMOSFET showed a switching loss of 6.82 mJ per cycle when controlling 120 kW of power (600 V, 200 A). The experimental results show that the SiC DMOSFETs are suitable for high speed, high power applications, which can revolutionize the power electronics industry.

ACKNOWLEDGMENT

Authors are grateful to Cree, Inc., and Army Research Laboratory for generous support of the project.

REFERENCES

[1] R. J. Callanan et al., IEEE Industrial Electronics 34th Annual Conference – IECON 2008, pp. 2885 – 2890, Nov. 2008

[2] http://www.cree.com/press/press_detail.asp?i=1295272745318

[3] S. Ryu, U.S. Patent 7,074,643

[4] B. A. Hull et al., Materials Science Forums Vols. 615-617 (2009), pp. 749-752

[5] M. Holz, J. Hilsenbeck, R. Otremba, A. Heinrich, P. Turkes, and R. Rupp, Materials Science Forums Vols. 615-617 (2009), pp. 613 - 616

[6] http://www.cree.com/products/power_docs2.asp

[7] L. Balogh, http://focus.ti.com/lit/ml/slup169/slup169.pdf

Proceedings of the 23rd International Symposium on Power Semiconductor Devices & IC's
May 23-26, 2011 San Diego, CA

High-Voltage GaN SBD on Si Substrate by Suppressing Metal Spikes

Min-Woo Ha, Cheong Hyun Roh, Hong Goo Choi,
Jun Ho Lee, Hong Joo Song, Ogyun Seok[*],
and Cheol-Koo Hahn

Compound Semiconductor Devices Research Center
Korea Electronics Technology Institute, Seongnam, Korea
[*] Seoul National University, Seoul, Korea
Email: isobar@keti.re.kr

Abstract—**We have successfully fabricated high-voltage GaN Schottky barrier diodes (SBDs) on Si substrate by suppressing metal spikes under ohmic contacts. The breakdown voltage of GaN SBDs is 450 V with superior device-to-deice uniformity. Metal spikes are suppressed by low-temperature annealing at 700 °C. The low contact resistance of 0.6 ohm-mm is also achieved due to ohmic contacts on the doped GaN. The diffusion of Ti/Al/Mo/Au into GaN is analyzed by Auger electron spectroscopy and scanning electron microscope. The depth and the number of metal spikes are proportional to the annealing temperature of ohmic contacts. Metal spikes in GaN power devices should be suppressed for the low power loss and the high breakdown voltage.**

I. INTRODUCTION

GaN devices are promising for high-voltage switching applications due to a wide bandgap, a high critical field, a high electron mobility, a high saturation velocity and a low intrinsic carrier generation [1-2]. Recently, GaN epitaxy on Si substrate has been attracted considerable attentions due to a large diameter to 8-inch. Additionally, Si substrate is not expensive compared with widely used sapphire and SiC substrate [3]. GaN Schottky barrier diodes (SBDs) have been investigated to replace Si diodes due to their high breakdown voltage and low dynamic loss. A floating-metal ring [4] and a low-temperature GaN cap [5] have been reported to suppress the leakage current of AlGaN/GaN SBDs. However, effects of ohmic contacts on reverse characteristics of GaN SBDs on Si substrate have hardly been reported.

The purpose of our work is to report high-voltage GaN SBDs on Si substrate by suppressing troublesome metal spikes. We have fabricated, measured and analyzed high-voltage GaN SBDs using a doped GaN/unintentionally-doped (UID) GaN on Si substrate. Metal spikes are generated during annealing of ohmic contacts and those are evidenced by measuring Auger electron spectroscopy (AES) and scanning electron microscope (SEM) after stripping metals. The suppression of metal spikes achieves the low buffer leakage current and the high breakdown voltage of GaN SBDs.

This research is funded by the Korea Government Ministry of Knowledge Economy under project no. 101030002B.

II. FABRICATION

The doped GaN/UID GaN was grown on 4-inch Si (111) substrate by a metal-organic chemical vapor deposition. The thickness of the doped and UID GaN were 200 nm and 1 μm, respectively. The doped GaN is a thick channel and that is suitable for high current operation. The doping concentration and the electron mobility of the doped GaN were 4.2×10^{17} cm^{-3} and 318.1 cm^2/Vs by Hall measurements.

A cross-sectional view of the proposed GaN SBD is shown in Fig. 1. The 1 μm-deep mesa was formed to define active regions. BCl_3 and Cl_2 were used to etch GaN in inductively coupled plasma etcher. A lift-off method was used to define metal patterns. The 6:1 buffered oxide etchant removes the native oxide before evaporation. Ohmic contacts of Ti/Al/Mo/Au were evaporated by electron beam and annealed under N_2 ambient. The annealing temperature of ohmic contacts was between 700 °C and 900 °C under N_2 ambient. The annealing time of ohmic contacts was fixed to 30 s. Schottky regions were etched for a recessed anode. A depth of the recess was around 200 nm. Plasma damages at recessed regions were cured by annealing at 500 °C under N_2 ambient. Schottky contacts of Ni/Au were directly evaporated on the recessed surface. Schottky contacts were annealed at 500 °C for stable contacts. Conventional devices having a planar anode were also fabricated.

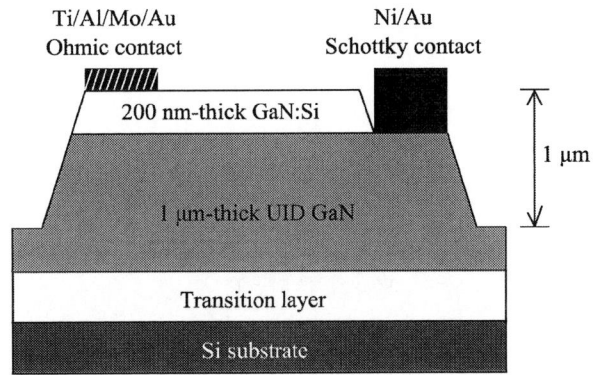

Figure 1.   Cross-sectional view of the proposed GaN SBD

III. EXEXPERIMENTAL RESULTS

The high resistivity of UID GaN buffer is critical for the large depletion at the reverse voltage. A test structure was fabricated to measure the buffer leakage current between two ohmic contacts which were isolated by mesa. Fig. 3 shows the

978-1-4244-8425-6/11 $26.00 © 2011 IEEE

test structure to measure the buffer leakage current. A width of the test structure was 100 μm. A distance between two ohmic contacts was 25 μm. The buffer leakage current was measured at 100 V. When ohmic contacts are annealed at 700 °C and 800 °C, the measured buffer leakage current are 9.7 nA and 782.0 nA, respectively. The high-temperature annealing of ohmic contacts induces the high buffer leakage current.

Fig. 4 shows the measured leakage current of planar GaN SBDs. Ohmic contacts of the devices were annealed at 700 °C and 800 °C, respectively. The leakage current of GaN SBD after annealing ohmic contacts at 800 °C is 13.45 A/cm^2 at -100 V while that of GaN SBD after annealing ohmic contacts at 700 °C is 0.37 A/cm^2 at -100 V. The high-temperature annealing of ohmic contacts finally induces the highly reverse current.

Metal diffusion into GaN buffer may decrease the resistivity of GaN buffer and cause the leakage current [6]. The diffusion of ohmic metals was measured by AES at accelerating voltage of 5 keV. Fig. 5 shows the measured AES-depth profiles of Ti/Al/Mo/Au on the epitaxial GaN before and after annealing. Annealing at 700 °C exhibits the diffusion depth of 324 nm (Au), 324 nm (Ti) and 367 nm (Mo) while annealing at 800 °C shows the diffusion depth of 458 nm (Au), 583 nm (Ti) and 458 nm (Mo). The higher annealing temperature results in the deeper diffusion of ohmic metals into GaN buffer. Especially, Mo does not block the interdiffusion of Au during annealing of ohmic contacts. Deep spikes of Mo and Au, TiN through dislocations [7], the change of mesa-isolated region and oxygen are responsible for the leakage current.

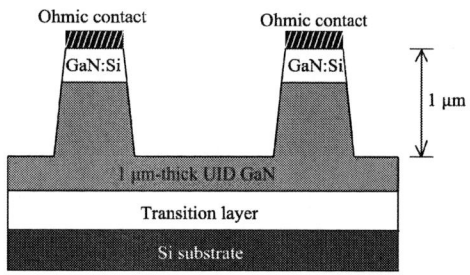

Figure 2. Cross-sectional view of test structure to measure the buffer leakage current

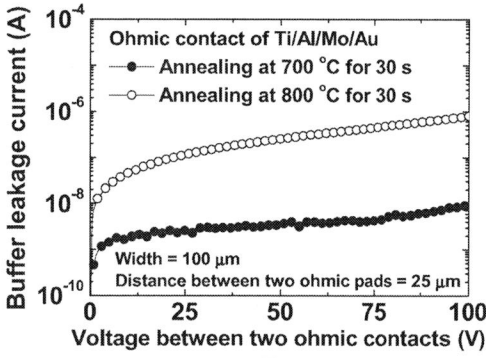

Figure 3. Measured buffer leakage current between two ohmic contacts which are isolated by mesa

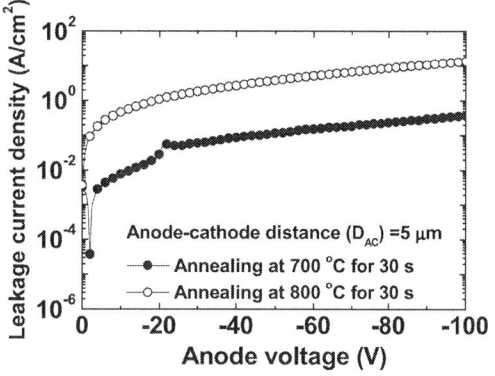

Figure 4. Measured leakage current of planar GaN SBDs after annealing

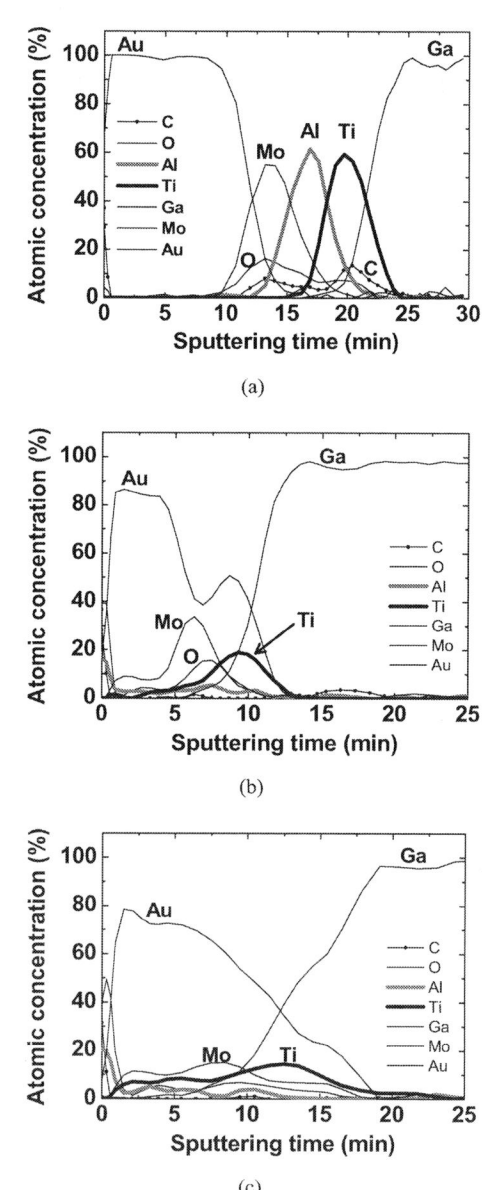

Figure 5. Measured AES-depth profiles of Ti/Al/Mo/Au on GaN before (a) and after annealing at (b) 700 °C and (c) 800 °C. Measured diffusion depth of Au are 324 nm at 700 °C and 458 nm at 800 °C, respectively.

When alloy metals on epitaxial GaN were stripped by wet etchant, metal spikes were evidenced by pits of surface measurement. Ti/Al/Mo/Au was evaporated on the epitaxial GaN after etching the native oxide. Samples were annealed at 700 °C, 800 °C and 900 °C for 30 s under N_2 ambient, respectively. The samples were etched away by 3:1 HCl : HNO_3 solution for 900 s. Etched samples were fully rinsed with the distilled water. Etched surfaces were measured by SEM as shown in Fig. 6. The sample without any annealing exhibits metal residues and any pit was not found. A small number of shallow pits were found in the sample after annealing at 700 °C. However, considerable amount of pits were found in the sample after annealing at 800 °C and 900 °C. Annealing for ohmic contacts at high temperature induces extensive and deep metal spikes. Metal spikes under ohmic contacts should be suppressed because they are plausible for the leakage path of GaN power devices.

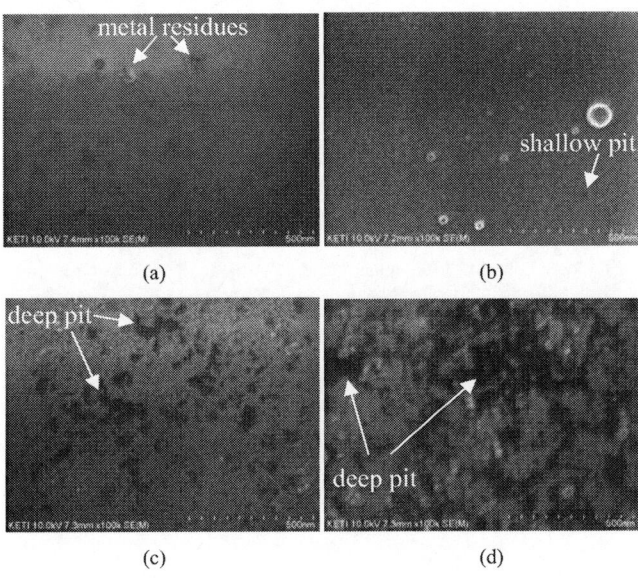

(a)

(b)

(c)

(d)

Figure 6.   Measured SEM micrographs of Ti/Al/Mo/Au on GaN after stripping metals by HCl and HNO_3 solution for 900 s. Sample (a) is not annealed. Sample (b), (c) and (d) are annealed at 700 °C, 800 °C and 900 °C, respectively.

We annealed ohmic contacts on the doped GaN at 700 °C to suppress deep metal spikes. Annealing temperature for ohmic contacts of planar and recessed devices is identical. Annealing of ohmic contacts at 700 °C also achieves the low contact resistance of 0.6 Ω-mm because the doping of GaN makes ohmic contacts easier. The recessed anode is adopted for the high Schottky barrier and the low tunneling current. We measured electric characteristics of planar and recessed devices. Fig. 7 shows the measured leakage current of planar and recessed devices. When an anode-cathode space (D_{AC}) is 5 μm, the measured leakage current of planar and recessed devices are 0.37 A/cm² and 0.05 A/cm² at -100 V, respectively. The high Schottky barrier on UID GaN buffer further decreases the leakage current.

Fig. 8 shows the measured forward I-V of GaN SBDs of which ohmic contacts are annealed at 700 °C. When D_{AC} is 5 μm, the measured forward voltage drop of planar and recessed devices are 1.05 V and 2.13 V at 100 A/cm², respectively. The forward voltage drop of recessed devices is rather increased due to the increase of Schottky barrier. However, an on-resistance of recessed devices is not considerably degraded compared with that of planar devices. The measured on-resistances of planar and recessed devices are 4.65 mΩ-cm² and 5.66 mΩ-cm², respectively.

Fig. 9 shows the measured breakdown voltage of GaN SBDs of which ohmic contacts are annealed at 700 °C. Each 21 devices having D_{AC} of 20 μm show the uniform breakdown voltage around 450 V. The planar device after annealing of ohmic contacts at 800 °C has the low breakdown voltage of 350 V at D_{AC} of 20 μm. The breakdown voltage of planar and recessed devices has no difference when ohmic contacts are annealed at 700 °C. The results suggest that 1 μm-thick GaN buffer is fully depleted at the breakdown. The thick GaN buffer on Si substrate should be required for high voltage operation. The suppression of metal spikes, the low ohmic contact resistance and the thick GaN buffer are necessary for the low power loss and the high voltage operation.

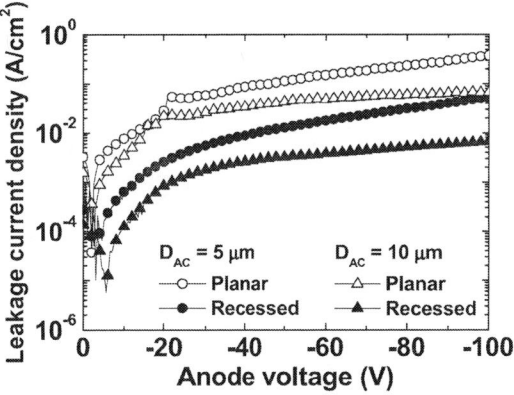

Figure 7.   Measured leakage current of planar and recessed GaN SBDs of which ohmic contacts are annealed at 700 °C

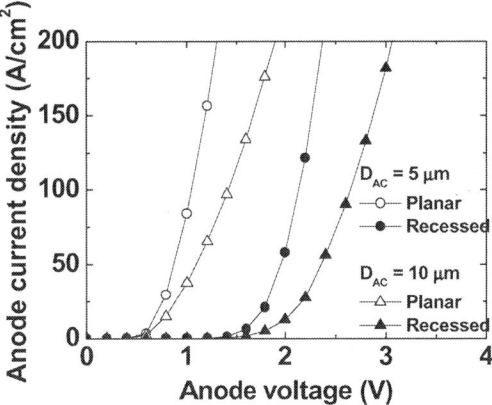

Figure 8.   Measured forward I-V of planar and recessed GaN SBDs of which ohmic contacts are annealed at 700 °C

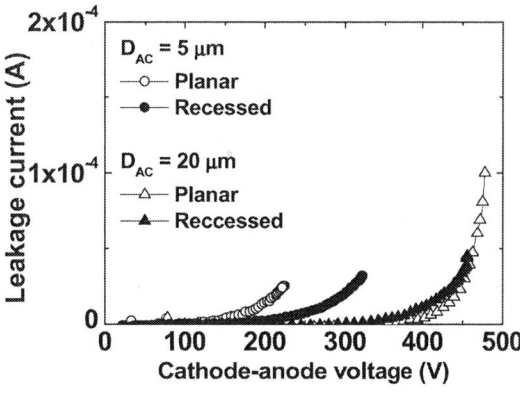

(a)

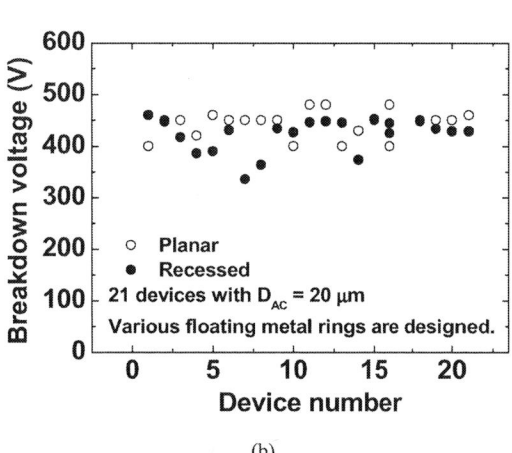

(b)

Figure 9. Measured breakdown voltage of planar and recessed GaN SBDs of which ohmic contacts are annealed at 700 °C

ACKNOWLEDGMENT

This work was supported by the "Power Generation and Electricity Delivery" of the Korea Institute of Energy Technology Evaluation and Planning (KETEP) grant funded by the Korea Government Ministry of Knowledge Economy. (Project No. 101030002B)

REFERENCES

[1] T. P. Chow and R. Tyagi, "Wide Bandgap Compound Semiconductors for Superior High-Voltage Unipolar Power Devices", IEEE Trans. Electron Devices, vol. 41, no. 8, pp. 1481-1483, August 1994.

[2] J. Millan, "Wide Band-gap Power Semiconductor Devices", IET Circuits Devices Syst., vol. 1, no. 5, pp. 372-379, October 2007.

[3] N. Ikeda, S. Kaya, J. Li, Y. Sato, S. Kato, and S. Yoshida, "High Power AlGaN/GaN HFET with a High Breakdown Voltage of Over 1.8 kV on 4 inch Si Substrates and the Suppression of Current Collapse", Proc. 20th ISPSD, pp. 287-290, 2008.

[4] S.-C. Lee, M.-W. Ha, J.-C. Her, S.-S. Kim, J.-Y. Lim, K.-S. Seo, and M.-K. Han, "High Breakdown Voltage GaN Schottky Barrier Diode Employing Floating Metal Rings on AlGaN/GaN Hetero-junction", Proc. 17th ISPSD, pp. 247-250, 2005.

[5] A. Kamada, K. Matsubayashi, A. Nakagawa, Y. Terada, and T. Egawa, "High-voltage AlGaN/GaN Schottky Barrier Diodes on Si Substrate with Low-Temperature GaN Cap Layer for Edge Termination," Proc. 20th ISPSD, pp. 225-228, 2008.

[6] Y. Dora, A. Chakraborty, S. Heikman, L. McCarthy, S. Keller, S. P. DenBaars, and U. K. Mishra, "Effect of Ohmic Contacts on Buffer Leakage of GaN Transistors", IEEE Electron Device Lett., vol. 27, no. 7, pp. 529-531, July 2006.

[7] L. Wang, F. M. Mohammed, and I. Adesida, "Dislocation-induced Nonuniform Interfacial Reactions of Ti/Al/Mo/Au Ohmic Contacts on AlGaN/GaN Heterostructure", Appl. Phys. Lett., vol. 87, pp. 141915, September 2005.

IV. CONCLUSION

We have fabricated GaN SBDs on Si substrate for high-voltage by suppressing metal spikes. Metal spikes are generated during annealing of ohmic contacts. Those are evidenced by measuring AES and SEM. Deep metal spikes are responsible for the leakage current of GaN SBDs. Annealing of ohmic contacts on the doped GaN at 700 °C achieves the low contact resistance as well as the low buffer leakage current. When ohmic contacts are annealed at 700 °C and 800 °C, the buffer leakage current per the width of 100 μm are 9.7 nA and 782.0 nA, respectively. The recessed structure of Schottky contacts further decreases the leakage current from 0.37 A/cm^2 to 0.05 A/cm^2 at -100 V. The GaN SBD having the anode-cathode space of 5 μm achieves the forward voltage drop of 2.13 V at 100 A/cm^2, the on-resistance of 5.66 mΩ-cm^2 and the breakdown voltage of 300 V. When the anode-cathode space is increased to 20 μm, the breakdown voltage of GaN SBD is further increased to 450 V. Metal spikes in GaN SBDs should be suppressed for the low power loss and the high voltage operation.

978-1-4244-8425-6/11 $26.00 © 2011 IEEE

Effect of Oxygen Annealing Temperature on AlGaN/GaN HEMTs

Ogyun Seok, Young-Shil Kim, Jiyong Lim, and Min-Koo Han
School of Electrical Engineering
Seoul National University
Seoul, Korea
ogseok@emlab.snu.ac.kr

Abstract—We have investigated an effect of oxygen annealing temperature on the leakage current and breakdown voltage of AlGaN/GaN HEMTs. The breakdown voltage of 830 V and a low drain leakage current of 1.2 nA/mm at V_{DS}= 50 V and V_{GS}= -5 V are exhibited by employing oxygen annealing at 550 °C. The blocking characteristics are improved with increasing annealing temperature up to 550 °C due to high density of deep traps generated by oxygen annealing. However, the blocking characteristics of the annealed device were degraded when the annealing temperature exceeds 550 °C due to thermal damage on the surface of AlGaN/GaN HEMTs.

INTRODUCTION

AlGaN/GaN high electron mobility transistors (HEMTs) have attracted considerable attention for high power applications due to its high critical electric field, low on-resistance by 2 dimensional electron gas (2DEG) and high saturation velocity [1-2]. Furthermore a low intrinsic carrier concentration ($\sim 10^{-11}$ 1/cm^3) induced by wide bandgap (3.4 eV) of GaN supports a stable operation at high temperature condition [2].

Recently, high voltage and large current AlGaN/GaN HEMTs have been reported [3-4]. However, the surface leakage current is rather large and the on/off current ratio should be improved for high efficiency power systems. The surface traps induced by a lattice mismatch between a GaN buffer layer and substrates, the electron trapping through the surface traps may be a dominant mechanism of the surface leakage current and soft breakdown characteristics of AlGaN/GaN HEMTs [5]. Also, the Schottky barrier height lowering caused by defects at Schottky/GaN interface is critical problem of high performance AlGaN/GaN HEMTs [6].

Surface states have the shallow traps below conduction band of GaN so that electrons can trap and de-trap into those states [7-9]. It is well known that the shallow trap level is related to the nitrogen vacancies and the non-combined electron states [10]. The electron trapping through the shallow

traps aggravates the off state blocking property of AlGaN/GaN HEMTs.

Meanwhile, oxygen acts as the deep trap within forbidden bandgap of GaN and screens the shallow trap [11]. Oxygen based treatment or various oxygen based passivation has been reported [12-13]. Oxygen annealing may be simple and effective method to inject oxygen into GaN and reduce surface leakage current of AlGaN/GaN HEMTs. It doesn't require any masks or long time process. We have already reported the improved blocking characteristics of AlGaN/GaN HEMT employing oxygen annealing [14]. The deep trap states also influence on RF dispersion so that optimizing proper density of oxygen states is important [15]. Also, the nitrogen can vaporize at high temperature process so that annealing temperature should be optimized [16].

The purpose of the work is to investigate the effect of oxygen annealing temperature on the AlGaN/GaN HEMTs, also analyze their electrical properties such as leakage current, breakdown voltage and pulsed I-V characteristics. The oxygen annealing temperature was varied from 250 °C to 650 °C at 100 °C increments. We obtained low leakage current of 1.2 nA/mm and high breakdown voltage of 830 V by emplaoying oxygen annealing at 550 °C.

DEVIECE STRUCTURE AND FABRICATION

The cross sectional view of the fabricated AlGaN/GaN HEMT is shown in Fig 1. AlGaN/GaN heterostructure was grown on 4H-SiC substrate by metalorganic chemical vapor deposition (MOCVD). The nucleation layer was grown, followed by the 3 μm-thick Fe GaN buffer layer. The 25 nm-thick unintentionally doped (UID) $Al_{0.26}Ga_{0.74}N$ layer and the 10 nm-thick UID GaN capping layer were grown in sequence.

The 270 nm-depth mesa isolation was performed by inductively coupled plasma (ICP) etching. An Ohmic contact for source and drain, Ti/Al/Ni/Au (20/80/20/100 nm), were deposited using an e-gun evaporator and annealed at 870 °C

This work was supported by the 'Power Generation & Electricity Delivery' of the Korea Institute of Energy Technology Evaluation and Planning (KETEP) grant funded by the Korea Government Ministry of Knowledge Economy.

for 30 sec under N_2 ambient. After the ohmic contact formation, we annealed AlGaN/GaN HEMTs by furnace as shown in Table Ⅰ. Finally, a Schottky contact, Ni/Au/Ni (50/150/50 nm), were deposited using e-gun evaporator and defined a pattern by lift-off method. The gate length (L_G), gate width (W) and gate-drain length (L_{GD}) of the fabricated AlGaN/GaN HEMT were 5 µm, 50 µm, 20 µm, respectively.

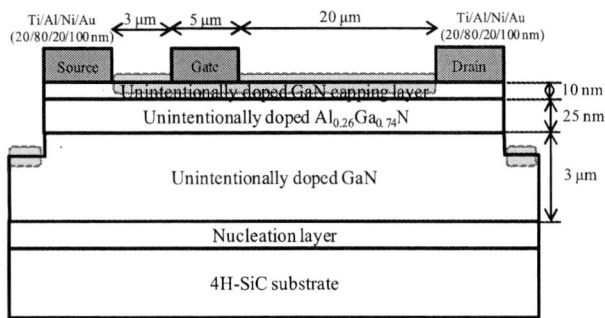

Fig. 1: Cross-sectional view of the fabricated AlGaN/GaN HEMT

TABLE I.   OXYGEN ANNEALING CONDITION

	Condition
Cleaning	30:1 BOE, 30 sec
Oxygen flow	4 SLPM
Process temperature	250 °C, 350 °C, 450 °C, 550 °C, 650 °C

EXPERIMENTAL RESULTS AND DISCUSSION

The drain leakage current of the fabricated AlGaN/GaN HEMT is shown in Fig 2. The leakage current of the conventional device without oxygen annealing was 621 µA/mm at V_{DS} of 50 V and V_{GS} of -5 V. The leakage current of the oxygen annealed devices at 250 °C, 350 °C, 450 °C, 550 °C and 650 °C were 12 µA/mm, 71 nA/mm, 14 nA/mm, 1.2 nA/mm and 2.7 nA/mm, respectively. The leakage current of the oxygen annealed devices was decreased with increasing annealing temperature up to 550 °C. The oxygen annealed AlGaN/GaN HEMTs at high temperature has high density of oxygen states within forbidden bandgap and low probability of de-trapping into conduction band so that the leakage current was successfully suppressed. Also the trapped electron into deep trap level near the gate may be attributed to relaxation of field distribution. The leakage current of the oxygen annealed device at 650 °C were slightly increased. We believe that the density of oxygen was saturated and the surface of GaN was affected by thermal damage at higher temperature than 550 °C.

The breakdown voltage was shown in Fig. 3. The conventional device (BV=180 V) and the oxygen annealed devices at low temperature show soft breakdown characteristics, while those of high temperature show improved blocking characteristics. The breakdown voltage by defined 1 mA/mm of the oxygen annealed devices at 250 °C, 350 °C, 450 °C, 550 °C and 650 °C were 300 V, 600 V, 700 V, 830 V and 800 V, respectively. The oxygen annealed devices at high temperature at over 450 °C showed low leakage current of 100 µA/mm at breakdown voltage. Low leakage current and high breakdown voltage is very important to achieve high quality of power switch. It is difficult to anneal at higher temperature than 650 °C due to thermal damage on GaN surface such as large amount of nitrogen vacancies generation and influence on ohmic metal layer [17].

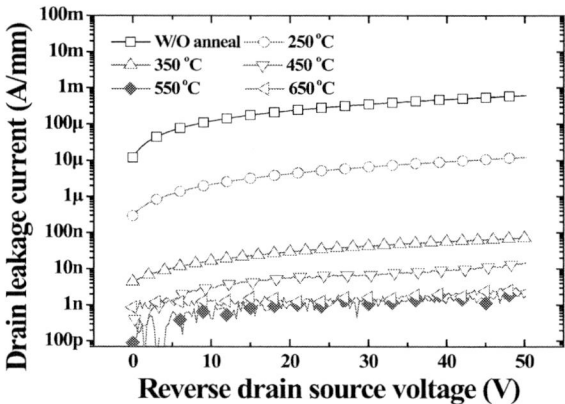

Fig. 2: Drain leakage current of the conventional device and the oxygen annealed devices

Annealing is an effective method to inject oxygen impurities into GaN, suppresses surface leakage current and obtain high breakdown voltage of AlGaN/GaN HEMTs. Furthermore, it doesn't require additional mask so that we can easily apply for AlGaN/GaN HEMTs process compare with any other edge termination structure such as metal field plate.

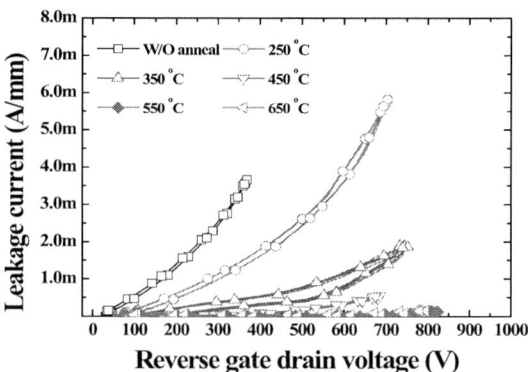

Fig. 3: Breakdown voltage of the conventional device and the oxygen annealed devices

DC forward I-V characteristics of the conventional device and the oxygen annealed devices were shown in Fig. 5. The drain current was measured with a gate voltage sweep, 1 V to -3 V with -2 V/step. The threshold voltages of all samples were -2.3 V. oxygen annealing didn't have a significant influence on DC forward I-V characteristics of AlGaN/GaN HEMT. The average drain current at V_{DS}= 20 V, V_{GS}=1 V was 268.8 mA/mm.

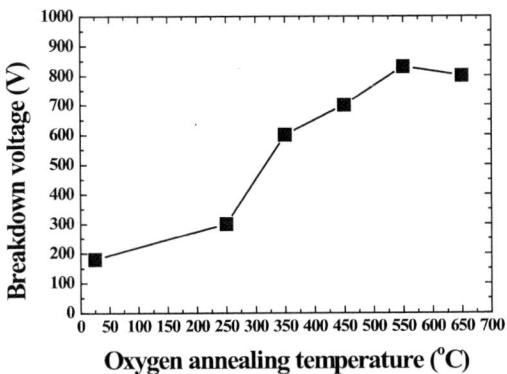

Fig. 4: Breakdown voltage with annealing temperature variation

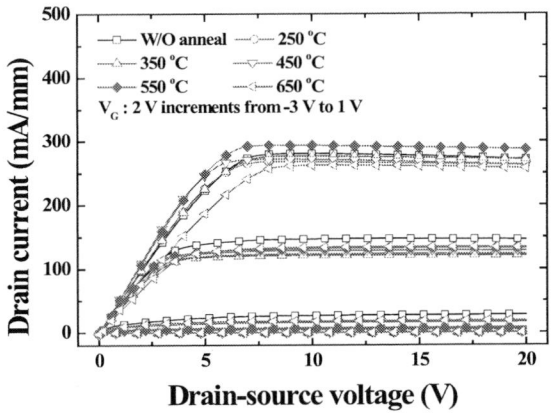

Fig. 5: DC forward I-V characteristics of the conventional device and the oxygen annealed devices

We investigated pulsed I-V characteristics of the conventional device and the oxygen annealed devices in order to analyze the effect of oxygen annealing temperature on AlGaN/GaN HEMTs. We can intuitionally understand the trap level which exist within forbidden bandgap and its role such as rf dispersion. The measured pulsed I-V characteristics are shown in Fig 6. The pulse width and interval were 5 μs and 1 ms, respectively. Gate pulse based on -4 V and rose to 1 V for turn on/off AlGaN/GaN HEMT. Drain based on 0 V, the pulse level was increased to 20 V with 1V/step. The pulsed drain current ($I_{DS,Pulse}$) / the DC drain current ($I_{DS,DC}$) was figured out in Fig 6. The $I_{DS,Pulse}$/$I_{DS,DC}$ of the conventional device without oxygen annealing was 0.94. The ratio was decreased with increasing annealing temperature until 350 °C. That of

350 °C was 0.61 which is the lowest value among the various annealing temperature condition. It is due to amount of oxygen impurities injected by annealing process under oxygen ambient and/or Ga-O bonding which produces Ga vacancies. The trapping and de-trapping through deep trap level is long time process so that pulsed input couldn't respond to it. However, the $I_{DS,Pulse}$/$I_{DS,DC}$ was increased from 0.62 to 0.83 at high annealing temperature over 450 °C. A high temperature process may influence on thermal damage such as vaporization of nitrogen, which produces the shallow traps. The oxygen annealed AlGaN/GaN HEMTs at over 450 °C has the mixed trap level with the shallow and deep traps. In spite of increased $I_{DS,Pulse}$/$I_{DS,DC}$ at 550 °C, the oxygen annealed AlGaN/GaN HEMT at 550 °C showed a lowest leakage current and a highest breakdown. We concluded the optimum annealing temperature which showed good blocking characteristics and pulse response is 550 °C.

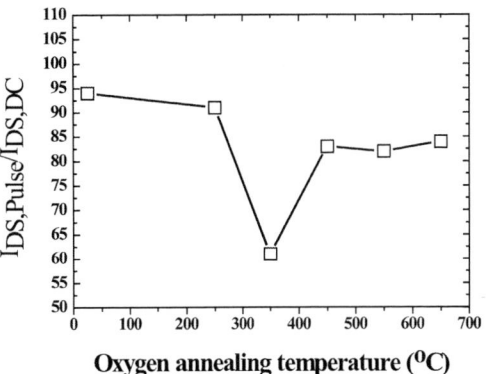

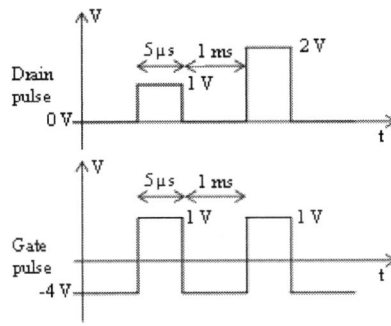

Fig. 5: Pulsed I-V characteristics of the conventional device and the oxygen annealed devices

CONCLUSION

We have investigated the effect of oxygen annealing temperature on AlGaN/GaN HEMTs. The experimental results was summarized in table II. We obtained a low drain leakage current of 1.2 nA/mm and high breakdown voltage of 830 V by oxygen annealing at 550 °C. The blocking capability was improved with increasing oxygen annealing temperature

978-1-4244-8425-6/11 $26.00 © 2011 IEEE

up to 550 °C due to amount of deep traps generation. In the case of $I_{DS,Pulse}/I_{DS,DC}$, the annealed device at 350 °C showed a lowest value and $I_{DS,Pulse}/I_{DS,DC}$ was increased at over 350 °C. Our experimental results showed that the optimum oxygen annealing temperature for AlGaN/GaN HEMT is 550 °C. The oxygen annealed AlGaN/GaN HEMT at 550 °C had high breakdown voltage and high $I_{DS,Pulse}/I_{DS,DC}$.

TABLE II.   SUMMARY OF EXPERIMENTAL RESULTS

Oxygen annealing temperature	BV*	Leakage current**	$I_{DS,Pulse}/I_{DS,DC}$***
W/O annealing	180 V	621 μA/mm	0.94
250 °C	300 V	12 μA/mm	0.91
350 °C	600 V	71 nA/mm	0.61
450 °C	700 V	14 nA/mm	0.83
550 °C	830 V	1.2 nA/mm	0.82
650 °C	800 V	2.7 nA/mm	0.84

* defined by 1mA/mm

** V_{DS} = 50 V, V_{GS} = -5 V

*** V_{DS} = 20 V, V_{GS} = 1 V

REFERENCES

[1] S. J. Pearton, J. C. Zolper, R. J. Shul, and F. Ren J, "GaN: processing, defects, and devices," J. Appl. Phys., vol 86, pp. 1-78, January 1999.

[2] J. L. Hudgins, G. S. Simin, E. Santi, and M. A. Khan, "An assessment of wide bandgap semiconductors for power devices," IEEE Trans. Power Electronics, vol. 18, pp. 907-914, May 2003.

[3] H. Kambayashi, Y. Satoh, S. Ootomo, T. Kokawa, T. Nomura, S. Kato, T. P. Chow, "Over 100 A oeration normally-off AlGaN/GaN Hybrid MOS-HFET on Si substrate with high-breakdown voltage", Sold-state Electron., vol 54, p. 660-664, January 2010.

[4] M. Yanagihara, Y. Uemoto, T. Ueda, T. Tanaka, and D. Ueda, "Recent advanced in GaN transistors for future emerging applications," Physics Status Solidi A, vol. 206, pp. 1221-1227, June 2009.

[5] B. Gil, "Group III Nitride Semiconductor Compounds – Physics and Applications," Clatendon Press Oxford,1998, Chapter 2.

[6] S.-C. Lee, J. Lim, M.-W. Ha, J.-C. Her, C.-M Yun, and M.-K. Han, "High performance AlGaN/GaN HEMT switches employing 500° C oxidized Ni/Au gate for very low leakage current and improvement of uniformity," in Proc. of Int. Symp. on Power Semiconductor Devices and ICs, Italy, June 4-8, 2006, pp. 177-180.

[7] M. Faqir, M. Bouya, N. Malbert, N. Labat, D. Carisetti, B. Lambert, G. Verzellesi, F. Fantini, "Analysis of current collapse effect in AlGaN/GaN HEMT: Experiments and numerical simulations", Microelectron. Reliab., vol. 50, pp. 1520-1522, 2010.

[8] M. L. Nakarmi, N. Nepal, J. Y. Lin and H. X. Jiang, "Unintentionally doped n-type $Al_{0.67}Ga_{0.33}N$ epilayers", Appl. Phys. Lett., vol. 86, 261902, 2005.

[9] K. H. Ploog and O. Brandt, "Doping of group III nitrides", J. Vac. Sci. Technol., vol. A 16, pp. 1609-1614, 1998.

[10] H. Kim, J. Lee, D. Liu, and W. Lu, "Gate current leakage and breakdown mechanism in unpassivated AlGaN/GaN high electron mobility transistors by post-gate annealing," Appl. Phys. Lett., vol. 86, 143505, March 2005.

[11] J. Oila, J. Kivioja, V. Ranki, and K. Saarinen, D. C. Look, R. J. Molnar, S. S. Park, S. K. Lee, and J. Y. Han, "Ga vacancies as dominant intrinsic acceptors in GaN grown by hydride vapor phase epitaxy", Appl. Phys. Lett., vol. 82, pp. 3433-3435, 2003.

[12] M.-W Ha, S.-C Lee, J.-H. Park, J.-C.l Her, K.-S. Seo, and M.-K. Han, "Silicon Dioxide Passivation of AlGaN/GaN HEMTs for High Breakdown Voltage", in Proc. of Int. Symp. on Power Semiconductor Devices and ICs, Italy, June 4-8, 2006, pp. 169-173.

[13] J. W. Chung, J. C. Roberts, E. L. Piner, and T. Palacios, "Effect of Gate Leakage in the Subthreshold Characteristics of AlGaN/GaN HEMTs", IEEE Electron Device Lett., vol. 29(11), pp.1196-1198, 2008.

[14] Y.-H. Choi, J. Lim, Y.-S Kim, O. Seok, M.-K Kim, and M.-K. Han, "High Voltage AlGaN/GaN High-lectron-Mobility Transistors (HEMTs) Employing Oxygen Annealing", in Proc. of Int. Symp. on Power Semiconductor Devices and ICs, Hiroshima, June 6-10, pp. 233-236, 2010.

[15] P. B. Klein, J. A. Freitas, S. C. Binari, and A. E. Wickenden, "Observation of deep traps responsible for current collapse in GaN metal–semiconductor field-effect transistors," Appl. Phys. Lett., vol. 75(25), pp. 4016–4018, 1999.

[16] W. Saito, M. Kuraguchi, Y. Takada, K. Tsuda, I. Omura, and T. Omura, "Influence of surface defect charge at AlGaN/GaN-HEMT upon Schottky gate leakage current and breakdown voltage," IEEE Trans. Electron Devices, vol. 52, no. 2, pp. 159–164, 2005.

[17] Q. Feng, L.-M. Li, Y. Hao, J.-Y Ni, J.-C. Zhang, "The improvement of ohmic contact of Ti/Al/Ni/Au to AlGaN/GaN HEMT by multi-step annealing method", Solid-State Electron., vol. 53 pp. 955–958, 2009.

Proceedings of the 23rd International Symposium on Power Semiconductor Devices & IC's
May 23-26, 2011 San Diego, CA

Normally-off High-Voltage p-GaN Gate GaN HFET with Carbon-Doped Buffer

O. Hilt, F. Brunner, E. Cho, A. Knauer, E. Bahat-Treidel and J. Würfl

Ferdinand-Braun-Institut, Leibniz Institut fuer Hoechstfrequenztechnik
Gustav-Kirchhoff-Strasse 4, 12489 Berlin, Germany
hilt@fbh-berlin.de

Abstract—**Normally-off GaN transistors for power applications in p-type GaN gate technology with a modified carbon-doped GaN buffer are presented. A combination of an AlGaN back-barrier with the carbon-doped buffer prevents early off-state punch-through. Simultaneously, the on-state resistance could be kept low and the threshold voltage with 1.1 V high enough for secure normally-off operation. 1000 V breakdown strength has been obtained for devices with 6 μm gate-drain spacing. The resulting breakdown scaling slope is 170 V/μm gate-drain distance. The on-state resistance is 7.4 Ωmm. The resulting V_{Br}-to-$R_{ON}A$ ratio (1000 V, 0.62 mΩcm^2) is beyond so far reported ratios for normally-off GaN transistors. Modifications of the p-type GaN layer have shown to additionally increase the threshold voltage by 0.4 V without paying a price in the on-state resistance of the device.**

I. INTRODUCTION

AlGaN/GaN HEMTs are generally promising candidates for switching power transistors due to their high breakdown strength and the high current density in the transistor channel giving a low on-state resistance, R_{ON} [1]. However, their inherent normally-on behavior would exclude them from most power-electronic applications. Recent attempts to convert AlGaN/GaN HEMTs into normally-off devices, using gate recess [2] or fluorine incorporation [3] showed limited applicability for power electronics due to their low threshold voltages $V_{th} < +1$ V, and their low gate swing of ~2 V. Realizing a high threshold voltage in normally-off GaN transistors is essential for gaining acceptance of GaN devices in power-electronic applications.

A p-type doped semiconductor as gate is able to deplete the transistor channel when unbiased, thus yielding a normally-off device. P-type gate GaN transistors combine the high-mobility 2DEG transistor channel known from AlGaN/GaN HEMTs with secure normally-off operation, as required for applications in power electronics. This concept has been realized with gates made of Mg-doped GaN, AlGaN [4] and also (p-type) nickel oxide [5]. However, the required $V_{th} > +1$ V is often achieved by a low Al-concentration in the AlGaN barrier, giving a reduced electron density in the 2DEG of the transistor channel and compromising R_{ON}.

Normally-off p-GaN gate transistors using an AlGaN buffer for a high V_{th}/R_{ON} have recently been presented [6]. These devices with $V_{th} = +1.25$ V and $V_{Br} = 870$ V showed a competitive V_{br}/R_{on} ratio for normally-off GaN transistors, see also Fig. 5. The breakdown strength scaled with ~50 V/μm gate-drain distance. This is significantly below the GaN material limit and normally-on GaN HEMTs with higher breakdown strengths have been demonstrated with Fe- [1] and C-doped [7] GaN-buffers. Buffer punch-through leakage [8] of the AlGaN buffer was identified as V_{Br} limitation.

We recently demonstrated normally-on GaN-HEMTs with $V_{Br} > 1000$ V with a V_{Br}/L_{GD} slope of > 160 V/μm by using a carbon-doped GaN-buffer (GaN:C) [9]. However, simple replacement of the AlGaN-buffer by a GaN:C-buffer in a p-GaN gate device is not of favour. The polarization charges at the interface between the GaN channel and the AlGaN back-barrier act as virtual p-type doping. Their presence close to the transistor channel is needed for good normally-off performance [6]. But for a GaN:C-buffer, deep traps associated with the C-doping may give rise to increased dispersion, if they are located too close to the 2DEG of the transistor channel [9].

For improving the breakdown strength of the p-GaN gate devices, we introduced a 3000 μm thick GaN:C-buffer with 4e19 cm^{-3} C-concentration and placed a 60 nm thick AlGaN back-barrier between the buffer and the 40 nm thick GaN channel. This structure, as shown in Fig. 1, gives sufficient positive potential to the 2DEG for normally-off performance, suppresses punch-through in the off-state and should give unhindered current-flow in the on-state.

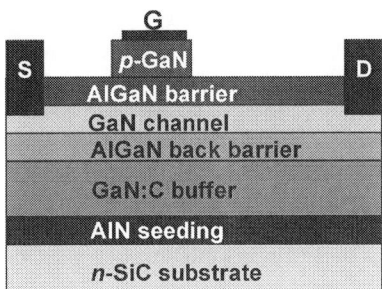

Figure 1. Schematic cross section of the p-GaN gate GaN-transistor with carbon-doped GaN buffer and AlGaN back-barrier on a conductive substrate.

This work was supported by the German Aerospace Center (DLR) under contract Nr. 50 PS 0704 – UAN.

978-1-4244-8425-6/11 $26.00 © 2011 IEEE

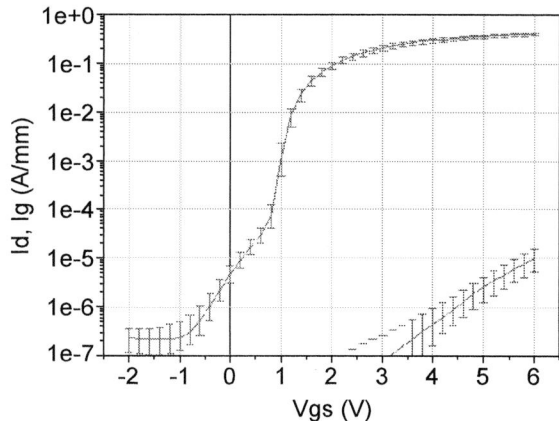

Figure 2. Transfer characteristic (red) and gate current (blue) (median and 25%/75% percentiles for both) for 1.62 mm wide devices with $l_{GD} = 6$ µm on a 3" wafer. $V_{DS} = 10$ V.

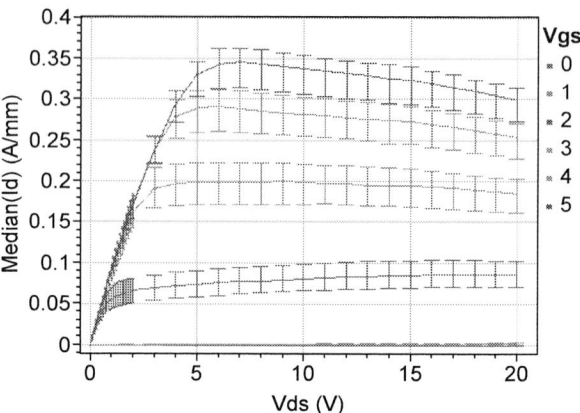

Figure 3. Output characteristic (median and 25%/75% percentiles) for 1.62 mm wide devices with $l_{GD} = 6$ µm. V_{GS} from +5 V in steps of $\Delta V = 1$ V.

To increase the threshold voltage of p-GaN gate devices, the electron density in the transistor channel can be reduced by i.e. a thinner AlGaN barrier or by a lower Al-content of it. But there is a trade-off between the on-state resistance (or maximum drain current) and V_{th} for p-GaN gate GaN-transistors. To minimize this trade-off, an AlGaN back-barrier has been introduced in [6]. Sanken used a gate-recess in combination with the p-type NiO_x-gate [5]. Here, we show that a positive shift of the threshold voltage by ~0.4 V was achieved by modifying the growth of the p-GaN layer. Devices with $V_{th} = 1.8$ V have been obtained.

II. EXPERIMENTAL

The devices have been processed on 3" n-type doped 4H-SiC wafers with a resistivity of ~30 mΩcm. The MOCVD-grown p-GaN/AlGaN/GaN heterostructures consist of a wetting layer, a 3 µm GaN:C buffer with 4e19 cm^{-3} carbon concentration, a 60 nm thick $Al_{0.05}Ga_{0.95}N$ back-barrier, a 40 nm thick uid GaN channel and a 15 nm $Al_{0.23}Ga_{0.77}N$ barrier. The Mg doped p-GaN was 110 nm thick with an effective doping of 3e17 cm^{-3}. The gate is 1.4 µm long and 1.62 mm wide and processed with optical lithography. The source-gate distance is 1 µm and the gate-drain distance $L_{GD} = 6$ µm for the 1.62 mm wide device. Additionally, smaller devices with $w_G = 0.25$ mm and scaled gate-drain spacings between 2 and 18 µm have been processed for the breakdown voltage scaling tests.

Ohmic source and drain contacts consists of a Ti/Al/Mo/Au metallization, annealed at 830 °C. The p-GaN gate was metalized with a Ni/Au ohmic contact. The p-GaN epitaxial layer was selectively plasma-etched, except for the gates. RTP annealing of any etch damage has been done at 500 °C. The devices were isolated by nitrogen implantation and are passivated with PECVD-deposited SiN.

III. DC-CHARACTERISTICS

Fig. 2 shows the median of the transfer characteristics and of the gate current for devices with 1.62 mm gate width. To underline a certain maturity of the process technology, the data are displayed as median with 25% / 75% error bars of 80 devices across the 3" wafer. The threshold voltage is $V_{th} = +1.1$ V. The sub-threshold leakage current drops close to the threshold voltage with a slope of 0.2 V/decade. This slope reduces to 0.7 V/decade below a drain leakage current level of ~50 µA/mm which is belived to be due to shallow traps beneath the p-GaN gate. The off-state drain leakage current is 4 µA/mm at $V_{GS} = 0$ V and thus almost 5 orders of magnitude below the on-state current. The gate current in the on-state (defined as $V_{GS} = +5$ V) is 3 µA/mm and thus more than 5 orders of magnitude below the on-state drain current. A gate bias > 5 V still gives a higher drain current (and lower R_{ON}) but on the expense of an increased gate current. Fig. 3 shows the median output characteristics. For $V_{GS} = +5$ V, I_{dsmax} is 0.35 A/mm or 0.56 A absolute. R_{ON} is 10.2 Ωmm. The devices are in off-state for $V_{GS} = +1$ V.

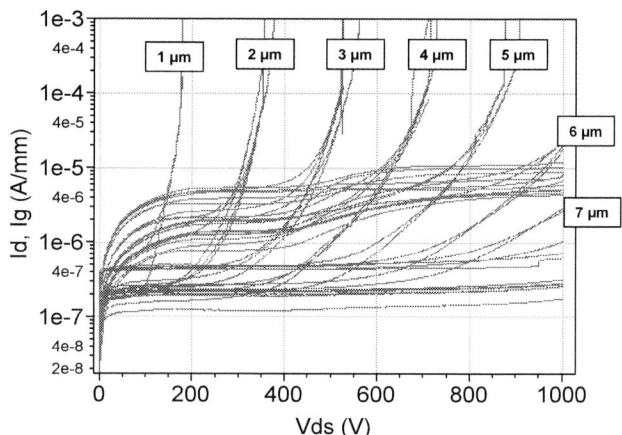

Figure 4. Off-state drain- (red) and gate- (blue) leakage currents for test transistors with different gate-drain spacings $L_{GD} = 1...18$ µm, as indicated in the figure. $w_G = 0.25$ mm and $V_{GS} = 0$ V. 1000 V was the highest available bias from the measurement system.

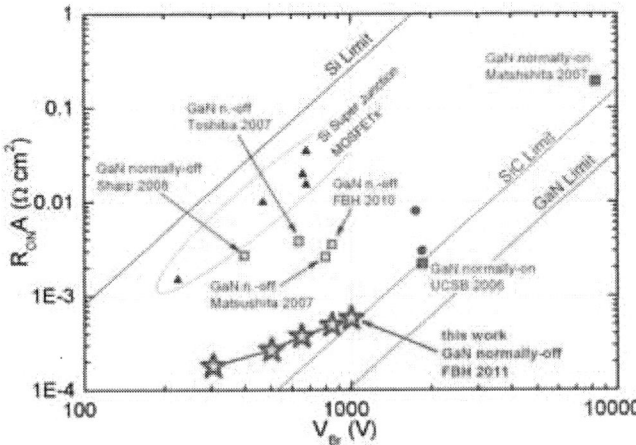

Figure 5. Specific on-state resistance versus breakdown voltage for normally-off and normally-on GaN transistors. Si- and SiC-based devices are also included. The stars represent transistors of this work with $w_G = 0.25$ mm. The gate lengths were 2, 3, 4, 5, and 6 μm, with increasing V_{Br}. FBH 2010 data point from [6].

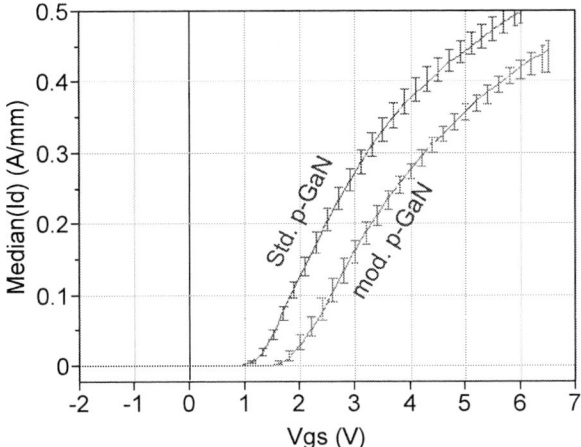

Figure 6. Transfer characteristics (median and 25% / 75% percentiles) for p-GaN gate transistors with standard growth of the p-GaN layer (blue) and with a modified growth scheme (red). $L_{GD} = 2$ μm, $w_G = 0.1$ mm and $V_{DS} = 10$ V.

IV. OFF-STATE CHARACTERISTICS

The breakdown voltage, V_{Br}, of the devices is determined with the drain leakage-current reaching 1 mA/mm. The off-state drain- and gate-leakage current for different gate-drain spacings is displayed in Fig. 4. V_{Br} (defined at $I_D = 1$ mA/mm) increases almost linearly with the gate-drain spacing L_{GD}. It is 176 V for $L_{GD} = 1$ μm, 875 V for $L_{GD} = 5$ μm and > 1000 V for $L_{GD} = 6$ μm. 1000 V is the highest available bias of the measurement set-up and no breakdown was observed for devices with $L_{GD} = 6, 7, 8, 10, 12, 15$ and 18 μm. The slope of the scaling up to $L_{GD} = 6$ μm is $V_{Br}/L_{GD} = 170$ V/μm. The off-state leakage currents before the first signatures of starting breakdown are < 10 μA/mm for the drain current and < 1 μA/mm for the gate current, indicating that the drain leakage is not determined by the gate current. Closer to the breakdown, gate- and drain currents approach each other, indicating that the breakdown is finally triggered by the gate-drain field with an electron flow from gate to drain.

The On-state resistance of these devices at $V_{GS} = 5$ V is $R_{ON} = 4.0, 4.8, 5.7, 6.1$ and 7.4 Ωmm for $L_{GD} = 2, 3, 4, 5$ and 6 μm, respectively. These devices are benchmarked in the R_{ON}-vs.-V_{Br} graph (Fig. 5) against other reported normally-off GaN transistors. Top performance, as it has so far only been reported for normally-on GaN devices is demonstrated for $L_{GD} = 6$ μm with $R_{ON}A = 0.62$ mΩcm^2 and $V_{Br} = 1000$ V. Comparing these data with the performance of p-GaN gate transistors with an AlGaN buffer ($R_{ON}A = 3.52$ mΩcm^2 and $V_{Br} = 870$ V for $L_{GD} = 18$ μm, [6]) show a significant improvement in the power figure-of-merit from $V_{Br}^2/R_{ON}A = 0.22$ GW/cm^2 for devices with AlGaN buffer to 1.6 GW/cm^2 for devices with the new GaN:C buffer combined with the AlGaN back-barrier. The cause for this is that the breakdown strength for a given L_{GD} is increased by a factor 3.4 while the on-state resistance stayed constant.

V. R_{ON}-V_{TH} TRADE-OFF

To escape the relation between the on-state resistance and the threshold voltage for p-GaN Gate GaN-transistors (given by the electron density in the 2DEG) towards higher V_{th} and lower R_{ON}, special attention was paid to the details of the p-GaN layer epitaxy that forms the gate. In particular the first nanometers of the p-GaN gate need a high acceptor density, since their field has most impact on depleting the transistor channel. In contrast, Mg-doping of GaN in MOCVD growth is known to give a delayed incorporation.

We did variations in the p-GaN growth scheme and could increase the threshold voltage by 0.4 V without reducing the on-state performance. Fig. 6 shows two transfer curves, the left one (blue) with the standard p-GaN and the right one (red) with the modified p-GaN and the shift in V_{th} is obvious. All other epitaxial layers have been kept the same. The relation

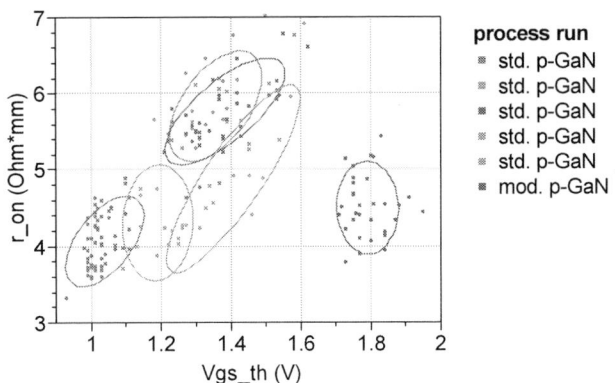

Figure 7. (a) On-state resistance and threshold voltage of reference test transistors from different p-GaN gate process runs. The 5 runs with the standard p-GaN group along a line, while 6th run with the modified p-GaN (purple) is set off towards higher V_{th} or lower R_{ON}. $L_{GD} = 2$ μm, $w_G = 0.1$ mm

978-1-4244-8425-6/11 $26.00 © 2011 IEEE

between R_{ON} and V_{th} for cross-reference test transistors from 6 different p-GaN gate transistor process runs is displayed in Fig. 7. Their epitaxy of the buffer, the GaN channel and the AlGaN barrier partly deviates. All devices of the first 5 process runs, labeled with "std. pGaN", accumulate along a line from $(R_{ON} / V_{th}) = (3 \text{ m}\Omega / 0.9 \text{ V})$ to $(7 \text{ m}\Omega / 1.6 \text{ V})$. This line represents the trade-off between R_{ON} and V_{th}. The 6th process run with the modified p-GaN growth scheme (purple) shows a significant off-set from this line towards a higher threshold voltage $V_{th} = 1.8$ V.

VI. CONCLUSION

A normally-off GaN transistor for power applications with low on-state resistance and high breakdown strength was presented. The used GaN:C-buffer in combination with the AlGaN back-barrier gave a very high breakdown strength to the devices that scales with 170 V/μm gate-drain distance. 1000 V breakdown strength has been achieved for devices with $L_{GD} = 6$ μm. Simultaneously, the on-state resistance was kept low. The combination of 1000 V breakdown strength and $R_{ON}A = 0.62 \text{ m}\Omega\text{cm}^2$ gives a power figure-of-merit of $V_{Br}^2/R_{ON}A = 1.6 \text{ GW/cm}^2$ and is one of the best ratio reported for normally-off GaN transistors so far.

Realizing a high threshold voltage in normally-off GaN transistors is essential for gaining acceptance of the GaN technology in power-electronic applications. By engineering the p-GaN growth scheme, the threshold voltage could get shifted to 1.8 V without simultaneously increasing R_{ON}.

REFERENCES

[1] Y. Dora, A. Chakraborty, L. McCarthy, S. Keller, S. P. DenBaars, and U. K. Mishra, "High breakdown voltage achieved on AlGaN/GaN HEMTs with integrated slant field plates", IEEE Electron Device Lett. Vol. 27 No. 9, pp. 713-715, 2006.

[2] J W. Saito, Y. Takada, M. Kuraguchi, K. Tsuda and I. Omura, "Recessed-gate structure approach towards normally off high-voltage AlGaN/GaN HEMT for power electronics applications", IEEE Trans. on Electron Devices Vol. 53 No. 2, pp. 356-362, 2006.

[3] Y. Cai, Y. Zhou, K.J. Chen and K.M. Lau, "High performance enhancement-mode AlGaN/GaN HEMTs using fluoride-based plasma treatment", IEEE Electron Device Lett. Vol. 26 No. 7, pp. 435-437, 2005.

[4] Y. Uemoto, M. Hikita, H. Ueno, H. Matsuo, H. Ishida, M. Yanagihara, T. Ueda, T. Tanaka and D. Ueda, „Gate injection transistor (GIT) - A normally-off AlGaN/GaN power transistor using conductivity modulation", IEEE Trans. on Electron Devices Vol. 54 No.12, p. 3393, 2007.

[5] N. Kaneko, O. Machida, M. Yanagihara, S. Iwakami, R. Baba, H. Goto, A. Iwabuchi: "Normally-off AlGaN/GaN HFETs using NiOx Gate with Recess", Proc. ISPDS 2009, pp 25-28.

[6] O. Hilt, A. Knauer, F. Brunner, E. Bahat-Treidel and J. Würfl "Normally-off AlGaN/GaN HFET with p-type GaN Gate and AlGaN Buffer", Proc. ISPSD 2010, pp. 347-350.

[7] N. Ikeda, L. Jiang, and S. Yoshida, "Normally-off operation power AlGaN/GaN HFET," in Power Semiconductor Devices and ICs, 2004. Proceedings. ISPSD 2004, pp. 369-372.

[8] E. Bahat-Treidel, O. Hilt, F. Brunner, J. Würfl, and G. Tränkle "Punchthrough-voltage enhancement of Al-GaN/GaN HEMTs using AlGaN double-heterojunction confinement" IEEE Trans. on Electron Devices, Vol. 55 No. 12, pp. 3354-3359, 2008.

[9] E. Bahat-Treidel, F. Brunner, O. Hilt, E. Cho J. Würfl and G. Tränkle, "AlGaN/GaN/GaN:C Back-Barrier HFETs With Breakdown Voltage of Over 1 kV and Low Ron x A", IEEE Transactions on Elec-tron Devices Vol. 57, No. 6, pp. 3050-3058, 2010.

Proceedings of the 23rd International Symposium on Power Semiconductor Devices & IC's
May 23-26, 2011 San Diego, CA

Safe Operating Area of AlGaAs/InGaAs/GaAs HEMT Power Transistors

Vipindas Pala, Mona Hella and T. Paul Chow

Center for Integrated Electronics
Rensselaer Polytechnic Institute
Troy, New York, U.S.A. 12180
[palav,hellam,chowt]@rpi.edu

Abstract—**A study of the physical phenomena leading to second breakdown of AlGaAs/InGaAs/GaAs power HEMTs in high voltage and high current conditions is presented. The boundary of the safe operating area (SOA) is measured in both DC and pulsed conditions. The effect of gate de-biasing and triggering of the parasitic bipolar transistor are identified as reasons for deterioration of the SOA. A model for these effects is also presented.**

I. INTRODUCTION

Recently, Enhancement Mode AlGaAs/InGaAs/AlGaAs p-HEMTs have been shown to have superior FOM for high frequency switching [1] due to their low ON resistance and gate charge. Indeed, using this p-HEMT technology, we have demonstrated a 4.5V-to-3.3V 100MHz dc-dc converter [2] with a peak efficiency of 87%. In these devices, the OFF-state reverse voltage is limited by the maximum tolerable drain-source leakage current, which is usually lower than the avalanche breakdown voltage limit. However, during turn-ON or turn-OFF transients, the device momentarily operates in the high-voltage, high-current regime whence impact ionization can limit the safe operating area (SOA) of the transistor. This work studies the second breakdown of GaAs/InGaAs/AlGaAs power HEMTs and focuses on the power limiting mechanisms of the device by experimentally determining the boundary of the maximum sustaining voltage/conduction current when biased in DC and pulsed conditions.

The schematic cross-section of the pHEMT structure is shown in Fig. 1. The transistor used for characterization was fabricated using a commercial MMIC process [3]. The conducting channel consists of a 2DEG formed in an $In_{0.18}Ga_{0.82}As$ layer sandwiched between two $Al_{0.24}Ga_{0.76}As$ layers, all lattice matched to a semi-insulating GaAs substrate. The channel is controlled using a *0.5µm* long, Schottky-type recessed gate. The recess depth is controlled to make the transistor normally OFF, with a threshold voltage of +0.36V. The maximum ON state voltage at the gate is about +0.85V, at which point the gate Schottky junction starts to get heavily forward biased.

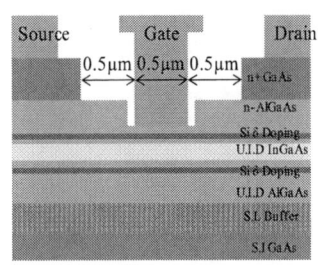

Figure 1. Structure of the pHEMT Device

II. SOA MEASUREMENTS

At high drain bias, the voltage is sustained between the *0.5µm* lateral distance between the gate edge and drain when the device is OFF (when gate is biased to 0V). The peak drain voltage is usually limited by the reverse gate current. When there is a large voltage supported between the gate and drain, the current through the Schottky junction rises due to the increase in electric field at the drain edge of the gate. In the OFF condition, the breakdown can be thought of as a two terminal junction breakdown. However when there is appreciable current through the channel at high gate bias voltages this model is not valid, and three terminal effects start to become important. The safe operating area of the transistor increasingly gets affected by the avalanche process that is initiated by the electrons in the 2DEG channel. It has been shown that the holes generated in the process are transported to the gate Schottky electrode [5]. The presence of a high reverse gate gate current can thus be thought of as an indication of the avalanche process.

The initiation of avalanche process in pHEMTs is followed by a sharp increase in drain and gate current, followed by current snapback and catastrophic failure due to thermal runaway. To study the limits imposed by the second breakdown on ON state operation, a *100µm* wide, single finger pHEMT was taken deep in to the saturation regime at various gate bias voltages till the device was destroyed. The V_{DS} -I_{DS} locus at failure is determined as the safe operating area of the transistor. Under DC conditions, when the device is

This work was supported by TriQuint Semiconductor Inc.

978-1-4244-8425-6/11 $26.00 © 2011 IEEE   243

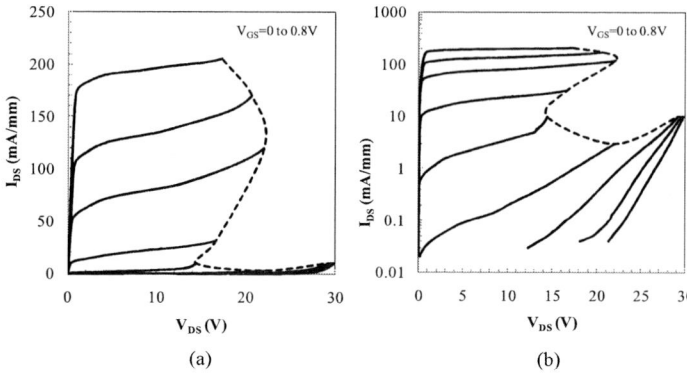

Figure 5. Safe Operating area of the device under DC conditions in a) linear scale, b) log scale.

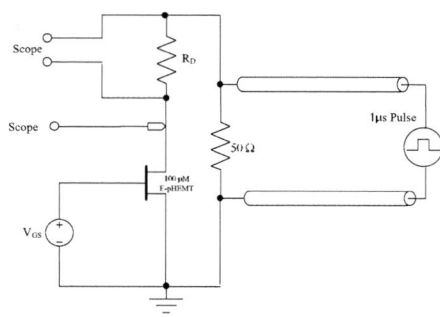

Figure 2. Set-up for pulsed SOA measurement.

allowed to self-heat, the SOA of is shown in Fig.2. The maximum operating voltage of the device depends strongly on the gate bias. The shape of the SOA can be seen as formed by three parts. In the OFF state (V_{GS}=0V), the device can sustain close to 30V before destructive breakdown. However the peak operating voltage falls sharply to as low as 13V (V_{GS} = 0.4V) before it increases again to 22V (V_{GS} = 0.6V) and then eventually falls to 17V (V_{GS}=0.8V).

To eliminate the effect of self-heating, the measurement was repeated in pulsed conditions with short durations and low duty cycles. In the first measurement, with a fixed gate bias, a pulsed drain bias was applied to the transistor as shown in Fig. 3. A limiting resistor was used to prevent the destruction of the device due to stray oscillations. The width of the drain pulse was fixed at *1μs* and the duty cycle was varied to adjust the extent of self-heating in the pHEMT. In the first experiment, the duty cycle was fixed at 0.1%. In this case, self heating is assumed to play a negligible role, so that the SOA is measured under isothermal conditions. For the isothermal SOA, shown in Fig. 4, two major differences can be seen in comparison to the DC case. First, the region of safe operation increases in area. Second, the drastic reduction in the operating bias voltage between V_{GS}=0V and V_{GS}=0.4V observed in the DC case is absent. When the duty cycle was increased to 50%, the SOA observed is shown in Fig. 5. As show in Fig. 6, the area of safe operation promptly falls in between the DC case and the isothermal case.

We also observed that the SOA is affected greatly by the presence of gate resistance. A reverse gate current, when generated due to impact ionization has the effect of raising the

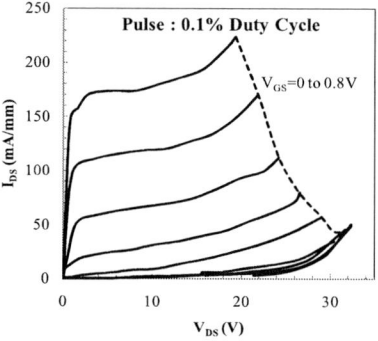

Figure 4. The pulsed safe operating area of the transistor when the duty cycle is 0.1%.

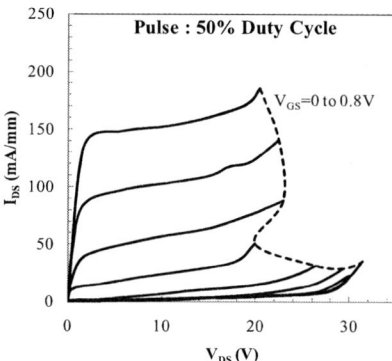

Figure 3. The pulsed safe operating area of the transistor when the duty cycle is 50%.

potential of the Schottky gate. Therefore, we would expect the drain current to increase due to this effect. Our experiments validate this behavior, as shown in Fig. 7. In presence of an extrinsic gate resistance of *2KΩ*, the isothermal SOA deteriorates considerably.

III. DISCUSSION

If the safe operating area is limited by impact ionization, then the observation that self heating deteriorates the SOA seems counter-intuitive, since we expect the impact ionization rate to decrease with increasing temperature. This behavior

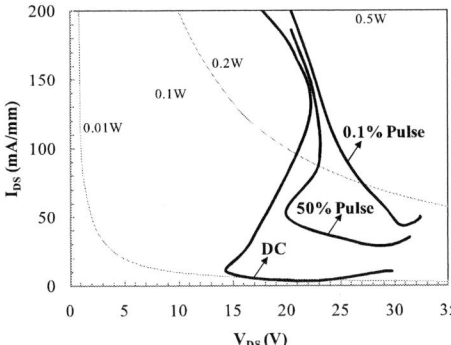

Figure 6. The SOA of the transistor as a function of duty cycle. Also shown are constant power contours for comparison.

can only be explained by a three terminal model for ON state breakdown.

In three-terminal operation, the breakdown of the pHEMT can be explained by the presence of a feedback mechanism due to a current gain. When the device is turned OFF, the channel carrier concentration is very small, and the voltage limit is decided by the drain-channel-gate n+-i-n+ structure. The breakdown mechanism is similar to that of an open base bipolar transistor [6]. This structure exhibits snapback when the avalanche process causes enough carriers to be generated so that the background doping concentration can be exceeded. This point is close to where the impact ionization multiplication factor M reaches large value, where M is given by,

$$M = \frac{1}{1 - (V / BV_{DGO})^n} \qquad (1)$$

After snapback, the conductivity of the structure increases due to the formation of an electron-hole plasma, and the voltage settles to the self sustaining voltage, BV_{DSO}. This limit is decided by the positive feedback that ensues, which is given by

$$BV_{DSO} = \frac{BV_{DGO}}{1 + \beta^{\frac{1}{n}}} \qquad (2)$$

where β is the current gain of the structure.

The current gain can come from a parasitic bipolar transistor which is formed due to the accumulation of holes generated by the impact ionization process, either in the channel or in the semi-insulating substrate as observed by many workers [6]-[9]. In a TCAD simulation of the device, the ON state breakdown voltage was seen to be heavily dependent on the hole lifetime in the device, as shown in Fig. 8. As the hole lifetime in the device is increased in the simulation, the gain of the parasitic bipolar device increases, which leads to an increased gain and hence reduced SOA. A bipolar mechanism has a positive temperature coefficient, because the gain of the parasitic bipolar device increases in presence of self-heating. The internal device temperature thus continues to rise indefinitely resulting in the destruction of the device.

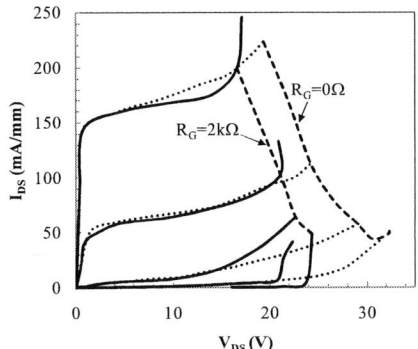

Figure 7. The deterioration of the isothermal safe operating area of the transistor in presence of a gate resistance.

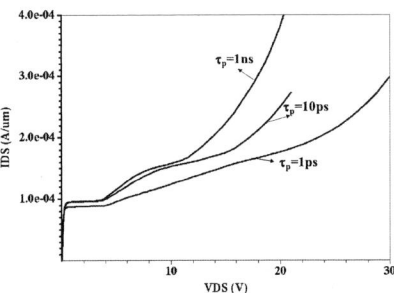

Figure 8. TCAD simulation of the variation of drain current at $V_{GS}=0.7V$ for various hole lifetimes in the device.

The second mechanism of current gain is through gate de-biasing by the reverse gate current. The de-biasing effect can come through the accumulation of holes in the channel, effectively raising the quasi Fermi level in the channel, which decreases the threshold voltage of the device [10], further increasing the current. The increase in current leads to more holes being generated due to impact ionization, culminating in positive feedback. De-biasing can also be caused by extrinsic gate resistances that are present. The reverse hole current raises the Schottky gate potential, which in turn increases the current in the channel. A simple model for these effects can be described as below.

Assuming that there is a constant lateral electric field between the drain and gate, the electric field is given by

$$E_{LAT} = V_{DS} / L_{GD} \qquad (3)$$

where L_{GD} is the spacing between the gate and drain electrodes. Using Fulop's approximation, if the impact ionization parameters are given by

$$\alpha_n = \alpha_p = A e^{-B/E_{LAT}} \qquad (4)$$

The ionization integral is written as,

$$I_{II} \approx \int_0^{L_{GD}} \alpha_n dx = A_n L_{GD} e^{-\frac{L_{GD} B_n}{V_{DS}}} \qquad (5)$$

The impact ionization multiplication factor is given by:

978-1-4244-8425-6/11 $26.00 © 2011 IEEE   245

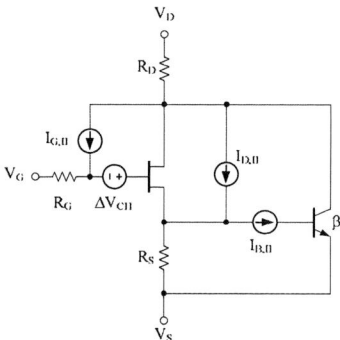

Figure 8. The circuit model for HEMT second breakdown

$$M = \frac{1}{1 - I_{II}} = \frac{1}{1 - A_n L_{GD} e^{-\frac{L_{GD} B_n}{V_{DS}}}} \quad (6)$$

Each electron flowing from source to drain will thus produce *M-1* additional electrons and holes. The hole concentration in the channel increases, which in turn lowers the quasi-fermi potential. To ensure charge balance, this increase in potential can be shown to be [10]:

$$\Delta V_{CH} = \Delta E_{Fn} = \frac{kT}{q} \ln\left(\frac{p_0 + p_{II}}{p_0}\right) \quad (7)$$

where p_0 is the equilibrium hole concentration and the p_{II} is net increase in concentration due to impact ionization, which is proportional to *M-1*. If X_1 is a model parameter, equation (7) can be written as,

$$\Delta V_{CH} = \frac{kT}{q} \ln\left[1 + X_1 (M-1) I_{DS0} (V_{GS} - V_T)\right] \quad (8)$$

where I_{DS0} is the drain current in absence of impact ionization. If a fraction of the impact ionization current flows through the gate, the additional de-biasing due to the gate resistor can be written as:

$$\Delta V_G = X_2 (M-1) I_{DS0} R_G \quad (9)$$

The final drain current through the FET can then be written as,

$$I_{DS,FET} = M \times I_{DS0} \quad (10)$$

$$I_{DS0} = g_m \left(V_{GS} + \Delta V_G + \Delta V_{CH} - V_T\right) \quad (11)$$

If a parasitic n-p-n bipolar transistor is present, with a base current proportional to the impact ionization generated hole current,

$$I_{DS,BJT} = X_3 \beta (V_{GS}, T)(M-1) I_{DS0} \quad (12)$$

The total current is then,

$$I_{DS} = I_{DS0} \left(M + X_3 \beta (V_{GS}, T)(M-1)\right) \quad (13)$$

A circuit schematic of this model is shown in Fig. 9. In conjunction with an empirical model to represent the drain current without impact ionization in pHEMT [11], the model was fitted to the experimental data as shown in Fig. 10.

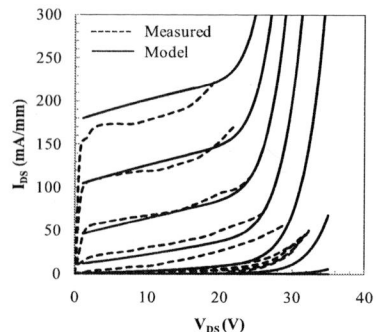

Figure 9. Comparison between the device model and experiment in the isothermal case.

IV. SUMMARY

The safe operating area of power pHEMTs is measured in both pulsed and DC conditions. It is determined that the SOA is a strong function of duty cycle, indicating that self heating plays a major role. SOA is also degraded by the presence of parasitic gate resistance, indicating that gate debiasing effects are crucial. Based on these observations a simple circuit model is developed to explain the electrothermal breakdown of the device in deep saturation.

REFERENCES

[1] V. Pala, K. Varadarajan and T.P Chow, "GaAs pseudomorphic HEMTs for low voltage high frequency DC-DC converters," *ISPSD 2009*, pp. 120-123.

[2] V. Pala, H. Peng, M. Hella and T.P Chow, "Application of GaAs pHEMT technology for efficient high-frequency switching applications," *ISPSD 2010*, pp. 65-68.

[3] *TQPED Process Datasheet*, TriQuint Semiconductor Inc., www.triquint.com

[4] R. Menozzi, "Off-State breakdown of GaAs pHEMTs: Review and new data," *IEEE Trans. Device and Material Reliability*, vol. 4, pp. 54-62, March 2004.

[5] M.H Somerville, R. Blanchard, J.A del Alamo, K.G Duh and P.C Chao, "On-State breakdown in power HEMTs: Measurements and modeling," *IEEE Trans. Electron Devices*, vol. 46, pp. 1087-1093, June 1999.

[6] V.A Vashchenko, V.F Sinkevitch and Y.B Martynov, " Physical limitation on drain voltage of power PM HEMT," *Microelectron. Reliab.*, vol. 37, pp. 1137-1141, 1997.

[7] G. Meneghesso, A. Chini, M. Maretto and E. Zanoni, "Pulsed measurements and circuit modeling of weak and strong avalanche effects in GaAs MESFETs and HEMTs," *IEEE Trans. Electron Devices*, vol. 50, pp. 324-332, February 2003.

[8] A. Di Carlo, L. Rossi, P. Lugli, G. Zandler, G. Meneghesso, M. Jackson and E. Zanoni, "Monte Carlo Study of dynamic breakdown effects in HEMT's," *IEEE Electron Device Lett.*, vol. 21, pp. 149-151, April 2000.

[9] Q. Cui, S. Parthasarathy, J.A. Salcedo, J. J. Liou, J.H. Hajjar and Y. Zhou,"Snapback and postsnapback saturation of pseudomorphic high-electron mobility transistor subject to transient overstress," *IEEE Electron Device Lett.*, vol. 31, pp. 425-427, May 2010.

[10] T. Suemitsu, H. Fushimi, S. Kodama, S. Tsunashima and S. Kimura, "Influence of hole accumulation, kink effect and on-state breakdown of InP based high electron mobility transistors: light irradiation study," *Jpn. J. Appl. Phys.* Vol. 41, pp. 1104-1107, February 2002.

[11] R.B. Hallgren and P.H Litzenberg, "TOM3 capacitance model: linking large and small-signal MESFET models in SPICE," *IEEE Trans. Microwave Theory and Techniques*, vol. 47, pp. 556-561, May 1999.

978-1-4244-8425-6/11 $26.00 © 2011 IEEE

Proceedings of the 23rd International Symposium on Power Semiconductor Devices & IC's
May 23-26, 2011 San Diego, CA

A New Vertical GaN SBD Employing in-situ Metallic Gallium Ohmic Contact

Jiyong Lim, Ogyun Seok, Young-Shil Kim and Min-Koo Han
Department of Electrical Engineering
Seoul National University
Seoul, Korea
mkh@snu.ac.kr

Minki Kim
Convergence Compaonents and Materials Reaserch
Laboratory
Electronics and Telecommunications Research Institute
Daejeon, Korea
mkk@etri.re.kr

Abstract—We proposed and fabricated new vertical GaN Schottky barrier diodes (SBDs) employing in-situ metallic gallium (Ga) ohmic contacts which increase the forward current of a vertical GaN SBD considerably. Highly conductive metallic Ga was formed in-situ at the bottom of n+ GaN substrate due to a high thermal budget during n- epi layer growth so that the ohmic contact was well-formed due to the metallic Ga. The forward current density of the proposed device was 625 A/cm^2 at 2 V while that of the conventional device was 300 A/cm^2. We also employed the floating metal ring and field plate to achieve the high breakdown voltage. The breakdown voltage of the proposed and conventional device was 880 V and 850 V respectively.

I. INTRODUCTION

GaN is attractive in a wide range of applications, including radio frequency (RF) to high-voltage electric power, hybrid electric vehicles and commercial lighting where high-frequency, high power and high voltage are needed in combination. Schottky rectifiers are a key element of power modules because of their high switching speeds and low switching losses, which are important for improving the efficiency of inductive motor controllers and power modules [1].There have been many researches to improve the electric characteristics of GaN Schottky diodes, such as surface treatment [1], floating metal ring [2].

However, most of researches on vertical GaN devices have been concentrated on the reverse electric characteristics of vertical GaN devices. Not only reverse electric characteristics but also forward electric characteristics are important. In order to increase the forward current of GaN power devices such as AlGaN/GaN HEMTs and vertical GaN SBDs, various ohmic metals, such as WSi$_x$ [3] and V/Al/Ni/Au [4], and additional treatments such as an ion implantation before the ohmic contact formation [5], have been reported. However, vertical GaN SBDs employing in-situ ohmic contact formation have been reported scarcely.

The purpose of our work is to report a novel vertical GaN SBD employing in-situ metallic Ga ohmic contact to increase the forward current of GaN SBDs without sacrificing other characteristics. In-situ metallic gallium had been formed during n-epi growth. The forward current of the proposed device was increased to 625 A/cm^2 at 2 V, while that of the conventional GaN SBD was 300 A/cm^2.

II. EXPERIMENTAL RESULTS AND DISCUSSIONS

We fabricated the various types of vertical GaN SBDs on the freestanding GaN substrate with and without metallic Ga layer. Backside-view of GaN substrate with and without metallic Ga is shown in Fig.1. Metallic Ga was formed during the epitaxial growth process and the resistivity was 10 mΩcm which was lower than that of the conventional n+ GaN by 3 times. The thickness of GaN substrate was 400 um and the doping concentration was 2×10^{18} /cm^3. GaN substrate was grown by metal-organic chemical vapor deposition on sapphire and seperated by using laser. 7 um n- GaN epitaxial layer was grown on GaN substrate and the doping concentration was 1×10^{16} /cm^3. The top-view and schematic cross-sectional view of the proposed and the conventional vertical GaN SBDs are shown in Fig. 2.

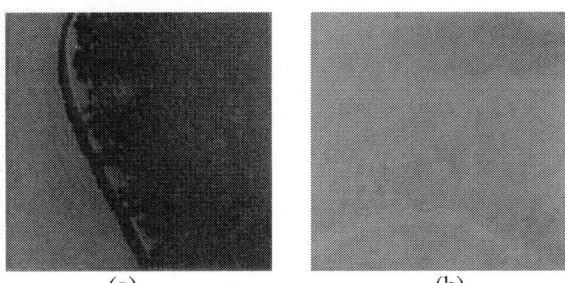

Fig. 1: The backside-view of GaN wafer with metallic Ga (a) and without metallic Ga (b)

This work was supported by the IT R&D program of the MKE/KEIT[10035171, Development of High Voltage/Current Power Module and ESD for BLDC Motor].

978-1-4244-8425-6/11 $26.00 © 2011 IEEE

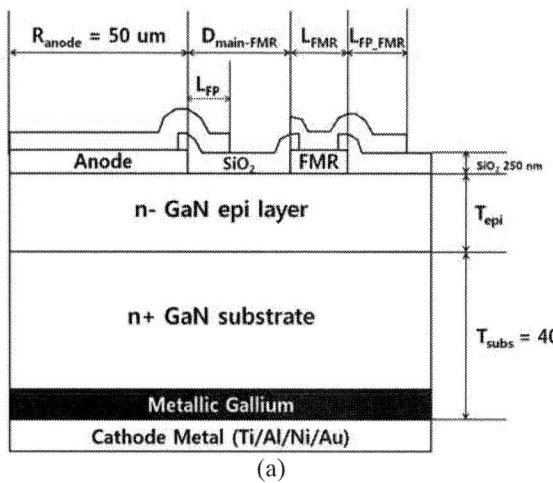

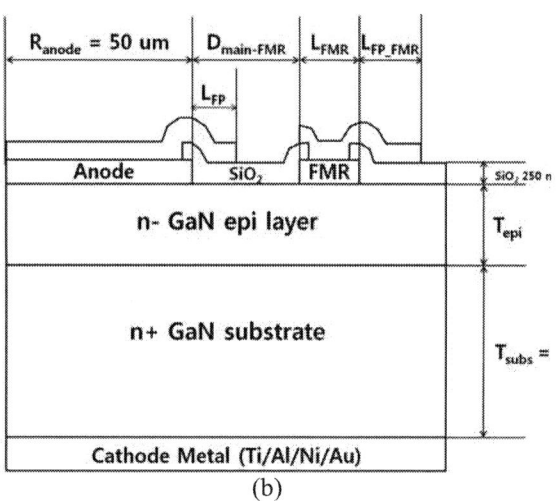

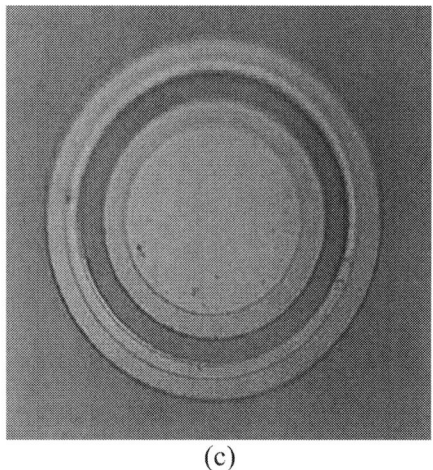

Fig. 2: Schematic cross-sectional view of the proposed (a), conventional (b) vertical GaN Schottky barrier diode with floating metal ring and field plate and top-view of the fabricated device with FMR and FP (c)

Ohmic metal of Ti/Al.Ni/Au (20/80/20/100 nm) was deposited on the backside of GaN substrate with an e-gun evaporator and annealed at 810 °C and 870 °C under N_2 ambient for 30 s by using RTA. Schottky metal of Ni/Au/Ni (50/250/50 nm) was deposited on the front-side of GaN substrate and patterned by a lift-off technique. Schottky metal was annealed at 500 °C for 5 minute under N_2 ambient. We passivated the device with 250 nm thick SiO_2 layer by PE-CVD. Then, the FP, which was extended to the anode and the FMR, was formed with e-gun evaporator and defined by lift-off technique. The design parameters and electric characteristics of the GaN SBD without edge termination structure and GaN SBD employing both FMR and FP are shown in Table. 1. Both the proposed GaN SBD employing metallic Ga and the conventional GaN SBD without metallic Ga were fabricated with or without FMR and FP.

TABLE I.   DESIGN PARAMETERS OF THE PROPOSED AND CONVENTIONAL VERTICAL GaN SBDs

Epi thickness T_{epi}	7 um	
Design	Conv	FP+FMR
R_{anode}	50 um	
L_{FP}	NA	2 um
$D_{main-FMR}$	NA	15 um
L_{FMR}	NA	10 um
L_{FP_FMR}	NA	2 um

Measured forward current of the proposed and conventional device without edge termination structure are shown in Fig. 3. We also optimized the annealing condition of ohmic contact. The forward current of the proposed device, which had been annealed at 870 oC, was 625 A/cm² at 2 V, while that of the conventional device was 300 A/cm² at the same condition. The proposed GaN SBD employing a metallic Ga layer showed superior current capability.

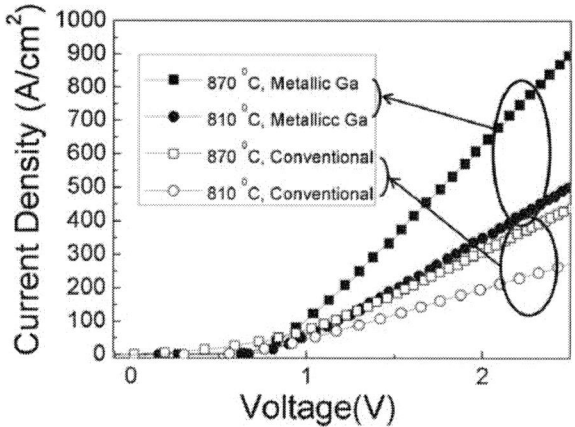

Fig. 3: The backside-view of GaN wafer with metallic Ga (a) and without metallic Ga (b)

We fabricated ohmic to ohmic pads on the backside of GaN substrates with metallic Ga and without metallic Ga to verify that the metallic Ga contributed the decrease of ohmic

978-1-4244-8425-6/11 $26.00 © 2011 IEEE   248

contact resistance. Metallic Ga between two pads was etched by HCl to reduce the difference of between the current path of two pads with metallic Ga and that of two pads without metallic Ga. The distance between two pads was 12 um and the width of two pads was 300 um. The resistance between two pads with metallic was 5.88 Ω while that of the two pads without metallic Ga was 6.84 Ω.

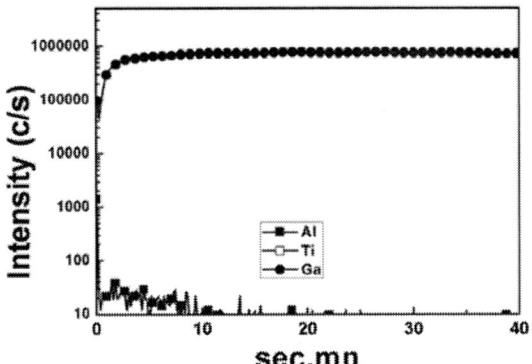

Fig. 4: SIMS measurement result of bare GaN wafer with metallic Ga

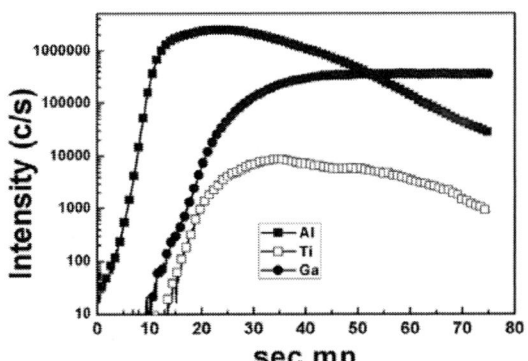

Fig. 5: SIMS measurement result of annealed Ti/Al metal system on GaN wafer without metallic Ga

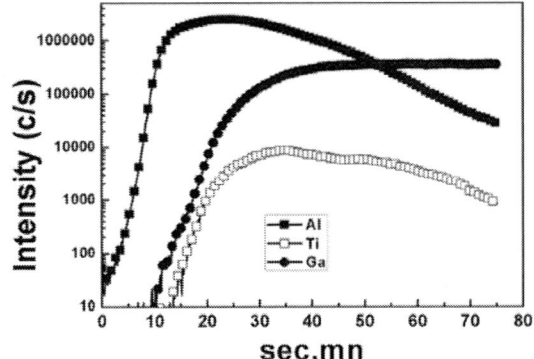

Fig. 6: SIMS measurement result of annealed Ti/Al metal system on GaN wafer with metallic Ga

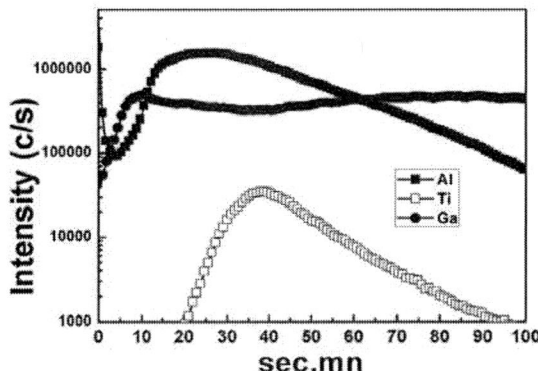

Fig. 7: SIMS measurement result of non-annealed Ti/Al metal system on GaN wafer with metallic Ga

To confirm that the black-spot of the backside is metallic Ga, the backside of the substrate was analyzed by SIMS measurement. The results of SIMS measurement are shown in Figure 4. Ti/Al/Ni/Au ohmic metal system was deposited on the backside of GaN wafers with or without metallic Ga. Among those metals and substrate, only Al, Ga and Ti were observed by using dynamic SIMS due to the limitation of the number of observed elements. Fig. 4 shows the SIMS results of bare GaN wafer with metallic Ga. This result shows that metallic Ga and GaN region are not different from each other in the density of Ga. Therefore, it could be assumed that metallic Ga was not formed by out diffusion of Ga but formed by evaporation of N. Fig. 5-7 supports this assumption. Fig. 5 shows SIMS result of annealed Al/Ti/GaN without metallic Ga. In comparison Ti and Ga, Ti diffused slightly into GaN due to annealing. It is well known that diffusion of Ti during the annealing process is one of the most important factor of ohmic contact formation on GaN [6-8]. Fig. 6 and Fig. 7 show the SIMS results of annealed and non-annealed samples with metallic Ga. In Fig. 7, Ti diffused deeper than that of Fig. 5 although the sample of Fig. 7 had not been annealed. Moreover, in Fig. 6 and 7, there is Ga layer above Ti. It is well known that N in GaN and Ti of deposited ohmic metal system forms TiN so that N vacancies are formed during the annealing process [9-10]. In case of backside of GaN substrate with metallic Ga, there are many N vacancies in metallic Ga region. Therefore, Ti could diffuse into GaN well due to porus surface of n-face GaN which had sufficient N vacancy. In comparison Fig. 6 and 7, it is shown that Ti of Fig. 6 diffused deeper that of Fig. 7 due to annealing process. Therefore, TiN of Fig. 6 was formed in deeper region that that of Fig. 7 so the nitrogen vacancies were.

It is noted that metallic Ga layer is actually Ga rich (N vacancy rich) GaN so that the diffusion of Ti is enhanced due to the nitrogen vacancy in metallic Ga layer. Therefore, TiN could be formed in deeper region of GaN backside so that more nitrogen vacancies were formed during the annealing process. The diffusion of Ti and nitrogen vacancy formation due to TiN formation are well known as two major factors of ohmic contact formation of Ga. Therefore, it is concluded that nitrogen vacancy formation during n- GaN layer epitaxial growth donated the improvement of the forward electric characteristics of vertical GaN SBDs.

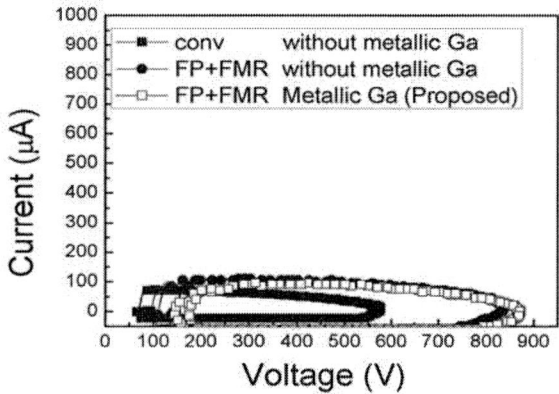

Fig. 8: Measured breakdown voltage of the proposed and conventional GaN SBDs

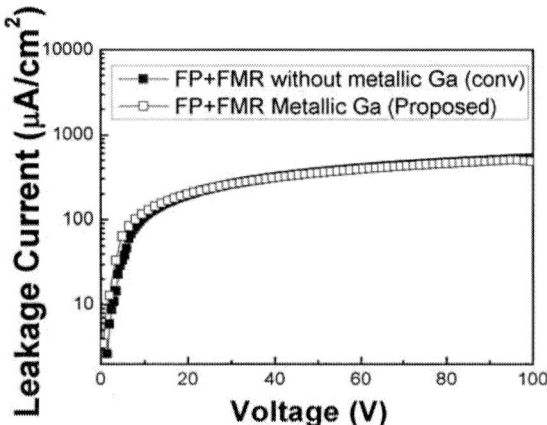

Fig 9: Measured leakage current of the proposed and conventional GaN SBDs with both FP and FMR

Fig. 8 and 9 shows the measured breakdown voltage and leakage current of the proposed and conventional vertical GaN SBDs. The breakdown voltage of the vertical GaN without metallic GaN and edge termination method was 600 V. Both the proposed vertical GaN SBD (with metallic Ga) and the conventional vertical GaN SBD (without metallic Ga) employing FMR and FP shows good reverse characteristics. The breakdown voltage of the proposed device and the conventional device employing FMR and FP was 850 V and 800 V, respectively. The leakage current at -100 V of the proposed and conventional device was 490 uA/cm^2 and 520 uA/cm^2, respectively.

It should be noted that, the metallic Ga layer increased the forward current capability, successfully without sacrificing the reverse electric characteristics.

III. CONCLUSION

We designed and fabricated the vertical GaN SBDs employing in-situ metallic Ga layer. Metallic Ga layer was formed during the epitaxial growth process. Metallic Ga layer increased the forward current of the proposed vertical GaN SBDs successfully due to the increase of nitrogen vacancy.

The forward current density of the proposed device was 625 A/cm^2 at 2 V while that of the conventional device was 300 A/cm^2 at the same condition. We also employed the floating metal ring and field plate to achieve the high breakdown voltage. The breakdown voltage of the proposed and conventional device was 880 V and 850 V respectively. It should be noted that, the metallic Ga layer increased the forward current capability, successfully without sacrificing the reverse electric characteristics

REFERENCES

[1] S. Arulkumaran1, T. Egawa1, H. Ishikawa1, T. Jimbo1, and Y. Sano, "Surface passivation effects on AlGaN/GaN high-electron-mobility transistors with SiO2, Si3N4 and silicon oxynitride", Applied Physics Letters., vol. 84,pp. 613-615, 2004.

[2] S. –C. Lee, J. –C. H, S. –S. Kim, M. –W. Ha, K. –S. Y. –I. C, and M. – K. Han, "A New Vertical GaN Schottky Barrier Diode with Floating Metal Ring for High Breakdown Voltage", ISPSD, 18th, pp.177-180, 2006.

[3] S.J. Pearton, J. C. Zolper, R. J. Shul, et al., "GaN: Processing, defects, and devices", Journal of Applied Physics, vol. 86, no. 1, pp. 1-78, 1999.

[4] K. O. Schweitz, P. K. Wang, S. E. Mohney, and D. Gotthold , "V/Al/Pt/Au Ohmic contact to n-AlGaN/GaN heterostructures", Applied Physics Letters, vol. 80, no. 11, pp. 1954-1956, 2002.

[5] Recht, F, et. Al., "Nonalloyed ohmic contacts in AlGaN/GaN HEMTs by ion implantation with reduced activation annealing temperature", Electron Device Letters, Vol. 26, pp.283-285, 2005.

[6] O. Ambacher, J. Smart, J. R. Shealy, N. G. Weimann, K. Chu, M. Murphy,W. J. Schaff, L. F. Eastman, R. Dimitrov, L. Wittmer, M. Stutzmann, W.Rieger, and J. Hilsenbeck, "Two-dimensional electron gases induced by spontaneous and piezoelectric polarization charges in N- and Ga-face AlGaN/GaN heterostructures", Journal of Applied Physics, vol. 85, pp. 3222-3233, 1999.

[7] S. Ruvimov, Z. Liliental-Weber, J. Washburn, K. J. Duxstad, E. E. Haller,Z. F. Fan, S. N. Mohammad, W. Kim, A. E. Botchkarev, and H. Morkoc, "Microstructure of Ti/Al and Ti/Al/Ni/Au Ohmic contacts for n-GaN ", Applied Physics Letters, vol. 69, pp. 1556-1558, 1996.

[8] C. J. Lu, A. V. Davydov, D. Josell, and L. A. Bendersky, "Interfacial reactions of Ti/n-GaN contacts at elevated temperature", Journal of Applied Physics, vol. 94, pp. 245-253, 2003.

[9] J. S. Foresi and T. D. Moustakas, "Metal contacts to gallium nitride", Applied Physics Letters, vol. 62, pp. 2859-2861 , 1993

[10] M. E. Lin, Z. Ma, F. Y. Huang, Z. Fan, L. H. Allen, and H. Morkoc, "Low resistance ohmic contacts on wide band-gap GaN", Applied Physics Letters, vol. 64, pp. 1003-1005, 1994.

Proceedings of the 23rd International Symposium on Power Semiconductor Devices & IC's
May 23-26, 2011 San Diego, CA

High Breakdown Voltage AlGaN/GaN HEMT by Employing Selective Fluoride Plasma Treatment

Young-Shil Kim, Jiyong Lim, O-Gyun Seok and Min-koo Han

School of Electrical Engineering
Seoul National University
Seoul, Korea
yskim@emlab.snu.ac.kr

Abstract— We proposed and fabricated AlGaN/GaN HEMT with high stable reverse blocking characteristics employing fluoride plasma treatment using CF_4 gas. The plasma treatment with various rf power was performed selectively on drain-side gate edge region where electric field was concentrated. Unlike normally-off process, fluoride plasma treatment with attenuated RF power expanded gate depletion region in the direction of drain electrode. Expansion of depletion was confirmed by the change of measured off-state gate-drain capacitance. Expanded gate depletion spread E-field more uniformly with reducing peak of field intensity and prevented from drastic surface potential drop at the gate edge under large reverse bias condition. By the mitigation of field concentration and gradual potential change due to plasma treatment, was leakage current reduced and high breakdown voltage achieved. The breakdown voltage of plasma treated device with optimized rf power was1400 V while that of untreated sample was 900 V. The leakage current of plasma treated device was 9.5 nA .

I. INTRODUCTION

Gallium nitride has been a subject for intensive investigation and emerged as an attractive material for a high power and high frequency applications. Its high energy band gap is reflected into a very high breakdown field, piezoelectric and spontaneous polarization effects within AlGaN/GaN hetero structure result in two-dimensional electron gas (2DEG) of high density above 10^{13} cm^{-2} without any doping of the barrier layer. Saturation and overshoot velocity are around 3×10^7 cm/s, with relatively good electron mobility value (1200 cm^2/Vs) [1]. Due to its high breakdown voltages originated from energy band gap, GaN-based HEMTs can operate at a voltage substantially higher than those which devices with other semiconductor material cannot readily sustained.

It is well known that a significant degradation takes place when the device is under a high bias in the off-state in the power systems [2-3]. During two-terminal reverse biasing between gate and drain electrode, very high electric field was induced within the active area of the device. A large portion of them is concentrated at the depletion formed below the Schottky junction. Leakage current is due to the electron injection through rectifying contact by various emission mechanisms [4-6], which is mainly attributed to high electric field dumped into the gate depletion. By the existence of the higher electric field beyond which gate depletion can sustain, catastrophic increase in gate leakage current take place [7]. The leakage current of active area deteriorates blocking capability of the device by causing power loss during off-state operation and also trigger a premature breakdown which limits the performance of high power application [8-9].

In order to suppress the leakage current and increase breakdown voltage, various methods to mitigate the phenomena identified as culprits for a device failure have been proposed. To control the surface states causing the virtual gate and transient phenomenon, passivation technique with various dielectric materials have been used [10-12]. Edge termination employing T-shaped off-set gate, Γ-shaped field plate and multi-layer field plate have been proposed to control of internal electric field of the device [13-16]. The field plate, however, may induce an additional gate capacitance between metal plate and AlGaN layer [17].

The purpose of our work is to propose a fluoride plasma treatment as a feasible edge termination technique to improve the reliability of the device by the simple plasma treatment and to give an account for field modulation caused by plasma treatment. Several studies on the fluoride plasma treatment to AlGaN/GaN HEMTs were reported [17-20]. However, most of them focused on the normally-off process. The increase of breakdown voltage on AlGaN/GaN HEMTs by fluoride plasma treatment was scarcely reported. In this paper, we have successfully fabricated AlGaN/GaN HEMT with high breakdown voltage by employing selective fluoride plasma treatment

---

This work was supported by the 'Power Generation & Electricity Delivery' of the Korea Institute of Energy Technology Evaluation and Planning (KETEP) grant funded by the Korea Government Ministry of Knowledge Economy.

978-1-4244-8425-6/11 $26.00 © 2011 IEEE

II. DEVICE FABRICATION

The cross-sectional view and microscopic image of fabricated AlGaN/GaN HEMT is shown in Fig.1. AlGaN/GaN heterostructure was grown on SiC substrate by MOCVP. A 30-nm-thick unintentionally doped $Al_{0.3}Ga_{0.7}N$ and n 3-μm-thick GaN buffer formed 2DEG channel of AlGaN/GaN HEMT then undoped GaN capping layer was grown. The mesa structure with the thickness of 140 nm was formed for device isolation by ICP (Inductively Coupled Plasma Reactive Ion etching) RIE apparatus. For source and drain electrode, Ti/Al/Ni/Au (20/80/20/100 nm) based Ohmic contact was e-gun evaporated then annealed at 870℃ for 30 sec under N_2 ambient. Ni/Au (30/150 nm) based Schottky contact was also formed by an e-gun evaporation. All these metallization patterns were defined by standard lift off method. Then the device was annealed at 500℃ for 5min under O_2 and N_2 ambient. Fluoride-based plasma treatment using CF_4 CCP (Capacitive Coupled Plasma) RIE apparatus was performed for 120 s with various RF power from 30 W to 60 W employing photo resist mask. The flow rate of CF_4 was 20 sccm at a pressure of 50mT.

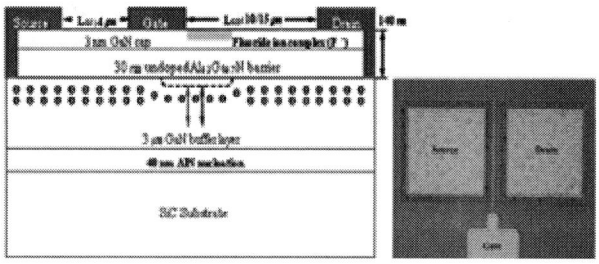

Fig.1. Cross-sectional view and Microscopic image of fabricated AlGaN/GaN HEMT

III. RESULT AND DISCUSSION

AES (Auger Electron Spectroscopy) measurement is shown in Fig.2. Plasma treatment was performed with attenuated RF power to implant fluoride ions into the AlGaN barrier layer with minimizing the damage during plasma treatment. From the measured AES data, it was confirmed that fluorine atoms were implanted within 5 nm from surface of the AlGaN layer considering total profiling time of AlGaN layer. Concentration of fluoride atoms was decreased along the depth of AlGaN barrier layer. Most fluoride atoms were distributed near the surface of AlGaN Layer in which negative polarization charge was localized [21]. The implanted fluoride ions changed the polarization field induced by a dipole system of the AlGaN layer. The modification of the electric field generated an alteration of depletion contour. All these changes caused by plasma treatment was confirmed by the measured electrical characteristics

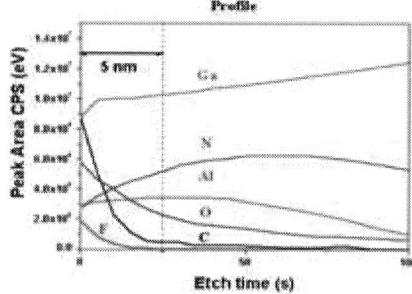

Fig.2. Auger Electron Spectroscopy measurement (AES) of the fluoride plasma treated AlGaN/GaN HEMT .

We investigated leakage and breakdown characteristics of plasma treated AlGaN/GaN HEMT. The leakage current was measured under the reverse bias condition where the gate and drain voltage were -7 V, 50 V respectively. The measured leakage current of conventional device was increased in proportional to the bias while that of proposed device remained at a low level regardless of drain bias as shown in Fig. 3. The measured leakage current of conventional device was 0.28 µA while that of proposed device was 9.5 nA. This was as low as 3.4% of the leakage current of untreated sample. From the measured leakage current, it was found that a selective fluoride plasma treatment performed on the gate edge region suppressed the leakage current efficiently.

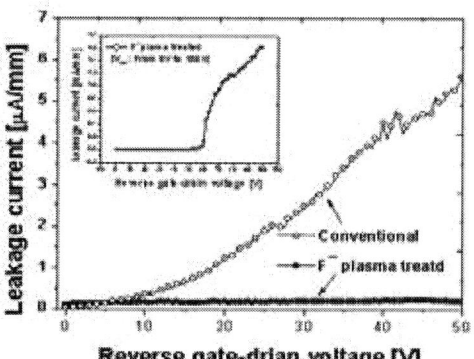

Fig.3. Measured leakage current of AlGaN/GaN HEMT

The suppression of leakage current would be explained by a decrease in peak intensity of electric field dumped into the gate depletion. Negatively charged fluoride ions implanted at the AlGaN surface screened negative localized polarization charges by interacting with donor–like states or with positive polarization charge localized at AlGaN/GaN interface. As shown in Fig.4, an electrical screening caused by fluoride ions disrupted a dipole system in the AlGaN layer decreasing polarization-induced field at the interface between GaN and AlGaN. Reduction of polarization field depleted carriers (2DEG) in the channel making depletion region more extended under reverse bias condition. By the expanded depletion region, electric field was distributed more uniformly. Weakening of electric field at the gate edge region led a suppression of leakage current.

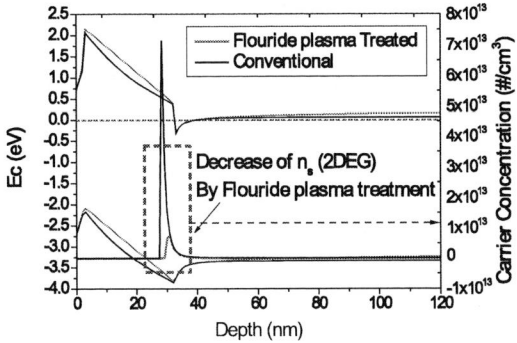

Fig.4. Simulated energy band diagram and carrier concentration of fluoride plasma treated AlGaN/GaN HEMT

The C-V characteristics of AlGaN/GaN HEMT was measured and analyzed to verify the alteration of depletion contour by fluoride plasma treatment. We adopted a simplified equivalent capacitance model of plasma treated AlGaN/GaN HEMT. The gate-drain capacitance under the reverse bias condition is the parallel connection between the capacitance of GaN capping layer and the series of two AlGaN capacitances and a channel capacitance. The gate-drain capacitance, C_{GD} becomes as follows. (1)

$$C_{GD} = C_{(Cap)} + C_{ch} C_{AlGaN} / (2C_{ch} + C_{AlGaN}) \quad (1)$$

When the gate voltage is decreased below the pinch-off voltage, C_{ch} means the depletion capacitance. C_{GD} can be approximated as follows. (2)

$$C_{GD} = C_{(Cap)} + C_{dep}, \quad C_{dep} = \varepsilon_{GaN} / t_{dep} \cdot W \cdot L_{exp} \quad (2)$$

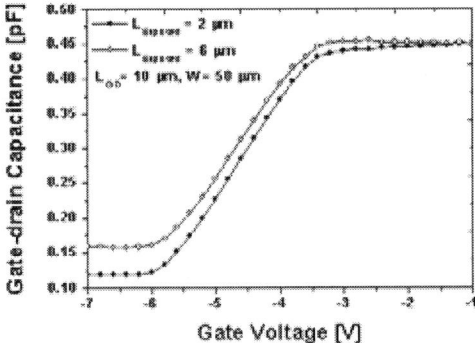

Fig.5. Measured C-V curve of plasma treated AlGaN/GaN HEMT with different exposed area.

The measured capacitance of the plasma treated AlGaN/GaN HEMT is shown in Fig. 5. From the data extracted measured off-state C_{GD} curve, Off-state C_{GD} of conventional device was 0.117 pF (not shown in the Fig.5) while that of plasma treated device with 2μm, 4μm exposed area were 0.119 pF and 0.16 pF respectively. Because the capacitance of GaN capping layer was fixed, Increase in gate-drain capacitance largely depends on the increase of the depletion capacitance. Electrostatic interaction between negatively charged fluoride ions and localized polarization charges ($\pm\sigma_{PZ}$) reduces polarization-induced field at the AlGaN/GaN interface with introducing extra gate-drain capacitance. As an area exposed to fluoride plasma was increased, depletion region was expanded laterally rather than vertically. Laterally expanded depletion distributed electric field and moderated the gradient of potential along the drift region.

The measured off-state breakdown voltage of AlGaN/GaN HEMT was investigated. As shown in Fig.7, The breakdown voltage was defined at drain current of 1mA/mm. Breakdown voltage of fabricated device with fluoride plasma treatment was considerably increased. The measured breakdown voltage of conventional device with L_{GD} of 10 μm was 495 V and that of plasma treated device was 520 V (inset of Fig.5). Breakdown voltage of fabricated device with L_{GD} of 20 μm is much higher than that of the device with 10μm gate-drain length. Breakdown of the plasma treated and untreated device with L_{GD} of 20 μm occurred at 1400 V, 900 V respectively.

The difference in breakdown voltage between plasma treated and untreated sample was increased from 7% (25V) to 55% (500V) as gate-drain length was increased. Leakage current of plasma treated device measured right before the breakdown with L_{GD} of 20 μm were 0.96 mA/mm while that of untreated device was 0.168 mA/mm.

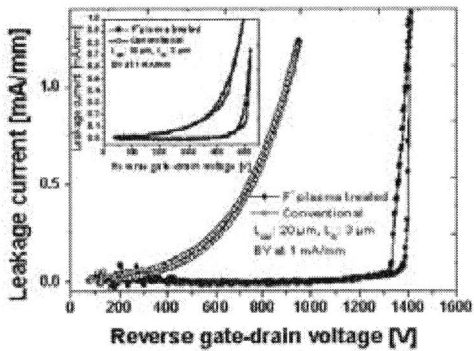

Fig.7. Measured breakdown voltage of F⁻ plasma treated AlGaN/GaN HEMT.

Considering the trade-off relationship between current loss (Fig.9) and improvement of breakdown characteristics, the stable breakdown characteristics of plasma treated device with minimized degradation of output current was obtained at an the optimized rf power of 30 W. Breakdown voltage of the device plasma treated with rf power lower than 30 W was similar to that of untreated one. Breakdown voltage of the plasma treated device with rf power higher than 30 W was increased considerably, but its degree of blocking was not stable as that shown by the device treated with 30 W. The plasma treatment over the optimized rf power would induce the severe damage on the surface of the AlGaN layer. The side effect cause by the damage may overwhelm the enhancement of the blocking capability by field modulation.

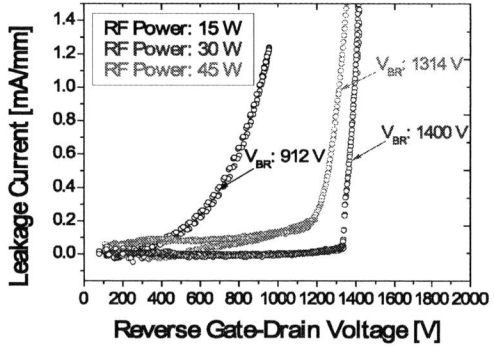

Fig.8. Breakdown characteristics of AlGaN/GaN HEMT with different RF power (15/30/45 W)

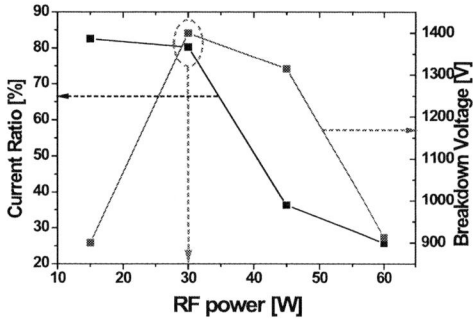

Fig.9. Dependency of the current loss and breakdown voltage on rf power

Contrary to improved reverse characteristics, there was some degradation in forward characteristics. Measured transfer characteristic of AlGaN/GaN HEMT with plasma treatment is shown in Fig.10. The Saturation output current was also decreased from 492.5 mA/mm to 412 mA/mm.

Specific on-resistance was slightly increased from 1.62 mΩ·mm to 1.94 mΩ·mm.

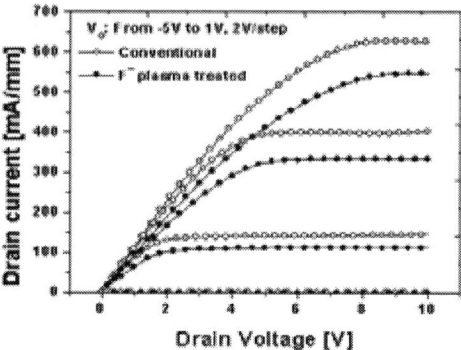

Fig.10. Measured I-V characteristics of fluoride plasma treated AlGaN/GaN HEMT

Degradation in output current is due to the field modulation. Implanted fluoride ions deplete the channel carriers by weakening polarization-induced field at the interface between GaN and AlGaN. The decrease in channel carrier density is directly led to the degradation of forward characteristics.

IV. CONCLUSION

A fluoride plasma treatment for the AlGaN/GaN HEMT with stable blocking capability was proposed. It was confirmed by the experimental data that fluoride plasma treatment is a promising tool to suppress the leakage current and increase the breakdown voltage with simple process. Fluoride plasma treatment expanded gate depletion region laterally through the field modulation effect cause by plasma treatment . Expansion of depletion region weakened a peak of the electric field concentrated at the gate edge by distributing electric field more uniformly. The breakdown voltage of fluoride plasma treated device can achieve high breakdown voltage with a slight degradation in forward characteristics. Measured breakdown voltage of plasma treated AlGaN/GaN HEMT with optimized rf power (30 W) was 1400 V while that of untreated sample was 900 V. The leakage current of the proposed device was 9.5 nA

ACKNOWLEDGMENT

This work was supported by the 'Power Generation & Electricity Delivery' of the Korea Institute of Energy Technology Evaluation and Planning (KETEP) grant funded by the Korea Government Ministry of Knowledge Economy.

References

[1] Mishra U.K, Shen L, Kazior T.E, Wu Y.F, "GaN-based rf power devices and amplifiers." Proc. IEEE, 96 (2008) pp.287-305.

[2] Nakajima S, "State of the art performnace for high power & high efficiency GaN HEMTs." Proc. of European Workshop on Compound Semiconductroe Device and Intergrated Circuits, (2007) pp. 323.

[3] www.nitronix.com.

[4] Hideki Hasegawa, and Tamotsu Hashizume, "Mechanism of current collapse and gate leakage currents in AlGaN//GaN heterostructure field effect transistors." J. Vac. Sci. Technol. B 21(4) (2003), pp. 1844.

[5] Dawei Yan, Hau Lu, and Youdou Zheng, " On the reverse gate leakgea curretn of AlGaN/GaN high electron mobility transistors" Appl. Phys. Lett. Vol. 97 (2010), pp 153503.

[6] Shreepad Karmalkar, naresh Satyan and D.Mahaveer Sathaiya. "On the Resolution of the Mechanism for Reverse Gate leakage in AlGaN/GaN HEMTs ." IEEE Electron Dev. Lett, Vol. 27 (2006), pp 87

[7] Joh J, del Alamo J. A, "Critical voltage for electric degradation of GaN high-electron mobility transistors." IEEE Electron Dev. Lett, Vol. 29 (2008), pp. 287-289

[8] A. Khaligh, A. Emadi, J. Electr Eng. Technol, pp 63, 2006

[9] Gaudezio Meneghesso, Enrico Zanoni "Realiability issue of Gallium Nitride High Electron Mobility Transistors." International Jouranl of Microwave andd Wireless Technol, 2(1), pp.35 (2010).

[10] M.W Ha, Y.H Chio and M K Han, "Silicon Dioxide passivation of AlGaN/GaN HEMTs for High Breakdown voltage." Proc. Int. Sym. om Power Semiconductor Devices & IC's. pp ,2006

[11] M. L. Huang, Y. C. Chang and P. Cheng "Surface passivation of compound semiconductors using atomic layer depositioin grown Al_2O_3" Appl. Phys. Lett Vol. 87 (2005), pp. 252104.

[12] Tamotus Hashizume, Sinya Ootomo and hideki Hasekawa, "Suppressoin of Current collapse in insulated Gate AlGaN/GaN heterostructure field-effect transistors using ultra thin Al_2O_3 dielectric." Appl. Phys. Lett Vol. 83 (2003), pp. 2952.

[13] Y. Ando, Y. Okamoto, H. Miyamoto and M. Kuzuhara. "10-W/mm AlGaN/GaN HFET with a field modulation plate." IEEE. Electron Dev. Lett. Vol. 24 (2003), pp. 289

[14] Y. F. Wu, M. Moore, T. Wisleder, U. K. Mishra and P. Parikh. "High-gain microwave GaN HEMTs with Source-terminated field-plates." IEDM Tech. Dig (2004), pp. 1078

[15] Y. Ando, A. Wakejima, Y. Okamoto and H. Yamanoguchi. "Novel AlGaN/GaN dual-field-plate FET with high gain increased linearity and stability " IEDM Tech. Dig (2005), pp. 576.

[16] Huili Xing, Y. dora, A. S. Keller and U. K . Mishra "High Breakdown voltage AlGaN-GaN HEMTs Achievde by Multiple Field Plates." IEEE. Electron. Dev. Lett. Vol. 25 (2004), pp. 161.

[17] D. Song, J. Liu, Z. cheng and K. J. Chen "Normally off AlGaN/GaN low-density drain HEMT (LDD-HEMT) with enhanced breakdown voltage and reduced current collapse" IEEE Elctron. Dev. Lett. Vol. 28 (2007), pp. 189

[18] Shou Jia, Yong CAi, Deliang Wang and Kevin J. Chen. "Enhancement-Mode AlGaN/GaN HEMTs on Silicon Substrate." IEEE. Trans. Electr. Dev. Vol. 53 (2006), pp. 1474

[19] Yong CAi, Yugang Zhou, Kevon J. Chen and May Lau "High-Performance Enhancement-Mode AlGaN/GaN HEMts Using Fluoride-Based Plasma Treatment." IEEE Electron Dev. Lett. Vol. 26 (2005), pp. 435.

Proceedings of the 23rd International Symposium on Power Semiconductor Devices & IC's
May 23-26, 2011 San Diego, CA

Design and Characterization of a 3D Half-Bridge Semiconductor Power Module in a DFN3x3 Package for DC-DC Buck Converter Application

Yi Su, Anup Bhalla, Daniel Ng, Fei Wang, Jonathan Xue, and Ji Pan

Alpha & Omega Semiconductor. Sunnyvale, California, USA

Email: ysu@aosmd.com

Abstract—**This paper presents a 30V half-bridge 3D semiconductor power module (SPM) in a DFN3x3 package for DC-DC buck converter applications. The 3D Half-Bridge (HB) SPM is compared with a 2D side-by-side HB SPM through a synchronous buck converter at the test conditions of input voltage Vin=19V, output voltage Vo=1.8V, output current Io=12A, and a switching frequency of 300kHz. Both of the HB SPMs are based on shielded gate trench power MOSFET technologies. The DC-DC results show that the proposed 3D HB SPM has better performance than the 2D HB SPM in three categories: (1) 0.9% higher DC-DC efficiency, (2) 3 °C degree lower junction temperature, and (3) 1.7V lower gate spikes.**

I. INTRODUCTION

Integration of semiconductor power devices at a package level provides high efficiency, more reliability, low cost, and overall small footprints for power electronics [1]. A 2D side-by-side die arrangement is usually used [2] due to its easy integration. But it requires small die size for the available footprint size of the package. To overcome the limitation of the 2D approach, we present a 3D stack dual HB SPM [3]. The 3D HB SPM approach has more flexibility, better module performance and higher efficiency as compared to the 2D HB SPM approach. It also allows a much larger low side MOSFET in the same package, and therefore can be scaled to much higher power levels in DC-DC applications.

II. 3D DEVICE STRUCTURE AND FABRICATION

The 3D HB SPM includes two power trench MOSFETs. Both of them (low-side and high-side) are fabricated by a shielded gate trench power MOSFET technology. The key fabrication process for the 3D HB SPM is schematically shown in Figure 1. First, the low-side trench power MOSFET is made with its first metal layer (A and B in Figure 1). Then, a passivation layer is deposited and patterned with openings for gate and source metal connections of the low-side MOSFET. The purpose of this passivation layer is for

isolation between the first metal and the second metal layer. A second metal layer is deposited and patterned for the high-side MOSFET drain connection. The second metal layer is at input voltage node of the half-bridge SPM. The source terminal (top surface) of the low-side MOSFET is at power ground. Finally, a high-side MOSFET is stacked on the top of the low-side MOSFET. A top view of the fabricated HB SPM in a DFN 3x3 package and its schematic circuit are shown in Figure 2. The high-side MOSFET is the small one which is stacked on the top of the large low-side MOSFET. For comparison, a 2D side-by-side HB SPM with equivalent device performance in a DFN 3x3 package is used. The 2D HB SPM is shown in Figure 3.

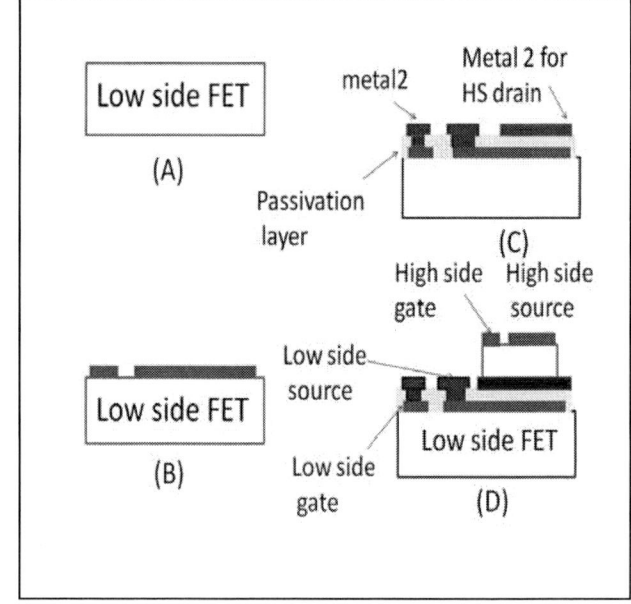

Figure1. A schematic major process flow for the 3D half-bridge SPM.

978-1-4244-8425-6/11 $26.00 © 2011 IEEE

(A)

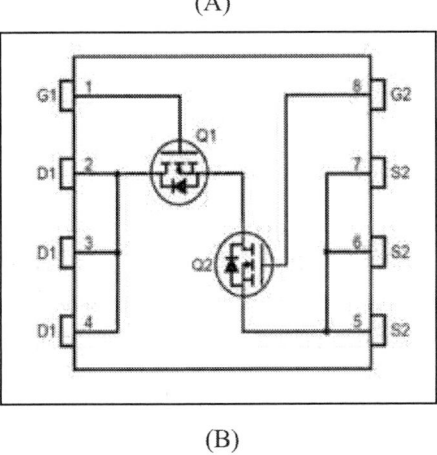

(B)

Figure 2. (A) A top view of the fabricated 3D HB SPM.

(B) Its schematic circuits.

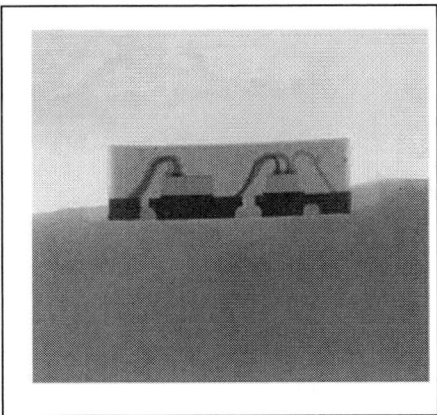

Figure 3. A cross section image of the 2D side-by-side HB SPM.

III. DEVICE CHARACTERIZATION

A comparison of device parameters of the 2D and 3D half-bridge SPMs in a DFN 3x3 package is listed in the Table I. All these trench power MOSFETs have the same body diode forward voltage drop of 0.7V. All low-side MOSFETs and high-side MOSFETs have equivalent gate resistance, Rg. The output capacitance, Coss from the 2D HB SPM is much lower than those from the 3D HB SPM. High Coss contributes high switching loss due to the charging and discharging of the output capacitor in DC-DC buck converter application. The breakdown voltage BVdss from the 3D HB SPM is much higher than 30V device requirement. This indicates that there is still a room for epi optimization. The low-side MOSFET from the 3D HB SPM has lower Rdson due to large die size in the 3D approach. It is preferred and will have low power loss at the DC-DC buck converter application.

TABLE I THE DEVICE PARAMETERS OF THE 2D AND 3D HB SPMs

System in Package	3D HB SPM (AON7900)		2D HB SPM	
Device type	High-side	Low-side	High-side	Low-side
BVdss (V)	39	39	33	33.7
Vgsth (V)	1.8	1.8	2.2	1.8
Rdson(10V) mohm	16.9	5.2	13.5	6.4
Rdson(4.5V) mohm	21.2	6.8	23.7	8.7
Vfsd (V)	0.70	0.70	0.70	0.70
Rg (ohm)	1.5	1.4	1.4	1.4
Ciss at 15V (pF)	600	1800	450	1120
Coss at 15V (pF)	380	865	180	412
Crss At 15V (pF)	13	64	26	44

IV. EXPERIMENTAL RESULTS AND DISCUSSIONS

A DC-DC demo board with a 5V gate driver ISL6207 is used to evaluate the 2D and the 3D HB SPM. The DC-DC test conditions are: 19V input voltage (Vin), 1.8V output voltage (Vo), 12A output current (Io), and 300kHz switching frequency. Figure 4 shows the DC-DC efficiency vs the output current Io. The tested result shows that the 3D approach has higher DC-DC efficiency from light load to full load. The 3D HB SPM has 0.9% higher DC-DC efficiency than that of the 2D HB SPM at the output current of 12A.

978-1-4244-8425-6/11 $26.00 © 2011 IEEE

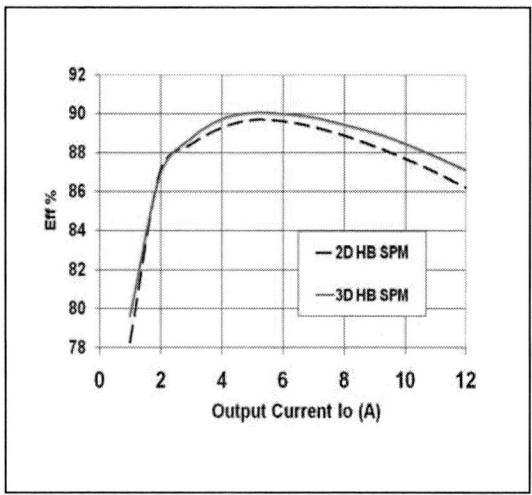

Figure 4. The tested DC-DC efficiency vs output current from the 2D and 3D HB SPMs.

Figure 5 and 6 show phase node voltage waveform at the high-side MOSFET turn-on and turn-off, respectively. Both the 2D and 3D HB SPM have equivalent phase node voltage peak. But the 2D approach shows high ringing frequency. The net loop parasitic inductance can be calculated based on the following equation.

$$\omega^2 = 1/(L \cdot Coss\_ls) \qquad (1)$$

Where, ω is the phase node voltage waveform frequency; L is the net loop parasitic inductance; $Coss\_ls$ is the output capacitance from the low-side MOSFET. From Figure 5 and device parameters in the Table I, the calculated net loop parasitic inductance is 0.71nH and 0.91 nH for the 2D and 3D HB SPM, respectively.

Figure 5 also shows shoot-through in the 2D approach. It is because the gate spikes of the low-side MOSFET is higher than its threshold voltage, when the high-side MOSFET is turn-on and low-side MOSFET is turn-off. Figure 7 shows 2.42 V gate spikes voltage in the 2D HB SPM, which is much higher than its threshold voltage of 1.8V. The gate spike is only 0.74V from the 3D HB SPM. The high gate spike is not desired, since it causes additional power loss at the switching for the DC-DC buck converter application.

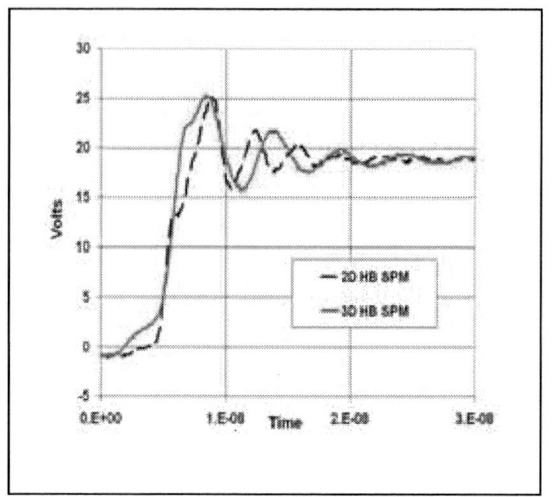

Figure 5. The phase node voltage at the high-side MOSFET turn-on.

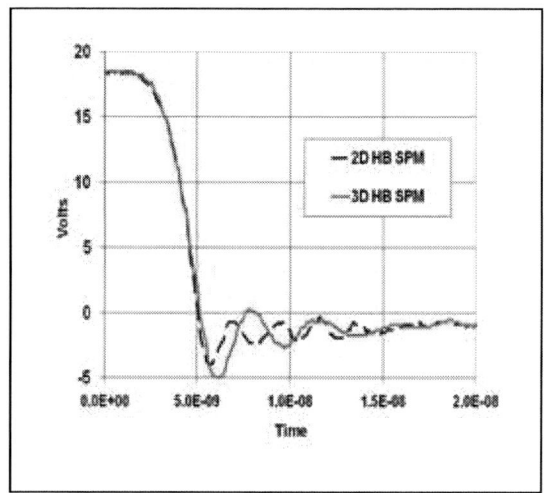

Figure 6. The phase node voltage at the high-side turn-off.

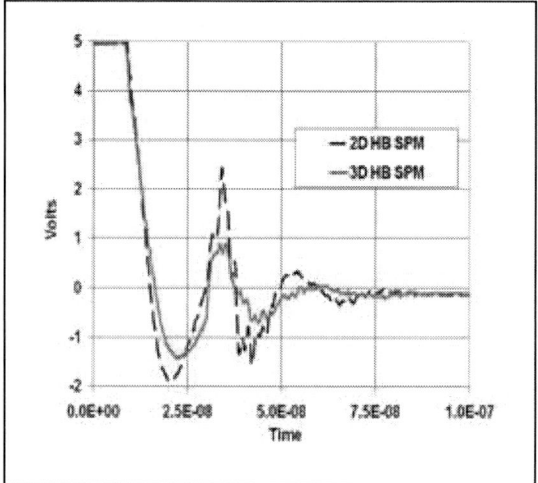

Figure 7. The gate spikes from the low side at the high-side turn-on.

978-1-4244-8425-6/11 $26.00 © 2011 IEEE   258

Figure 8 and 9 show the junction temperature at a full load output current of 12A during DC-DC buck converter testing. The junction temperature is 75.3 °C and 92.6 °C degree for the high- side MOSFET and the low-side MOSFET of the 2D HB SPM, respectively, while it is 84.9 °C and 80 °C degree, respectively for the 3D HB SPM at the test condition of input voltage of 19V, output voltage of 1.8V, output current of 12A, and switching frequency of 300kHz. The thermal measurement shows that the 3D HB SPM has better thermal conduction.

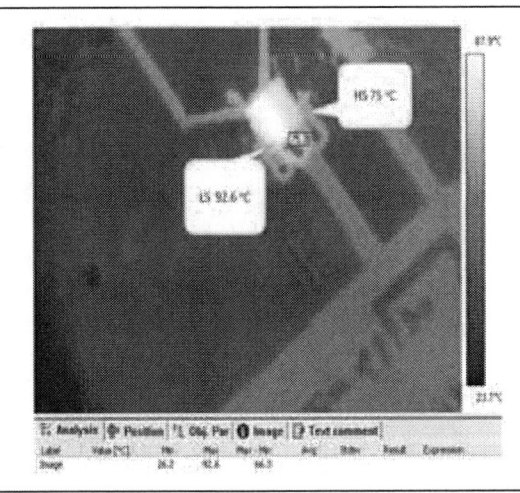

Figure 8. The junction temperature from the 2D HB SPM at Vin=19V, Vo=1.8V, Io=12A, 300kHz.

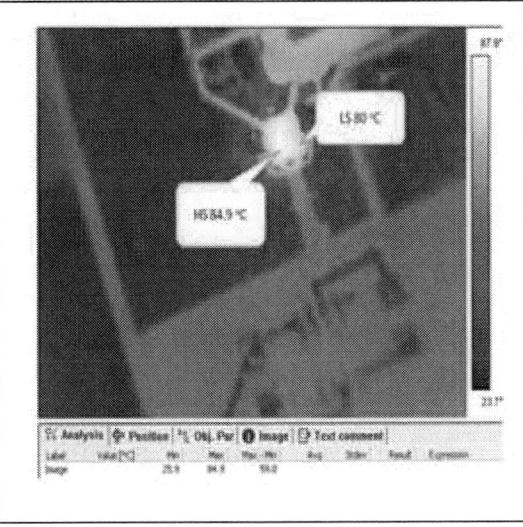

Figure 9. The junction temperature from the 3D HB SPM at Vin=19V, Vo=1.8V, I0=12A, 300kHz.

V.   CONCLUSION

A 3D stacked dual die half-bridge semiconductor power module (product AON7900) has been successfully developed in a DFN 3x3 package for DC-DC buck converter application. The 3D HB SPM has better DC-DC performance as compared to the 2D approach. The tested results show that the 3D approach has higher DC-DC efficiency, lower junction temperature, lower gate spikes, and equivalent phase node voltage ringing.

ACKNOWLEDGMENT

We would like to thank Chang Hong, Limin Weng, Wenjun Li, Hailin Zhou and Jun Lu for manufacturing the devices, and developing the package. We also want to thank Chung Yee for the device characterization and DC-DC testing.

REFERENCES

[1] Yong Liu, Scott Irving, Timwah Luk and Dan Kinzer, "Trends of power electronic packaging and modeling," The 10th Electronics Packaging Technology Conference, pp1-11, 2008.

[2] Laurent Dupont, Stephane Lefebvre, Mounira Bouaroudj, Zoubir Khatir, Jean-Claude Fraugieres, Francis Emorine, "Ageing test results of low voltage MOSFET modules for electrical vehicles," European Conference on Power Electronics and Application, PP1-10, 2007.

[3] AOS US patent application #12/819,111; #12/534,057.

Proceedings of the 23rd International Symposium on Power Semiconductor Devices & IC's
May 23-26, 2011 San Diego, CA

Reliability Study of Au-In Transient Liquid Phase Bonding for SiC Power Semiconductor Packaging

Brian Grummel[1,2], Habib A. Mustain[1,3], Z. John Shen[1], and Allen R. Hefner[2]

[1] Dept. of Electrical Engineering and Computer Science
University of Central Florida
Orlando, FL USA,
bgrummel@mail.ucf.edu

[2] Semiconductor Electronics Division
Physical Measurement Laboratory
National Institute of Standards and Technology
Gaithersburg, MD USA

[3] Cree, Inc.
Durham, NC USA
(Current Affiliation)

Abstract—**Transient liquid phase (TLP) bonding is a promising advanced die-attach technique for wide-bandgap power semiconductor and high-temperature packaging. TLP bonding advances modern soldering techniques by raising the melting point to over 500 °C without detrimental high-lead materials. The bond also has greater reliability and rigidity due in part to a bonding temperature of 200 °C that drastically lowers the peak bond stresses. Furthermore, the thermal conductivity is fractionally increased 67 % while the bond thickness is substantially reduced, lowering the thermal resistance by an order of magnitude or more. It is observed that Au-In TLP bonds exude excellent electrical reliability against thermal cycling degradation if designed properly as experimentally confirmed in this work.**

I. INTRODUCTION

As SiC and other wide bandgap power semiconductor devices gain increasing acceptance in the marketplace, it becomes increasingly important to develop high temperature packaging technologies that allow for higher operating temperatures for applications in extreme environments and to reduce the size, weight, cost, and complexity of thermal management systems. Traditional lead-based die-attach solders have a melting temperature below 200°C and are incapable of supporting the high-temperature operation of SiC MOSFETs that have been demonstrated to operate up to 500 °C for an extended period [1].

An ideal wide-bandgap die-attach material system needs to have a high melting temperature, high thermal conductivity, low mechanical stress, and high tensile and yield strength [2]. Transient liquid phase (TLP) bonding is a novel die-attach method for power devices capable of providing these qualities. Several papers, primarily on proof-of-concept studies, have looked into the TLP bonding technique, but have not investigated its electrical reliability or attempted to characterize multiple bond structures [3]. This work expands the prior work in the field of TLP bonding. It is intended to characterize multiple TLP bond compositions and to investigate bond reliability over temperature cycling conditions by monitoring changes in thin film resistivity and surface morphology of specially fabricated test structures. The

influence of indium composition on reliability in the gold-indium (Au-In) TLP bond with thermal cycling is also studied.

II. TRANSIENT LIQUID PHASE BONDING

A TLP bond is composed primarily of two materials, a base layer which has a high melting point and an interlayer which has a much lower melting point and a high diffusion rate into the base layer. To utilize TLP bonding as a die-attach method, the base material is typically deposited on a substrate or a semiconductor die while the interlayer is deposited on the adjoining piece. The two are then put in contact with applied pressure at a temperature above the interlayer melting point causing it to quickly melt and diffuse into the base layer. The bond becomes stable by being held at the melting temperature until the interlayer is completely consumed into the base; this is known as isothermal solidification [4]. An example of our TLP bonding experiment is shown in Fig. 1 where a SiC Schottky rectifier die is TLP bonded to a silicon nitride DBC substrate. Fig. 2 shows the I-V characteristics of the packaged SiC rectifier.

The advantage of this technique is that the resulting compound of the base and interlayer materials become one solid TLP bond with a melting temperature significantly higher than that of the original interlayer material but is joined at a much lower temperature. This reduces the total mechanical stress applied to the die-attach and device throughout its operation over a wide temperature range and increases reliability, one of the greatest concerns for die-attach techniques. This is a large improvement over alternative high-temperature brazes.

III. EXPERIMENT

While the complete SiC device package shown in Fig. 1 represents the final solution, it is difficult to observe the variations in thin film electrical properties and surface morphology in the TLP bonds over thermal cycling with these structures. To this end, we have designed and fabricated several special TLP bond test structures with all TLP and barrier metal layers sequentially deposited on a glass substrate as opposed to forming an actual bond of a substrate on a

---

This work supported by the National Institute of Standards and Technology; not subject to copyright.

978-1-4244-8425-6/11 $26.00 © 2011 IEEE

Figure 1.   Figure 2.   SiC Schottky rectifier TLP bonded to a silicon nitride DBC substrate with copper metallization.

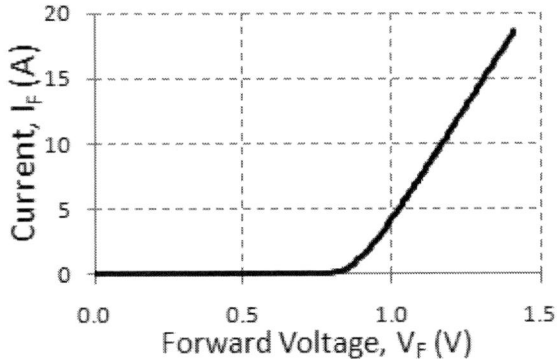

Figure 2.   I-V characteristics of a SiC Schottky rectifier TLP bonded on a silicon nitride DBC substrate.

semiconductor die. Glass is chosen as our substrate due to its electrical insulation along with its coefficient of thermal expansion (CTE) comparable to that of SiC and silicon nitride. While this test structure is not a true representation of the real TLP device package shown in Fig. 1, it does offer an opportunity to measure the thin film resistivity of the Au-In bond and its variation over temperature cycling as well as observe any changes in surface morphology. The material stack is shown in Fig. 3.

A base layer of gold with a melting point of 1064 °C along with an indium interlayer with a melting point of 156 °C is used as this has been shown to be a workable material pair for TLP bonding [5-7].

As an extension over the prior TLP work, we investigated

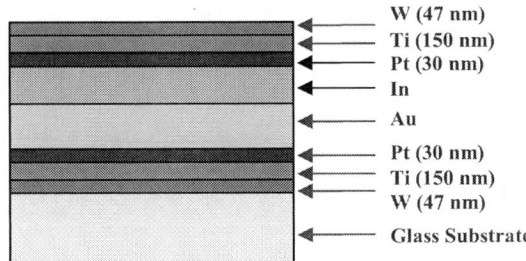

Figure 3.   Material stack of TLP samples emulating actual TLP bonds. Au and In thicknesses vary by sample to create different material ratios seen in Table 1.

TABLE I.   TLP SAMPLE PARAMETERS

Sample	t_{In} (µm)	t_{Au} (µm)	Indium Mass Fraction, w_B	Indium Mole Fraction, χ_B
A	1.261	1.0	0.323	0.45
B	3.087	1.0	0.538	0.67
C	3.087	0.223	0.84	0.90

three different Au-In TLP bonds with three different mole fractions of In. Bond TLP-A has a relatively high concentration of Au in the bond with an In mole fraction of 0.45 while bond TLP-B is proportioned to form $AuIn_2$ compound throughout the bond as opposed to forming multiple Au-In material phases, this occurs at an In mole fraction of 0.67. TLP-C has the highest In composition with an In mole fraction of 0.90. In this particular sample, there may be pure In remaining after reacting with the available Au to form Au-In compounds, possibly causing instability in the final bond. The three samples are summarized in Table 1 and identified on the Au-In phase diagram in Fig. 4 [8].

Each of the three Au-In TLP bonds are sandwiched within a stack of barrier layers of tungsten for adhesion, titanium as the primary diffusion barrier, and platinum to prevent undesirable Au-Ti intermetallics [9], [10]. All barrier layers are added to both the glass substrates along with the exposed face of the TLP bond to completely emulate the entire material stack that is present when utilizing TLP as a bond, as in Fig. 1.

All samples are fabricated by first electron beam evaporating 47 nm of W followed by 150 nm of Ti and then 30 nm of Pt. The TLP-A samples are then deposited with 1.261 µm In while the B and C substrates receive 3.087 µm of In. All samples are then e-beam evaporated with 50 nm of Au as a temporary oxidation barrier before transferring the samples out of the e-beam system and immediately into a sputtering system. TLP-A and B substrates are then sputtered

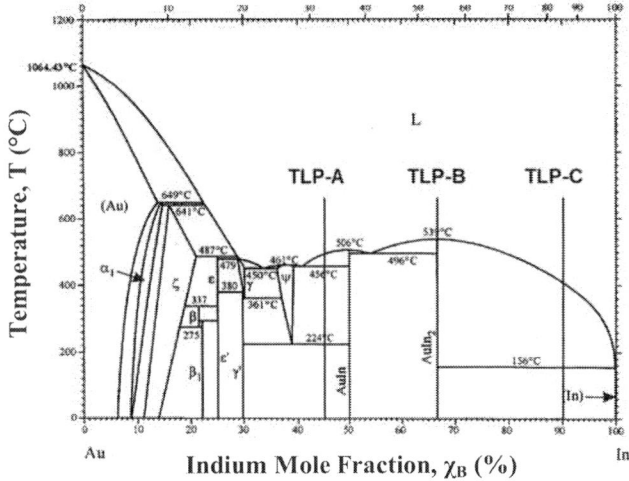

Figure 4.   Binary phase diagram of the Au-In system with three test sample concentrations denoted at mole fractions of indium 0.45, 0.67, and 0.90 within samples TLP-A, B, and C, respectively.

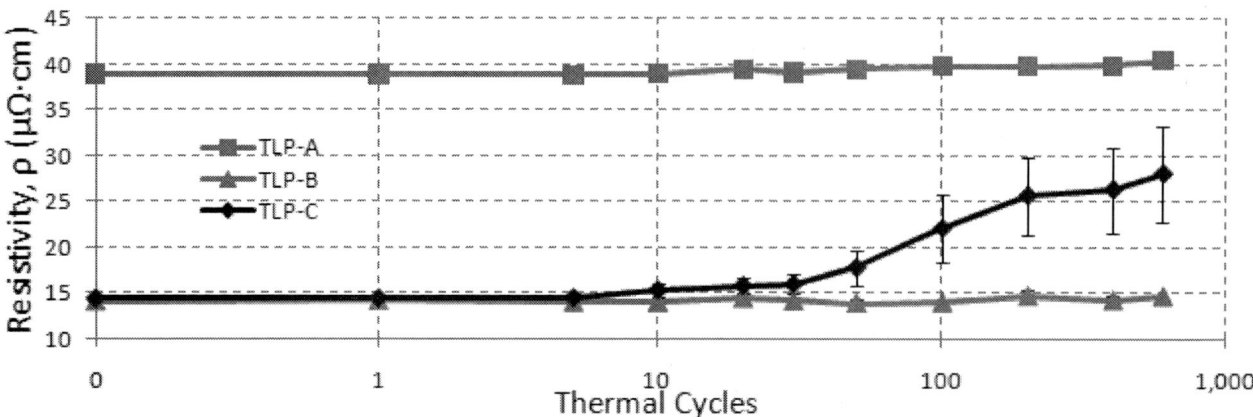

Figure 5. Logarithmic graph of resistivity measurements of samples TLP-A, B, and C versus thermal cycling degradation showing excellent reliability in higher Au concentration samples A and B while the sample, C, containing high In exuded poor reliability with increasing resistivity. Standard deviation error graphed with n = 20.

with 777 nm of Au before the C samples are placed into the chamber. All samples are then completed by sputtering an additional 173 nm of Au followed by the stack of barrier layers again but in reverse order of 30 nm of Pt, 150 nm of Ti, then 47 nm of W to complete the test structure. Though Au is deposited in two or three depositions, the total Au thickness is calculated to create the desired In mole fraction in each of the samples. The slides are then loaded into a 200 °C oven and immediately purged with nitrogen and then vacuum pumped to 1.44 kPa for 15 min. They are then removed and allowed to cool in air to complete the TLP samples.

To characterize the electrical reliability of the TLP bonds, the samples are temperature cycled in air to simulate the operating conditions that power electronics experience from self-heating, extreme-temperature environments, or both. The thermal cycles ramp between 25 °C and 200 °C at a rate of 20 °C/min with a dwell at 25 °C and 200 °C for 5 min each for a total cycle of approximately 30 min.

IV. RESULTS

A. Resistivity

Each sample is tested for its thin film resistivity before and after thermal cycling and is shown in Fig. 5. All resistivity measurements are taken using a four point probe with probe currents of both 10.0 mA and 9.0 mA measured at five specified locations uniformly distributed over the sample surface during each characterization. The measurements are then averaged to create a mean resistivity value. Initially, the high Au concentration sample, TLP-A, has a resistivity of $\rho_A = 38.9 \, \mu\Omega \cdot cm$ while the samples TLP-B and TLP-C hold lower resistivity values of $\rho_B = 14.067 \mu\Omega \cdot cm$ and $\rho_C = 14.356 \, \mu\Omega \cdot cm$.

After 100 thermal cycles, there is no observation of change in resistivity in samples TLP-A and TLP-B with higher Au concentrations, while TLP-C with the highest In concentration shows a large increase in resistivity indicating degradation of this Au-In TPL bond. After only 200 thermal cycles, the resistivity of TLP-C has fractionally increased 78 % to $\rho_C = 25.66 \, \mu\Omega \cdot cm$ and by 83.23 % after 800 cycles to $\rho_C = 26.3 \, \mu\Omega \cdot cm$. The large resistivity increase is partially due to in-homogeneity of resistivity over the surface of the sample with some areas rising greatly while others remain lower at $\rho \cong 20 \, \mu\Omega \cdot cm$. This non-uniformity is the primary cause for the increased standard deviation seen in the graph as well. Over the same period of 800 cycles, TLP-A has fractionally risen only 2.43 % and TLP-B just 1.38 % indicating little change within the TLP bond.

There is also a highly visible change in the color of the TLP-C surface in response to the thermal cycling which, as seen in Fig. 6, has gone from the original brown color created primarily by the exterior W layer to a gray color indicative of In presence. The color of the TLP-A and TLP-B samples remain brown and only slightly increase in darkness with thermal cycling, with TLP-B again showing the most reliable results of little color change.

B. Surface Microscopy and Diffusion

SEM inspection of the samples reveals no morphological changes in the fine structure surface of the samples as a result of thermal cycling on all three samples, thus appearing identical to their un-degraded counterparts, as seen in Fig. 7. This is an indicator of isothermal solidification during the sample preparation as this unchanged morphology would

Figure 6. Surface color change of TLP samples due to thermal cycling. Original sample surface color displayed above surface color after 400 thermal cycles on samples TLP-A, B, and C.

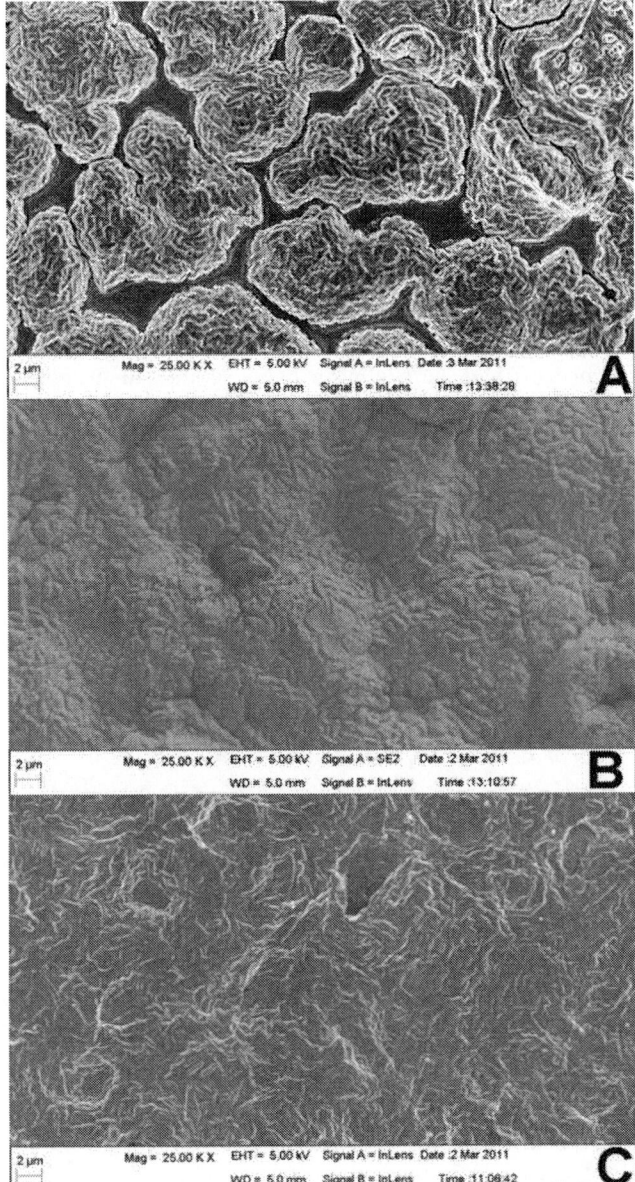

Figure 7. SEM microscopy of surface morphologies of degraded samples at 25 000 X magnification. Appearance of these samples is identical to the original morphologies prior to thermal cycling (*not pictured*).

likely not occur if a layer of pure In were in existence within the bond and was melting with each thermal cycle. It therefore is also an indicator of adverse In diffusion out of the TLP bond in sample TLP-C.

Energy-dispersive X-ray spectroscopy of the samples confirms this hypothesis and reveals diffusion within the bonds which also explains the change in resistivities. The TLP-A and B samples, exhibiting little resistivity variation, show only a small increase in the In concentration near the surface which is expected with continued homogenization of the TLP bond during cycling. The In concentration near the TLP-C surface has increased more significantly and has affected the TLP barrier layers due to indium's high

diffusivity showing high volatility in the sample and the cause of the increase in measured resistivity.

V. CONCLUSION

It has been demonstrated that TLP bonding is a viable method for high-temperature die-attach for SiC and other wide-bandgap power semiconductors in terms of its electrical reliability. Samples TLP-A and B, with optimum Au-In compositions, have shown a negligible rise in resistivity in response to thermal cycling while also maintaining a stable surface morphology and predictable diffusion. Alternatively, sample TLP-C contains an In concentration exceeding what is necessary for proper Au-In bonding and has demonstrated poor thermal cycle reliability. For this reason, Au-In TLP bonds below an In mole fraction of 0.67 remain a very promising die-attach method for SiC devices. It is worthwhile to continue electrical reliability testing of the TLP bonds that have shown promising results in this work while also expanding to perform a robust mechanical reliability investigation utilizing SiC devices and silicon nitride substrates as well.

ACKNOWLEDGEMENTS

The authors thankfully note that research performed in part at the NIST Center for Nanoscale Science and Technology.

REFERENCES

[1] P. G. Neudeck et al., "6H-SiC Transistor Integrated Circuits Demonstrating Prolonged Operation at 500 C," in IMAPS International Conference and Exhibition on High Temperature Electronics (HiTEC 2008), 2008.

[2] B. Grummel, R. McClure, L. Zhou, A. P. Gordon, L. Chow, and Z. J. Shen, "Design Consideration of High Temperature SiC Power Modules," in Industrial Electronics, 2008. IECON 2008. 34th Annual Conference of IEEE, pp. 2861–2866, 2008.

[3] P. O. Quintero, T. Oberc, and F. P. McCluskey, "High Temperature Die Attach," presented at the High Temperature Electronics (HiTEC 2008), Albuquerque, New Mexico, p. 207, 2008.

[4] W. F. Gale and D. A. Butts, "Transient Liquid Phase Bonding," Science and Technology of Welding & Joining, vol. 9, no. 4, pp. 283–300, 2004.

[5] Y. C. Chen and C. C. Lee, "Indium-copper Multilayer Composites for Fluxless Oxidation-free Bonding," Thin Solid Films, vol. 283, no. 1, pp. 243–246, 1996.

[6] H. A. Mustain, W. D. Brown, and S. S. Ang, "Transient Liquid Phase Die Attach for High-Temperature Silicon Carbide Power Devices," Components and Packaging Technologies, IEEE Transactions on, no. 99, p. 1, 2010.

[7] C. Lee, C. Wang, and G. Matijasevic, "Au-In Bonding Below the Eutectic Temperature," Components, Hybrids, and Manufacturing Technology, IEEE Transactions on, vol. 16, no. 3, pp. 311-316, 1993.

[8] H. Okamoto, "Au-In (gold-indium)," Journal of Phase Equilibria and Diffusion, vol. 25, no. 2, pp. 197–198, 2004.

[9] Y. Kitaura, "Long-term Reliability of Pt and Mo Diffusion Barriers in Ti-Pt-Au and Ti-Mo-Au Metallization Systems for GaAs Digital Integrated Circuits," Journal of Vacuum Science & Technology B: Microelectronics and Nanometer Structures, vol. 12, no. 5, p. 2985, 1994.

[10] H. A. Mustain, W. D. Brown, and S. S. Ang, "Tungsten Carbide as a Diffusion Barrier on Silicon Nitride Active- Metal-Brazed Substrates for Silicon Carbide Power Devices," Journal of Electronic Packaging, vol. 131, no. 3, p. 034502, 2009.

978-1-4244-8425-6/11 $26.00 © 2011 IEEE

Proceedings of the 23rd International Symposium on Power Semiconductor Devices & IC's
May 23-26, 2011 San Diego, CA

Thermal Impedance Spectroscopy of Power Modules during Power Cycling

Alexander Henlser, Daniel Wingert,
Christian Herold, Josef Lutz
Chemnitz University of Technology
Chemnitz, Germany
Email: alexander.hensler@etit.tu-chemnitz.de

Markus Thoben
Infineon Technologies AG
Warstein, Germany

Abstract—**The presented thermal impedance spectroscopy of power modules simplifies significantly the failure analysis of power modules. It enables online observation of degradation within the cooling path with detailed information about failure mechanisms. The degradation of certain layer within the power module is detected by observation of Z_{th} parameters. Several tests results are compared with analysis of the scanning acoustic microscope.**

I. INTRODUCTION

The reliability of power modules at power cycling load is determined besides bond wire lift-off by deterioration of different layers within heat flow path (Fig. 1). In order to detect this failure during reliability tests, today the quasi steady state thermal resistance between junction and heat sink or coolant is monitored. This measurement, however, delivers no information which layer of the power module is degraded. Thus for detailed information about specific failure mechanism, subsequent time consuming and often destructive failure analyses are needed, e.g. metallographic preparation. Thermal impedance spectroscopy of power modules promises a simpler and faster method for the failure localization.

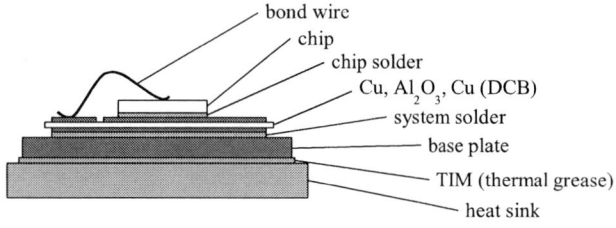

Figure 1: Typical power module with base plate

II. METHOD DESCRIPTION

A. Definition of Thermal Impedance Spectrum

The transient thermal behavior of power modules is usually given between the junction and a reference temperature T_{ref} (case, heat sink or ambient). This function is determined by means of step response. The power device is exposed to an active power pulse P_V generated with load current. After the steady state is reached, the load current is

switched off and the cooling behavior is measured until the junction temperature is equal to the reference temperature. From this the Z_{th} function is extracted (Eq. 1). Today this function is often determined by simulations.

$$Z_{th} = \frac{T_j - T_{ref}}{P_V} \qquad (1)$$

In applications this function is used to estimate the junction temperature of the power device depending on power losses and pulse time. This Z_{th} function is approximated with an equivalent Foster network as shown in Fig. 2. The mathematical description is given in Eq. 2.

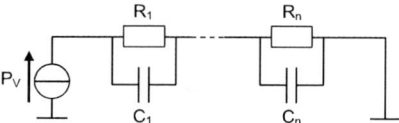

Figure 2: Equivalent Foster network

$$Z_{th}(t) = \sum_{i=1}^{n} R_i \left(1 - e^{-\frac{t}{\tau_i}} \right); \; \tau_i = R_i \cdot C_i \qquad (2)$$

In most cases few RC elements are sufficient to describe the thermal system properly. E.g. in data sheet for the thermal impedance between junction and case (bottom side of the base plate) four elements are used. For direct liquid-cooled power modules the approximation with five elements is standard.

For further considerations the correlation of Eq. (2) is defined as the thermal impedance spectrum. Time constants τ_i are on the x-coordinate, magnitudes R_i are on the y-coordinate. For example the thermal impedance spectrum of a power module is depicted in Fig. 3. This spectrum describes the thermal impedance between junction and case (Z_{thjc}).

978-1-4244-8425-6/11 $26.00 © 2011 IEEE

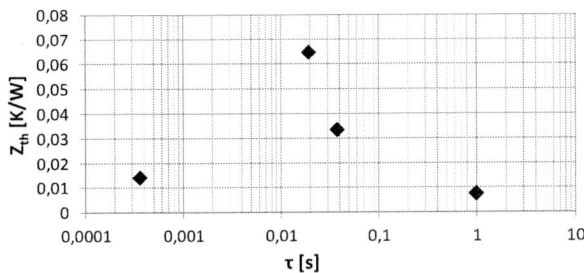

Figure 3: Thermal impedance spectrum of a power module

B. Approach for Failure Localization

It is assumed that different failures within the heat flow path of the power module lead to different change of the thermal impedance spectrum. Partial thermal resistances R_i with corresponding low time constant τ_i represent the heat flow area near to the chip. An increase is expected in this R_i part, if the chip solder layer degradation determines the increase of the thermal resistance. RC elements with higher time constants describe the material layers which are further away from the chip.

For this simplified approach the failure localization is implemented with the following schedule:

- Periodical measurement of the $Z_{th}(t)$ function during power cycling test

- Extraction of the equivalent FOSTER network with an approximation method (e.g. [1])

- Comparison of the thermal impedance spectrum with the initial state

III. EXPERIMENTAL APPLICATION

A. Separation of chip solder degradation from thermal grease effect

First verification of the thermal impedance spectroscopy was performed during a power cycling test. The device under test was a standard power module of Infineon Technologies AG, Fig. 4.

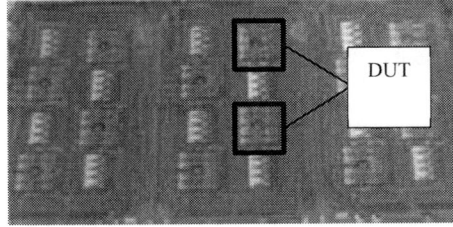

Figure 4: DUT (IGBT of standard power module)

During the reliability test the DUT was loaded with active thermal cycles induced by cycling direct current. The temperature cycle of the junction is depicted in Fig. 5.

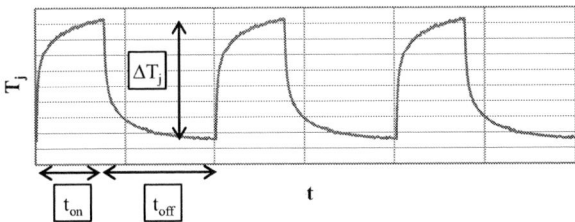

Figure 5: Junction temperature during power cycling test

Test parameters in detail are:

I_{Load}=400A; t_{on}=0,6s; t_{off}=3,1s; T_{jmax}=175°C, ΔT_j=105K, P_V=340/cm²

The steady state thermal resistance R_{thja} between junction and coolant was measured corresponding to the conventional method of the aging monitoring. The trend of the R_{th} is shown in Fig. 6. In the beginning of the test there is a significant decrease of the R_{thja} and in the range of 30.000 cycles a low increase. Thereby the failure criterion (20% increase of the initial value) was not reached.

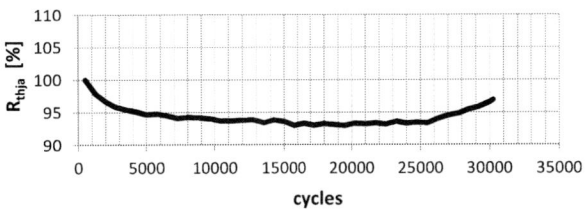

Figure 6: Trend of steady state thermal resistance R_{thja}

Parallel to the conventional method the thermal impedance spectrum was monitored. The $Z_{thja}(t)$ was measured every 250 cycles. For this measurement the cycling test was interrupted. After that a load pulse was applied with 250A load current, 40s heating and 40s cooling time. The $Z_{thja}(t)$ was recorded during the cooling phase.

With the thermal impedance spectroscopy different effects can be separated. In Fig. 7 the change of the spectrum at 30.000 cycles is shown in comparison to the initial state.

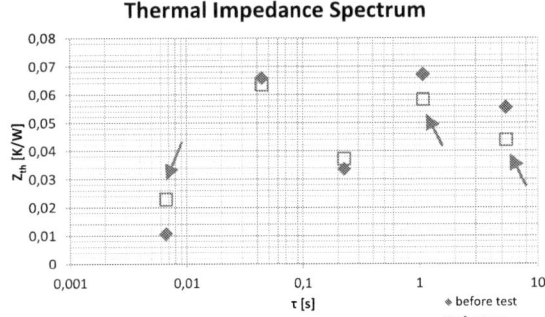

Figure 7: Thermal impedance spectrum before and after test

As shown in Fig. 7 the R_i parts of RC elements with higher time constants decrease. It should be the influence of the

978-1-4244-8425-6/11 $26.00 © 2011 IEEE

thermal grease. The partial R_i corresponding to the lowest time constant show a significant increase. The RC elements in the middle of the spectrum show no change. According to the simplified approach for the failure localization first RC element should indicate chip solder layer failure, elements with time constants greater than 1s show thermal grease effect and elements in-between contain the influence of the system solder layer. The trend of the thermal impedance spectrum of Fig. 7 is shown in Fig. 8.

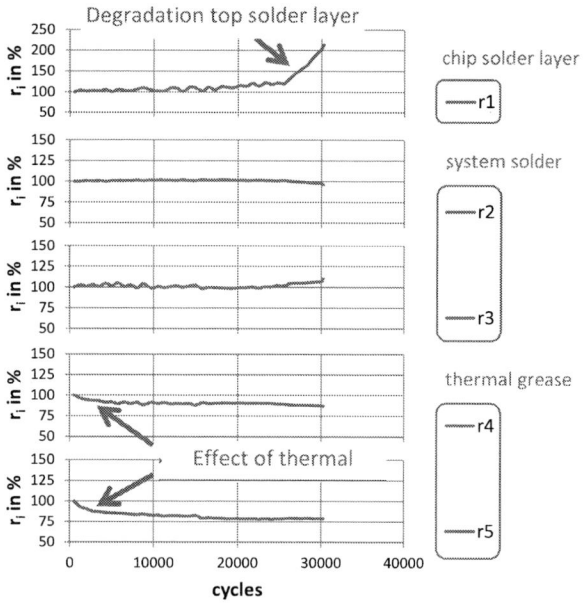

Figure 8: Trend of thermal impedance spectrum

Corresponding to this analysis the power module should have degradation within the chip solder layer and no degradation within the system solder layer. It was verified with scanning acoustic microscopy. The analysis is shown in Fig. 9. In the middle image bright regions of the chip solder layer show clearly the degraded chip solder layer, whereas the system solder layer has no degradations. It is conform to the result of the thermal impedance spectroscopy.

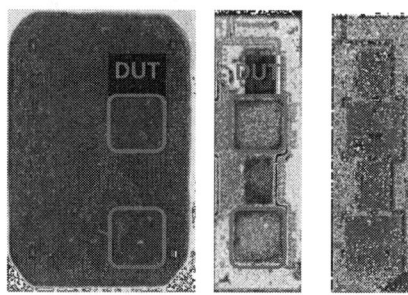

Figure 9: left: system solder layer after test without degradation, middle: degraded chip solder layer after test, right: unstressed chip solder layer

B. Localization of system solder layer failure

For the verification of this method for localization of the system solder layer failure, a superimposed power cycling test was performed. The DUT was the power module type "HybridPACK1" as well.

The power module was mounted on a liquid-cooled heat sink. With an external heating/cooling station the power module was heated and cooled passively. During the heating phase power, cycles were superimposed. The temperature trend is shown in Fig. 10 Test parameters are listed in Table 1. These test parameters were chosen to activate the system solder layer degradation.

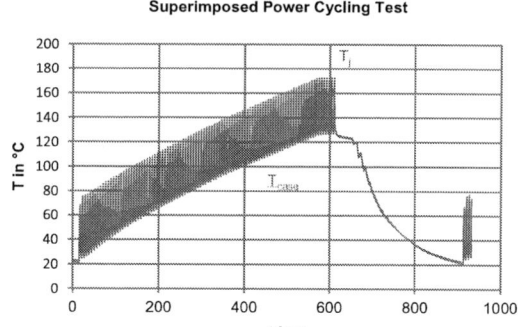

Figure 10: Temperature trend of superimposed power cycling test

Table 1: Test parameters of superimposed power cycling test

$T_{coolant\_min}$	22°C
$T_{coolant\_max}$	122°C
$t_{heating}$ (passive cycle)	10min
$t_{cooling}$ (passive cycle)	5min
I_{Load}	220A
t_{on} (power cycling)	2s
t_{off} (power cycling)	4s
P_V	140 W/cm²
Power cycles per passive cycle	100
T_{jmax}	175°C
ΔT_j (power cycling)	53K

The thermal impedance was measured periodically every 20 passive cycles. For this measurement the cycling test was interrupted at the lowest coolant temperature at the end of the cooling phase. After that a load pulse was applied with 250A load current, 20s heating and 20s cooling time. During the cooling phase the Z_{th} was measured between junction and case as shown in Fig. 11.

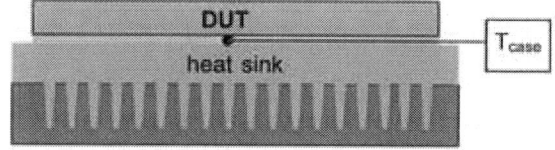

Figure 11: Test set-up for Z_{thjc} measurement

The thermal impedance spectrum was monitored for the localization of the failure mechanism. The partial thermal resistances are shown in Fig. 12. In this power cycling test the parameter r2 has a significant increase. It indicates the system solder layer degradation. The effect of the thermal grease is shown in the parameter r4 and can be clearly separated from other failure mechanisms.

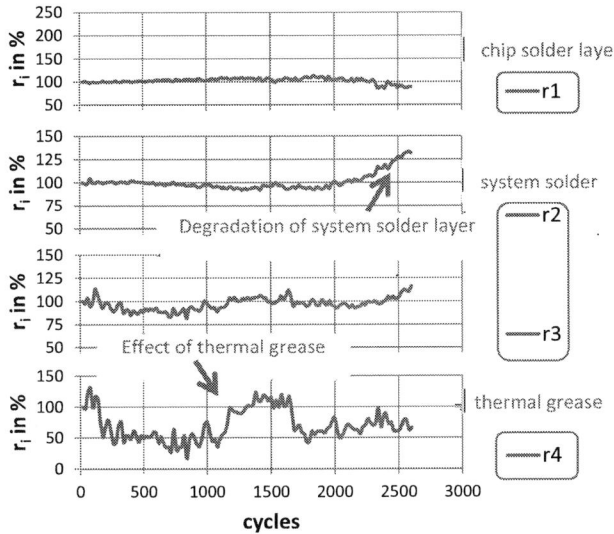

Figure 12: Trend of thermal impedance spectrum

The subsequent failure analysis with the scanning acoustic microscopy confirmed the failure of the system solder layer. Bright region beneath the chip is the delaminated solder area, which caused the increase of the thermal resistance.

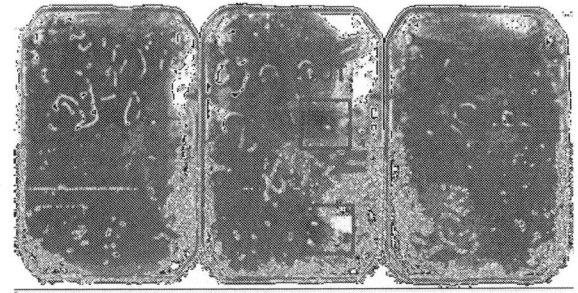

Figure 13: System solder layer after superimposed power cycling test

IV. CONCLUSION

The thermal impedance spectroscopy is an appropriate non-destructive failure analysis method for power modules. This method provides the separation of partial thermal resistances and enables an online monitoring of different failures. With this method typical failures can be clearly identified in power cycling tests. With experimental tests the degradation of the chip solder layer, the failure of the system solder and the effect of the thermal grease could be distinguished and localized.

An experimental proof for power modules is published here for the first time. A further improvement of this method is conceivable corresponding to the high resolution evaluation method described in [2]. Also the recursively rapid Foster-Cauer circuit transformation described in [3] is a further tool for more detailed conclusions from thermal impedance spectra. This is the subject for further investigations.

ACKNOWLEDGEMENT

The work was supported by grants of the Federal Ministry of Economics and Technology (BMWi).

REFERENCES

[1] C. L. Lawson and R. J. Hanson, "Solving Least Squares Problems," p. 161, 1974.

[2] V. Szekely, "A new evaluation method of thermal transient measurement results," *Microelectronics Journal*, no. 28, pp. 277-292, 1997.

[3] Y. C. Gerstenmaier, W. Kiffe, and G. Wachutka, "Combination of Thermal Subsystems Modeled by Rapid Circuit Transformation," in *THERMINIC*, Budapest, 2007.

[4] A. Wintrich, U. Nicolai, W. Tursky, and T. Reimann, "Applikationshandbuch Leistungshalbleiter," pp. 82-85, 2010.

[5] J. Lutz, H. Schlangenotto, U. Scheuermann, and R. De Doncker, *Semiconductor Power Devices*. 2011.

[6] U. Scheuermann and R. Schmidt, "Investigations of the VCE(T)-Method to Determine the Junction Temperature by Using the Chip Itself as Sensor," in *Proceedings of PCIM Europe 2009*, Berlin, 2009.

Proceedings of the 23rd International Symposium on Power Semiconductor Devices & IC's
May 23-26, 2011 San Diego, CA

Application driven integrated design of a half-bridge power switch

Adane Solomon, Alberto Castellazzi

Department of Electrical & Electronic Engineering
University of Nottingham
NG7 2RD Nottingham, UK
alberto.castellazzi@nottingham.ac.uk

Abstract— **This paper presents the assembly of a half-bridge switch taking into account the actual current commutations encountered in power converter topologies. So, rather than optimizing the connection between an active switch (e.g., IGBT) and its anti-parallel freewheeling diodes, a novel approach is proposed which favors compact interconnection of the high-side transistor with the low-side diode and vice-versa. The result is a very low-inductive, double-sided cooled power switch, suitable for advanced integration of power conversion equipment.**

I. Introduction

The anti-parallel connection of a controllable device (e.g., an IGBT) with a freewheeling diode (Fig. 1a) is the basic building block of most commercial power modules (in the following, this arrangement is referred to as the *basic switch*). This is mainly due to the structural characteristics of the devices, mostly of vertical type, and to the widespread use of wire-bond technology for the surface interconnection, which make it easy to assemble basic switches of this kind. Fig. 1b shows a typical assembly of an elementary switch as per Fig. 1 a) in standard commercial packaging technology: the devices backside (i.e., IGBT collector an diode cathode) is soldered onto a ceramic DCB (Direct Copper Bonded) substrate and the interconnections to the remaining terminals (i.e., IGBT gate and emitter and diode anode) are implemented by means of aluminum wires bonded to the device surface and to the DCB copper tracks, properly patterned. In the example of Fig. 1 b), more devices are connected in parallel for current rating requirements, with a typical ratio of 2 IGBTs per 1 diode.

Bond-wire technology exhibits some limits, which the power electronics community has long wished to see overcome: to name but a few, bond-wire lift-off and degradation of the contact metallization due to thermal and power cycling [1]; relatively high parasitic inductance causing power sharing unbalance at higher operating frequencies [2];

single side (backside) cooling of the devices, while heat-generation mainly takes place at the surface. All of these issues significantly affect the performance of the switches and are of major concern in guaranteeing a given operational lifetime. To eliminate some of these drawbacks, recently, alternative sandwich packaging and interconnection concepts have been proposed, which, based on the use of solid posts as opposed to bond-wires, enable designs with better electro-magnetic and electro-thermal performance and the potential for either higher power densities for a given lifetime or a longer lifetime for a given power density [3,4].

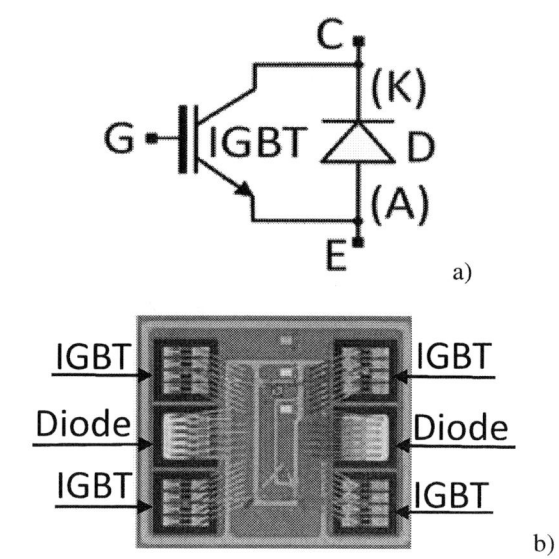

Fig. 1: Schematic of a anti-parallel IGBT-Diode basic switch, a), and example of standard commercial assembly, b).

978-1-4244-8425-6/11 $26.00 © 2011 IEEE

Hitherto presented novel approaches still aimed at the integration the active switches with the anti-parallel freewheeling diode. However, in the application, two switches as per Fig. 1 need to be connected in series (high-side and low-side switch) to build a half-bridge power switch configuration, as extensively required, for instance, by synchronous rectified DC-DC converters and by DC-AC converters (i.e., inverters). During operation within such power converter topologies, current commutations always take place between a high-side transistor and a low-side and, vice-versa, between a low-side transistor and a high-side diode. Current commutation between anti-parallel transistor-diode pairs only takes place at zero voltage in the case of synchronous rectified DC-DC converters, and at zero current in the case of inverters and is thus by definition non critical in both cases (i.e., it does not imply power dissipation or potentially destructive voltage overshoots). So, as pointed out in [5] and illustrated in Fig. 2, for improved switching performance, packaging design should aim at integrating and minimizing not the high-side or low-side basic switches, but rather the *positive* and *negative switching cells*, that is, the high-side transistor/low-side diode and the low-side transistor/high-side diode pairs, respectively (the terminology positive and negative cell is derived from the sign of the load current during inverter operation).

substrates and interconnected by means of copper posts soldered between their top metallization, treated with silver to be made solderable, and the other DCB. This solution still results in a sandwich type assembly, but with the novel feature of having a basic switch top-flipped with respect to the other. This feature has the twofold advantage of enabling more compact designs, with improved layout from an electro-magnetic point of view, as discussed above, and also of providing a better utilization of the cooling surfaces. In fact, even in the case of double-sided cooling, the favorite path for heat removal remains through the backside; the estimated heat-removal via the device surface is around 30-35% in the best cases. Conventional sandwich packaging approaches have all the chips mounted backside down on the same surface and thus imply unbalanced cooling conditions. Here, on the other hand, thermal symmetry is ensured. In this example, bond-wires were still used here for the connections to the IGBT gate and emitter terminals used for driving the transistor, but this was just for ease of laboratory assembly: all interconnections could be easily implemented with solid posts provided properly patterned DCBs are available. The load current flow is however entirely via the bumps and substrate copper tracks. Moreover, as is clearly visible in Fig. 3 b), which shows the final switch assembly, this integration approach enables to easily keep separate driving and power loops, positioning the relative electrical interconnections on opposite sides of the switch; it is suitable for double-sided (liquid) cooling of the

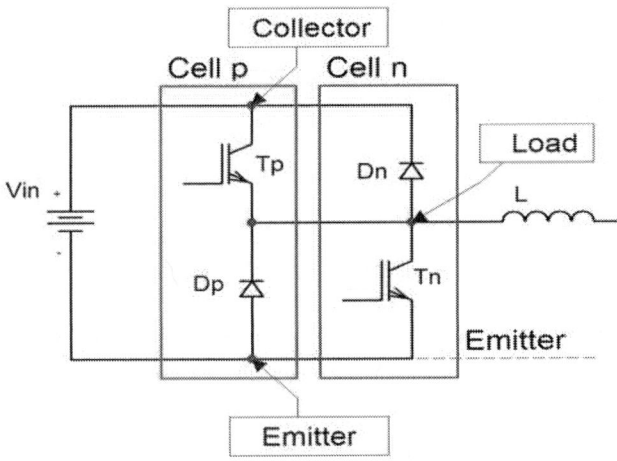

Fig. 2: Indication of application driven integration requirement.

Taking into account the above considerations, this paper presents an advanced, application-driven integration approach. In particular, in trying to optimize the switching performance of the half-bridge switch, use is made of power bump technology, so as to keep all of the positive features of sandwich packaging approaches, such as double-sided cooling and reduced stray inductance.

II. ADVANCED INTEGRATION OF THE HALF-BRIDGE SWITCH

Here, IGBTs are used as the controllable devices. As shown in Fig. 3 a), the chips were soldered on separate DCB

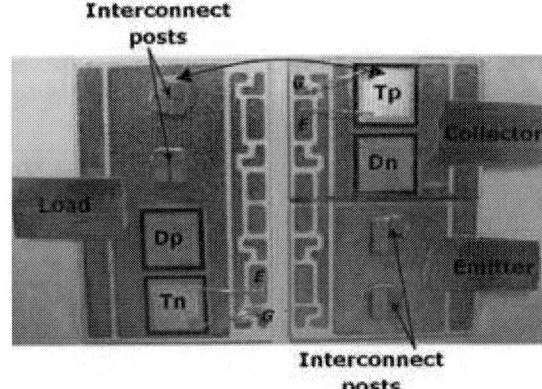

a)

b)

Fig. 3: Open view (i.e., before assembly) of the half-bridge switch, a), and integrated power switch, b).

devices and for the possible additional integration of passives and gate-drivers on the top and bottom substrates.

III. SWITCH CHARACTERISATION

The developed switch was characterized electro-magnetically and electro-thermally, employing structural numerical analysis tools, and tested for functional performance. Fig. 4 shows the 3D model mesh for the lector-magnetic characterization of the assembly, which was performed with FastHenry [6]. The values of parasitic inductance and resistance between all terminal pairs (Collector, *C*, Emitter, *E*, and Load, *L*, see Fig. 3) were extracted and provided very encouraging indications. For instance, the estimated parasitic inductance between C and E was only 6.4 nH at a frequency of 10 kHz .

Fig. 4: 3D structural model mesh of the switch assembly for the extraction of equivalent electro-magnetic elements.

Fig. 5 shows the results of numerical steady-state thermal simulations for different packaging schemes [6]: a), silicon chip dissipating 100 W and being effectively cooled only via the backside (representative of standard bond-wire technology); b), effect of having surface interconnections connected to a cooling surface (i.e., conductive boundary conditions on top of bumps, representative of 2D integration approaches, as in [3,4]). Here, the chip dimensions are 10 x 10 x 0.5 mm; the initial temperature is 300K in all cases; boundary conditions on the cooling surfaces emulate conduction (i.e., very high heat-transfer coefficient), on the free surfaces natural convection; the bumps are copper, 2 mm thick. These results clearly show that surface bumps can be effectively used for partial heat-removal via the surface. Fig. 6 shows the results of thermal simulation for the whole assembly and with a more realistic value of heat-transfer coefficient, corresponding to forced convection liquid cooling conditions. In particular, the total symmetry of operation for the two cooling surfaces can be observed.

Finally, the prototype assembly was tested in a classic double-pulse test circuit (see [2], for example). Fig. 7 shows a photograph of the test circuit and the experimental results. The photograph of the hardware well demonstrates the possibility for advanced integration of passives and drivers on the bottom and top DCBs (used here only for cooling an mechanical assembly), additional connotative features of the sandwich packaging approaches presented in [3,4]. Though still at conservative voltage and current levels, the experimental results show good performance, with barely any ringing. The voltage overshoot at turn-off is due to the need of inserting an

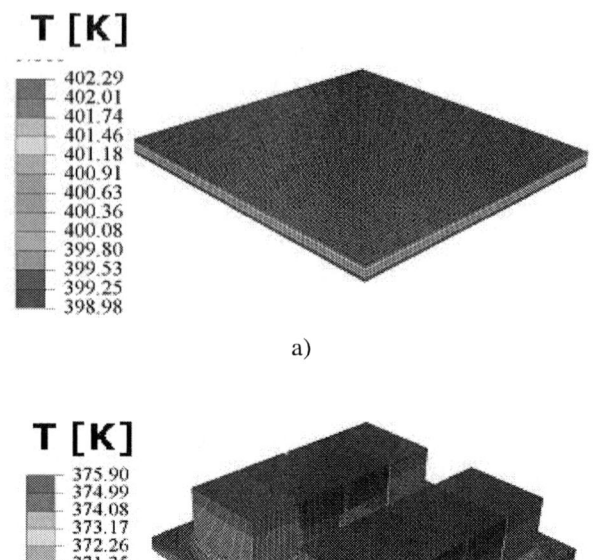

a)

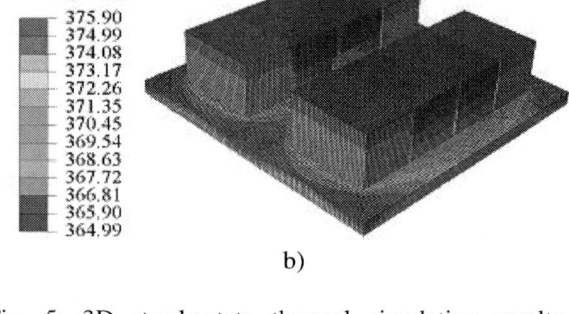

b)

Fig. 5: 3D steady-state thermal simulation results of various chip packaging concepts: a), single silicon chip cooled only via the backside; b), effect of using surface bump interconnections connected to a cooling surface.

overlong conduction path to enable insertion of a current probe, also visible in Fig. 7.

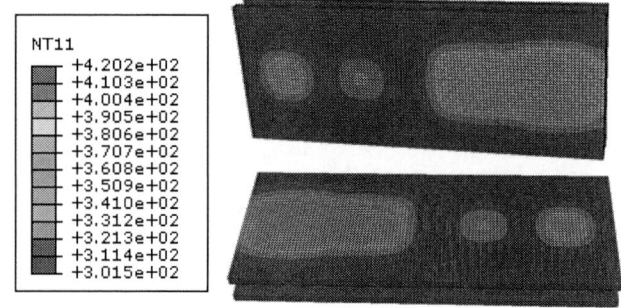

Fig. 6: Simulated temperature distribution for the same applied power dissipation in the assembly built here. The temperature scale is in K.

IV. CONCLUSION AND OUTLOOK

This work has presented the development of a novel integration approach for a widespread basic building block of power electronics conversion, the half-bridge power switch. The novel approach takes into consideration not only chip structural aspects, but also application related functional issues and optimizes the interconnection among single chips accordingly. Preliminary results show the potential for an interesting advancement of the state-of-the-art, represented by recently proposed sandwich package concepts based on solid bump or post interconnections and double-sided cooling, in particular by ensuring a more even temperature distribution within the power module and by further minimizing parasitic inductance.

Presently, the optimization of the half-bridge switch layout (i.e., positioning of IGBTs and diodes) is being investigated: an interleaved arrangement of the devices, that is, with the locations of Dp and Dn exchanged as compared to Fig. 3a), provides enhanced characteristics, both in terms of parasitic inductance and thermal performance. A sketch of the interleaved half-bridge switch is shown in Fig. 8 a). Fig. 8 b) shows the results of thermal simulations for this arrangement, under the same load conditions used in producing Fig. 6: a comparison between the two clearly shows the gain in performance that can be achieved with the interleaved concept.

ACKNOWLEDGMENT

We are grateful to Dr. Munaf Rahimo of ABB Semiconductors, Lenzburg, Switzerland, for providing the IGBT and Diode chips used in this work. We also wish to acknowledge the support received by the Faculty of Engineering, University of Nottingham, UK, in the framework of the High Flier research placement scheme.

REFERENCES

[1] M. Ciappa, *Selected failure mechanisms of modern power modules*, Microelectronics Reliability 42 (2002), 653-667.

[2] A. Castellazzi, M. Ciappa, W. Fichtner, G. Lourdel, M. Mermet-Guyennet, *Compact modeling and analysis of power-sharing unbalances in IGBT-modules used in traction applications*, Microelectronics Reliability 46 (2006), 1754 – 1759.

[3] M. Mermet-Guyennet, *New structure of Power Integrate Module*, in Proc. of the 4th Conference on Integrated Power Electronics Systems (CIPS 2006), Naples, Italy, 2006.

[4] C.M. Johnson et al., *Compact Double-Side Liquid-Impingement-Cooled Integrated Power Electronic Module*, ISPSD 2007, Jeju, Korea, pp. 53-56.

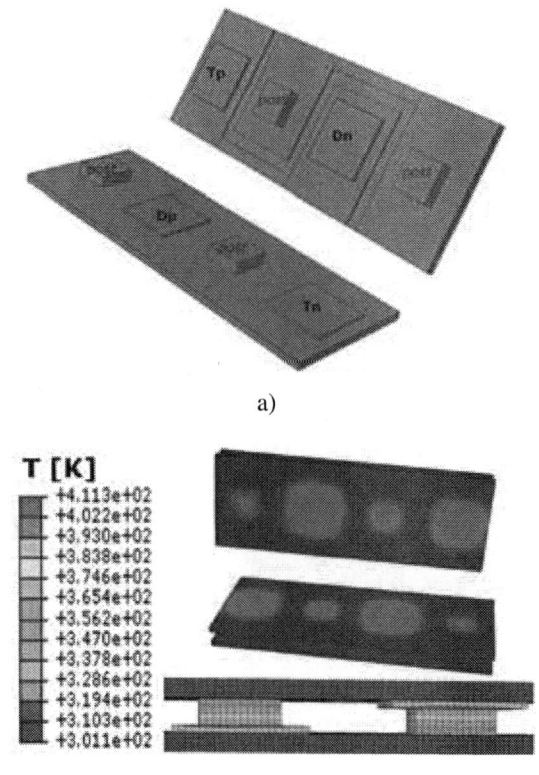

a)

b)

Fig. 8: *Interleaved* integration approach, a), simulated temperature distribution for the same load conditions applied in the case above, c).

[5] F.Z. Peng, *Revisit Power Conversion Circuit Topologies–Recent Advances and Applications*, in Proc. of the 6th IEEE International Power Electronics and Motion Control Conference (IPEMC-ECCE Asia 2009), Wuhan, China, May 2009.

[6] http://www.fastfieldsolvers.com/.

[7] http://www.simulia.com/products/abaqus_fea.html

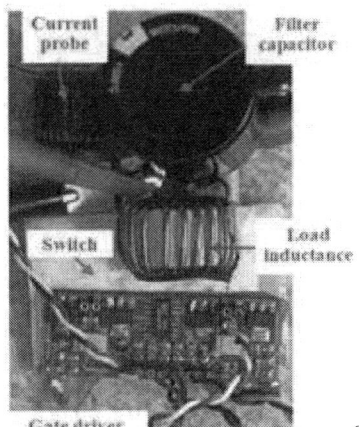

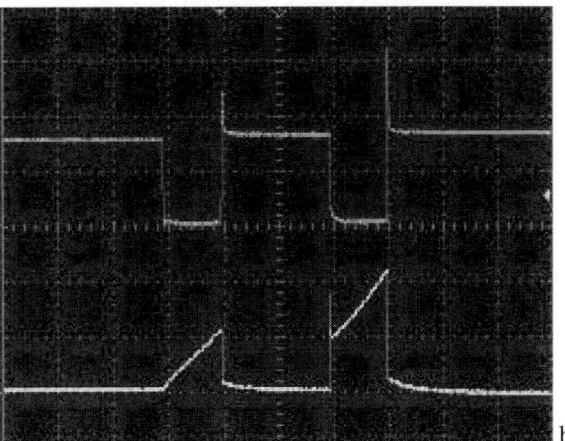

a)

b)

Fig. 7: Double-pulse test circuit, a), and experimental current and voltage waveforms, b). The time scale is 25us/div; the current (bottom waveform) is 50A/div; the voltage (top waveform) is 10V/div.

Proceedings of the 23rd International Symposium on Power Semiconductor Devices & IC's
May 23-26, 2011 San Diego, CA

Investigation on Wirebond-less Power Module Structure with High-Density Packaging and High Reliability

Yoshinari Ikeda[1], Yuji Iizuka[1], Yuichiro Hinata[1], Masafumi Horio[1], Motohito Hori[2] And Yoshikazu Takahashi[2]

[1]Energy and Environmental System Research Center
Fuji Electric Holdings Co., Ltd., Matsumoto, Nagano, Japan
ikeda-y@fujielectric.co.jp

[2]Semiconductor Development Center,
Fuji Electric Systems Co., Ltd., Matsumoto, Nagano, Japan

Abstract—**A newly developed module with wirebond-less structure is investigated. This structure has multi-pin attached interconnection structure implanted into power circuit board with connecting line between chips and other elements inside the power module. Additionally, heat-spreader-like copper blocks bonded to ceramic insulated substrates performing high thermal conductivity, enable to realize high current capability operations effectively. Moreover, full molded resin package performs higher reliability comparing with the conventional module, shows the package structure is one of the potential candidates of power module for the high power applications including Wide Band Gap (WBG) devices.**

I. INTRODUCTION

Power modules are widely used in the power conversion applications, such as industrial, household, hybrid electric vehicles, and electric vehicles. Recent demands for power modules require more and more high power density than ever, especially in the area of renewable energy related application, such as wind power, photo voltaic, and power transformer for power grid. From the both views of gaining power efficiency and density, WBG semiconductor devices such as Silicon Carbide (SiC) devices and Gallium Nitride (GaN) devices are very promising. One of the biggest benefits is their wide band gap associated characteristics, high dielectric breakdown field and high thermal conductivity. In order to take advantages of their superior characteristics, packaging will be one of the most important keys to enable them to achieve their high power performances in higher temperature, for the reason that the lifetime of power module gets shorter in such a condition.

The mainstream of power module packaging is known as wirebonding structure, widely utilized for power modules to construct current path inside the package until now. However, it has drawbacks, which requires additional bonding areas on the substrates to make interconnection path, and causes high-density packaging development to be difficult. Recently, aiming to overcome the bottleneck point described above, several works concerning with alternative structure of power modules have been reported [1][2].

In this paper, we investigate newly developed module without wirebond structure, which has multi-pin attached interconnection structure implanted into power circuit board. The multi-pin are mounted on each chip electrode as

alternative interconnection technology for appropriate use of high power application, with newly added two parts to realize high-density packaging and high-reliability, simultaneously.

II. NEW PACKAGE STRUCTURE

In order to realize high-density packaging and high reliability, new structure has three different parts from conventional structure. Fig.1 shows the schematic view.

A. High-Density Packaging

Newly developed structure has power circuit board locating upper side with patterned multi-layer conduction lines. Power circuit board has copper pin inserted into through-holes. Copper pin attaches each chip electrode as alternative elements to wirebond.

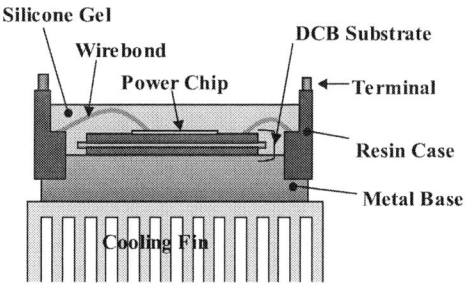

a. Conventional wirebond structure

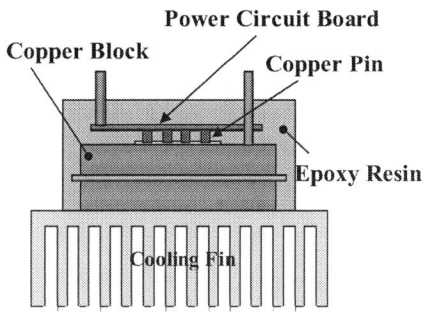

b. Newly developed structure

Figure 1.   Sectional view of package structure

978-1-4244-8425-6/11 $26.00 © 2011 IEEE

These copper pin connections on the surface of chip allow higher current capability than conventional wirebond structure, as its lower electrical resistance of copper material and much shorter electrical path into power circuit board facing to chips.

It will be noted following chapter, developed short connection structure decreases stray inductances inside the package, acts switching performance comparing with conventional wirebond structure.

B. Low-Thermal Resistanse

Lower structure side includes heat-spreader-like copper block with insulation substrate bonded to Silicon-Nitride (Si_3N_4), or Alumina (Al_2O_3). This structure affects the heat conduction improvements of transferring heat flow to the cooling fin. Under the power chip, heat diffusion via block spreading all over the mounting area, widening heat flux before the heat flow reaches to the insulator, which defines thermal resistances by its own poor conductivities. Fig. 2 shows the results by FEA. Thermal resistance (Rth) improvements of newly developed structure are 37%, comparing with traditional wirebond structure. Additionally, for the case of SiN insulator substrate structure shows 55% decreasing of Rth.

Typically, traditional wirebond makes the pattern layouts of normal Direct Copper Bonding (DCB) substrates to be complicated, not only for the requirements of bonding area in the DCB, but also making the connections between chips and copper pattern, to form the inner circuit lines inside the package. On the contrary, newly developed structure has power circuit board, as interconnecting lines between chips and terminals, which makes the lower substrates pattern topology to be simple. Considering ideal thermal conduction, the best way is to use plane without pattern edges. Additional effect to refine thermal properties of structure arises, because new structure allows simple pattering layout by the reason described above.

C. High-Reliability

Conventional package has soft silicone-gel filled structure mainly for protecting metal surfaces and joining interfaces from degradations. Silicone-gel has so small rigidity, with no effects on the structure as stress transferring material. A newly developed structure has additional function to improve the reliabilities by its own properties, rigidly epoxy-molded structure, which differs from silicone-gel filled structure. Fig.3 shows FE model and results, about 30% decreasing of strain amplitude per cycle for newly developed structure with epoxy molded resin in power-cycling-mode simulation.

By the simple inspection of the results, stress amplitudes in the joining layer reduced, showing the smoothing effects to the soft metal layer without prominent stress concentrations, which causes the reduction of amplitude of strain generated in the soft-solder layer, for the reason that high-rigidity molding resin activate as strain dispersion structure.

Derived strain amplitude specifies lifetime shortening effects by the mechanical fatigue accumulation in bonding layer, having relationship in the form of anti-proportional formula [3]. According to the previously derived reducing

strain results of module, it is expected lifetime improvements will be the benefit of newly developed structure.

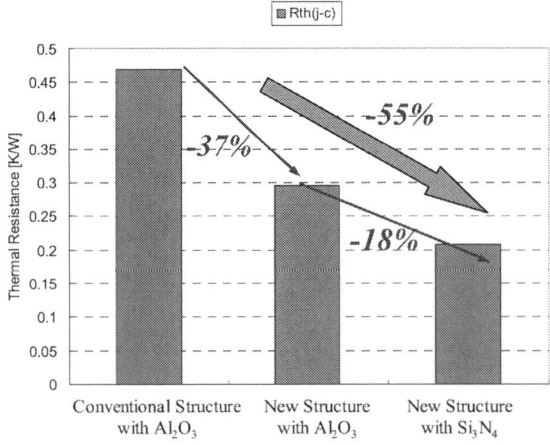

Figure 2.   FE Analysis: Comparison of package thermal resistances

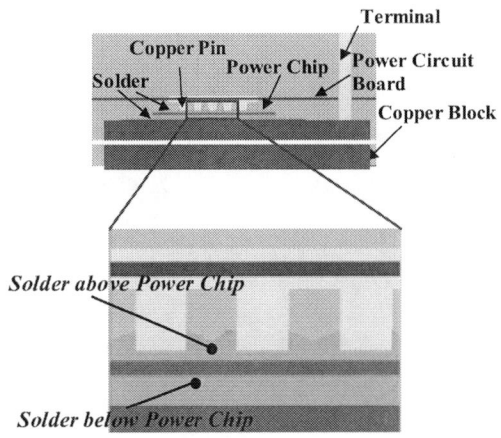

a. FE Model (Stress Analysis)

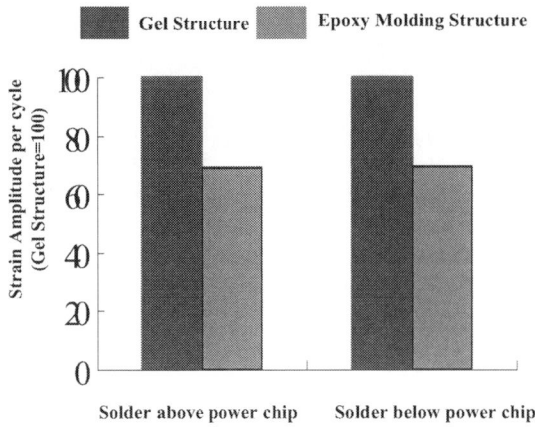

b. Strain Amplitude in Power Cycling Mode

Figure 3.   FE Analysis: Strain amplitude of power cycling mode

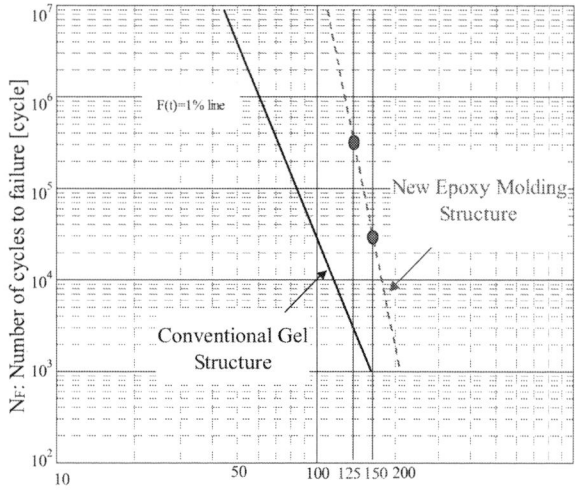

Figure 4.   Results of power cycling reliability test

Hybrid/All-SiC Module   **SiC SBD Module**

Figure 5.   Photos of newly developed structure modules, 1200V SiC devices embedded.

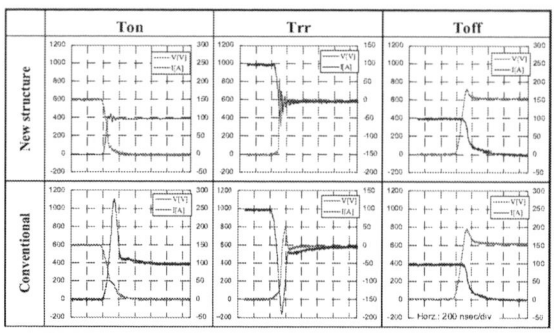

Figure 6.   Switching waveforms comparison of two type samples, new structure module with Si-IGBT & SiC-SBD (Upper side), conventional structure module with Si-IGBT & Si-FWD (Lower side)

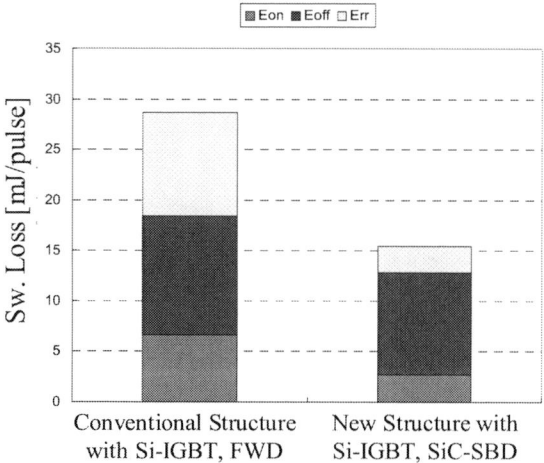

Figure 7.   Switching loss comparison in Fig.6 Conventional, new structure module

D. Reliability-Tests

Fig.4 shows the results of power-cycling test in the condition of starting temperature 25deg C, proving newly developed structure has longer lifetime more than conventional gel-filled structure. In the condition of junction temperature swing 125[K], N_F of new structure is about 100 times same as conventional structure. In ISPSD2010, we demonstrated that the conventional wirebonding structure modules are dominated by the withstanding capabilities of bonding interfaces, including electrodes of the chips, to thermal aging effects by electrical operations [4]. For the case of newly developed structure, physical observations indicate that failure modes are dominated by joining material fracture concerning with fatigue in the sense of cyclic accumulated, representing Rth increasing failure as module.

In the large temperature swings region, these lifetime curves seem to have cross point at the neighborhood of 200[K], in the test condition maximum temperature reaches melting point of joining materials. Swinging temperature increase causes acceleration effect of fatigue, monotonically. In the point of this, joining material choice is the key issue for the further improvement of the newly developed structure.

III.   APPLICATION TO WBG DEVICES

Previous investigations on newly developed structure gave insight into its structural features such as high-density packaging and high reliability. Finally, we investigated the validity of newly developed structure for the case, application to WBG device, from the point of views of electrical operations, comparing with conventional package. Fig.5 shows photos of them, left picture is half-bridge-type package sample (100A/1200V), and right picture is diode module in double series connection type (400A/1200V).

A. Hybrid-type Module (Si-IGBT & SiC-SBD)

Fig.6 shows switching waveforms of sample modules. In the turn-on phase including reverse recovery, it is obvious that for the case of newly developed structure embedding SiC-SBD, extreme reduction of surge current peak, owing to the unipolar-type SiC-SBD characteristics. Another feature of improvements motivated by structure are confirmed by the waveforms of turn-off, showing lower surge voltage. As mentioned previously, high-density packaging structure gives the effects on them, internal stray inductance reduction.

978-1-4244-8425-6/11 $26.00 © 2011 IEEE   274

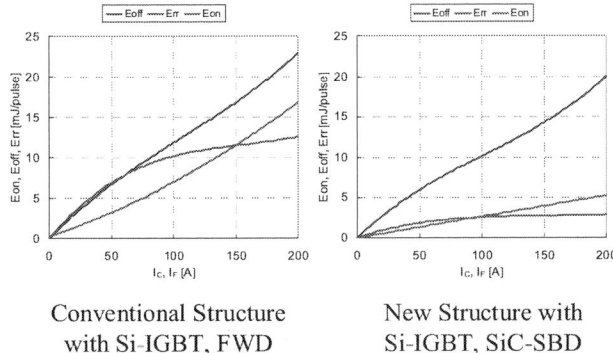

Conventional Structure
with Si-IGBT, FWD

New Structure with
Si-IGBT, SiC-SBD

Figure 8. Switching loss dependence on currents
(Comparison of two-type structure)

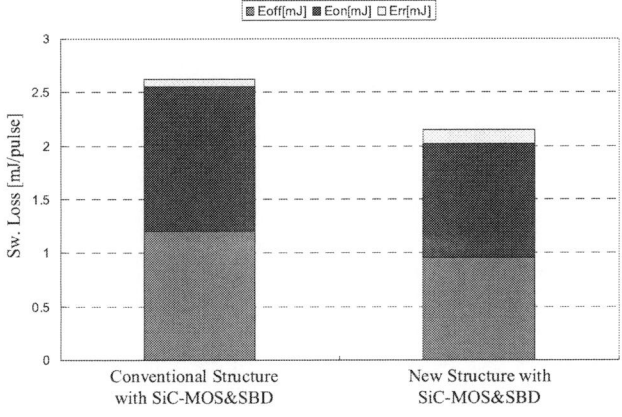

Figure 9. Switching losses: All-SiC embedded modules: conventional
structure versus new structure

Fig. 7 shows switching loss of both types module, conventional structure with Si-IGBT & Si-FWD, and newly developed structure with Hybrid chip-set (Si-IGBT & SiC-SBD). The results show total switching loss reduction is 46%, comparing with conventional module, almost half of that. Fig.8 shows each component of switching loss, dependency on current. As described previously, SBD related switching loss, Eon, Err, decrease obviously. And IGBT oriented switching loss, Eoff, decreasing about 20% is depicted as structural improvement effects motivated by stray inductance reduction. These switching loss reductions effectively act on the high current condition, these results prove the fact that newly developed structure is well fit with high current purpose.

B. SBD Module (SiC-SBD)

Further investigations are carried out to make confirmation of validity of the newly developed structure, for the case of passive device activation in the high current density conditions. Reverse recovery tests perform the sample operation to be stable, within the condition of forward current 400A. Considering about normal operation, conventional structure requires about 4 times as large mounting area as developed structure does [5].

C. All-SiC Module (SiC-MOS & SiC-SBD)

In addition to previous section, including unipolar-type active device, SiC-MOS & SiC-SBD embedding structure is investigated. In this case, not only stray inductance through main current path, but also gate control path optimization effectively acts on switching behavior, reducing switching loss about 20% shown in Fig. 9, comparing with conventional structure with the same device. Moreover, comparison with conventional structure with bipolar chipset (Si-IGBT & Si-FWD), achieve about 70% reduction of switching loss.

IV. CONCLUSIONS

A newly developed structure package has been developed and investigated. It has superior thermal properties by its high conductive structure, and reliability tests proved validity of full molded package structure, as high reliability structure. Implementation of WBG devices shows additional benefits of new structure, low loss characteristics suitable for high power applications, which is oriented by high-density packaging and high reliability structure.

ACKNOWLEGEMENTS

The authors would like to thank Mr. Okuma, Mr. Fujimoto, Mr. Matsubara, and Mr. Takubo for their technical support during electrical investigation and their useful advices.

REFERENCES

[1] M.Otsuki, H.Kanemaru, Y.Ikeda, K.Ueno, M.Kirisawa, Y.Onozawa, Y.Seki, "Advanced thin wafer IGBTs with new thermal management solution", Proc. of ISPSD., pp.144-147, 2003

[2] Y.Ikeda, Y.Iizuka, T.Asai, T.Goto, Y.Takahashi, "A study on the reliability of the chip surface solder joint", Proc. of ISPSD., pp.189-192, 2008.

[3] Kariya, Journal of electrical Material, Vol.27, No.7, p.866, 1998.

[4] Y.Ikeda, H.Hokazono, S.Sakai, T.Nishimura, Y.Takahashi, "A study of the bonding-wire reliability on the chip surface electrode in IGBT", Proc. of ISPSD., pp.289-292, 2010.

[5] Y.Ikeda, N.Nashida, M.Horio, H.Takubo, Y.Takahashi, "Ultra Compact, Low Thermal Impedance and High Reliability Module Structure with SiC Schottky Barrier Diodes", IEEE APEC Technical Session, 2011.

Proceedings of the 23rd International Symposium on Power Semiconductor Devices & IC's
May 23-26, 2011 San Diego, CA

A Novel Normally-off GaN Power Tunnel Junction FET

Li Yuan, Hongwei Chen, Qi Zhou, Chunhua Zhou, and Kevin J. Chen

Department of Electronic and Computer Engineering, Hong Kong University of Science and Technology
Clear Water Bay, Kowloon, Hong Kong SAR, CHINA
Tel. +852-23588530   Email: ylxac@ust.hk

Abstract— We demonstrate AlGaN/GaN tunnel junction FETs (TJ-FET) featuring a metal-2DEG Schottky junction at the source. The TJ-FETs exhibit normally-off operation in an otherwise normally-on as-grown sample owing to a current controlling scheme different from the conventional FETs. The high 2DEG density in AlGaN/GaN heterostructure results in a thin tunnel barrier whose effective thickness is controlled by an overlaying gate electrode. A positive gate bias results in a nanometer-thick barrier with high tunneling current, while a zero gate bias leads to a thicker barrier that effectively blocks the current flow. High drive current (326 mA/mm), low off-state leakage current (10^{-8} mA/mm) and high I_{ON}/I_{OFF} ratio (10^{10}) at a drain voltage of 50 V, and high off-state breakdown voltage (557 V) are obtained on a standard GaN-on-Si platform featuring a 1.8 μm buffer.

I. INTRODUCTION

The Schottky source/drain FETs have attracted much attention and been implemented in Si and Ge based MOSFETs [1-6], primarily as a promising candidate for beyond CMOS logic implementation. This device structure is capable of delivering favorable characteristics including highly-scalable low-resistance contacts, suppressed drain induced barrier lowering (DIBL) effect and low off-state leakage current. One challenge of the Schottky source/drain MOSFET is the low current drive capability limited by the inefficient carrier injection at the Schottky tunnel junction. By lowering the Schottky barrier height (SBH) with low work function silicides (e.g. ErSi$_2$, YbSi$_2$, for NMOS), the drive current can be improved, but at the price of weaker reverse blocking capability. The Schottky source/drain concept has also been implemented in GaN MOSFET [7], with a current drive less than 5 mA/mm as a result of inefficient tunneling process.

The wide-bandgap AlGaN/GaN heterostructures, on the other hand, possess unique characteristics that favors the realization of high-performance Schottky source tunnel FET. Most importantly, the strong spontaneous and piezoelectric charge polarization in AlGaN/GaN heterostructures yield 2-dimensional electron gas (2DEG) with high mobility (~2000 cm^{2}V^{-1}s^{-1}) and carrier density (~ 10^{20} cm^{-3}), without any intentional doping (or impurities). When metal is in direct contact to this 2DEG, a Schottky junction with relatively high

SBH but very small Schottky barrier width (SBW) could be formed, making it possible to achieve high tunnel current [8].

In this work, we demonstrate the metal-2DEG tunnel junction FETs (TJ-FET) using a baseline AlGaN/GaN HEMT sample grown on 4-inch (111) Si substrate and present detailed device characterizations in the context of power electronics application. The source features a metal-2DEG tunnel junction controlled by an overlapping gate electrode. Since the current turn-on/off is mainly controlled by the tunnel junction instead of the 2DEG channel, positive threshold corresponding to normally-off operation is realized in the as-grown normally-on structure. This method of realizing normally-off operation is fundamentally different from the previous approaches [9-17] that require a positive pinch-off voltage in the 2DEG channel. Furthermore, since the Schottky junction at the source electrode is naturally reverse biased at the OFF-state with large drain voltage, low leakage current and high I_{ON}/I_{OFF} ratio could be obtained.

II. DEVICE STRUCTURE AND OPERATION PRINCIPLE

A schematic cross-section view of the proposed metal-2DEG TJ-FET is shown in Fig. 1. The sample used in this work is a baseline AlGaN/GaN HEMT structure grown on a 4-inch Si (111) substrate. The epi-structure includes a 1.8 μm undoped buffer, a 20 nm barrier (18 nm AlGaN and a 2 nm GaN cap) with an AlN interface enhancement layer (~1 nm).

The device fabrication started with the active region isolation by mesa etching using Cl$_2$/He inductively coupled plasma reactive-ion etching (ICP-RIE). Then the ohmic drain electrode is formed by e-beam deposition of Ti/Al/Ni/Au followed by rapid thermal annealing in N$_2$ ambient at 850 °C

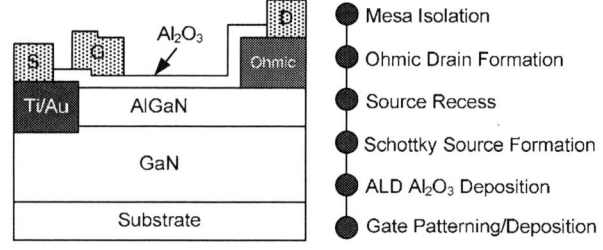

Fig. 1. The schematic cross section and the process flow of AlGaN/GaN TJ-FETs.

This work is supported by Hong Kong Innovation Technology Fund under grant ITS/122/09FP and Hong Kong RGC under GRF grant 611610.

978-1-4244-8425-6/11 $26.00 © 2011 IEEE

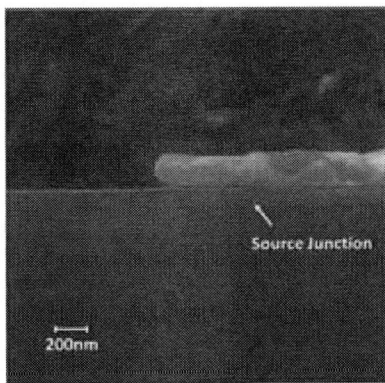

Fig. 2. SEM cross section of a metal-2DEG tunnel junction with an overlapping gate. The overlapping length is 0.5 μm in this testing structure, while the nominal overlapping length in the fabricated TJ-FETs is 0.25 μm.

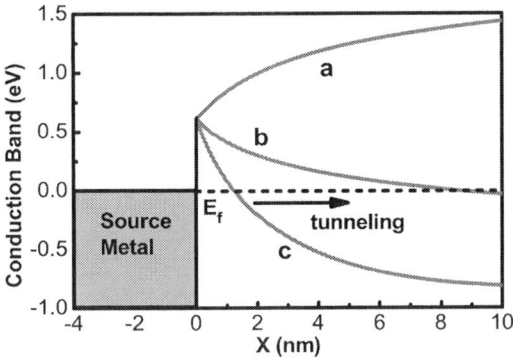

Fig. 3. The simulated conduction band distribution profiles of a TJ-FET. Label a-c represent three different bias conditions: a) V_{GS} = -3 V, V_{DS} = 10 V, the 2DEG channel is cut off; b) V_{GS} = 0 V, V_{DS} = 10 V, the 2DEG channel is turn on but the SBW is ~10 nm, the TJ-FET is still at the "OFF" state; c) V_{GS} = 3 V, V_{DS} = 10 V, the SBW is ~ 1 nm, turning the device to the "ON" state.

for 35 sec. In the next step, the source electrode region was recessed by Cl$_2$/He ICP-RIE etching. The recessed source region was then covered with Ti/Au (30 nm/5 nm) by electron beam deposition and lift-off. The source recess is about ~ 22-25 nm deep. The source metal is in direct contact to the sidewall of the 2DEG channel, as shown in the SEM cross section view of a testing structure fabricated on the same sample (Fig. 2). A 10 nm Al$_2$O$_3$ layer was then grown on the sample by atomic layer deposition (ALD) followed by gate electrode (Ni/Au) patterning, deposition and lift-off. The gate electrode was designed to overlap with the Schottky source contact by 0.25 μm.

The operating principle is illustrated by the conduction band profiles (Fig. 3) simulated by the 2-D Dessis-ISE device simulator. At zero gate bias, the 2DEG channel is fully turned on. However, the effective SBW at the source metal-2DEG junction is still about 9 nm. This thick tunneling barrier yields a negligible tunneling current and effectively blocks the current flow, keeping the device at the "OFF" state. As the gate bias becomes more positive, the 2DEG density is raised and the conduction band is pulled down, effectively reducing

SBW. At V_{GS} = 3 V (curve c in Fig. 3), the SBW is about 1 nm and the tunneling current is significantly higher, turning the device to the "ON" state. Thus, it can be understood that the current controlling mechanism in the TJ-FET is fundamentally different from that of HEMTs, suggesting a new way of realizing normally-off AlGaN/GaN transistors. It should be noted that the conduction band profiles in Fig. 3 only qualitatively explains the operating principle and more accurate simulation requires taking into account the other relevant mechanisms such as the barrier lowering effect by the image force of carriers. Another beneficial byproduct is the inherent reverse blocking capability of the Schottky junction at the source. This junction is naturally reverse biased at large drain bias, suppressing the leakage through the buffer layer.

III. DEVICE CARACTERISTICS

TJ-FETs with a 2 μm gate length (L$_G$) and 2 to 15 μm gate-drain spacing (L$_{GD}$) were fabricated. The measured I$_D$-V$_{GS}$ profiles of TJ-FET with a 2 μm L$_{GD}$ are shown in Fig. 4 and 5. In Fig. 4, the transfer characteristic of a conventional HEMT fabricated on the same wafer is also shown as a

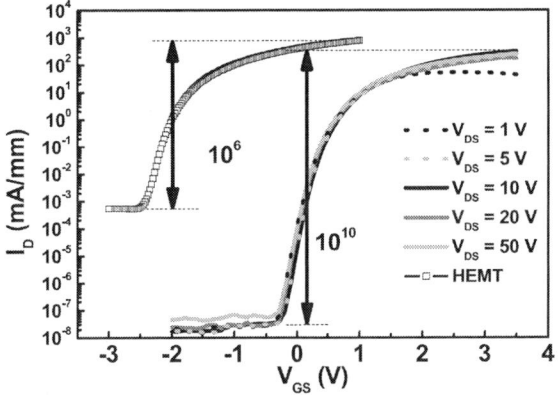

Fig. 4. The experimentally measured I$_D$-V$_{GS}$ transfer characteristics of a TJ-FET with L$_G$ = 2 μm and L$_{GD}$ = 2 μm at V$_{DS}$ = 1 to 50 V. The transfer characteristics of a conventional HEMT fabricated on the same wafer are measured at V$_{DS}$ = 10V and plotted for reference.

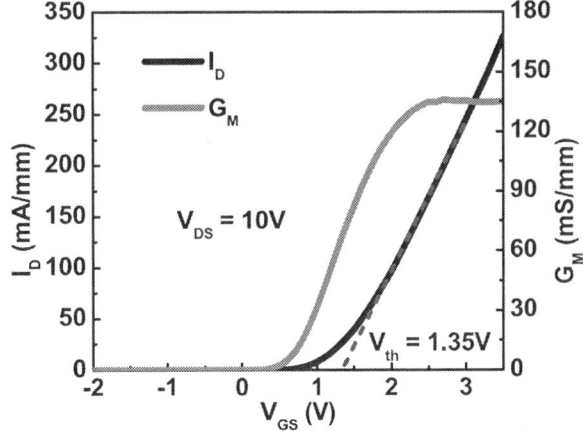

Fig. 5. The experimentally measured I$_D$-V$_{GS}$, G$_m$-V$_{GS}$ profiles of TJ-FET with L$_G$ = 2 μm and L$_{GD}$ = 2 μm at V$_{DS}$ = 10 V.

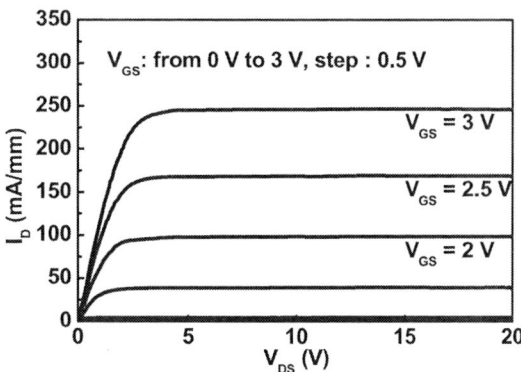

Fig. 6. The experimentally measured I_D-V_{DS} profiles of TJ-FET with L_G = 2 μm and L_{GD} = 2 μm.

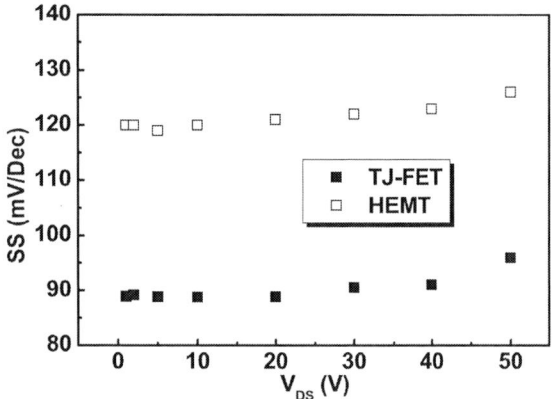

Fig. 7. The experimentally measured SS-V_{GS} profiles of TJ-FET and HEMT at V_{DS} = 10 V.

reference. Owing to the unique current control mechanism, the TJ-FET exhibits an I_{ON}/I_{OFF} ratio of 10^{10} at drain from 1 V to 50 V, 10^4 larger than the conventional HEMT. Especially, with the benefit of reverse-biased Schottky source blocking, the TJ-FET's leakage can be limited to 10^{-8} mA/mm even at a drain voltage of 50 V.

As shown in Fig. 5, the threshold voltage of TJ-FET is about +1.35 V, presenting the TJ-FET as a normally-off transistor. The drain current density is 326 mA/mm at V_{GS} = 3.5 V. The transconductance G_m is stabilized at 130-135 mS/mm in the gate bias range of 2-3.5 V, indicating good linearity in the input-output transfer relation.

Fig. 7 shows the sub-threshold swing (SS)-V_{DS} relation of a TJ-FET and a HEMT fabricated on the same wafer. The TJ-FET shows significantly lower SS than the conventional HEMT, due to the sensitive tunnel current controlling mechanism by the effective barrier thickness of the source Schottky junction as discussed in section II.

The three-terminal off-state characteristics of a TJ-FET and conventional HEMT with an L_{GD} of 2 μm fabricated on the same wafer is plotted in Fig. 8. Since the source/buffer junction is a naturally reverse biased Schottky junction in the off-state, the leakage current can be significantly smaller than that in the conventional HEMT that features no blocking

junction between the ohmic source and the unintentionally doped buffer layer. At V_{DS} < 70 V, the drain leakage current of TJ-FET originates from the leakage at the reverse biased source/buffer Schottky junction. At V_{DS} > 70 V, the drain leakage of TJ-FET is dominated by the gate leakage through the Al_2O_3 layer.

The off-state breakdown characteristics of the TJ-FETs with L_{GD} from 2 to 15 μm are summarized in Fig. 9-11. Using

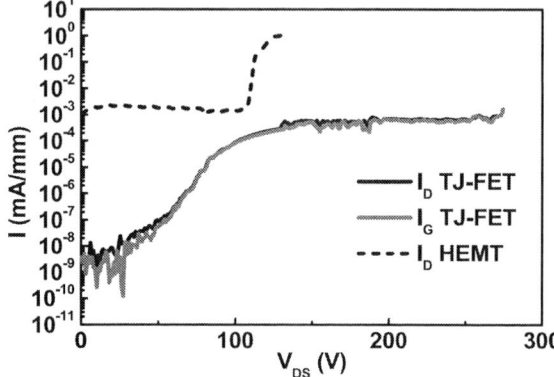

Fig. 8. The experimentally measured off-state breakdown characteristics of TJ-FET and HEMT at V_{GS} = -1 V.

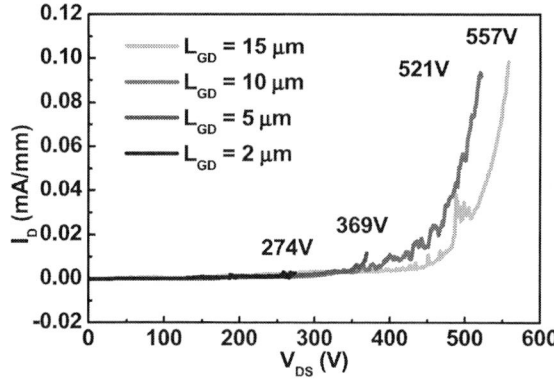

Fig. 9. The experimentally measured three-terminal off-state breakdown characteristics of TJ-FET with L_G = 2 μm and L_{GD} = 2 to 15 μm at V_{GS} = -1 V, indicating off-state breakdown voltage of 274 V to 557 V at I_D = 0.1 mA/mm.

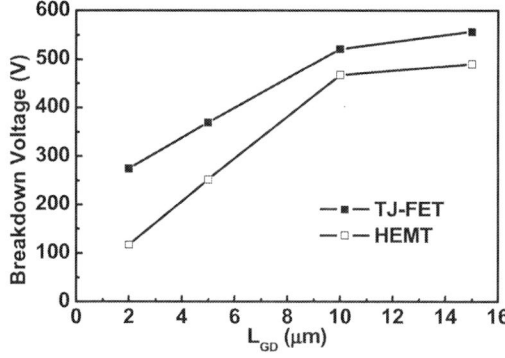

Fig. 10. The off-state breakdown voltage comparison between TJ-FET and HEMT fabricated on the same substrate with gate-drain distance L_{GD} = 2 to 15 μm.

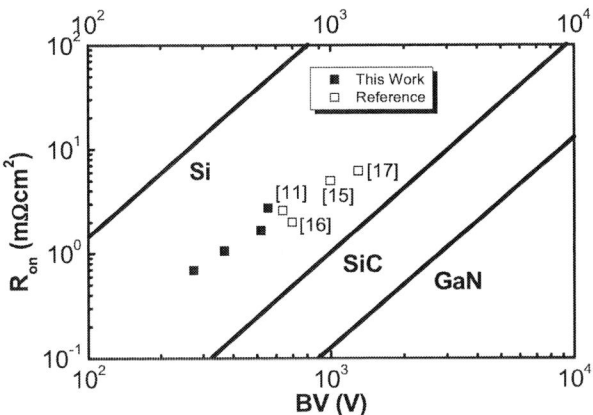

Fig. 11. Specific on-resistance and off-state breakdown voltage in the AlGaN/GaN TJ-FETs and recently reported normally-off AlGaN/GaN HEMTs, together with the theoretical Si-, SiC- and GaN-limits.

a 0.1 mA/mm drain leakage current limit, the OFF-state breakdown voltage (BV) of the TJ-FETs and HEMTs are compared in Fig. 10. The TJ-FETs consistently show larger BV, especially in devices with shorter gate-drain distance L_{GD}. With $L_{GD} = 2$ μm, the TJ-FET delivers a BV more than 2 times of that in the HEMT. The trade-off characteristics between the specific on-resistance (R_{on}, including the area of the ohmic drain electrode) and BV are plotted in Fig. 11. The R_{on} of TJ-FETs with $L_{GD} = 2, 5, 10$ and 15 μm (corresponding to BV of 274, 369, 521, and 557 V) are 0.69, 1.06, 1.66 and 2.74 mΩcm², respectively. The device performance comparison between TJ-FET and conventional HEMT is provided in table 1.

Table 1. The device performance summarise of AlGaN/GaN TJ-FET and HEMT fabricated on the same type substrate.

	TJ-FET	HEMT
I_{OFF} (mA/mm)	4×10^{-8} @ $V_{DS} = 50$ V	6×10^{-4} @ $V_{DS} = 10$ V
I_{ON} (mA/mm)	326	800
I_{ON}/I_{OFF}	10^{10}	10^{6}
SS (mV/Decade)	89	119
V_{th} (V)	+1.35	-2.1
BV (@0.1mA/mm) $L_{GD}=2$μm	274	117

IV. CONCLUSION

AlGaN/GaN metal-2DEG tunnel junction FETs are demonstrated for the first time, using a standard AlGaN/GaN HEMT structure grown on silicon substrate. The current flow is controlled by a gate-controlled metal-2DEG tunnel junction, allowing normally-off operation. The metal-2DEG tunnel junction at the source can deliver highly efficient carrier injection via quantum tunneling, enabling high current drive (326 mA/mm), low off-state leakage (10^{-8} mA/mm), and high I_{ON}/I_{OFF} ratio (10^{10}).

REFERENCES

[1] J. R. Tucker, C. Wang, P. S. Carney, "Silicon field-effect transistor based on quantum tunneling," *Appl. Phys. Lett.*, vol. 65, pp. 618-620, May 1994.

[2] J. Kedzierski, P. Xuan, E. H. Anderson, J. Bokor, T.-J. King, C. Hu, "Complementary silicide source/drain thin-body MOSFETs for the 20 nm gate length regime," *IEDM Tech. Dig.*, pp. 57-60, Dec. 2000.

[3] J. Guo and M. S. Lundstrom, "A computational study of thin-body, double-gate, Schottky barrier MOSFETs," *IEEE Trans. Electron Devices*, vol. 49, pp. 1897-1902, Nov. 2002.

[4] M. Nishisaka, S. Matsumoto, and T. Asano, "Schottky source/drain SOI MOSFET with shallow doped extension," *Jpn. J. Appl. Phys.*, vol. 42, pp. 2009-2013, Apr. 2003.

[5] S. Zhu, J. Chen, M. –F. Li, S. J. Lee, J. Singh, C. X. Zhu, A. Du, A. Chin, and D. L. Kwong, "N-Type Schottky barrier source/drain MOSFET using ytterbium silicide," *IEEE Electron Device Lett.*, vol. 25, No. 8, pp. 565-567, 2004.

[6] J. M. Larson and J. P. Snyder, "Overview and status of metal S/D Schottky-barrier MOSFET technology," *IEEE Trans. Electron Devices*, vol. 53, pp. 1048-1058, May 2006.

[7] H. –B. Lee, H. –I. Cho, H. –S. An, Y. –H. Bae, M. –B. Lee, J. –H. Lee, and S. –H. Hahm, "A normally-off GaN n-MOSFET with Schottky-barrier source and drain on a Si-auto-doped p-GaN/Si," *IEEE Electron Device Lett.*, vol. 27, No. 2, pp. 81-83, Feb. 2006.

[8] L. Yuan, H. Chen, and K. J. Chen, "Normally-off AlGaN/GaN metal-2DEG tunnel-junction field-effect transistors," *IEEE Electron Device Letters*, vol. 32, No. 3, pp. 303-305, 2011.

[9] W. B. Lanford, T. Tanaka, Y. Otoki and I. Adesida, "Recessed-gate enhancement-mode GaN HEMT with high threshold voltage," *Electron. Lett.*, vol. 41, pp. 449-450, Mar. 2005.

[10] Y. Cai, Y. G. Zhou, K. J. Chen, and K. M. Lau, "High-performance enhancement-mode AlGaN/GaN HEMTs using fluoride-based plasma treatment," *IEEE Electron Device Lett.*, vol. 26, pp. 435-437, July 2005.

[11] Y. Uemoto, M. Hikita, H. Ueno, H. Matsuo, H. Ishida, M. Yanagihara, T. Ueda, T. Tanaka, and D. Ueda, "A normally-off AlGaN/GaN power transistor using conductivity modulation," *IEEE Trans. Electron Devices*, vol. 54, pp. 3393-3399, Dec. 2007.

[12] R. Wang, Y. Cai, C. -W. Tang, K. M. Lau, and K. J. Chen, "Enhancement-mode Si₃N₄/AlGaN/GaN MISHFETs," *IEEE Electron Device Letters*, vol. 27, No. 10, pp. 793-795, Oct. 2006.

[13] D. Song, J. Liu, Z. Cheng, W. C. -W. Tang, K. M. Lau, and K. J. Chen, "Normally Off AlGaN/GaN Low-Density Drain HEMT (LDD-HEMT) With Enhanced Breakdown Voltage and Reduced Current Collapse," *IEEE Electron Device Lett.*, Vol. 28, No. 3, pp. 189-191, Mar. 2007.

[14] M. Kanamura, T. Ohki, T. Kikkawa, K. Imanishi, T. Imada, A. Yamada, and N. Hara, "Enhancement-mode GaN MIS-HEMTs with n-GaN/i-AlN/n-GaN triple cap layer and high-k gate dielectrics," *IEEE Electron Device Lett.*, vol. 31, pp. 189-191, Mar. 2010.

[15] K. Ota, K. Endo, Y. Okamoto, H. Ando, H. Miyamoto, H. Shimawaki, "A normally-off GaN FET with high threshold voltage uniformity using a novel piezo neutralization technique," *IEDM Tech. Dig.*, pp. 153-156, Dec. 2009.

[16] J. Derluyn et al., "Low leakage high breakdown E-Mode GaN DHFET on Si by selective removal of in-situ grown Si₃N₄," *IEDM Tech. Dig.*, pp. 157-160, Dec. 2009.

[17] K.S. Boutros, S. Burnham, D. Wong, K. Shinohara, B. Hughes, D. Zehnder, and C. McGuire, "Normally-off 5A/1100V GaN-on-silicon device for high voltage applications," *IEDM Tech. Dig.*, pp. 161-163, Dec. 2009.

Proceedings of the 23rd International Symposium on Power Semiconductor Devices & IC's
May 23-26, 2011 San Diego, CA

GaN Based Super HFETs over 700V Using the Polarization Junction Concept

Akira Nakajima, Mahesh H. Dhyani and E. M. Sankara Narayanan

Department of EEE, University of Sheffield
Mappin Street, Sheffield S1 3JD, UK
a.nakajima@sheffield.ac.uk

Yasunobu Sumida and Hiroji Kawai

POWDEC K.K.
1-23-15 Wakagi-cho, Oyama, Tochigi, 323-0028, JAPAN
sumida@powdec.co.jp

Abstract—GaN Super Heterojunction Field Effect Transistors (Super HFETs) based on the polarization junction (PJ) concept are demonstrated on Sapphire substrates. These Super HFETs were fabricated from a GaN/Al$_{0.23}$Ga$_{0.77}$N/GaN hetero structure with 2D hole and electron gas densities of 1.1×10^{13} and 9.7×10^{12} cm^{-2} at the respective hetero-interfaces. The Super HFETs show breakdown voltage above 700 V with on-resistances of 15 $\Omega\cdot$mm. In addition, the super HFETs have inherent body diodes and its reverse conducting characteristics are demonstrated.

I. INTRODUCTION

GaN based power devices are emerging candidates for the next generation of power integrated circuits due to their intrinsic high electric field strength. In particular, heterojunction field effect transistors (HFET) are expected to achieve lower power losses and higher switching performances than their silicon counterparts. However, electric field management has been a key issue for the GaN HFETs. Fig. 1 shows a simplified cross-section of a conventional HFET with a Schottky gate contact. In the absence of a field plate (FP), during the off-state, electric field is concentrated at the gate edge as illustrated in Fig. 1(b) due to positive polarization charges at the AlGaN/GaN hetero-interface. Generally, FP technologies are used to suppress the electric field crowding [1-4]. FP technologies enhance breakdown voltage of GaN HFETs, suppress gate leakage current and reduce hot electron generation inducing current collapse [2].

Polarization-junction (PJ) concept was proposed in 2006 for further improvement of area-specific on-resistance ($R_{on}\cdot A$) and the breakdown voltage (BV) in GaN power devices [5]. The PJ concept is a novel field shaping technology utilizing unique polarization properties of group-III nitride semiconductors. The PJ is made from a double hetero-structure of nitride semiconductors (e.g. GaN/AlGaN/GaN), where positive and negative polarization charges exist at heterointerfaces. Due to the inherent polarization properties, these positive and negative charge quantities are equal and a charge balance condition can be obtained.

Therefore, the PJ has similar effect as a RESURF or super-junction (SJ) in silicon or GaN [6-8], but, *without* intentional donor and acceptor doping. With the combination of the PJ concept and the high electric field strength, GaN devices are near ideal candidates for next-generation power devices.

Recently, we have developed novel GaN wafers for PJ devices [9] and reported the first demonstration of GaN-HFETs based on the PJ concept, named Super HFETs [10]. In this paper, we report more details of the device physics and analysis of GaN based Super HFETs.

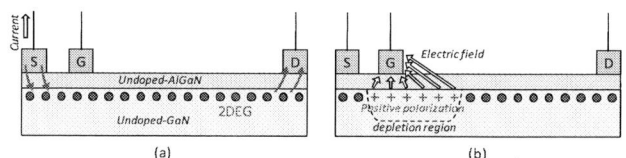

Fig. 1. Schematic of a conventional HFET in (a) the ON-state and (b) the OFF-state. There are three electrodes which are source (S), gate (G) and drain (D).

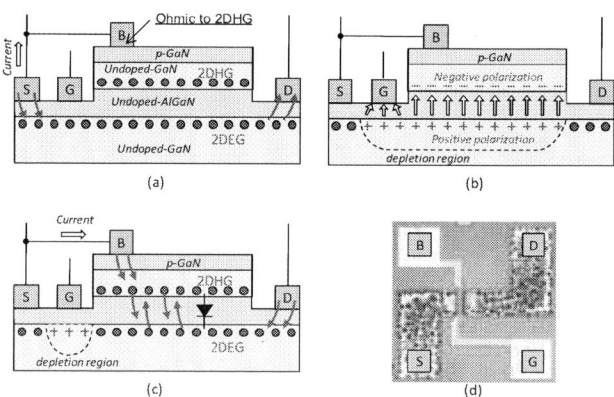

Fig. 2. Schematic of a Super HFET in (a) the ON-state, (b) the OFF-state and (c) the reverse biased condtion. There are four electrodes which are source (S), gate (G), drain (D) and base (B). (d) Microscope picture of a fabricated Super HFET.

The work is financially supported by the Royal Society of London, British Academy and the Royal Academy of Engineering under the Newton International Fellowship Scheme.

978-1-4244-8425-6/11 $26.00 © 2011 IEEE

II. DEVICE STRUCTURE AND PHYSICS

Fig. 2 shows a simplified cross-section of a normally-on Super HFET. The layer structure consists of an undoped double-hetero GaN/AlGaN/GaN structure with a p-GaN cap layer. The upper GaN(000-1)/AlGaN(0001) and the bottom AlGaN(000-1)/GaN(0001) interfaces have negative and positive polarization charges and they induce two dimensional hole gas (2DHG) and electron gas (2DEG), respectively. The Super HFET has four electrodes (source, gate, drain and base). However, the base can be electrically connected to the gate electrode as reported in [10] or to the source electrode to form three-terminal devices, as explained in this study. When the base is connected with the source, the device can provide an inherent body diode and therefore, these super HFETs can show reverse conducting characteristics.

The structural differences between conventional HFETs (Fig. 1) and the Super HFET (Fig. 2) are the additional base electrode and presence of the 2DHG induced by the negative polarization. The base makes an ohmic contact to the 2DHG through the top p-GaN layer.

In the on-state, current flows from the drain to source through the 2DEG (Fig. 2(a)), in a manner similar to the conventional HFETs. 2DEG at AlGaN/GaN interfaces has high mobility of about 1000 cm²/Vs due to low impurity scattering and, therefore low on-resistance can be obtained as same with conventional HFETs.

During the turn-off period, the 2DHG and 2DEG are discharged through the base and drain electrodes respectively, and the drift region between the base and drain is depleted at low drain voltages. With further increase in drain voltages, the drift region acts as an 'intrinsic semiconductor region' and a flat field distribution to maximize the breakdown voltage can be obtained as illustrated in Fig. 2(b). As a result, the breakdown capability of Super HFETs can be drastically enhanced.

In addition, unlike conventional HFETs, the Super HFET has an inherent pn-junction body diode formed by the 2DHG and 2DEG. In the reverse bias condition, holes and electrons are injected from the base and drain electrodes respectively and current flows from the base to the drain (Fig. 2(c)). Therefore, the Super HFET has reverse conducting characteristics.

III. DEVICE FABRICATION

The GaN/AlGaN/GaN double-heterostructure has been grown by metal organic chemical vapor deposition on 3" Sapphire substrates. The epitaxial growth starts with a 1-μm-thick undoped GaN layer on a GaN buffer layer. Subsequently, a 47-nm-thick undoped AlGaN layer with an Al composition of 23 %, a 10-nm-thick undoped GaN layer and a 30-nm-thick Mg doped (3×10^{19} cm⁻³) p-GaN layer have been grown. 2DHG and 2DEG are formed at the upper GaN/AlGaN and bottom AlGaN/GaN interfaces respectively and their sheet densities are 1.1×10^{13} and 9.7×10^{12} cm⁻² respectively as determined by Hall effect measurements at room temperature. The detailed layer design of the GaN wafers to obtain the high density 2DHG is reported in [9]. The top p-GaN layer enables an ohmic contact to 2DHG. The small difference between the

2DHG and 2DEG densities arises due to p-type doping in the top p-GaN layer.

Conventional HFETs and Super HFETs have been made on the same wafer without field plates. The conventional HFETs have no base electrode or 2DHG. The conventional HFETs were made by removing top p-GaN/undoped-GaN layers by inductively coupled plasma (ICP) etching. The source and drain ohmic electrodes (Ti/Al/Ti/Au) have been deposited on the AlGaN surface and annealed under N_2 ambient at 800 °C. The gate and base electrodes have been formed using Ni/Au/Ni/Au on the AlGaN and p-GaN layers, respectively and annealed in air at 550 °C to decrease the contact resistance of the base electrode. Finally, a 160-nm thick SiO_2 deposited by PECVD system is used as a passivation layer.

Fig. 2(d) shows a microscope picture of a fabricated Super HFET which has four pads (source, drain, gate and base). The base and source pads are electrically isolated each other and base and source currents can be measured separately. The gate width of the both conventional and Super HFETs are 50 μm. The source-gate distance, the gate length and the gate-drain distance of the both types of HFETs are same dimensions and are 3 μm, 3 μm and 13 μm respectively. The gate-base distance, base contact length and the distance between the p-GaN/i-GaN layers and the drain of the Super HFETs are 3 μm, 4 μm and 3 μm respectively.

IV. RESULTS AND DISCUSSION

A. On-state performance

Fig. 3 shows typical I_d-V_{gs} characteristics of fabricated conventional HFETs and Super HFETs at $V_{ds} = 10$ V. In this measurement, the base electrode is electrically connected with the source (base-source voltage $V_{bs} = 0$ V). When the gate voltage is less than -2 V, both conventional and Super HFETs have an identical drain current curve and shows the same gate-source threshold voltage of -4 V. When the gate voltage of more than -2 V, the transfer characteristics of the conventional and Super HFETs differ as shown in Fig. 3. In conventional

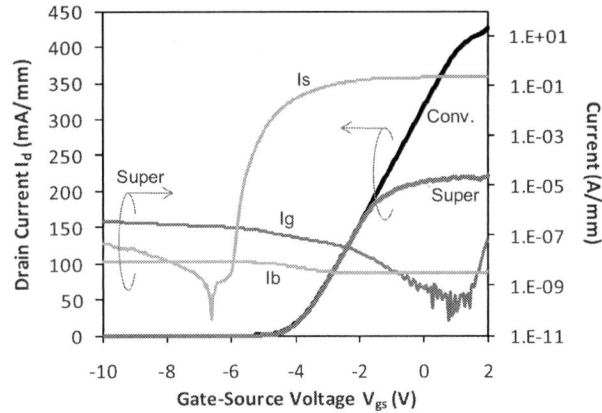

Fig. 3.   Transfer characteristics of fabricated conventional and Super HFETs at $V_{ds} = 10$ V. The base of the Super HFET is electrically connected with the source. The low saturation current of Super HFETs enhances the short-circuit capability. Source (I_s), gate (I_g) and base currents (I_b) of the Super HFET are also plotted.

HFETs, the drain current linearly increases with the gate voltage. On the other hand, Super HFETs show drain current saturation, when the V_{gs} is more than -2V. Source, gate and base currents of the Super HFET are also plotted in Fig. 3. The gate and base currents are less than 1 μA/mm. Therefore, in the on-state of Super HFETs, current flows from the drain to source through the 2DEG as illustrated in Fig. 2 (a).

Fig. 4 shows typical I_d-V_{ds} characteristics of fabricated Super HFETs. To understand the differences in the transfer characteristics between conventional and Super HFETs, we measured in conditions of the base electrode shorted with the source (*GND*) and also base floating (*FLT*). In the both conditions, Super HFETs show a same on-resistance of 15 Ω·mm evaluated at V_{ds} = 0.1 V and the calculated $R_{on}·A$ is 3.3 mΩ·cm². As can be seen in Fig. 4, drain current of *FLT* condition is higher than that of *GND* when the V_{gs} is more than -2 V. The maximum drain currents were 220 and 280 mA/mm in *GND* and *FLT* conditions at V_{gs} = 0 V respectively. These results clearly indicate that the drain current suppression of Super HFETs shown in Fig. 3 is due to the fact that the base contact maintains source potential at the drain side of the drift region.

The on-state behavior of Super HFETs can be explained as follows. Super HFETs have a body diode formed by 2DHG and 2DEG, where the base and drain act as an anode and cathode, respectively. The 2DHG and 2DEG densities decrease with increasing the drain voltage because a reverse bias, which corresponds to the V_{ds}, is applied across the body diode. As can be seen in the capacitance-voltage measurement results in [9], the 2DHG and 2DEG are completely depleted with V_{ds} of only 5 V in the case of the fabricated HFETs of this study. The depletion induces drain current suppression in the V_{gs} of more than -2 V as shown in Fig. 3. Therefore, Super HFETs have lower saturation current than conventional HFETs. The low saturation current is an advantage of Super HFETs in power converter applications because short-circuit capability is enhanced by reducing heat generation.

B. Reverse characteristic

In Fig. 5 is shown measured reverse characteristics of conventional and Super HFETs in the off-state conditions (V_{gs} = -10 V) as the base electrode of the Super HFET is connected with the source. Drain current of conventional HFETs was less than 1 μA/mm at the V_{ds} = -3.5 V. On the other hand, drain current of Super HFETs exponentially increases depending on the drain voltage although gate voltage is less than the threshold voltage. The Super HFETs have reverse conduction characteristics independent on the gate voltage. Source, gate and base currents of the Super HFET are also plotted in Fig. 5. As can be seen, the source and gate currents are negligible (less than 0.1 μA/mm) and the current mainly flows through the body diode between the base and drain.

In power converter applications, the body diode can be utilized as a freewheeling diode. Minority carrier lifetimes in GaN are inherently shorter than that of Si because GaN is a direct semiconductor. Lifetimes of less than 20 nsec have been reported [11]. Therefore, a fast recovery characteristic of the body diode can be expected.

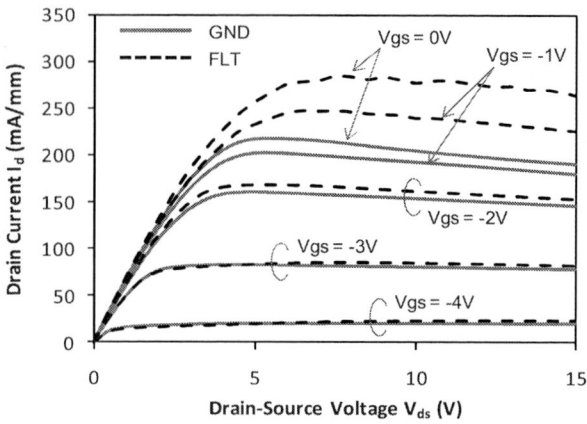

Fig. 4. I_d-V_{ds} characteristics of a fabricated Super HFET measured with the base connected to the source (*GND*) and base floating (*FLT*).

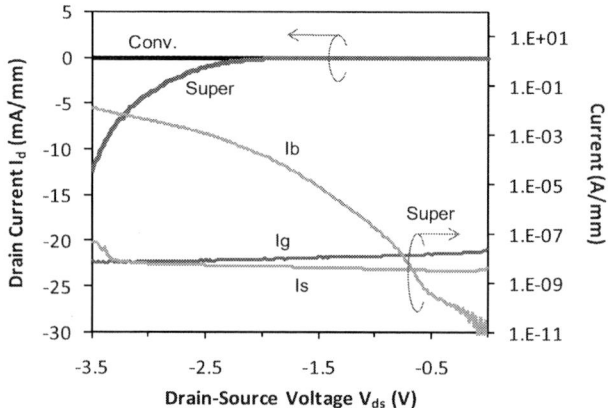

Fig. 5.   Reverse characteristics of conventional and Super HFETs at V_{gs} = -10 V. The base of the Super HFET is electrically connected to the source. Source (I_s), gate (I_g) and base currents (I_b) of the Super HFET are also plotted.

C. Forward blocking

Fig. 6 shows typical off-state characteristics of conventional and Super HFETs measured at V_{gs} = -10 V. It can be seen that the gate leakage current of fabricated conventional HFETs gradually increases to about 10 μA/mm with increasing the drain voltage and then rises up sharply at around V_{ds} of 100 V. The leakage current can be attributed to tunneling current of the Schottky gate contact due to electric field crowding at the gate edge as illustrated in Fig. 1(b). Moreover, AlGaN surface damage by ICP etching to remove the top p-GaN and undoped GaN layers may have contributed to the increased leakage current of the conventional HFETs [12]. As shown in Fig. 6, Super HFETs show a smaller gate leakage current less than 1 μA/mm independent of the drain voltage although the gate is also deposited on the etched AlGaN surface. These results imply field crowding at the gate edge is effectively suppressed by the charge compensation effect between positive and negative polarization charges.

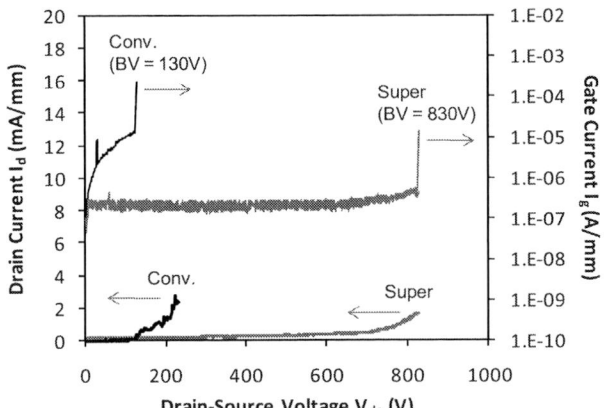

Fig. 6. Typical off-state characteristics of conventional and Super HFETs at V_{gs} = -10 V. The base of the Super HFET is electrically connected to the source. The BV is calculated at I_g reaching to 10 μA/mm.

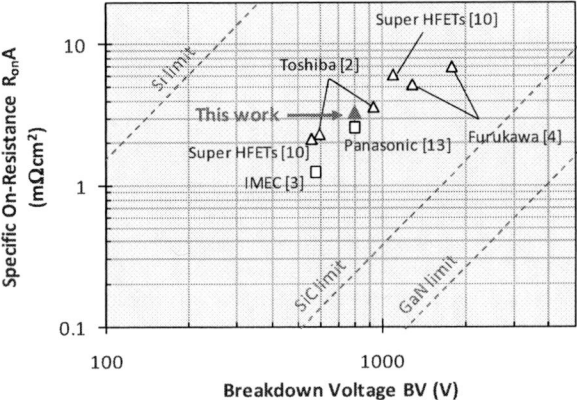

Fig. 7. Specific on-resistance versus breakdown voltage of GaN devices with D-mode (triangle) and E-mode operation (square). The dashed lines are theoritical material limits of Si, SiC and GaN unipolar devices.

Breakdown voltage of the fabricated devices is calculated when the gate current reaches a value of 10 μA/mm. The evaluated BV of conventional HFETs was less than 200 V and typically about 100 V. The BV is lower than that of previously reported conventional GaN HFETs [1-4], because no field plate was included in this study. On the other hand, BV of Super HFETs was more than 700 V and typically about 800 V without field plate. The BV of Super HFETs is drastically enhanced due to PJ effect.

From the on-state and off-state characteristics, a relationship between $R_{on} \cdot A$ and BV of the fabricated Super HEMTs is plotted in Fig. 7. In the same plot, relationships of the other reported conventional GaN-based HFETs with FP technologies [2-4], gate injection transistors reported in [13], and Super HFETs with base connected to gate [10] are shown.

V. CONCLUSIONS

In this paper, we have demonstrated the GaN based Super HFETs using the PJ concept and explained the device

operation and behaviors. Breakdown capability of the Super HFETs is drastically enhanced by the charge compensation effect between positive and negative polarization charges. The measured BV is over 700 V with R_{on} of 15 Ω·mm, show that the performance of the fabricated Super HFET is almost comparable with that of previously reported conventional HFETs with FP technologies as shown in Fig. 7. Further improvements can be expected by optimizing the device structure, in addition to using FP technologies. PJ GaN devices such as the Super HFETs are promising candidates for next generation ultra-low loss power devices beyond the GaN material limit. As lateral devices, they are ideally suited for ultra-high efficiency power ICs with small chip size.

REFERENCES

[1] N.–Q. Zhang, S. Keller, G. Parish, S. Heikman, S. P. DenBaars, and U. K. Mishra, "High Breakdown GaN HEMT with Over lapping Gate Structure," *IEEE Electron Device Lett.*, vol. 21, no. 9, pp. 421-423, Sep. 2000.

[2] W. Saito, T. Nitta, Y. Kakiuchi, Y. Saito, K. Tsuda, I. Omura, and M. Yamaguchi, "Suppression of Dynamic On-Resistance Increase and Gate Charge Measurements in High-Voltage GaN-HEMTs With Optimized Field-Plate Structure," *IEEE Trans. Electron Devices*, vol. 54, no. 8, pp. 1825-1830, Aug. 2007.

[3] F. Medjdoub, J. Derluyn, K. Cheng, M. Leys, S. Degroote, D. Marcon, D. Visalli, M. Van Hove, M. Germain, and G. Borghs, "Low On-Resistance High-Breakdown Normally Off AlN/GaN/AlGaN DHFET on Si Substrate," *IEEE Electron Device Lett.*, vol. 31 no. 2, pp. 111-113, Feb. 2010.

[4] N. Ikeda, Y. Niiyama, H. Kambayashi, Y. Sato, T. Nomura, S. Kato, and S. Yoshida, "GaN Power Transistors on Si Substrates for Switching Application," in *Proc. IEEE*, vol. 98, no. 7, pp. 1151-1161, Jul. 2010.

[5] A. Nakajima, K. Adachi, M. Shimizu, and H. Okumura, "Improvement of unipolar power device performance using a polarization junction," *Appl. Phys. Lett.*, vol. 89, no. 19, pp. 193501, Nov. 2006.

[6] K. Tang, Z. Li, T. P. Chow, Y. Niiyama, T. Nomura, and S. Yoshida, "Enhancement-mode GaN Hybrid MOS-HEMTs with Breakdwon Voltage of 1300 V," in *Proc. ISPSD*, 2009, pp. 279 – 282.T. Fujihira, "Theory of Semiconductor Superjunction Devices," *Jpn. J. Appl. Phys.*, vol. 36, pp. 6254-6262, Oct. 1997.

[7] T. Fujihira, "Theory of Semiconductor Superjunction Devices," *Jpn. J. Appl. Phys.*, vol. 36, pp. 6254-6262, Oct. 1997.

[8] L. Lorenz, G. Deboy, A. Knapp, and M. März, "COOLMOSTM – a new milestone in high voltage power MOS," in *Proc. ISPSD*, 1999, pp. 3-10.

[9] A. Nakajima, Y. Sumida, M. H. Dhyani, H. Kawai, and E. M. S. Narayanan, "High Density Two-dimensional Hole Gas Induced by Negative Polarization at GaN/AlGaN Heterointerface," *Appl. Phys. Express*, vol. 3, no. 12, p. 121004, Dec. 2010.

[10] A. Nakajima, Y. Sumida, M. H. Dhyani, H. Kawai, and E. M. S. Narayanan, "GaN Based Super Heterojunction Field Effect Transistors Using the Polarization Junction Concept," will be published in *IEEE Electron Device Lett.*

[11] Z. Z. Bandić, P. M. Bridger, E. C. Piquette, and T. C. McGill, "The values of minority carrier diffusion lengths and lifetimes in GaN and their implications for bipolar devices," *Solid-State Electron*, vol. 44, pp. 221-228, Feb. 2006.

[12] H. Hasegawa, T. Inagaki, S. Ootomo, and T. Hashizume, "Mechanisms of current collapse and gate leakage currents in AlGaN/GaN heterostructure field effect transistors," *J. Vac. Sci Technolo. B* vol. 21, pp. 1844-1855, Jul./Aug. 2003.

[13] Y. Uemoto, M. Hikita, H. Ueno, H. Matsuo, H. Ishida, M. Yanagihara, T. Ueda, T. Tanaka, and D. Ueda, "Gate Injection Transistor (GIT)-A Normally-Off AlGaN/GaN Power Transistor Using Conductivity Modulation," *IEEE Trans. Electron Devices*, vol. 54, no. 12, pp. 3393-3399, Dec. 2007.

Proceedings of the 23rd International Symposium on Power Semiconductor Devices & IC's
May 23-26, 2011 San Diego, CA

Over 1.7 kV normally-off GaN hybrid MOS-HFETs with a lower on-resistance on a Si substrate

Nariaki Ikeda, Ryosuke Tamura, Takuya Kokawa, Hiroshi Kambayashi, Yoshihiro Sato, Takehiko Nomura and Sadahiro Kato

Advanced Power Device Research Association
Yokohama City, Kanagawa, Japan

Abstract—In this study, normally-off GaN hybrid MOS-HFET devices on 4-inch Si substrates were fabricated, and the device characteristics were examined. As a result, the breakdown voltage (Vb) was improved using a combination of a high-resistive carbon-doped back barrier layer and a thin channel layer of 50 nm. The specific on-resistance (RonA) was estimated to be less than 7.1 mΩcm^2 for Lgd = 12 μm, and Vb was estimated to be over 1.71 kV for Lgd = 18 μm. To our knowledge, these values are the best results ever reported for normally-off GaN-based MOSFETs.

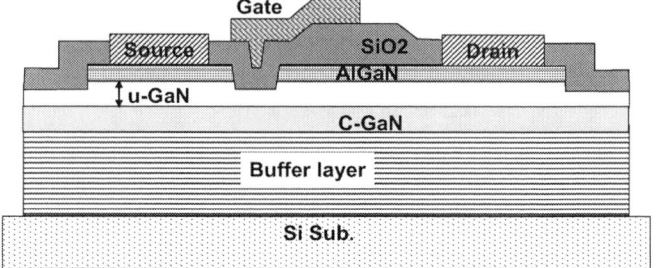

Figure. 1 Schematic structure of an AlGaN/GaN hybrid MOS-HFET unit

I. INTRODUCTION

GaN-based field effect transistors (FETs) can be operated under high-power, high-frequency and high-temperature conditions, producing lower loss and higher power switching characteristics [1] – [4]. Using an AlGaN/GaN HFET structure, a two-dimensional electron gas (2DEG) with high mobility and a high carrier density is generated at the hetero-interface. Conventional HFET devices usually operate in normally-on mode. However, as this state is not suitable for power application due to fail-safe mode issues, a normally-off operation transistor is necessary in the case of application for a switching device.

In the past, attempts to use normally-off GaN-FETs have been reported. Several device structure types have been demonstrated, including a p-type (Al)GaN semiconductor under the gate [5], a recess-gate structure [6], [7] and a fluorine-treated device [8]. MOSFET structures provide promising candidates, as there is a proven method for them in Si-based electronics. One such candidate is a GaN-based RESURF-MOSFET structure created using an ion implantation technique [9]. However, this type of device has a significant problem related to higher on-resistance, which is considered a result of increased resistance caused by the lower activation ratio of the RESURF portion.

In recent years, several researchers have demonstrated a novel hybrid MOS-HFET [10], [11]. This type of structure has several advantages over the conventional GaN-based RESURF-MOSFET structure.

We previously reported on normally-off GaN hybrid MOS-HFETs on Si substrates [12]. The breakdown voltage was 600 V and the RonA value was 10 mΩcm^2. However, the Vb value was lower than that of GaN-based HFETs. The thickness of the epitaxial layer is an important parameter in maximizing Vb values. Although increasing the gate-to-drain distance (Lgd) is one method of releasing the electric field from the gate to the drain, access resistance due to the large Lgd causes deterioration of the on-resistance. In the case of GaN-HFET structures, the use of a thin undoped (u)-GaN layer as a channel layer on a highly resistive carbon-doped GaN layer has been reported as an effective method of maximizing Vb and minimizing Lgd as much as possible, resulting in an improved trade-off between RonA and Vb [13].

In the case of Si devices, an SOI (silicon-on-insulator) structure was introduced to improve Si-MOSFET structure characteristics [14]. In the same manner, a thin u-GaN channel layer over a highly resistive GaN layer can be operated like an SOI structure. Accordingly, such a combination was introduced in this study to improve the Vb for a GaN hybrid MOS-HFET, and the thickness dependence of the u-GaN layer was examined.

978-1-4244-8425-6/11 $26.00 © 2011 IEEE

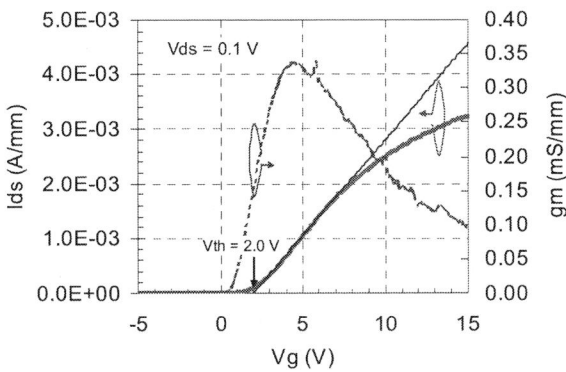

Figure 2 Transfer characteristics of GaN hybrid MOS-HFETs with an Lgd value of 12 μm

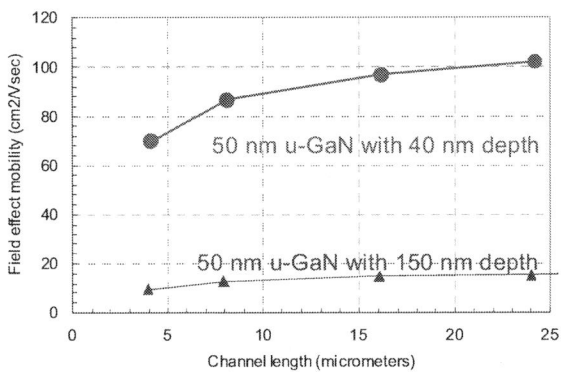

Figure 3 Field effect mobility for hybrid MOS-HFET devices with a recess etching depth of 40 nm compared to those with a depth of 150 nm

II. EXPERIMENTAL CONDITIONS

An AlGaN/GaN hetero-structure then a highly resistive buffer layer were formed on a 4-inch Si substrate using the MOCVD method. A thick epitaxial layer was used to create a total thickness of over 7.3 μm.

Figure 1 shows the GaN hybrid MOS-HFET device structure. A mesa structure was formed for electrical isolation, and the electrode deposits were made using the sputtering method. State-of-the-art Ti/AlSi/Mo-based ohmic electrodes were formed on the AlGaN layers as source and drain electrodes to reduce the contact resistance. SiO_2 layers deposited using the P-CVD method were used as gate insulator films and field plate films. A 40-nm gate insulator film was deposited, and a shallow recess etching structure was then formed. Ti/Au was used as the gate electrode material, including for the gate field plate structure. After the device fabrication process was finished, the DC characteristics were examined and the current collapse characteristics were investigated.

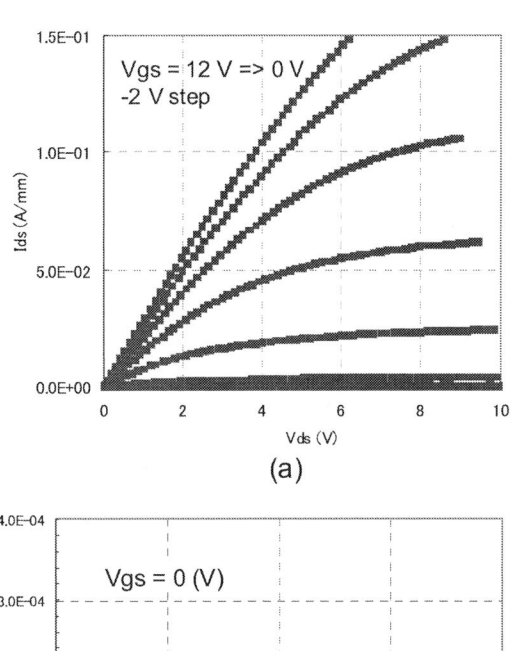

(a)

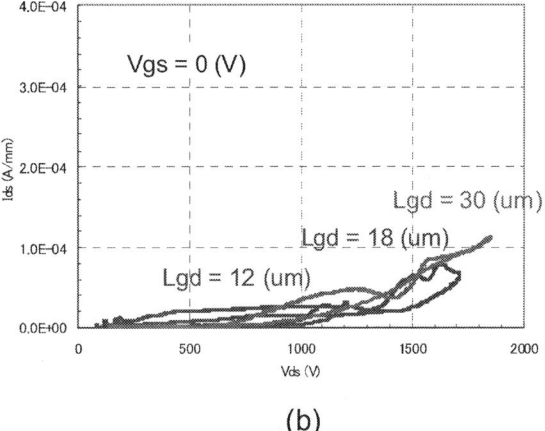

(b)

Figure 4 Drain current versus drain voltage characteristics for (a) on-state for GaN hybrid MOS-HFETs with an Lgd value of 12 μm, and (b) off-state for GaN hybrid MOS-HFETs with different Lgd values

III. RESULTS AND DISCUSSION

Hybrid MOS-HFET devices with several Lgd values were examined. Figure 2 shows the transfer characteristics for an Lgd value of 12 μm. The threshold voltage (Vth) was extrapolated as 2.0 V, resulting in the achievement of normally-off operation. Figure 3 shows the field effect mobility (μ_{FE}) values for the channel layer from gradual channel approximation. The μ_{FE} values were compared with the recess etching depths. In the case of a depth of 50 nm, μ_{FE} was estimated to be 102 cm^2/Vsec, while for 150 nm it was estimated to be 15 cm^2/Vsec. The lower-mobility carbon-doped layer affected the μFE value of the channel layer, and this was considered to be the reason for the difference. It is very important to control the etching depth under the gate to achieve lower on-resistance. Figure 4 (a) shows the drain current (Ids) versus the drain voltage (Vds) characteristics for a device. The thickness of the u-GaN channel layer was 50 nm. In this case, the recess etching depth under the gate was 40 nm. The gate length (Lg), the distance from the source electrode to the gate electrode (Lsg) and the Lgd value were 1.0 μm, 3.0 μm and 12.0 μm, respectively. Moreover, the peripheral width in relation to current diffusion around the edges of the ohmic electrodes was assumed to be 2 μm. A distance of 20 μm from the source to the drain (Lsd) and a gate width (Wg) of 1.0 mm were defined to calculate the active area, which was 2.0 x 10⁻⁴ cm² for this device. Its RonA value was estimated to be 7.1 mΩcm².

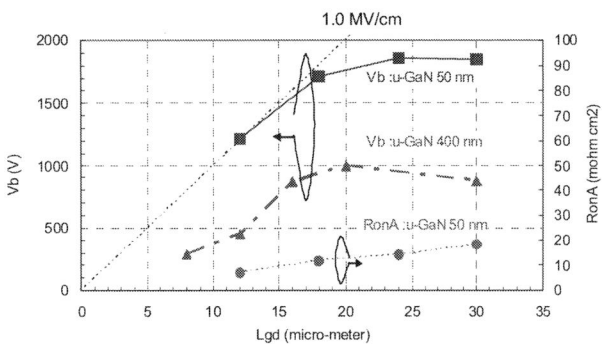

Figure 5 Vb versus Lgd for different u-GaN layer thicknesses, and RonA versus Lgd for a u-GaN value of 50 nm

Figure 4 (b) shows the off-state characteristics for a Vgs value of 0 V. During the measurements, a substrate bias was maintained in a floating condition. Several devices with different Lgd lengths were examined. The Vb value was increased along with that of Lgd, resulting in a Vb of 1.21 kV for an Lgd of 12 μm and over 1.71 kV for an Lgd of over 18 μm.

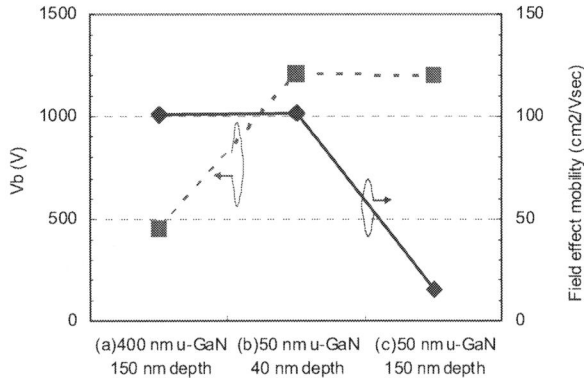

Figure 6 Comparison of Vb and field effect mobility between different u-GaN thicknesses and recess etching depths for a device with an Lgd of 12 μm

Figure 5 shows the Vb and RonA values versus the gate-to-drain distance (Lgd) for a u-GaN thickness of 50 nm. For a value of 50 nm, Vb was improved in comparison with a thickness of 400 nm. These outcomes indicate that using a thin channel layer on a highly resistive carbon-doped GaN layer was effective in obtaining a high Vb value, resulting in fully depleted channel layers. As a result, it can be concluded that a highly resistive GaN layer plays important roles such as dielectric insulation of Si SOI structures.

Figure 6 shows a comparison of Vb and μ_FE values for different u-GaN thickness and recess etching depths. In the case of (a) and (b), the μ_FE values were almost the same, but the Vb values were different, while for (b) and (c), the Vb values were almost the same but the μ_FE values were different. As a result, case (b) can be seen as the best combination to improve both Vb and μ_FE simultaneously.

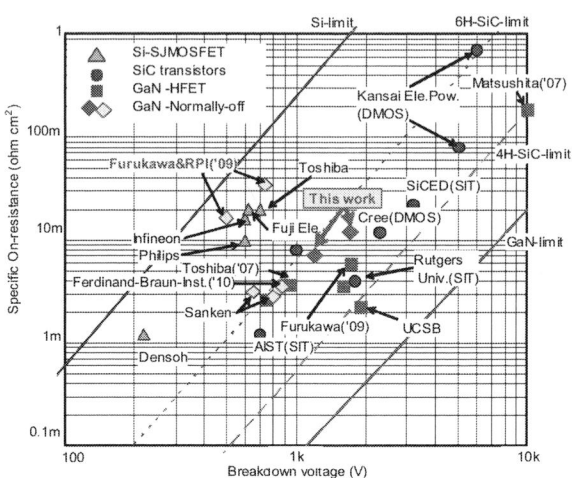

Figure 7 Specific on-resistance (RonA) versus breakdown voltage (Vb) for Si, SiC and GaN devices

Figure 7 shows the characteristics of RonA versus Vb. The RonA value for our normally-off hybrid MOS-HFETs was 7.1 mΩcm² and the Vb was 1.21 kV for an Lgd of 12 μm, while the corresponding values were 11.9 mΩcm² and 1.71 kV for an Lgd of 18 μm. These values are plotted in the figure between the limits of the 4H-SiC line and the 6H-SiC line (labeled "This work"), and are improved compared with our previously reported results [12]. To our knowledge, these are the best results ever reported for normally-off GaN-based MOSFETs.

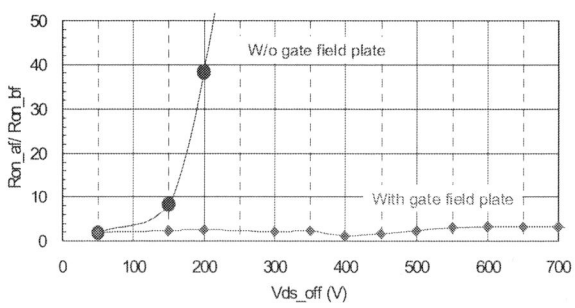

Figure 8 Comparison of current collapse ratios for hybrid MOS-HFETs with and without a gate field plate structure

Figure 8 shows a comparison of several current collapse results obtained when the stress bias was changed. The current collapse phenomena were observed using our own on-wafer measurement methods [15]. These results are for an Lgd value of 18 μm. Without a field plate structure, the current collapse

978-1-4244-8425-6/11 $26.00 © 2011 IEEE

ratio increases in a linear fashion; with one, it is maintained at a constant value despite an increase in the Vds_off stress bias.

IV. CONCLUSION

This paper reported on normally-off GaN hybrid MOS-HFETs with high Vb and lower RonA values. A thin u-GaN channel layer on a highly resistive carbon-doped layer was found to be effective in improving the RonA-Vb trade-off. As a result, RonA was estimated to be less than 7.1 mΩcm^2 for an Lgd value of 12 μm, and Vb was estimated to be over 1.71 kV for an Lgd of 18 μm. These outcomes indicate that a structure with a highly resistive carbon-doped back barrier layer and a thin channel layer in combination is promising as a normally-off GaN-FET for power switching application.

ACKNOWLEDGMENT

The authors would like to thank Professor T. P. Chow of the Rensselaer Polytechnic Institute for his constructive input on this study.

REFERENCES

[1] T. P. Chow, R. Tyagi, "Wide bandgap compound semiconductors for superior high-voltage unipolar power devices." IEEE Trans Electron Devices, Vol. 41, pp. 1481 – 1483, 1994.

[2] O. Akutas, Z. F. Fan, S. N. Mohammad, A. E. Botchkarev, H. Morkoç, "High temperature characteristics of AlGaN/GaN modulation doped field effect transistors." Appl Phys Lett, Vol. 69, pp. 3872 – 3874, 1996.

[3] S. Yoshida, J. Li, T. Wada, H. Takehara, "High-Power AlGaN/GaN HFET with a Lower On-state Resistance and a Higher Switching Time for an Inverter Circuit." In Proc. 15th ISPSDs, pp. 58 – 61, 2003.

[4] W. Saito, T. Nitta, Y. Kakiuchi, Y. Saito, K. Tsuda, I. Omura, M. Yamaguchi, "A 120-W Boost Converter Operation Using a High-Voltage GaN-HEMT." IEEE Electron Device Letters, Vol. 29, No. 1, 2007, pp. 8 – 10.

[5] Y. Uemoto, M. Hikita, H. Ueno, H. Matsuo, H. Ishida, M. Yanagihara, T. Ueda, T. Tanaka, D. Ueda, "Gate Injection Transistor (GIT)—A Normally-Off AlGaN/GaN Power Transistor Using Conductivity Modulation." IEEE Transactions on Electron Devices, Vol. 54, No. 12, 2007, pp. 3393 – 3399.

[6] M. Kanemura T. Ohki, T. Kikkawa, K. Imanishi, T. Imada and N. Hara, "High current operation of enhancement-mode GaN MIS-HEMTs with triple cap structure using atomic layer deposited Al2O3 gate insulator." DRC2009 proc., pp. 165 – 166.

[7] K. Ota K. Endo, Y. Okamoto, Y. Ando, H. Miyamoto, H. Shimawaki, "Normally-off GaN FET with High Threshold Voltage Uniformity Using a Novel Piezo Neutralization Technique." IEDM2009 proc., pp. 153 – 156.

[8] C. Zhou, W. Chen, E. L. Piner, K. J. Chen, "Self-Protected GaN Power Devices with Reverse Drain Blocking and Forward Current Limiting Capabilities." In Proc. 22nd ISPSDs, pp. 343 – 346, 2010.

[9] Y. Niiyama, S. Ootomo, J.Li, H. Kambayashi, T. Nomura, S. Yoshida, K. Sawano and Y. Shiraki, "Si Ion Implantation into Mg-Doped GaN for Fabrication of Reduced Surface Field Metal–Oxide–Semiconductor Field-Effect Transistors." Japanese Journal of Applied Physics, Vol. 47, No. 7, 2008, pp. 5409 – 5416.

[10] W. Huang, T. P. Chow, Y. Niiyama, T. Nomura, S. Yoshida, "Enhancement-mode gan hybrid MOS-HEMTs with ron,sp of 20 mΩ-cm^2." In Proc. 20th ISPSDs, pp. 295 – 298, 2008.

[11] T. Oka, T. Nozawa, "AlGaN/GaN Recessed MIS-Gate HFET With High-Threshold-Voltage Normally-Off Operation for Power Electronics Applications." Electron Device Letters, 29, 2008, pp. 668 – 670.

[12] H. Kambayashi, Y. Satoh, S. Ootomo, T. Kokawa, T. Nomura, S. Kato, T. P. Chow, "Over 100 A operation normally-off AlGaN/GaN hybrid MOS-HFET on Si substrate with high-breakdown voltage." Solid-State Electronics 54, No. 6, 2010, pp. 660 – 664.

[13] N Ikeda, S. Kaya, J. Li, T. Kokawa, M. Masuda, S. Kato, "High-power AlGaN/GaN MIS-HFETs with field-plates on Si substrates." Proc. 21st ISPSDs, pp. 251 – 254, 2009.

[14] T. Letavic, E. Arnold, M. Simpson, R. Aquino, H. Bhimnathwala, R. Egloff, A. Emmerik, S. Wong, S. Mukherjee, "High performance 600 V smart power technology based on thin layer Silicon-on Insulator." In Proc. 9th ISPSDs, pp. 49 – 52, 1997.

[15] N. Ikeda, Y. Niiyama, H. Kambayashi, Y. Sato, T. Nomura, S. Kato, S. Yoshida, "GaN Power Transistors on Si Substrates for Switching Applications." Proceedings of the IEEE Vol. 98, No. 7, July 2010, pp. 1151 – 1161.

Proceedings of the 23rd International Symposium on Power Semiconductor Devices & IC's
May 23-26, 2011 San Diego, CA

Low On-Resistance 1.2 kV 4H-SiC MOSFETs Integrated with Current Sensor

A. Furukawa, S. Kinouchi, H. Nakatake, Y. Ebiike, Y. Kagawa, N. Miura,
Y. Nakao, M. Imaizumi, H. Sumitani and T. Oomori[*]
Advanced Technology R&D Center, Mitsubishi Electric Corporation
8-1-1 Tsukaguchi-Honmachi, Amagasaki, Hyogo 661-8661/Japan
E-mail: Furukawa.Akihiko@df.MitsubishiElectric.co.jp
[*]Power Device Works, Mitsubishi Electric Corporation

Abstract — **4H-SiC MOSFETs integrated with a current sensor have been fabricated for the first time. The MOSFET shows superior characteristics with a specific on-resistance of 3.7 mΩcm^2 and a blocking voltage of 1.4 kV. The deviation of the current ratio (*Imain/Isense*) stays within 10% in the temperature range between 25°C and 175°C, which is desirable for the current sensor of high power devices. Furthermore, the main current shut-off operation at an over-current detected using the current sensor has been demonstrated successfully.**

I. INTRODUCTION

Intelligent Power Module (IPM) has been widely used in power systems such as industrial motor drives and PV control system [1]. Continuous improvement of silicon IGBT and diode performance in terms of power loss reduction, higher power density and higher power handling capability have been demanded for such applications, but these devices are considered to be approaching the limit in terms of performance improvement. Silicon carbide (SiC) is an attractive semiconductor for power devices since it exhibits exceptional material properties compared with silicon, such as wider bandgap, higher critical field, and higher thermal conductivity. SiC power devices are expected to save energy and miniaturize power electronics systems. We demonstrated a successful drive of a 3.7 kW/400 V induction motor using a full SiC module consisting of MOSFETs and Schottky barrier diodes and presented marked power loss reduction [2]. In practical use, IPMs equipped with SiC MOSFETs are required because of their usability. Recently operation of a full SiC IPM was demonstrated in a high temperature region [3]. An IPM needs current sensing to protect itself from an over-current. In case of a normal IPM, external current transformers are used for sensing currents. On the other hand, an on-chip current-sensor using integrated MOSFETs enables instant detection of an over-current (e.g. short circuit current). To date, small size SiC JFETs integrated with current sensor have already been reported [4]. However, SiC MOSFETs integrated with current sensor have not been reported yet. In this paper, we present the electrical characteristics of 1.2 kV class SiC MOSFETs that implement current sensing.

II. DEVICE STRUCTURE

SiC vertical MOSFETs have been fabricated on 4H-SiC substrate. The net donor concentration (N_D-N_A) and thickness of the drift epilayer are 1×10^{16} cm^{-3} and 12 μm, respectively. Ion implantation was carried out at multiple energies with a photo-resist mask for the source and p-well regions. The implantation on the contact region in the p-well regions was performed at high temperature [5]. Subsequent activation annealing was carried out in an Ar atmosphere at temperatures above 1600°C using a shuttle annealing technique. The field oxide layer was deposited and patterned, and the gate dielectric layer was then formed by the pyrogenic thermal oxidation and nitridation. After the gate electrode formation, ohmic contacts for the source and backside were formed by Ni alloy. An Al layer was deposited, patterned, and finally, covered by polyimide film. Figure 1 shows the top view photograph of the fabricated SiC MOSFETs. The gate pad area and the current sense-source pad area are arranged on the upper center and on the right corner, respectively.

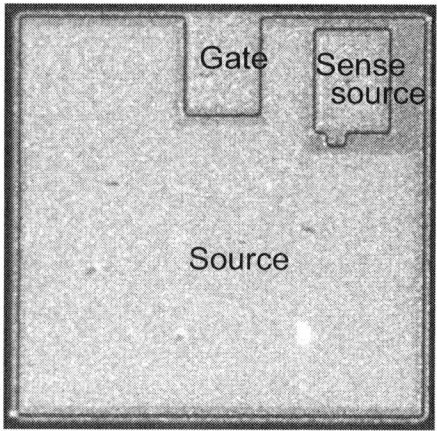

Figure 1. Top view of fabricated SiC MOSFET.

978-1-4244-8425-6/11 $26.00 © 2011 IEEE

Figure 2 shows an equivalent circuit of the current sensing scheme using the SiC MOSFET integrated with current sensor. The MOSFETs consist of parallel conjunctions of 10-μm square unit cells, and the cells for current sensor and the main cells are identical, having a common gate and drain and separate sources. The current sensor is currently arranged on the edge of the sense-source pad. In this study, we used an external shunt resistor to sense the current.

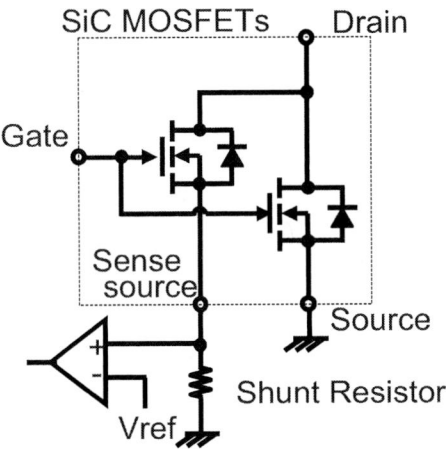

Figure 2. Equivalent circuit of the integrated SiC MOSFET current sensor and current sensing with an external shunt resistor.

Figure 3 shows the schematic of the cross-sectional structure between the main cell and the current sense cell. Each of the two MOSFET cell arrays, namely main and sense cell arrays, is surrounded by a p-well non-active area, which is connected to the corresponding source pad.

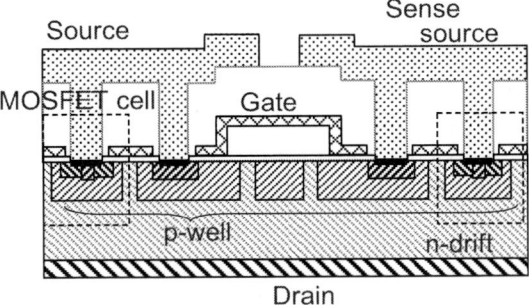

Figure 3. Schematic of the cross-sectional structure of the main MOSFET cell, the current sense MOSFET cell and the isolation region.

III. RESULTS AND DISSCUTION

Figure 4 (a) shows the typical drain output characteristics of the main MOSFET at room temperature. The specific on-resistance ($R_{on,sp}$) is estimated to be 3.7 mΩcm^2 at a drain current density of 100 A/cm^2 and a gate bias of 15 V. Figure 4 (b) shows the drain output characteristics of the current sense MOSFET. Small difference observed on the I-V curves in Fig.4, is possibly due to the self-heating effect on the main MOSFET. Nevertheless the sense current almost follows the main current.

The temperature dependence of the static behavior of the main MOSFET is investigated. The threshold voltage of the MOSFET is set to suppress the temperature dependence of the drain current density and to obtain a negative temperature coefficient. The $R_{on,sp}$ characteristics is plotted against temperature in Fig. 5. A very low $R_{on,sp}$ value of 6.9 mΩcm^2 is obtained maintained even at 230°C. The increase in the $R_{on,sp}$ is due to the reduction of the electron mobility in the channel and drift layer.

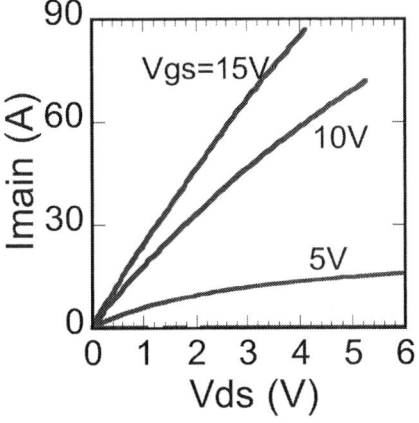

(a) Main SiC MOSFET

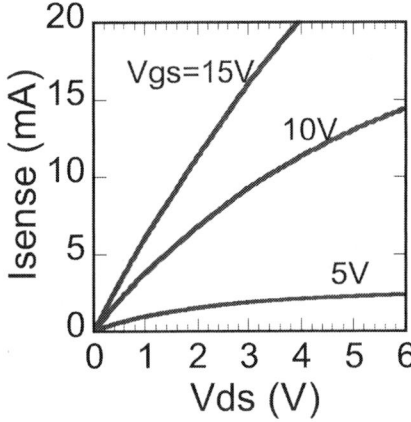

(b) Current sense SiC MOSFET

Figure 4. Drain output characteristics at room temperature.

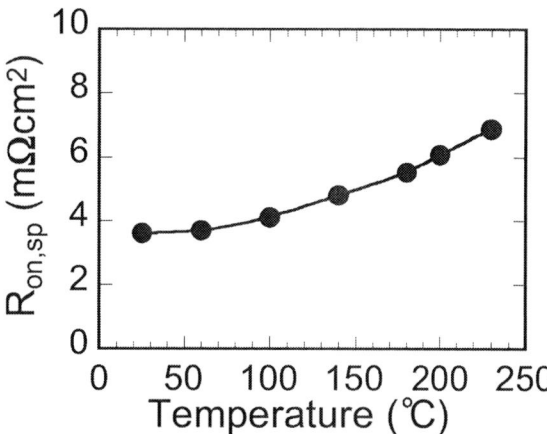

Figure 5. Temperature dependence of the specific on-resistance.

Figure 6 shows the typical blocking characteristics at a short configuration between the sense-source terminal and the source terminal. A stable avalanche breakdown of 1.4 kV is obtained at a gate bias of -5 V.

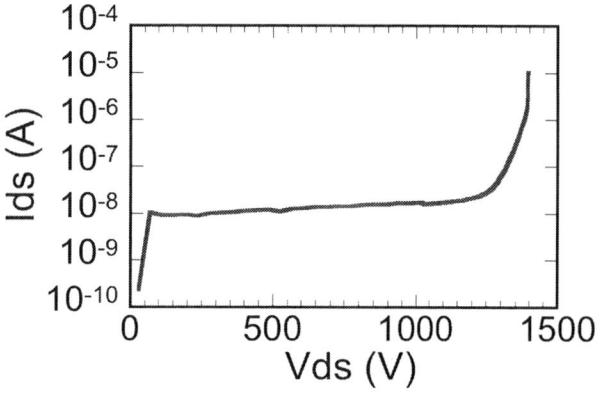

Figure 6. Blocking characteristics of SiC MOSFET.

Figure 7 shows the isolation characteristics between the sense-source treminal and the source terminal. The OFF state is selected as the measurement condition to investigate the isolation. From the analysis using the equivalent circuit model, the leak current of region A is corresponds to the diode forward current of the current sensor, and the current of region B is corresponds to the subthreshold leak current of the main MOSFET (see Fig. 3). No adverse effect was observed.

Next, the temperature dependence of the current ratio (*Imain/Isense*) has been investigated. *Imain* corresponds to the drain current, and *Isense* corresponds to the source current of the sense MOSFET. The value of *Isense* is estimated by using the shunt resistor connected to the ciruit configuration as shown in Fig. 2. The value of the shunt resistor is varied as 4.7, 51, and 110 Ω, and for each value, the current through the sense cells is measuered. Then the intercept of the current line plotted against shunt resistance is taken to be the value of *Isense*.

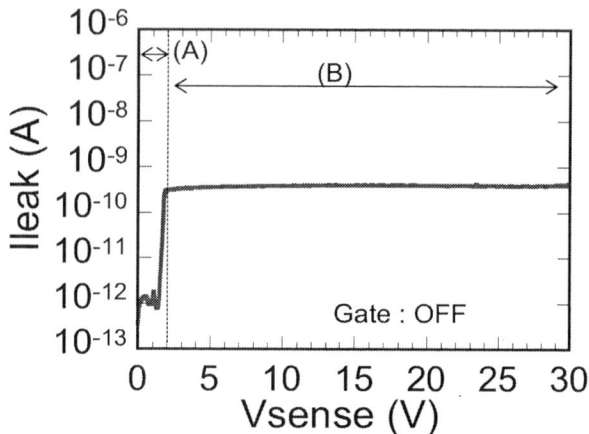

Figure 7. Isolation characteristics between sense-source and source terminals. Source terminal was set at 0V, drain terminal was set at floating.

Figure 8 shows the current ratio as a function of the main current. The current ratio stays almost constant at 4400 in a wide range of the main current at 25°C. This value is a little lower than that of the cell area ratio. Furthermore, the deviation of the current ratio is within 10% from 25°C to 175°C. This is due to the on-resitance having a small temperature coefficient (see Fig. 5).

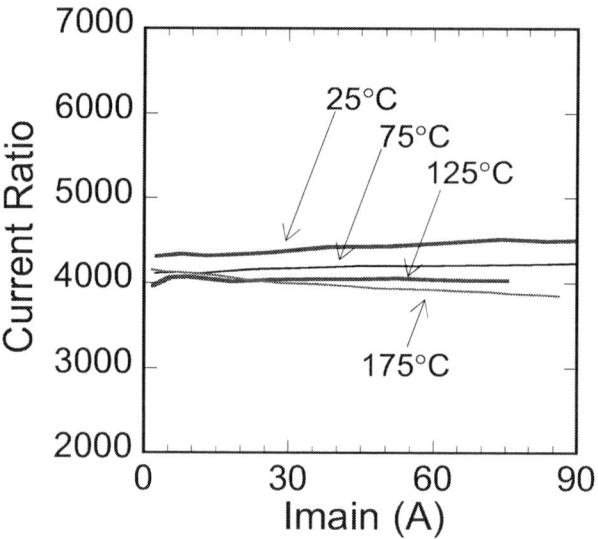

Figure 8. Measured current ratio (*Imain/Isense*) as a function of main current (*Imain*).

Finally, dynamic electrical characterization was perfomed on the SiC MOSFET integrated with current sensor. Figure 9 presents the dynamic waveforms with a shut-off feedback from the current sensor at 25°C. The measurement was performed under the following condition; *Vgs* = 15 V at ON state, -10 V at OFF state, and *Vcc* (*Vds*) = 600 V. The sense voltage increases in proportion to the main current. Safe shut-off operation at 56 A (600 A/cm²) is demonstrated using its current sensing. The comparable result was also obtained at 175°C.

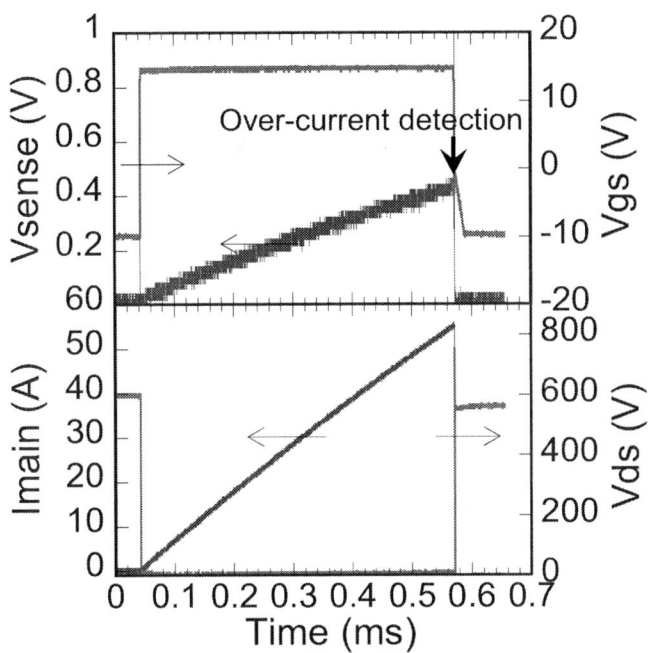

Figure 9. Dynamic shut-off waveforms with inductive load.

IV. CONCLUSION

4H-SiC MOSFETs integrated with current sensor are successfully developed for 1.2 kV class. The current sensing ratio stays almost constant for wide temperature and forward current ranges. Safe turn-off operation with inductive load is achieved by its current sensing through an external shunt resistor.

ACKNOWLEDGMENT

We would like to acknowledge Takeshi Oi and Shuhei Nakata for their continuous support. We also thank Tomokatsu Watanabe, Masayuki Furuhashi, Shiro Hino, Toshikazu Tanioka, Keiko Fujihira, Rina Kakimi and Tsuyoshi Kawakami for their valuable discussions.

REFERENCES

[1] G. Majumdar, "Power Module Technology for Home power Electronics", Proceedings on IEEE International Power Electronics Conference (2010) pp773-777.

[2] N. Miura, K. Fujihira, Y. Nakao, T. Watanabe, Y. Tarui, S. Kinouchi, M. Imaizumi and T Oomori, "Successful development of 1.2 kV 4H-SiC MOSFETs with the very low on-resistance of 5m$\Omega \cdot$cm2", Proc. ISPSD (2006) pp261-264.

[3] T. Nakamura, M. Sasagawa, Y. Nakano, T. Otsuka and M. Miura, "Large current SiC power devices for automobile applications", Proceedings on IEEE International Power Electronics Conference (2010) pp1023-1026.

[4] D. Tournier, M. Vellvehi1, P. Godignon, J. Montserrat1, D. Planson and F. Sarrus, "Current sensing for SiC Power Devices", Mater. Sci. Forum. 527-529 (2006) pp1215-1218.

[5] T. Watanabe, S. Aya, R. Hattori, M. Imaizumi and T. Oomori, "Effects of implantation temperature on sheet and contact resistance of heavily Al implanted 4H-SiC", Mater. Sci. Forum. 645-648 (2010) pp705-708.

Proceedings of the 23rd International Symposium on Power Semiconductor Devices & IC's
May 23-26, 2011 San Diego, CA

4H-SiC Bipolar Junction Transistors with Record Current Gains of 257 on (0001) and 335 on (000-1)

Hiroki Miyake[1], Tsunenobu Kimoto[1,2], and Jun Suda[1]

1) Department of Electronic Science and Engineering, Kyoto University
2) Photonics and Electronics Science and Engineering Center (PESEC), Kyoto University
A1-303, Kyotodaigakukatsura, Nishikyo, Kyoto 615-8510, JAPAN
miyake@semicon.kuee.kyoto-u.ac.jp

Abstract—We demonstrate 4H-SiC bipolar junction transistors (BJTs) with record current gains. Improved current gain was achieved by utilizing optimized device geometry as well as optimized surface passivation and continuous epitaxial growth of the emitter-base junction, combined with an intentional deep-level-reduction process based on thermal oxidation to improve the lifetime in p-SiC base. Current gain (β) of 257 was achieved for 4H-SiC BJTs fabricated on the (0001)Si-face. The gain of 257 is twice as large as the previous record gain. We also demonstrate, for the first time, BJTs on the (000-1)C-face that showed the highest β of 335 among the SiC BJTs ever reported.

I. INTRODUCTION

Power bipolar junction transistors (BJTs) based on 4H-SiC are recognized as very attractive candidates for high-power switching devices because of such characteristics as high breakdown voltage, low on-resistance, and high current density. For the best use of the advantages of SiC BJTs, they can be used at elevated temperature because they are free of gate-oxide reliability issues observed in metal-oxide-semiconductor field effect transistors. Therefore, the current gain at both room and elevated temperatures should be high enough for practical applications to reduce the current at the base-drive circuit. Although high-performance BJTs with low on-resistance and high breakdown voltage have been reported, the current gain is typically in the range of 60-70 [1, 2] at room temperature, and only a few groups have achieved current gain in the range of 102-134 [3-5], indicating that improving current gain remains difficult.

The improvements of current gain are reported by using optimized device geometry [1], optimized surface passivation [4, 5], and continuous epitaxial growth [1, 2]. For improved current gains in SiC BJTs, further reduction is crucial of the recombination current at the emitter-mesa sidewall, the base surface, and the emitter-base junction. In addition, poor lifetime in p-SiC is thought to be responsible for limiting the current gain [3, 6].

In this paper, we employ optimized device geometry as well as optimized surface passivation to suppress the surface

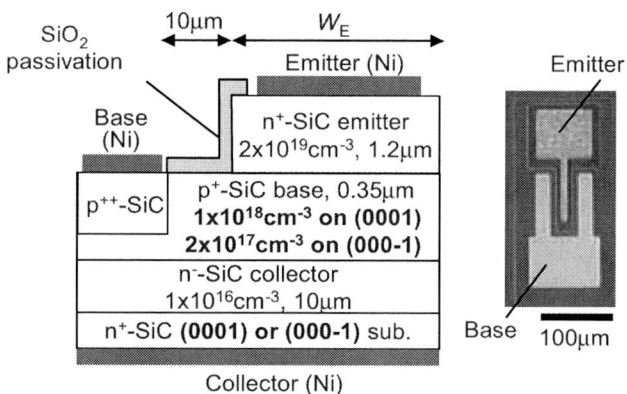

Fig. 1 Schematic cross section of a fabricated 4H-SiC BJT and optical image of fabricated single finger BJT.

recombination at the emitter-mesa sidewalls, and continuous epitaxial growth of the emitter junction to reduce interface states. To enhance the lifetime in p-SiC, we have employed an intentional deep-level-reduction-process (DLR-process) based on thermal oxidation (1150°C, 5 h × 2) [7, 8]. With these techniques, we successfully demonstrate 4H-SiC BJTs with a record current gain (β) of 257 fabricated on the (0001)Si-face. We also utilize SiC (000-1)C-face because lower interface state density can be obtained for SiO$_2$/SiC(000-1) structures adequately processed [9], which may reduce the surface recombination of BJTs and enhance the current gains. We demonstrate BJTs on the (000-1)C-face that showed the highest β of 335 among the SiC BJTs ever reported.

II. DEVICE FABRICATION

Fig. 1 shows the schematic structure of a fabricated BJT. The BJT structure was grown on both n-type 4H-SiC (0001) and (000-1) substrates (8° off-oriented toward [11-20]) with a 10-μm-thick N-doped n-SiC collector. A 0.35-μm-thick Al-doped p$^+$-SiC base and a 1.2-μm-thick N-doped n$^+$-SiC emitter

978-1-4244-8425-6/11 $26.00 © 2011 IEEE

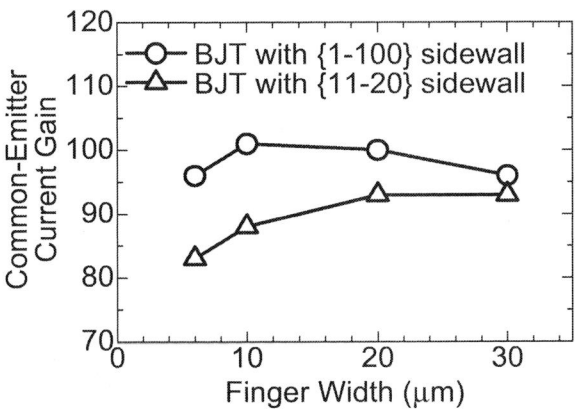

Fig. 2 Summary of current gain of BJTs with various device geometries fabricated on (0001). Current gain increases with finger width. BJTs with {1-100} sidewalls exhibited higher current gain compared with those with {11-20} sidewalls.

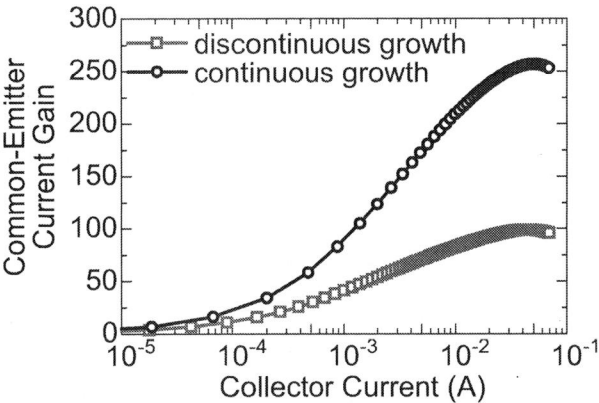

Fig. 3 Current gains as function of collector current for BJTs with 20-μm-wide fingers and {1-100} sidewalls grown by separated or continuous growth run .

for both BJTs were grown continuously in the same reactor by chemical vapor deposition. For reference, we also fabricated BJTs prepared by separated growth runs for the base and emitter layers. The impurity concentration of the base layer was 1×10^{18} cm^{-3} on (0001) and 2×10^{17} cm^{-3} on (000-1). Multiple Al$^+$ ion implantations at 500°C with an impurity concentration of 3×10^{20} cm^{-3} were carried out to form p^{++} base contact regions.

To maximize the lifetime in the p-SiC base layer, we have employed DLR-process based on thermal oxidation for BJTs on (0001). First, we utilized thermal oxidation in O$_2$ at 1150°C for 5 h before activation annealing of implants to the p-SiC base contact region. This process contributes to the reduction of major deep levels such as $Z_{1/2}$ and EH$_{6/7}$ observed in as-grown epi-layer [7], and gives rise to the HK0 center associated with CF$_4$-based RIE [10]. Then, we employed the activation annealing of implants at 1800°C for 10 min. This will regenerate $Z_{1/2}$ and EH$_{6/7}$, whereas the HK0 center that anneals out at 1550°C will disappear [10]. To re-eliminate $Z_{1/2}$ and EH$_{6/7}$, we utilized thermal oxidation in O$_2$ at 1150°C for 5 h once again. This process may reduce the deep levels in p-SiC base layer and implanted layer [11]. A carbon-cap used for activation annealing was removed during this oxidation process. For BJTs on (000-1), we utilized the same process except for the oxidation time (15 min) to obtain similar oxide thickness as on (0001).

Surface passivation by 80-nm-thick deposited oxide nitrided in NO was employed to minimize the surface recombination for both BJTs on (0001) and on (000-1). The deposited oxide (SiO$_2$) was prepared by plasma-enhanced chemical vapor deposition at 400°C using tetraethoxysilane and O$_2$. The oxides were then nitrided by 10%-diluted NO in N$_2$ at 1300°C for 30 min followed by N$_2$ annealing at 1300°C for 30 min. The details of surface passivation were described in [5].

Ni ohmic contacts were formed on the emitter and base surface using a standard lift-off technique, respectively. Rapid thermal annealing was performed at 950°C in Ar.

All the BJTs discussed in this letter have a single finger, 100-μm long, and the distance between the emitters and the base contact regions (L_{EB}) is 10 μm. Emitter fingers with variable width and different directions having a {11-20} sidewall or a {1-100} sidewall were fabricated to evaluate the effect of the surface recombination.

III. RESULTS AND DISCUSSION

Fig. 2 summarizes the current gain of BJTs with various device geometries fabricated on (0001). In this section, the BJTs grown in separated growth runs for emitter junctions are used. Fig. 2 clearly shows that the current gain depends on the finger width and direction. The current gain increases with the finger width because a wider BJT can minimize the effect of surface recombination at the sidewalls (emitter size effect), and the gain saturated for finger width over 20 μm. BJTs with {1-100} sidewalls showed higher current gains than BJTs with {11-20} sidewalls, indicating a lower surface recombination on {1-100} sidewalls. Yano et al. pointed out that MOS interface properties of the trench sidewalls fabricated on SiC 8° off-axis substrates strongly depend on the crystal planes [12]. According to their report, (11-20) 8° off channel plane showed a lower interface state density (D_{it}), whereas (-1-120) -8° off channel plane exhibited a higher D_{it} compared with {1-100}, due to the large off-angle of the substrate. As a result, the sum of the D_{it} at SiO$_2$/SiC{1-100} is smaller than that at SiO$_2$/SiC{11-20}. Since the D_{it} value has correlation with the current gain in BJTs [5, 13], we infer that the high current gain in the BJTs with {1-100} sidewalls is attributed to the smaller D_{it} at {1-100} planes that may suppress the surface recombination. Using 10-20-μm-wide fingers and {1-100} sidewalls, we achieved a current gain of 100.

Fig. 3 shows the current gain as a function of collector current for the BJTs with 20-μm-wide fingers and {1-100} sidewalls grown by separated or continuous growth. A higher current gain in the entire collector current range was obtained in a BJT using continuous growth. A record current gain of 257 in the BJT grown by a continuous growth run was achieved (Fig. 4), whereas a current gain of 100 in the BJT grown by discontinuous growth run was obtained. This drastic improvement in current gain is attributed to the reduction of

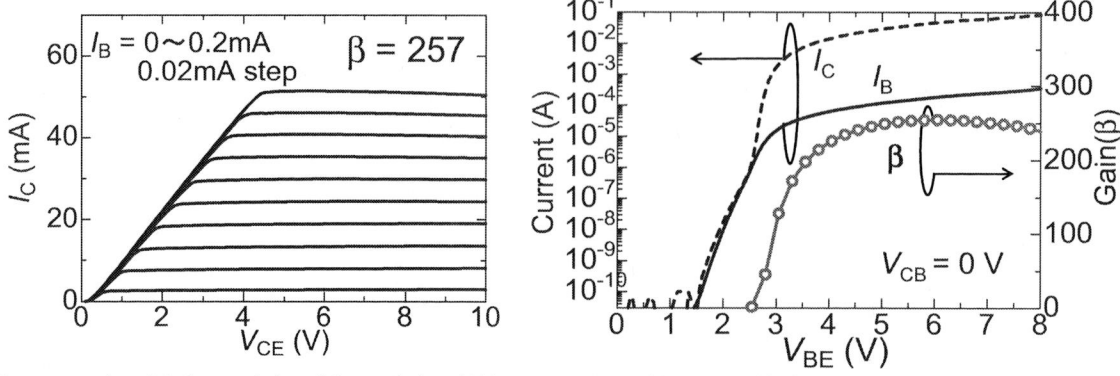

Fig. 4 Common-emitter *I-V* characteristic and Gummel plot of SiC BJT on (0001) with 20-μm-wide finger and {1-100} sidewall. Record current gain of 257 was achieved in BJT.

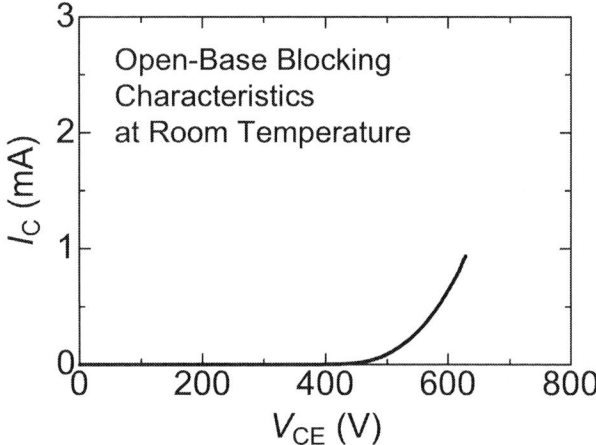

Fig. 5 The open-base blocking characteristics of the high-current-gain BJTs are shown

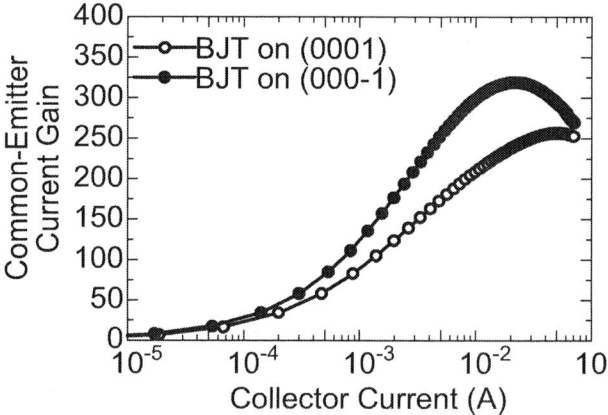

Fig. 6 Current gains as a function of collector current for BJTs with 20-μm-wide fingers and {1-100} sidewalls fabricated on (0001) and (000-1). A very high current gain of 335 was obtained in BJT fabricated on (000-1).

the recombination current at the emitter junction due to the reduction of the interface states [1, 2]. The current gain of 257 is twice as large as previous record current gain, although epi-structure is almost similar to the BJTs reported by other groups. We believe that well-optimized surface passivation [5] combined with an intentional DLR-process described in this letter contributed to such high current gains. The open-base blocking characteristics of the high-current-gain BJTs are shown in Fig. 5. Although these BJTs did not have edge termination, the leakage current was negligible at V_{CE} less than 400V.

We also demonstrate BJTs fabricated on (000-1). Fig. 6 shows the current gains as a function of collector current for the BJTs with 20-μm-wide fingers and {1-100} sidewalls fabricated on (0001) and (000-1). Since the doping concentration of the base layer on (000-1) is smaller than that on (0001), the BJT on (000-1) enters a high injection mode at a lower collector current compared with BJTs on (0001); thus it showed a peak current gain at a lower collector current. As shown in Fig. 7, a very high current gain of 335 was obtained in the BJT fabricated on (000-1), which is the highest value ever reported among the SiC BJTs. It should be noted that we applied the surface passivation process developed for SiC(0001), and that DLR-process did not work well on

current gain. We believe that optimization of the device fabrication process such as surface passivation, DLR-process, growth process for C-face BJTs will further enhance the current gain.

IV. CONCLUSION

We demonstrated 4H-SiC BJTs with a record current gain of 257 on the (0001)Si-face, which is twice as large as the previous record gain. Usage of {1-100} sidewalls, surface passivation with deposited oxides annealed in NO, and continuous epitaxial growth of the emitter-base junction, combined with deep-level-reduction process based on thermal oxidation dramatically improved the current gain. We also demonstrated the operation of 4H-SiC BJTs on the (000-1)C-face with the highest β of 335 among the SiC BJTs ever reported.

ACKNOWLEDGMENT

This work was supported in part by the Ministry of Education, Culture, Sports, Science and Technology, Japan, through the Global Center of Excellence (G-COE) Program (C09). The work of H. Miyake was supported by the Japan Society for the Promotion of Science through the Grant-in-Aid for Research Fellow and Scientific Research Grant 21226008.

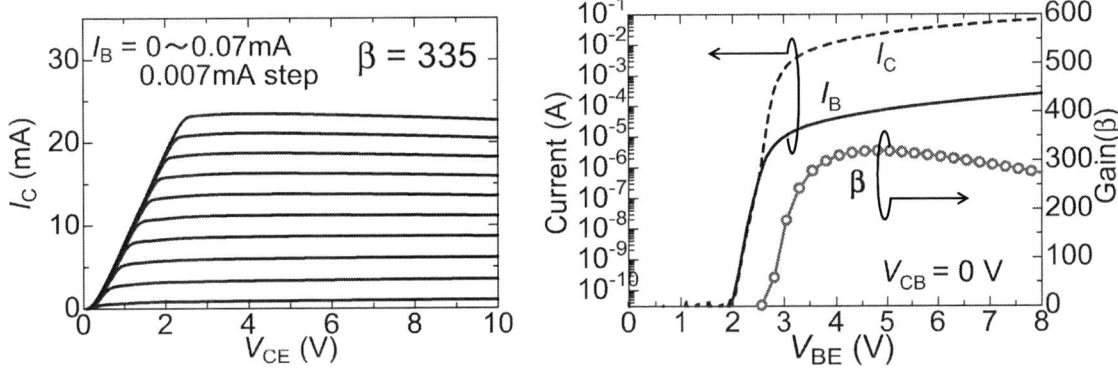

Fig. 7 Common-emitter *I-V* characteristic and Gummel plot of SiC BJT on (000-1) with 20-μm-wide finger and {1-100} sidewall. Record current gain of 335 was achieved in BJT.

REFERENCES

[1] M. Domeij, H.-S. Lee, E. Danielsson, C.-M. Zetterling, M. Östling, and A. Schöner, "Geometrical effects in high current gain 1100-V 4H-SiC BJTs," *IEEE Electron Device Lett.*, vol. 26, no. 10, pp. 743-745, Oct. 2005.

[2] J. Zhang, X. Li, P. Alexandrov, L. Fursin, X. Wang, and J. H. Zhao, "Fabrication and characterization of high-current-gain 4H-SiC bipolar junction transistors," *IEEE Trans. Electron Devices,* vol. 55, no. 8, pp. 1899-1906, Aug. 2008.

[3] Q. Zhang, A. Agarwal, A. Burk, B. Geil, and C. Scozzie, "4H-SiC BJTs with current gain of 110," *Solid-State Electronics,* vol. 52, no. 7, pp. 1008-1010, July 2008.

[4] K. Nonaka, A. Horiuchi, Y. Negoro, K. Iwanaga, S. Yokoyama, H. Hashimoto, M. Sato, Y. Maeyama, M. Shimizu, and H. Iwakuro, "A new high current gain 4H-SiC bipolar junction transistor with suppressed surface recombination structure; SSR-BJT," *Mater. Sci. Forum,* vol. 615-617, pp. 821-824, 2009.

[5] H. Miyake, T. Kimoto, and J. Suda, "Improvement of current gain in 4H-SiC BJTs by surface passivation with deposited oxides nitrided in N₂O or NO," *IEEE Electron Device Lett.,* vol. 32, no. 3, pp. 285-287, Mar. 2011.

[6] B. Buono, R. Ghandi, M. Domeij, B. G. Malm, C.-M. Zetterling, and M. Ostling, "Modeling and characterization of current gain versus temperature in 4H-SiC power BJTs," *IEEE Trans. Electron Devices*, vol. 57, pp. 704–711, 2010.

[7] T. Hiyoshi and T. Kimoto, "Reduction of deep levels and improvement of carrier lifetime in n-type 4H-SiC by thermal oxidation," *Appl. Phys. Express*, vol. 2, no. 4, p. 041101, 2009.

[8] T. Hayashi, K. Asano, J. Suda, and T. Kimoto, "Impacts of Thermal Oxidation and Surface Passivation on Carrier Lifetimes in p-type and n-type 4H-SiC Epilayers," presented at ECSCRM2010, Oslo, Norway, Aug. 29–Sep. 2, 2010, We1-4

[9] M. Noborio, J. Suda, S. Beljakowa, M. Krieger, and T. Kimoto, "4H-SiC MISFETs with nitrogen-containing insulators," *Phys. Stat. Sol. (a),* vol. 206, no. 10, pp. 2374-2390, 2009.

[10] K. Danno, and T. Kimoto, "Deep level transient spectroscopy on as-grown and electron-irradiated p-type 4H-SiC epilayers," *J. Appl. Phys.,* vol. 101, no. 10, p. 103704, 2007.

[11] K. Kawahara, J. Suda, G. Pensl, and T. Kimoto, "Reduction of deep levels generated by ion implantation into n- and p-type 4H-SiC," *J. Appl. Phys.,* vol. 108, no. 3, p. 033706, 2010.

[12] H. Yano, H. Nakao, H. Mikami, T. Hatayama, Y. Uraoka, and T. Fuyuki, "Anomalously anisotropic channel mobility on trench sidewalls in 4H-SiC trench-gate metal-oxide-semiconductor field-effect transistors fabricated on 8° off substrates," *Appl. Phys. Lett.,* vol. 90, no. 4, pp. 042102-042104, 2007.

[13] R. Ghandi, M. Domeij, R. Esteve, B. Buono, A. Schöner, J. Han, S. Dimitrijev, S. A. Reshanov, C.-M. Zetterling, and M. Östling, "Experimental evaluation of different passivation layers on the performance of 3kV 4H-SiC BJTs," *Mater. Sci. Forum,* vol. 645-648, pp. 661-664, 2010.

Proceedings of the 23rd International Symposium on Power Semiconductor Devices & IC's
May 23-26, 2011 San Diego, CA

5kV class 4H-SiC PiN Diode with Low Voltage Overshoot during Forward Recovery for High Frequency Inverter

S. Ogata, Y. Miyanagi, K. Nakayama, A. Tanaka, and K. Asano

Power Engineering R&D Center, The Kansai Electric Power Company,
3-11-20 Nakoji, Amagasaki, Hyogo 661-0974, Japan

Abstract— **Forward recovery characteristics have been reported in a 5 kV class SiC pin diode used for a high frequency inverter. The 5 kV class SiC pin diode obviously has low forward voltage overshoot and an extremely small voltage shift along with a higher forward current increase rate or junction temperature as compared to the Si fast diode. The minority carrier lifetime has also been evaluated from the forward recovery characteristics, and its dependence on temperature has been investigated. Next, the relation between the minority carrier lifetime and the forward voltage drop were investigated. Even at a higher junction temperature, it was confirmed that the calculated relations between the drift region thickness and the ambipolar diffusion length approximated the best values to maintain low forward voltage drop.**

I. INTRODUCTION

Si power devices are facing limitations in terms of their physical properties. The development of SiC devices has been welcomed, because SiC has physical properties superior to those of Si. For example, the 6.2 kV class 4H-SiC pin diode [1], the SiC Commutated Gate turn-off Thyristor (SiCGT) [2], IGBT [3], and BJT [4] have been reported. Furthermore, some applications using SiC devices have been developed, and we have reported on the 100 kVA SiC inverter using 5 kV class SiCGTs and 5 kV class SiC pin diodes for connection between the battery and the distribution system [5].

When the inverter is operated at high frequency, the cost and size of the electric power conversion system can be reduced. To operate the inverter at high frequency, the switching device with fast switching speed is required. A pin diode used as a free wheeling diode (FWD) is also connected in anti-parallel with the switching device, and must switch at high speed when a fast switching device is used in the voltage type of electric power conversion system. These switching devices and FWDs are installed in the upper and lower arms of the inverter, respectively, and the switching devices are alternately turned on and off. This results in a state where current flows to one specific arm or a state where current flows to the opposite arm exclusively. When the switching device of one arm turns off, the current flows to the FWD of the opposite arm. When the turn-off speed of the switching device of one arm quickens, the current flows faster to the opposite arm. That is, the forward current increase rate (d*i*/d*t*) of the opposite arm becomes higher. The forward voltage overshoot of the FWD becomes larger at a high d*i*/d*t* than overshoot at a lower d*i*/d*t*. This is because more current flows without conductivity modulation at large part of the drift region at a high d*i*/d*t* [6]. And there is an anxiety that the switching device is destroyed if the forward voltage overshoot

of this FWD exceeds the reverse blocking voltage of the switching device connected in anti-parallel.

As for the SiC pin diode, the thickness of the drift region is thinner and the impurity concentration is higher than those of the Si diode. Therefore, the forward voltage overshoot of the SiC pin diode is expected to be smaller than that of the Si diode. In this paper, we report on d*i*/d*t* dependence and temperature dependence of the forward recovery characteristics of the 5 kV class SiC pin diode, and the minority carrier lifetime estimated from the forward recovery characteristics.

II. DEVICE AND PACKAGE STRUCTURE

This time, the device that we evaluated was the 5 kV class SiC pin diode with mesa Junction Termination Extension (JTE) [1]. The pin diode was fabricated using a 4H-SiC wafer with an n-type drift layer about 75 μm thick and an impurity concentration of $1-5 \times 10^{14}$ cm^{-3}. The chip is 8 mm×8 mm in size, and was fabricated by CREE.

To use the SiC pin diode at high temperature, a metal package was fabricated [5]. Because the diode is connected in anti-parallel and used with the switching device, insulation is required. Therefore, the SiC pin diode was mounted on a Direct Bond Copper Ceramic Substrate in the package.

III. MEASUREMENTS AND CONSIDERATIONS

A. Forward Recovery Characteristics

Figure 1 shows the test circuit used to measure the forward recovery characteristics. Measurements were performed by using the chopper circuit and the 5 kV class SiCGT was installed in the upper arm as a switching device; the 5 kV class SiC pin diode was installed in the lower arm. To decrease the inductance and make the flow of current faster, the capacitor was arranged between the DC power supply and the SiC pin diode. The 2 kV 100 A class Si fast diode was investigated for comparison with the SiC pin diode.

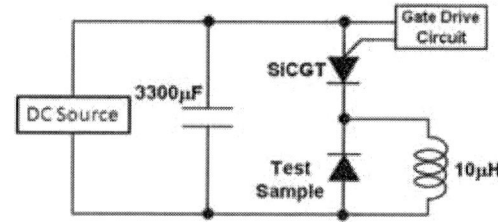

Figure 1. Test circuit for measuring forward recovery characteristics of diode.

978-1-4244-8425-6/11 $26.00 © 2011 IEEE

B. Dependence on di/dt

The forward recovery waveforms of the SiC pin diode and Si fast diode at room temperature (RT) are shown in Fig.2 (a) and (b), respectively. The voltage supplied to the diode was set to 25 V, and the current applied to the diode was changed to 50 A, 75 A and 100 A for investigation. The SiC pin diode obviously has low forward voltage overshoot and an extremely small voltage shift with higher di/dt as compared to the Si fast diode. The SiC pin diode's forward recovery time is also shorter than the Si fast diode's time when the forward voltage overshoot becomes to a steady forward voltage drop.

Figure 3 shows the di/dt dependence of the forward voltage overshoot of the SiC pin diode and Si fast diode, based on the results shown in Fig.2. The forward voltage overshoot of the SiC pin diode increased only by 4.5 V though the overshoot of the Si fast diode increased by 25.7 V when di/dt increased from 100 A/μs to 300 A/μs. These results suggest that a larger proportion of the drift region of the SiC pin diode becomes conductivity modulated in a short time because the SiC pin diode apparently has a thinner thickness and a higher concentration of impurities of the drift region, despite having higher reverse blocking voltage than the Si fast diode. In the SiC pin diode, for the reasons described on, the forward voltage overshoot is hardly affected by the value of di/dt as compared to the Si fast diode.

By using the SiC pin diode as the FWD, we consequently found that the high-speed switching device connected in anti-parallel with the FWD will not be destroyed, and that the inverter can operate at high frequency.

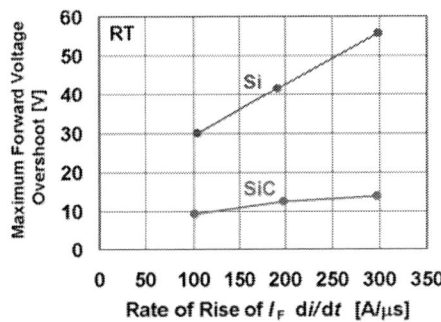

Figure 3. di/dt dependence of maximum forward voltage overshoot.

C. Dependence on Temperature

We then investigated the temperature dependence of the forward recovery characteristics. The voltage supplied by the DC Source was set to 25 V, and the current applied to the diode was fixed at 100 A. The temperature was changed to RT, 100 °C, 200 °C and 300 °C for SiC pin diode (and to RT, 75 °C, 100 °C and 125 °C for Si fast diode).

Figure 4 shows the observed waveforms when the temperature was changed. The SiC pin diode obviously has low forward voltage overshoot and an extremely small voltage shift along with higher temperature as compared to the Si fast diode. And the SiC pin diode's forward recovery time when the forward voltage overshoot becomes to a steady forward voltage drop is shorter than that of the Si fast diode.

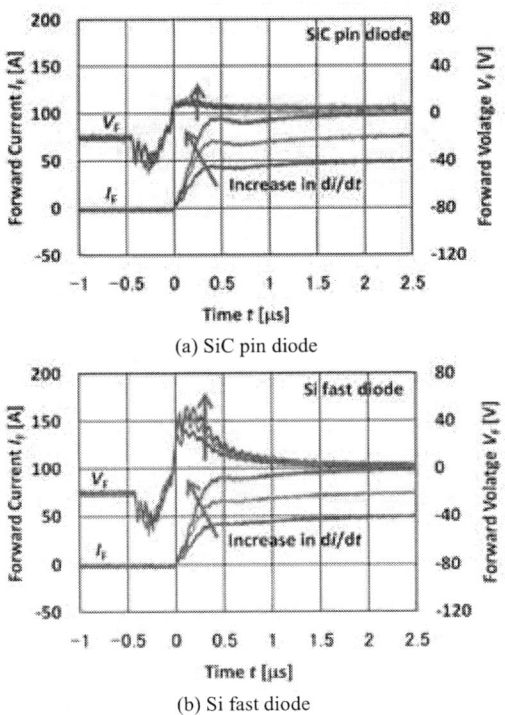

Figure 2. Forward recovery waveforms of diodes at various di/dts.

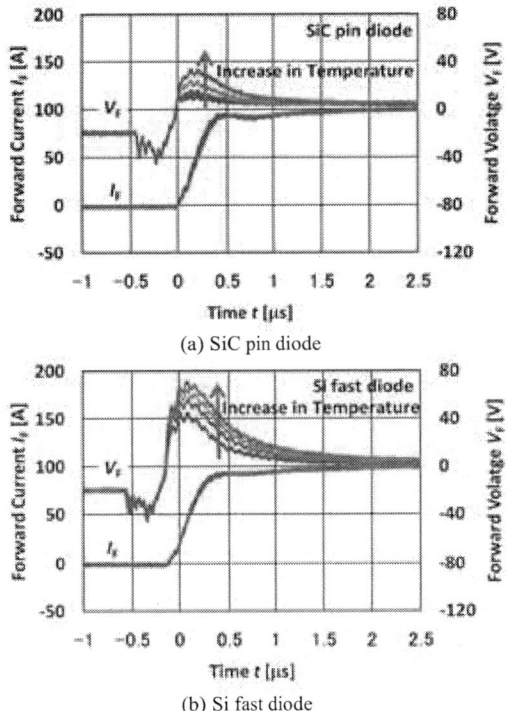

Figure 4. Forward recovery waveforms of the diodes at various temperatures.

Figure 5 shows the junction temperature dependences of forward voltage overshoot. The forward voltage overshoot increases as the temperature rises, however, an increase in the forward voltage overshoot of SiC is less than that of Si. Even if the SiC pin diode heats up to 300 °C, the forward voltage overshoot of the SiC pin diode was found to be 11.5 V which was smaller than that of the Si fast diode at RT. These results suggest that the increase in bulk resistance according to the temperature rise (which negates a decrease in resistance caused by extending the carrier lifetime in the drift region of the SiC pin diode) is smaller than that of the Si fast diode.

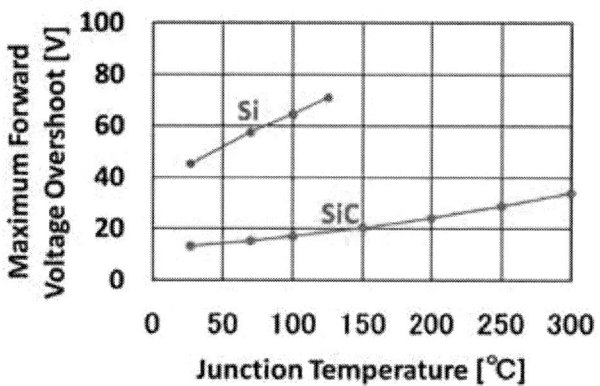

Figure 5. Junction temperature dependences of maximum forward voltage overshoot.

D. Minority Carrier Lifetime

The higher the junction temperature of the diode in the inverter, the longer the minority career lifetime of the diode and the larger the charge accumulated in the drift region layer. As this charge increases, the switching time of the diode becomes long, and the reverse-recovery loss increases, so that the inverter efficiency decreases. However, the effect of conductivity modulation also decreases when the minority carrier lifetime of the diode is excessively shortened by lifetime control, coupled with an increase in on-resistance and on-loss. Therefore, when diode performance is considered to improve inverter efficiency, it is important to understand this trade-off relation and the characteristics of the minority carrier lifetime.

We then confirmed the method of measuring the minority carrier lifetime. There is another method of estimation from the reverse recovery characteristics in addition to the method of estimation from the forward recovery characteristics. It is known that the minority carrier lifetime calculated from the reverse recovery characteristics is a lower value than that calculated from the forward recovery characteristics. There are some reports that this has been attributed to the very short minority carrier lifetime in a potential barrier of the metal/P^+ - anode interface of the SiC pin diode [7] [8].

Although there is no precision technique for extracting the minority carrier lifetime from the transient characteristics, it is clear from simple physical reason that the forward recovery time is equivalent to the minority carrier lifetime [9]. The forward recovery time is then extracted based on the standard of the diode [10]. The method of extracting the minority

carrier lifetime from the forward recovery characteristics is as follows. Specifically, the forward recovery time is from the time when the forward voltage drop rose to $0.1V_F$ after the forward current began flowing to the time when the voltage dropped to $1.1V_F$ after the voltage reached the maximum value (with V_F denoting the steady forward voltage drop).

Figure 6 shows the temperature dependence of the minority carrier lifetime of the SiC pin diode estimated from data on the forward recovery characteristics based on the above-mentioned definition. The voltage supplied to the diode was set to 25 V, and the current thrown into the diode was 100 A. The temperature was changed from RT to 300 °C. The minority carrier lifetime at RT was about 1 µs; the minority carrier lifetime depends $T^{1.1}$ as a function of temperature (T).

The results of examining the relation between the minority carrier lifetime and the forward voltage drop are presented below. It is important to understand this relation as noted above. In general, the voltage drop across the drift region (V_M) is shown as follows [11]:

$$\frac{V_M}{kT/q} = \frac{8b}{(b+1)^2} \frac{\sinh(d/L_a)}{\sqrt{1-B^2 \tanh^2(d/L_a)}} \cdot \arctan\left[\sqrt{1-B^2 \tanh^2(d/L_a)} \sinh(d/L_a)\right] + B \ln\left[\frac{1+B \tanh^2(d/L_a)}{1-B\tanh^2(d/L_a)}\right] \quad (1)$$

where, q is the elementary charge, k is Boltzmann's constant, T is the absolute temperature, d denotes the half of the drift region thickness, L_a is the ambipolar diffusion length, and b $= \mu_n/\mu_p$, B $= (\mu_n-\mu_p)/(\mu_n+\mu_p)$.

For silicon, the temperature dependence of the mobility for electrons (μ_n) and holes (μ_p) [6]:

$\mu_n = 1360(T/300)^{-2.42}$ cm²/Vs, $\mu_p = 495(T/300)^{-2.20}$ cm²/Vs

For 4H-SiC,

$\mu_n = 1140(T/300)^{-2.70}$ cm²/Vs, $\mu_p = 120(T/300)^{-2.70}$ cm²/Vs

We then investigated the relation between the total forward voltage drop (V_{ON}) and current density (J_T). In general, this relation is shown as follows [11]:

$$J_T = \frac{2qD_a n_i}{d} F\left(\frac{d}{L_a}\right) e^{\frac{qV_{ON}}{2kT}} \quad (2)$$

$$F\left(\frac{d}{L_a}\right) = \frac{(d/L_a)\tanh(d/L_a)}{\sqrt{1-0.25\tanh^4(d/L_a)}} e^{-\frac{qV_M}{2kT}} \quad (3)$$

where, D_a is the ambipolar diffusion coefficient, n_i is the intrinsic carrier density.

$$V_{ON} = \frac{2kT}{q} \ln\left[\frac{J_T d}{2qD_a n_i F(d/L_a)}\right] \quad (4)$$

(4) is obtained from (2). Fig.7 shows the extracted V_{ON} at RT and 300 °C. From Fig.7, it is understand that V_{ON} is minimized when $d/L_a = 1.5$. Thus, as d is the half of the drift region thickness, in order to minimize the V_{ON}, the minority carrier lifetime should be adjusted until the ambipolar diffusion length equals one-third of the drift region thickness. Moreover, V_{ON} increases drastically when the value of $d/L_a =$

4 exceeds. V_{ON} increase rapidly when the ambipolar diffusion length is less than one-eighth of the drift region thickness. In contrast, the value of d/L_a (calculated based on the minority carrier lifetime shown in Fig.6) fell within the range from 1.48 at RT to 1.78 at 300 ^{O}C. Even at a higher junction temperature, the calculated values of d/L_a were confirmed as approximating the best values to maintain low V_{ON}. The temperature dependences of μ_n and μ_p at $T^{-2.7}$ while that of the minority carrier lifetime is $T^{1.1}$. As a decrease in the ambipoler diffusion length denies an increase in minority carrier lifetime by the temperature rise, the value of L_a hardly changed. Therefore, there was little change in d/L_a, even at higher temperatures.

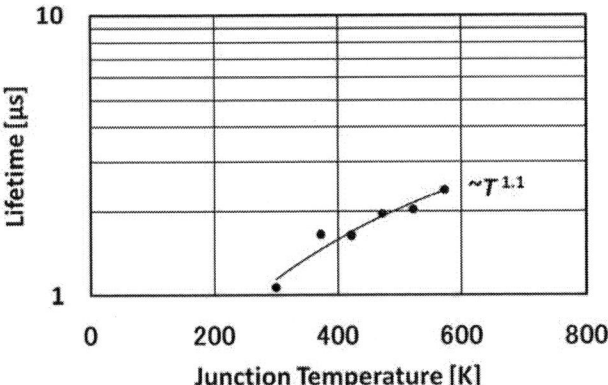

Figure 6. Temperature dependence of minority carrier liftime of the SiC pin diode.

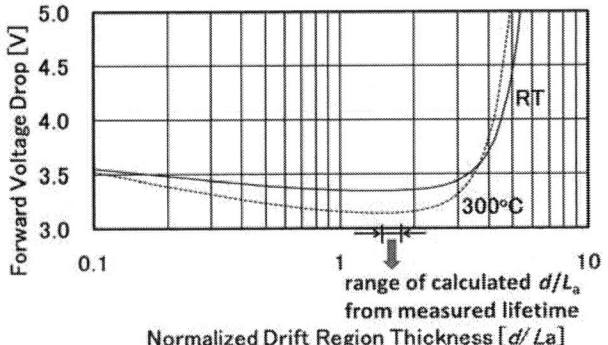

Figure 7. Forward voltage drop for the SiC pin diode.

IV. CONCLUSION

The forward recovery characteristics of the 5 kV class SiC pin diode were observed and the results were compared to characteristics of the Si fast diode. The forward voltage overshoot value of the 5 kV class SiC pin diode was determined as being smaller than that of the Si fast diode, even when di/dt or the junction temperature was high. Therefore the use of the 5 kV class SiC pin diode as a FWD in an inverter is therefore expected to prevent destruction of the fast switching device connected in anti-parallel with the diode. Moreover, the minority carrier lifetime was estimated from the forward recovery characteristics and its dependence on junction temperature was investigated. The minority carrier lifetime at room temperature is about 1 μs and its temperature dependence is $T^{1.1}$. Even at a higher junction temperature, the calculated values of d/L_a were confirmed as approximating the best values to maintain low forward voltage drop.

ACKNOWLEDGMENT

The Japan Society for the Promotion of Science (JSPS) supported this research through its "Funding Program for World-Leading Innovative R&D on Science and Technology (FIRST Program).

The authors gratefully acknowledge Dr. A. Agarwal and Dr. J. W. Palmour of CREE Inc. for the fabrication of SiC device chips.

REFERENCES

[1] Y. Sugawara, K. Asano, R. Singh, and J. W. Palmour. "6.2kV 4H-SiC pin Diode with Low Forward Voltage Drop," Proceedings of ICSCRM'99, pp.170, 1999.

[2] Y. Sugawara, D. Takayama, K. Asano, A. Agarwal, S. Ryu, J. W. Palmour, and S. Ogata. "12.7kV Ultra High Voltage SiC Commutated Gate Turn-off Thyristor: SICGT," Proceedings of ISPSD'04, pp.365-368, 2004.

[3] Q. Zhang, M. Das, J. Sumakeris, R. Callanan, and A. Agarwal. "12-kV p-Channel IGBTs With Low On-Resistance in 4H-SiC," IEEE Electron Device Letters, pp.1027-1029, Vol. 20, No. 9, 2008.

[4] S. Balachandran. C. Li, P. A. Losee, I. B. Bhat, and T. P. Chow. "6kV 4H-SiC BJTs with Specific On-resistance Below the Unipolar Limit using a Selectively Grown Base Contact Process," Proceedings of ISPSD'07, pp.293-296, 2007.

[5] Y. Sugawara, S. Ogata, T. Izumi, K. Nakayama, Y. Miyanagi, K. Asano, A. Tanaka, S. Okada, and R. Ishii. "Development of a 100kVA SiC Inverter with High Overload Capability of 300kVA," Proceedings of ISPSD'09, pp.331-334, 2009.

[6] B. Jayant Baliga. "Fundamentals of Power Semiconductor Devices," Springer, pp.236-243, 2008.

[7] N.V. Dyakonova, P.A. Ivanov, V.A. Kozlov, M.E. Levinshtein, J.W. Palmour, S.L. Rumyantsev, and R.Singh. "Steady-State and Transient Forward Current-Voltage Characteristics of 4H-Silicon Carbide 5.5kV Diodes at High and Superhigh Current Densities," IEEE Transactions on Electron Devices, Vol. 46, No. 11, 1999, pp.2188-2194.

[8] M.E. Levinshtein, T.T. Mnatsakanov, P. Ivanov, J.W. Palmour, S.L. Rumyantsev, R. Singh, and S.N. Yurkov. "Paradoxes of Carrier Lifetime Measurements in High-Volatge SiC Diodes," IEEE Transactions on Electron Devices, Vol. 48, No. 8, 2001, pp.1703-1709.

[9] J. G. Kassakian, M. F. Schlecht, and G. C. Verghese. "Principles of Power Electronics," AddisonWesley, pp.486, 1991.

[10] Standard of the Japanese Electrotechnical Committee,"JEC-2402 Rectifier Diodes," Denkishion, 2002.

[11] S.K. Ghandhi. "Semiconductor Power Devices," pp.112-128, John Wiley and Sons, 1977.

Proceedings of the 23rd International Symposium on Power Semiconductor Devices & IC's
May 23-26, 2011 San Diego, CA

A SiC Static Induction Transistor (SIT) Technology For Pulsed RF Power Amplifiers

Francis K. Chai, Bruce Odekirk, Ed Maxwell, Mar Caballero[†], Terri Fields, Mike Mallinger[†] & Dumitru Sdrulla

Microsemi Corp.

307 SW Columbia St, Bend, OR. 97702; [†]3295 Scott Blvd., Suite 150, Santa Clara, CA. 95054

Tel: (541) 382-8028; email: fchai@microsemi.com

Abstract - **A Static Induction Transistor (SIT) technology capable of delivering output power over 2200W at VDD of 125V in UHF band of 406MHz-450MHz is presented. The transistor technology is built on 3" 4H-SiC epitaxial wafers. L-band performance is presented based on the prototype transistor unit cells. An improved transistor architecture is proposed to further boost frequency and power performance.**

I. INTRODUCTION

It is well known that 4H-SiC has important qualities inherent with wide band-gap materials such as capability of high-temperature, high-power density operation, high critical breakdown field allowing the use of more heavily doped epitaxy reducing transistor conduction loss and high saturated electron drift velocity beneficial for power applications [1]. Together with basic material properties including high thermal conductivity and low thermal expansion, 4H-SiC has been widely regarded as a natural candidate in power switching and RF applications.

However, quality and cost of starting materials has been a barrier to the commercialization of 4H-SiC until recent years. The progress made in SiC MOSFET and Schottky diodes driven by the rapidly growing markets of electrical vehicles and alternative energy not only pushed the envelope of process and device technologies, but has been coupled with a clear positive trend in both the quality and cost of SiC starting materials. This trend initiated by the demand in the power switching market is expected to benefit the adoption of SiC transistors in power RF applications.

Since its invention in the early 1900's, radar has found wide applications in air traffic control, air and marine navigation, weather monitoring, remote sensing, law enforcement and highway safety, beyond the original use in military and aerospace. With the typical transmitter output peak power levels up to mega Watts and an average of several kilo Watts, system designers are in a perpetual quest for simplification of power amplifier blocks with reduced weight and size as well as lowered cost. The timely readiness of the overall SiC technology is finding itself in an important role in the evolution of the transmitter block of radar systems toward solid-state amplifiers.

Power amplifiers based on SITs have been reported in application frequency spans from MF up to S-band [2-4] with transistor technologies built both on silicon and SiC platforms.

Despite the high output power projected for SiC PA into the 10kW range, the reported power levels have been under 1kW. The incentive to boost the output power of amplifier building blocks is apparent as the power transmitter combining all blocks would be significantly simplified.

A 4H-SiC SIT technology targeted for power RF applications is presented in this paper. With the output power over 2.2kW at **VDD** of 125V from 406MHz to 450MHz, this is the highest output power in UHF band demonstrated to date by a solid-state amplifier package to the knowledge of the authors.

A. SIT Transistor process

RF SIT devices reported here were fabricated on 3" SiC wafers consisting of four epitaxial layers on n-type 4H-SiC substrates. Guard ring trenches and source mesas were patterned by reactive ion etching (RIE) using fluorine based chemistry. Sidewall spacers were then formed by PECVD of SiO_2, followed by anisotropic RIE. Al ion implantation was then done at elevated wafer temperature (600°C) to form a highly doped box implant profile ($>1\times10^{19}$ cm^{-3} peak doping), with PECVD SiO_2 serving as the implant mask. This implant forms both the p-type gate region as well as the p-type guard rings surrounding the transistor active area. After implantation, the SiO_2 layer was removed and a carbon film deposited by PECVD to suppress step bunching during the implant activation anneal, which was carried out at 1675°C in Ar in a Centrotherm HV100 vertical furnace.

After the implant anneal, the carbon film was removed by O_2 plasma; BPSG was then deposited by PECVD and flowed in a conventional horizontal furnace. Contact openings to both source mesas and gate regions were then patterned by RIE. Nickel silicide ohmic contacts were simultaneously formed to source, gate and drain regions by reacting a thin Ni film at elevated temperature. The drain contact is on the back of the wafer. A gold overlay metal was then patterned on the gate fingers by lift-off to reduce the gate resistance. An inter-layer dielectric film consisting of ~1μm PECVD SiO_2 was then deposited, and contact vias patterned and etched by RIE. Tungsten was then deposited by CVD after reactive sputter deposition of a TiW/TiWON barrier stack. The W metallization is patterned by RIE. Bond pads consisting of Ti/Pt/Au films were patterned by liftoff. A final passivation layer of SiO_xN_y was deposited by PECVD, and bond pad openings formed by RIE. After die sort testing was completed,

978-1-4244-8425-6/11 $26.00 © 2011 IEEE

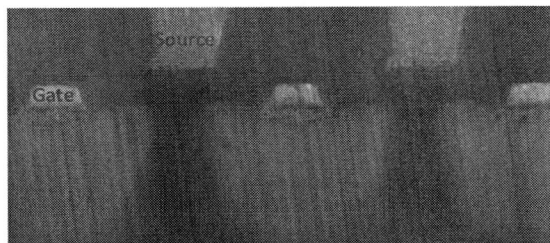

Fig. 1. SEM x-section image of the power RF SiC static induction transistor.

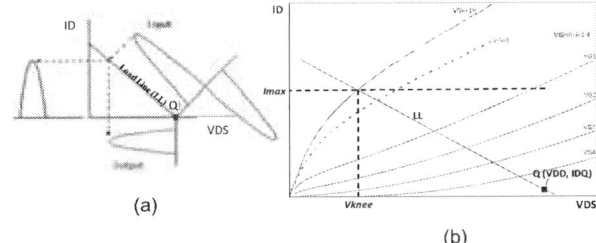

Fig. 3. A simple conceptual view illustrating the role of **VP** in amplifier linearity. **(a)** Current and voltage swing as transistor is driven by RF power. **(b)** Simulated SIT output characteristic overlaid with load-line.

a sputtered silicon layer was deposited on the backside of the wafers to facilitate Au/Si eutectic die attach. The final prototype unit cell transistor has the single-mesa architecture as described in [5]; the SEM image of a transistor x-section is shown in **Fig. 1**.

B. Optimization of RF power performance

There are two areas of focused efforts to engineer a SIT for optimized power RF performance, namely, transistor frequency response and power capability. In order to fulfill the need of power amplification for UHF frequencies and beyond, transistor frequency response is carefully optimized. This includes the optimization of transconductance, **gm** and reduction of total gate parasitic capacitance, **Cgg**. For applications in UHF band, the gate implant is engineered to achieve an effective channel length to deliver sufficient **gm** while ensure **Cgg** remains in check. Aside from the gate doping profile, device geometrical parameters, such as source mesa and spacer (gate implant mask) widths, and source mesa height, greatly influence the transistor RF power performance. These geometrical parameters consequently determine the critical DC electrical parameters such as **VP**, the pinch voltage, which in turn, plays a significant role in both RF power gain and amplifier power capability as explained next.

2D TCAD process and device simulations using Silvaco software were performed in the course of the prototype unit cell development. Key transistor structural parameters such as source mesa width were set by the desired pinch voltage (**VP**) based on simulations as shown in **Fig. 2**, where **VP80** refers to **VP** at **VDD**=80V. Such design curves greatly facilitate technology development. It is clear from **Fig. 2** that transistor blocking voltage gain (BV gain=∂**VDS**/∂**VGS** at a given **ID**) improves as **VP** becomes less negative. Sufficient amplifier power gain requires a high transistor BV gain. However, there is a lower bound in the pursuit of a minimized -**VP** to boost

BV gain. The lower bound of -**VP** is set by amplifier linearity requirement, or the 1dB gain compression output power, **P1dB**. To optimize the output power of a transistor, it is necessary to maximize **Imax** (max. drain current when gate/source junction is slightly turned on) and minimize **Vknee** (**VDS** when **Imax** occurs). As illustrated in **Fig. 3(a)**, transistor current swings from quiescent current **IDQ** and **VDS** from quiescent voltage **VDQ** as input is driven by RF power (a class AB configuration, pure sinusoidal current and voltage waveforms are assumed for simplicity). As RF power increases, gate/source RF voltage swing eventually starts to turn on the gate/source junction. At this point, the transistor full current swing, **Imax**-**IDQ**, is reached while the maximum voltage swing is equal to **VDD**-**Vknee**, as shown in **Fig. 3(b)**. Further gate swing does not increase the current drive as the gate starts to conduct and both **ID** and **VDS** swings begin to be clipped when the gain compression sets in. As device **VP** becomes less negative (i.e., channel is more readily pinched), **Imax** degrades and **Vknee** increases. Consequently, less input drive power is necessary to push the amplifier into compression (measured by **P1dB**). This is demonstrated by the measured gain versus output power for various transistor pinch voltages as shown in **Fig. 4**. In summary of the discussion above, the floor of -**VP** is set by **P1dB** and ceiling by the power gain.

Meanwhile, **Imax** and **Vknee** are strong functions of the transistor on-resistance, **RDSON**, which is determined by the resistivity of epitaxial layers and trades off with device breakdown voltage. The resistivity of the epitaxy and the

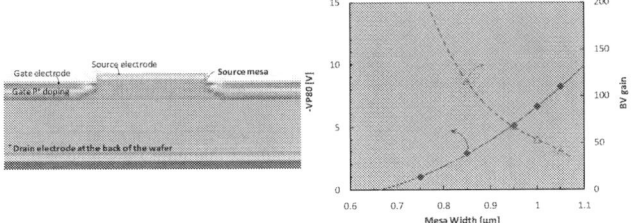

Fig. 2. Simulated transistor structure using Athena and design curves using Atlas relating SIT structural parameters to electrical performance.

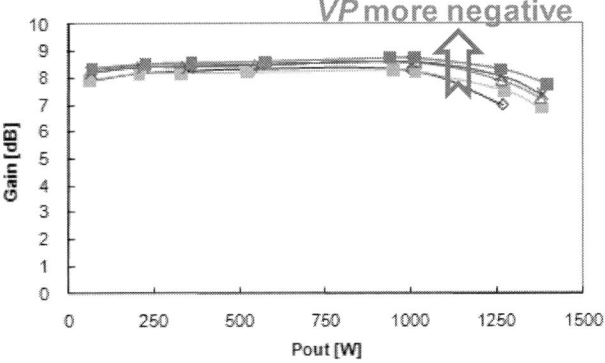

Fig. 4. Measured pulsed RF power gain at 450MHz illustrates the important correlation between pinch voltage and transistor RF power capability.

978-1-4244-8425-6/11 $26.00 © 2011 IEEE

transistor guard rings are optimized such that the breakdown voltage is at least 2× of **VDD** and additionally, the transistor is RF rugged. A minimum **RDSON** value is set as a result and cannot be further optimized. Consequently, it is critical not to over rate transistor breakdown voltage so that power capability of the final power amplifier is not compromised.

Besides the frequency response, achieving the high output power levels required by pulsed radar applications mandates significant up-scaling of the prototype transistor cell size. However, this increase in transistor power-delivering periphery does not readily translate into amplifier output power scaling unless the chip floor plan is optimized to account for severe heat generation as a consequence of the extreme output power. Experiments with design factors such as transistor cell size, chip floor plan as well as chip thickness were evaluated to arrive at a combination such that the measured output power approaches the theoretical transistor periphery scaling factor. The sensitivity of measured **P1dB** to transistor unit cell spacing and chip thickness, as shown in **Figs. 5** and **6**, illustrates the importance of such optimization to achieve the expected power scaling.

II. RESULTS AND DISCUSSION

A prototype UHF transistor cell built on 4H-SiC 3" epitaxial wafers was first demonstrated and serves as a

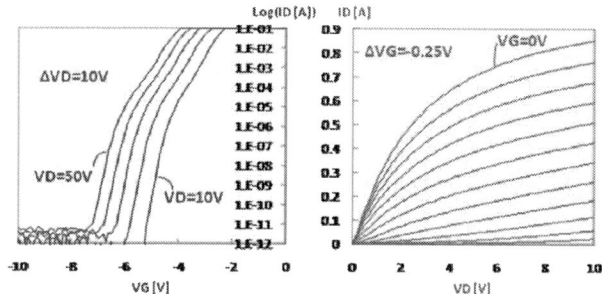

Fig. 7. Measured **ID-VG** and **ID-VD** characteristics of a prototype transistor unit cell.

building block to exercise critical design rules such as transistor channel width and length. **Fig. 7** shows the measured DC **IV** characteristics of a prototype transistor cell.

A source pulser board, shown **Fig. 8**, is used in conjunction with the RF test fixture to establish the desired DC quiescent point for RF performance evaluation. Automated load- and source-pull allows quick determination of optimized output/input matching as prototype unit cells are combined to scale up the output power of the amplifier. Load and source power contours of an amplifier part are shown in **Fig. 9**.

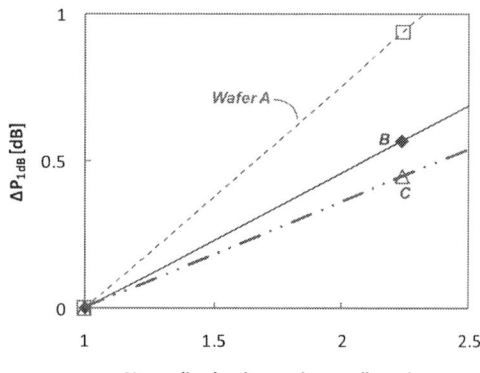

Fig. 5. Increase in **P1dB** versus unit cell spacing for three different materials.

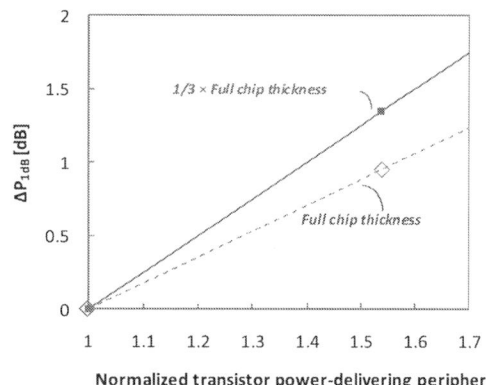

Fig. 6. Increase in **P1dB** versus transistor power-delivering periphery scaling at two different chip thicknesses.

Fig. 8. The source pulser board and the RF test fixture used for RF performance evaluation.

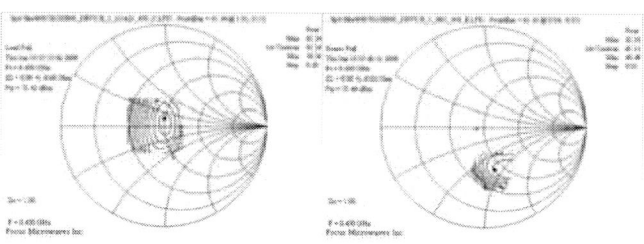

Fig. 9. Load (left) and source power contours of a power amplifier part.

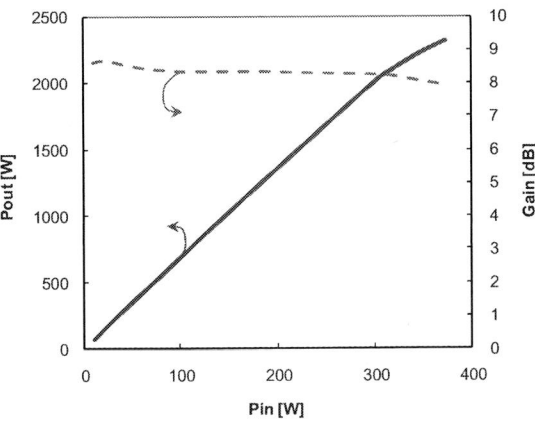

Fig. 10. Measured RF drive-up characteristic of the SiC SIT-based UHF power amplifier at 450MHz.

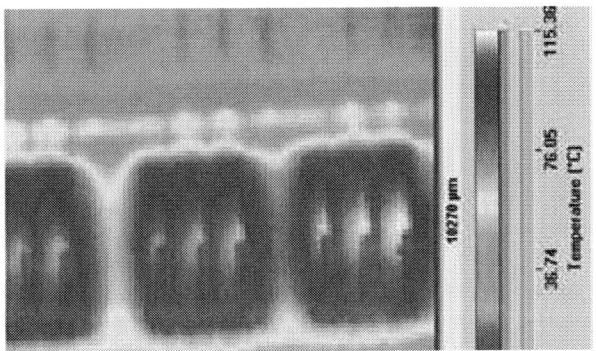

Figure 11. Partial image of an IR scan of a 21-cell power amplifier.

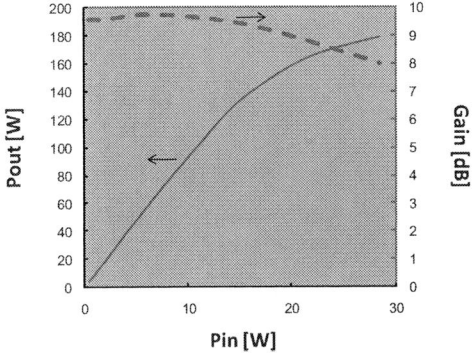

Figure 12. The drive-up characteristic of a 3-cell power amplifier evaluated at L-band frequency of 1215MHz.

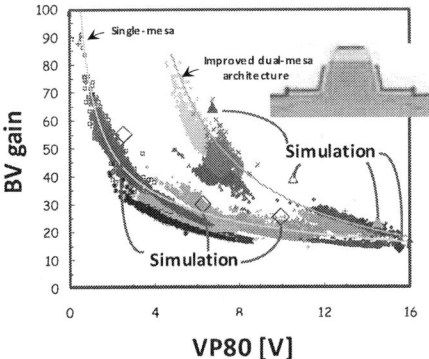

Figure 13. Measured blocking voltage gain (BV gain) versus pinch voltage at 80V drain bias (*VP80*). Symbols represent various lots.

The drive-up characteristic of a pulsed power amplifier delivering over 2.2kW output power is shown in **Fig. 10**. This corresponds to a power density of 13W/cm (120kW/cm^2) at **VDD**=125V, a pulse width of 300µs and 6% duty cycle. **Fig. 11** shows the result of IR scan of a 21-cell power amplifier terminated at 10:1 load mismatch while delivering 480W of output power and dissipating over 2100W. A temperature rise of 115°C from the base plate temperature of 70°C is observed. A *thermally-limited* die with less-than-optimal chip floor plan shows 30°C more temperature rise at identical dissipated power density.

Finally, a power amplifier based on proto type cells discussed so far has been evaluated in L-band frequencies from 960MHz to 1215MHz. The drive-up characteristic of a 3-cell amplifier at 1215MHz, **VDD**=75V under the pulse condition of 300µs/2%, is shown in **Fig. 12**. Experimental designs incorporating a new SIT structure, termed dual-mesa, has further boosted transistor *gm* while reducing *Cgg* using a novel gate ion implant [6-7]. This structure demonstrates significantly improved BV gain for a given *VP*, as shown in **Fig. 13**, predicted accurately by simulation. The boost in BV gain demonstrated by the dual-mesa transistor is expected to translate into improvement in both frequency and power performance.

III. CONCLUSION

A solid-state power amplifier package delivering over 2.2kW output power in UHF band frequencies up to 450MHz has been demonstrated based on transistors manufactured on a 3" 4H-SiC technology platform. The frequency capability of the technology is further demonstrated in L-band RF amplifier performance with additional enhancement in power based on an improved dual-mesa device architecture.

ACKNOWLEDGMENT

The authors would like to acknowledge the support of L-band technology development by US Air Force under the contract FA 8650-07-C-5400 monitored by Mr. John Blevins.

REFERENCES

[1] C. E. Weitzel, Digest of 2000 GaAsMANTECH, pp. 197-200.

[2] De Salvo et. al., International Journal of High Speed Electronics and Systems, vol. 14, no. 3 (2004), pp. 906-908.

[3] Merrett et. al., Materials Science Forum, vols. 527-529 (2006), pp. 1223-1226.

[4] Clarke et. al., Proceedings of 2000 IEEE/Cornell Conference on High Performance Devices, pp. 141-143.

[5] Siergiej et. al., United States Patent No. 5,903,020, May 11th, 1999.

[6] Bruce Odekirk et. al., United States Patent Application Publication No.US 2011/0049532 A1, March 3rd, 2011.

[7] Bruce Odekirk et. al., Provisional United States Patent Application Serial No. 61/424,912, December 20th, 2010.

Proceedings of the 23rd International Symposium on Power Semiconductor Devices & IC's
May 23-26, 2011 San Diego, CA

UltiMOS : A Local Charge-Balanced Trench-Based 600V Super-Junction Device

P Moens, F. Bogman, H. Ziad, H. De Vleeschouwer, J. Baele, M. Tack

Corporate R&D

ON Semiconductor, Oudenaarde, Belgium

Peter.moens@onsemi.com

G. Loechelt, G. Grivna, J. Parsey, Y. Wu, T. Quddus, P. Zdebel

Corporate R&D

ON Semiconductor, Phoenix, Arizona, USA

Abstract—This paper for the first time reports on a novel "local" charge balanced trench-based super junction transistor. The local charge balance is achieved by selectively growing thin highly-doped n-type and p-type layers in a deep trench structure. The final charge-balanced trench structure is finished with an oxide-sealed airgap. Devices rated at 10A with V_{bd}=730V and a Ron=23 mΩ.cm^2 are demonstrated.

I. INTRODUCTION

The super-junction (SJ) concept allows achieving device performance well below the so-called silicon 1D-limit through full depletion of alternating P-N columns [1]. The SJ concept becomes attractive for higher voltage ratings, with a sweet spot in the 500-900V range. SJ-MOSFETs are showing increased market penetration as main power switching devices for various power supply application circuits. Today, two different approaches to achieve super-junction action are widely studied. The first concept is using multi-epitaxy/multi-implant steps to create the alternating n- and p-type pillars. Since the introduction of CoolMOS in 1998 [2], this concept has been dominating the SJ market. To achieve further shrinking of the SJ device, a lot of research is currently ongoing to use deep trench etching and epi re-growth for making the alternately doped pillar structure [3-7]. Although both approaches are quite different, they have in common that the thick planar base epi layer (typically n-type) serves to counterbalance the doping charge in the columns (typically p-type), being created either by implant or by deep trench etch and refill. Hence, both approaches can be categorized as "global" charge balance, meaning that the base planar epi contributes to the electrical performance of the device.

This paper reports for the first time on a novel "local" charge balanced trench-based SJ transistor exceeding 700V. The local charge balance is achieved by selectively growing both n-type and p-type epitaxial layers in deep trenches. Instead of fully sealing the deep trench by selective epi growth, the trench structure contains an airgap and is sealed by an oxide plug. This makes the concept much less dependent on the tapering angle of the deep trench.

The "local" charge balance approach is different from the above described "global" charge balance approach. So far, only one paper reports on local charge balance [8] showing only diode performance up to 120V, but no transistor data has been reported. This paper shows MOSFET device performance of V_{bd} =730V and Ron=23 mΩ.cm^2, for a 10A rated device. In addition, the UIS and switching performance of the device is presented, along with reliability data.

II. DEVICE CONCEPT

Figs. 1 and 2 show SEM cross-sections of a few unit cells of the transistor. The starting wafer consists of a highly doped n-type substrate on which a thick lowly doped epi (i-epi) is grown with resistivity > 50 Ω.cm. Deep trenches with an aspect ratio > 1/10 are etched using the Bosch process and an oxide hard mask. The deep trenches penetrate into the highly doped substrate. Subsequently, thin but highly doped n-and p-type layers are selectively grown on the sidewalls of the deep trenches. No epi is grown on the oxide hard mask. To separate the n- and p-charge, a thin intrinsic layer (I_l) is grown in between the n- and p-layers, which is not shown in Fig.2 for clarity reasons. However, the intrinsic layer is clearly visible in the SCM of Fig. 3. The trench structure is not filled completely with epi. Instead, an "airgap" structure is made in which the trench silicon surface after epi-growth is first passivated by a high quality liner oxidation, after which the trench is sealed with an oxide plug. This approach does not rely on a tight control of the trench taper angle which is required for complete epi fill. In the latter, a taper angle of at least 1° is required for a seamless epi fill, resulting in an n-rich structure at the bottom (as a result of the tapering the spacing between trenches is larger at the bottom than at the top). It is verified by μ-Raman that the airgap structure and oxide seal yield no strain in the structure as is shown in Fig.4.

The n- and p-type layers need to be perfectly charge balanced, taking into account the contribution from the i-epi (<3%) and charges in the liner oxide ($N_{it} < 2x10^{10}$ cm^{-2}). Note that the intrinsic bulk epi layer does not provide a conduction path for current in the on-state, and is basically only present for mechanical support.

The gate structure consists of a trench gate. An n-type implant is performed to connect the channel end of the MOS structure with the n-type charge balanced layer, see also Fig. 2. The termination structure of the devices is a standard p-ring racetrack structure.

This work is partially sponsored by the IWT project "GreenFETs"

978-1-4244-8425-6/11 $26.00 © 2011 IEEE

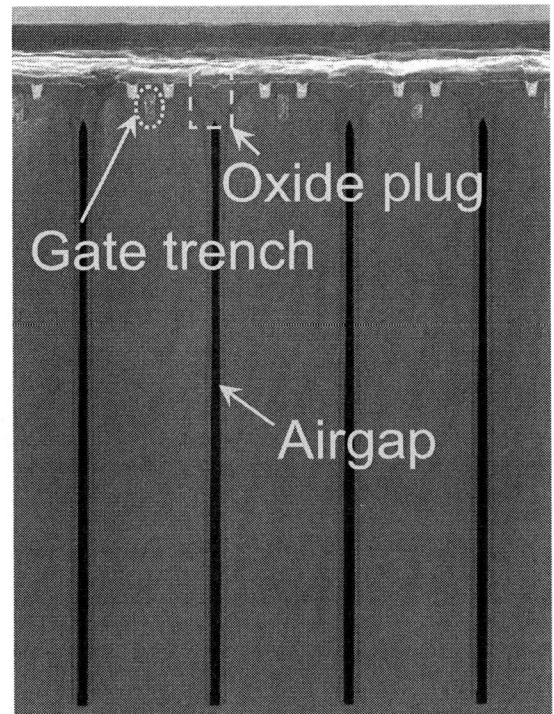

Figure 1 : Schematic cross-section of the device. The structure consists of an air-gap which is sealed by an oxide plug. The gate structure consists of a gate trench and connection implants to the SJ structure.

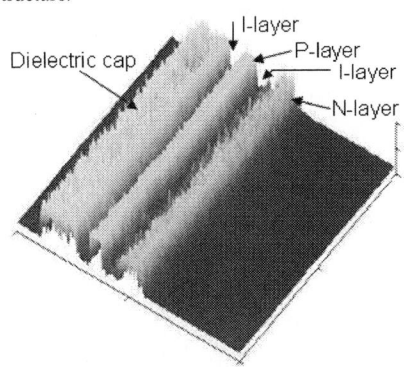

Figure 3 : Scanning Capacitance Microscopy (SCM) of the charge balanced selectively grown epi (SEG) inside the trench. A dielectric cap is present inside the trench structure. The intrinsic layer separating the N- and P-type layers is clearly visible.

Table 1 : Typical Results obtained on a 10A rated device.

parameter	Unit	value	Remark
V_{bd}	V	>600	RT, I_{ds}=250 µA
Ron	mΩ.cm^2	23	RT, I_{ds}=10A
I_{dss}	nA	<100	RT, V_{ds}=600V
I_{gss}	nA	<1	RT, V_{gs}=25V
V_{th}	V	3.8	I_{ds}=100 mA
UIS	J/cm^2	4.5	I_{ds}=10A, V_{dd}=50V
Q_{gtot}	nC/cm^2	280	V_{gs}=10V, V_{dd}=300V
Q_{rr}	µC/cm^2	30	
t_{rr}	ns	200	

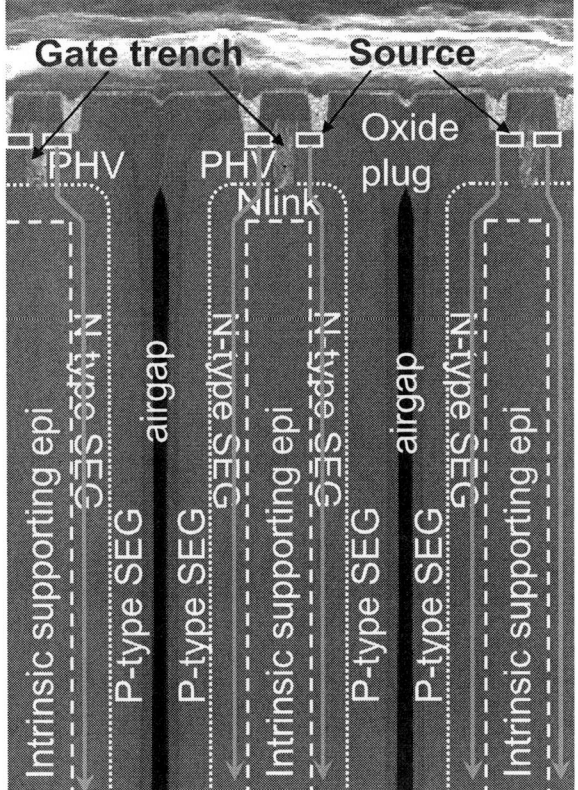

Figure 2: Detail of the SJ structure. The n-and p-type epi layers (selectively grown, SEG), are indicated. Nlink is the connection implant to connect the MOS channel to the n-type SEG layer. The current flow in on-state is shown by green lines and arrows. The I_l layer (separating N-and P-type SEG), is not shown for clarity of the figure).

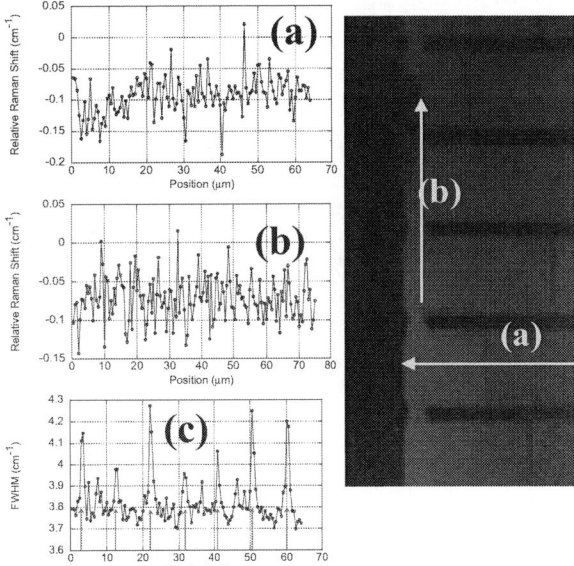

Figure 4 : µ-Raman measurements on a device cross-section. (a) Data parallel (in between two trenches) and (b) perpendicular to the SJ deep trench structure, (c) shows the Full-width at Half Maximum (FWHM—a measure of the crystallinity of the measurement point) for measurement (b), clearly showing the presence of the oxide plug denoted by arrows).

III. EXPERIMENTAL RESULTS

Transistors with an active area ranging from 0.4 to 15 mm^2 are fabricated. For reverse recovery, UIS and Ron testing at high currents, the parts are packaged in either DPAK or TO220FP. Bonding is done using thick Al wires. All other parameters are tested at wafer level.

Fig. 5 shows the off-state leakage current as a function of ambient temperature. Bona fide diode leakage behavior is observed. No additional leakage from the trench airgap seal is present. Fig. 6 shows the Ron versus V_{bd} for variations in the n-type dose (CN) and separating intrinsic layer thickness (I_1). The data indicate the median, 5 percentile and 95 percentile of statistical distributions.

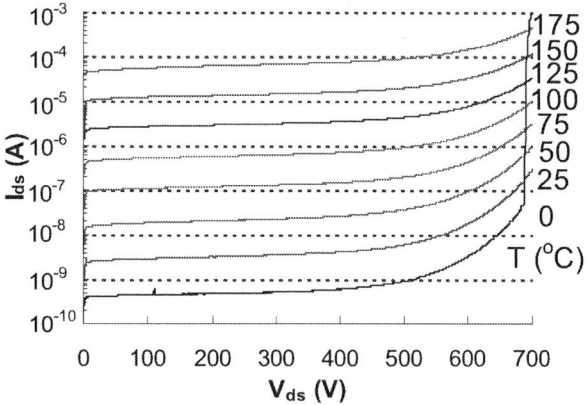

Figure 5 : Off-state leakage current as a function of temperature, The maximum of the measurement system (chuck isolation) is 700V.

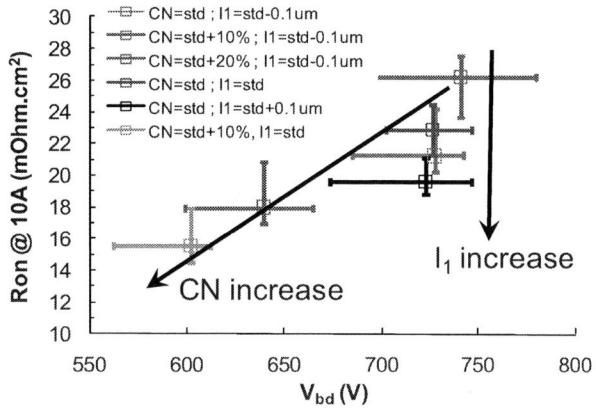

Figure 6: Ron (at 10A) versus V_{bd} for different process splits. Boxes denote the median of the distribution, error bars correspond to the 5 and 95 percentiles. Data for optimum CB. Increasing the n-type dose (CN) or the spacing between N-and P-type layers (I_1) has a different effect on the Ron/V_{bd} trade-off.

Increasing I_1 spaces the p- and n-type layers further apart and hence avoids inter-diffusion of n-and p-type charge. As a result, the conduction efficiency of the n-type layer is increased, thus lowering the Ron. The effect of the inter-

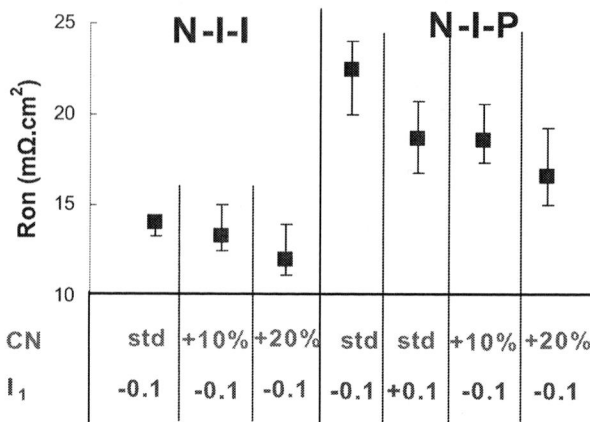

Figure 7 : Effect of inter-diffusion of N- and P-layers on the Ron (measured at 10A), as from Resistor structures : N-I-I structure only has n-type Selective Epi growth (SEG) in the trench ; N-I-P structure has both n-and p-type SEG.

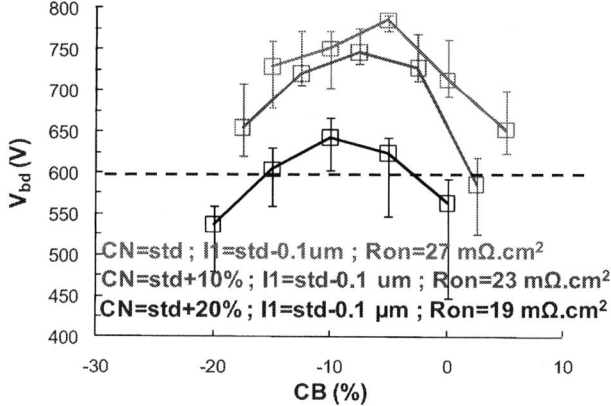

Figure 8: V_{bd} versus charge balance of the N-type and P-type layers. Boxes denote the median of the distribution, error bars correspond to the 5 and 95 percentiles of the distribution.

diffusion of the p-type layer into the n-type layer is assessed through special resistor structures without a gate trench (n+ on top), in which the p-type is left floating. The results are shown in Fig. 7. The presence of the floating p-type layer increases the Ron due to the close proximity and inter-diffusion of the two layers.

As can be noticed, Ron can be substantially lowered if CN is increased (as expected), but also if I_1 is increased. The latter increases the conduction efficiency for the same dopant concentration. As a result, increasing I_1 is not impacting the maximum charge balance too much, see Fig. 6 and can be used as an extra design parameter to improve the Ron/V_{bd} tradeoff. The impact of CN on the CB window is shown in Fig. 8. For each given process split, the n-type doping is kept constant whereas the p-type concentration is varied.

Figs. 9(a) shows the benchmarking of Ron versus V_{bd} using the median of the statistical distributions, against competitor parts which are available on the market today. On the other hand, Fig. 9(b) compares our best devices against published data from literature. The local charge balanced device is competitive with the best parts on the market.

978-1-4244-8425-6/11 $26.00 © 2011 IEEE

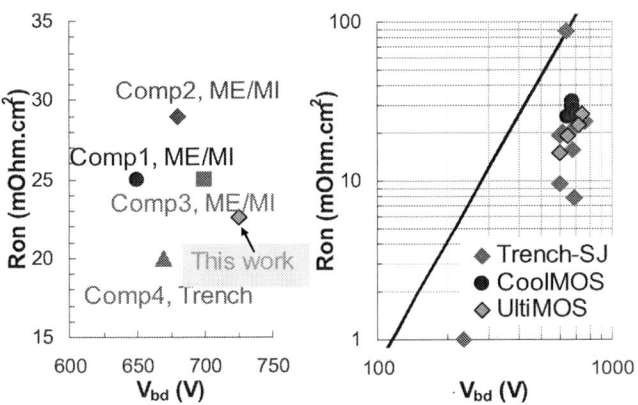

Figure 9 : (a) Benchmarking of the median of the statistical distribution for Ron and V_{bd} with competitor devices on the market today [CoolMOS-type i.e. multi-epitaxy/multi-implant (ME/MI) and trench-based] (b) Benchmarking of best devices from our process splits with data from publications.

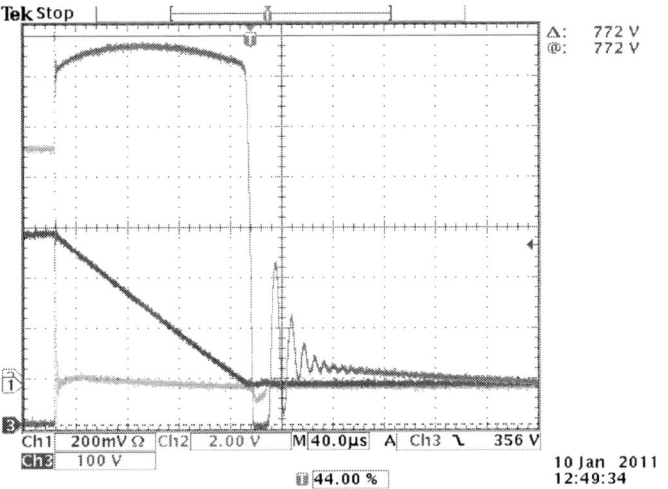

Figure 10 : Typical UIS characteristic measured on the device. I_{pulse}=12A, V_{dd}=50V, L=10 mH.

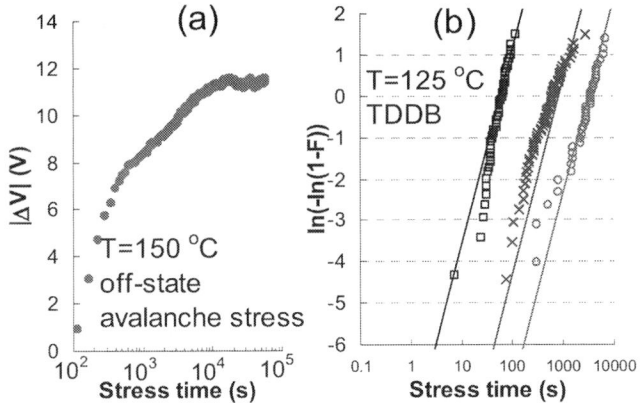

Figure 11 : (a) Shift in V_{bd} during avalanche stressing at T=150°C. Shift is limited, completely saturates after $\sim 10^4$s; (b) gateox TDDB data at T=125 °C. Data obtained when stressing the transistors in accumulation at three different gate fields.

Fig. 10 shows a typical UIS characteristic, measured at I_{pulse}=12A, L=10 mH and V_{dd}=50V. For these conditions, the avalanche energy is approximately 4.5 J/cm^2. Taking into account all parameters (V_{bd}, Ron, UIS, manufacturing window, etc), typical results for the transistor parameters are shown in Table 1. Note that the switching parameters are excellent.

Fig. 11 shows some initial reliability data, HRTB at T=150°C and TDDB of the trench-gate. For the TDDB, no extrinsic failures are observed.

IV. CONCLUSIONS

In this paper a novel "local" charge balanced trench-based super junction transistor is reported. The local charge balance is achieved by selectively growing both n-type and p-type doped thin layers in a deep trench structure. The trench structure consists of an oxide-sealed airgap structure, making the concept independent of the trench taper angle and epi-filled induced stress. The thick planar epi in which the trenches are etched, is lowly doped and does not contribute to the electrical performance of the device. The spacing between the charge-balanced n- and p-type layers is an important design parameter allowing to improve the Ron versus V_{bd} trade-off. The gate structure of the devices consists of a trench gate with a dedicated n-type connection implant to connect the end of the MOS channel with the n-type selective epi layer. Devices rated at 10A with a breakdown voltage of 730V and a Ron of 23 mΩ.cm^2 and good UIS performance are fabricated. The switching characteristics of the devices are excellent.

REFERENCES

[1] T. Fujihira, "Theory of Semiconductor SuperJunction Devices", Jpn J. Appl. Phys. **36** (10), pp6254-6262 (1997).

[2] G. Deboy, M. Marz, J.P. Stengl, H. Strack, J. Tihany and H. Weber, "A New Generation of High Voltage Power MOSFETs breaks the Limit line of Silicon", IEDM 1998, pp683-685.

[3] C. Rochefort and R. Van Dalen, "A Scalable Trench Etch Based Process for High Voltage Vertical RESURF MOSFETs", ISPSD 2005, pp35-38.

[4] K. Takahashi, H. Kuribayashi, T. Kawashima, S. Wakimoto, K. Mochizuki and H. Nakazawa, "20 mΩ.cm2—660V Super Junction MOSFETs Fabricated by Deep Trench Etching and Epitaxial Growth", ISPSD 2006, pp297-300.

[5] W. Saito, I. Omura, S. Aida, S. Koduki, M. Izumisawa, H. Yoshioka, H. Okumura, M. Yamaguchi and T. Ogura, "A 15.5 mΩ.cm2—680V SuperJunction MOSFET : Reduced On-Resistance by lateral Pitch Narrowing", ISPSD 2006, pp293-296.

[6] J. Sakakibara, Y. Noda, T. Shibata, S. Nogami, T. Yamaoka and H. Yamaguchi, " 600V Class Super Junction MOSFET with High Aspect Ratio P/N Columns Structure", ISPSD 2008, pp299-302.

[7] S. Ono, L. Zhang, H. Ohta, M. Watanabe, W. Saito, S. Sato, H. Sugaya and M. Yamaguchi, "Development of 600V Class Trench-Filling SJ-MOSFET with SSRM Analysis Technology", ISPSD 2009, pp303-306.

[8] M. Rub, D. Ahlers, J. Baumgartl, G. Deboy, W. Friza, O. Haberlen and I. Steinigke,"A Novel Concept for the Fabrication of Compensation Devices", ISPSD 2003, pp203-206.

Proceedings of the 23rd International Symposium on Power Semiconductor Devices & IC's
May 23-26, 2011 San Diego, CA

Vertical Charge Imbalance Effect on 600 V-class Trench-Filling Superjunction Power MOSFETs

T. Tamaki, Y. Nakazawa, H. Kanai, Y. Abiko, Y. Ikegami,
M. Ishikawa, E. Wakimoto, T. Yasuda, and S. Eguchi
Renesas Electronics Corporation
111 Nishiyokote, Takasaki 370-0021, JAPAN
Email: tomohiro.tamaki.fz@renesas.com

Abstract—**600V-class superjunction (SJ) MOSFETs fabricated by trench-filling process are investigated by analytical and numerical solutions with experimental results. The careful consideration on the effects of trench taper and p-column profile is given for accurate charge control. The breakdown voltage (V_b), specific on-resistance (R_{onAa}), and gate-to-drain charge (Q_{gd}) of 736 V, 16.4 mΩ-cm^2, and 6 nC, respectively, have been achieved for the fabricated SJ-MOSFET.**

I. INTRODUCTION

Superjunction MOSFET (SJ-MOSFET) has been drawing attention as a switching power device due to its superiority of the low specific on-resistance, *RonAa*, compared to conventional power MOSFETs. The SJ structure formed by alternately aligned p and n-columns makes it possible to go beyond the Si-limit. Since the breakdown voltage of SJ devices is determined by the charge compensation principle [1], the control of the charge balance (C. B.) becomes a critical issue. The breakdown voltage of SJ devices has been considered to be increased proportionally to the column depth in the C. B. state for the simplest, i.e. vertical sidewall trench SJ (VSJ) structures [2][3] as shown in Fig. 1 (a). Since tapered sidewall trenches are preferable both for void-less and for higher growth rate silicon trench-filling epitaxial growth, the tapered sidewall trench SJ (TSJ) as shown in Fig.1 (b) must be considered in such case. Also, the dopant profile in p-columns is another critical parameter [4], and is often graded due to the thermal diffusion during epitaxial growth to fill trenches.

In this paper, the effects of tapered trenches and p-column dopant profiles have been investigated both theoretically and experimentally. A simple analytical solution shows that tapered trenches and p-column dopant profiles have identical effects on the breakdown voltage in terms of the charge compensation. Experimentally, SJ-MOSFETs have been fabricated by 50 um-deep trench process with the 10-µm cell pitch design under our optimized process conditions. Results indicate that the breakdown voltages agree well with theoretical predictions.

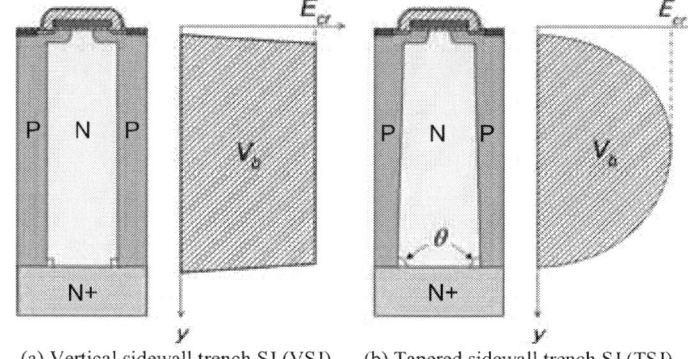

(a) Vertical sidewall trench SJ (VSJ)   (b) Tapered sidewall trench SJ (TSJ)

Figure 1.   Electric field distribution in the blocking state.

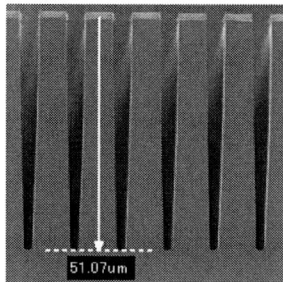

Figure 2.   XSEM image of 51 µm-deep tapered trenches.

II. FABRICATION OF TRENCH-FILLING SJ

The n on n+ epitaxial silicon wafers are chosen for SJ fabrication process. The hard mask of oxide is deposited on Si surface for deep Si etching. Non-Bosch type dry etching process with a microwave plasma etching equipment is employed so that trenches are nearly 50-µm in depth with a tapered sidewall angle as shown in Fig. 2. The angle of sidewalls is controlled by varying gas flow ratio, stage

978-1-4244-8425-6/11 $26.00 © 2011 IEEE   308

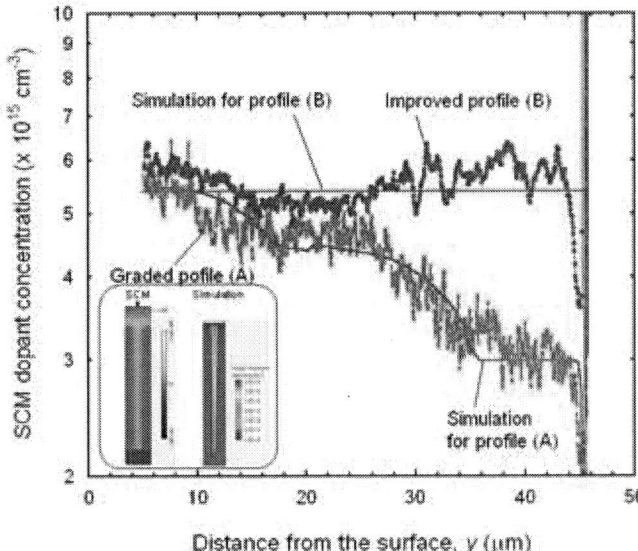

Figure 3. SCM profiles of the graded Profile (A) and the improved Profile (B) along the middle line of the p-column with bold lines by simulations. The simulation is calibrated for the Profile (A), and no calibration is done for the Profile (B)

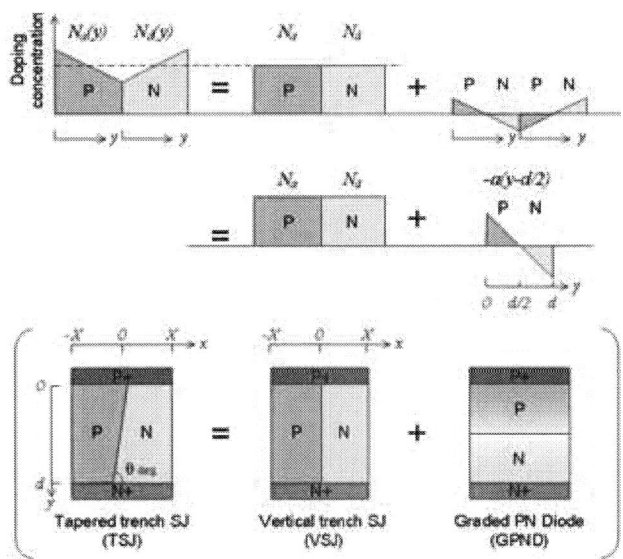

Figure 4. Charge superposition principle for a tapered trench SJ (TSJ) in the C. B. state. A TSJ is decomposed into a vertical sidewall trench SJ (VSJ) and a graded PN diode (GPND) in terms of charge component.

temperature, and RF bias [5]. The alternately aligned p/n columns are formed by the boron doped silicon epitaxial growth to fill trenches, followed by planarization of the over-growth silicon layer. After the SJ structure is formed, the edge termination and the active cells of planar DMOSFET are fabricated. A detailed discussion regarding the edge termination structure is beyond the scope of the present paper.

Scanning capacitance microscopy (SCM) technique is applied to observe p-column dopant depth profiles as shown in Fig. 3. The SCM image embedded in Fig. 3 shows a graded profile (profile (A)). Since the graded profile affects the C. B. condition, process simulation is calibrated to agree with the profile. Also the silicon epitaxial growth condition to fill trenches is improved to obtain a flatter depth profile (profile (B)).

III. CHARGE SUPERPOSITION PRINCIPLE OF TAPERED SJ

A. Formulation using charge superposition principle

A TSJ can be decomposed into a VSJ and a linearly graded pn diode (GPND) in terms of charge component as shown in Fig. 4. For simplicity, a flat p-column dopant profile, an equal width of p and n-columns with the same doping, and a fully depleted condition in the C. B. state are assumed. The taper angle θ is expressed by a gradient a of the net dopant concentration in the GPND as follows:

$$\theta = \tan^{-1}(N/aX), \tag{1}$$

where N is the dopant concentration of each column and X is a half column width.

In case of the TSJ in which the p-column dopant profile is linearly graded with a gradient a', the effective dopant gradient a_{eff} is defined by:

$$a_{eff} = a + a'. \tag{2}$$

The electric field sustained by the GPND is given by:

$$E_{GPND}(y) = \frac{6V_t}{d^3}\left[(d'/2)^2 - (y - d'/2)^2\right], \tag{3}$$

where d' is the depletion width of the GPND and V_t is the voltage drop across the junction:

$$V_T = \frac{qd'^3}{12\varepsilon}\left(\frac{N}{X\tan\theta} + a'\right). \tag{4}$$

Also, the electric field across the VSJ reported in [1] is given by:

$$E_{bal,VSJ}(y)\big|_{x=X} = -2\frac{V_D y}{d^2} + \frac{V_R + V_D}{d} + \sum_{n=1}^{\infty}\frac{K_n\gamma_n}{2}\frac{\cos(K_n y)}{\cosh(K_n X)}, \tag{5}$$

where $V_D \equiv (qN/2\varepsilon_s)d^2$, V_R is the applied reverse voltage, $K_n = n\pi/d$, and $\gamma_n = 8V_D/(n\pi)^3[(-1)^n - 1]$. Combining (3) and (5) gives the electric field across the TSJ as shown below:

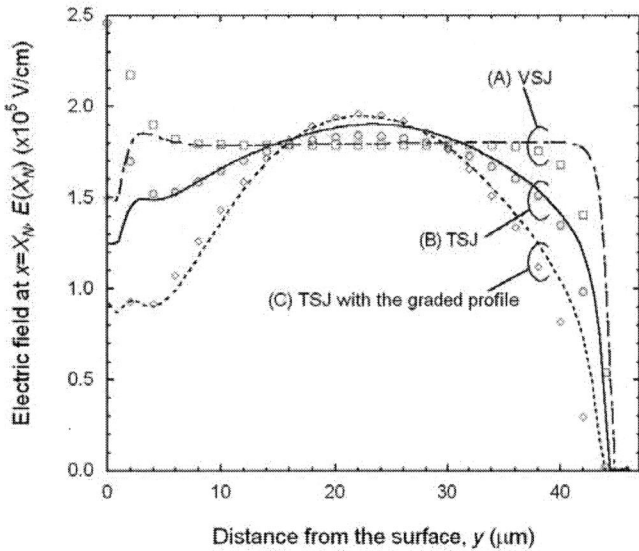

Figure 5. Analytical (as plots) and numerical (as lines) solutions of electcric field distribution along the middle line of the n-column in the C. B. state for VSJ, TSJ, and TSJ with the graded profile (A).

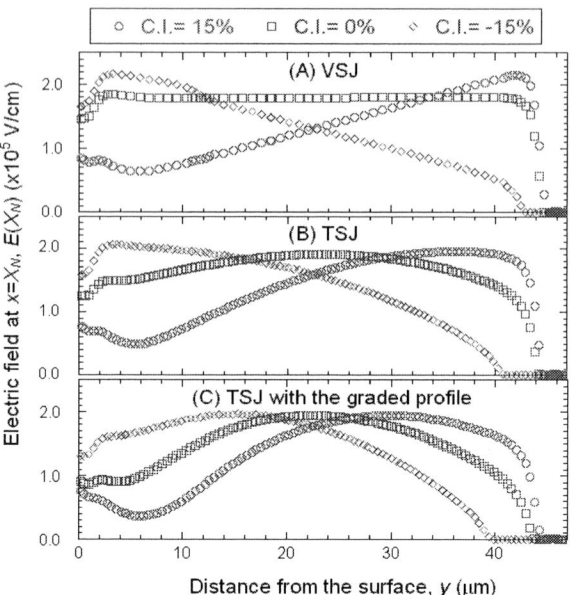

Figure 6. Electric field distributions for C. I. ratio of -15% ($Q_P<Q_N$), 0% (C. B. state), and +15% ($Q_P>Q_N$) by TCAD simulation.

$$E_{bal,TSJ}(y)\big|_{x=X} = E_{bal,VSJ}(y)\big|_{x=X} + E_{GPND}(y)$$

$$= -2\frac{V_D y}{d^2} + \frac{V_R + V_D}{d} + \sum_{n=1}^{\infty}\frac{K_n \gamma_n}{2}\frac{\cos(K_n y)}{\cosh(K_n X)}$$

$$+ \frac{6V_t}{d^3}\left[(d/2)^2 - (y-d/2)^2\right], \quad (5)$$

which peaks at $y=0$ or $y=d/2$, depending on the V_t term. If the junction gradient a_{eff} is small enough to fully deplete the junction, the depletion width d' in (3) and (4) is replaced with the column depth d.

B. Simulation of the electric field of the tapered SJ

The analytical solutions by (5) are compared with simulation results for VSJ, TSJ, and TSJ with the gradient profile (A), respectively. Synopsys® TCAD software is used to simulate SJ-MOSFETs. In our analytical model, the dopant profile (A) is approximated by the linear gradient a' of 7.1×10^{17} cm^{-4}, which is large enough to leave the quasi-neutral region in the GPND. Thus the depletion width d' of 36.6 μm is used for plot (C), whereas the other two plots are calculated by d' of 45 μm (=d) due to full depletion.

In the C. B. state, every simulation result supports the validity of the analytical prediction except for the surface region, as shown in Fig. 5. The difference of the electric field near the surface is due to the JFET region in simulation, whereas there is neither MOSFET nor JFET in the analytical modeling. In the C. B. state, the VSJ is fully depleted, thus the charge of columns is compensated perfectly such that the electric filed in VJS is constant. However the electric field of the TSJ shows a peak at the middle of columns where the junction of GPND is formed, whereas the charge in columns is not compensated perfectly near the surface and bottom region of columns even in the C.B. condition.

In case that the charge imbalance (C. I.) occurs due to the process variation, the breakdown voltage is degraded due to the deformed electric field distribution as shown in Fig. 6. A VSJ in the C. I. state is decomposed into the C. B. SJ and the PiN diode [3], thus the peak electric field occurs at the surface region of p+n junction for $Q_P<Q_N$ and at the bottom region of pn+ junction for $Q_P>Q_N$, respectively. The peak electric field of TSJs in the C. I. state moves upwards for $Q_P<Q_N$ and downward for $Q_P>Q_N$ depending on the excess charge amount because the junction depth in the GPND moves so.

Since the area of the electric field is the breakdown voltage, the possible smallest a_{eff} is suitable for higher breakdown voltage, but the degradation of the breakdown voltage caused by the C. I. is less sensitive to the C. I. ratio for larger a_{eff} as expected from the area of the electric field.

IV. Device Performance

The breakdown voltages and their dependencies on the C.I. ratio for fabricated SJ-MOSFETs are shown in Table I and Fig. 7 with simulation results. Notice that as the effective gradient a_{eff} increases, the breakdown voltage decreases as expected from the analytical and simulation results. The disagreement of the breakdown voltage between experimental results and simulations is mainly due to the edge termination structures because only the active cell region is simulated. Also the optimum design of the edge termination is applied to Device (A) and (B), whereas Device (C) has a poor edge termination structure.

TABLE I. FABRICATED SJ-MOSFETs

Device #	Process Condition		Breakdown Voltage	
	Trench Taper Angle[a]	P-Column Profile	$V_b@1uA$ (V)	C.I. Margin over 600V (%)
Device (A)	Taper (A)	Profile (B)	800	-7.5, +11.3
Device (B)	Taper (B)	Profile (B)	730	-9.8, +12.7
Device (C)	Taper (C)	Profile (A)	513	N/A

a. Angle :(A) > (B) ~ (C)

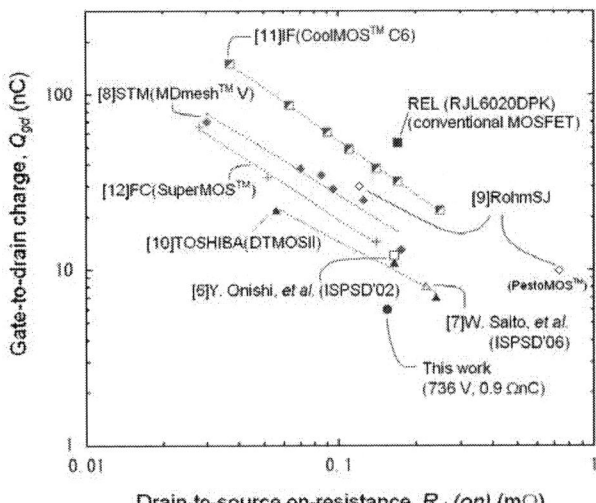

Figure 8. Trade-off between $R_{ds(on)}$ vs. Q_{gd} for 600V-class SJ MOSFETs (some plots are from supplier's data sheets) with the reference of conventional MOSFET.

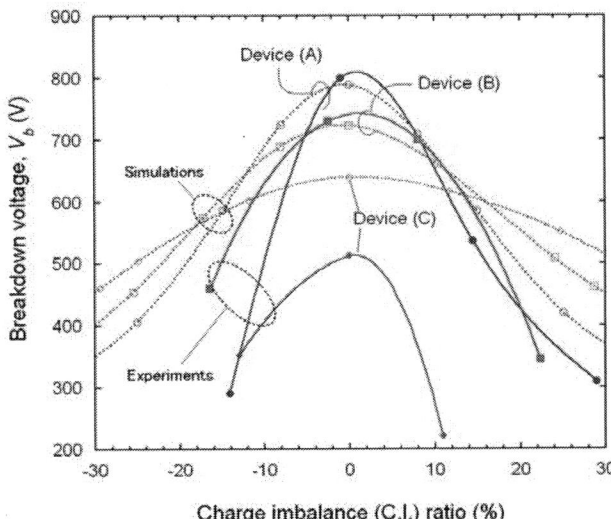

Figure 7. Breakdown voltage dependence on the C.I. ratio for SJs with different trench tapers (angle: A>B~C) and p-column profiles.

Among three fabricated devices, Device (B) yields the best balance between the breakdown voltage and its margin against the C. I. ratio. The breakdown voltage (V_b) of 730 V and the excellent charge imbalance margin of -9.8 to +12.7% to obtain over 600 V have been achieved for Device (B). The fabricated SJ-MOSFET of Device (B) in a TO-3PSG package shows the drain-to-source on-resistance of 150 mΩ corresponding to the specific on-resistance ($R_{on}A_a$) of 16.7 mΩ-cm^2 and the gate-to-drain charge (Q_{gd}) of 6 nC. The tradeoff relationship between $R_{ds(on)}$ and Q_{gd} is compared with previously reported 600 V-class SJ MOSFETs and competitors' data sheets in Fig. 8.

CONCLUSIONS

The analytical modeling of SJ structures having the tapered sidewall trench and the graded p-column profile has been described in comparison to numerical simulations and experimental results. The tapered trench and the graded p-column profile degrade the breakdown voltage, whereas the breakdown margin against the C. I. ratio increases. By shrinking the cell pitch and utilizing optimized process conditions with accurate charge balance control, a good balance between the device performance and the process margin has been achieved at a high level. The fabricated device performance of $R_{ds(on)}Q_{gd}$ of 0.9 ΩnC has ranked a top level among previously reported 600 V-class SJ MOSFETs.

ACKNOWLEDGMENT

The authors would like to thank process engineers of Takasaki Factory, Renesas Electronics Corporation. Especially, the support on the device fabrication from Mr. H. Ideyama and other technical staffs are appreciated. Also, many thanks to Mr. S. Kubo and Ms. R. Takada for their arrangement on XSEM and SCM. Furthermore, the authors would like to thank Mr. T. Kanazawa for his encouragement on this project.

REFERENCES

[1] A. G. M. Strollo and E. Napoli, IEEE, Trans. Electron Devices, vol. 48, pp.2161-2167, 2001.

[2] T. Fujihira, Jpn. J. Appl. Phys. vol. 36, pp. 6254-6262, 1997.

[3] H. Wang, E. Napoli, and F. Udrea, IEEE Trans. Electron Devices, vol. 56, pp. 3175-3183, 2009.

[4] S. Ono, L. Zhang, H. Ohta, M. Watanabe, W. Saito, S. Sato, H. Sugaya, and M. Yamaguchi, Proc. ISPSD'09, pp. 303-306, 2009.

[5] T. Maruyama, T. Narukage, R. Onuki, and N. Fujiwara, J. Vac. Sci. Technol. B 28, pp. 854-861, 2010.

[6] Y. Onishi, S. Iwamoto, T. Sato, T. Nagaoka, K. Ueno, and T. Fujihira, Proc. ISPSD'02, pp. 241-244, 2002.

[7] W. Saito, I. Omura, S. Aida, S. Koduki, M. Izumisawa, H. Yoshioka, H. Okumura, M. Yamaguchi and T. Ogura, Proc. ISPSD'06, pp. 293-296, 2006.

[8] STMicroelectronics, Data Sheet, "MDmesh V series", 2009.

[9] Rohm, Data Sheet, R6025ANZ(TO-3PF) and R6008FNX(TO-220FM), 2009.

[10] Toshiba, Data Sheet, "DTMOS II series", TO-3P(N) package, 2010.

[11] Infineon, Data Sheet, "600V CoolMOS C6 series", TO-247 package, 2010.

[12] Fairchild, Date Sheet, "SuperMOS series", TO-247 package, 2010.

Proceedings of the 23rd International Symposium on Power Semiconductor Devices & IC's
May 23-26, 2011 San Diego, CA

Energy limits for Unclamped Inductive Switching in High-Voltage Planar and SuperJunction Power MOSFETs

J. Roig, P. Moens, J. McDonald*, P. Vanmeerbeek, F. Bauwens, M. Tack

Power Technology Centre - Corporate R&D, ON Semiconductor, Oudenaarde (Belgium)
* Automotive and Power Group, ON Semiconductor. Phoenix (USA)
jaume.roigguitart@onsemi.com

Abstract— **In this work the maximum UIS energy capability (E_{as}) for High-Voltage (600V-900V) Planar and SuperJunction (SJ) power MOSFETs is analyzed through experiment, TCAD simulation and analytical modeling. A new theoretical approach considering a buried heat source is presented to accurately predict E_{as} values in a wide range of voltage capability and load inductor values.**

I. INTRODUCTION

The Power MOSFET robustness during a single UIS event is a critical success criterion for many applications. This robustness is normally quantified by the maximum energy capability (E_{as}) in the datasheets of commercially available devices. In a standard procedure, E_{as} is calculated from the maximum allowed current (I_{av}*Area) delivered by the series load inductor (L) in a similar UIS test setup as in Fig. 1a. Former works [1-3] provide detailed information for UIS measurements in Low-Voltage (LV) Power MOSFETs (<200V), however, the test conditions and the device area for a given E_{as} are barely specified for High-Voltage (HV) Power MOSFETs. As a result, it is extremely difficult to compare and predict E_{as} at different voltage ranges and HV technologies (SJ and Planar). This work provides own measurements carried out in different HV technologies under different test conditions. In order to predict the measured I_{av} and E_{as}, a new analytical model is developed for both SJ and Planar power MOSFETs with BV up to 900V. Differently from previous theoretical models [2-6], the dynamic self-heating process and the critical temperature (T_c) are related to the device voltage capability (BV). Besides, TCAD simulations are performed to give physical insight into some aspects of the model and the experimental observations.

II. THEORETICAL MODEL

This model proposes an electro-thermal assessment of the dynamic avalanche during the UIS event, like other reported models [2-6]. Based on empirical observations, the V_{ds} and I_d curves for destructive and non-destructive UIS tests are

represented in Figs. 1b and 1c. These observations are valid for both SJ and Planar power MOSFETs. In our real test, the external applied voltage (V_{dd}) and the gate voltage (V_g) are set to 50V and 12V, respectively, whereas the load current is progressively increased at different UIS events. The load current at a fixed L is increased by enlarging load time until reaching the failure event. The last test avoiding the failure during the UIS time (T_{av}) determines I_{av}, which is normally expressed as

$$I_{av} = C_{th} \cdot \Delta T^{2/3} \cdot L^{-1/3} \cdot BV_{inc}^{-1/3} \cdot \left(1 - V_{dd} / BV_{inc}\right)^{1/3} \quad (1)$$

where ΔT is the maximum temperature increment ($\Delta T = T_c - T_{amb}$), BV_{inc} is an average V_{ds} value and C_{th} is a thermal coefficient with the dimensions $K^{-2/3} J^{1/3} W^{1/3}$. An original derivation of C_{th}, T_c and BV_{inc} is presented and discussed in the following subsections.

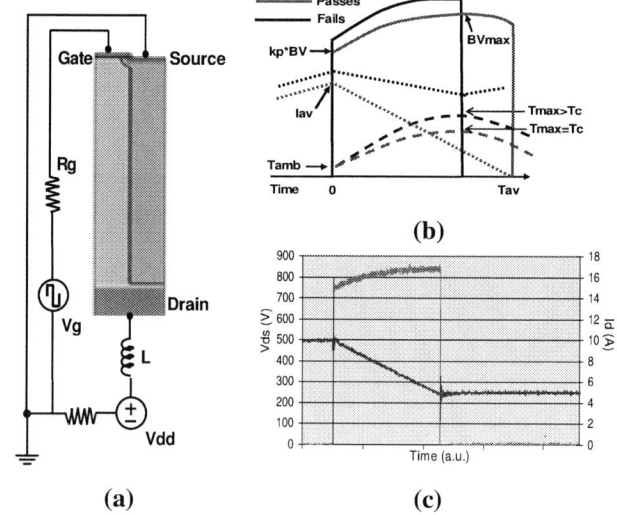

Figure 1. (a) Schematic description of the setup used to perform UIS measurements and mixed-mode simulations. (b) Dynamic evolution of V_d, I_d and T_{max} during a destructive and non-destructive UIS event. (c) Example of destructive UIS measurement in a SJ power MOSFET with BV~650V.

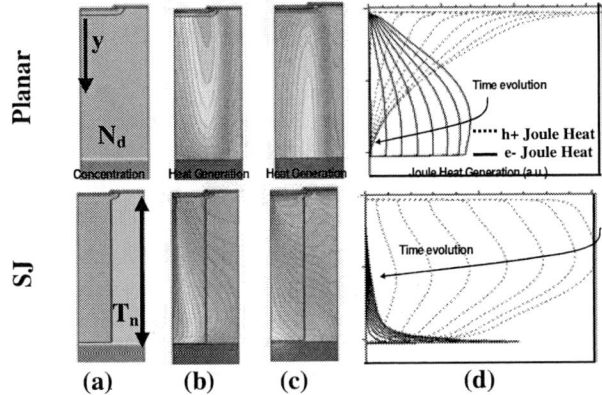

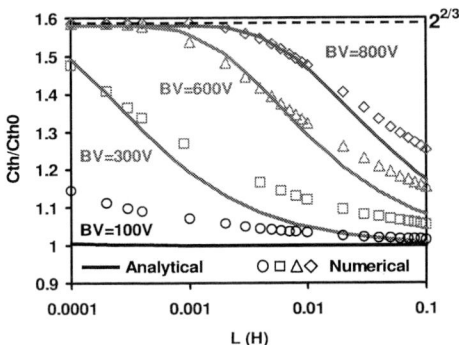

Figure 2. (a) Doping concentration, (b) h+ and (c) e- Joule heat generation for SJ and Planar structures. (c) Profiles of h+ and e- Joule heat at different times between 0 and T_{av} during the UIS event.

Figure 3. C_{th}/C_{th0} vs. L for different BV values. The analytical values are extracted from (4) while the numerical values are calculated from $(\Delta T_1/\Delta T_0)^{2/3}$. C_{th} tends to C_{th0} at small BV and high L (or large T_{av}).

A. Thermal analysis with a buried heat source

In LV devices, C_{th} is normally inferred by solving the one-dimensional heat flow equation in a semi-infinite Silicon slab with a plane heat source at the surface. In this way, the ΔT dependence on the total dissipated power (P) is simplified to $\Delta T_0 = 2.C_{th0}^{3/2}.T_{av}^{1/2}.P$, where C_{th0} (or C_{th} in LV devices) is

$$C_{th0} = (2R/\pi C)^{-1/3} \qquad (2)$$

being C and R the device effective thermal resistance and capacitance. A similar 1D thermal approach can be applied to HV devices considering: (a) a device active area covering the complete die area, (b) a negligible heat evacuation at top, (c) a T_{av} short enough to minimize the thermal influence of the package. Contrarily, it has been already observed [7] that a surface heat source is inaccurate to determine ΔT in HV devices. Indeed, the large drift length (T_n) suggests that a buried heat source must be defined in our thermal system. A relevant Joule heat generated inside the Silicon is identified in Fig. 2 from UIS mixed-mode simulations. This effect, more accentuated in SJ power MOSFETs, implies a less severe heating for $T_{av} < T_n^2/4D_t$, where $D_t = Area/RC$. The buried heat source assumption implies solving the heat flow equation [8]

$$\Delta T_1 \alpha \int_0^{T_{av}} P(t) \cdot (1/\sqrt{t-t'}) \cdot \exp(-(y-Tn)^2/4D_t(t-t'))dt' \qquad (3)$$

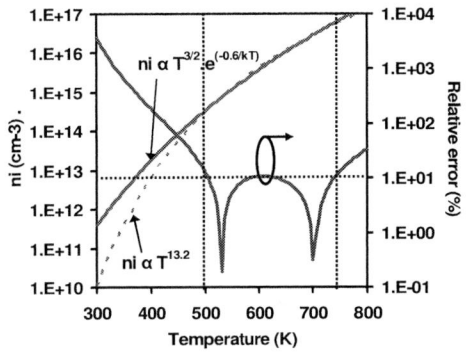

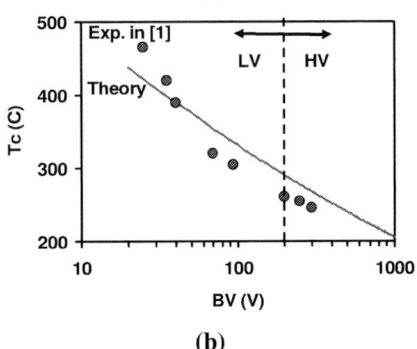

Figure 4. (a) Comparison between conventional and simplified $n_i(T)$ with a relative error < 10% in the n_i range of interest. (b) Measured [1] and analytical T_c vs. BV in Planar power MOSFETs.

with $P(t) = BV_{inc}.(I_{av}.t.(BV_{inc}-V_{dd})/L)$. When $T_n=0$, ΔT_1 is reduced to ΔT_0. However, when $T_n>0$, then (3) must be solved by numerical methods. Therefore the thermal problem is preferably tackled by defining C_{th} as

$$C_{th} = C_{th0} \cdot (2^{2/3} + (1-2^{2/3}) \cdot \exp(-\tau/T_{av})) \qquad (4)$$

where τ/T_{av} is $k_t.BV^3.L^{-2/3}$ in a first order approach. A good adjustment respect to the exact $(\Delta T_1/\Delta T_0)^{2/3}$ solution is obtained by fixing the k_t parameter to 1.5×10^{-10} $H^{2/3}V^{-3}$ (see Fig. 3). Note that C_{th} becomes C_{th0} for LV devices and large L values.

B. T_c dependence on device voltage capability

Based on a pure thermal failure criterion, the device destruction occurs when the maximum temperature (T_{max}) surpasses the intrinsic temperature of the Silicon (T_c) at a certain point of the active area. When $T_{max}>T_c$, then the intrinsic carrier density (n_i) reaches the lowest doping concentration level ($n_i=N_d$ at the drift region) triggering the filamentation effect. Any other failure mechanism related to bipolar activation [9,10], gate propagation delay [11] or termination avalanche [12] results in E_{as} below the thermal one. Unlike other models, a relation between ΔT and BV is found for Planar and SJ power MOSFETs. Hence, N_d is analytically related to T_c in

$$T_c = (N_d/9 \cdot 10^{-22})^{1/13.2} \qquad (5)$$

after replacing the classical $n_i(T)$ dependence $T^{3/2}.e^{-0.6/kT}$ by $T^{13.2}$ (see Fig. 4a). The relative error of this approach is less than 10% when N_d ranges from $2x10^{14}$ to $7x10^{16}$ cm^{-3}. The dependence of T_c with BV is straightforward by using the well known ideal relations between N_d and BV

Planar $\qquad N_d = 1.9 \cdot 10^{18} \cdot BV^{-4/3}$ (6)

SJ $\qquad N_d = 6.3 \cdot 10^{11} \cdot (pitch/2)^{-7/6}$ (7)

The final T_c(BV) dependence for planar Power MOSFETs is validated in Fig. 4b by experimental data reported in [1]. Interestingly, T_c drops about 200K for 20V<BV<200V whereas less than 100K T_c reduction is observed for 200V<BV<1000V. Despite the lack of T_c measured data for SJ devices, a larger T_c is normally expected for a fixed BV value due to the superior N_d values.

C. Average V_{ds} and BV increment with T

The average V_{ds} value or BV_{inc} is extracted by geometrical considerations in the non-destructive UIS curve of Fig. 1b. If V_{ds} linearly increases during T_{av}, then BV_{inc} is approached by

$$BV_{inc} = k_p \cdot BV + 0.5 \cdot (BV_{max} - k_p \cdot BV)$$ (8)

where $k_p \geq 1$ is a coefficient that accounts for a boosted BV value due to the Egawa effect [13] and BV_{max} obeys the linear ΔT dependence

$$BV_{max} = k_p \cdot BV + Beta \cdot \Delta T$$ (9)

being Beta the BV increment with T, which is calibrated for Planar and SJ power MOSFETs by experimental data (see Fig. 5a). In both Planar and SJ devices, a constant Beta/BV = $9x10^{-4}$ K^{-1} is measured, which slightly differs from $7.8x10^{-4}$ K^{-1} in [4]. In Planar devices, Beta increases with BV due to the N_d diminishment. Notwithstanding N_d is higher in the SJ devices, Beta remains equal as in the Planar ones. A possible explanation of this fact is the counteracting effect of the dense and vertical super-abrupt PN junction [14]. Differently from Beta, k_p is a more difficult parameter to determine due to its multiple dependencies. On top of the I_{av} dependency, it is known from [13] that k_p is strongly dependent on N_d in Planar devices, thus tending to 1 for small N_d. This trend is corroborated in Fig. 5b by TCAD simulations as well as the charge balance (CB) dependence in SJ devices, tending to 1 for CB=0. In both cases, the snapback current (I_s) correlates to k_p, thus needing a k_p>1.2 or, what is the same, I_s>400A/cm^2 to avoid an electrical failure at very short times. Finally, k_p is fixed to 1.3 for the sake of comparison with LV devices [5].

III. EXPERIMENTAL RESULTS AND DISCUSSION

The theoretical and measured data for I_{av} in Planar and SJ power MOSFETs are compared in Figs. 6 and 7. The measured data for BV<120V is extracted from literature [2,3] whereas data for BV>600V corresponds to own measurements. It is inferred from Fig. 6 that the model data with $C_{th}=C_{th0}$ overestimates the I_{av} drop in HV Planar devices. The inability to mimic an accelerated I_{av} drop at large L in Fig. 7b is another limitation of using $C_{th}=C_{th0}$ in HV SJ and Planar devices. These trends are captured with the new analytical model including C_{th}(BV,L). Although the I_{av} vs. L trend is correctly described in LV devices (Fig. 7a), both models show mismatch in the I_{av} vs. BV trend due to T_c vs. BV variations in

Fig. 4. Note that C_{th0}, k_p, k_t and Beta/BV are identical for all cases to provide a universal model. Otherwise, the parameters C_{th0} and k_t of the new model can be readjusted to fit the measured I_{av} vs. L in a specific device. Contrarily, an adjustment of the model with $C_{th}=C_{th0}$ is not extensible to the complete L range of interest. It can be noticed in Fig. 7b that the good match between $C_{th}=C_{th0}$ model and experiment at high L is not suitable for low L. A comparative E_{as} analysis between HV Planar and SJ power MOSFETs is carried out by using the C_{th}(BV,L) model. The BV is fixed to 800 V whereas Ron is set to 650 mΩ by scaling the area of the devices (see

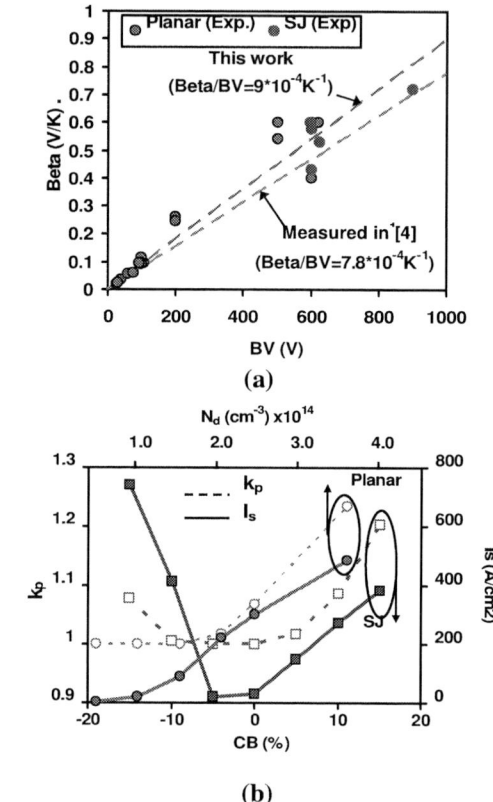

(a)

(b)

Figure 5. (a) Experimental and theoretical Beta vs. BV (Beta is the BV increment with T) for Planar and SJ power MOSFETs. (b) Simulated k_p and I_s dependence on N_d and CB for Planar and SJ devices, respectively.

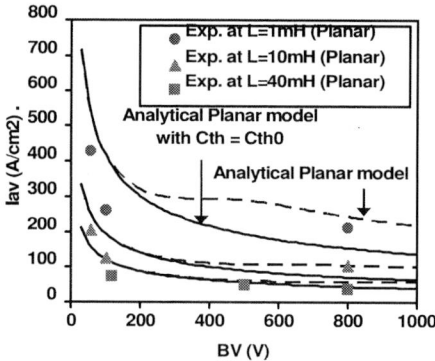

Figure 6. I_{av} vs. BV from theory and experiment for L=1, 10 and 40mH. Theoretical data is calculated from (1) by using $C_{th}=C_{th0}$ or C_{th}(BV,L).

Fig. 8). Eventually, E_{as} is extracted by

$$E_{as} = 0.5 \cdot L \cdot \left(I_{av} \cdot Area \right)^2 \cdot \left(BV_{inc} / BV_{inc} - V_{dd} \right) \qquad (10)$$

As it was pointed out in literature [15], SJ devices show superior I_{av} compared to Planar ones, however, their aggressive area reduction for a fixed Ron is predominant and the resulting E_{as} is degraded.

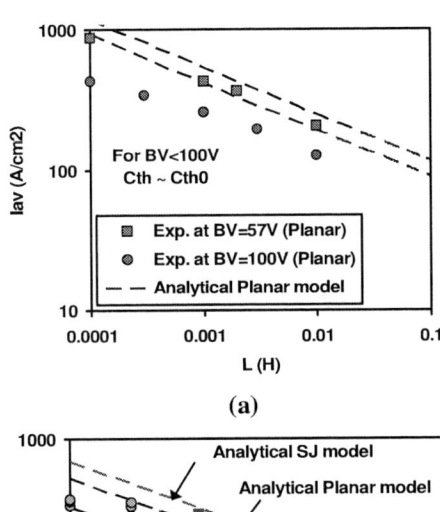

(a)

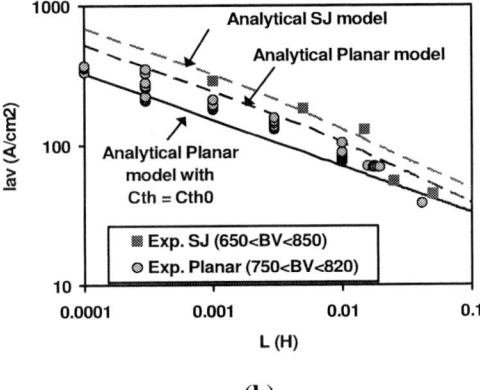

(b)

Figure 7.   (a) I_{av} vs. L from theory and experiment [2,3] for BV=57 and 100V Planar power MOSFETs. (b) I_{av} vs. L from theory and experiment for Planar and SJ power MOSFETs with 650V<BV<850V. The theoretical curves are extracted from (1) by using $C_{th}=C_{th0}$ or $C_{th}(BV,L)$.

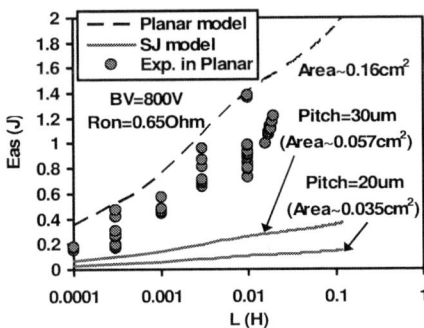

Figure 8.   Experimental and theoretical E_{as} vs. L values for a 0.65Ohm/800V Planar power MOSFET and prediction of E_{as} for a SJ power MOSFET with identical BV and Ron.

IV.   CONCLUSION

The UIS energy capability has been evaluated by experiment and theory in HV Planar and SJ power MOSFETs. A small I_{av} variation with BV is noticed for HV devices while different I_{av} vs. L decay slopes are identified at small or large L. To overcome the limitations of the previous models on replicating the aforementioned trends, a new analytical model with a more accurate thermal behavior is proposed, discussed and validated by experimental data. Finally, a practical E_{as} prediction case is assessed by using the new model.

ACKNOWLEDGMENT

This work is carried out in the frame of the GREENFETS project and financed by the IWT (Flanders, Belgium).

REFERENCES

[1]   D. Kinzer, "Advances in power switch technology for 40 V - 300 V applications", EPE, 2005, pp. 1-11.

[2]   R. R. Stoltenburg, "Boundary of Power-MOSFET, unclamped inductive-switching (UIS, avalanche current capability)", APEC, 1989, pp. 359-364.

[3]   E. Oxner and P. A. Dunning., "Unclamped Inductive Switching Rugged MOSFETs for Rugged Environments", AN601 SMP30N10, Feb. 1994.

[4]   G. A. M. Hurkx and N. Koper, "A physics-based model for the avalanche ruggedness of power diodes," ISPSD, 1999, pp. 169-172.

[5]   I. Pawel, R. Siemieniec, M. Rosch, F. Hirler, and R. Herzer, "Experimental study and simulations on two different avalanche modes in trench power MOSFETs," IET Circuits, Devices & Systems, vol. 5, pp. 341–6, 2007.

[6]   K. Chinnaswamy, P. Khandelwal, M. Trivedi, and K. Shenai, "Unclamped Inductive Switching Dynamics in Lateral and Vertical Power DMOSFETs," IAS, 1999, pp. 1085-92.

[7]   J. Roig, E. Stefanov, and F. Morancho, "Thermal behavior of a superjunction MOSFET in a high-current conduction", IEEE Trans. On Elec. Devices, vol. 53, pp. 1712-20, 2006.

[8]   R. Joy and E. S. Schlig, "Thermal properties of very fast transistors", IEEE Trans. On Elec. Devices, vol. 17, pp. 586-593, 1970.

[9]   D. Farenc, G. Charitat, P. Dupuy, T. Sicard, I. Pages, and P. Rossel, "Clamped inductive switching of LDMOST for smart power IC's," ISPSD, 1998, pp. 359-62.

[10]   A. Icaza-Deckelmann, G. Wachutka, F. Hirler, J. Krumrey, and R. Henninger, "Failure of multiple-cell power DMOS transistors in avalanche operation", ESSDERC, 2003, pp. 323-6.

[11]   I. Pawel, R. Siemieniec, M. Rösch, F. Hirler, C. Geissler, A. Pugatschow, and L. J. Balk, "Design of avalanche capability of Power MOSFETs by device simulation," EPE, 2007, pp. 1-10.

[12]   J. Roig, Y. Weber, J.-M. Reynes, F. Morancho, E.N. Stefanov, M. Dilhan,   and G. Sarrabayrouse, "Electrical and Physical Characterization of 150-200V FLYMOSFETs," ISPSD, 2006, pp. 1-4.

[13]   H. Egawa, "Avalanche characteristics and failure mechanism of high voltage diodes", IEEE Trans. On Elec. Devices, vol. 13, pp. 754-758, 1966.

[14]   C. Y. Chang, S. S. Chiu, L. P. Hsu, "Temperature dependence of breakdown voltage in silicon abrupt p-n junctions", IEEE Trans. On Elec. Devices, vol. 18, pp. 391-3, 1971.

[15]   J.-S. Lai et al., "Characteristics and utilization of a new class of low on-resistance MOS-gated power device," IEEE Trans. Ind. Appl., vol. 37, pp. 1282–9, 2001.

Proceedings of the 23rd International Symposium on Power Semiconductor Devices & IC's
May 23-26, 2011 San Diego, CA

Improvement of Switching Trade-off Characteristics between Noise and Loss in High Voltage MOSFETs

Wataru Saito, Satoshi Aida, Shigeo Koduki and Masaru Izumisawa

Semiconductor Company, Toshiba Corp.
Kawasaki, Japan
wataru3.saito@toshiba.co.jp

Abstract—A new MOS-gate structure was proposed and demonstrated to improve the switching trade-off characteristics between noise and loss in high-voltage MOSFETs. The lightly p-doped dummy base layer under the gate electrode modulates C_{gd}-V_{ds} curve due to the depletion under high applied voltage and the turn-off dV/dt can be suppressed even with high-speed switching. The fabricated device showed the surge voltage suppression of 50 V or the turn-off loss reduction of 20% in the turn-off switching test with an inductive load. In the flyback converter operation, it was also shown that the trade-off characteristics between the radiation noise and total power loss were improved by the proposed dummy base structure.

I. INTRODUCTION

Power electric systems are continuously required the efficiency improvement and those downsizing. As a solution, high speed switching operation has been employed due to the switching loss reduction and downsizing the passive component size by high frequency operation. Since the switching loss is mainly determined by the feedback capacitance [1], [2] the gate-drain charge Q_{gd} has been reduced for the high-speed operation [3]-[6].

High turn-off dV/dt, however, induces large voltage surge because of the stray inductance in the circuit and generates EMI noise. As a result, the efficiency degradation due to low dV/dt must be accepted for the noise reduction. In the previous works, the stray inductance was reduced by assemble processes, such as multi chip module and Cu lead [7], [8]. Therefore the high switching frequency operation has been approached by a combination of low Q_{gd} chip and low inductance package. In addition, it has been mainly studied in low voltage MOSFETs. The noise reduction has not been approached from the device design.

In this paper, a new MOS-gate structure is proposed to improve the switching trade-off characteristics. The proposed structure realizes suppressing the voltage surge at the turn-off switching with low switching loss by the feedback capacitance management. The fabricated 600 V-class MOSFET demonstrated a flyback converter operation and it is shown that the trade-off characteristics between the radiation noise and total power loss were improved.

Identify applicable sponsor/s here. *(sponsors)*

II. DEVICE STRUCTURE AND DESIGN

The dummy base structure is proposed to improve the switching trade-off characteristics by the feedback capacitance modulation. In the proposed structure, the dummy base layer is formed under the gate electrode and connected to the p-base layer as shown in Fig. 1. The dummy base layer is lightly p-doped and depleted under high applied voltage. Therefore the gate electrode area facing the drain is increased at high drain voltage V_{ds} and the feed-back capacitance C_{gd} is increased as

(a) Low V_{ds} (Before Depletion of Dummy Base)

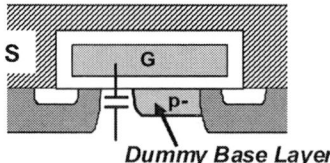

(b) High V_{ds} (After Depletion of Dummy Base)

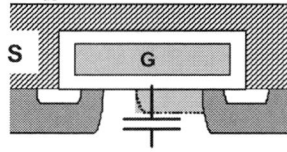

Fig. 1 Cross-sectional structure of proposed MOS-gate with dummy base layer for improvement of switching trade-off characteristics.

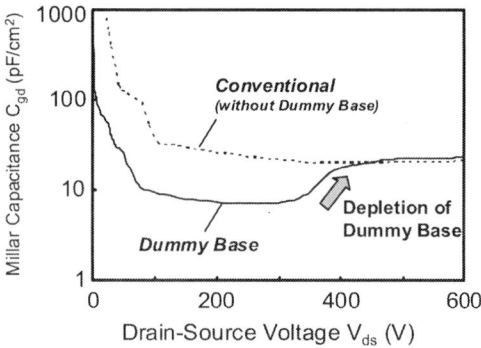

Fig. 2 Simulated C_{gd}-V_{ds} characteristics modulated by depletion of dummy base layer.

978-1-4244-8425-6/11 $26.00 © 2011 IEEE

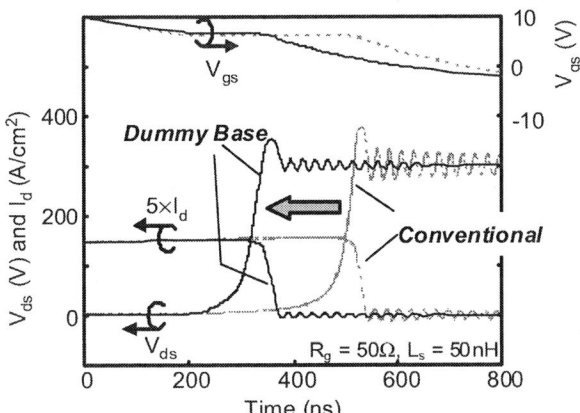

Fig. 3 Simulated turn-off switching waveform.

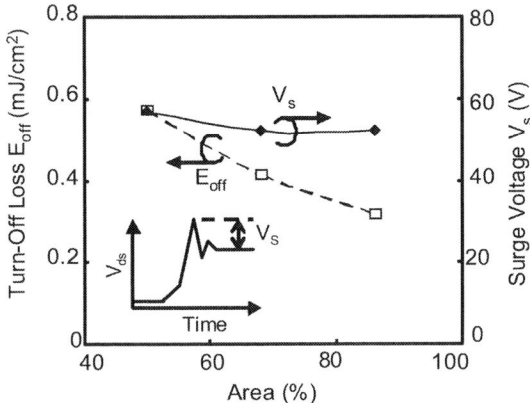

Fig. 4 Relation between turn-off loss, surge voltage and dummy base area.

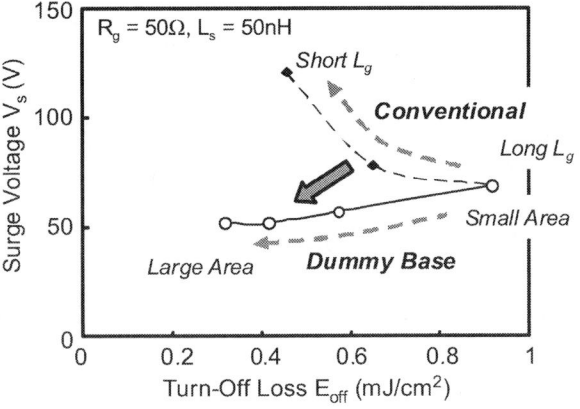

Fig. 5 Trade-off characteristics between turn-off loss and surge voltage comparing with proposed dummy base structure and conventional one.

shown in Fig. 2. Therefore C_{gd} at low V_{ds} can be reduced maintaining large C_{gd} at high V_{ds}. Since the dV/dt is inverse proportional to the C_{gd} at the turn-off switching due to the charge through the outer gate resistance, initial dV/dt at low V_{ds} can be increased for low loss switching and a final dV/dt at high V_{ds} can be suppressed for the noise reduction. As a

result, the switching trade-off characteristics between noise and loss can be improved by the proposed dummy base structure.

The effect of the proposed structure was estimated using the device simulator Dessis as follows. The stray inductance was assumed 50 nH in the simulation. An example of the simulated switching waveform is shown in Fig. 3. The turn-off time is decreased by the dummy base structure, although the voltage oscillation is slightly suppressed. These behaviors are determined by the C_{gd}-V_{ds} characteristics, which are modulated by the dummy base structure.

The design parameters of the dummy base structure are the area under the gate and the doping concentration. The area determines the C_{gd} at low V_{ds} and the C_{gd} change at high V_{ds}. The doping concentration determines the voltage of C_{gd} change due to the dummy base layer depletion. The dummy base layer was designed for the 600 V-class MOSFET. Fig. 4 shows a relation between the area, the loss and the surge voltage. In this work, the surge voltage was defined as the difference between the peak voltage and the supply DC voltage. Both of the loss and the surge voltage is reduced with the increase of the area due to small C_{gd} at low V_{ds} and large C_{gd} change.

The switching loss can be reduced only by the decrease of the C_{gd}. Therefore fine gate-pattern (short L_g) realizes high speed switching. The voltage surge, however, is enhanced by large dV/dt due to small C_{gd} at high V_{ds} and the relation between noise and loss becomes a trade-off at the conventional structure as shown in Fig. 5. On the other hand, at the proposed structure, the turn-off loss can be reduced by small C_{gd} at low V_{ds}. In addition, the voltage surge is suppressed by large C_{gd} at high V_{ds} due to the depletion of the dummy base layer. As a result, the switching trade-off characteristics are improved by the increase of the dummy base area as shown in the same figure.

III. TURN-OFF SWITCHING TEST

600 V-class MOSFETs with and without the dummy base layer were fabricated. The mesh-gate pattern was employed and the dummy base layer was formed under the cross point of the mesh-gate as shown in Fig. 6. It was because to avoid the degradation of the static characteristics. To suppress the resistance increase in JFET-n region, the n-doping concentration was increased comparing with the conventional structure. The breakdown voltage lowering, however, can be avoided by the relaxation of the electric field concentration at the p-base edge by the p-doped dummy base layer.

In this experiment, the area accounted for over 50% of JFET-n region under the gate and the doping concentration was designed to deplete under 350 V. The n-doping concentration of the JFET-n region was 25% higher than that of the conventional structure to suppress the on-resistance increase.

The fabricated MOSFETs achieved no degradation of static characteristics compared with the conventional one. The relation between the dummy base area and the static characteristics is shown in Fig. 7. The same on-resistance was achieved at the area of less than 67%. The breakdown voltage

978-1-4244-8425-6/11 $26.00 © 2011 IEEE

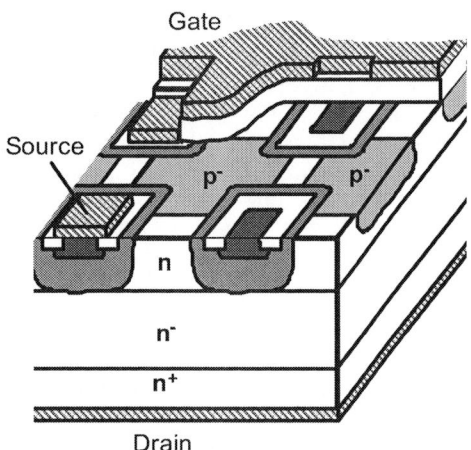

Fig. 6 Fabricated structure with dummy base layer under cross-point of mesh gate electrode.

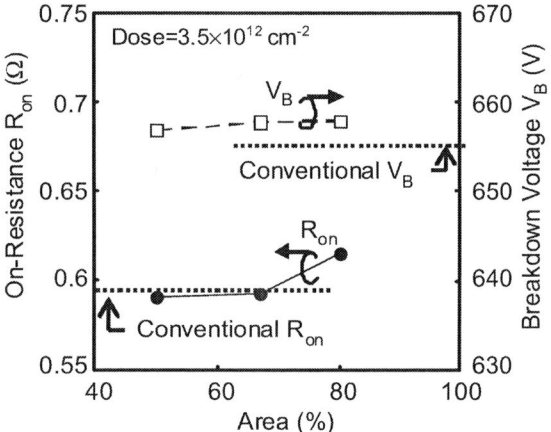

Fig. 7 Relation between static characteristics and dummy base area.

was slightly higher than that for the conventional structure due to the relaxation of the electric field concentration.

At a turn-off switching with the inductive load of 1 mH, the dummy base structure suppressed the voltage surge with high speed switching as shown in Fig. 8. Although the turn-off loss was decreased with the increase of the dummy base area, there was the optimum condition for the minimum surge voltage due to the depletion of the dummy base layer as shown in Fig. 9. From view points of the on-resistance and the surge voltage, the area of 67% is an optimal point in this case.

The trade-off characteristics between the surge voltage and the turn-off loss were measured by the varied outer gate resistance. The optimized structure improved the trade-off characteristics as shown in Fig. 10. The turn-off loss was reduced about 20% maintaining the surge voltage. On the other hand, the trade-off characteristics for turn-on switching were almost as same as those of conventional structure, because the turn-on behavior was determined by the recovery diode characteristics.

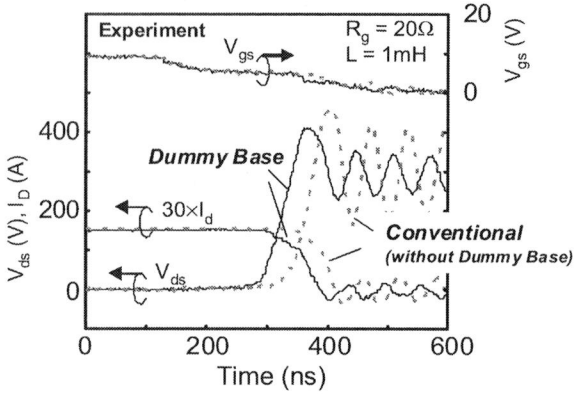

Fig. 8 Turn-off switching waveforms at inductive load of 1 mH for fabricated dummy base and conventional structures.

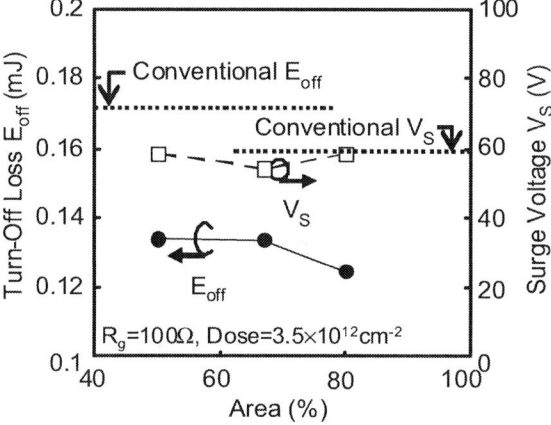

Fig. 9 Relation between switching characteristics and dummy base area.

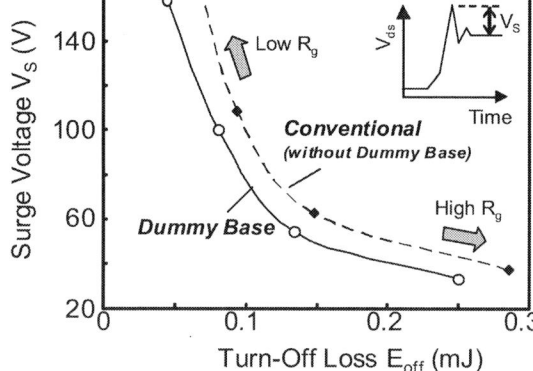

Fig. 10 Trade-off characteristics between surge voltage and turn-off loss improved by dummy base structure.

IV. FLYBACK COVERTER OPERATION

The fabricated device was tested as a switching device in the primary side of the flyback converter as shown in Fig. 11. The radiation noise from the circuit was measured using an electric field prove and a spectrum analyzer. The operating conditions were fixed to the input voltage of 150 V, the duty

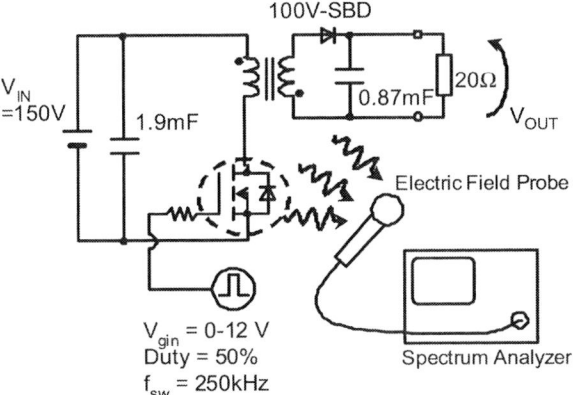

Fig. 11 Tested flyback converter circuit. Radiation noise was measured using an electric field probe.

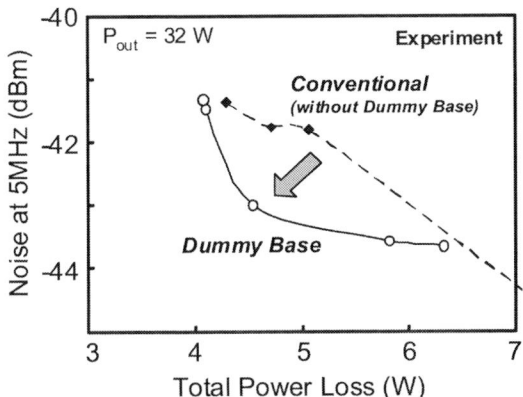

Fig. 12 Trade-off characteristics between radiation noise and total power loss improved by dummy base structure.

ratio of 50% and the switching frequency of 250 kHz. The output power was 32 W. The trade-off characteristics between the radiation noise and the total power loss were measured by the varied outer gate resistance of 20-120 Ω.

The fabricated dummy base structure obtained the improvement of the trade-off characteristics in the circuit operation. The total loss was reduced from 6 W to 4.5 W without the increase of the noise amplitude for 5 MHz as shown in Fig. 12. The switching loss was reduced without the noise enhancement and the conduction loss was maintained by the optimal dummy base design. Although there are other radiation noise peak at higher frequency of 15 and 24 MHz, the trade-off characteristics between these noises and the loss were also improved by the dummy base structure.

V. CONCLUSIONS

The dummy base structure was proposed and demonstrated to improve the switching trade-off characteristics between noise and loss. The dummy base layer formed under the gate modulates C_{gd}-V_{ds} characteristics due to its depletion at high V_{ds}. The switching loss is reduced by small C_{gd} at low V_{ds} and the noise is suppressed by large C_{gd} at high V_{ds}. A fabricated 600 V-class MOSFET achieved 20% turn-off loss reduction with the same voltage surge compared with the conventional structure at the inductive load switching test. In the flyback converter operation, the trade-off characteristics between the radiation noise and the total power loss were improved by the dummy base structure. These results show the candidate of increasing the switching frequency with low EMI noise by the feedback capacitance management in the switching device design.

ACKNOWLEDGMENT

The authors wish to thank Prof. I. Omura in Kyushu Institute of Technology for his fruitful discussion of this work. They are also grateful to A. Ono, T. Yoshihira, T. Ogura, T. Matsuda and N. Matsuura for their support and fruitful discussion of this work.

REFERENCES

[1] R. Sodhi, S. Brown Sr., and D. Kinzer; "Integrated design environment for DC/DC converter FET optimization," in Proceedings of ISPSD'99 pp.241-244, 1999.

[2] R.J.E. Hueting, E.A. Hijzen, A.W. Ludikhuize and M.A.A. in't Zandt, "Switching Performance of Low-Voltage N-Channel Trench MOSFETs," Proceedings of ISPSD'02, pp.177-180, 2002.

[3] T. Kobayashi, H. Abe, Y. Niimura, T. Yamada, A. Kurosaki, T. Hosen and T. Fujihira; "High-voltage power MOSFETs reached almost the silicon limite," in Proceedings of ISPSD'01, pp.435-438, 2001.

[4] M. Darwish, C. Yue, K. H. Lui, F. Giles, B. Chan, K. Chen, D. Pattanayak, Q. Chen, K. Terrill and K. Owyang; "A new power W-gated trech MOSFET (WMOSFET) with high switching performance," in Proceedings of ISPSD'03, pp.24-27, 2003.

[5] L. Ma, A. Amali, S. Kiyawat, A. Mirchandani, D. He, N. Thapar, R. Sodhi, K. Spring and D. Kinzer; "New trench MOSFET technology for DC-DC converter applications," in ISPSD'03, pp.354-357, 2003.

[6] P. Goarin, G. E. J. Koops, R. van Dalen, C. Le Cam and J. Saby; "Spilt-gate resurf stepped oxide (RSO) MOSFETs for 25 V applications with recored low gate-to-drain charge," in Proceedings of ISPSD'07, p.61-64, 2007.

[7] Y. Kawaguchi and T. Kawano, H. Takei, S. Ono and A. Nakagawa, "Multi chip module with minimum parasitic inductance for new generation voltage regulator," in Proceedings of ISPSD'05, p.371-374, 2005.

[8] T. Hashimoto, T. Uno, Y. Satou, M. Shiraishi, T. Kawashima and N. Matsuura, "Advanced power SiP with wireless bonding for voltage regulators," in Proceedings of ISPSD'07, p.125-128, 2007.

Proceedings of the 23rd International Symposium on Power Semiconductor Devices & IC's
May 23-26, 2011 San Diego, CA

300A 650V 70 um Thin IGBTs with Double-Sided Cooling

Hsueh-Rong Chang [1], Jiankang Bu [1], George Kong [2] and Ricky Labayen [3]

[1] Automotive Power Switches Development, [2] Temecula Manufacture Center, [3] Power Device Characterization Lab.
International Rectifier Corp. 101 N. Sepulveda Blvd, El Segundo, CA 90245
Phone: 310-726-8854, Email: hrchang1@irf.com

Abstract— **Large IGBTs with a current rating of 300A and a blocking voltage of 650V on ultra thin wafers have been successfully developed with double-sided cooling capability. The deposition of solderable metals on the front and back sides of the IGBT produced flat thin wafers with less than 2 mm warpage and good mechanical yield. A large reduction of on-state voltage drop 390 mV at 300A is achieved in a wirebond-less Cu-clip package. The combination of lower on-state voltage drop and larger heat exchange area increases the IGBT current carrying capability by 200%.**

I. INTRODUCTION

The demand for high voltage and high current in hybrid electric vehicles (HEVs) presents technical challenges for power conversion beyond those normally associated with vehicle electrical and electronic systems. An increasing demand for fuel efficiency warrants light weight and small volume power control unit (PCU) in HEV [1, 2]. Conventional IGBT power modules use wire bonds which is the common site of module failures due to coefficient of thermal expansion (CTE) difference, and is the limiting factor for module life. IGBTs with dual solderable metals on top and bottom of the die offer a wirebond-less assembly option with enhanced reliability and lower manufacturing cost [3, 4]. In addition, it allows double sided cooling with significantly larger heat exchange area, thus improving thermal management when compared with wire bonds used in traditional power modules.

In this paper, we report the development of large IGBTs with a current rating of 300A, a blocking voltage of 650V, and double-sided cooling capability on ultra thin wafers. A process for the deposition of solderable metals on the front and back sides of the IGBT was developed to produce flat thin wafers with less than 2 mm warpage and good mechanical yield. A wirebond-less Cu-clip package were used to demonstrate significant lower on-state voltage drop of the IGBTs with solderable front metals (SFM) and large increase in heat exchange area, thus resulting in higher current carrying capability by 200%.

II. PROCESS CHALLENGES AND SOLUTIONS

IGBT with field stop (FS) structure offers a superior trade-off between static and dynamic losses as compared with Punch-Through (PT) - IGBTs [5]. The typical thickness of 650V FS IGBT is in the range of 65 – 75 um. Thin wafers could be severely warped due to compressive or tensile metal stress, thus resulting in high mechanical breakage and poor yield [6].

The IGBTs reported in this paper were processed following a normal planar IGBT process first till the aluminum was sputtered on the front side. After the deposition of the passivation layer, the solderable front metal (SFM) was sputtered and patterned, followed by a ultra-thin-wafer (UTW) grinding process and an implantation to form the buffer-layer on the backside of the wafer. Collector was formed by the implantation of boron followed by back metal processes. The composition of the solderable metals includes Ti-Ni-Ag. Cu-clip was chosen for contacting emitter pad to eliminate wire bonds and to enable double-sided cooling. Fig. 1 shows the side view of the IGBT. The SFM coverage over passivation layer is shown in Fig. 2.

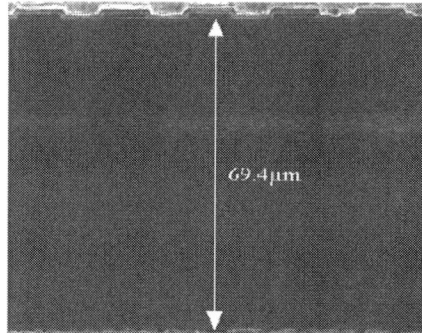

Fig. 1 Device structure.

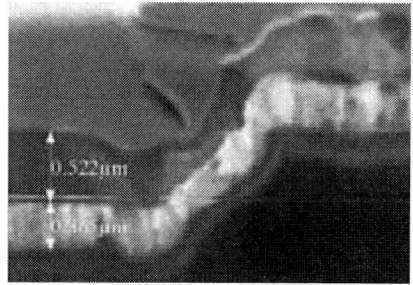

Fig. 2 SFM coverage over the passivation layer.

978-1-4244-8425-6/11 $26.00 © 2011 IEEE

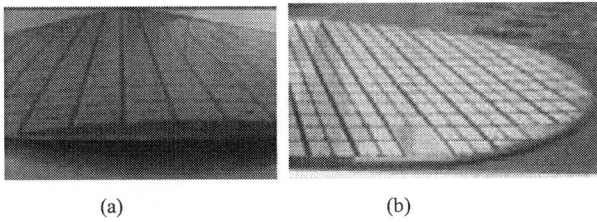

(a)   (b)

Fig. 3 Standard process could lead to (a) compressive or (b) tensile stressed wafers.

Fig. 4 Wafer with warpage less than 1 mm.

(a) Cu-clip IGBT.   (b). Wire-bonded IGBT
Fig. 5 Top views of IGBTs packaged in Cu-clip (a) and wirebond (b).

Fig. 6 A side view of Cu-clip IGBT

The metal Ni used in the solderable metal composition produces high surface stress on Si wafers, which poses tremendous challenges for the UTW process. Un-optimized metal deposition process could lead to high compressive or tensile stress, resulting in severely warped wafers, as shown in Fig. 3a and 3b. Wafers with warpage greater than 6 mm cannot be used in manufacturing process and electrical testing thereafter due to high risk on wafer breakage. The Zero-stress process is susceptible to metal peeling. To overcome these challenges, we have developed a novel stress control scheme to allow compressive stress balanced by the tensile stress at the end of the fabrication of the thin IGBT wafers. Fig. 4 shows the ultra thin IGBT wafer with near-zero-warpage.

Cu-clip package has been widely used in microelectronic components to eliminate wire bonds and to enhance device performance and reliability. But it has not been explored in the assembly of high voltage IGBTs. In this study, we use the wirebond-less feature of Cu-clip to evaluate the performance improvement of large IGBT die. Fig. 5 shows the top views of the IGBTs packaged in Cu-clip and in wirebond configuration. Cu-clip was designed to perfectly fit the emitter pad. A side view of the Cu-clip IGBT is shown in Fig. 6. The Cu-clip has a thickness of 2mm. It significantly reduces the current spreading resistance at the emitter, thus resulting in low on-state voltage drop, Vceon.

III. ELECTRICAL PERFORMANCE AND DISCUSSION

1. Static Performance

Fig. 7 shows the IGBT die layout. The output characteristic is shown in Fig. 8. A large reduction of 390 mV in Vceon at Ice=300A was achieved for the Cu-clip IGBTs as compared with the IGBTs packaged with wire-bonds, as shown in Fig. 9.

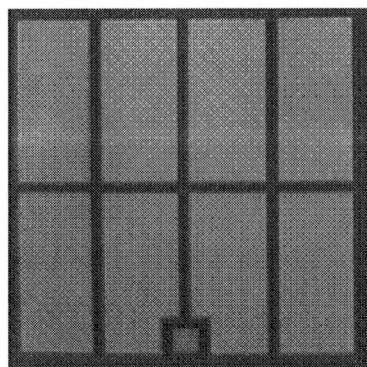

Fig.7 IGBT Die Layout.

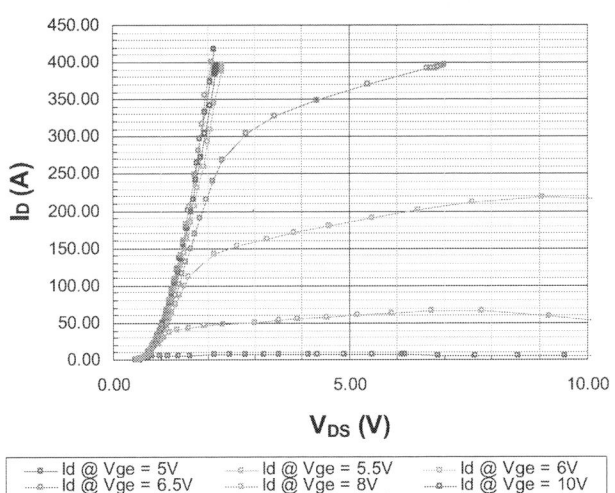

Id @ Vge = 5V	Id @ Vge = 5.5V	Id @ Vge = 6V
Id @ Vge = 6.5V	Id @ Vge = 8V	Id @ Vge = 10V
Id @ Vge = 15V	Id @ Vge = 18V	

Fig. 8 Output Characteristics at 25C at various Vge.

978-1-4244-8425-6/11 $26.00 © 2011 IEEE   321

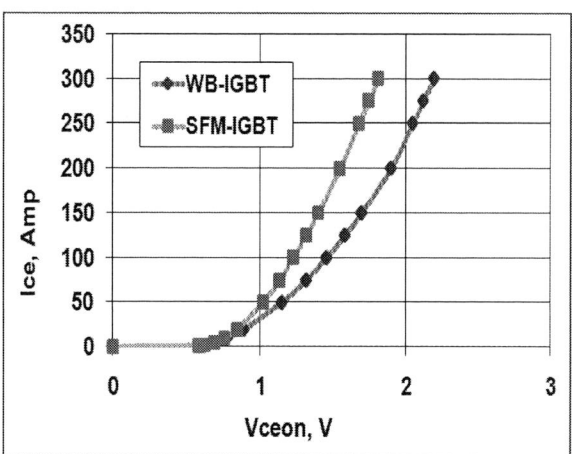

Fig. 9 I-V curves for WB-IGBT and SFM-IGBT.

Fig. 10 shows the breakdown voltage distribution of the ultra thin SFM IGBT wafers. The average breakdown voltage is 706V at 25C with 3 Sigma of 21V, which provides sufficient margin for 650V applications. The breakdown voltage is increased to above 800V at 150C, as shown in Fig. 11.

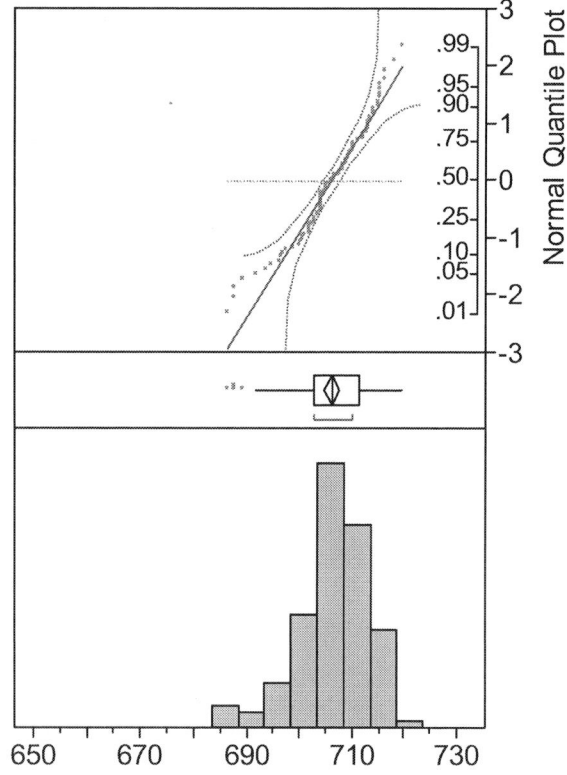

Fig. 10 Breakdown voltage distribution.

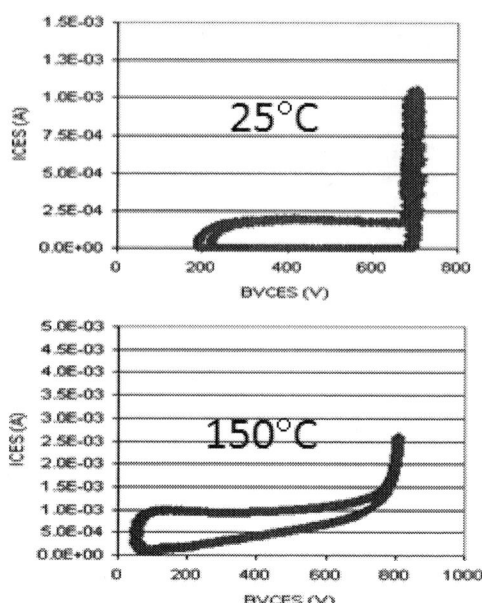

Fig. 11 Breakdown voltages at 150C for SFM- IGBTs with Cu clip.

2. Dynamic Characterization

Dynamic switching of the Cu-clip IGBTs was measured at Vcc=400V and Ic=300A. The current and voltage waveforms are shown in Fig. 12. The turn-off speed of the Cu-clip IGBTs is in the range of 200-215 ns at 25°C. Because no electron irradiation is used for lifetime control, the t_{fall} time does not vary much with temperature. The leakage current of the thin IGBTs is low at elevated temperatures, i.e., 4.5 mA at 600V and 175°C, enabling the operation at Tjmax of 175°C while the Tjmax is limited to 150C for the PT-IGBTs due to a rapid increase in the leakage current with increasing temperature.

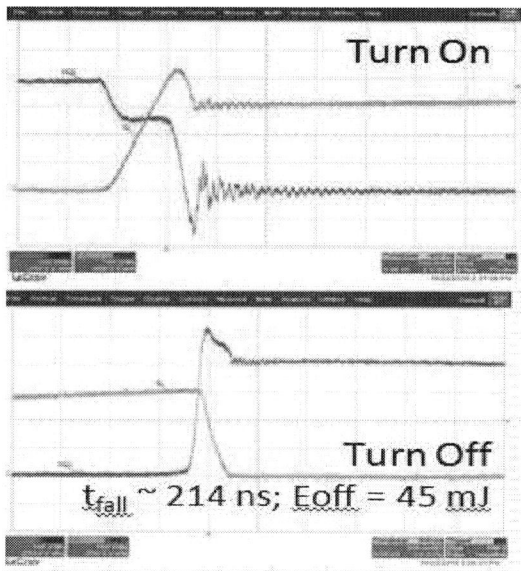

Fig. 12 Turn-on and turn-off waveforms of SFM - IGBT with Cu-clip. (Vcc=400V, Ice=300A, 25°C)

978-1-4244-8425-6/11 $26.00 © 2011 IEEE   322

The thin IGBTs with dual solderable metals on top and bottom of the die enable double-sided cooling for effective heat removal. The junction-to-case thermal resistance of the SFM-IGBTs, $R_{\theta jc\_SFM}$ can be calculated by Eqn.(1). It is equivalent to a WB-IGBT in parallel to a top SFM emitter.

$$\frac{1}{R_{\theta jc\_SFM}} = \frac{1}{R_{\theta jc\_WB}} + \frac{1}{R_{\theta jc\_TOP}} \qquad (1)$$

In this study, the SFM coverage on the emitter pad is 67% of the die area. This leads to 40% reduction of $R_{\theta jc\_SFM}$ as compared with that of WB-IGBT's., thus significantly increasing the power handling capability of the SFM-IGBT. In addition, the SFM IGBT has a much lower Vceon than WB-IGBT, as shown in Fig. 9. At Vceon=1.8V, the SFM-IGBTs would conduct 300A while only 175A for the WB-IGBT's. The combination of lower Vceon and double-sided cooling allows 200% improvement in the SFM-IGBT power handling capability as compared with conventional WB-IGBTs.

The wirebond-less IGBTs offer additional advantage in terms of reliability. Conventional IGBT power modules use wire bonds which is the common site of module failures due to CTE difference, and is the limiting factor for module life. The wirebond-less SFM IGBTs were evaluated in power cycling test with $\Delta T=100°C$. The preliminary results are very encouraging. The number of cycles is increased by 260% for the wirebond-less IGBTs as compared with the wire-bonded IGBTs.

IV. CONCLUSION

We have successfully developed high current, high voltage IGBTs on ultra thin wafers with double-sided cooling capability. Wirebond-less IGBTs with Cu-clip contacted emitter exhibit significant reduction of on-state voltage drop as well as enhanced reliability. The feature of double-side cooling allows 67% increase in the heat exchange area. The combination of lower Vceon and larger heat exchange area leads to 200% improvement in the IGBT power handling capability as compared with the conventional wire-bonded IGBTs.

ACKNOWLEDGMENT

Authors would like to acknowledge IR Temecula management and engineering team for their support.

REFERENCES

[1] T. Kikuchi, et. Al., EVS21, 2005

[2] R. Hironaka, et. Al., EVS22, 2006.

[3] T. Burress, International Energy Conversion Engineering Conference,Nashville, Tennesse, June 2010.

[4] "Power EV-HEV 2010 – power electronics in Electric and Hybrid Cars." Yole development. France. 2009.

[5] T. Laska, et al, "Short Circuit Properties of the trench/Field-Stop-IGBTs – Design Aspects for a Superior Robustness', Proc. ISPSD'03, 2003.

[6] T. Laska, M. Matschitsch, W. Scholz, "Ultra thin-wafer technology for a new 600 V-NPT-IGBT", ISPSD '97.

Proceedings of the 23rd International Symposium on Power Semiconductor Devices & IC's
May 23-26, 2011 San Diego, CA

SKiN: Double side sintering technology for new packages

Thomas Stockmeier, Peter Beckedahl, Christian Göbl, Thomas Malzer

SEMIKRON Elektronik GmbH & Co. KG
Sigmundstrasse 200
D-90431 Nuremberg

Abstract—**SKiN technology comprises the sintering of power chips to a substrate (i.e. DBC), a top side sintering of the power chips to a flexible circuit board, and the sintering of the substrate to a pin-fin heat sink. The resulting power device has a very low volume and weight and demonstrates unprecedented thermal, electrical, and reliability performance. Therefore, this technology is ideally suited to provide better power electronic solutions for a range of applications, i.e. electric vehicles, renewable energies and variable speed motor drives. The process to manufacture a 400 Amp, 600 V Dual IGBT, sintered to an aluminum pin-fin heat sink will be described in detail and electrical, thermal and reliability results will be shown.**

I. INTRODUCTION

Power Electronic Module packaging technology faces many challenges, as the width and depth of applications for power modules grows significantly. The latest power semiconductor switching devices such as MOSFETs and IGBT achieve very high power densities so that conventional thick aluminum wire bonding technology represents a bottleneck for load current capability and reliability. Recently, attempts have been made to improve this technology /1/. However, these attempts lead to comparably large efforts in the power device metallization and processing. Wide bandgap materials enlarge the usable temperature and switching frequency range so that low inductive packages beyond 200°C are required /2/. Environmental conditions such as bias, humidity, temperature, industrial gas atmosphere, and mechanical shock and vibration play an increasingly important role, for which off-shore wind turbines, electric vehicles and marine drives may serve as examples.

However, not only these technical and environmental challenges have to be faced, but also power electronic modules need to be reduced in size and weight for use in vehicles and have to be interfaced in a simple and reliable manner, even at high load currents.

II. SKiN TECHNOLOGY PROCESS FLOW

A. Chip to Substrate Sintering

Sintering of silver particles at low temperature and low pressure is a well established technology /3/ today and has started to replace soldering of chips to DBC substrates already in mass production. Thanks to its unprecedented reliability and thermal behaviour, this joint technology makes power modules better suited for higher temperatures and more demanding applications.

In the conventional silver sintering process, a sinter paste which contains silver particles or flakes (nm to μm size) is stencil printed on a metallised ceramic substrate (i.e. DBC). This is very similar to a solder paste printing process. Then power semiconductor dies, i.e. Silicon, SiC, GaN devices are picked from wafer or chip tray and placed onto the paste. Afterwards, this populated substrate is subjected to a heated mechanical press, where the silver sinter layer is formed, typically at 250 °C, 40 MPa, for 1 min. Some improved pastes allow today already a significant reduction of pressure and temperature. Fig. 1 shows a 5" x 7" DBC card which contains four substrates (before separation by laser cutting) with sintered chips, as used today in series production for power modules in automotive applications.

Fig. 1: 5" x 7" DBC card with four substrates (before laser cutting) with sintered chips, used in power modules for automotive series production.

Up to this point, the SKiN technology follows this well established process. However, rather than continuing in the process sequence, which would be Al-wire wedge bonding, electrical testing, laser cutting, and final automated optical inspection (AOI) as in conventional devices, the SKiN technology /4/ proceeds differently:

B. SKiN Design

Fig. 2 shows a schematic cross section of the new device. The key design element is a flexible circuit board (called "flex board") of polyimide with patterned metal tracks on both sides. The bottom metal, called "power side", exhibits a thick metal layer which will be able to carry the load current. Thus, depending on the material (Cu or Al) and on

978-1-4244-8425-6/11 $26.00 © 2011 IEEE   324

the required current, a thickness in the range of 100 μm is most suitable. The top side metal layer, called "logic side", needs only a relatively thin metallization (i.e. 30 μm Cu), as it carries only gate, auxiliary and sense signals. The material selection and thickness of the polyimide itself depends on the application conditions, such as intended operating temperature and voltage class. Typically, it is several tens of μm thick.

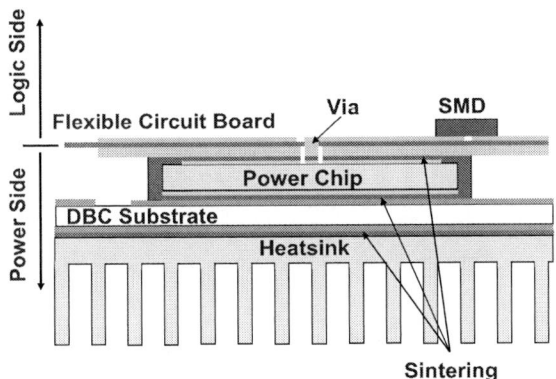

Fig. 2: Schematic cross section of the SKiN device. The device exhibits no solder layers, nor any wire bond interconnects and is integrated with a pin fin heat sink.

In order to bring the gate contacts of the power switches from the power side to the logic side, the polyimide layer exhibits through-hole contacts (vias) in the flex board which can be laterally located at any suitable position. One gate via is drawn schematically in the above figure.

C. SKiN Process: Power Part

Fig. 3 shows the flex board (top side), as it is prepared to be positioned and sintered within the next process step. In areas which will connect the chips with the flex board, the power side of the board is screen or stencil printed with silver paste. The auxiliary contacts, such as gate, emitter and temperature sensor contacts are carried out as tracks on the flex board, reaching out to the left and right side. The gate contacts are not visible as they are on the logic side.

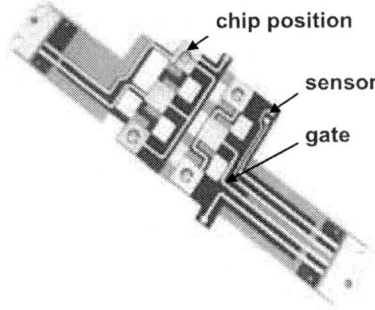

Fig. 3: Photo of a Flex board (top side) prior to sintering to the substrate. In this particular design, the auxiliary contacts, such as gate and emitter, are carried out as tracks on the flex board, reaching out to the left and right to be folded-in, later.

In order to protect the flex board from cracking at sharp edges (such as power chips) in the following sintering

process, a thin layer of organic material is dispensed or screen printed at the rim of chips on the populated DBC substrates. Also, silver paste is stencil printed in areas, where collector contacts and power terminals will be connected by sintering, later. Fig. 4 shows the photograph of a DBC substrate with chips, chip edge protection, and printed silver paste, prior to be merged with the flex board (comp. Fig. 3).

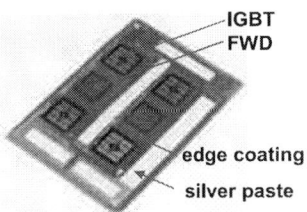

Fig. 4: Photograph of a substrate with IGBT and freewheeling diodes (FWD) chips, where the chip edges are protected by an organic material and silver paste is printed in power terminal and collector contact areas.

Now, the substrate and flex layer are aligned to each other and put into a conventional sinter press, where the flex board power side is now sintered to the top side of the chips and to the collector contact areas. The resulting power device is now finished and can be fully electrical tested, comprising static tests, dynamic tests, and insulation / partial discharge.

D. SKiN Process: Interconnect and Cooling

However, the SKiN technology doesn't stop at this stage. SKiN technology proceeds in providing thermal and electrical interfaces by sintering the power terminals and the heat sink to the power device. Fig. 5 shows the AC- and DC- power terminals, made from Ag-coated Al or Cu, as well as a pure Al pin-fin heat sink. On top of the heat sink, a defined area is stencil printed with silver paste where the power part will be attached.

Fig. 5: Photograph of DC- and AC- power terminals, as well as an Al pin-fin cooler with a silver paste area on the top side.

Now, the heat sink, the power terminals, and the power part are all assembled into a sintering jig and put again into a conventional sinter press, this time forming the sinter connections from heat sink to substrate and substrate to power terminals.

Afterwards, a simple plastic frame is added to provide guidance for the auxiliary terminals and to provide alignment for the use of this device in the power electronic inverter system. Fig. 6 shows, how all parts are assembled together to form a 400 A, 600V dual SKiN device. The device contains each two 200 Amp IGBTs and one 400

978-1-4244-8425-6/11 $26.00 © 2011 IEEE   325

Amp freewheeling diode for the upper and the lower switch, as well as a temperature sensor chip.

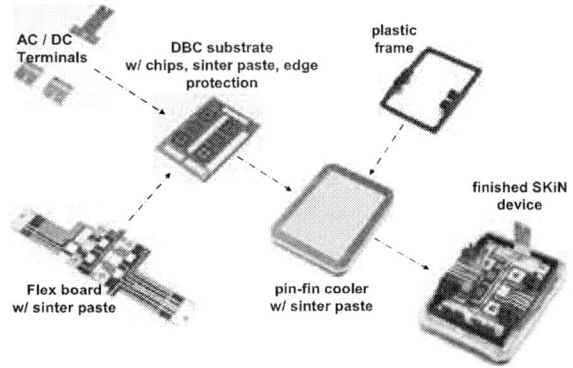

Fig. 6: Exploded view of a 400A, 600V dual IGBT device. The device comprises a heat sink, a DBC substrate with IGBT and FWD chips, a flex board and power terminals all being connected by silver sintering technology.

Fig. 7 shows the finished device, as it is in production right now. In this specific layout, the auxiliary contacts are formed as flex layer tracks, folded such that they can be connected to a printed circuit board (not shown). The auxiliary contacts of the two switches come out to the left and the right of the wide side of the heat sink, whereas the plus and minus and phase terminals come out on the narrow sides of the heat sink, opposing each other.

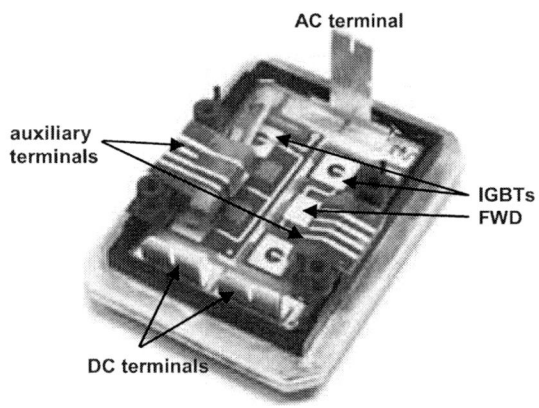

Fig. 7: Photo of a 400A, 600V dual IGBT SKiN device: 74 mm x 56 mm x 11 mm (L x W x H). The device weighs 0,095 kg.

The manufacturing process above was described as a sequence of three sinter steps: First, sintering the chips to the substrate, then sintering the flex layer to the chips, and finally sintering the power device to the cooler. However, this sequence is not mandatory. Our tests have shown that all sinter layers can be formed in one single sintering process, as well. It is a matter of process and yield optimization, as well as process modularity, which way the process should be set-up.

The 400 A, 600V dual IGBT SKiN device, including the heat sink, weighs 95 g and has an overall dimension of 74 x 56 x 11 mm³ (L x W x H). Therefore, the weight is only 40 % of a conventional power module. The size is

difficult to compare, as the SKiN device contains a part of the heat sink, already.

III. RESULTS

A. Electrical Test Results

The electrical properties of the SKiN device demonstrate the superiority of a continuous metal layer on top of the chips, compared to discrete bond wires. A detailed comparison of electrical test results of the SKiN device with flex board, as shown here, versus a SKiN device with the very same design, but with bond wires instead of flex board is given elsewhere /5/. Fig. 8 shows a typical turn-off waveform of the SKiN device. Here, the bottom switch is turned-off in unclamped inverter operation from 500 A at room temperature. The DC voltage is 400 V.

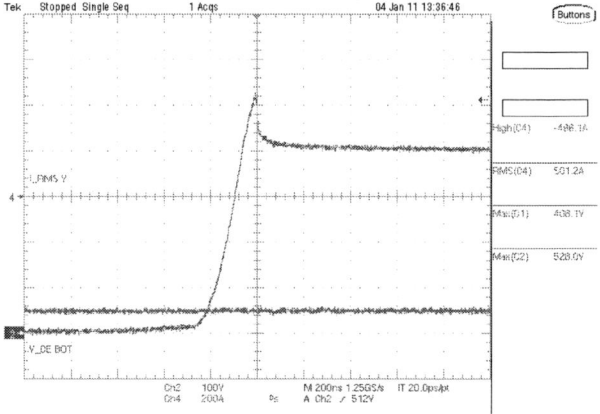

Fig. 8   Turn-off waveform of the BOT IGBT at 400V DC, 500 A. The peak voltage is 528 V.

The flex board leads to a significantly higher surge current capability of the free wheeling diodes than for wire bonded diodes.

The short circuit capability of IGBTs is extended also, as the flex board provides some heat sinking capability in the msec range, due to the thermal capacity of the power side. Insulation requirements of 1700V devices are fulfilled: 4000VAC, 50 Hz, 1 min.

The finished device was tested in a three phase inverter system. The system ran continuously at 300 Arms at a DC voltage of 450V. The water inlet temperature was 70 °C, with a flow rate of 5 l/min and pressure drop of 50 mbar. Fig. 9 shows a photograph of the inverter before test.

Fig. 9: Photograph of a 400 A, 600V inverter, based on SKiN devices.

B. Thermal Tests Results

The device has a thermal resistance of junction to ambient of just 0.44 Kcm²/W which is 35% lower than for conventional power modules. Therefore, the current rating can be increased, accordingly. Fig. 10 shows the device layout and Fig. 11 shows infrared images of the SKiN device under load current, for the IGBT and the FWD, respectively. As it is shown, the chip temperature is very uniform across the chip and from chip to chip. The spacing between the chips is designed such that there is almost no thermal overlap between the chips.

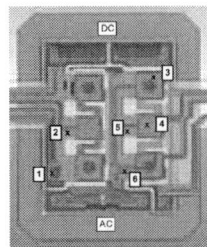

Fig. 10: Schematic view of the SKiN device. The IGBTs can be identified by having a gate contact in the middle of the chip.

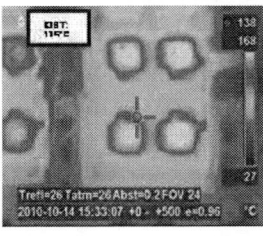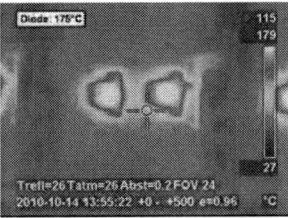

Fig. 11: Infrared image of the SKiN device. Left: the IGBTs are heated by a DC current of 400A. Right: the diodes are heated by the same load current. The spacing between the chips is optimized to provide no thermal cross talk between the chips.

C. Reliability Test Results

Fig. 12 shows power cycling results. In power cycling, the device is heated by the power dissipation during the conduction period and then cooled down when the maximum junction temperature is reached. The maximum number of cycles can vary orders of magnitude, depending on the temperature difference between the highest and lowest temperature, the medium temperature, the raise and fall times of the temperature, as well as the test being carried out under constant maximum power, constant maximum temperature, or constant maximum current. Therefore power cycling results are not comparable amongst different technologies and vendors, if the tests are not carried out in the very same way. However, if test are performed under comparable conditions, one can see that the SKiN device holds up to 500 kcycles (when cycling the junction from 40 °C to 150 °C within a 10 sec. cycle), whereas a conventional power module would exhibit a significant fatigue at 20 to 40 kcycles, already.

Passive temperature cycling is challenging for such an integrated SKiN device, as the thermal coefficients of heat sink, substrate, chips, and flex board are all different. Therefore, extensive temperature cycling tests were carried out, typically cycling the devices between -50 °C and +150°C, with a temperature rise and fall time of 3K/min.

Several hundred cycles up to a thousand cycles were achieved, before the DBC substrate started to delaminate from the heat sink. In various design of experiments it could be shown that thickness and morphology of the sinter layer between substrate and heat sink are key parameters to achieve a good passive thermal cycling performance.

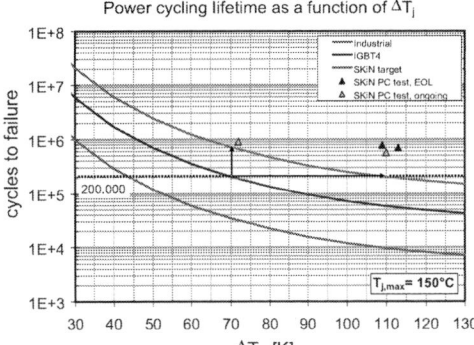

Fig. 12: Power cycling capability of SKiN as a function of ΔTj, compared to standard and improved conventional devices. 500 kcycles can be reached at $\Delta Tj = 110K$.

IV. SUMMARY AND OUTLOOK

In this paper, a new packaging technology, call SKiN technology was demonstrated for the first time, which eliminates the need for solder layers and wire bonds. The SKiN device contains power chips which are sintered on both sides. The chip top side is sintered to a special flexible circuit board. The bottom chip side is sintered to an insulating substrate, such as a conventional DBC substrate. The entire structure is sintered to an aluminium pin fin heat sink. Therefore the device exhibits a very low volume and weight, excellent thermal resistance and very good power cycling capability. The careful material selection makes the device usable for ambient temperatures beyond 150 °C.

As the silver sinter technology develops rapidly towards simpler processes, it will soon make SKiN device processing less demanding. The flex board which replaces the wire bonds has the capability to integrate passive and active components for gate drive, current and temperature sensing, and filtering purposes on the logic side, thus proceeding towards true 3D integration.

REFERENCES

[1] F. Hille, F. Umbach, T. Raker:
"Failure mechanism and improvement potential of IGBT´s short circuit operation"
Proc. ISPSD 2010, pp. 33, Hiroshima, 2010

[2] J.W. Kolar:
"Performance, Trends and Limitations of Power Electronic Systems",
Proc. CIPS 2010, pp.17, Nuremberg ,2010

[3] C. Göbl:
"Low Temperature Sinter Technology Die Attachment for Power Electronic Applications"
Proc. CIPS 2010, pp. 327, Nuremberg, 2010

[4] German Patent:
DE 10 2007 006 706

[5] P. Beckedahl, M. Hermann, M. Kind, M. Knebel, J. Nascimento, A. Wintrich:
"Performance comparison of traditional packaging technologies to a novel bond wire less all sintered power module"
Proc. PCIM 2011, Nuremberg, 2011

Proceedings of the 23rd International Symposium on Power Semiconductor Devices & IC's
May 23-26, 2011 San Diego, CA

A novel Power System in Package with 3D chip on chip interconnections of the power transistor and its gate driver

*Simonot Timothé, *Rouger Nicolas, *Crebier Jean-Christophe, *Gaude Victor, **Irène Pheng

*Grenoble Electrical Engineering Lab
UMR 5269 CNRS / Grenoble University
St Martin d'Hères, France

**CIME Nanotech
3 parvis Louis Neel
38000 Grenoble

Abstract— Currently, most industrial power modules, even IPEMs (Intelligent Power Electronics Modules), are interconnected in a planar way, and interconnections are made with bonding wires. This paper presents a three dimensional interconnection solution based on the idea to flip chip the gate driver directly on the surface of the power device, simplifying and optimizing the packaging and the interconnections among the two devices and improving the overall performances. Various approaches and interconnection solutions will be presented in this paper, and the advantages of the chosen approach will be discussed. Then the technological process for the realization of the interconnections will be explained, and practical realizations will be shown.

I. INTRODUCTION

To satisfy the need of higher performances, simple implementation and higher power densities in power electronics modules, innovative packaging assembly and interconnect solutions have been developed these past years. For example, in order to reduce the parasitic impedance between the power devices and their gate drivers, IPEMs (Intelligent Power Electronics Modules) have been developed [1]. These modules include the power dies and their associated drivers in the same package, reducing parasitic inductances due to interconnections that can disturb the control signal of power transistors. Nevertheless, the interconnection solution of most industrial IPEMs are still based on wire bonding technique, mostly because this interconnection technique is well known, cheap and reliable. Three dimensional (3D) interconnection solutions have been developed for power modules, mostly by the CPES (Center for Power Electronic Systems) at Virginia Tech [2, 3]. These 3D interconnection solutions suppress the wire bondings by realizing interconnections with solder bumps or electroplated metal pillars. The assembly of these modules is made in flip chip through a double sided flex or PCB layer. Even in this case, this layer can generate electrical (EMI) or thermo-mechanical issues, due to the stacking of different materials with different CTEs (Coefficient of Thermal Expansion).

The approach that will be described in this paper is to customize the design of the power device in order to flip chip its associated gate driver directly on the surface metallization of the power chip, in order to minimize parasitic impedances.

The objective is then to design the layout and the functions of the two dies in order to develop their functional interactions and to allow their assembly based on a flip chip interconnection and their fully autonomous implementation. Figure 1 presents the resulting P-SOC (Power System On Chip) where the power die surface metallization layout is designed to receive the flip chip of the gate driver die but also the storage capacitor of the gate driver supply and an additional couple of terminals for the control signal to be transmitted from the remote control to the isolated gate driver. The power and driver dies are not only electrically connected, but also from the mechanical and thermal point of views, making very attractive the global system.

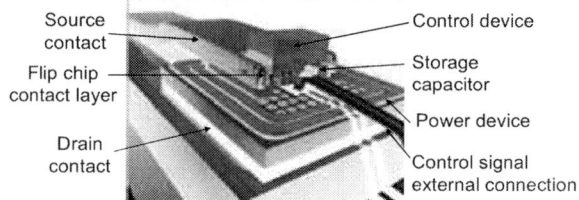

Figure 1. Structure of the imagined autonomous power switch

II. INTERCONNECTION SOLUTIONS

As stated before, several solutions can be employed to interconnect the power chip with its driver. The most common solution is a planar assembly with wire bondings, this solution being cheap and reliable since it is in use in the industry for many years. However this technique has known limitations, especially from the electrical point of view due to the strong parasitic impedance of the bonding wires generating inductive coupling and EMI issues [4]. In order to limit the length of wire bond, a first solution can be to stack both dies one on the other and to interconnect them with wire bonds. The reduction of wire bond length is a first improvement from the electromagnetic point of view. To overcome the limitation introduced by the use of wire bonds, solder bumps or electroplated pillars can be used, as they have a much lower parasitic inductance than wire bondings [5]. In figure 2 it can be seen that the flip chip of the driver die on top of the power die can be carried out in two ways: the first way that has been previously developed is to interpose a double sided flex or PCB layer between the two dies to realize interconnections.

978-1-4244-8425-6/11 $26.00 © 2011 IEEE

This is employed because the surface layout of the two chips is not designed for the direct flip chip of one chip on top of the other. The joint design of both dies, and especially the surface metallization layout customization of the power die, allows this direct assembly, suppressing the PCB layer and thus minimizing the parasitic impedance and thermo-mechanical issues. This will be shown later in this paper. Depending on technological accesses, the driver chip could integrate TSV to optimize the interconnection [6]. The drawback of the direct flip chip of the two dies one on top of the other is the apparition of dielectric constraints when control signal insulation is needed. It can be seen on figure 2 that two dielectric constraints are located in the module, one on the power die with the pads for the connection of the remote control signal with wire bondings, and one in the driver die where an insulation system is needed. These dielectric constraints must be studied in order to obtain a sufficient insulation voltage.

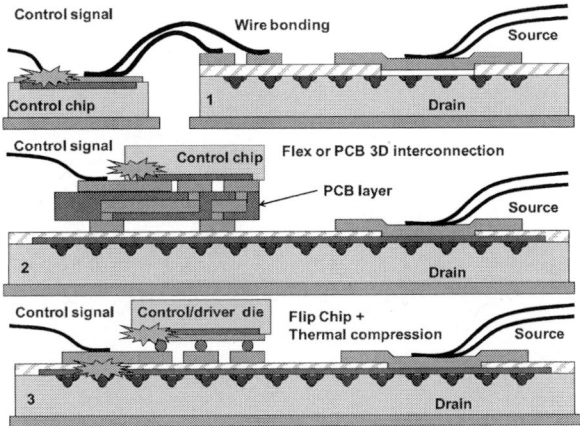

Figure 2. Possible interconnection solutions between the power die and its driver.

The advantage of former approach is that the driver and power parts can be associated one on top of the other with tight electrical connections. It is then possible for the designer to take advantage of both dies in order to develop the functionalities and performances of the global power switch. Especially, it becomes possible to mix functions from both dies as it is done today in microelectronics in multichip modules and chip on chip assemblies [7]. Figure 3 shows the layout view of the power chip, which includes a power MOSFET and two auxiliary MOSFETs for the supply system of the driver chip [8, 9]. The surface metallization of the power chip includes the footprints for electroplating the copper pillars for the flip chip of the driver die, and also the footprints for the soldering of the storage capacitor also needed for the self supply system.

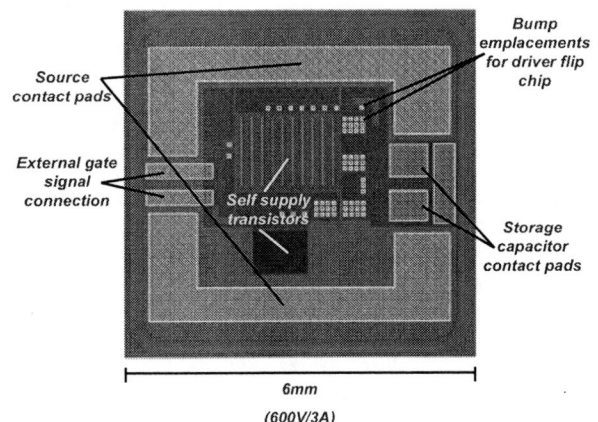

Figure 3. Layout view of the power chip with surface metallization for the flip chip of the driver die

III. ELECTRICAL AND THERMAL ASPECTS

The interconnection solution proposed offers several advantages, from both the electrical and the thermo-mechanical point of views. It also introduce thermal operational constraints. These points are addressed in this section.

A. Electrical aspects :

The reduction of parasitic impedances at the interconnection between the power device and its gate driver not only improves the efficiency and the performances of the switching device implementation [4], but also makes possible high frequency functional interactions between the driver and the power device and gives the opportunity to drive the power device near to its limits. Especially, electromagnetic couplings are reduced, allowing either to reduce the specifications on the driver or to operate the switch with a unipolar gate driving signal. This is currently very significant as the trend in power converters is to increase the switching speed of the power devices.

B. Thermal aspects

From a thermal point of view, it is important to notice that in this case, the heat dissipated by the driver chip has to be evacuated through the power device. The thermal model employed for this simulation is shown in figure 4 allows to carry out a simple investigation related to this issue.

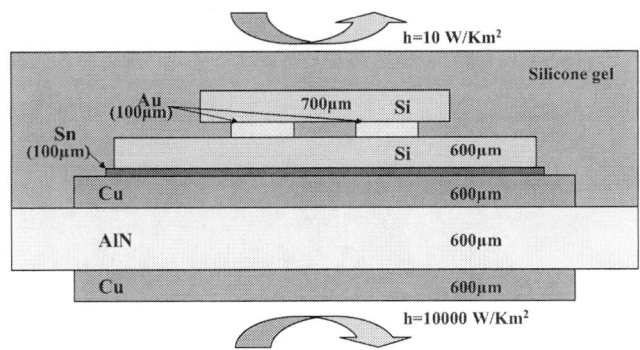

Figure 4. Structure of the thermal model used for the simulation

TABLE I.   VALUES OF THE THERMAL PARAMETERS OF THE PACKAGE

	Driver	Bump	Transi stor	Sn	Cu	AlN
Rth [K/W. m]	0.78	0.79	0.09	0.03	7e-3	0.02
Cth [J/K]	7e-3	3e-3	0.03	7e-3	0.1	0.2

The most important thermal parameters of the different materials in the package used for the thermal 1D simulation are listed in the table I below. The power to be dissipated is estimated to 70mW for the driver (including the dissipation of all functionalities) and about 10W for the power transistor, assuming that the commutation losses are equal to conduction losses and for a load current of 5A, a voltage drop across the transistor of 2V during the ON state and a duty cycle of 50%. The thermal resistance of the radiator is assumed to be 2 K.W-1.m-1 for a 25°C ambient temperature. Figure 5 shows the results of the simulation which has been carried out with the software Flotherm. It can be seen that the temperature difference between the driver and power chips is in the order of 0.8°C (the temperature of the driver chip is 52.3°C whereas the hottest point in the power chip is 51.5°C). This means that the power dissipated by the driver chip is well evacuated through the power die, and that the thermal coupling between the two dies is satisfactory. Since driver parts are usually low voltage devices compared to the power devices and since the temperature limitation are related to voltage ratings of semi-conductors, it appears that in any case, the driver part should be well cooled below its limits if the power device is also correctly cooled down. Another interesting issue is related to the small temperature difference between the two dies. This could be advantageously used to monitor the temperature of the power device directly from gate driver at no expense and with a correct estimation.

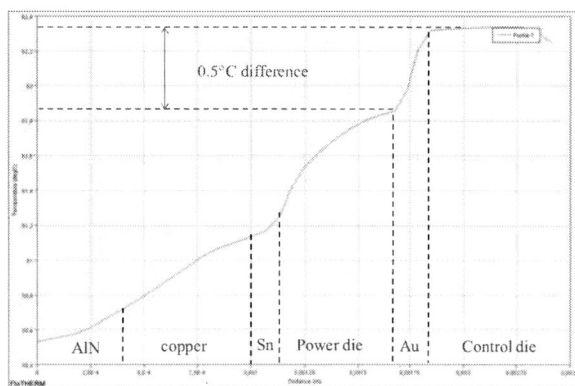

Figure 5.   Static thermal temperature profile simulation with Flotherm of the 3D module

C.   Thermo-mechanical aspects

Since the power and the control/driver dies are in silicon, they have identical thermal expansion coefficients. This is a good point for the reliability of the assembly because if their temperatures remain all the time close to each other, this means that reduced forces will be applied across the solder joints. If between the solder joints and the silicon, thermo-mechanical stresses are still possible, the thickness of the

solder joints can be optimized to improve the reliability of the assembly. Further more, since both silicon dies will have the same temperature and the same CTE, shear forces between solder joints will be extremely reduced. We have seen that in steady state operation, the temperature differences are below one degree which is very good. A first order simulation based on the 1D equivalent model presented above has been carried out in order to investigate this issue under dynamic conditions. The time domain temperature evolution of the two dies is given figure 17 below.

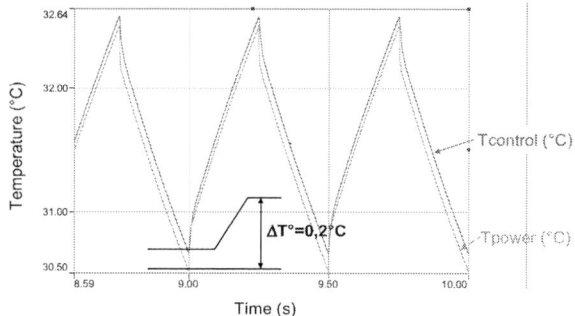

Figure 6.   1D dynamic thermal simulation of the package

It appears that under dynamic conditions, the temperature differences are remaining a similar range and again below 1°C. This analysis gives us the opportunity to affirm that the assembly of the control/driver chip on top of the power chip should offers very nice robustness with respect to thermo-mechanical stresses.

IV.   TECHNOLOGICAL REALIZATIONS

A.   Technological interconnection solutions

The realization of the interconnections can be made by two different ways: stud bumps or electroplated copper pillars. The first solution with gold stud bumps is a simple solution for prototyping as no additional steps are required in the power process, and this requires only a bumping machine. However this solution can be time consuming when a lot of bumps are needed. The second solution is electroplating of copper pillars to realize the interconnections. This solution can be seen at first more costly as it requires additional steps in the process (seed layer deposition, photo resist deposition and electroplating), but it can be applied at wafer level. Thus, this solution can be time saving if the process is in production at a higher level. Figure 7 shows the realized interconnections, either gold stud bumps deposited on the pads of the CMOS gate driver industrially fabricated [8], or the electroplating of copper pillars on the surface of the power chip.

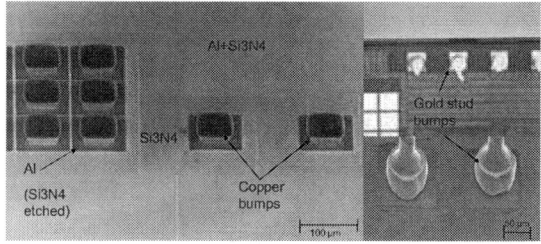

Figure 7.   Back end metallization of the power transistors made of thick electroplated copper layer forming copper pillars for the flip chip of the control die and gold stud bumps on the control die.

The electroplating process is a back end process, after the aluminum metallization of the power chip. A layer of silicon nitride (Si3N4) is deposited, and then etched to open the contacts of the back end interconnections. The electroplating process being an electrolytic reaction, a conductive layer has to be deposited on the whole wafer. Moreover, a layer has to be deposited to avoid the migration of gold or copper in aluminum. For this, successive layers of Ti/TiN/Cu are deposited to form the seed layer. Next a thick photo resist (>10μm) is deposited and developed, to form molds in which the copper pillars will be deposited. Then the electroplating is done and the photo resist is removed, forming the copper pillars. The objective of these realizations is to achieve the assembly of the two dies one flip-chipped on the other with the different proposed solutions, then to characterize the interconnections from the electrical, thermal and thermo-mechanical point of views. This is currently under realization and the proposed set of tests will be presented in the next section.

B. Interconnections characterization

To characterize the interconnections, a set of test patterns have been designed, in which the bottom chip having a 6x6mm^2 dimension, is interconnected in flip chip with a smaller chip having a 3x3mm^2 dimension. In these chips, daisy chained contact patterns are drawn, with various dimensions and spacing. This way, contact resistance of the interconnections can be extracted through electrical measurements. The proposed test experimental set up is shown in figure 8.

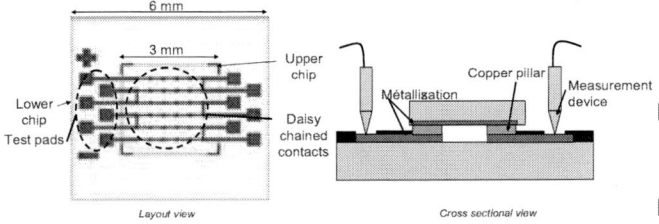

Figure 8. Interconnection contact resistance pattern and measurement experimental set up

These test chips have been realized, and the back end interconnections have been made with both gold stud bumps and electroplated copper pillars. The flip chip process is currently under development. As a matter of fact, a placement/soldering machine is needed to achieve the positioning of the of gold stud bumps on top the the pads, as pressure, heat and ultrasonic energy are needed to form an interconnection with gold. Another solution for stud bumps is to cover them with conductive adhesive, then only pressure has to be applied in order to form an interconnection. For copper pillars, a small tin coating can be deposited on top of the pillar, then the chip stacking has to be heated enough for tin to melt and form an interconnection. These solutions will be investigated in the close future. The daisy chain pattern allows also the interconnection to be tested from a thermo-mechanical point of view, which will also be investigated. The proposed test procedure is thermal cycling, then electrical test or SEM observation of a cross section of the chip stacking to see if all bumps are still correctly connected.

V. CONCLUSION

This paper has presented a three dimensional packaging technique in order to interconnect a power transistor and its gate driver. The proposed solution is an improvement of the state of the art, usually based on double sided flex or PCB layer between the two dies, reducing the parasitic impedances due to interconnections. This reduction of parasitic impedance offers an improvement of the device performance, and also a closer and more precise control of the device, allowing the development of precise control functionalities. Moreover, the proposed package suppresses the stacking of materials with different coefficients of thermal expansion, which is a good point for the thermal and thermo mechanical behaviors and thus the reliability of the package. Perspectives are now to validate the expected advantages introduced by this approach from the electrical side. Then, an important effort will have to be carried out focusing on the reliability of the package especially at the soldering level between the two dies. This will be investigated thanks to a set of electrical tests and thermal cycling on dedicated test chips.

ACKNOWLEDGMENT

The authors also acknowledge the French Research National Agency (ANR) for its founding (projects MOBIDIC ANR-06-BLAN-0204-03 and ECLIPSE ANR-09-BLAN-0036-01) and the french RTB Centers, PTA, LAAS, FEMTO-ST and CIME-Nanotech for their technological supports. This work has been performed with the help of the "Plateforme technologique amont" de Grenoble,

REFERENCES

[1] Stockmeier, T., "From Packaging to "Un"-Packaging - Trends in Power Semiconductor Modules," Power Semiconductor Devices and IC's, 2008. ISPSD '08. 20th International Symposium on , pp.12-19, 18-22 May 2008

[2] K. Siddabattula, Z. Chen, D. Boroyevich, "Evaluation of metal post interconnected parallel plate structure for power electronics building blocks," in proc. IEEE APEC 2000, vol. 1, pages 271-276, 6.-10. February, 2000.

[3] J. Z. Liang, J-D. V. Wyk, F-C. Lee, « Integrated packaging of a 1kW switching module using a novel planar integration technology », IEEE Transactions on Power Electronics, January 2004

[4] Cottet, D., Hamidi, A., "Numerical Comparison of Packaging Technologies for Power Electronics Modules," PESC '05. IEEE 36th , pp.2187-2193, June 2005

[5] Xingsheng Liu, Guo-Quan Lu, "Power Chip Interconnection: From Wirebonding to Area Bonding" , The International Journal of Microcircuits and Electronic Packaging, Volume 23, Number 4, Fourth Quarter, 2000

[6] www.dalsa.com

[7] M.G. Sage, D.R. Gross, "Trends in MCM and Microelectronics assembly", MCB Up Ltd.

[8] Simonot Timothé, Rouger Nicolas, Crebier Jean-Christophe, "Design and characterization of an integrated CMOS gate driver for vertical power MOSFETs," Energy Conversion Congress and Exposition (ECCE), 2010 IEEE, pp.2206-2213, 12-16 Sept. 2010

[9] Simonot Timothé, Rouger Nicolas, Crebier Jean-Christophe, "Design and characterization of a signal insulation coreless transformer integrated in a CMOS gate driver chip", Power Semiconductor Devices and IC's, 2008. ISPSD '08. 20th International Symposium on, in press.

978-1-4244-8425-6/11 $26.00 © 2011 IEEE

Proceedings of the 23rd International Symposium on Power Semiconductor Devices & IC's
May 23-26, 2011 San Diego, CA

Innovative heat removal structure for power devices - the drift region integrated microchannel cooler

Kremena Vladimirova*, Jean-Christophe Crebier*, Yvan Avenas*, Christian Schaeffer*, Stepahne Litaudon**

* Grenoble Electrical Engineering Lab (G2Elab)	**3 parvis Louis Néel
961 Houille Blanche	BP 257
38402 St Martin d'Hères, France	38016 Grenoble, France

e-mails: vladimirova@g2elab.grenoble-inp.fr, crebier@g2elab.grenoble-inp.fr, avenas@g2elab.grenoble-inp.fr

Abstract— **Liquid microchannel cooling is approved to be a compact and high-performance solution to deal with the thermal requirements of power devices and modules. This is due to the large heat exchange surface, the high heat transfer coefficient and the minimized thermal interfaces that microchannel coolers offer. This paper reports an original concept for efficient thermal management of power devices based on the integration of a microchannel cooler, including numerous parallel through wafer fluid vias, directly into the drift region of the power device. Simulation results proved that no negative side effects are resulting from this integration regarding the electrical performance of the device. The effectiveness of the proposed cooling technique was evaluated with hydraulic and thermal numerical models. Vertical power diodes with integrated microchannel cooler were fabricated and characterized to demonstrate the feasibility and the effectiveness of the concept.**

I. INTRODUCTION

Numerous research studies have been realized in order to develop and to evaluate the interest of liquid microchannel cooling for planar integrated circuits [1, 2] and for 3D chip stacks configurations [3, 4]. Although it has been demonstrated that reducing the hydraulic diameter of channels allows to greatly increase the heat transfer coefficient and the ratio surface/volume of the heat exchanger, the resulting significant increase of the pressure drop values must also be taken into consideration. In addition, in standard power modules, the cooling fluid is flowing in lateral microchannels in the parallel plan of the module [5] which causes temperature gradients along the cooled area [6] due to the warming up of the fluid.

This paper presents an innovative concept for liquid microchannel cooling of power devices and modules providing a high performance heat removal and a uniform temperature distribution whereas drastically reducing the pressure drop values. This new concept, called the Dift Region Integrated Microchannel cooler (DRIM cooler), is based on the integration of multiple microchannel fluid vias directly into the drift region of the power device. Vertical power diodes integrating the DRIM cooler were fabricated and

experimentally tested. Electrical tests were carried out to measure their direct and reverse biased characteristics. At the end of the paper we present the experimental setup allowing the measurement of the thermal and hydraulic performances of the devices with DRIM cooler.

II. DESIGN AND REALIZATION

A. Concept

The concept of the DRIM cooler is to etch directly into the active region of the power device numerous parallel microchannels which will serve as fluid vias enabling the circulation of a dielectric cooling fluid. Fig. 1 shows a schematic view of the principle using a vertical power diode as example.

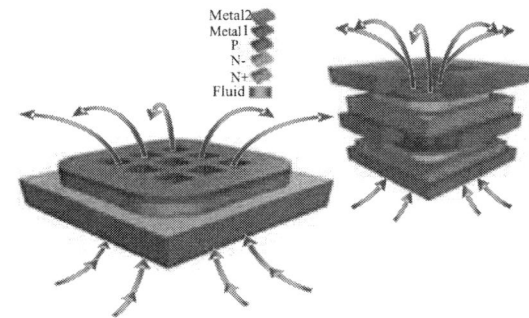

Figure 1. Schematic view of the cooling concept with the microchannel cooler integrated in the drift region of a vertical power PIN diode and application of the cooling concept for 3D chip stack configuration.

As it can be seen in fig. 1 the voltage handling capability of the device is assured by a peripheral deep trench termination [7]. The multiple parallel microchannels (MC) with small length, take advantage of this edge termination technique to offer an effective management of the electric field at the interface between the drift region of the power device and the dielectric cooling fluid. MC length is equal to the thickness of the substrate, providing very low pressure drop values. Besides, the numerous MC with small hydraulic diameter

---

This work is supported in part by the French Research National Agency (ANR) with projects MOBIDIC ANR-06-BLAN-0204-03 and ECLIPSE ANR-09-BLAN-0036-01.

978-1-4244-8425-6/11 $26.00 © 2011 IEEE

(50μm - 400μm) offer a high heat transfer coefficient (more than 10000W/m²/K) and a greatly expanded exchange surface (e.g. for a device of 1cm2 with 500μm thickness of the substrate where 10% of the total surface is occupied by square microchannels of 100x100μm² the exchange surface is equal to 2cm²). Another important profit of this cooling concept is the resulting homogeneous temperature of the semiconductor chip due to the homogenous spread of parallel MC and the high thermal conductivity of silicon.

A key benefit of the concept is the possibility to distinguish thermal and electrical "exchange" surfaces, thus enabling advanced design packaging for power modules. Fig. 1 shows that the wire bonds can be replaced by massive copper connections surrounding the MC region. It allows the series implementation of multiple power devices in a 3D Power Chip on Chip configuration [8]. In this particular case the electrical contact can be realized only at the periphery of the device where a special area is reserved not occupied by the channels and where the electrical connection can be realized. The homogenous repartition of the current must then be assured by a thick metal layer at the electrodes of the devices.

B. Theoretical results

A detailed analysis with 2D finite element simulations (Silvaco) was realized in order to examine the possible effects considering the electrical performances of a vertical power diode with DRIM cooler. The obtained results were compared with the optimum case of the infinite plane junction for a device having the same physical parameters. In fig. 2 is depicted the electric field distribution contours and the equipotential lines at a reverse voltage of 800V for a vertical PIN diode with 50x50μm² fluid via passing through the drift region of the diode and filled with a dielectric material.

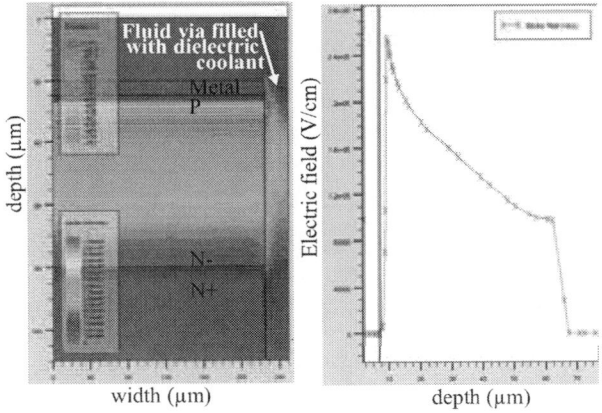

Figure 2. Electric field distribution contours and equipotential lines for 800V reverse biased PIN diode with 50x50μm² width microchannel fluid via filled with a dielectric material and passing through the drift region of the device; Electric field cross sectional view at the interface Si/dielectric fluid. The simulated structure has the following characteristics – 90μm thickness of the Si substrate where 25-28μm is the depth of the P region with 3×10¹⁷/cm3 concentration, 28-80μm is the drift N region with 2×10¹⁴/cm3 concentration and 80-110μm is the N+ region with 1×10¹⁹/cm3 concentration.

The results show that the integration of the microchannel fluid vias into the drift region does not affect the voltage

handling capability of the device. Fig. 2 shows a conformal and close to ideal electric field distribution. Nevertheless, voltage and current ratings are not affected by the integration of the MC. Especially, the reduction of the active area cross section is greatly compensated by the enhanced heat removal.

Fig. 3 shows one of the key benefits of the proposed cooling technique (square vertical microchannels) compared with another cooling approach (square lateral microchannels) which relies on the circulation of the cooling fluid in the parallel plan of the module or the PN junction through the realized on the backside of the semi-conductor device or the module microchannels.

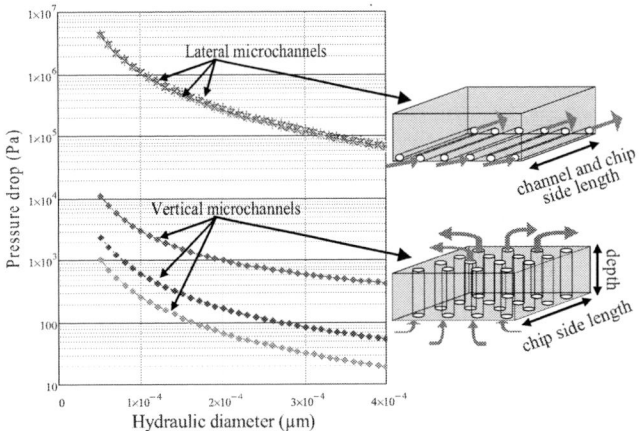

Figure 3. Pressure drop values in function of the hydraulic diameter for different chip side lengths - 5mm red curve, 10mm blue curve, 15mm green curve

The pressure drop values represented in fig. 3 were calculated as function of the hydraulic diameter of the channels for a constant fluid flow rate (0.12L/min determined so that the temperature difference of the cooling fluid, water in this case, from the inlet to the outlet of the channels is 50°C and the power losses are 400W) and fixed exchange surface (10% of the total surface of the chip is occupied by MC) when varying the chip side length (5mm, 10mm, 15mm for 500μm thickness). With these parameters the flow regime is determined to be laminar. In the case of vertical MC the variation of the chip side length causes a variation of the number of channels in parallel but their length remains constant and equal to the chip thickness (500μm in our case) and thus the pressure drop values decrease. In the case of lateral MC the variation of the chip side length causes a variation of the number of channels in parallel but also determines the length of the channels. Thus the pressure drop values remain identical. As it can be seen in fig. 3 for all hydraulic diameters the pressure drop values in the case of vertical MC are significantly lower compared with those for lateral microchannels.

Fig. 4 shows the numerical evaluation of the thermal performances of the DRIM cooler. The simulation represents the temperature distribution throughout the wall of a square MC of 50μm hydraulic diameter with 500μm length with dielectric coolant flowing through it with total flow rate of 0.27L/min (the temperature difference of the cooling fluid from the inlet to the outlet of the channel is 50°C and the total power heat is 400W). The heat transfer coefficient is in that case $1.2*10^4 W/m^2/K$ and the thermal resistance 0.08K/W.

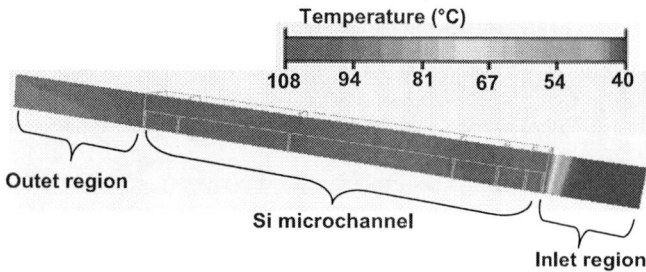

Figure 4.   Temperature distribution at the wall of a silicon MC with inlet and outlet region of the dielectric coolant.

As it can be seen on fig. 4 the temperature throughout the silicon microchannel is rather homogenous.

C. Realization

In order to evaluate the performance of this new cooling concept vertical power diodes with DRIM cooler were fabricated. In our case 500μm N+ type (100) substrate plus 55μm lightly doped (20 Ω cm) epitaxial layer was used for the realization of the prototypes. After the deposition of the 3μm thick aluminum layer on the both sides of the silicon substrate a double side lithography step was realized to distinguish the different patterns. Top and bottom patterns are different because of the surface peripheral deep trench termination needed to ensure the voltage handling capability of the devices. The through wafer fluid vias were realized by Deep Reactive Ion Etching (DRIE) performed on both sides of the silicon substrate. We used the aluminum metal layer as masking material. No post processes for surface or wall treatment were performed after the silicon etching step. Fig. 5 shows pictures of the realized prototypes.

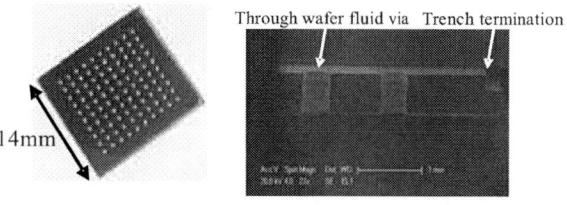

Figure 5.   Photograph and cross sectional SEM image of the realized power diode with DRIM (200μm diameter of the fluid vias) cooler.

III.   PRACTICAL RESULTS AND DISCUSSION.

A. Electrical characterization of the device with DRIM cooler

For the experimental validation of the electrical performances, the fabricated power diodes with DRIM cooler were packaged and passivated with silicone oil. Fig. 6 shows the realized package.

The periphery of the cathode not occupied by the microchannels was soldered on a pcb track especially designed in order to optimize the power interconnections and to allow the use of Kelvin probe measurement technique. The anode was wire bonded with multiple bond connections spread all over the surface periphery of the device in order to improve the homogenous distribution of the current.

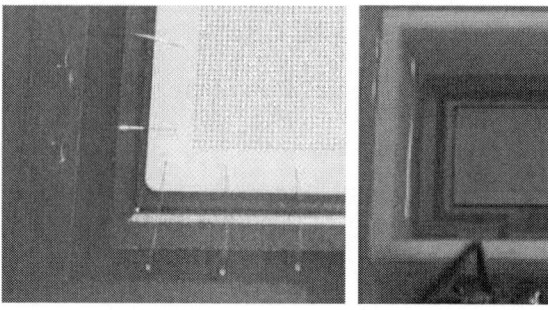

Figure 6.   Photograph of the realized package.

The packaged prototypes were experimentally tested with a static characteristic power curve tracer HP 371A. Fig. 7 shows the reverse biased static characteristic of the diode with DRIM cooler (prototype fig. 6). The device achieved a breakdown voltage of 300V with a leakage current of 1μA.

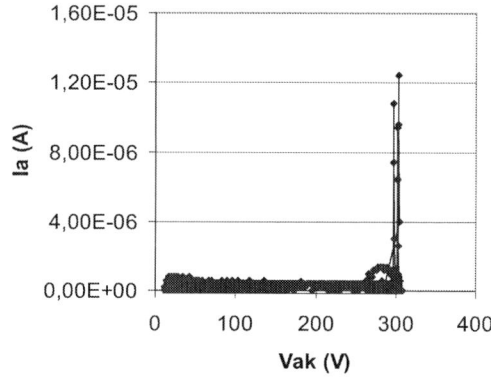

Figure 7.   Reverse biased characteristic of the diode with DRIM cooler.

This result is only partial because the device was initially designed for a breakdown voltage of 600V and it is supposed to handle such voltage breakdown levels. In our case the passivation of the device and the trench terminations was realized in a non protected atmosphere which may justify this result. Additional experimental measurements should be carried out in order to identify the reason of the premature breakdown of the device. Fig. 8 represents the forward biased characteristics of the same diode. Although the Kelvin probe measurement this result is below our expectations. Indeed the forward voltage drop was expected to be 1.5V at I_{AK}=20A instead of 3V obtained in practice. This large difference can not be associated to the reduction of active surface since it only represents 10% of the active surface. Again additional work will have to be carried out to investigate this issue. Nevertheless both static results are greatly encouraging and allow to continue the thermal and fluidic characterizations.

978-1-4244-8425-6/11 $26.00 © 2011 IEEE   334

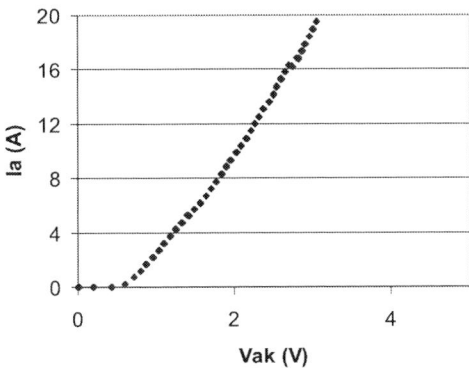

Figure 8. Forward biased characteristic of the diode with DRIM cooler.

B. Thermal characterization of the device with DRIM cooler

Fig. 9 shows a picture of the experimental setup designed for the hydraulic and thermal characterization of the DRIM cooler concept which is based on the measurement of the junction temperature of the device with thermosensitive electrical parameters and the measurement of the pressure drop values needed for the identification of the pumping power necessary to cool the device. A vacuum pump is used for evacuating the air from the hydraulic circuit before its filling and for degassing the fluid. The goal of this procedure is to reduce the vapor or the air bubbles in the circuit and especially in the DRIM cooler. The diode with DRIM cooler (device under test) is fixed between two electrodes in aluminum. The electrical contact between the periphery of the device and the electrodes is made by applying a pressure force. In both electrodes, a circular duct is machined for the fluid circulation. Two holes allow the measurement of the pressure drops by a differential pressure sensor. The flow rate is measured with an ultrasonic flowmeter.

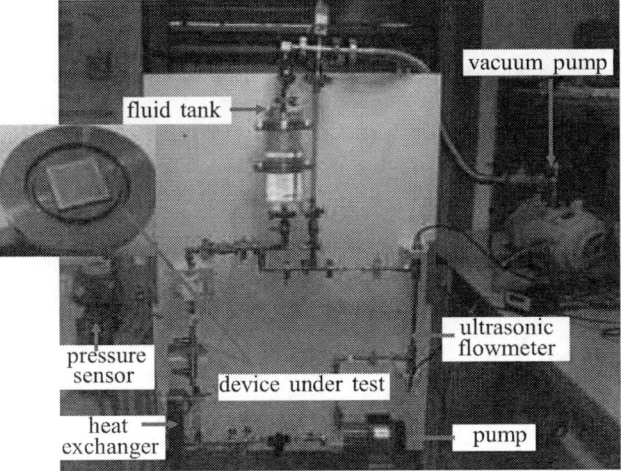

Figure 9. Experimental setup for the validation of the DRIM cooler concept.

The first step of the DRIM cooler characterization is the measurement of the pressure drops. Because no current circulates in the device, water was used in order to compare first experimental and simulation results. Fig. 10 shows a good conformity of theoretical and practical results.

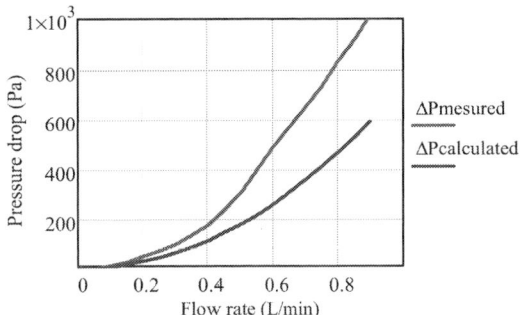

Figure 10. Measured and calculated pressure drop values versus flow rate for a diode with DRIM ($400 \times 400 \mu m^2$ MC with 15% occupation) cooler.

IV. CONCLUSION

An original concept for efficient direct cooling of power electronic devices was presented. It is based on the integration of numerous microchannel fluid vias directly into the drift region of the power device. The small hydraulic diameter of the fluid vias offers a high heat transfer coefficient and thus excellent thermal performances are expected. The exchange surface is greatly developed with the multiple parallel channels with small length which are also providing low pressure drop values. The practical characterization of the fabricated prototypes proves that the integration of microchannels in the drift region of the device is feasible but must be improved.

ACKNOWLEDGMENT

The authors would like to thank CIME-nanotech and RTB centers PTA and FEMTO-ST for their technical support.

REFERENCES

[1] D.B. Tuckerman, and R.F. Pease, "High-performance heat sinking for VLSI", IEEE Electron Device Letters, vol.2, no.5, pp. 126-129, 1981.

[2] S.A. Solvitz, L.D. Stevanovic, and R. Beaupre, "Microchannels take heatsinks to the next level", Power Electronics Technology Mag., 2006

[3] K. Matsumoto, S. Ibaraki, M. Sato, K. Sakuma, Y. Orii, and F. Yamada, " Investigations of cooling solutions for three-dimensional (3D) chip stacks", in Proc. IEEE SEMITHERM 2010, pp. 25 - 32

[4] B. Dang, M. Bakir, D. Sekar, C. King, and J. Meindl, "Integrated microfluidic cooling and interconnects for 2D and 3D Chips", IEEE Trans. on advanced packaging, vol.33, no.1, February 2010

[5] T. Steiner, and R. Sittig, "IGBT module setup with integrated micro-heat sinks", in Proc. IEEE, ISPSD 2000, pp. 209 – 212

[6] G. Hetsroni, A. Mosyak, and Z. Segal, "Nonuniform temperature distribution in electronic devices cooled by flow in parallel microchannels", IEEE Trans. on Components and Packaging Technologies,.2001, vol. 24, pp. 16 – 23

[7] K. Vladimirova, J.C. Crebier, Y. Avenas, C. Schaeffer and T. Simonot, "Single die multiple 600V power diodes with vertical voltage terminations and isolation", IEEE ECCE 2010, 2010, pp. 2200-2205.

[8] E. Vagnon, P.O. Jeannin, J.-C. Crebier, and Y. Avenas, "A Bus-Bar-Like Power Module Based on Three-Dimensional Power-Chip-on-Chip Hybrid Integration,"IEEE Trans. IAS, vol.46, no.5, 2010.

Proceedings of the 23rd International Symposium on Power Semiconductor Devices & IC's
May 23-26, 2011 San Diego, CA

Interface Charge Trapping and Hot Carrier Reliability in
High Voltage SOI SJ LDMOSFET

M. Antoniou*[1], F. Udrea[1], E. Kho Ching Tee[2], Yang Hao[2], S. Pilkington[2], Kee Kia Yaw[2], D. K. Pal[2], A. Hoelke[2]

[1]Department of Engineering, University of Cambridge, UK
[2]X-FAB Sarawak Sdn. Bhd. Kuching, Sarawak, Malaysia
*Tel: +44 (0)1223 748 311, Fax: +44 (0)1223 748 348, E-mail: ma308@cam.ac.uk

Abstract— **This paper demonstrates and explains the effects of hot carrier injection and interface charge trapping correlated with impact ionization under normal on-state conditions in a highly dense low-resistance Super-Junction LDMOSFET. The study is done through extensive experimental measurements and numerical simulations using advanced trap models. The introduction of the SJ structure in the drift region of the LDMOSFET allows a shorter length and significantly higher drift doping both of which result in very low on-state resistance for a given breakdown voltage 170V. However careful design and optimization of the Super-junction layers is needed to avoid the combined effects of parasitic JFET effect, impact ionization and charge trapping. The paper discusses these complex phenomena and gives solutions to increase robustness against instability problems.**

I. INTRODUCTION

THERE is currently a lot of interest in high voltage transistors that are implemented in sub-micron CMOS processes. The fact that these devices are CMOS process compatible makes them very popular and cost efficient. The Super-Junction technology has enabled for shorter drain lengths and higher doping levels. As a result, the on-state resistance and breakdown voltage ratings of the CMOS based high voltage devices have been improved.

In this paper we are investigating the effects of interface charge trapping and hot carrier injection in ultra dense, high performance SOI SJ LDMOSFET [1] and give effective solutions to address them. This is the first time that the effects of interface traps on the performance and reliability of a high voltage device under pre-stress conditions are demonstrated and explained; Silicon is tetrahedrally bonded with each Si atom bonded to four Si atoms in the wafer bulk. When the Si is oxidized, the bonding configuration at the surface is such that most Si atoms are bonded to oxygen. An interface trapped charge, often called 'interface trap', is due to trivalent Si atoms each with one unsaturated (unpaired) valence electron at the SiO2/Si interface. The defects result from the naturally occurring mismatching – induced stress at the SiO2/Si interface during oxide growth.
Silicon-oxide interfaces therefore exhibit interface traps that may alter the electric field properties of the semiconductor. These traps are electrically active defects with an energy

distribution throughout the Si band gap. The surface potential dependence of the occupancy of interface traps is illustrated in Fig. 1. Interface traps at the SiO2/Si interface are acceptor-like in the upper half and donor-like in the lower half of the band gap. Hence, as shown in Fig. 1, at flatband, with electrons occupying states below the Fermi energy, the states in the lower half of the band gap are neutral (occupied donors designated by ''0''). Those between midgap and the Fermi energy are negatively charged (occupied acceptors designated by ''_''), and those above EF are neutral (unoccupied acceptors). As a result the amount of charge arising from the interface traps is a function of the surface potential. Depending on the level of traps a depletion region is formed below Si/SiO2 interface with its depth depending on the doping of the semiconductor underneath. If a positive bias on the n doped region is applied, as shown in Fig. 1(b), the number of occupied interface traps between mid gap and the Fermi level is enhanced.

It is also well known that lateral devices operating at high electric fields suffer from hot carrier injection. This phenomenon refers to the trapping of charge at the semiconductor –insulator interface and generation of interface states under high electric field conditions. An immediate effect includes the shifting or distortion of the transfer (drain current versus gate voltage) characteristic given the electric field direction and conducting materials doping. The injection rate and the related formation of defects are non-uniform along the oxide of the drift layer of the LD MOSFET (Fig. 2) because they depend on the local values of the carrier energy and vertical component of the electric field. In contrast to the case of uniform degradation induced by irradiation or Fowler-Nordheim injection, the defects generated by hot carriers are concentrated within the pinch-off region in the case of low voltage MOSFETs and the transition between the poly gate and poly field plates for standard LD MOSFETs. In our case due to the presence of the Super-junction, the drain side is also exposed to hot carrier injection.

Under certain operating conditions, the impact ionization rate and substrate current I, are very large. The direction of the

978-1-4244-8425-6/11 $26.00 © 2011 IEEE

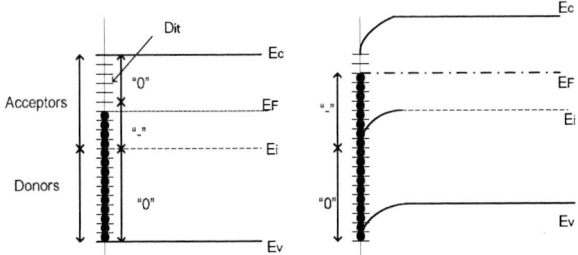

Figure 1: Band diagrams of an Si n- doped layer showing the occupancy of interface traps and the charge polarities with (a) negative interface trap charge at flatband and (b) positive voltage applied.

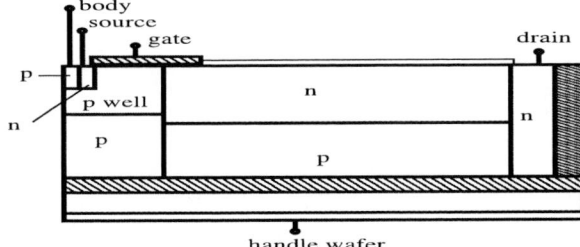

Figure 2: The SJ LDMOS structure.

electric field in the oxide determines the hot carrier injection type and intensity. The affected area has therefore a high concentration of interface states. Therefore given the channel/ accumulation/drift region parameters the injection of holes or electrons will directly affect the properties of the device.

II. THE DEVICE PERFORMANCE

Fig. 2 shows the investigated SJ LDMOS structure. The device consists of a channel region (p well) the super-junction type drift region above a buried oxide layer and a top-surface oxide layer. The electrons current path under on-state conditions is therefore the gate channel/ accumulation layer and the n layer of the drift region. The breakdown rating of this device is 170V.

The body current versus gate voltage (Fig.3) shows a first peak at low values of Vg which is related to the increase of the hole current due to the impact ionization at the gate side p-well /n layer junction region [2]. As the gate voltage (Vg) increases the reduction of the electric field at the gate side (p-well/n layer junction) leads to a reduction in the body current. As we further increase Vg impact ionization at the drain side increases leading to a second peak. This can be partly explained by Kirk effect which in turn is more prominent due to the presence of the multi layer super-junction structure.

In the case of the SJ LDMOSFET structure due to the fact that all the current flows through the n layer, (where high electric field is also expected) higher e-/h+ pairs generated by impact ionization are also expected. Under simulation conditions and the use of appropriate initial interface trap density and the capture cross section of the interface states, the initial density and the rate of the trapped electrons builds up (before any significant impact ionization (pre-stress) conditions) is found to be significant. Hence the trapped electrons concentration

cannot be ignored in our simulations. The actual interface trap density and hot carrier degradation mechanism can be simulated using fitting parameters on the Sentaurus simulation.

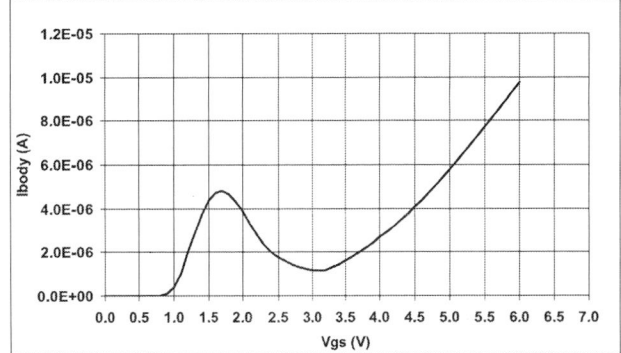

Figure 3: The Ibody versus Vdrain characteristic (experimental results).

This locally increased interface state density of the Si/SiO_2 interface at the device surface (electrons trapped), results in an increased surface depletion under the surface oxide and, consequently leads to an increase in the on-state resistance (this effect can be seen in Fig.4). A depletion region appears along the interface of Si/SiO_2, therefore the nature and location of these traps is very important. Obviously, if the device is stressed under prolonged high field further degradation will occur [3-7].

Ramping the drain voltage from low to high, the depletion region is expanding under the oxide (as already explained above) because of the increase in the negative interface charge. A depletion region also appears along the reverse bias junction of the SJ structure, – the combination of both depletion regions creates a JFET effect in the n layer where the flow of current is constrained, increasing the on-state resistance. As expected, this results into current densities that are significantly lower than the values we expect if the existence of such a depletion region is ignored.

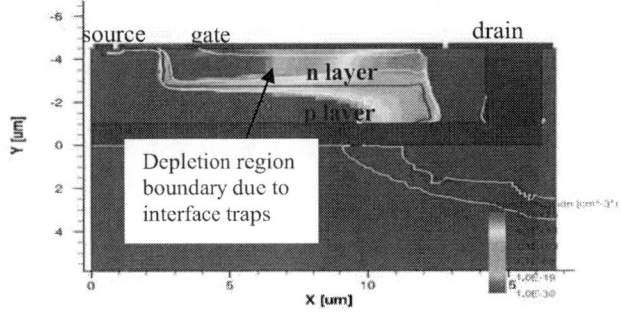

Figure 4: Impact Ionisation (Vg=5V, Vd=25V)

From Fig. 5 (simulation results – these are very similar to the experimental results), as we continue to increase the drain current what we observe is a kink in the on-state characteristic. This kink can be explained by inspecting the behavior of the impact ionization induced holes; the holes produced at the drain side of the device. Due to increased impact ionization, the holes gain enough energy to be injected towards the oxide and neutralize the interface trapped electrons, therefore collapsing the surface depletion region and allowing the current to flow into a wider area (Fig.6). Note that depending on the interface trap level, an accumulation layer may be formed at the surface. In any case, at high drain voltages the depletion region is significantly reduced.

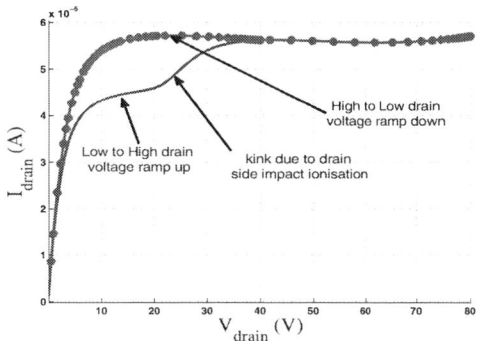

Figure 5. Vg=5V (a) Vdrain ramped from 0 to 80V (black line) (b) Vdrain ramped from 80 to 0V (blue line) – simulation results for Dit=3x10^{11} cm^{-2} σe=1x10^{18} cm^{2}

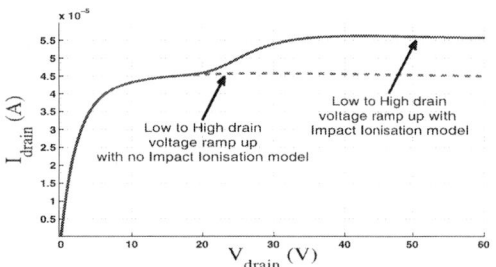

Figure 6: Vg=5V, Idrain versus Vdrain with and without impact ionisation model– **simulation results.**

This kink can also be verified in simulations by turning on or off the impact ionization parameter; when this parameter is absent no kink is observed (Fig.6). Therefore as long as these holes remain at the Si/SiO2 interface the trapped charge is virtually neutralized and sweeping the DC voltage from high to low or low to high results in no hysteresis behaviour (identical characteristic).

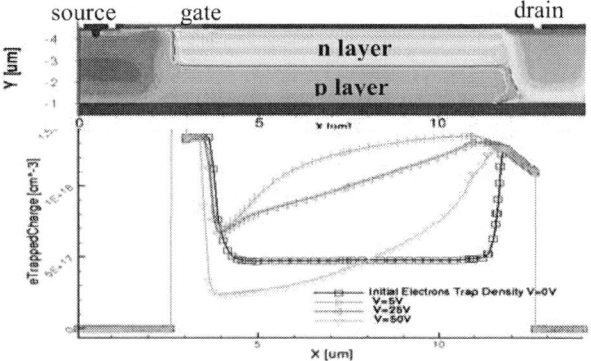

Figure 7: The interface electron trapped density, V_d=0V (initial), 5, 25 and 50V.

Fig.7 shows the interface electron trapped charge for Vg equal to 5V and V_{drain} varying from 0 to 50V. Following the changes in the electron trapped charge we can clearly see that as the drain voltage increases, the electron injection is very dominant at the drain side. This is because the bands close to the drain bend more than those away from the high voltage contact. Hence, at V_{drain} equal to 5V the interface charge has already reached its maximum value at the drain side whereas regions away from the drain see a relatively smaller increase in the trapped charge. As expected, as we further increase the voltage more interface traps get occupied.

For V_{drain} greater than 25V the trapped electron concentration begins to drop (Fig.7) both at the gate and drain side indicating trapped electron neutralization from impact ionization due to holes injected across the Si/SiO2 interface. Part of the holes (generated by impact ionization) is collected

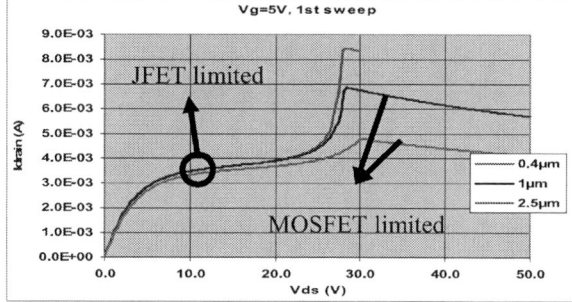

Figure 8: Vg=5V, Idrain versus Vdrain - low to high ramp– **experimental measurements** for different **channel lengths.**

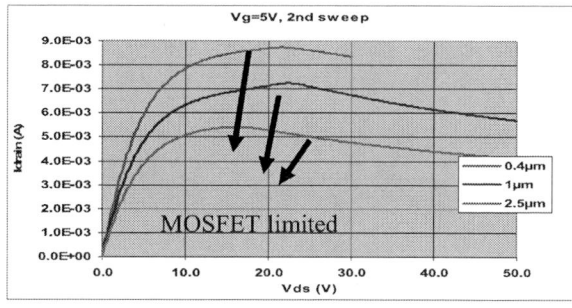

Figure 9: Vg=5V, Idrain versus Vdrain - high to low ramp– **experimental measurements** for different **channel lengths.**

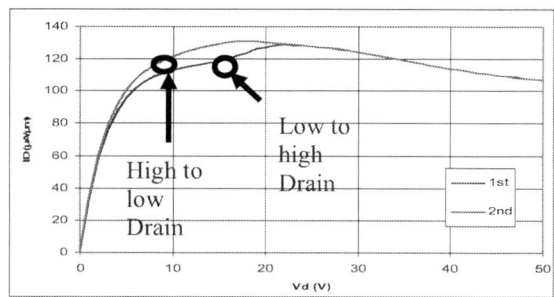

Figure 10: The on state output for an SJ LDMOSFET for Vg=5V and V_{drain} varying at Temp=300K, employing drain field plates – experimental results.

by the body contact at the source side and part of these

compensates some of the negative charge trapped at surface states between gate and drain.

Fig. 8 and 9 show the result of varying the channel length before and after the impact ionization takes place confirming the existence of the JFET effect. In Fig.8 the current is strongly limited by the JFET effect in spite of the channel resistance being reduced (through decreasing the channel length). After impact ionization takes place the JFET is virtually eliminated and the current varies significantly with the channel length, as expected (Fig. 9).

III. SOLUTIONS

Benefits can be achieved by extending the metallization – field plates from the drain towards the drift region. In fact, the introduction of field plates helps to reduce the electric field and hence (i) the impact ionization and (ii) reduction of the electron injection at the drain side. Fig.11 shows the device featuring field plates at the drain side. The JFET effect is still present but its impact is reduced as we extend the field plate towards the drift region. Therefore the effectiveness of the field plate at the drain side requires careful optimization. On the other hand, the reduced impact ionization reduces the levels of the interface trapped charge hence the difference between low to high and high to low voltage ramping is almost eliminated. Fig 10 shows the experimental results of the on state output characteristics for the SJ LDMOSFET for Vg=5V at Temp=300K, employing drain field plates. The experimental results confirm the effectiveness of the optimized field plates. The device on-state resistance difference due to hysteresis is now reduced to less than 8%.

In order to suppress the JFET effect a relatively thin and more highly doped n layer is placed at the surface (Fig.12). This layer is extended from the drain side towards the edge of the accumulation region. The layer helps to minimize the depletion region at the silicon close to the oxide interface due to the increase in doping. The doping charge of this layer needs to be carefully optimized to minimize the disturbance in the breakdown of the device.

The on-state output characteristic of this optimized device is shown in Fig.12. The JFET is eliminated hence the saturation is higher compared to an optimized field plate -without the top n surface layer- structure. At the same time, the low to high and high to low drain voltage sweeps follow identical paths.

IV. CONCLUSION

In this paper we have showed and explained the effect of interface charge trapping and impact ionization under normal operating conditions in a highly dense super-junction structure. The study is based on both numerical simulations and experimental results. Due to the concomitant presence of interface traps, high current densities and the depletion region formed around the super-junction, impact ionization could occur at the drain side leading to hole carrier injection. This leads to instability in the on-state characteristics. The solution proposed in this paper, based on surface field plates and the

addition of a more highly doped n type surface layer overcomes the instability effect.

REFERENCES

[1] Hoelke, A.; Pal, D.K.; Yang Hao; Kee Kia Yaw; Kho, E.; Kittler, G.; Kuniss, U.; Gessner, J.; , "A 200V Partial SOI 0.18μm CMOS technology," *Power Semiconductor Devices & IC's (ISPSD), 2010 International Symposium on.*, pp.257-260.

[2] Perez-Gonzalez, J.; Sonsky, J.; Heringa, A.; Benson, J.; Chiang, P.Y.; Yao, C.W.; Sua, R.Y.; , "HCI reliability control in HV-PMOS transistors: Conventional EDMOS vs. Dielectric RESURF and lateral field plates," *Power Semiconductor Devices & IC's, 2009. ISPSD 2009.* pp.61-64, 14-18 June 2009.

[3] Moens, P.; Van den bosch, G.; , "Characterization of Total Safe Operating Area of Lateral DMOS Transistors," *Device and Materials Reliability, IEEE Transactions on* , vol.6, no.3, pp.349-357, Sept. 2006

[4] Moens, P.; Bauwens, F.; Nelson, M.; Tack, M.; , "Electron trapping and interface trap generation in drain extended pMOS transistors," *Reliability Physics Symposium, 2005. Proceedings. 43rd Annual. 2005 IEEE International* , pp. 555- 559, 2005

[5] Varghese, D.; Moens, P.; Alam, M.A.; , "on-State Hot Carrier Degradation in Drain-Extended NMOS Transistors," *Electron Devices, IEEE Transactions on* , vol.57, no.10, pp.2704-2710, Oct. 2010

[6] I. Cortes, J. Roig, D. Flores, J. Urresti, S. Hidalgo and J. Rebollo, "Analysis of hot-carrier degradation in a SOI LDMOS transistor with a steep retrograde drift doping profile", *Microelectron Reliability* 45 (2005) (3), pp. 493–498.

[7] Chen, J.F.; Kuen-Shiuan Tian; Shiang-Yu Chen; Kuo-Ming Wu; Liu, C.M.; , "On-Resistance Degradation Induced by Hot-Carrier Injection in LDMOS Transistors With STI in the Drift Region," *Electron Device Letters, IEEE* , vol.29, no.9, pp.1071-1073, Sept. 2008

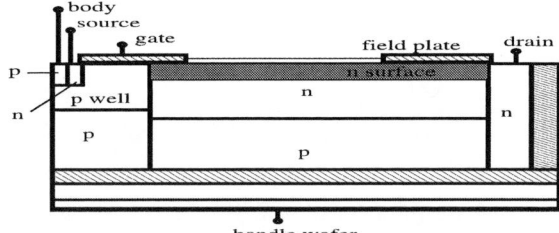

Figure 11: The SJ LDMOSFET, employing drain field plates and n injector layer.

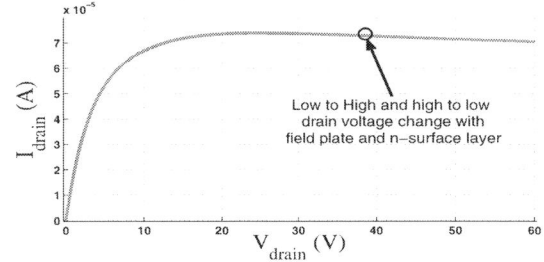

Figure 12: The on state output for an SJ LDMOSFET for Vg=5V and Vdrain varying @ Temp=300K, employing a field plate (lower curves) and employing a field plate and n injector (upper curve).

Proceedings of the 23rd International Symposium on Power Semiconductor Devices & IC's
May 23-26, 2011 San Diego, CA

Reliability and Performance Optimization of 42V N-channel Drift MOS Transistor in Advanced BCD Technology

A.Molfese, P. Gattari, G. Marchesi, G.Croce

Technology R&D, STMicroelectronics
Via C. Olivetti 2, 20041 Agrate Brianza (Milano) Italy
antonio.molfese@st.com

G. Pizzo, F. Alagi, F. Borella

Technology R&D, STMicroelectronics
Via Tolomeo 1, 20010 Cornaredo (Milano) Italy

Abstract— Optimization flow for a 42V N-channel drift MOS in an advanced BCD technology in terms of performance and stability is described. The origin of the very fast on state resistance (R_{on}) degradation detected during reliability tests under off state on the starting device has been identified in borderless silicon nitride used as stop layer during contact etch. The final solution including a process step introduction, device geometry modification and drain doping profile optimization improves performance and addresses both voltage capability and reliability requirements.

I. INTRODUCTION

Smart-Power applications require medium to high voltage (typically 8 to 100 V) field-effect transistors (FETs) integrated in standard Low Voltage CMOS technology platforms [1][2]. Low cost process together with high and reliable performance is required. Low cost process is generally achieved trying to embed high-voltage transistors in a standard CMOS process flow with the minimum number of additional dedicated process steps. High performance, within the target voltage capability, is translated into low drain/source on-state resistance and area consumption.

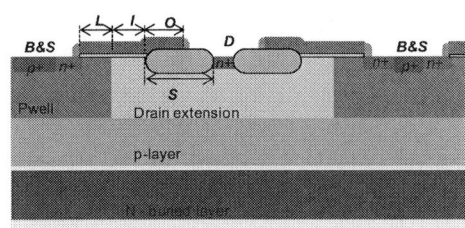

Figure 1: Drift NMOS cross section and geometrical parameters
(I=1.25 a.u. O=1 a.u. S= 2.25 a.u.)

Drift MOS transistors (Fig. 1) are widely used to realize high voltage NMOS in advanced BCD technology since they are both cost-effective, sharing the body implant/mask with the available CMOS one, and flexible, being the channel length a parameter that can be tuned for performance optimization.

In this paper 42V N-channel drift MOS performance and reliability optimization is reported. A preliminary device which showed fast R_{on} degradation during reliability tests under off- state is presented. The degradation reason has been firstly identified, then ruggedness and performance

improvements obtained through process and geometrical modifications are presented. Finally it is experimentally demonstrated that hot carrier degradation has been reduced on the final structure.

II. DEVICE DESCRIPTION

Silicon high Voltage N-channel Drift MOS schematic cross section is shown in Fig. 1. The key factors in the device development are:

1) the dedicated drain extension doping profile definition;

2) the geometrical parameters definition and specifically: the total drift length in the field oxide region of the device (S), the drift length in the source/body active area (I) and the poly-silicon gate overlap on the field oxide (O), as shown in Fig. 1.

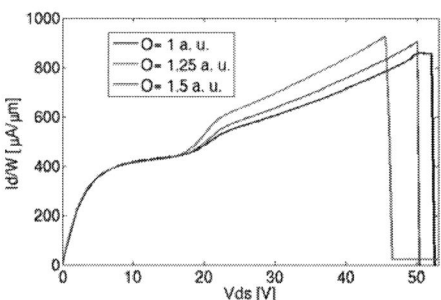

Figure 2: Pulsed output characteristic ($V_{gs} = V_{gs,max}$) up to on state breakdown voltage (BV_{on}) and its dependence on geometrical parameter O

State of the art device performance ($Ron \cdot Area = 32$ m$\Omega \cdot$mm^2; $BV_{dss} = 54$V) was originally achieved fixing geometrical parameters and drain doping profiles to ensure the requested device voltage capability both in off-state and in on-state with no limitation in electrical Safe Operating Area (SOA). In Fig. 2 the output characteristics and on-state breakdown voltage at V_{gsmax} are shown for different O.

Thanks to the existing CMOS body implant doping level and profile, the intrinsic npn parasitic bipolar (source/body/drain) base resistance is quite low hence retarding the bipolar turn on: therefore snapback occurs at large drain voltage. This important feature enables higher electric fields to be reached when impact ionization is occurring at the drain (large V_{gs}) and the practical result is an extension of the drain curves to higher drain voltages, revealing an increase and subsequent saturation

978-1-4244-8425-6/11 $26.00 © 2011 IEEE

of the current at high drain bias. This behavior has been already observed in similar devices [3]. This kind of devices can be schematized as an intrinsic NMOS with a non linear resistance in series (R_d) which takes in account drift region contribution. A "quasi-saturation" is initially observed caused by the saturation of the electron drift velocity and related increase of drain resistance. The effective Vds of intrinsic MOS is small and the intrinsic transistor is not working in saturation region. At larger Vgs and Vds, device experiences Kirk effect so electric field in region close to drain active area heavily increases causing impact ionization and consequent increase of conduction up to allow intrinsic MOS to reach saturation condition.

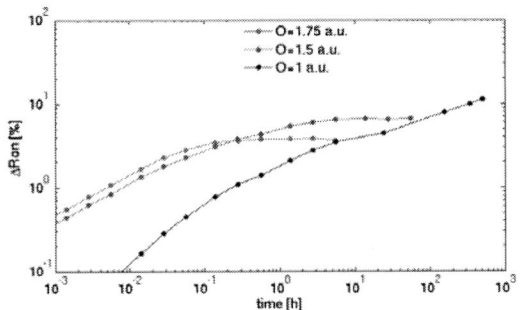

Figure 3 Ron degradation during HTRB dependence on O (HTRB condition V_{ds} = 46V T = 175°C)

Devices with larger O present premature current increase (Fig. 2) due to larger electric field in the region closer to drain active area caused by poly-silicon gate limiting the device electrical Safe Operating area.

III. OFF-STATE RELIABILITY RESULTS

A. High Temperature Reverse Bias (HTRB) stress

High Temperature Reverse Bias (HTRB) stress test is typically done on elementary high voltage devices during technology qualification phase to avoid product failure caused by device electrical parameters degradation over time [4]. In this test devices have been biased at V_{ds} = 46 V and V_{gs} = 0 at T=175°C. Large and very fast Ron degradation has been detected (Fig. 3). It has been also observed that when O is increased R_{on} degradation is faster but with a clear saturating trend to a lower asymptotic value.

B. Degradation Mechanism Hypothesis

In advanced CMOS technology an amorphous silicon nitride (a-Si$_x$Ni$_y$; defined as "borderless nitride") has been introduced as first layer in the pre-metal dielectric (PMD) stack. It is used as stop layer during contact etch to ensure minimum active area overlay on contact [5]. In more advanced technology the intrinsic layer stress have been used to improve MOS transistors performance [6]

In previous works [7][8][9], it was proven the borderless nitride negative impact on high voltage devices reliability behavior. This layer is in straight contact with silicided poly-silicon gate, spacer nitride layer, field oxide and silicided drain active area regions.

Conduction through nitride layers has been extensively studied [10] [11]: electric field, temperature and trap concentration are playing the major role in the carrier transport

(Poole-Frenkel emission). The current in this kind of dielectric film is given by the following equation [10][11]

$$ I = A \exp\left(\frac{\beta \cdot E^{\frac{1}{2}} - \phi}{kT} \right) \quad (1) $$

E is the electrical field and A, β and ϕ are constants at a given temperature and k is Boltzmann constant. For Poole-Frenkel effect which is field enhanced thermal excitation of electron from the shallow traps, the constant ϕ is the energy level from the conduction band for these traps.

During HTRB stress, a meaningful electric field in the borderless nitride layer is generated by the large voltage applied between drain and gate (V_{dg} = 46 V). Due to the high temperature present during the stress, a small current starts to flow in the layer.

Neutral traps are located at the interface between field oxide and borderless. These traps can be electrically activated (filled by electrons) by the electrons transport occurring in the upper nitride layer. These trapped charges modulate the conductivity of drain extension region not covered by poly-silicon gate overlap on the field oxide. The saturation trend occurs when all available traps are occupied. Traps amount and nature may be changed when different material is used for borderless nitride or when different process treatments are performed.

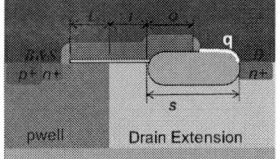

O [a.u.]	ΔR$_{on}$ SIM. [%]	ΔR$_{on}$ After HTRB [%]
1	10.6	11.2
1.25	6	6.4
1.75	3.8	3.8

Figure 4: TCAD simulations results with fixed density charge q = -1e12 cm^{-2} at interface nitride/field oxide

Experimental R_{on} asymptotic degradation level (Fig. 3) and dependence on O dimension are well predicted by TCAD simulations placing the same negative charge density (q = -1e12 cm^{-2}) at borderless/field oxide interface (Fig. 4).

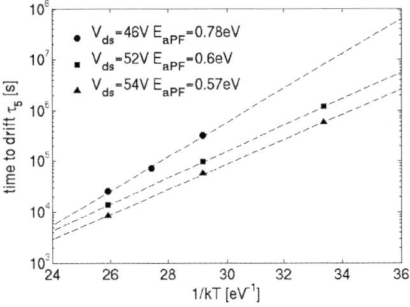

Figure 5: Time-to-drift (τ_S) dependence on $1/kT$ and applied voltage.

A similar result has been obtained placing the same amount of charge in the nitride layer bulk. It's therefore very difficult to exclude a partial (or total) contribution of the nitride bulk properties to the experimental degradation results.

To exclude that traps were located at the interface between the two present nitride layers (spacer and borderless) [8], the

978-1-4244-8425-6/11 $26.00 © 2011 IEEE

same method has been used by placing a density charge at that interface. The density charge value needed to reproduce the electrical degradation was very large (~1e13 cm^{-2}) and it has not been possible to replicate the degradation trend with O parameter.

Increasing O the starting degradation rate is larger since the same voltage drop is applied to a smaller region (S-O) increasing the electric field in the nitride layer. On the other hand the charged traps impact on saturating R_{on} degradation value is minimized since the drain region coverage by the poly-silicon gate is larger.

C. Temperature dependence investigation

It has been assumed that R_{on} degradation is proportional to interface trapped charge.

Trapped charge is proportional to neutral traps density and to the current flowing in the nitride:

$$\Delta R_{on} \approx B \cdot q \approx C \cdot I \cdot t \qquad (2)$$

τ_5 is defined as the time to reach 5 % degradation level:

$$\tau_5 = \frac{\Delta R_{on}\big|_{5\%}}{C \cdot I} \approx D \exp\left(\frac{\phi - \beta \cdot E^{\frac{1}{2}}}{kT}\right) = D \exp\left(\frac{E_{aPF}}{kT}\right) \qquad (3)$$

To derive equation (3), it has been implicitly assumed that Poole Frenkel mechanism is the main responsible for carrier transport in the film.

Measurements at different temperature and three different supply voltages were performed and results are summarized in Fig. 5. Experimental results confirm that the degradation activation energy (E_{aPF} in Fig. 5) is a function of the applied stress voltage and it decreases increasing the supply voltage. Therefore Poole-Frenkel like conduction mechanism seems to play a major role in the observed phenomena.

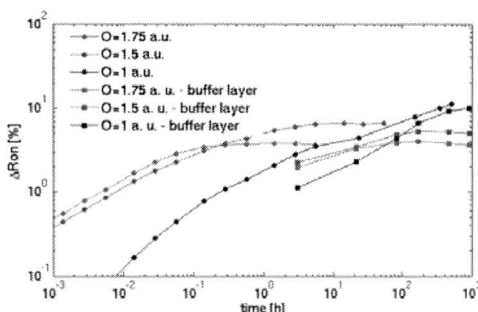

Figure 6: Ron degradation without and with buffer layer.

D. Buffer layer introduction

From the previous considerations, device robustness may be increased by reducing the traps density at analyzed interface (or in the bulk) or by minimizing electrical conduction in the nitride layer. Buffer layer added before borderless nitride should address all these goals: the interface is modified and nitride will not be directly electrically connected to silicided polysilicon gate and drain active area.

R_{on} degradation is drastically delayed even if the saturating drift value, occurring after large stress time, is not changed

(Fig. 6). Buffer layer introduction increases time-to-drift, thanks to the reduced electrical coupling but does not reduce the final drift since the impact on the total number of trap seems minimum.

From Fig. 6, it has been identified the dimension needed to minimize this effect: O = 1.6 a.u. obtaining ΔR_{on} < 5% after 1000h at T = 175°C, without the fast degradation detected before buffer layer introduction.

IV. DEVICE OPTIMIZATION

It has been shown that device optimization with larger O is necessary to guarantee R_{on} stability after HTRB stress. This geometrical parameter has a detrimental impact on the device SOA (BV_{on} at large V_{gs} as in Fig. 2).

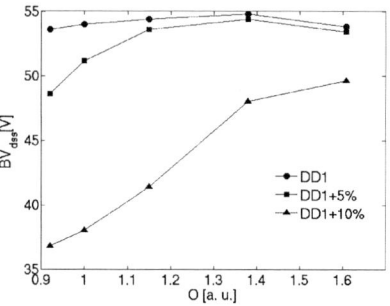

Figure 7: BV_{dss} as function of O for different drain dose doping: starting dose DD1 and two increased dose drain doping DD1+5% and DD1+10%.

BV_{on} increase may be achieved by changing geometrical parameters or by modifying drain doping profile. Drain doping profile modification in order to delay Kirk effect and consequent multiplication has been the solution adopted.

The other important aspect to take into account with increasing drain doping is off-state voltage capability: BV_{dss} dependence on O and drain doping is shown in Fig. 7. 5% drain doping increase allows working with larger O with minimum impact on voltage capability.

A device with increased drain dose doping profile and larger O has been finally defined ensuring the full electrical SOA (Optimized device in table I). HTRB stress has been performed on several devices with the final architecture from different diffusion lots. It was confirmed that R_{on} degradation level stay below 5% with a clear saturating trend even after 2500h (Stress conditions: T = 175°C; V_{ds} = 46V) (Fig. 8).

A complete comparison of key parameters is summarized in Table I.

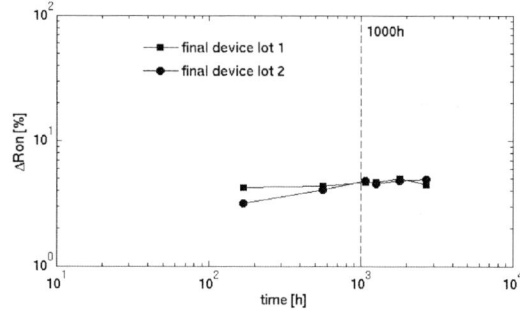

Figure 8: R_{on} degradation on final device after HTRB stress test on two lots. .

TABLE I.   COMPARISON BETWEEN STARTING AND OPTIMIZED DEVICE

Parameter	Unit	Starting Device	Optimized Device
O	a.u.	1	1.6
$BVdss$	V	54	53.6
$BVon$	V	51	49
$R_{on} \cdot Area$	mΩ·mm^2	32	31
ΔR_{on} after 1000h HTRB T=175°C	%	10	4.6

V.   HOT CARRIER DEGRADATION

High Voltage N-channel Drift MOS transistors Hot Carrier Injection typical reliability characterization is based on the well known CMOS tests. The device is submitted to a DC stress, by applying a constant voltage to gate, source and drain contacts. Stress bias conditions are fixed by application constraints. For power stages, hot carrier injection occurs during transistor commutation phases. Maximum drain to source voltage is kept as far as the gate to source voltage is close to the threshold voltage when the transistor is turned on. Generally this condition (high V_{ds}, moderate gate to source over-drive) is coincident with the worst case.

R_{on} is the electrical parameter which shows the fastest degradation (increase) after hot carrier stress test. The time-to-drift is implicitly defined by the time needed to reach 10% degradation level.

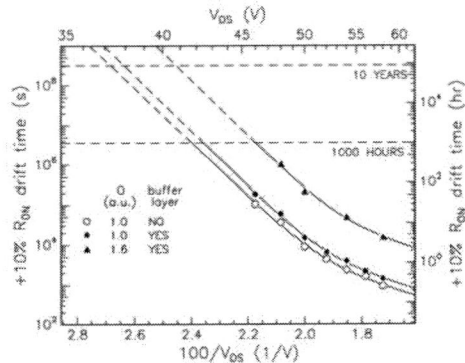

Figure 9: Time-to-drift as function of the drain voltage for original device (O = 1 a.u., without and with buffer layer) and optimized device (O = 1.6 a.u. with buffer layer and increased drain doping)

Time-to-drift at V_{gs}-V_{th} = 0.9 V is reported in Fig. 9 and shows that the initial device (O =1 a.u., no buffer layer) ensure safe DC operations at maximum V_{ds} (42 V) for about 1000 hours. Experimental results (Fig. 9) shows that by increasing the nominal layout parameter O from 1 to 1.6 a.u. (and modifying slightly the drain doping profile), a time-to-drift improvement of more than one order of magnitude is observed in all the experiments and it is extrapolated at the maximum operating voltage (V_{ds} = 42 V). No relevant change has been detected due to buffer layer introduction.

Two main reasons are responsible for the observed improvement:

1) Poly-silicon gate overlap on the field oxide (O) increase reduces electric field in the source/body drift region (I) where hot carrier injection occurs under moderate overdrive conditions[12][13]

2) drain doping increase makes the device more robust vs the injected charge.

VI.   CONCLUSIONS

Optimization flow for 42V N-channel MOS with state of art performance and good stability has been described. On starting device fast and unexpected Ron degradation was observed during High Temperature Reverse Bias stress. Electrons injected and trapped during stress through the borderless nitride layer have been recognized as responsible physical mechanism. The introduction of a buffer layer to decouple borderless from silicided poly-silicon gate and drain active area produced large improvements on degradation time but did not impact total degradation level. A device with optimized drain doping profiles, buffer layer and larger O (poly-silicon gate overlap on the field oxide) has been finally developed fulfilling both voltage capability and reliability requirements.

REFERENCES

[1]   D. Riccardi, A. Causio, I. Filippi, A. Paleari, A. Pregnolato, P Galbiati and C. Contiero, "BCD8 from 7V to 70V: a new 0.18μm technology platform to address the evolution of applications towards smart power ICs with high logic contents", Proceedings of. ISPSD '07, pp 73-76, May 2007

[2]   S. Pendharkar, R Pan, T Tamura, B. Todd and T. Efland "7 to 30V state-of-art power device implementation in 0.25 μm LBC7 BiCMOS-DMOS process technology"; Proceedings. ISPSD '04, pp. 419-422, 2004.

[3]   S.Reggiani, E. Gnani. A .Gnudi, G. Baccarani, M. Denilson, S. Pendharkar, R. Wise and S. Seetharaman "Investigation on saturation effect in the rugged LDMOS transistor", Proceeding of ISPSD '09, pp. 208-211 June 2009

[4]   S. Bach, F. Borella, J. Cambieri, G. Pizzo, A. Causio , L. Atzeni, D. Riccardi, L. Zullino, G. Croce and A. Nannipieri, "Simulation of off-state degradation at high temperature in high Voltage NMOS transistor with STI architecture", Proceedings of. ISPSD '10, pp. 189-192, June 2010

[5]   Quirk M, Serda J. Semiconductor manufacturing technology. Prentice-Hall Inc. pp. 214.

[6]   H.S.Yang et al, "Dual stress liner for high performance sub45nm gate length SOl CMOS" Electron Devices Meeting, 2004. IEDM Technical Digest. IEEE International, pp. 1075-1077, 2004.

[7]   G. Beylier, S. Bruyere, D. Benoit and G. Ghibaudo "Refined electrical analysis of two charge states transition characteristic of ''borderless'' silicon nitride", Microelectronics Reliability, vol 47, Isuue 4-5, pp. 743–747, 2007.

[8]   G. Marchesi, J. Cambieri, A. Dundulachi, G. Pizzo, F. Pozzobon, M. Annese, A. Andreini and G. Croce "High voltage P-channel MOS breakdown voltage instability during high temperature gate stress induced by pre-metal nitride layers", Proceedings of. ISPSD '08, pp. 275-278, May 2008

[9]   D. Lachenal, Y. Rey-tauriac, L. Boissonnet and A. Bravaix, "Reliability investigation of NLDEMOS in 0.13μm SOI CMOS", Proc. 25th International Conference on Microelectronics, pp.555-558, May 2006.

[10]   B. Swaroop and P.S. Schaffer, "Conduction in silicon nitride and silicon-nitride-oxide films" J. Phys. D: Appl. Phys. Vol.3, n. 3, 1970.

[11]   S. Manzini, "Electronic processes in silicon nitride", J. Appl. Phys., vol. 62, is. 8, pp.3278-3284, 1987

[12]   P. Moens, J. Mertens, F. Bauwens, P. Joris, W. De Ceuninck and M. Tack, "A comprehensive model for hot carrier degradation in LDMOS transistors", Reliability physics symposium, pp. 492-497, April 2007

[13]   M. Annese, S. Carniello, and S. Manzini, "Design and optimization of a hot-carrier resistant high-voltage nMOS transistor", IEEE Transaction on Electron Devices, vol. 52, n. 7, July 2005 pp 1634-1639

Proceedings of the 23rd International Symposium on Power Semiconductor Devices & IC's
May 23-26, 2011 San Diego, CA

Practical Approaches to Improve Thermal SOA for Smart Power IC

T. Nitta, A. Omichi, S. Yanagi, Y. Yoshihisa, T. Kuroi,
K. Hatasako, and S. Maegawa
Renesas Electronics Corp.
4-1, Mizuhara, Itami, Hyogo, 664-0005, Japan
e-mail: tetsuya.nitta.xd@renesas.com

K. Furuya
Renesas Semiconductor Engineering Corp.
4-1, Mizuhara, Itami, Hyogo, 664-0005, Japan

Abstract— **In this paper, approaches to improve thermal SOA (T-SOA) of LDMOS have been presented. We show three important points for T-SOA based on experimental data. Firstly, improvement of thermal stability, which is expressed by simple index "α"; a ratio of drain current at 200°C divided by that at 25°C. Secondly, suppression of parasitic NPN action that causes device destruction and the correlation between failure energy and off-state leak current is studied. Thirdly, reduction of the thermal impedance. We examined an effect of Cu plate on wafer surface and a thinning effect of wafer thickness, and found thinner wafers were quite effective for long pulse energy.**

I. INTRODUCTION

LDMOS has been widely used in BiC-DMOS technology, and that is a key element for mixed signal applications such as automotive, display driver and digital audio. Continuous progress of lower resistance for LDMOS has been made, and it has contributed to shrink output drivers on smart power ICs. On the other hand, it has caused an increase in power density of output drivers, and the importance of thermal SOA (or electro thermal SOA) has been increasing. For automotive applications such as airbags, ABS and powertrains, large T-SOA is generally required, and the improvement of T-SOA is as significant as that of on-resistance.

While the stress power of T-SOA is much lower than that of the electrical SOA (E-SOA), the stress pulse is generally longer for T-SOA (Fig.1), and accordingly the self heating effect has a major impact on device destruction.

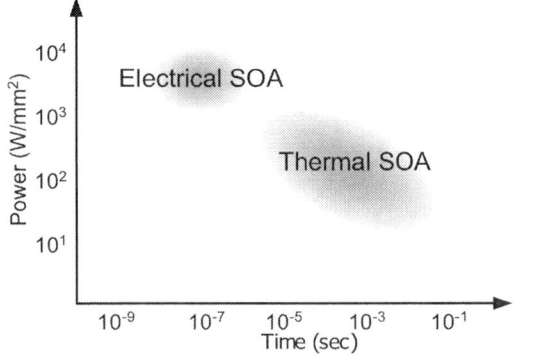

Figure 1.   Typical regime of Electrical SOA and Thermal SOA.

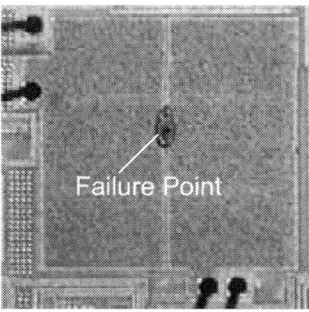

Figure 2.   An example of T-SOA failure point.This device is an output driver of a powretrain solenoid driver.

A typical failure point on T-SOA evaluation is shown in Fig. 2. The destruction occurred at the device center, which was the most severe point of thermal dispersion. The thermal stability (thermal planarization) is one of the most important points on T-SOA. The characterization and the analysis of T-SOA for LDMOS were previously reported [1-5], and what is responsible for T-SOA has been studied. However the practical discussions to improve T-SOA without sacrificing on-resistance are not quite sufficient. In this paper, we discuss practical methods to optimize T-SOA of LDMOS, in other words, ways to improve both T-SOA and on-resistance.

II. WAFER LEVEL EXPERIMENT

A. Setup

Wafer level T-SOA measurements were done for various type LDMOS fabricated on 0.25μm BiC-DMOS process. The device size is 0.5mm². The power stress was applied with constant drain voltage (Vds) and constant gate voltage (Vgs). The measurements were done for various Vds, and the failure time was controlled to about 1ms by adjusting Vgs for each Vds.

B. Analysis of Thermal Stability

In order to quantitatively clarify the relationship between thermal stability [3] and failure energy, we defined an index "α" (Fig.3). The α is a ratio of drain current (Ids) at 200°C divided by that at 25°C. The α was measured with small size LDMOS. Fig.4 shows examples of measured drain current of T-SOA measurement. For cases of large α, such as Fig. 4(a),

978-1-4244-8425-6/11 $26.00 © 2011 IEEE

the Ids increase with self-heating and positive feedback occurs at the heat spot. For cases of small α, such as Fig.4(b), the Ids is suppressed by self-heating and the power distribution may tend to be homogeneous.

Fig.5 shows the measured relationship between α and failure energy by changing Vds for "same type" structure A1, A2 and A3. These structures have same the diffusion layers and have a different cell pitch (A3 > A2 > A1). The value of α is lower for lower Vds conditions because a high gate bias is needed to apply the same energy (Fig. 3). The failure energy has a very good correlation to α even for variant structures in Fig.5. We can say that the degree of thermal stability is well expressed by index α, and the root cause of drain bias dependence of failure energy [4] is the difference in thermal stability.

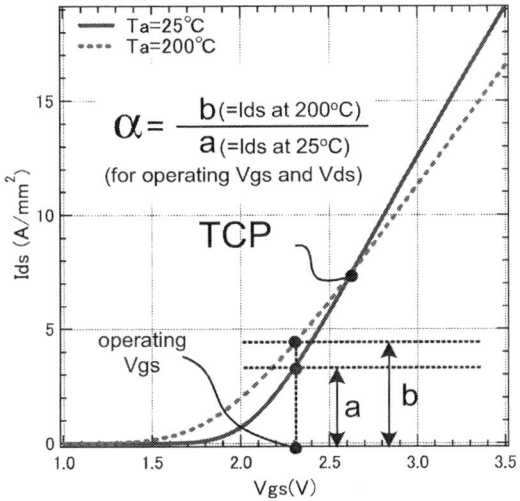

Figure 3. Vds-Ids characteristics of LDMOS (structure A1) and the definition of α. The crossing point (α =1) is called Temperature Compensation Point (TCP). The α generally becomes lower at higher gate voltages.

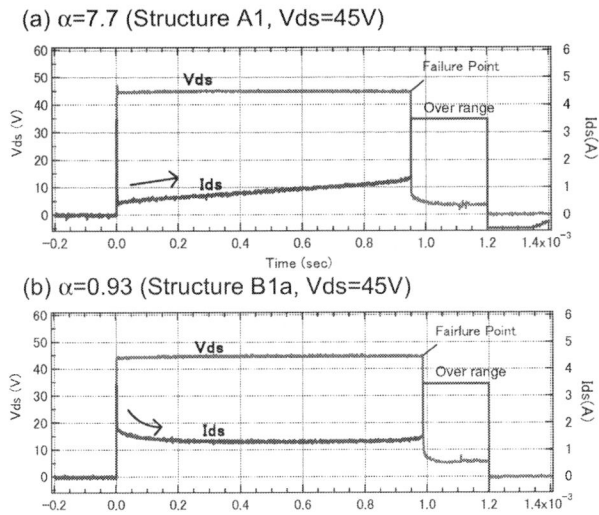

Figure 4. Measured drain current shifting during stress for two cases.

As α is easily calculated by device simulation, and the measurement does not need large size devices, it is a very helpful index for device structure design with optimizing T-SOA and other characteristics, such as on-resistance. As the power density of LDMOS has been increasing with progress in on-resistance, this method to estimate T-SOA by simulation stage or at early stages of device experimentation has great importance.

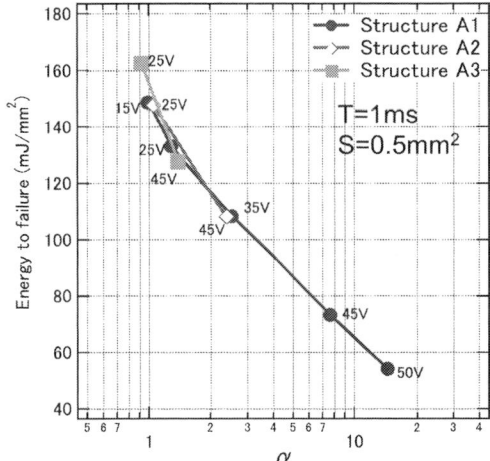

Figure 5. The relationship between α and failure energy for different drain voltages for structure A1, A2 and A3. The failure energy was measured with 0.5 mm^2 sample, and the α was measured with small sample size.

C. Influence of Parasitic NPN

We investigated the universality of index α by measuring other type structures, B1a, B1b, and B1c. These structures have not only different cell pitches but also different diffusion layers as shown in Table I.

TABLE I.   THE FEATURE OF EACH STRUCTURE

A1, A2, A3		High doping level P body layer (dedicated layer) is used.
B1	a	Pwell is used for P body.
	b	Pwell is used for P body, and the doping of Noffset is higher than others.
	c	Optimized Pwell is used for P body (doping level is lower than A type).

Fig. 6 shows the measured relationship between α and failure energy for these structures. We can see that unlike the result of Fig. 5, the relationship between α and failure energy is different for each structure. This result shows the failure energy is not determined by only thermal stability.

Another important parameter for failure energy is critical temperature (Tcrit), and these structures may have different Tcrit. Although the Tcrit is important for T-SOA, it is difficult to measure directly. As Tcrit is associated with parasitic NPN transistor in LDMOS (Fig. 9) [6], the hole current has an important role on turning on the NPN Tr. For typical T-SOA conditions, the drain current is not large enough to cause the Kirk effect, and the electric field is close to that of off-state. So the off-state leak current (Ioff) should be a good index to evaluate hole current aspect on T-SOA.

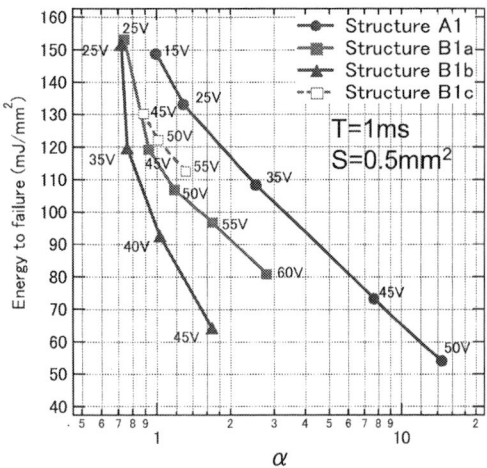

Figure 6. The relationship between α and failure energy for different drain voltages for structure A1, B1a, B1b and B1c.

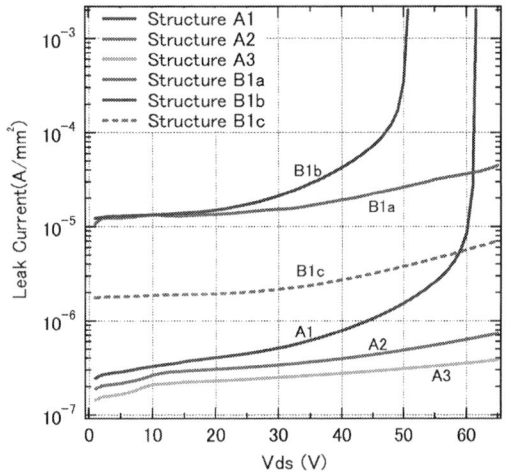

Figure 7. Off-state leak current for each structure at 150℃.

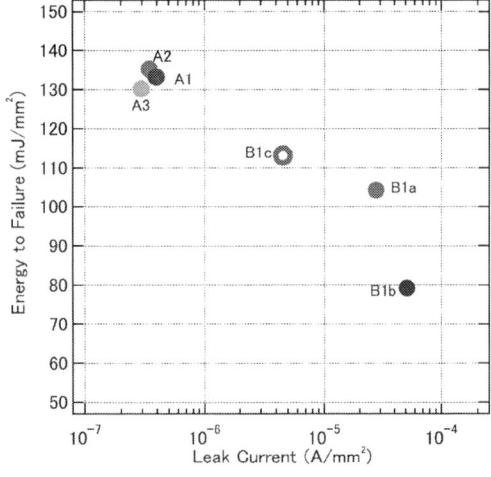

Figure 8. The relationship between off-state leak current and the failure energy at the condition of α=1.3. (The drainvoltage is different for each structure. ex. Vds=25V for structure A1, Vds=51V for structure B1)

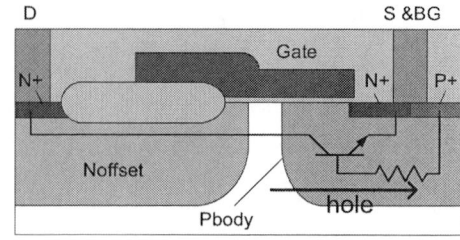

Figure 9. Schematic cross section of LDMOS. A parasitic NPN transistor consists of N+Source, P-body and N-Drain.

Fig. 7 shows Ioff at 150°C for the structures evaluated in Fig. 5 and Fig. 6. Fig. 8 shows the dependency between Ioff at the condition of α=1.3 and failure energy at the condition of α=1.3. It is clear that the failure energy can be increased by reducing the leak current. From this result, we can say that Ioff expresses the difference of Tcrit and it can be used as an index of T-SOA estimation. It should be mentioned that as both α and Ioff are general parameters and easy to simulate or measure directly, optimizing these parameters is very practical for improving T-SOA.

III. REDUCTION OF THERMAL IMPEDANCE

While improving T-SOA by optimizing device structure and/or diffusion profile is important, a method to improve T-SOA without optimizing device structure is very useful. One such method is reduction of thermal impedance (Zth). We investigated two methods to lower thermal impedance. One was using Cu plate [7] on the device surface (thickness=6μm), and the other was to introduce thinner wafer grinding (wafer thickness after grinding was 150μm). Fig.10 shows the simulated Zth compared to reference condition; wafer thickness=400μm, w/o Cu plate. The simulation result shows that the Cu plate is effective for short pulse, and the thin wafer is effective for long pulse (Fig. 10). The thermal capacitance of Cu plate is saturated at an early stage, and the heat conduction needs about 1ms to reach wafer backside (Fig. 11). The package level experimental results for the pulse width from about 1ms to 10ms are shown in Fig. 12.

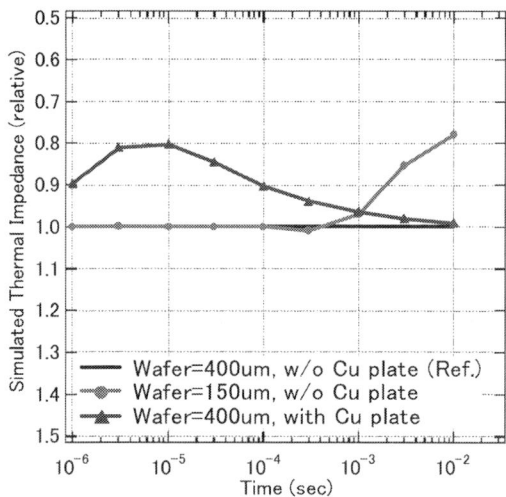

Figure 10. Simulated 1D thermal impedance with die frame and epoxy mold. (Ref. condition =1.0)

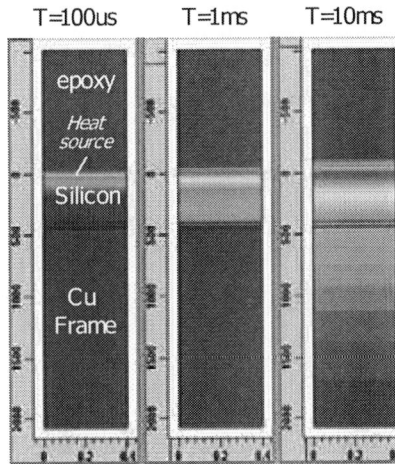

Figure 11. Simulated heat dispersion for typical T-SOA pulse widths.

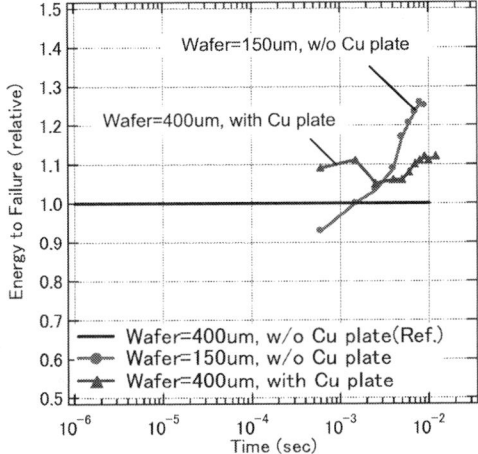

Figure 12. Measured failure energy for thin wafer (150um) or applying Cu plate. (Ref. condition =1.0)

The failure energy of thin wafer (150um) compared to that of thick wafer (400um) is clearly higher as the pulse is longer. The improvement is over 20% for a pulse of 10ms. Taking measurement error into consideration, the effect of Cu plate is not clear for these time ranges. To confirm the effect of Cu plate, further investigation of short pulse is needed.

IV. DISCUSSION

In general device scaling, on-resistance has a trade-off relationship with T-SOA [4], and pursuing low on-resistance structure without considering T-SOA may be deadlocked by T-SOA specification. Although further work is needed to fully understand T-SOA, we can say that it is possible to realize large T-SOA while keeping lower on-resistance based on the above discussion. Table. II shows an example of device optimization. The structure D, which applied novel scaling methods considering T-SOA, achieves 20% lower Ron,sp and 30% larger T-SOA than structure C. That is enabled by introducing a wider cell pitch, optimized P-body profile and optimized RESURF effect (No Zth lowering techniques were applied in this comparison). Such optimization can be efficiently achieved by considering the α, Ioff or other indexes.

TABLE II.   A DEMONSTRATION OF IMPRVING T-SOA

	Cell Pitch (Relative)	BVoff (V)	Ron,sp ($m\Omega cm^2$)	Energy to Failure (mJ/mm^2) (*1)
Structure C	1.00	61	0.94	120
Structure D	1.12	78	0.76	160

(*1) Vds=50V, Pulse=1ms, S=1mm^2

V. CONCLUSION

We investigated practical methods to improve thermal SOA for LDMOS. Measurements showed that the thermal stability of LDMOS was expressed by an index α; a ratio of drain current at 200°C divided by that at 25°C, and had good correlation with T-SOA. Measurements also showed that the critical temperature can be estimated by measuring off-state leak current at 150°C. As both α and Ioff are easily simulated or measured, optimizing these parameters is a very useful method for improving T-SOA. We also studied the influence of thermal impedance for T-SOA, and found that a thin wafer was effective for long pulse stress (>1ms) by experimentation, and Cu plate was effective for short pulse stress (< 100ns) by simulation. Investigations discussed in this paper are helpful and necessary for further scaling of LDMOS overcoming T-SOA specification limits.

ACKNOWLEDGMENT

The authors would like to thank K. Senda and K. Minagawa for analyzing data and also thank A. Uenisi, S. Kikuyama, T. Kawahara and K. Nishimoto for fruitful discussions.

REFERENCES

[1] Vishnu Khemka, Vijay Parthasarathy, Ronghua Zhu, Amitava Bose, and Todd Roggenbauer, "Experimental and Theoretical Analysis of Energy Capability of RESURF LDMOSFETs and Its Correlation With Static Electrical Safe Operating Area (SOA)", IEEE Trans. on Electron Devices, vol. 49, No. 6, pp. 1049-1058, 2002.

[2] Vishnu Khemka, Vijay Parthasarathy, Ronghua Zhu, Amitava Bose, and Todd Roggenbauer, "Detection and Optimization of Temperature Distribution Across Large-Area Power MOSFETs to Improve Energy Capability", IEEE Trans. on Electron Devices, vol. 51, No. 6, pp. 1025-1032, 2004.

[3] Marie Denison, Martin Pfost, Klaus-Willi Pieper, Stefan Märkl, Dieter Metzner, Matthias Stecher, "Influence of Inhomogeneous Current Distribution on the Thermal SOA of Integrated DMOS transistors", Proc. ISPSD2004, pp. 409-412, 2004

[4] G. Van den bosch, D. Wojciechowski, B. Elattari, P. Moens and G. Groeseneken, "Characterization of dynamic SOA of power MOSFETs limited by electro thermal breakdown", Proc. ESSDERC2005, pp. 465-468, 2005.

[5] Philip L Hower and Sameer Pendharkar, "Short and long-term safe operating area considerations in LDMOS transistors", Proc. IRPS2005, pp. 545-550, 2005.

[6] Vishnu Khemka, Vijay Parthasarathy, Ronghua Zhu, and Amitava Bose, "A Novel Technique of Decouple Electrical and Thermal Effect in SOA Limitation of Power LDMOSFET", IEEE Electron Device Letters, vol. 25, No. 10, pp. 705-707, 2004.

[7] Young S. Chung, Terry Willett, Veronique Macary, Steve Merchant, Bob Baird, "Energy Capability of Power Devices with Cu Layer Integration", Proc. ISPSD1999, pp. 63-66, 1999.

Proceedings of the 23rd International Symposium on Power Semiconductor Devices & IC's
May 23-26, 2011 San Diego, CA

Hot carrier degradation of HV-SOI devices under off- and on-state current conditions.

R. van Dalen, S. Dhar[#], A. Heringa[#], M.J. Swanenberg, A.B. van der Wal, P.W.M. Boos, V. Braspenning-Girault

NXP, Wafer Technology & Foundry Organisation (WT&FO), Nijmegen, The Netherlands
[#]NXP Research, Eindhoven, The Netherlands

Abstract— **Optimisation of High Voltage (HV) devices is typically governed by life-time considerations, most notably requirements for sufficient immunity against Hot Carrier Injection (HCI). Based on insight in the different degradation mechanisms in HV-SOI, we have identified distinctive acceleration methodologies for on- and off-state stress conditions.**

I. INTRODUCTION

The degradation of high voltage devices is typically determined by their susceptibility to hot carrier degradation. Proper acceleration or extrapolation techniques are essential to be able to guarantee long-term reliability without the need to resort to many time-consuming stress tests.

Extensive work on sub-100V LDMOS devices has resulted in so-called "universal" degradation models for both off- and on-state degradation [1,2], which allow the extraction of the life-time for arbitrary operating conditions based on limited short-term stress data. Acceleration is achieved by stressing at elevated gate and/or drain bias. In both cases, the degradation was found to occur at the accumulation region and LOCOS bird's beak with the dominant parameter shift occurring in the specific on-resistance.

In this work we investigate hot carrier degradation in NXP's in-house EZ-HV technology, being a thin layer SOI high voltage (650V) process with record $R_{ds,on}$-BV_{ds} performance [3]. Such technology features an extended drift region that is simultaneously depleted by the top field plate and the bottom handler wafer, resulting in peak electric fields that occur well within the drift region (Fig.1). Thus the hot carrier degradation is expected to differ markedly from low voltage and/or bulk technology, as indeed shown in [4] for (sub-threshold) on-state stress:

- Hot hole injection occurs in the drift region.

- The injected holes disturb the RESURF balance, such that the location of charge injection can shift during stress.

- The dominant parameter shift is in the multiplication, ultimately leading to an excessive leakage current.

These deviating mechanisms require a renewed study into applicable acceleration methodologies, now particularly for SOI technology. Such accelerated tests should obviously mimic the actual degradation mechanism during operation. Several publications have shown that such assumptions are not necessarily valid even when comparing very similar stress conditions [5] and need detailed investigation.

II. PROBING ELECTRIC FIELDS AND CHARGE INJECTION

HV-SOI features a unique depletion such that location and magnitude of local lateral field peaks can be directly extracted from the multiplication curve (Fig.1). As injected charge disturbs the RESURF balance and thus affects the potential distribution throughout the device, changes in this extracted field distribution can be effectively used to monitor hot carrier degradation [4].

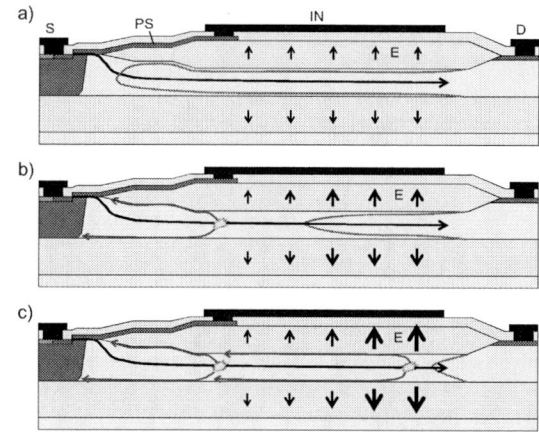

Figure 1. Schematic on-state current flow with increasing V_{ds} (a→c) in HV-SOI. Top field plates and handler wafer push the electron flow towards the center of the SOI layer. Gradual lateral depletion through local lateral field peaks results in additional multiplication (taken from [4]).

III. MECHANISMS GOVERNING HOT CARRIER INJECTION IN HV-SOI

Fig.2 illustrates the mechanisms governing hot hole generation and injection showing a strong dependence on the local lateral and vertical field components. Note that both components change oppositely during HCI.

978-1-4244-8425-6/11 $26.00 © 2011 IEEE

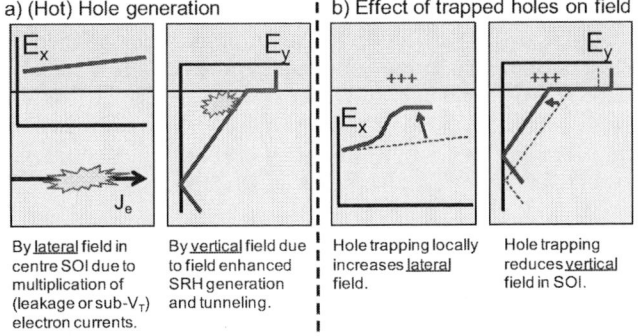

a) (Hot) Hole generation

| E_x | E_y |

By <u>lateral</u> field in centre SOI due to multiplication of (leakage or sub-V_T) electron currents.

By <u>vertical</u> field due to field enhanced SRH generation and tunneling.

b) Effect of trapped holes on field

| E_x | E_y |

Hole trapping locally increases <u>lateral</u> field.

Hole trapping reduces <u>vertical</u> field in SOI.

Figure 2. Degradation in HV-SOI devices is a complex interplay between the lateral and vertical fields; a) Several hole generation mechanisms can be identified with different field dependencies; b) Resulting hot carrier injection affects field components oppositely.

IV. OUTLINE OF THE EXPERIMENTS

A. Objective and set-up of the experiments

Main objective of this study was to establish the role of the lateral and vertical field components in the HCI degradation. To identify the dominant contribution, we use the fact that there are several means to enhance/suppress the individual mechanisms sketched in Fig.2, either by process changes or by varying measurement conditions.

A good example of a process change that has a very large impact on the fields is the <u>choice of dielectric thickness</u>. Increasing the dielectric thickness *decreases the vertical field* (as the same applied voltage is now dropped across a thicker layer) whereas it *increases the lateral field* (as it suppresses the RESURF action of the field plates). In our particular case, due to field-fringing effects, the impact on the lateral field far outweighs that of the vertical component. Similarly, changing the <u>extension dose</u> is only expected to significantly affect the *lateral field component*.

It is well known that HCI is strongly affected by the temperature. Increasing the temperature however not only decreases the capture rate but at the same time influences the amount of hot carriers that are being generated. Avalanche multiplication of sub-threshold (on-state) current decreases at higher temperatures, but Shockley-Read-Hall generated (off-state) leakage currents increase dramatically. The typical statement that HCI is dominant at low temperatures will therefore prove to be not necessarily valid in HV-SOI.

In this work we have examined the degradation of our EZ-HV HVndmos as a function of:

- Two EZ-HV process settings with slightly varying top oxide thickness (Process A vs. B, the former featuring an enhanced lateral field component as shown in Fig.3).

- Extension doping (again affecting the lateral field component, Fig.4).

- Temperature (impacting both trapping efficiency as well as the hot-carrier generation rate).

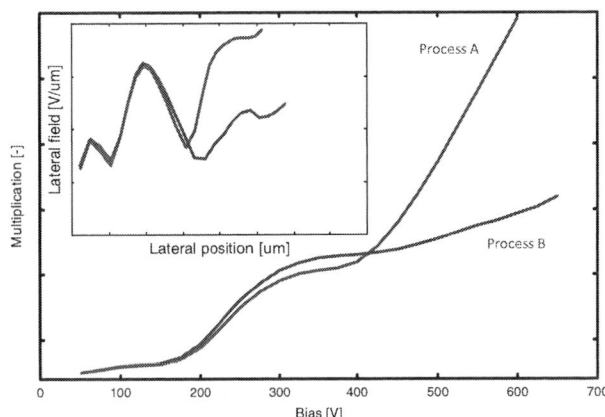

Figure 3. Multiplication and extracted lateral field distribution (inset, derived using the technique described in [4]) for two process settings with different lateral field in the drain extension but overall similar build and implant settings.

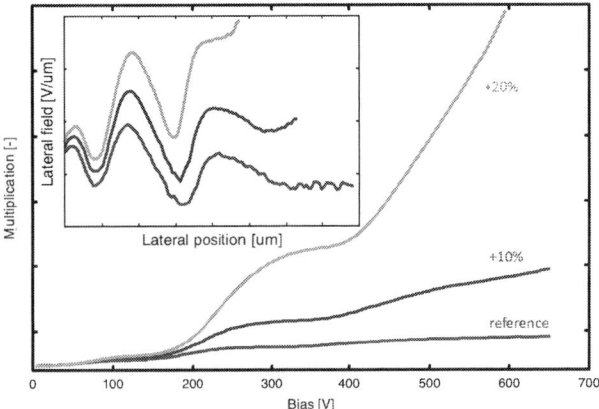

Figure 4. Multiplication and extracted lateral field distribution (inset) as a function of drain extension doping (Process A).

Note that process settings well outside the typical process-window were used to induce significant degradation within a reasonable time-scale.

B. Measurement procedure

All life tests were performed on-wafer with the exception of the room-temperature off-state tests that, because of their duration, were done on packaged samples.

To obtain insight in the degradation mechanism, the on-wafer stress measurements were periodically interrupted to perform an off-state leakage and a multiplication measurement. It was verified that these characterisation measurements did not significantly add to the overall degradation. Although the latter can be easily satisfied for sub-threshold stress tests, simply by limiting the currents during the multiplication measurement below the stress currents, it is clearly much less obvious when performing an off-state stress measurement. Limiting the currents during characterisation as well as their duration proved to result in comparable time-to-fail, defined by a sharp increase in multiplication and/or 'eakage current, with or without characterisation.

978-1-4244-8425-6/11 $26.00 © 2011 IEEE

V. EXPERIMENTAL RESULTS

A. Sub-threshold (on-state) stress

The dominant role of the lateral electric field in on-state HCI degradation is shown in Fig.5, revealing up to three orders of magnitude difference in on-state degradation for the two different process settings. At higher lateral fields (higher in process A than in process B) less HCI is needed before the critical failure point is reached. The latter is also shown by the extension dose dependence, the lateral field scaling with the dose. The temperature dependence of the hot carrier degradation during sub-threshold on-state stress was already shown in [4] to correspond to typical HCI: a higher temperature reduces the amount of HCI.

A typical evolution of the multiplication during on-state stress is shown in Fig.6. The extracted corresponding lateral fields (inset) indicate a broad region within the drift region affected by HCI.

B. Off-state stress

Comparing the evolution of the multiplication and electric field distribution during on- and off-state stress clearly indicates two completely different signatures (Fig.6,8). While on-state stress degradation exhibits a broad region affected by HCI, off-state stress results in a very localised injection at the very end of the depletion region where the vertical field is highest.

The dominance of the vertical field during off-state stress is further supported by the fact that off-state degradation in both processes A and B occurs after similar stress duration (Fig.7) as well as its limited dependence on the extension dose (Fig.9). Finally, the off-state life-time exhibits a very strong inverse temperature dependence (Fig.10). All these observations are in sharp contrast to that observed earlier for on-state stress.

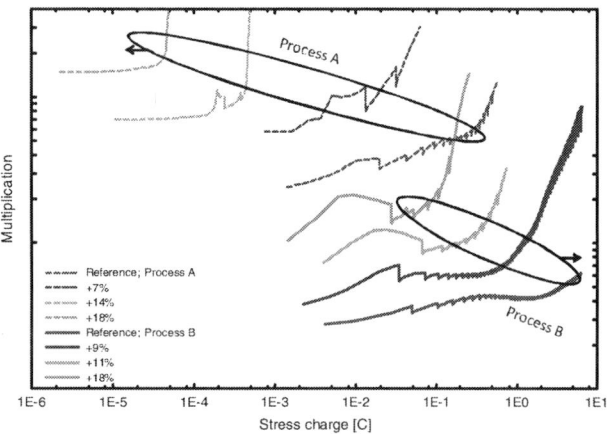

Figure 5. Evolution of the multiplication during on-state stress for various extension doses. As the degradation scales linearly with the applied stress current, the results are plotted versus the stress charge (integrated stress-current over time). Note the hugely reduced robustness for process A.

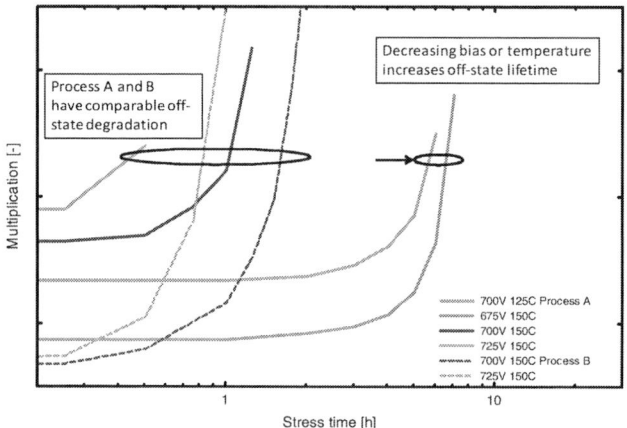

Figure 7. Evolution of the multiplication during off-state stress. Characterisation and stress correspond to the same drain bias.

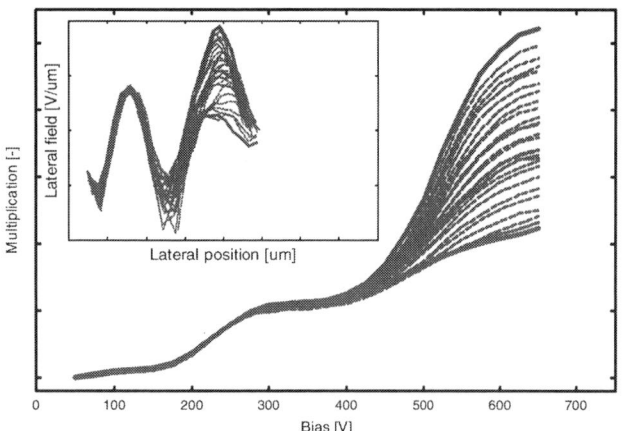

Figure 6. Typical evolution of the multiplication as a function of bias during on-state stress, inset shows the corresponding extracted lateral field distribution.

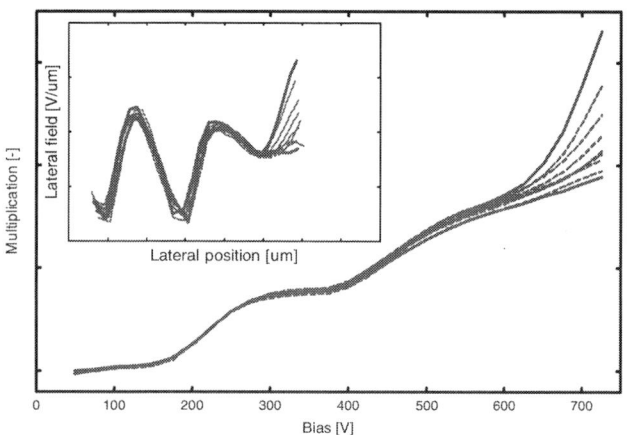

Figure 8. Typical evaluation of the multiplication as a function of bias and extracted lateral field during off-state stress. Note the different signature as compared to that found for on-state stress (see Fig.6).

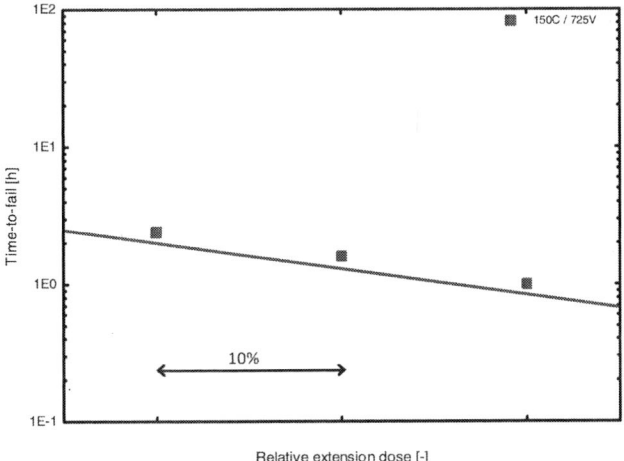

Figure 9. Off-state life-time versus extension dose.

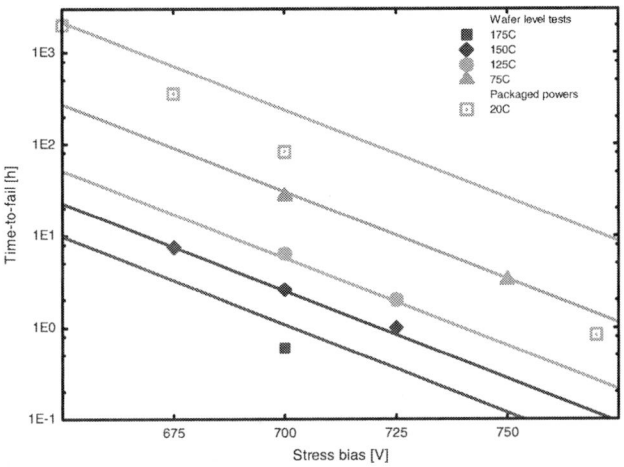

Figure 10. Measured off-state life-times as function of bias and temperature. The solid yellow line corresponds to an extrapolation to room temperature based on a simple fit to the results obtained from the wafer level tests at elevated temperatures.

Finally, our off-state wafer level tests (WLT) at elevated temperatures were extrapolated to typical off-state qualification measurements at room-temperature on large (packaged) power devices (Fig.10). Note that the packaged devices are of a slightly different design, with a typically lower BV_{ds} as compared to the designs used during WLT at the extension doses involved. Despite the latter, predictions from the WLT measurements are reasonably in line at stress biases close to their nominal voltage rating (650V).

CONCLUSIONS

From the previous paragraphs, it is clear that the degradation mechanisms for on- and off-state stress are dominated by different mechanisms, exhibit distinct HCI signatures and feature dissimilar dependencies. Most notably off-state degradation is greatly enhanced at increased temperature, in contrast to on-state stress.

Optimisation of a HV-SOI device therefore depends strongly on the application and the stress profile that it is subjected to throughout its lifespan. Devices operating at high voltages in (semi-) on-state greatly benefit from a relaxed design in which the lateral fields and multiplication are minimised. Operation at high temperatures beneficially increases the lifetime. In contrast, devices that essentially only experience high bias in off-state current conditions, such as power convertors that utilise zero-voltage switching, allow a much more aggressive design with lower $R_{ds,on}$ due to the limited dependence on extension dose. The resulting reduction in dissipation helps to limit the device's operating temperature, thereby partially compensating the de-rating with temperature.

REFERENCES

[1] D. Varghese et al., "Off-state degradation in drain-extended NMOS transistors: Interface damage and correlation to dielectric breakdown", IEEE Trans. Elec. Dev. 54, 2007, pp. 2669-2678.

[2] P. Moens, D. Varghese, M. Alam, "Towards a universal model for Hot Carrier Degradation in DMOS transistors", Power Semiconductor Devices & IC's (ISPSD), 2010, pp. 61-64.

[3] T. Letavic et al., "600 V power conversion system-on-a-chip based on thin layer silicon-on-insulator", Power Semiconductor Devices and IC's (ISPSD), 1999, pp. 325-328.

[4] R. van Dalen, A. Heringa, P.W.M. Boos, A.B. van der Wal, M.J. Swanenberg, "Using multiplication to evaluate HCI degradation in HV-SOI devices", Power Semiconductor Devices & IC's (ISPSD), 2010, pp. 89-92.

[5] S. Poli et al., "Investigation on the temperature dependence of the HCI effects in the rugged STI-based LDMOS transistor", Power Semiconductor Devices & IC's (ISPSD), 2010, pp. 311-314.

Proceedings of the 23rd International Symposium on Power Semiconductor Devices & IC's
May 23-26, 2011 San Diego, CA

A Novel Silicon-Embedded Coreless Transformer for Isolated DC-DC Converter Application

Rongxiang Wu and Johnny K.O. Sin
Department of Electronic and Computer Engineering
Hong Kong University of Science and Technology
Clear Water Bay, Kowloon, Hong Kong
Email: eewurx@ust.hk; eesin@ust.hk

S.Y. (Ron) Hui
Department of Electronic Engineering
City University of Hong Kong
Kowloon, Hong Kong

Abstract— In this paper, a novel silicon-embedded coreless transformer (SECT) is proposed and demonstrated for isolated dc-dc converter applications. By embedding two interleaved thick Cu coils in the bottom layer of the Si substrate, the designed 2 mm^2 SECT can achieve a maximum transformer efficiency of 85% at 50 MHz. Compared to the on-silicon coreless power transformer reported earlier for 0.5 W isolated dc-dc conversion at 170 MHz with a transformer efficiency of over 70%, the much lower operating frequency of the SECT allows the power losses of the power MOSFETs and Schottky diodes to be reduced by around 50%, leading to a converter loss reduction of 38%. Since only 4 vias are opened at the top layer of the substrate, most of the top layer of the substrate can be used for power IC implementation to achieve efficient monolithic integration. Experimental results show that the SECT provides a maximum transformer efficiency of 73% at 50 MHz, which is lower than expected and due to the non-optimized isolation oxide used.

I. INTRODUCTION

Monolithic power converters are highly desirable for the rapidly increasing portable electronic applications [1]. As a result, monolithic integration of the related power electronic components has attracted a lot of research interest recently. Fig. 1 shows the block diagram of a basic isolated dc-dc converter. For monolithic isolated dc-dc conversion, a monolithic power transformer is needed. Several monolithic magnetic-core power transformers have been reported previously [2, 3]. However, these transformers have large sizes (in the order of 10 mm^2) as the poor high frequency performance of the magnetic core prevents their size to be reduced by increasing the operating frequency. On the other hand, coreless transformers have been developed and commercialized for various power electronic applications [4, 5]. A 2 mm^2 on-silicon coreless power transformer has been reported earlier for 0.5 W isolated dc-dc conversion with an estimated transformer efficiency of over 70% [6, 7]. However, the small coil inductance and large coil resistance of the on-silicon coreless power transformer requires the converter to be operated at a very high frequency of 170 MHz in order to provide acceptable transformer performance. This high

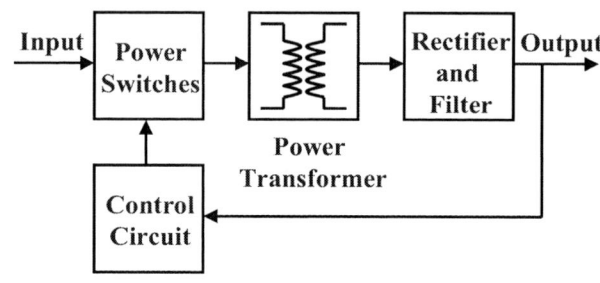

Figure 1. Block diagram of a basic isolated dc-dc converter.

operating frequency results in significant power MOSFETs gate drive loss and Shottky diodes reverse recovery loss, leading to a poor converter efficiency of 33%.

In this paper, a novel silicon-embedded coreless transformer (SECT) will be proposed and demonstrated for isolated dc-dc converter application. Having larger coil inductance and smaller coil dc resistance than the on-silicon coreless power transformer, the SECT allows a much lower operating frequency and reduced converter loss for power transfer, making it very promising for monolithic isolated dc-dc conversion.

II. TRANSFORMER DESIGN

In order to operate at lower frequency, monolithic coreless power transformer is required to have a larger coil inductance in order to maintain an acceptable coil reactance. This in turn requires the coil to be made with more turns within the limited device area. This leads to a larger coil resistance due to the reduced coil track width and increased total coil track length, resulting in a lossier transformer. To solve this problem, thicker coil tracks are required to reduce the coil resistance. By embedding two interleaved thick Cu coils in the bottom layer of the Si substrate, the SECT is realized monolithically, allowing a smaller Cu track width and more turns to be applied for a larger inductance, while keeping a small coil resistance. Fig. 2 shows the schematic 3-D view of the SECT structure. Since only 4 vias are opened at the top layer of the

---

This work was partially supported by the City University of Hong Kong under the Strategy Research Grant 7002218.

978-1-4244-8425-6/11 $26.00 © 2011 IEEE

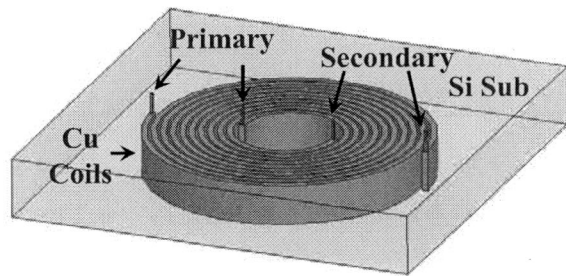

Figure 2. Schematic 3-D view of the SECT.

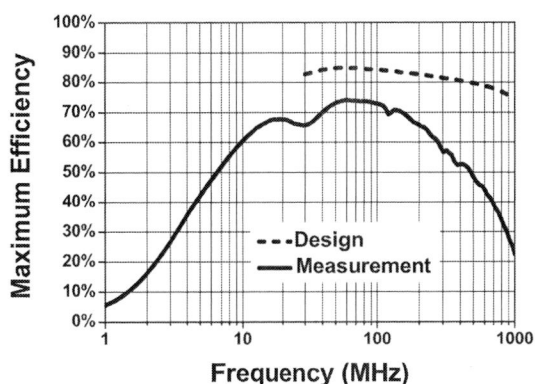

Figure 3. Maximum transformer efficiency of the SECT.

substrate, most of the top layer of the substrate can be used for power IC implementation to achieve efficient monolithic integration.

For 0.5 W isolated dc-dc conversion, a 2 mm² SECT was designed with 5 turns for each coil, with an isolation oxide thickness of 1 μm, and with Cu track thickness, width and spacing of 100 μm, 15 μm and 10 μm, respectively. The performance of the SECT was simulated using Ansoft's HFSS (high frequency structure simulator). Assuming optimal load is used, the maximum transformer efficiencies η_{MAX} achievable at different operating frequencies are calculated from the simulated two-port s-parameters using the following equations [8], and as shown in Fig. 3:

$$\eta_{MAX} = |S_{21}/S_{12}| \left(K - \sqrt{K^2 - 1} \right), \qquad (1)$$

$$K = \frac{1 - |S_{11}|^2 - |S_{22}|^2 + |S_{11}S_{22} - S_{12}S_{21}|^2}{2|S_{12}S_{21}|}. \qquad (2)$$

It can be seen that a maximum transformer efficiency of 85% could be achieved at 50 MHz. The simulated performance of the SECT is compared with the on-silicon coreless transformer [6, 7] in Table I. It can be seen that the SECT shows a much larger coil inductance of 36 nH (4.5x) and a lower coil dc resistance of 0.46 Ω (-43%, assuming an electroplated Cu resistivity of 4 Ω-cm), allowing similar reactance and quality factor to be achieved at a much lower frequency of 50 MHz (0.3x).

By assuming that the power MOSFETs, power transformer, and Schottky diodes (main sources of power loss [6]) have equal power losses, the total power loss of the converter using the designed SECT for 0.5 W isolated dc-dc conversion [6, 7] is estimated. For the Schottky diodes, the power losses include the conduction loss $P_{C,D}$ and the reverse recovery loss $P_{RR,D}$ as given below:

$$P_{LOSS,D} = P_{C,D} + P_{RR,D} = V_{on,D} I_{on,D} D_D + K_{RR,D} f, \qquad (3)$$

where $V_{on,D}$, $I_{on,D}$, and D_D are the on-state voltage (estimated to be 0.4 V), on-state current (0.1 A), and duty ratio of the diodes, respectively, and $K_{RR,D}$ is a frequency-independent

TABLE I
SECT AND ON-SILICON CORELESS [6, 7] TRANSFORMER PERFORMANCE COMPARISON

	On-Silicon Coreless [6, 7]	SECT Design	Measure
Area (mm²)	2	2	2
R_{DC} (Ω)	0.8	0.46	0.44
f (MHz)	170	50	50
L (nH)	8	36	36
X (Ω)	8.5	11.4	11.2
Q	10.7	9.2	4.1
k	N/A	0.97	0.95
Transformer η	over 70%	85%	73%

parameter. For the power MOSFETs, the switching loss and body diode reverse recovery loss are eliminated using resonant switching techniques. The remaining conduction loss $P_{C,M}$ and gate drive loss $P_{GD,M}$ of the power MOSFETs are estimated as:

$$
\begin{aligned}
P_{LOSS,M} &= P_{C,M} + P_{GD,M} = K_{C,M}/A_M + K_{GD,M}A_M f \\
&\geq 2\sqrt{K_{C,M}K_{GD,M}}\sqrt{f}
\end{aligned}
, \qquad (4)
$$

where $K_{C,M}$, and $K_{GD,M}$ are area-independent and frequency-independent parameters, and A_M and f are the power MOSFET area and operating frequency, respectively. For the power transformer, the DC loss $P_{DC,T}$ and AC loss $P_{AC,T}$ are estimated as:

$$
\begin{aligned}
P_{LOSS,T} &= P_{DC,T} + P_{AC,T} = I_S^2 \left(R_{DC,P} + R_{DC,S} \right) + I_M^2 R_{AC,P} \\
&= I_S^2 \left(R_{DC,P} + R_{DC,S} \right) + K_T / X_P Q_P
\end{aligned}
, \qquad (5)
$$

where I_S, $R_{DC,P}$, $R_{DC,S}$, I_M, $R_{AC,P}$, X_P and Q_P are the secondary current, primary dc resistance, secondary dc resistance, magnetizing current, primary ac resistance, primary reactance, and primary quality factor, respectively, and K_T is a parameter not related to the transformer characteristics. The calculated

978-1-4244-8425-6/11 $26.00 © 2011 IEEE   353

total power loss of the converter using the designed SECT is compared with that of the converter using the on-silicon coreless counterpart [6, 7] as shown in Table II. Results show that a total converter loss reduction of 38% can be obtained by using the designed SECT.

III. EXPERIMENTAL RESULTS AND DISCUSSION

The SECT is fabricated based on the silicon-embedded coreless power inductor technology reported previously [9]. The fabrication was carried out with a 315 μm-thick, 17 Ω-cm, double-side-polished p-type silicon wafer. Fig. 4 illustrates the schematic cross-sections of the SECT at major fabrication steps. Trenches for vias and trenches for coils are etched from the two sides of the Si wafer, deposited with low-temperature-oxide (LTO), and filled with electroplated Cu. Polishing is then performed to remove the over-plated Cu, and pads are formed by masked wet etching.

The SECT was characterized by on-wafer measurements. Using the four point probe method, the dc resistances of the primary and secondary coils of the SECT were measured to be 0.44 Ω and 0.45 Ω, respectively, corresponding to an electroplated Cu resistivity of 3.9 μΩ-cm. Using Rohde & Schwarz's ZVB8 vector network analyzer, one-port s-parameter of the SECT primary coil was measured while the secondary coil is open. After converting the measured one-port s-parameter to one-port z-parameter, the primary coil inductance L_P and quality factor Q_P are calculated and shown in Fig. 5. The low frequency primary coil inductance is found to be 35.4 nH at 1 MHz. A peak quality factor of 5.8 is achieved at 25 MHz, with an inductance of 34 nH. A resonant peak is observed at 70 MHz, with an inductance of 36.3 nH. The self-resonant frequency is around 280 MHz. Two-port s-parameters of the SECT were also measured with one port added to the primary coil and the other port added to the secondary coil. Fig. 6 shows the coupling factor k of the SECT calculated from the two-port z-parameters which are converted from the measured two-port s-parameters. It can be seen that a very good coupling factor of 0.95 is achieved up to 100 MHz. Assuming optimal load is used, the maximum transformer efficiencies η_{MAX} achievable at different operating frequencies are calculated by (1) and (2) using the measured two-port s-parameters, and compared to the designed performance in Fig. 3. It can be seen that a maximum transformer efficiency of 73% is achieved at 50 MHz, which is much lower than the designed value of 85%. The measured performance of the SECT is also summarized and compared to the designed performance in Table I. It can be seen that the lower measured transformer efficiency is mainly due to the lower quality factor. This is caused by the non-optimized isolation oxide deposition process (poor conformity) used, which leads to a significant substrate loss through the capacitive coupling between the coils and the substrate; although 3 μm of LTO were deposited from both sides of the wafer for isolation, the equivalent oxide thickness inside the trenches is only around 30 nm. This is verified by the simulation results obtained in Fig. 7. This poor isolation oxide thickness has also been confirmed by the measured DC breakdown voltage of only around 20 V between the two coils,

TABLE II
CALCULATED CONVERTER POWER LOSS FOR CONVERTERS USING THE DESIGNED SECT AND THE ON-SILICON CORELESS COUNTERPART [6, 7]

	Calculated on-silicon coreless [6, 7]	Designed SECT simulation
Schottky Diodes	0.34 W	0.16 W (-53%)
Conduction	0.08 W	0.08W
Reverse Recovery	0.26 W	0.08 W (-69%)
Power MOSFETs	0.34 W	0.18 W (-47%)
Power Transformer	0.34 W	0.29 W (-15%)
DC conduction	0.02 W	0.01 W (-50%)
AC conduction	0.32 W	0.28 W (-13%)
Total	1.02 W	0.63 W (-38%)

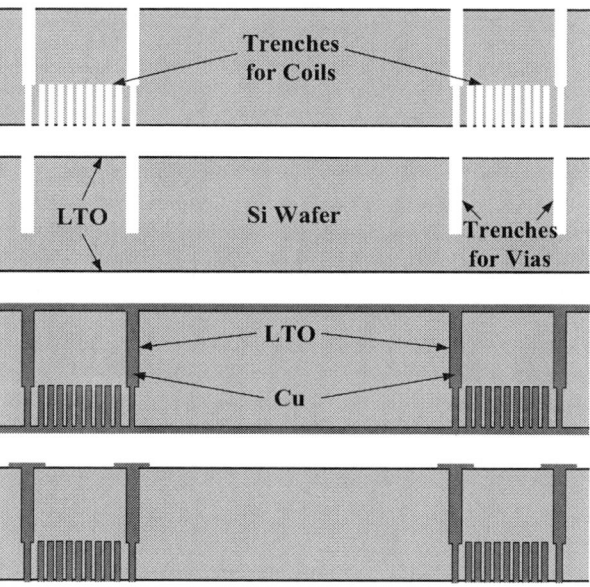

Figure 4. Schematic cross-sections of the SECT at major fabrication steps.

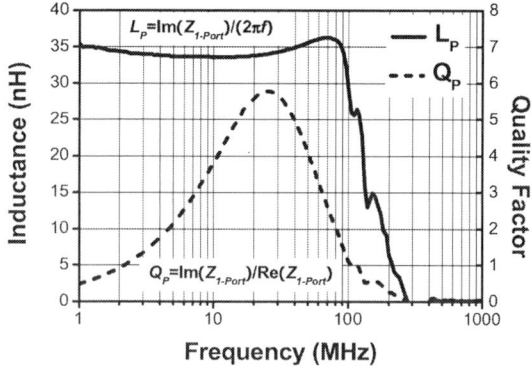

Figure 5. Experimental primary coil inductance and quality factor of the SECT.

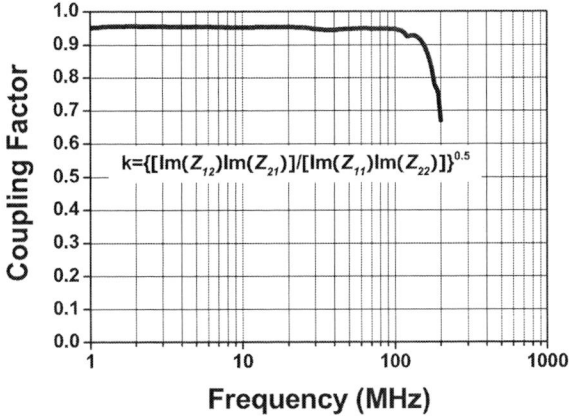

Figure 6. Experimental coupling factor of the SECT.

$$k=\{[\mathrm{Im}(Z_{12})\mathrm{Im}(Z_{21})]/[\mathrm{Im}(Z_{11})\mathrm{Im}(Z_{22})]\}^{0.5}$$

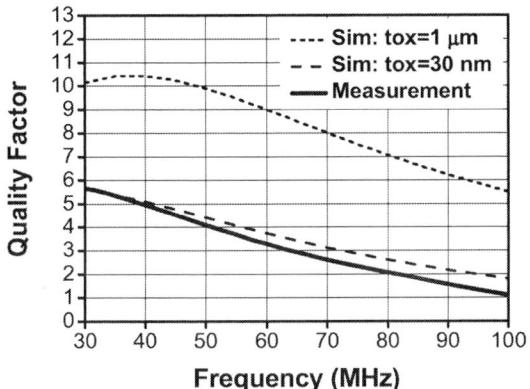

Figure 7. Comparison of the simulated and measured quality factors of the SECT.

considering that a breakdown strength of 600 V/μm has been reported earlier for LTO [10].

IV. CONCLUSION

A novel silicon-embedded coreless transformer (SECT) is proposed and demonstrated for isolated dc-dc converter applications. The designed 2 mm^2 SECT shows a maximum transformer efficiency of 85% at 50 MHz. Compared to the on-silicon coreless power transformer reported earlier for 0.5 W isolated dc-dc conversion at 170 MHz with a transformer efficiency of over 70%, the much lower operating frequency of the SECT allows the power losses of the power MOSFETs

and Schottky diodes to be reduced by around 50%, leading to a total converter loss reduction of 38%. Experimental results show that the SECT provides a maximum transformer efficiency of 73% at 50 MHz, which is lower than expected and due to the non-optimized isolation oxide deposition. The SECT shows great potential for monolithic isolated dc-dc converter appplications.

ACKNOWLEDGMENT

The authors would like to thank the staff of the Nanoelectronics Fabrication Facility and the Semiconductor Product Analysis and Design Enhancement Centre at the Hong Kong University of Science and Technology for their excellent support.

REFERENCES

[1] R. Foley, F. Waldron, J. Slowey, A. Alderman, B. Narveson, and S.C. O'Mathuna, "Technology roadmapping for power supply in package (PSiP) and power supply on chip (PwrSoC)", Applied Power Electronics Conference and Exposition, Palm Springs, USA, pp. 525-532, February 2010.

[2] C.R. Sullivan and S.R. Sanders, "Measured performance of a high-power-density microfabricated transformer in a dc-dc converter", Power Electronics Specialists Conference, Baveno, Italy, Vol. 1, pp. 287-294, June 1996.

[3] M. Brunet, T. O'Donnell, L. Baud, N. Wang, J. O'Brien, P. McCloskey, and S.C. O'Mathuna, "Electrical performance of microtransformers for dc-dc converter applications", IEEE Transactions on Magnetics, Vol. 38, No. 5, pp. 3174-3176, 2002.

[4] S.C. Tang, S.Y. (R.) Hui, and H.S.-H. Chung, "A low-profile low-power converter with coreless PCB isolation transformer", IEEE Transactions on Power Electronics, Vol. 16, No. 3, pp. 311-315, 2001.

[5] M. Munzer, W. Ademmer, B. Strzalkowski, and K.T. Kaschani, "Coreless transformer a new technology for half bridge driver IC's", International Exhibition and Conference for Power Electronics Intelligent Motion Power Quality (PCIM), Nuremburg, Germany, May 2003.

[6] B. Chen, "Fully integrated isolated dc-dc converter using micro-transformers", Applied Power Electronics Conference and Exposition, Austin, USA, pp. 335-338, February 2008.

[7] B. Chen, "Fully integrated isolated dc-to-dc converter and half bridge gate driver with internal power supply", 1st International Workshop on Power-Supply-on-Chip, Cork, Ireland, pp.40-41, September 2008.

[8] D.M. Pozar, Microwave Engineering, 3rd edition, New York: Wiley, pp.536-553, 2005.

[9] R. Wu and J.K.O. Sin, "A novel silicon-embedded coreless inductor for high frequency power management applications", IEEE Electron Device Letters, Vol. 32, No. 1, pp. 60-62, 2011.

[10] S.W. Nam, Y.W. Chang, S.B. Lee, and N.J. Kim, "Investigation on the LPCVD LTO thin film as a new dielectric layer for the future ULSI devices", Key Engineering Materials, Vol. 345-346, pp. 1549-1552, 2007.

978-1-4244-8425-6/11 $26.00 © 2011 IEEE

Proceedings of the 23rd International Symposium on Power Semiconductor Devices & IC's
May 23-26, 2011 San Diego, CA

Integrated Low Power and High Bandwidth Optical Isolator for Monolithic Power MOSFETs Driver

Nicolas Rouger, Jean-Christophe Crébier and Olivier Lesaint

Grenoble Electrical Engineering Lab (G2Elab)

Centre National de la Recherche Scientifique, Grenoble-INP, UJF-Grenoble university 1, Saint Martin d'Hères, France

Email: nicolas.rouger@g2elab.grenoble-inp.fr

Abstract—An integrated solution for the galvanic isolation between power transistors and their control unit is presented in this paper. This solution is based on a monolithic integration of a photodetector within a power MOSFET without any modification of its fabrication process. This photoreceiver can be associated with a monolithic driver to drive high side switches. Exhaustive characteristics for several integrated photodetectors are presented and discussed: quantum efficiency, step response, small signal analysis and sensitivity to the High Voltage MOSFET's Drain. The results of this analysis are photoreceivers with a Full Width at Half Maximum above 300MHz and a responsivity above 0.15A/W at a wavelength of 500nm. This leads to an integrated low power and high bandwidth optical isolation.

Index Terms—Monolithic integration, Power ICs, integrated optical sensor, power transistor gate driving circuits.

I. INTRODUCTION

SiC or GaN based power transistors are very attractive and start to exhibit outstanding performances [1]. However, the Silicon platform still has numerous advantages, most significantly its availability, reliability and performances. The heterogeneous and monolithic integration within Silicon power transistors is still a great way to improve not only the transistor itself, but the whole power switch function in power converters [2], [3]. Although several integrated solutions for driver supplies or gate drivers are available, only a few solutions have been presented to date for the integration of the required isolation between the power switch and its external control driver [4], [5]. However, these solutions remain difficult to monolithically integrate within power transistors. Considering integration related issues, the optical isolation is the perfect candidate for power transistors [6], annihilating Electro Magnetic Interferences between the remote control circuit and the power transistor. The most advanced and attractive solution so far for an integrated optical isolation has been recently presented in [7], [8]. This solution, however, requires more than one optical Watt for the power transistor's turn ON and is based on a heterogeneous assembly of a GaAs optical detector and a power transistor. Following our preliminary results [6], we propose hereinafter a different approach than in [7], [8]. The monolithically integrated photo receiver acts as an optocoupler and delivers a small current to a monolithically integrated or flip chipped gate driver [9]. The gate driver will then use an efficient and integrated supply [10] to deliver the energy for the switching of the power transistor. In this paper, we will demonstrate the benefits and possible drawbacks of an integrated photodetector.

II. INTEGRATION APPROACH

In order to optimize the efficient operation of power transistors, their associated drivers must be located as close as possible to them. Indeed, any parasitic inductance between the power transistor and the driver circuit can drastically reduce the switching performances dynamics and can even introduce undesired firing. The monolithic integration of all the required driving functions looks therefore attractive. Our goal is here to provide a generic power switch function that can be used either as a high side or low side switch, while integrating the required isolation circuit with the highest isolation capability. Our integration approach in this context is to monolithically integrate a light-sensitive detector within the power VDMOS without any modification of its fabrication process. As previously demonstrated, other required functions can also be monolithically integrated within VDMOS (loss free gate driver supply [3]). In the view of improving the driver's intelligence (advanced sensors and protection circuits), a flip chipped CMOS driver die [9] can be preferably used. The resulting approach is presented in figure 1.

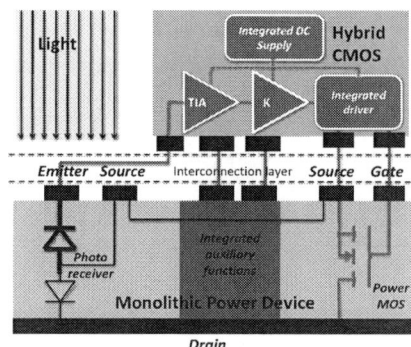

Fig. 1. Integration approach: a monolithically integrated photodiode, connected to a flip-chipped readout circuit (TransImpedance Amplifier + K voltage gain), a gate driver and an efficient self driver supply.

The power die monolithically integrates the power transistor cells, the photodetector and other required functions for the driver self supply technique [11]. The CMOS die integrates the required circuits for the photodiode readout circuit, the

978-1-4244-8425-6/11 $26.00 © 2011 IEEE   356

self supply operation and the gate current amplification. Since the switching of power MOSFET usually requires at least hundreds of mW, transferring this power by light will definitely not be efficient (limited by the light emitter efficiency, transmission channel attenuation and detector responsivity). The integrated photoreceiver acts here as a sensor while converting optical power into a generated low current which is then filtered, amplified and detected by the CMOS circuits. The required energy for the power transistor's switching transitions is then taken from a storage capacitor. This integrated Power System On Chip would be very attractive since it will be autonomous, externally driven by a low power optical signal. However, this solution will be possible only with a highly sensitive and fast photodetector, which must not be affected by the common High Voltage drain. The next sections will focus on these requirements.

III. DESIGN AND NUMERICAL SIMULATION OF INTEGRATED PHOTODETECTORS

A. Photodetector structure

The integrated photo receiver takes benefit of the parasitic NPN bipolar structure built in power VDMOS figure 2. Additional P+ highly doped base regions can be required, depending on the P- well concentration. The Collector backside electrode is merged with the High Voltage VDMOS Drain electrode. While connecting the Base region to the VDMOS lowest potential (Source), the vertical Collector-Base N++/Nν/P- junction will always be reverse biased. As a consequence, the N+/P- Emitter-Base junction can be used as an integrated diode, as long as the Emitter-Base junction is not forward biased. Since this Emitter-Base junction will act as the photo sensitive junction and should not be affected by the Drain/Collector potential, the Emitter-Base junction must always be reverse biased and the vertical bipolar transistor must not be activated neither during Drain voltage transients nor steady state or dynamic photo generation within the vertical device.

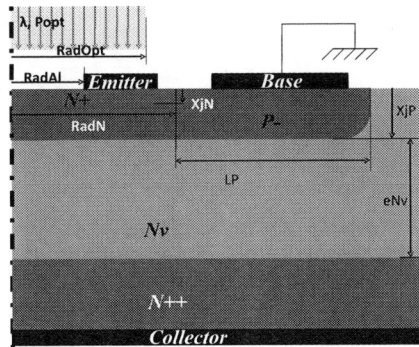

Fig. 2. Cross section of the integrated photodiode and its parameters. The center vertical axis represents the axis of revolution corresponding to the cylindrical structure of the photodiode.

Considering the proposed structure, the Base current will be the sum of the Emitter-Base leakage current and the Collector-Base leakage current. In this structure, the Emitter current will be used as the optically generated electrical current. One can notice that this structure with diffused P and N regions is definitely not optimized for a photodetector. Usually, fast and efficient optical detectors are made of an intrinsic/lightly doped region and two thin highly doped contact regions. Such classical designs maximize the drift phenomenon of the free carrier generated by light absorption within a wide space charge region. Diffusion currents are kept to a minimum due to the quantum efficiency and electrical bandwidth reductions by recombination and diffusion. In our design, diffusion currents generated by optical absorption are not negligible relatively to drift currents. Long current paths (recombination) must be avoided between the Emitter contact and the free carriers generation regions in order to increase the responsivity of the device. However, shadow effects and strong optical absorption of the Emitter contacts must also be prevented. The control of carrier lifetimes is also challenging in this context and for fast integrated photodetectors.

B. DC modeling

Since doping profiles of the integrated Emitter-Base junction are not uniform and close to Gaussian profiles, an analytic modeling is therefore impossible especially in 3D. Numerical simulations were conducted in the Silvaco software Suite with the following assumptions:

- The epi $N\nu$ drift layer is designed for a 600V voltage breakdown capability. Accordingly, the epi region doping is $2 \cdot 10^{14} cm^{-3}$ with $eN\nu = 50\mu m$,
- The Base thickness X_{jP} is $5\mu m$ with a surface concentration of $5 \cdot 10^{16} cm^{-3}$,
- The Emitter surface concentration is $1 \cdot 10^{20} cm^{-3}$,
- The optical beam power P_{opt} is uniformly distributed to the photodetector sensitive area (collimated beam). Therefore, the Irradiance I_{opt} is constant within the optical beam with the relationship $P_{opt} = I_{opt} \cdot \pi Rad_{Opt}^2$,
- The complex refractive index of Silicon is wavelength dependant,
- For faster simulations, only $5\mu m$ of the backside N++ contact region are simulated,
- Since intrinsic performances are first to be demonstrated, the electrical contacts are assumed ohmic, and aluminium contacts' thickness infinitely thin,
- Shockley-Read-Hall and Auger recombination processes are activated during simulations, as well as band-gap narrowing, concentration and field dependent mobilities,
- Impact ionization is also activated for high Collector voltages.

Figure 3 presents the total Quantum Efficiency (QE) of several different devices, where the N+ Emitter thickness X_{jN} was varied in the range of $0.7\mu m$, $1\mu m$ and $1.2\mu m$. The QE is calculated accordingly to $QE(\lambda) = \frac{I(\lambda)}{S \cdot \frac{q \cdot I_{opt} \cdot \lambda}{h \cdot c}}$ for each current. The denominator in the QE equation is commonly called the source photocurrent, as it represents the maximum possible current that can be generated by the detector, i.e. a 100% QE optical detector. Knowing that the light absorption coefficient

α is reduced for higher wavelengths, free carriers will be generated deeper in the Silicon device for higher wavelengths. For wavelength below $400nm$, α is higher than 10^5 cm^{-1}, corresponding to an absorption depth below $100nm$. As a result, most of the free carriers generated close to the sensitive surface are recombining through diffusion. Moreover, Silicon is highly dispersive at short wavelengths, leading to increased reflections below $400nm$. The overall QE is therefore limited when short wavelengths are used. The highest QE for the Emitter current is 40% at $\lambda = 500nm$ whereas the highest QE for the Base current is 60% at $\lambda \approx 700nm$. Considering the Air/Silicon optical interface, more than 33% of the incident optical power is lost due to reflections.

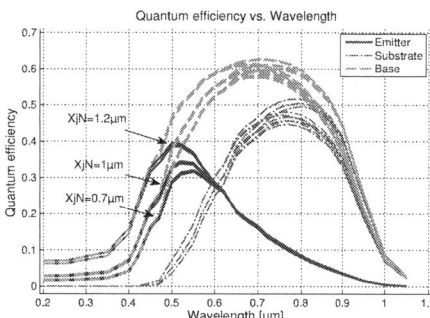

Fig. 3. Quantum Efficiency computed with a 3D Finite Element Analysis. Device parameters: $Rad_{Opt} = Rad_N = 4.5\mu m$, $P_{Opt} = 10; 20; 100\mu W$, $Rad_{Alu} = 0; 2.5\mu m$ and $X_{jN} = 0.7; 1; 1.2\mu m$.

Since both surface concentration of the N+ Emitter region and P- Base region are fixed, and both doping profiles are Gaussian, increasing the N+ Emitter region's depth X_{jN} will reduce the doping charge density around the P-/N+ junction. As a result, the space charge region at zero Emitter-Base bias will be wider with larger values of X_{jN}. Therefore, the QE of deeper Emitter regions will be increased due to reduced recombination levels. This is clearly shown by figure 3. However, N+ implantation dose limitation and width reduction of the P- pinched region reduces the maximum possible X_{jN}. The relationship between the QE and the responsivity of the photodetector is $R_\lambda(\lambda) = QE(\lambda) \cdot q\frac{\lambda}{h \cdot c}$. For the best detector with $X_{jN} = 1.2\mu m$ the responsivity is around $0.16A/W$ at $500nm$. No significant effects have been noted with small modifications of the Aluminium contact nor the optical power, considering our assumptions. It was also verified that the generated current does not increase significantly the base potential to guarantee the Emitter-Base reverse bias - Figure 4.

C. Small signal and large signal modeling

Both the transient response and the small signal analysis must be investigated. Figure 5 presents the emitter current transient response for different optical detectors where the optical beam and emitter radii have been modified as well as the wavelength, while keeping a same optical power step. All the simulated designs exhibit a step response below $3ns$, even

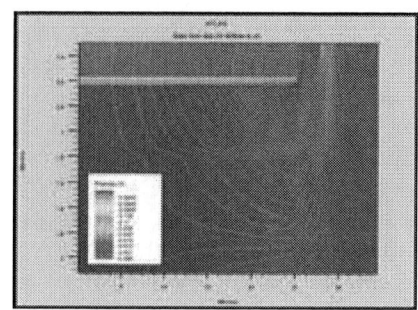

Fig. 4. Potential distribution with a $100\mu W$ - $\lambda = 500nm$ - $Rad_{Opt} = 25\mu m$ optical beam. Locally, the potential is increased by less than 0.05V. Current flow lines are also plotted.

for larger optical devices. One can also notice that when the optical beam radius is much higher than the emitter radius, the QE of the device is clearly decreased due to increased recombination. Since smaller devices also have smaller capacitances and smaller current paths, the best transient response is obtained with $Rad_{Opt} = 4.5\mu m$ and $Rad_N = 4.5\mu m$. This analysis is confirmed by the small signal analysis - figure 6. The dynamic behavior of the transient response is confirmed by the small signal analysis. The fastest detector has a 3dB bandwidth above $1GHz$ whereas the slowest is still above $300MHz$. Small devices are definitely attractive when illuminated by a $500nm$ or $550nm$ wavelength optical beam.

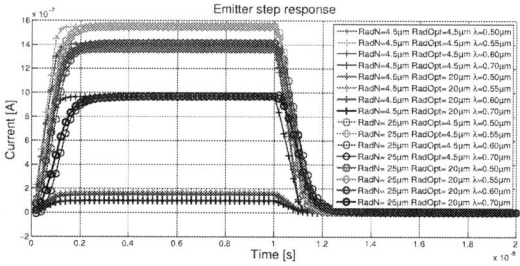

Fig. 5. Emitter current transient response to a $10\mu W$ optical power step. Device parameters: $Rad_{Opt} = 4.5; 20\mu m$, $Rad_N = 4.5; 25\mu m$, $\lambda = 500; 550; 600; 700nm$ and $X_{jN} = 1\mu m$.

D. Effects of the high voltage drain

When the Drain/Collector voltage is increased, several phenomena are affecting the isolated Emitter-Base photodetector. First, the Collector-Base space charge region is extending within the Base region, therefore reducing the Emitter QE. Numerical simulations show that the Emitter QE is reduced by 12% at $\lambda = 600nm$ and 3% at $\lambda = 500nm$ when the Collector is biased at 400V. Second, a high leakage current is created when a positive dV/dt is applied to the Drain/Collector. This Collector-Base capacitive current locally increase the potential of the base region therefore locally activating the vertical bipolar transistor. This is also more important when a high optical power is injected at the center of the photodetector. Applying a $400V/50ns$ Collector transient dV/dt within $50ns$

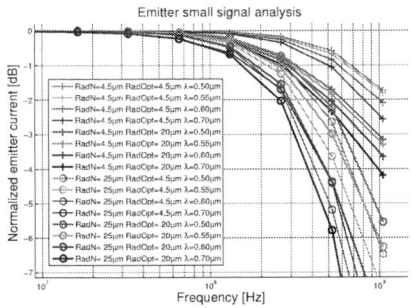

Fig. 6. Emitter current small signal analysis. Bias optical power is $10\mu W$ and small signal optical power amplitude is $1\mu W$. Device parameters: $Rad_{Opt} = 4.5; 20\mu m$, $Rad_N = 4.5; 25\mu m$, $\lambda = 500; 550; 600; 700nm$ and $X_{jN} = 1\mu m$. The emitter current was normalized relatively to the DC current for each design.

and under a $100\mu W$ optical beam did not locally forward bias the Emitter-Base junction. This dynamic self shielding was observed for every design with reduced Emitter and Base radii.

IV. DEVICE CHARACTERIZATION

An electro-optical setup has been realized - figure 7. Since low power coherent optical sources at $500nm$ are difficult to obtain, our first experimental characterizations use a $2mW$ - $640nm$ HeNe laser. The laser optical beam is spatially filtered and focused to the packaged photodetector. The laser was directly modulated. The radius of our devices was $Rad_N = 50\mu m$ and the best transient response observed was $0.2\mu s$. At such long wavelength, the substrate voltage also has a significant effect. However, this first experimental result demonstrates that an integrated photodetector is capable of a sufficient bandwidth, higher than 1MHz. The highest measured responsivity was 0.08A/W at 640nm. The differences between the numerical modeling and the experimental results are explained by larger devices (more than 4 times larger than a $25\mu m$ radius emitter region) and required parameters fitting.

V. DISCUSSION AND FUTURE WORKS

A detailed numerical analysis has been presented. Shorter wavelengths, smaller detectors and thicker emitter regions must be used in order to optimize both the DC and AC performances of the devices. Although the numerical analysis investigated key parameters, temperature effects must be taken into account. A deeper characterization of both DC and AC performances of the integrated photodetectors is also required. We are currently improving the electro-optical setup in order to characterize the QE of integrated detectors for a wide range of wavelengths. DC and AC performances for large detectors and long wavelength have been demonstrated as a first proof of concept.

ACKOWLEDGEMENT

The authors thank Asier Saenz de Urturi for his contribution to the characterization. Part of this work was supported by the French Research Agency under grant # ANR 2010 JCJC 0907 01 SiPowLight.

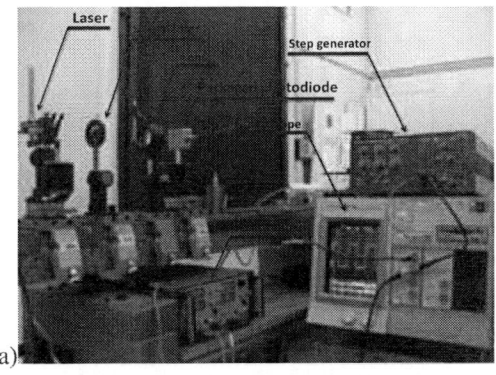

a)

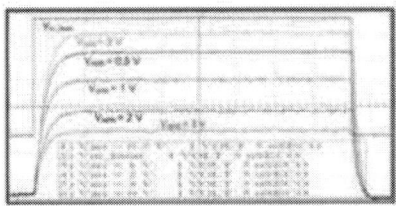

b)

Fig. 7.   a) Electro optical setup. b) Characterized substrate effect and photodiode step response.

REFERENCES

[1] K. Matocha *et al.*, "1400 volt, 5 mΩ /cm2 sic MOSFETs for high-speed switching," in *Power Semiconductor Devices IC's (ISPSD), 2010 22nd International Symposium on*, 2010, pp. 365 –368.

[2] F. Capy *et al.*, "New self-controlled and self-protected igbt based integrated switch," in *Power Semiconductor Devices IC's, 2009. ISPSD 2009. 21st International Symposium on*, 2009, pp. 243 –246.

[3] N. Rouger *et al.*, "Fully integrated driver power supply for insulated gate transistors," in *Power Semiconductor Devices and IC's, 2006. ISPSD 2006. IEEE International Symposium on*, 2006, pp. 1 –4.

[4] C. Batard *et al.*, "Wireless transmission of igbt driver control," in *Applied Power Electronics Conference and Exposition, 2009. APEC 2009. Twenty-Fourth Annual IEEE*, 2009, pp. 1257 –1262.

[5] S. Brehaut and F. Costa, "Gate driving of high power igbt by wireless transmission," in *Power Electronics and Motion Control Conference, 2006. IPEMC 2006. CES/IEEE 5th International*, 2006, pp. 1 –5.

[6] N. Rouger and J.-C. Crebier, "Integrated photoreceiver for an isolated control signal transfert in favour of power transistors," in *Power Semiconductor Devices and IC's, 2008. ISPSD '08. 20th International Symposium on*, May 2008, pp. 213 –216.

[7] S. Mazumder and T. Sarkar, "Optically-activated gate control of power semiconductor device switching dynamics," in *Power Semiconductor Devices IC's, 2009. ISPSD 2009. 21st International Symposium on*, 2009, pp. 152 –155.

[8] T. Sarkar and S. K. Mazumder, "Epitaxial design of a direct optically controlled gaas/algaas-based heterostructure lateral superjunction power device for fast repetitive switching," *Electron Devices, IEEE Transactions on*, vol. 54, no. 3, pp. 589 –600, 2007.

[9] T. Simonot, J. Crebier, N. Rouger, and V. Gaude, "3d hybrid integration and functional interconnection of a power transistor and its gate driver," in *Energy Conversion Congress and Exposition (ECCE), 2010 IEEE*, 2010, pp. 1268 –1274.

[10] J.-C. Crebier and N. Rouger, "Loss free gate driver unipolar power supply for high side power transistors," *Power Electronics, IEEE Transactions on*, vol. 23, no. 3, pp. 1565 –1573, May 2008.

[11] T. Simonot, N. Rouger, and J. Crebier, "Design and characterization of an integrated cmos gate driver for vertical power mosfets," in *Energy Conversion Congress and Exposition (ECCE), 2010 IEEE*, 2010, pp. 2206 –2213.

978-1-4244-8425-6/11 $26.00 © 2011 IEEE

Proceedings of the 23rd International Symposium on Power Semiconductor Devices & IC's
May 23-26, 2011 San Diego, CA

Design and characterization of a signal insulation coreless transformer integrated in a CMOS gate driver chip

Simonot Timothé, Rouger Nicolas, Crebier Jean-Christophe
Grenoble Electrical Engineering Lab
UMR 5269 CNRS / UJF-Grenoble University 1/Grenoble-INP
St Martin d'Hères, France

Arnould Jean-Daniel
IMEP LAHC
UMR 5130 CNRS / UJF-Grenoble University 1/Grenoble INP/Université de Savoie
Grenoble, France

Abstract— **With the development of multi-level, multiphase or network converters requiring the implementation of numerous distinct power transistor gate drivers, the control signal insulation is becoming more and more important in power converters. This paper presents an isolation technique based on a coreless transformer integrated in a CMOS silicon die together with the gate driver and other required functions. The associated demodulation circuit will also be presented, as the control signal must be modulated at a high frequency through the coreless transformer. The chosen design methodology will be explained and experimental results will be shown in order to validate the functionality.**

I. INTRODUCTION

An insulation system is often required to carry and bring the control signal of power transistors at their reference potential. Indeed, the voltage difference between their reference potential and the external remote command can be brought to high floating voltages. Insulation systems in most power electronics circuits are level shifters [1], optocouplers [2] or discrete magnetic or piezoelectric transformers [3, 4]. The transformers' benefits over other insulated control signal transfer systems are that they are not as sensitive to EMI and voltage ratings as level shifters, they are faster and less consumer than optocouplers [5]. Moreover, they offer a fairly high dielectric insulation range (up to several kV), depending on the thickness and dielectric strength of the insulation material between the two windings. Initially, magnetic insulation was carried out with discrete pulse transformers [3], then with coreless PCB transformers [6] and now with silicon integrated coreless transformers [7]. The main benefits of this integration is that the insulation system can be directly implemented within a CMOS gate driver with no extra technological steps in the fabrication process, while consuming a reasonable silicon area, presenting a large bandwidth and being directly interconnected to the gate driver. The use of a CMOS process also allows the integration of complex logic functions, which can be required for the demodulation of the gate control signal, and other protection and driving circuits. Finally transformers can be used to transfer data in both directions to send back useful information to the control unit for example.

This paper presents the design methodology of an integrated coreless transformer and its associated demodulation circuit for power electronics transistor driving purposes. First, a geometrical layout of the transformer is presented, based on geometrical design considerations deduced from analytic formulas and technological constraints coming from the foundry characteristics. A simple model of the transformer is then established using analytic formulas and finite-elements simulations with the Flux^R 2D software. This

a standard CMOS technology. Finally, the design is validated with a practical implementation, based on Austria MicroSystem's C35B4M3 technology, and experimental results and transformer characteristic curves are shown.

II. OPERATING PRINCIPLE AND DESIGN CONSIDERATIONS

A. Operating principle

Most of HF silicon integrated transformers have no magnetic core to guide the flux lines, which means that the coupling is severely reduced at low frequencies. The resonant frequency of this type of transformers is generally high, as the magnetizing inductance of the integrated transformer is small (around 100nH). Therefore, the remote control signal has to be modulated at a high frequency (hundreds of MHz) in order to minimize the losses inside the transformer. A modulation technique has to be chosen, depending on which signal is sent into the transformer. The PWM signal is generated in the external remote circuit. Corresponding to the ON state, a modulated high frequency sine wave carrier is applied to the transformer. When the OFF state is desired, a zero DC voltage difference is applied to the primary winding. This modulation technique is appropriate and simple. The carrier frequency can be set in a wide frequency range (between 90 and 500MHz) and the transmitted envelope is a square wave corresponding to the transistor triggering signal (PWM).

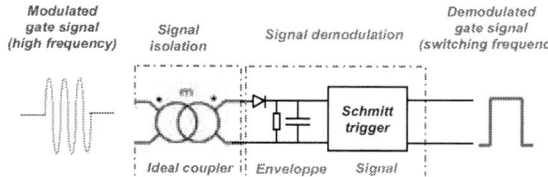

Figure 1: Control signal modulation principle

Figure 1 shows the principle of the insulation system. The sine wave signal is transferred through the transformer, then an envelope detection circuit at the secondary side winding of the transformer recovers the low frequency square wave modulating envelope. A Schmitt trigger is introduced to avoid false triggering of the transistor due to parasitic signals. The signal is finally demodulated and sent into the gate driver amplification circuit.

B. Design considerations

The design's main objective is to obtain the best coupling while using a silicon surface as small as possible. The technological data, conductor and isolator thicknesses and layout constraints, are then used to determine the geometry of the transformer. The chosen transformer topology is a stacked ransformer, using the multiple metal layers available in the

978-1-4244-8425-6/11 $26.00 © 2011 IEEE

selected CMOS technology to maximize the voltage insulation capability. As shown in figure 2, one can distinguish geometrical parameters such as outer and inner diameters, conductor width and spacing between the turns, and vertical parameters such as the conductor and oxide thicknesses between the windings, and oxide thickness between the windings and the substrate. The geometrical parameters are critical to determine the leakage and magnetizing inductances as well as the winding series resistance, and the coupling coefficient. The thickness of the conductors will modify the resistance of the windings, and the oxide thickness between the conductors and the substrate will change the capacitive coupling between the windings and to the substrate. The oxide thickness between the two windings is critical when insulation is needed, as the breakdown voltage is directly proportional to the thickness of the oxide according to equation (1)

$$Tox*DS=BV \qquad (1)$$

(with Tox : oxide thickness, DS dielectric strength and BV breakdown voltage)

The oxide used in IC fabrication is often low temperature silicon dioxide (SiO2), which usually has a dielectric strength below 10MV/cm. As a result, the breakdown voltage of the oxide is around 1kV for a 1μm-thick oxide. One must notice that the designer can not modify the vertical parameters since they are technology dependant in standard CMOS processes (set by the foundry). The design is then focused on the geometrical parameters to obtain the desired inductance, resistance and coupling coefficient, while minimizing the parasitic elements which reduce the coupling.

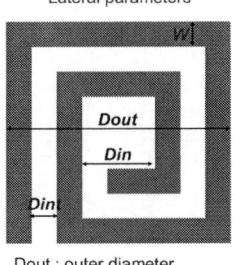

Figure 2: Geometrical parameters of the integrated transformer

Dout : outer diameter

Din : inner diameter

Dint : distance between the turns

W : conductor width

T : conductor thickness

Tox : oxide thickess between the two windings

Tsub : oxide thickness between the conductor and the substrate

In the next section, it will be shown how the lateral parameters alter the typical values of the integrated transformer, and which compromises must be optimized.

III. DESIGN GUIDELINES

The typical values of an integrated transformer are the windings inductance and resistance, the mutual inductance between the windings, and the coupling coefficient. The parasitic elements which deteriorate the coupling are at first the leakage inductance and winding resistance and then the capacitances between the windings, the capacitances between the windings and the substrate, and the substrate capacitance and resistance. Analytic formulas found in refs [8] and [9] describe these elements as a function of the geometric

parameters of the transformer. These formulas will be employed in this paper to see the evolution of the elements of the transformer depending on the geometry of the transformer in a simple manner

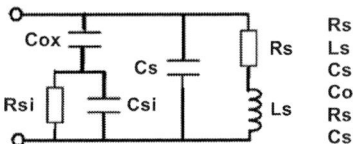

Rs : series resistance
Ls : series inductance
Cs : spiral capacitance
Cox : oxide capacitance
Rsi : substrate resistance
Csi : substrate capacitance

Figure 3: Physical model of an inductor on silicon

Figure 3 shows the equivalent model of an inductor on silicon, with the series inductance, resistance, and capacitance, and the parasitic elements. From this model, one can express the equivalent impedance of the parasitic elements Cox, Csi, Rsi of the spiral. The impedance of the spiral can be expressed with Rs, Cs and Ls. For high frequencies in the range of hundreds of MHz, it can be assumed that the term $(Ls*\omega)^2$ will be predominant over the term R^2 and $1/(C*\omega)^2$ in the equation of the spiral impedance, so the impedance of the spiral is only a function of the inductance of the spiral. Equation (2) expresses the equivalent parasitic impedance Zparas, and equation 3 the impedance of the inductance Xs.

$$Zparas := \left[\cfrac{1}{\cfrac{1}{\cfrac{1}{Rsi^2} + (Csi \cdot \omega)^2} + \cfrac{1}{(Cox \cdot \omega)^2}} \right]^{0.5} \qquad (2)$$

$$Xs := \left[(L \cdot \omega)^2 \right]^{0.5} \qquad (3)$$

From equations (2) and (3), the ratio Xs/Zparas can be expressed. It is a good parameter for the designer to see at which point the impedance of the spiral is predominant over the impedance of the parasitic elements for a given frequency. Another significant parameter that can be used to design the transformer is the coupling coefficient of the transformer. This parameter is the ratio of the spirals mutual inductance divided by their series inductance, as shown in equation (2), where Lp and Ls are respectively the primary winding and secondary winding inductances, and M the mutual inductance. In the case of a stacked transformer, the windings have identical geometries, so that Lp=Ls=L. One can then obtain the mutual inductance M=k*L. The coupling coefficient can be expressed as a function of the geometrical parameters with equation (4), where ds is the center to center distance of the turns, and davg the average diameter [8].

$$k := 0.9 - \frac{ds}{davg} \qquad (4)$$

The equations stated above show both the effect of parasitic impedance over spiral inductance and the coupling between the spirals as a function of the geometry of the transformer. On this basis, the design of the transformer's geometry can be made by searching the largest ratio of inductance impedance over parasitic impedance, while optimizing the coupling coefficient k.

A. Influence of the outer and inner diameters

First, the outer diameter is varied while setting the inner diameter as a the portion of the outer diameter (e.g.

978-1-4244-8425-6/11 $26.00 © 2011 IEEE

din=dout/4). As a result, the number of turns N will then increase when dout increases. The other geometrical parameters which are the conductor width and the distance between the turns are set respectively to 10μm and 2μm. The frequency is set to 100MHz.

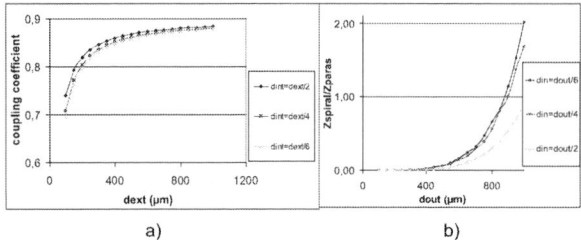

Figure 4: a) Ratio Zspiral/Zparas as a function of the outer diameter b) Coupling coefficient as a function of the outer diameter

As one can see on figure 4-a, the ratio Xs/Zparas is close to zero for outer diameters below 500μm, showing that parasitic impedance is predominant, and starts to increase exponentially for outer diameters above 500μm. The ratio is lower when the inner diameter is half the outer diameter, but is roughly the same when it is a quarter or a sixth of the outer diameter. Figure 4-b shows a weak coupling coefficient for small outer diameters, and coupling above 0.85 for outer diameters larger than 500μm. Figure 4-b also introduces that there is no significant influence of the inner diameter on the coupling coefficient. Then, to achieve a satisfactory coupling and low parasitic influence on the spiral inductance, the outer diameter should be larger than 500μm for the chosen frequency and spiral geometry. A compromise has then to be made between the silicon space used by the transformer and the desired ratio.

B. Influence of the spiral width

To see the influence of conductor width and spacing between the turns, these parameters will be wept while setting the outer and inner diameters respectively to 1mm and 250μm, and keeping a same frequency. As it can be deducted from equations, (2), (3) and (4), both the ratio Xs/Zparas and the coupling coefficient decrease when the width of the conductor increase, showing that the conductor width should be kept as small as possible. The width can then be chosen as a function of the skin depth, which should be the smallest value of the conductor width for the chosen design frequency. From the same equations, the influence of the spacing between the turns on both the impedance ratio and coupling coefficient is comparable to the influence of the conductor width, it should also be chosen as small as possible. The smallest value of this spacing is determined by the design rules of the technology used for the fabrication of the transformer.

IV. EXPERIMENTAL VERIFICATION

After having set these design guidelines, the transformer can be designed in the chosen technology and for the chosen frequency. We chose the technology AMS C35B4M3 from Austria MicroSystems, with 4 metal layers available. The chosen topology of the transformer is to use the upper metal layer for the primary winding and the lower metal layer for the secondary winding. The largest oxide thickness between the two windings is therefore achieved. The chosen geometry for the spirals is 600μm for the outer diameter and 100μm for the

inner diameter, resulting in a 1/6 ratio to maximize the spiral inductance. The chosen design frequency is 100MHz, so the conductor width is set to 16μm which is two times the aluminum skin depth at this frequency. The spacing between the turns is set to 2μm, which is the minimum spacing between two thick metal tracks in the C35B4M3 technology.

A. Analytic and Finite Elements simulation comparison

The resulting values of the transformer elements are calculated from the formulas previously used from [8] and [9], and compared to the values computed with the finite elements simulation software Flux 2D. The resulting values are shown in table I.

TABLE I.   COMPARISON OF TRANSFORMER MODEL VALUES

Transformer element	Calculated values	
	Analytic	Finite Elements
Series spiral inductance	90 nH	85 nH
Mutual inductance	76.9 nH	76,5 nH
Coupling coefficient	0.85	0.9
Primary winding series resistance (DC)	23.2 Ω	25.8 Ω
Secondary winding series resistance (DC)	97.7 Ω	107.5 Ω

The values in Table I show small differences between the analytic values and the values obtained with simulations. It can then be assumed that values computed from analytic equations are accurate enough over a wide range of parameter values. The series resistances of the windings are frequency dependant due to skin and proximity effects. In this case only signal is transmitted, so the current needed is only the magnetizing current of the transformer. As the period of the modulated signal is small, it can be assumed that the current needed will be small, so these effects can be neglected even at the high operating frequency of the transformer. This has been confirmed with numerical simulations. These values will be compared to the fabricated transformer measurements.

B. Analytic model and experimental measurements comparison

The characterization of the transformer at frequencies from 40MHz to 40GHz has been performed using an ANRITSU ME7808C vector network analyzer. Figure 5 shows the comparison between the model and the measurements for both the primary and secondary winding impedance real and imaginary parts. The model curve fits the measurements up to the first resonant frequency around 500MHz, but differ after this frequency due to other capacitive and inductive couplings. Figure 6 shows that the transmission parameter S21 model fits the measurements up to 5GHz, except for the two resonant frequencies around 500MHz. From this curve one can deduct that the optimal operating point of the transformer is around 490MHz, since the transmission coefficient S21 is maximum at this frequency and starts to decrease after 500MHz. From these measurements it can be said that the transformer model is incomplete, as model curves don't fit the measurements above 500MHz. However, for our frequency range of operation, it is sufficient to have a good idea of the transformer's impedance and coupling behavior, as the design has been made for a 100MHz operating frequency.

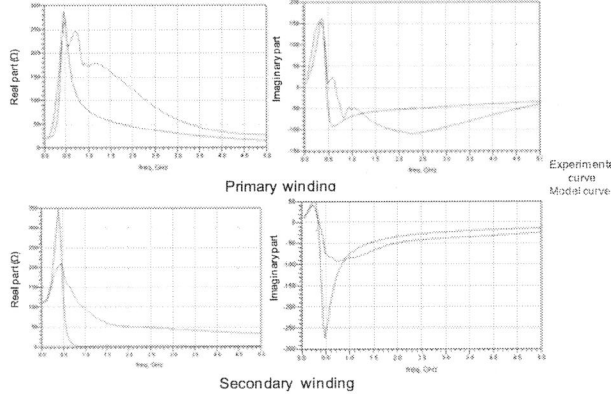

Figure 5: Transformer real and imaginary parts model and measurements comparison

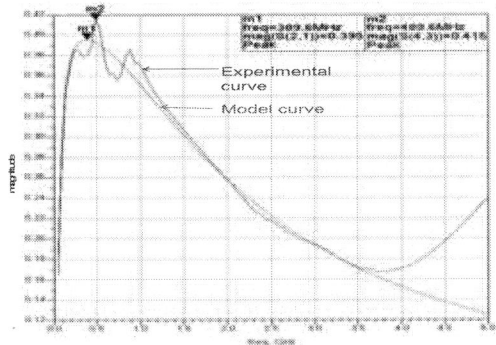

Figure 6: Transformer S21 parameter model and measurement comparison

C. Demodulation circuit and breakdown voltage experimental measurements

Experimentations of the transformer and its demodulation circuit showed good results with a modulation of the carrier at a 90MHz frequency and for a 12V peak to peak amplitude signal at the primary winding (figure 7). A 10ns response delay was observed, which is fast compared to standard optocouplers. The isolation voltage capability of this transformer was also tested and the dielectric breakdown of the oxide was observed for a voltage of 1200V between the two windings, making this transformer suitable for many applications.

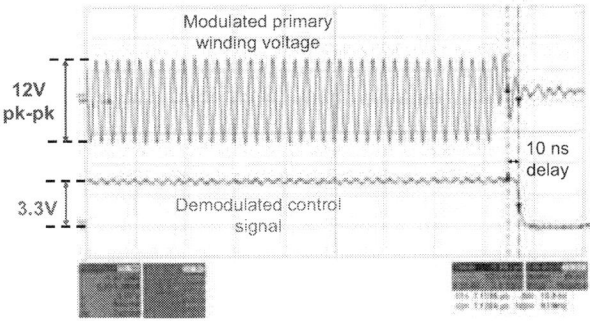

Figure 7: Waveforms of the modulated primary winding voltage and demodulated control signal

V. SUMMARY AND CONCLUSIONS

The design of an integrated transformer and its associated demodulation circuit has been presented in this paper. The purpose of this transformer is the insulation of the remote control signal of a power MOSFET, which can be integrated within a power MOSFET gate driver processed in a standard CMOS technology at no extra cost. To help the designer to achieve an efficient design, design guidelines taking into account the parasitic impedance and the coupling coefficient of the transformer as a function of its size and geometry have been given. Therefore, a good coupling can be obtained while minimizing the transformer's size and parasitic elements. Experimental results of an integrated transformer fabricated using Austria Microsystems AMS C35B4M3 standard CMOS technology have been shown. The analytical model used for the design showed good accuracy in the desired frequency range of operation compared to the measurements. The fabricated transformer was capable to withstand a voltage of up to 1200V, while exhibiting a small propagation delay of 10 ns.

ACKNOWLEDGMENT

The authors would like to gratefully acknowledge the contributions of N. Corrao from IMEP-LAHC (Grenoble - France) for the characterization of the coreless transformer. We would also like to thank O. Deleage for his contribution to the modeling of the coreless transformer.

REFERENCES

[1] Yongcheol Choi, Changki Jeon, Minsuk Kim, "Design and process considerations for 1200V HVIC technology," ISPSD 2009. 21st International Symposium on , pp.311-314, 14-18 June 2009

[2] Mazumder, S.K., Sarkar, T., "Optically-activated gate control of power semiconductor device switching dynamics," ISPSD 2009. 21st International Symposium on pp.152-155, 14-18 June 2009

[3] Herzer, R.; Pawel, S.; Lehmann, J., "IGBT driver chipset for high power applications" Power Semiconductor Devices and ICs, 2002. Proceedings of the 14th International Symposium on, pp. 161- 164

[4] Vasic, D.; Costa, F.; Sarraute, E.; , "Piezoelectric transformer for integrated MOSFET and IGBT gate driver," Power Electronics, IEEE Transactions on , vol.21, no.1, pp. 56- 65, 2006

[5] Munzer, M. et al, "Insulated signal transfer in a half bridge driver IC based on coreless transformer technology," PEDS 2003. The Fifth International Conference on, pp. 93- 96 Vol.1, 17-20 Nov. 2003

[6] Pawel, S., Thalheim, J, "1700V Fully Coreless Gate Driver with Rugged Signal Interface and Switching-Independent Power Supply," ISPSD '08. 20th International Symposium on , pp.319-322, 2008

[7] Baoxing Chen; , "Isolated half-bridge gate driver with integrated high-side supply," Power Electronics Specialists Conference, 2008. PESC 2008. IEEE , pp.3615-3618 2008

[8] Hasaneen, E.-S.A.M.; , "Modeling of on-chip inductor and transformer for RF integrated circuits," Power Systems Conference, 2006. MEPCON 2006. Eleventh International Middle East , pp.65-69, 2006

[9] del Mar Hershenson, M.; Mohan, S.S.; Boyd, S.P.; Lee, T.H.; , "Optimization of inductor circuits via geometric programming," Design Automation Conference, 1999. Proceedings. 36th , pp.994-998, 1999

Proceedings of the 23rd International Symposium on Power Semiconductor Devices & IC's
May 23-26, 2011 San Diego, CA

Reduction of Conducted Electromagnetic Interference in SMPS using Programmable Gate Driving Strength

A. Shorten, A.A. Fomani, W.T. Ng
The Edward S. Rogers Sr. Electrical and Computer
Engineering Department
University of Toronto
Toronto, Canada
ngwt@vrg.utoronto.ca

H. Nishio, and Y. Takahashi
Fuji Electric Systems Co. Ltd.,
4-18-1, Tsukama, Matsumoto,
Nagano 390-0821, Japan

nishio-haruhiko@fujielectric.co.jp

Abstract— A gate driver IC with programmable driving strength to reduce conducted electromagnetic interference (CEMI) in SMPS is presented in this paper. The solution presented is to dynamically adjust the gate driving strength (output resistance R_{out}) at the arrival of each gate pulse to minimize CEMI while maintaining low switching loss. Dynamically adjusting R_{out} is not possible with conventional gate driver designs. A segmented gate driver is designed and fabricated in the AMS 0.35μm 40V HVCMOS process. Unlike snubber circuits, the proposed method does not require extra discrete components or wasted energy. Experimental results indicate up to a 7dBμV improvement in peak CEMI between 20 MHz and 30 MHz.

I. INTRODUCTION

Despite the tremendous benefit in efficiency that SMPS can achieve over other energy conversion techniques, SMPS can exhibit a significant penalty in terms of undesired switching noise [1]. This noise takes the form of both conducted and radiated electromagnetic interference and can cause circuits near (or connected to) the converter to malfunction. This phenomenon necessitates either expensive, bulky shielding, filtering or novel techniques to reduce this unwanted interference.

Traditional SMPS design strategies encourage the selection of a gate driver with the best rise and fall times. This characteristic is achieved with a low gate driver output resistance R_{out} which is often described as a high driving strength. However, with a low R_{out}, the gate voltage of the power MOSFET has a tendency to ring at the rising or falling edge of the input signal. This ringing also appears at the output switching node, generating undesired conducted electro-magnetic interference (CEMI). The amount of ringing is heavily influenced by R_{out}, as it sets the damping ratio of the *RLC* circuit formed by the gate driver, the PCB trace and the power MOSFET (see Figure 1). By setting R_{out} to be very large, CEMI can be reduced. However, a large R_{out} will also increase the time required for the gate capacitance C_g to charge and discharge; increasing the T_{on} and T_{off} times of the power MOSFET and subsequently increasing the switching

loss due to shoot through current. A trade-off exists between switching loss, dead time and CEMI in SMPS [1]. This work is aimed at providing a new degree of freedom for this trade-off.

In this paper, an adjustable gate drive strength technique is implemented which utilizes a segmented gate driver IC in order to reduce CEMI. Figure 1 presents the topology of the gate driver circuit, which is composed of seven identical gate drivers connected in parallel that can be individually enabled or disabled digitally on-the-fly. Through this structure, R_{out} can be controlled in real time. During a gate transition, a reduction in CEMI is realized with little reduction of efficiency by initially driving the power transistors with a low R_{out} and then increasing R_{out} once the transition has passed. Furthermore, this adjustable driving strength approach is well suited for digitally controlled DC-DC converters.

Section II describes the topology and the signal timing of the gate driver. The experimental test setup is discussed in Section III. Experimental results are presented in Section IV. Conclusions are discussed in Section V.

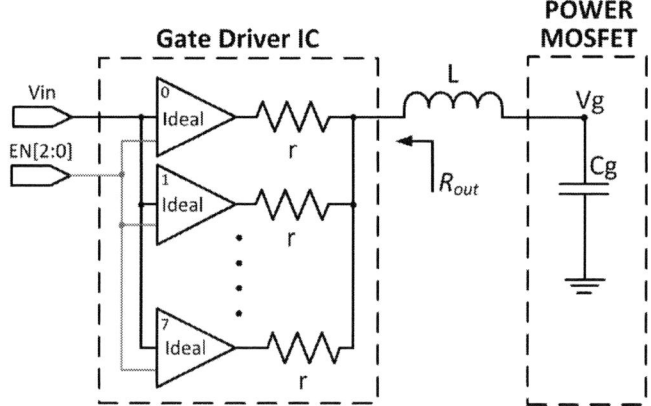

Fig. 1. Topology of the proposed segmented gate driver IC.

---

* This work was supported in part by the Natural Science and Engineering Research Council of Canada, Auto21, and Fuji Electric Systems Co. Ltd.

978-1-4244-8425-6/11 $26.00 © 2011 IEEE

II. CEMI Reduction Technique

A. Gate Driver Design

The gate driver consists of seven identical driver segments, connected in parallel as shown in Figure 1. Each segment is identical with an internal topology as shown in Figure 2. The segments are designed such that they can be digitally enabled or disabled independently. When a segment is disabled, the output of the gate driver goes into a high impedance state [2]. Thus, a disabled segment provides no conduction path to source or sink current from adjacent segments. A disabled segment therefore increases the R_{out} of the overall gate driver [2]. It should be noted that for simplicity, the current work only configures the gate driver such that either one or all of the segments are enabled.

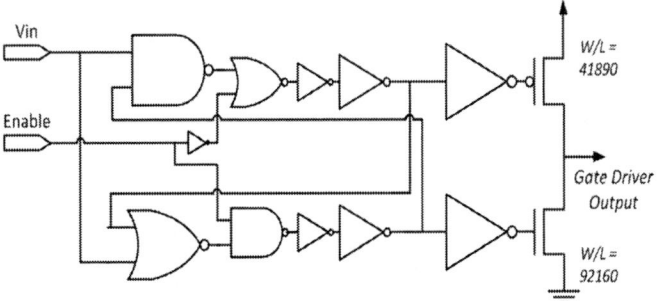

Fig. 2. Topology of an individual gate driver segment.

B Theory of Operation

As shown in Figure 1, the output resistance of the gate driver (R_{out}), the parasitic inductance from the PCB trace between the gate driver IC and the power MOSFET (L) and the gate capacitance of the power MOSFET (C_g), form a series *RLC* circuit. From the analysis of a simple *RLC* circuit, the amount of ringing is determined by the dampening factor, ζ as depicted by Equation 2. It can be seen that the value of ζ can be controlled by adjusting the output resistance of the gate driver IC, R_{out}.

$$\frac{Vg}{Vi} = \frac{1}{1 + R_{out}C_g s + LC_g s^2} \qquad (1)$$

$$\omega_0 = \frac{1}{\sqrt{LC_g}}, \quad \zeta = \frac{R_{out}}{2}\sqrt{\frac{C_g}{L}} \qquad (2)$$

$$\omega_{osc} = \omega_0 \sqrt{1 - \zeta^2}$$

$$\omega_{osc} \cong \omega_0 \qquad (3)$$

$$\omega_{osc} \cong \frac{1}{\sqrt{LC_g}}$$

By changing R_{out} according to the timing diagram in Figure 3, the T_{on} and T_{off} times of the power MOSFETs can be kept small while reducing the ringing at the switching node after the gate transition has passed. With this approach, two variables are defined in order to describe the operation of the gate driver. The first is T_{pre}, the time before the gate transition that R_{out} becomes low. The second is T_{post}, the time that R_{out} stays low after the gate transition. By adjusting T_{post} and T_{pre} appropriately, a reduction in CEMI is realized without a significant reduction in efficiency.

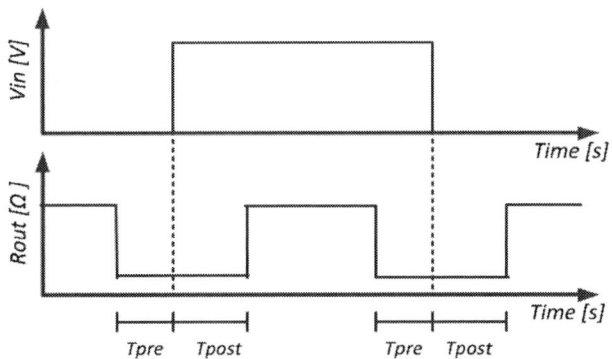

Fig. 3. Gate driver IC signal timing diagram.

III. Experimental Test Setup

In order to validate the CEMI reduction technique, an IC was fabricated containing an identical pair of segmented gate drivers. This IC was incorporated into a discrete synchronous buck converter with the pair of gate drivers used to drive the high side (HS) and low side (LS) power MOSFET switches. The current drawn by the converter was probed using a *Tektronix TCPA300/TCP312* current probe system. The signal from the current probe was analyzed using a *Hewlett Packard 8591E* spectrum analyzer. The spectrum measured was limited from 150 kHz to 30 MHz as this is the domain over which conducted EMI standards are defined by the FCC [4]. An overview of the test setup is seen in Figure 4.

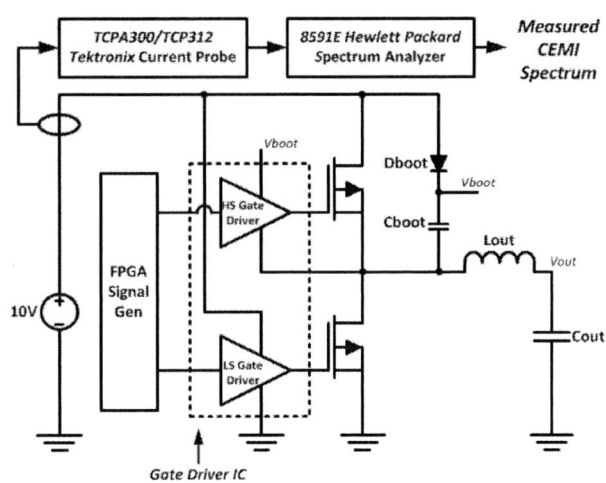

Fig. 4. Test setup for the proposed segmented gate driver technique.

978-1-4244-8425-6/11 $26.00 © 2011 IEEE

IV. EXPERIMENTAL RESULTS

The gate driver IC was fabricated using a 0.35μm 40V HVCMOS process from AMS. The final IC design allows the R_{out} of each gate to be varied between 1.8 to 13 Ω during pull up. During pull down, R_{out} can be varied between 0.65 to 4.5 Ω. A micrograph of the die is as shown in Figure 5. The die was packaged and mounted onto a PCB as part of a synchronous buck converter (Figure 6) with *International Rectifier IRF8707GTRPBF* power MOSFETs as the HS and LS switches. The conditions under which the system was tested can be found in Table I. The gating signals for the gate driver IC were generated using an *Altera Cyclone II FPGA*.

T_{post} was selected as 8ns by matching its value to the rise time of the gate switching node. It was observed that a short T_{post} resulted in converter behavior similar to the operation with high R_{out}. Furthermore, a long T_{post} resulted in converter behavior similar to the low R_{out} operation. The value of T_{pre} was found to have less bearing on the performance of the converter than T_{post}. The only constraints found for T_{pre} were that its value needed to be greater than ~3ns and be substantially less than $D \times T_s$, where D is the duty ratio and T_s is the switching period. A value of 5ns was selected for T_{pre}. Bringing the enable signals high early to set T_{pre} was simple, as the controller was implemented digitally and the gating pulses are generated with a DPWM.

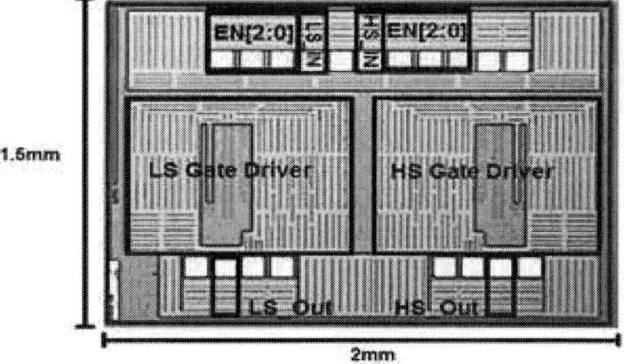

Fig. 5. Micrograph of the gate driver IC implemented in a 0.35μm 40V HVCMOS. Die size is 1.5mm × 2mm.

Fig. 6. A synchronous buck converter implemented using

TABLE I. TESITNG CONDITIONS FOR CEMI REDUCTION

Parameter	Value
V_{in}	10 V
V_{out}	1.2 V
C	20 μF
L	1 μH
I_{out}	2 A
f_{sw}	1 MHz

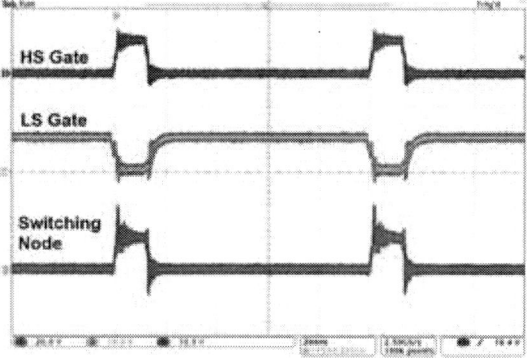

Fig. 7. Voltage waveforms for high gate driver R_{out} (one segment enabled).

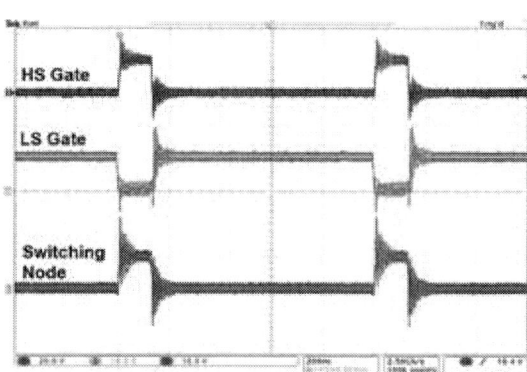

Fig. 8. Voltage waveforms for low gate driver R_{out} (all segments enabled).

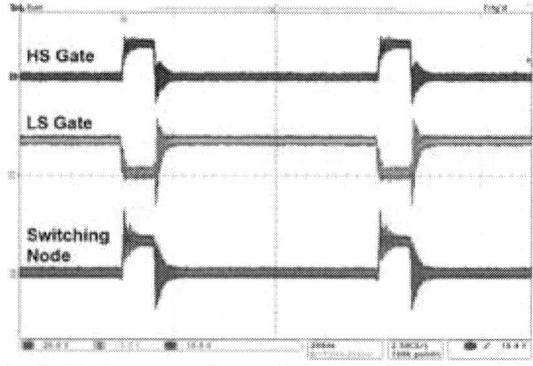

Fig. 9. Voltage waveforms for adjustable gate driving strength operation.

Voltage waveforms for the HS gate node, LS gate node and switching node of the converter for high R_{out}, low R_{out} and the proposed adjustable driving strength operation can be seen in Figures 7, 8, and 9, respectively. The behavior in these diagrams confirms that the ringing at the switching node was directly influenced by the output resistance of the gate drivers. The adjustable driving strength operation has demonstrated the capability to reduce ringing at the switching node. The converter efficiency of these three operating modes can be found in Table II. The output load during each efficiency measurement was 2A, the same as the CEMI measurement.

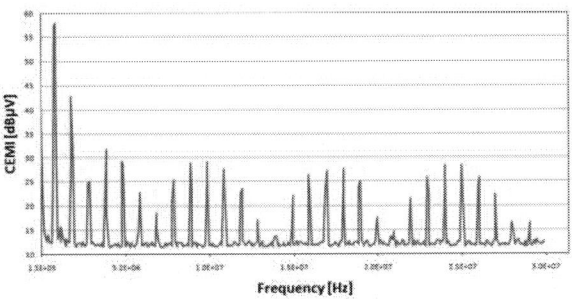

Fig. 10. EMI spectrum for high gate driver R_{out} (one segment enabled).

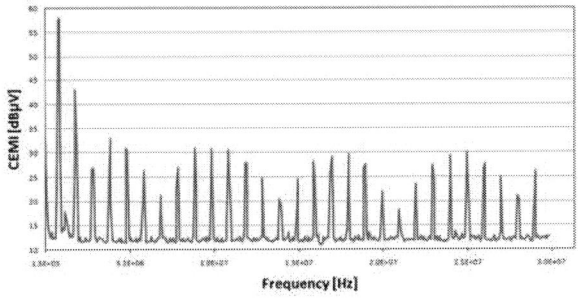

Fig. 11. EMI Spectrum for low gate driver R_{out} (all seven segments enabled).

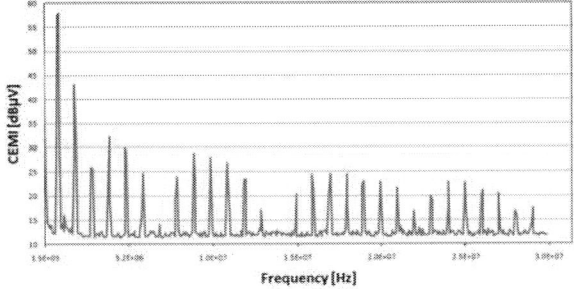

Fig. 12: EMI spectrum for adjustable gate driving strength operation.

The CEMI measurements for high R_{out}, low R_{out} and the proposed adjustable driving strength operation can be seen in Figures 10, 11 and 12, respectively. These figures confirm that the proposed technique reduced CEMI while Table II confirms that this technique did not significantly reduce the efficiency of the converter.

TABLE II. EFFICIENCY FOR DIFFERENT MODES OF OPERATION

Operation Mode	Converter Efficiency
High Output Resistance	74.3 %
Low Output Resistance	76.6 %
Adjustable Driving Strength	76.5 %

In this work, the adjustable driving strength operation exhibited an improvement of 7dBμV in CEMI between 20MHz and 30MHz from the situation where the gate driver R_{out} is held continuously low. This reduction in CEMI is obtained with only a small decrease in efficiency of 0.1%. In contrast to the situation where the gate driver R_{out} is held continuously high, only a small additional CEMI improvement is observed but at the expense of a 2.3% reduction in efficiency. Furthermore, it was observed that the CEMI improvement diminished as the load current was increased.

An interesting observation is that the frequency of the ringing at the switching node was measured to be approximately 70 MHz. The implication of this is that the bulk of the CEMI reduction is likely outside the FCC CEMI range of interest (150 kHz to 30 MHz) [4]. It would therefore be prudent to explore the impact of this technique on the radiated emissions of the converter as these emissions have an FCC spectrum which is defined above 30 MHz [4]. Additionally, future work should investigate whether or not using more than two settings for R_{out} can provide further improvements in CEMI.

V. CONCLUSIONS

In this work, a segmented gate driving technique was developed and shown to reduce conducted EMI in SMPS without a significant reduction in efficiency. An improvement in CEMI of 7dBμV between 20 MHz and 30 MHz was realized at an output load of 2 A. This technique is suitable for digitally controlled DC-DC converters implemented with discrete power MOSFETs.

REFERENCES

[1] Galluzzo, A.; Melito, M.; Belverde, G.; Musumeci, S.; Raciti, A.; Testa, A.; , "Switching characteristic improvement of modern gate controlled devices," *Power Electronics and Applications, 1993., Fifth European Conference on* , pp.374-379 vol.2, 13-16 Sep 1993.

[2] Fomani, A.A.; Ng, W.T.; "A segmented gate driver with adjustable driving capability for efficiency optimization," *Power Electronics Conference (IPEC), 2010 International*, pp.1646-1650, Sapporo, 21-24 June 2010.

[3] Siniscalchi, P.P.; Hester, R.K.; , "A 20W/channel Class-D amplifier with significantly reduced common-mode radiated emissions," *Solid-State Circuits Conference - Digest of Technical Papers, 2009. ISSCC 2009*. pp.448-449, San Francisco, 8-12 Feb. 2009.

[4] Federal Communications Commission (2009, Oct 1). PART 15— RADIO FREQUENCY DEVICES [Online]. Available: http://wireless.fcc.gov/index.htm?job=rules_and_regulations

[5] Strydom, J.T.; de Rooij, M.A.; van Wyk, J.D.; , "A comparison of fundamental gate-driver topologies for high frequency applications," *Applied Power Electronics Conference and Exposition, 2004. APEC '04. Nineteenth Annual IEEE* , vol.2, no., pp. 1045- 1052 vol.2, 2004.

978-1-4244-8425-6/11 $26.00 © 2011 IEEE

Proceedings of the 23rd International Symposium on Power Semiconductor Devices & IC's
May 23-26, 2011 San Diego, CA

Self-heating Analysis of Power MOSFET Module during Burn-in Test

Evgueniy N. Stefanov
RASG/Research Technology Group
FREESCALE Semiconductor
31023-Toulouse, FRANCE
este01@freescale.com

Rene Escoffier
On leave FREESCALE Semiconductor for
D2NT-L2MA
CEA-LETI
38054 Grenoble, FRANCE

Gael Blondel and Blaise Rouleau
R&D Department
VALEO VES
94017 Creteil, FRANCE

Abstract—**The paper deals with the thermal behavior for paralleled MOSFET's module during accelerated cycling burn-in test in harsh ambient and current conditions. The aim of the work is to optimize the key parameters acting on the self-heating in order to avoid undesirable failures resulting from overheating. An electro-thermal model is developed to simulate the device temperature during the test. Well calibrated to the experimental data for R_{on} and avalanche phases, our model allowed realistic thermal prediction. The impact of gate bias, pulse time, as well as disparities of breakdown voltage between the FETs was analyzed and the test conditions were optimized.**

I. INTRODUCTION

In order to achieve a high current rating parallel devices are commonly used in module or discrete realization for automotive switching applications [1]. The number of the needed devices is determined essentially by thermal considerations – the maximum operating temperature should not be exceeded even for short times of operation. Paralleling power chips has several benefits related to the conductive and switching power losses, the thermal load spread on the heat sink, or the lower cost due to the yield in respect to a single large die MOSFET.

However, the power MOSFETs parameters may exhibit some discrepancies which can be worse when the operating temperature varies. These discrepancies are often due to the statistical normal distribution of the device parameters (threshold voltage V_{th}, breakdown voltage BV, etc) and can lead to conduction current unbalance during both steady state and switching transients, resulting in power loss and corresponding thermal stress difference for different MOSFETs. The thermal dissymmetry during the turn-off period could also result in the destruction of the device because of thermal runaway. To overcome the problems related to the paralleled switches, the dynamic operation of the device must be studied in both, R_{DSon} and avalanche period. The static current sharing in conduction state for low-frequency applications depends on the resistance and also strongly on gate bias. The MOSFET having a lower R_{DSon} will conduct higher current to have a voltage drop equal to other parallel devices across its terminals and depends strongly on gate bias. As the MOSFET has a positive thermal coefficient a negative feedback between current flowing and

temperature balances the current sharing [2, 3]. The thermal balance is considerably improved when the paralleled dies are assembled in a module sharing the same heat sink. One papers is known treating the dynamics of current and thermal sharing during the avalanche for paralleled MOSFETs [4].

In this paper we analyzed the electro-thermal behavior of the module of VALEO micro-hybrid StARS™ starter-alternator system during accelerated burn-in test in harsh conditions. The module shown on Fig.1 consists of four planar MOSFETs $0.6m\Omega/25V$ per bridge, two connected in parallel for the Low Side (LS) and two for the High Side (HS) of the switch to sustain nominal driving current (400A) at low R_{DSon} maintaining lowest temperature T_J in the dies.

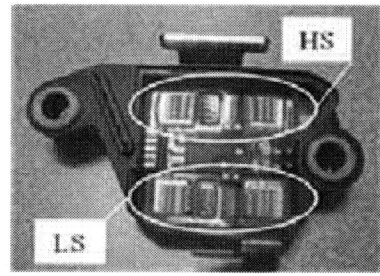

Figure 1. The Valeo I-star module

In Chapter II a short description of the burn-in test set-up conditions is presented as well as the motivation of the thermal analysis of the module. Chapter III presents the advanced dynamic electro-thermal models developed for the numerical analysis of the paralleled devices within the module. In Chapter IV the simulation results are presented for both the conductive and the avalanche modes of operation. Special emphasis is made on the calibration procedure to the experimental data. Special emphasis in the analysis is also put on the dynamic current sharing during turn-off phase and the dissymmetry of Joule heating in paralleled MOSFETs caused by breakdown discrepancy ΔBV between the paralleled devices during Unclamped Inductive Switch (UIS).

II. BURN-IN TEST SET-UP

The module is subject to accelerated cycling burn-in test at harsh conditions (the ambient temperature T_{amb} is set to

978-1-4244-8425-6/11 $26.00 © 2011 IEEE

150°C and driving current enhanced to 600A in respect to the nominal current 400A) in order to screen potentially defective devices, die attaching and wire bonding quality. R_{DSon} and UIS modes of operation are alternatively applied with a sequence of 10 pulses at low frequency f=1Hz and duty cycle signal 75-90% at the gates of both devices M1 and M2. The I_d current and V_{gs}, V_{ds} bias waves during burn-in are shown in Fig.2. Fig.3 shows the circuit of modeled HS branch under test including the load inductance L_{load}=4μH, the parasitic stray inductances L_s~50nH at the drain terminals and supply voltage V_{bat}=10V. The switch controlled by the Gate Driver circuit generates pulses V_{gs} with f=1HZ. The the gate access resistance R_G is identical for both M1 and M2, ensuring short turn-off times ~10μs.

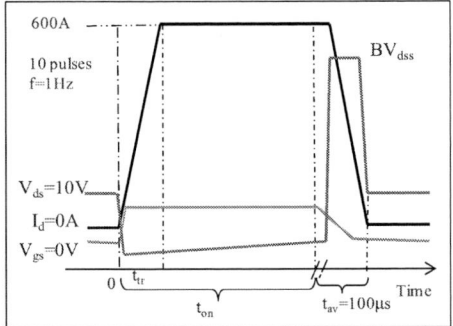

Figure 2.   Current and voltage bias waves during the burn-in test

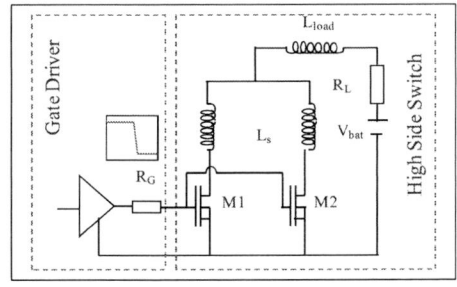

Figure 3.   The equivalent circuit of HS branch of the switch

During burn-in the MOSFETs are self-heating in both modes of operation, R_{DSon} (t_{on}~100-250ms) and UIS, due to the dissipated Joule heat. After turn-off, M1 and M2 enter in avalanche mode dissipating the energy for very short times t_{av}~100μs. Statistical discrepancies of BV in M1 and M2 cause unbalanced current flowing in the devices and then dissymmetry of the self-heating in the lowest BV device worsening dramatically the situation. The power dies gradually warm up with time and the number of pulses and can cause undesirable failures of "good" devices due to thermal overstress of tested dies.

A realistic thermal simulation of the burn-in test, based on reliable physical models of the device and the module stack, is the only alternative to predict the hot spot temperature rise and to optimize the acting critical parameters (R_{DSon} timing, gate bias V_{gs}, and disparities ΔBV between M1 and M2) in order to avoid damaged devices and so improving the module reliability and cost.

III.   ELECTRO-THERMAL MODELS

Advanced numerical electro-thermal models of the planar power MOSFETs and the layers composing the vertical structure of the inherent module stack is developed to simulate the self-heating during the test cycle and their hierarchy is shown in Fig.4.

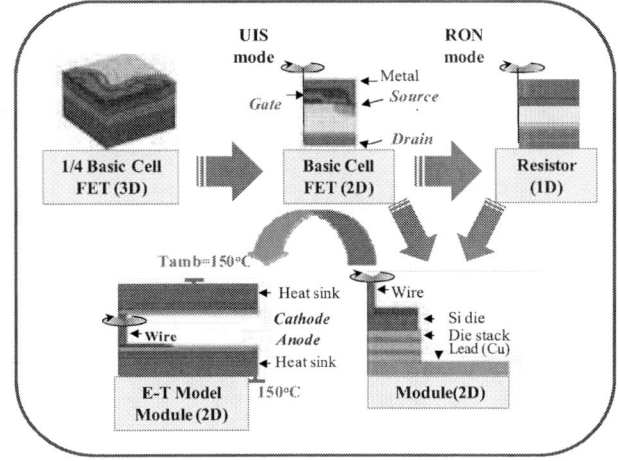

Figure 4.   Architecture of the electro-thermal models hierarchy

The **1st ET model** is approximated from the 3D device basic cell topology where the DC device characteristics are normalized to 2D concave circular FET cells. Total die and module stack area are split into 600000 parallel identical circular structures corresponding to the number of basic cells integrated in the power die. Each base cell is connected with individual wire to the source terminal, where the sum of all wires area corresponds to the real area of all bonding wires used in the module. The **1st ET model** allows coupled the simulation of both, R_{DSon} and UIS modes, but doesn't take into account the voltage drop along the top metal. The **2nd ET model** is approximated from the **1st ET model** of the device but the 2D doping structure is approximated by 1D resistor doping profile n+n-n+ normalized to the R_{DSon}(T) of the transistor. The total die and module stack area for **2nd ET model** are split and normalized to 8 (number of stitched bonding wires of 20 mils and 8mm) parallel identical circular structures for each device, corresponding to the area controlled by each wire bonded at the top metal. The simplified **2nd ET model** allows accurate separate simulation, respectively for the R_{DSon} mode of operation, taking into account the voltage drop along the Source top metal. The charge transport equations for the carriers are coupled by lattice-heat model accounting for the thermal transport and Joule heating, and solved with Synopsys-Sentaurus package [5]. The thermal conditions T_{amb}=150°C are applied at the heat sink terminals. Mixed-mode (lumped element SPICE & device) simulation is applied to account the test set-up conditions. The models allow accurate prediction of the transient performances, across the device and the module subjected to burn-in self-heating.

978-1-4244-8425-6/11 $26.00 © 2011 IEEE

IV. RESULTS AND DISCUSSION

A. Analysis of R_{DSon} mode of operation

Experimental verification of the **2nd ET model** to predict self-heating in the MOSFETs during R_{DSon} phase is done. Fig.5 shows the comparison between the measured dynamic R_{DSon} and simulation during the 1st and the last 10th pulse according the test set-up. The V_{gs} is 15V, the conduction time per pulse is t_{on}=200ms, T_{amb}=150°C, and current I_d=600A per switch. The experimental R_{DSon} was extracted from the measured V_{ds} drop between the terminals. The excellent fit achieved confirms the capability of the model to predict adequately the temperature distribution in the device and the module stack.

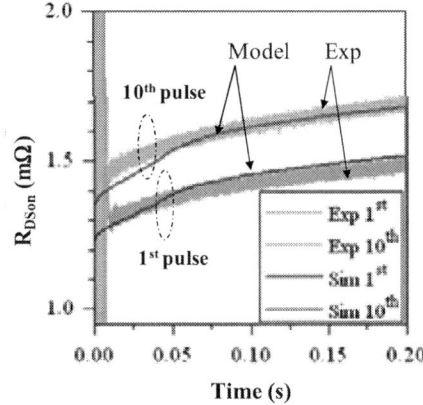

Figure 5. R_{DSon} mode calibration. Measured and simulated R_{DSon} vs. t_{on}.

The impact of both, gate bias V_{gs} and conduction time t_{on}, on the hot spot T_J is simulated for the whole cycle test and Fig.6 shows the ΔT_J evolution during the 1st and 10th on-state pulses. The simulated curves concern respectively 3 cases a)V_{gs}=10V, t_{on}=250ms; b)V_{gs}=15V, t_{on}=200ms; c)V_{gs}=15V, t_{on}=100ms. It is seen that V_{gs} and the on-state timing t_{on} play essential role on the Joule pre-heating of the MOSFETs before the UIS mode of operation.

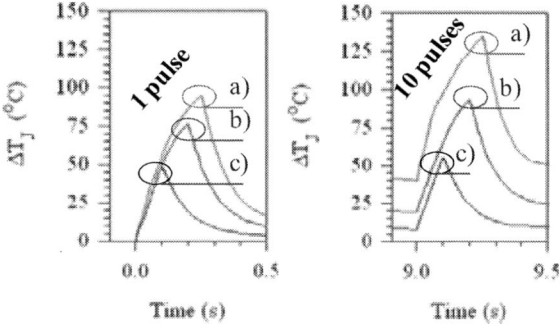

Figure 6.   Simulated ΔT_J during the 1st and 10th pulses

Fig. 7 shows the 1D temperature distribution along the bonding wires, Si die and module stack after the 1st (dashed line) and 10th (full line) on-state pulses for the same set-up conditions cited in Fig.6. It is seen that the maximum of T is located along the bonding wires because to their resistance and thermal conductivity limiting the dissipation of generated heat flow to the heat sink. The simulated T at the interface

metal/Si is relatively lower, but for high V_{ds} drop (V_{gs}=10V) and long on-state operation times (t_{on}=250ms) could be achieved high pre-heated Si (T_J=220°C after one pulse).

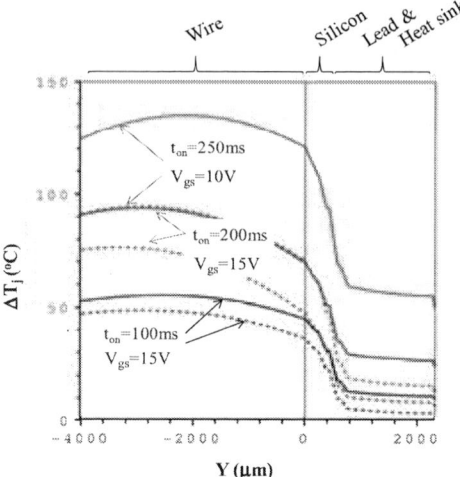

Figure 7.   Vertical ΔT_J distribution along the wire, die, and module stack

B. Analysis of UIS mode of operation

The modeling of UIS mode of single MOSFET assembled in TO3 package has been calibrated to the experimental data measured in both, mono- and multi-pulse (1000 pulses) regimes on a special purpose designed test bench. Comparison of simulation results during the avalanche with the measured T_J rise by means of on-chip integrated temperature sensor is shown on Fig.8. High value L_{load}=500µH was chosen in order to measure precisely T_J during relatively long energy dissipation times. T_{amb} is 25°C, and the current I_d is 43A. Very good fit to the experimental I_d, V_{ds} waves, and self-heating T_J rise .is demonstrated.

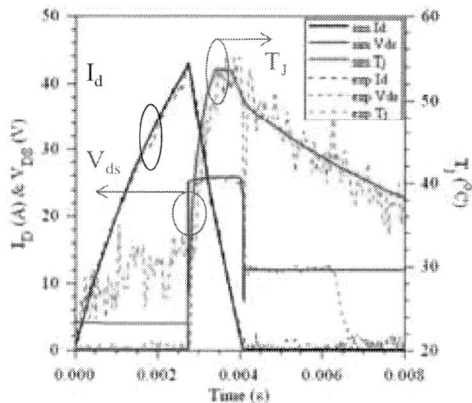

Figure 8.   Single device mono-pulse UIS calibration.

Next we proceeded with multi-pulse UIS calibration at f=20Hz for the single device. 1000 on/off pulses are generated and T_J was measured before the turn-off and after T_J rise during UIS phase. Fig.9 shows the comparison between experimentally measured and simulated T_J rise, where an excellent fit was achieved. Test set-up conditions are the same as for the mono-pulse experiment. These results confirmed the capability of the **1st ET model** to predict realistically the Joule self-heating in simulated MOSFETs.

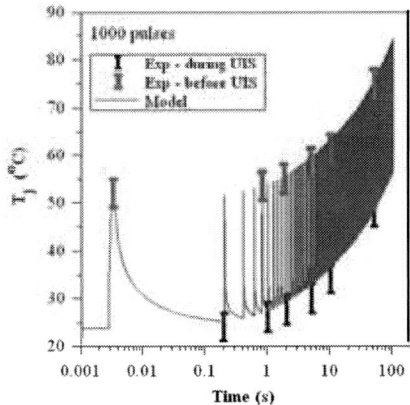

Figure 9.   Single device multi-pulses UIS calibration.

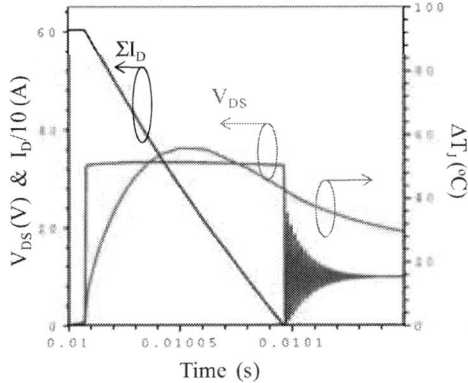

Figure 10. Simulated I_d,V_{ds} waves and ΔT_J vs. time with ΔBV=0V.

MaxT$_J$ of the pre-heated die is extracted from simulations generated by the **2nd ET model** during on-state operation. This temperature is imposed as initial thermal condition on the source top surface of the paralled MOSFETs before the UIS turn-off. Two cases are investigated. The first case concerns two MOSFETs with the same BV (ΔBV=0). Fig.10 shows the simulated Id, Vds waves, and Joule heating ΔT_J vs. time during the 1st UIS switch. The current I_d is 600A per switch, t_{on} is 200ms, V_{gs} is 15V, t_{av} is 100µs, T_{amb}=150°C and the initial die T_J is 200°C. ΔT_J is rising during the turn-off and reach maxT$_J$=250°C some 50µs from the beginning of the switch. Both MOSFETs having the same BV, share uniformly the current and reach the same maxT$_J$.

The worst treated case is when the discrepancy between the devices is ΔBV=1V. This BV shift is defined by the 6 sigma statistical distribution of BV in a fabricated lot. Fig.11 shows the simulated I_d, V_{ds} waves, and Joule heating ΔT_J vs. time during the 1st UIS switch. The applied test set-up conditions are the same as in previous case, and the terminal stray inductances are kept the same L$_s$=50nH.

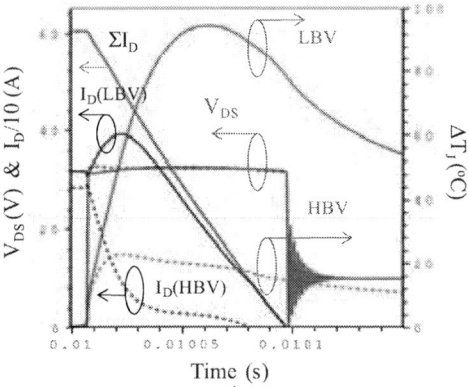

Figure 11.   Simulated I_d and V_{ds} waves and ΔT_J vs. time with ΔBV=1V.

The I_d waves on Fig.11 show when both MOSFETs are in avalanche the device with lowest BV (LBV) clamps the V_{ds} providing a low resistive path for the stored energy in the load. Most of the current tends to flow through the LBV device, causing higher power loss and respectively drastic rise of the temperature. Because to the positive BV(T) coefficient, for longer dissipation times T$_J$ of the LBV will increase until its BV(T) becomes equal to the BV of higher BV (HBV) device. Such balancing is impossible for short time of t_{av} for the burn-in set-up (L$_{load}$=4µH). In result, strong disparity of generated T$_J$ for both devices is observed. In this simulated case we conclude that the LBV device would reach T$_J$ about 290°C. Taking into account the posteriori self-heating during the cycling, the choice of this simulated test set-up risks to cause thermal overstress. Based on the proposed modeling approach we were able to define "safe" set-up parameters avoiding premature failures and damaging.

V.   CONCLUSIONS

Advanced electro-thermal model for the I-Star module is developed to simulate the thermal and current sharing in 2 paralleled MOSFETs during accelerated burn-in test at high ambient temperature and enhanced driving currents. Based on fine calibration to measured data for R$_{DSon}$ and UIS phases of operation the maxT$_J$ in result of Joule heating was predicted taking into account the test critical parameters. Safe test set-up conditions were established, allowing drastic reduction (>10 times) of the failed modules and power dies due to overstress. In result efficient improvement of the module and die yield is achieved, allowing further reduction of its cost.

REFERENCES

[1]   D.A. Grant and J. Glower, Power MOSFETs theory and applications, John Wiley & Sons, Ed. 1989.

[2]   T. Lopez and R. Elferich, "Static Paralleling of Power MOSFETs in Thermal Equilibrium", APEC'06, pp.841-7, 2006.

[3]   B. Abdi, A. et al. "Problems Associated with Parallel Performance of High Current Semiconductor Switches and their Remedy", SPPEDAM'08, pp. 1379-83, 2008.

[4]   J. Chen, S. Downer, A. Murray and D. Divins, "Analysis of Avalanche Behavior for Paralleled MOSFETs", SAE World Congress, 2004.

[5]   Sentaurus Device, version D-2010.03 Copyright © 1994-2010, Synopsys^R-Inc

Proceedings of the 23rd International Symposium on Power Semiconductor Devices & IC's
May 23-26, 2011 San Diego, CA

Design of an 80V-class high-side capable double-resurf JI L-IGBT

Hiroki Fujii, Shinichi Komatsu, Masaharu Sato and Toshihiko Ichikawa

Devices & Analysis Technology Division, Renesas Electronics Corporation

1120 Shimokuzawa Chuo-ku Sagamihara, Kanagawa, 252-5298 Japan

E-mail: hiroki.fujii.zj@renesas.com

Abstract— **This paper proposes a suitable design of an 80V-class high-side capable double-resurf lateral IGBT (L-IGBT) using our cost-effective HV-MOS process, which excludes SOI/DTI structures. This junction-isolated (JI) L-IGBT, unlike a drift-NMOSFET, suffered a large substrate leakage caused by a parasitic PNP BJT whose collector was a p- substrate. It also had a disadvantage in turn-off time due to the floated n- drift layer. We tried a unique, promising approach — the p+ collector was connected to the outmost enclosing n+ sinker internally by n- drift region resistor and externally by metallization, which needs no additional process steps. Our measurement results show that it successfully eliminated both problems without almost any sacrifice in performance of current drivability per area, breakdown voltage, HCI, and ESD endurance.**

I. INTRODUCTION

As a large number of automotive parts come under the control of an electrical system, the chip size of each IC needs to be minimized to be accommodated in a limited space in a car. This background makes an L-IGBT, whose current drivability is substantially greater than that of a drift-MOSFET, a powerful candidate for the automotive electrical system. L-IGBT can be fabricated just by replacing an n+ drain diffusion region of a drift-MOSFET by p+ one (this region is renamed the "collector" although it acts as an emitter for a PNP BJT section), which brings an advantage in current drivability because of a drastic reduction in the resistance of the n- drift region. Conductivity modulation caused by hole injection from the p+ collector is responsible for that, and this approach is confirmed to be highly effective for increasing current per area by measurement results for not only a several hundreds volt-class L-IGBT whose on-resistance is dominated by the n-drift region, but also a relatively low-voltage class, specifically, under 200V or 100V class L-IGBT [1] [2]. When an L-IGBT is integrated into logic circuits, it may cause a substrate leakage problem that negatively affects nearby logic MOSFETs, especially in a high-side configuration, because the parasitic PNP BJT whose collector is a p- substrate is formed. Although some solutions to this have been provided, such as enclosing the L-IGBT by deep trench isolation (DTI) combined with SOI [1] or by double p+/n+ sinker (an emitter-shorted p+ sinker and a floated n+ sinker) combined with double p+/n+ buried layer [2], their higher process costs do not fit the recent trend toward lower costs.

We have already reported the development of a cost-effective (SOI/DTI-free and single sinker/buried layer) HV-MOS process that can handle up to 80V such as a load dump surge in a generally used 12V car system [3] [4]. In this paper, we report that we have successfully incorporated an 80V-class high-side capable L-IGBT into this process without extra process steps.

II. DEVICE STRUCTURES

The new junction-isolated (JI) L-IGBT (Type K in Fig.1) has an enclosing n+ sinker at its outermost position, and the sinker is connected to the p+ collector externally by metallization and internally by an n- drift region resistor (R1a in Fig.1) that is inserted for both activating the intrinsic PNP BJT and pulling out residual holes during turn-off time. For comparison, we also fabricated the conventional n- drift floated L-IGBTs by placing no contacts on the n+ sinker (Type M) or replacing the n+ diffusion region in the n+ sinker with the p+ diffusion region (Type B). Another conventional L-IGBT whose p+ collector diffusion region was shared with n+ diffusion region (Type F) [5] and the drift-NMOSFET (Type N) were also tried. Note that each layout size is the same between all types except 'Type F' whose additional n+ diffusion region increases the cell pitch.

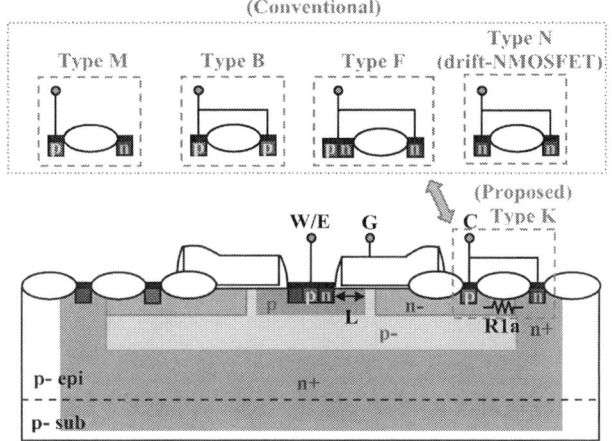

Fig.1 Cross-sectional view of high-side capable L-IGBT. L was set at 1μm. Gate finger width was fixed at 50μm.

III. EXPERIMENTAL RESULTS

A. DC characteristics

The proposed structure (Type K) obtained about twice as much saturation current (Ion) as the drift-NMOSFET (Type N) as did the conventional 'n- drift floated L-IGBTs' (Type M, B), though the Ion of the also conventional 'common p+/n+ diffusion collector L-IGBT' (Type F) remained almost the same as that of the drift-NMOSFET (Type N) (Fig.2). This

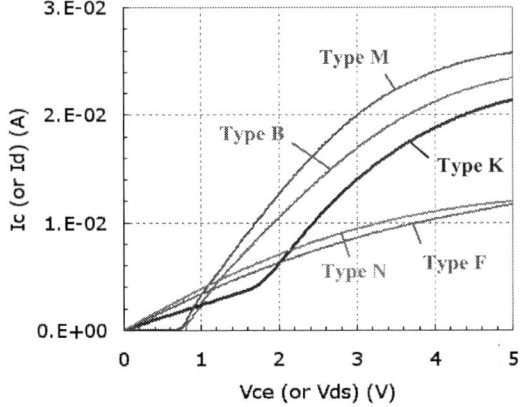

Fig.2 Measured Ic-Vce (or Id-Vds) characteristics at Vg=5V for various L-IGBTs and drift-NMOSFET.

means the 'R1a' addition was necessary for the PNP activation because the n- drift region optimized for the 80V double resurf condition could not provide enough base resistance to initiate the PNP BJT consisting of common p+/n+ diffusion and p- well. Despite its high Ion, the proposed structure (Type K) was demonstrated to have almost 80V on-state breakdown voltage (BVon) (Fig.3) due to the low base

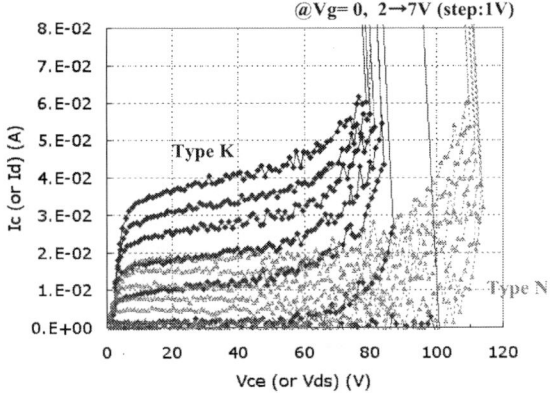

Fig.3 Measured pulsed Ic-Vce (or Id-Vds) characteristics for 'Type K' L-IGBT (solid line with filled marks) and drift-NMOSFET (dotted line with empty marks). Pulse width was set at 50ns.

resistance of the parasitic NPN, which was realized by the interdigited emitter-well layout. The off-state breakdown voltage (BVoff) also surpassed 80V as did the conventional ones (Fig.4). The punch-through between the p+ collector and the p- well was blocked until the collector bias reached as high as 90V due to the fully depleted p- well, which gave the

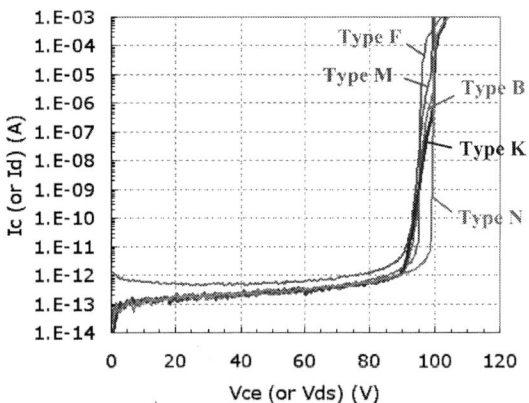

Fig.4 Measured off-state Ic-Vce (or Id-Vds) characteristics for various L-IGBTs and drift-NMOSFET.

proposed structure almost the same BVoff (monitored at Ic=1μA) as that of the drift-NMOSFET.

Among the L-IGBTs whose intrinsic PNP BJTs were found to be active, only the proposed structure (Type K) achieved almost the same extremely low substrate leakage as the drift-NMOSFET (Fig.5) due to the deactivation of the parasitic PNP BJT consisting of p+ collector, n+ sinker and p-substrate. It also had shorter turn-off time than the other PNP-active L-IGBTs (Type M, B) (Fig.6) because the residual holes in the n- drift region could be extracted via the n+ sinker.

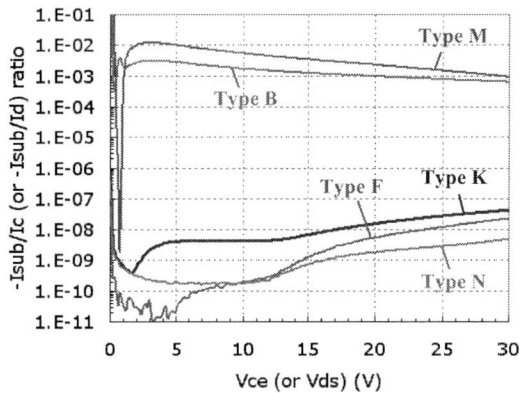

Fig.5 Measured Isub to Ic (or Id) ratio as functions of Vce (or Vds) at Vg=5V for various L-IGBTs and drift-NMOSFET.

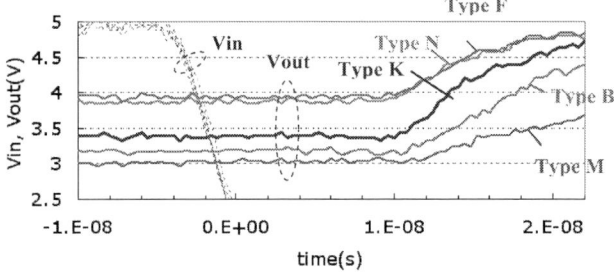

Fig.6 Measured turn-off characteristics with a 5V supply via 110ohm resistor to collector/drain for various L-IGBTs and drift-NMOSFET.

So far we have discussed the case in which the gate finger number 'Nf' is the minimum value of two. However, in many

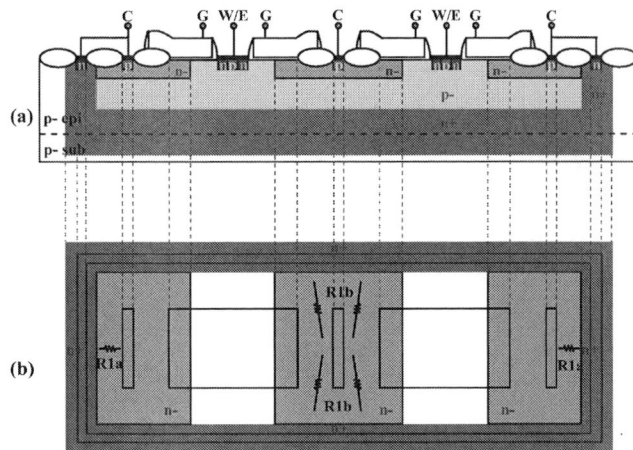

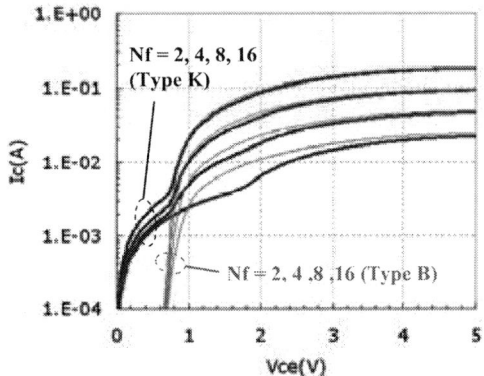

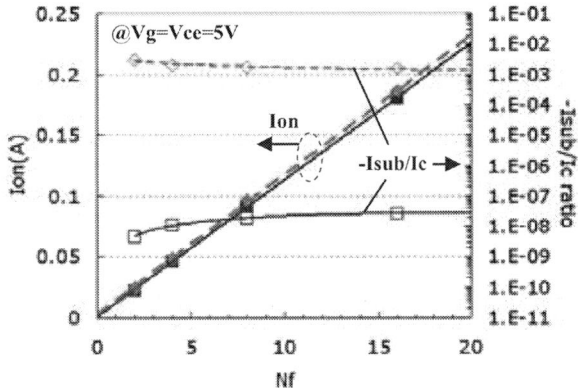

Fig.7 Cross-section (a) and top view (b) of multi-finger layout of 'Type K' L-IGBT.

applications such as a driver circuit, the 'Nf' needs to be increased to satisfy the demands for drive current. This new case produces an internal drift region, which is seemingly floated from the outmost n+ sinker in a cross-section along the L-direction (Fig.7(a)) but is actually shorted to the n+ sinker along the W-direction via an internal resistor (R1b in Fig.7(b)). The intrinsic PNP BJT of the proposed structure (Type K) came to initiate at a fixed collector voltage of about 0.7V

Fig.8 Measured Ic-Vce characteristics for 'Type K' and 'Type B' L-IGBTs with different gate finger number 'Nf' at Vg=5V.

Fig.9 Measured Ion and Isub/Ic ratio as functions of 'Nf' for 'Type K' (solid line with ■ or □) and 'Type B' (dashed line with ◆ or ◇) L-IGBTs.

(Fig.8) because the base resistance 'R1b' was much larger than the outer base resistance 'R1a'. Note that also the outmost part could be initiated at about 0.7V by separating its drift drain region from the n+ sinker along the L-direction at the cost of the area. However, it seemingly was not worth trying because the Ion already has stayed in proportionality with the 'Nf' (Fig.9). In addition, the substrate leakage was determined to be well suppressed even at the increased 'Nf' (Fig.9).

B. HCI characteristics

The Ion shift during hot carrier injection (HCI) for the proposed structure (Type K) was nearly the same as that for the drift-NMOSFET (Type N) at the gate voltage of 2V which gave the maximum well current in the drift-NMOSFET (Fig.10(a)). At higher gate bias, the initial Ion increase found in the n- drift floated structures (Type M, B) was almost totally diminished in the 'Type K' (Fig.10(b)). This suggests that shorting the n- drift region to the collector made the hole current from the p+ collector flow in the n- drift region somewhat away from the LOCOS where the hot electrons were injected.

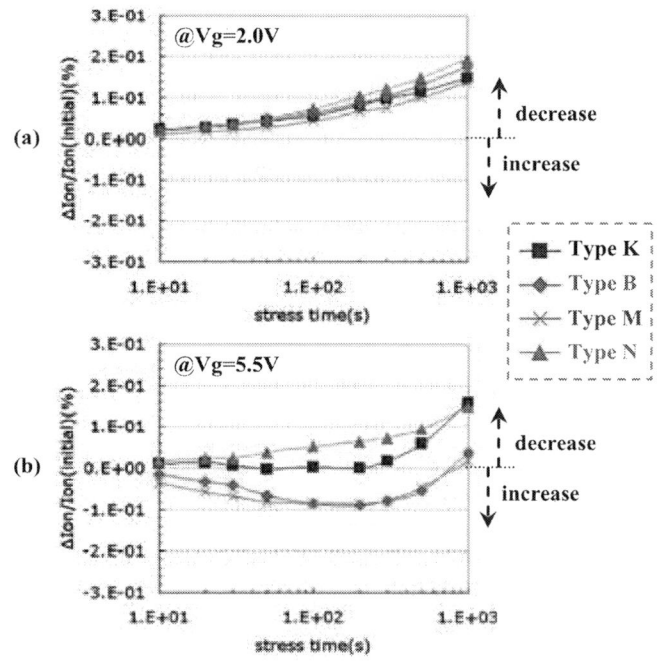

Fig.10 Measured time dependence of Ion degradation for various L-IGBTs and drift-NMOSFET. Stress collector bias was 40V while gate bias was set at Vg=2V (a) and Vg=5.5V (b).

C. TLP characteristics

The proposed 'Type K' structure as well as the conventional 'Type B' survived against a TLP surge to some level although the drift-NMOSFET was destroyed immediately after snapback (Fig.11). This was because the current dispersion, which means the collector currents of PNP and NPN flowed in different routes, decreased the temperature at the hot spot [6]. The buried n+ layer failed to draw much of the snapback current due to its deep location for achieving over 80V breakdown voltage, which is responsible for the poor performance of the drift-NMOSFET. This self-protective performance of 'Type K' stayed almost the same as that of 'Type B' even if the layout parameter of the interdigited emitter-well diffusion was changed in a way by which the parasitic NPN BJT came to be influential (Figs.12,13), namely, the thyristor action became accelerated. When the required BVon is much lower than 80V, the estimated ESD endurance can reach 6kV-HBM (It2 >4A) with over 5V holding voltage.

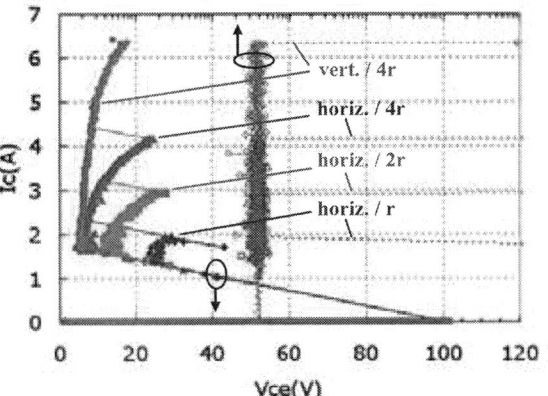

Fig.13. Measured pulsed Ic-Vce characteristics for 'Type K' L-IGBT with various emitter to well area ratios. Pulse width was set at 50ns. Gate was grounded.

IV. CONCLUSIONS

The proposed L-IGBT, which was easily integrated into our 80V-MOS process without any additional process steps, was found to have no problems of substrate leakage, large turn-off time, and high cost. Our measurement results also found that it made almost no sacrifice in performance of current drivability per area, breakdown voltage, HCI, and ESD endurance.

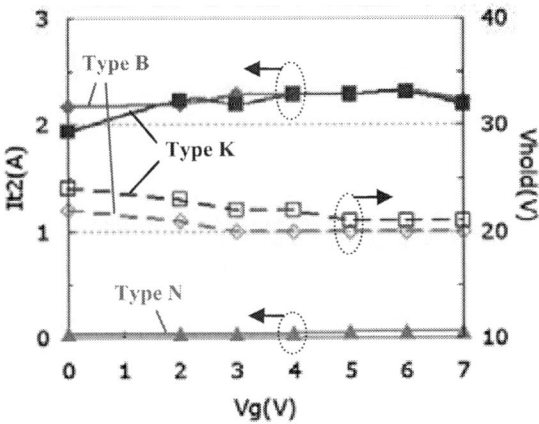

Fig.11 Gate bias dependence of secondary breakdown current (It2) and holding voltage (Vhold) obtained from 50ns TLP pulse measurement. Total channel width was set at 100μm. Note that for 'Type N', first breakdown current (It1) instead of 'It2' was plotted due to immediate destruction just after snapback.

REFERENCES

[1] T.Nitta et al, "Wide Voltage Power Device Implemention in 0.25μm SOI BiC-DMOS", Proc. of ISPSD'06, pp.341-344

[2] B.Bakeroot et al, "A 75V lateral IGBT for Junction-Isolated Smart Power Technologies", Proc. of ISPSD'06, pp.201-204

[3] H.Fujii et al., "A novel 80V-class HV-MOS platform technology featuring high-side capable 30V-gate-voltage drift-NMOSFET and trigger voltage controllable ESD protection BJT", Proc. of ISPSD'09, pp.323-326

[4] H.Fujii et al., "A study of p stopper effect on 30V-gate/80V-drain bi-directional NMOSFET and 80V ESD protection BJT", Proc. of ISPSD'10, pp.401-404

[5] Mark R. Simpson, "Analysis of negative differential resistance in the I-V characteristics of shorted-anode LIGBT's", IEEE Trans. on ED'91, pp.1633-1640

[6] Nils Jensen et al., "Coupled Bipolar Transistors as Very Robust ESD Protection Devices for Automotive Applications", Proc. of EOS/ESD Symposium'03,pp.01-06

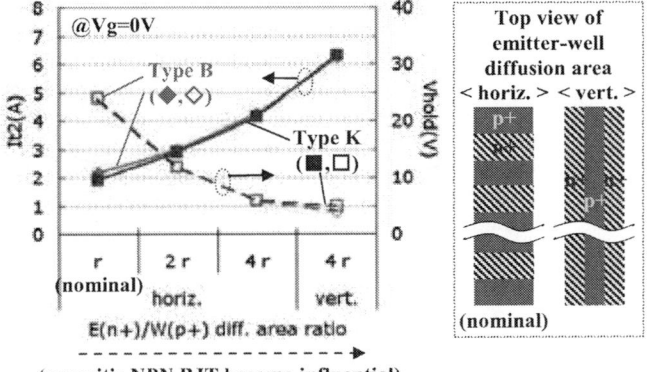

Fig.12 Emitter to well diffusion area ratio dependence of 'It2' and 'Vhold' obtained from 50ns TLP pulse measurement for 'Type K' and 'Type B' L-IGBTs. Meaning of parameter on x-axis is illustrated in right figure. Interdigited layout was named 'horiz'. Nominal ratio value was symbolized as 'r'. Total channel width was set at 100μm.

Proceedings of the 23rd International Symposium on Power Semiconductor Devices & IC's
May 23-26, 2011 San Diego, CA

150 V, 100 mΩ, SOI Power LDMOS with High Avalanche Current Capability for MHz Frequency Power Switching Applications

Patrick M. Shea and Z. John Shen

Department of Electrical Engineering and Computer Science
University of Central Florida
P.O. Box 162450, Orlando, FL 32816-2450, USA
Tel: 407-823-0379, Fax: 407-823-5835, E-mail: johnshen@mail.ucf.edu

Abstract— **To enable MHz frequency switching in high density power converters, power MOSFETs with very low gate charge and low on resistance are needed. In this paper, we report on the design and fabrication of a 150V, 100 mΩ, chip-scale SOI lateral power MOSFET. The power LDMOS transistor is based on the SOI RESURF principle and fabricated in a 0.35um CMOS foundry using a custom process. The device incorporates the "adaptive RESURF" principle as well as a P+ buried layer tied to the source in order to achieve high avalanche current capability. The adaptive RESURF technique is implemented by controlling the spacing between the N+ drain and the shallow trench isolation which runs across the drift region. This allows a drain buffer region to be formed without requiring any additional masks or ion implants. The source metal extends above the drift region as a field plate to reduce the electric field between drain and gate without increasing Miller capacitance. The device is packaged in a low-inductance, low profile flipchip package. The Qg×Rdson figure-of-merit for this device is considerably lower than state of the art commercial power MOSFETs.**

I. INTRODUCTION

Previous work on SOI LDMOS was mainly focused on optimizing the tradeoff between BVdss and Rdson based on the SOI Resurf principle [1-4]. Those works analytically defined the optimization of the lateral doping profile of the LDD region of the power MOSFET, where the heaviest LDD doping concentration occurs near the N+ drain, and the doping concentration is linearly decreased toward the source/drain p-n junction. This gives the theoretically optimal tradeoff between BVdss and Rdson for an SOI LDMOS. Those structures were fabricated on thin SOI device layers, generally on the order of 100-200 nm thick. This allows the N+ drain to diffuse all the way to the buried oxide layer, thereby reducing the optimization of the LDD doping profile to a 1-D problem in the lateral direction. While this linearly-graded thin SOI LDMOS structure provides for the lowest possible Rdson, the issues of gate charge, switching losses, and SOA were not discussed. It is thought that this type of device would not be very rugged, because the thin SOI device layer does not allow for a low resistance P+ body contact, thereby making the parasitic NPN bipolar transistor more susceptible to activation during avalanche.

Some designs were also reported which focused on using "adaptive Resurf" and other techniques in order to form a rugged LDMOS, however those works also did not include gate charge or switching efficiency as part of the design optimization equations [5-8]. In fact, the switching efficiency of many of these LDMOS structures is assumed to be quite low, due to high Miller capacitance. This is because the popular LDMOS design requires the poly gate to double as a field plate, overlapping the drain region and thereby reducing the peak electric field between the drain and the gate oxide. Normally a thick field oxide is present between the poly gate and the drain, which helps limit the Miller capacitance to some degree but does not eliminate the penalty entirely.

In this work, we focus on developing rugged LDMOS transistors on SOI substrate with very low gate charge which is optimized for MHz frequency switching applications. The concept behind this device is to take elements from the thin SOI Resurf devices for low Rdson, while incorporating other features to enhance avalanche ruggedness and minimize gate charge.

II. DEVICE CONCEPT

Unlike previously reported SOI Resurf devices, we use a relatively thick 1.0 μm p-type SOI device layer on top of a 1.0 μm buried oxide. The device features a P+ buried layer tied to the source and an N-buffer region surrounding the N+ drain, both of which are features intended to increase avalanche ruggedness. The STI oxide runs along a portion of LDD, and a source metal extension over the drift region acts a field plate to reduce the electric field between drain and gate. A cross-sectional drawing and SEM micrograph of the new SOI LDMOS are shown in figures 1 and 2 respectively.

978-1-4244-8425-6/11 $26.00 © 2011 IEEE

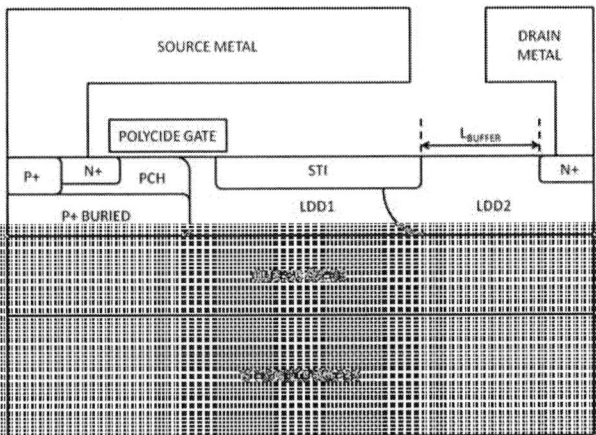

Figure 1. Cross-sectional view of the 150 V SOI LDMOS with low gate charge and P+ buried layer and N buffer to enhance avalanche ruggedness.

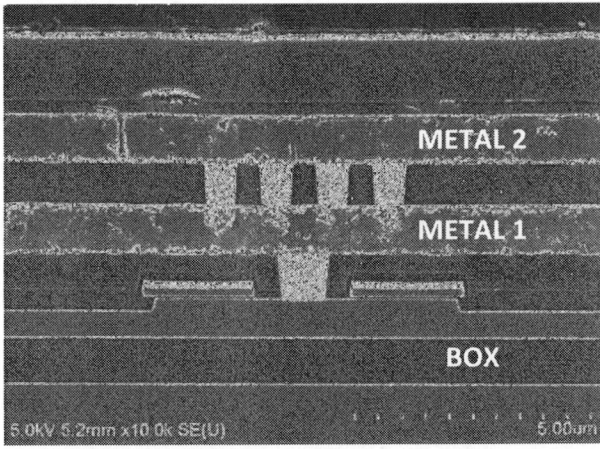

Figure 2. SEM micrograph showing the source and channel regions of the SOI power MOSFET.

The P+ buried layer is implanted at a depth near the buried oxide interface and tied to the P+ source using contacts spaced every 1 μm along the width of the transistor. This provides a very low resistance path to divert hole current away from the sensitive base-emitter junction of the parasitic NPN bipolar transistor. The SOI layer is sufficiently thick so that the vertical diffusion of the heavily doped P+ buried layer does not significantly affect the P-channel doping concentration near the surface. This allows the design of the P+ buried region to be decoupled from the MOSFET threshold voltage to such a degree that the P+ buried layer can be degenerately doped.

The LDD region is formed differently from previously reported SOI LDMOS. In previous works, the lateral grading of the LDD doping concentration was achieved by patterning successively larger LDD implant windows from source to drain. In this device, two discrete LDD regions are patterned and then formed using different phosphorus implant doses. The LDD1 implant covers the entire drift region, while the LDD2 implant is patterned only in the portion of the drift region near the drain side. The doping concentration of the LDD2 region is defined by the sum of the LDD1 and LDD2 implant doses. After implant, the two LDD implants are driven together by a high temperature furnace anneal. The balance of LDD implant doses and patterning, in conjunction with the LDD drive, allows the lateral diffusion of the LDD implants to form a near ideal linear grading of doping concentration in the lateral direction in accordance with the SOI Resurf principle.

The shallow trench isolation (STI) is formed after the LDD drive, so a portion of both LDD implants are consumed by the STI process. A drain buffer is formed by adjusting the space between the STI and the N+ drain, which leaves a region in which the entire LDD1 and LDD2 implant doses are preserved in the silicon. The buffer length and LDD ion implant doses must be tuned together to achieve the desired balance between Rdson, BVdss and ruggedness. Increasing L_{buffer} improves avalanche ruggedness and Rdson but can result in a decrease in BVDSS.

In order to minimize Miller capacitance, we try to minimize both the gate-drain overlap and also the LDD doping concentration near the gate. Instead of being self-aligned, the LDD1 implant is patterned some distance away from the poly gate and then diffused laterally during the aforementioned drive. Extensive process simulations were performed to determine the proper spacing between the LDD1 implant and the poly gate to ensure that the drain-source p-n junction is formed on the source side of the STI but with minimal overlap of the poly gate. The location of this junction is also dependent on lateral diffusion of the P+ buried layer during the source-drain anneal. This process results in a very low LDD doping concentration underneath the poly, according to simulation.

To further minimize Qgd, we avoid the use of the poly gate as a field plate over the drain, and instead we extend the source metal across a wide portion of the LDD. The source metal field plate affects BVdss differently depending on the ILD thickness and the length of the metal extension over the drift region, the mechanics of which are well described in previous literature. One effect of this design is that the STI is not necessary at all to achieve a high breakdown voltage, provided the LDD doses are tuned accordingly; however this precludes the formation of the N buffer and was shown to sacrifice ruggedness.

TCAD simulation was used extensively in the optimization of the design and process. Figure 3 shows the simulated potential contours of a typical design at 150 V. A variety of designs were incorporated into a test mask array to explore design and process tradeoffs.

III. RESULTS

Devices were fabricated in a 0.35 μm CMOS foundry on a SmartCut® SOI substrate. Device performance varied across a wide range of designs and process conditions. The devices with the best combination of DC performance and UIS

978-1-4244-8425-6/11 $26.00 © 2011 IEEE   377

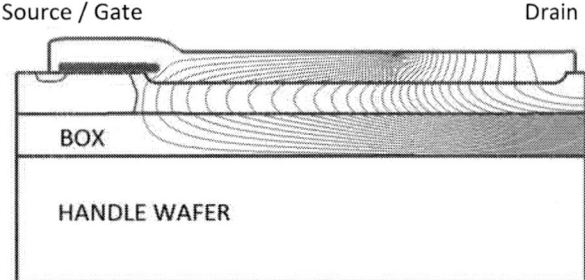

Figure 3. Simulated equipotential contours at VDS = 150 V for a design without drain buffer.

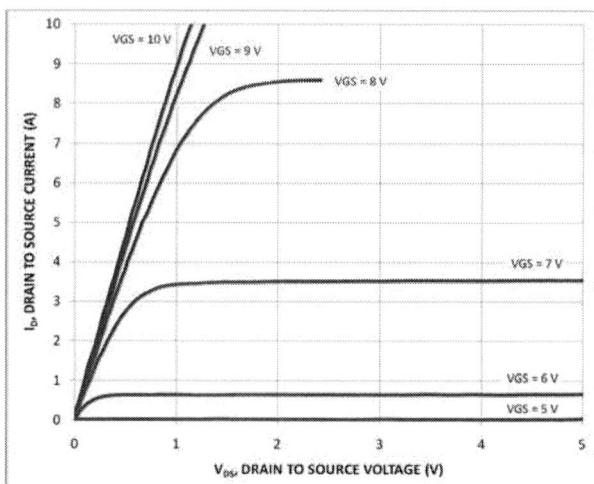

Figure 4. Measured forward IV characteristic of the 150 V SOI LDMOS. Vth measured at 250 µA = 4.8 V.

capability were mounted on evaluation boards for AC characterization.

Figure 4 shows the measured forward I-V characteristics of the device. Rdson values measured at Ids = 4 A and Vgs = 8V, 10V, and 12V were 115 mΩ, 109 mΩ, and 102 mΩ respectively. BVdss of this design, which included the drain buffer, was 165 V with Vth = 4.8 V. Figure 5 shows the BVdss curve for a design from the same wafer without drain buffer, with a measured BVdss = 180 V.

The device is designed with very low gate-to-drain charge and reverse recovery charge in order to improve switching efficiency at MHz switching frequencies. Figure 6 shows the typical gate charge waveforms of the SOI LDMOS design. Qgd for this design is only 0.8 nC measured at Ids = 4 A, Vds = 72 V and Vgs = 8 V. Figure 7 shows the reverse recovery current waveform. Table I summarizes the performance of this work in comparison with two state of the art discrete vertical trench power MOSFETs. While Rdson of the SOI power MOSFET is considerably higher than the trench, the very low Qg results in an overall improvement in switching efficiency figure-of-merit.

Device avalanche capability is an important parameter for many applications. Even with the inclusion of the P+ buried layer and drain buffer, ruggedness varied across a wide array of designs and processes. All of the devices discussed in this paper had the same drift length, defined as the spacing between

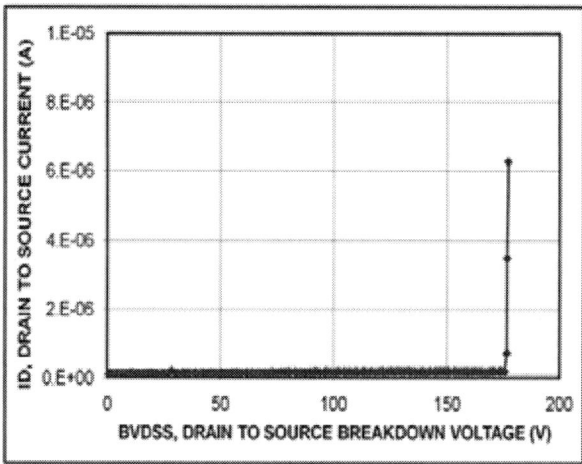

Figure 5. Measured BVdss curve for a design without drain buffer. BVdss = 180 V. The addition of the drain buffer reduces BVdss but significantly enhances the avalanche ruggedness of the device.

the N+ drain diffusion and the poly gate. Only the drain buffer length was varied, which is defined as the spacing between the N+ drain and the STI. Figure 8 shows the peak avalanche current as a function of drain buffer length for different LDD implant doses. In general, an increase in buffer length results in an improvement in device ruggedness. Devices with no STI showed weaker UIS capability for most LDD doses. In this series of tests, the self-driven current was limited to 45 A. For the chosen design and process split whose DC and AC characteristics are summarized in Table 1, the avalanche current capability was greater than 73 Amps at 230 V during subsequent constant-current avalanche tests.

TABLE I. COMPARISON OF KEY SWITCHING PARAMETRICS WITH STATE-OF-THE-ART 150V COMMERCIAL POWER MOSFETs

	This Work	Commercial A	Commercial B
Rdson (mΩ)	102	56	47
Qg total (nC)	4.6	8.9	25
Qgd (nC)	0.8	2.0	6.6
Trr (ns)	36	61	62
Qrr (nC)	32	71	164
FOM (Rdson * Qg)	469	498	1175
FOM (Rdson * Qgd)	82	112	310

978-1-4244-8425-6/11 $26.00 © 2011 IEEE   378

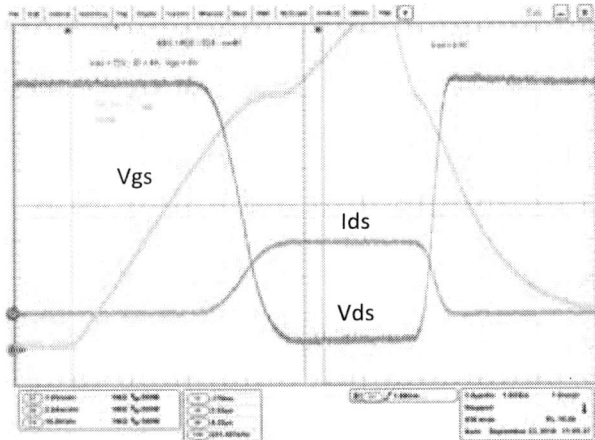

Figure 6. Measured gate charge waveforms for the representative design. At VDS = 72 V, VGS = 8 V, IDS = 4 A. Qtot = 4.6 nC, Qgs = 3.3 nC, Qgd = 0.8 nC.

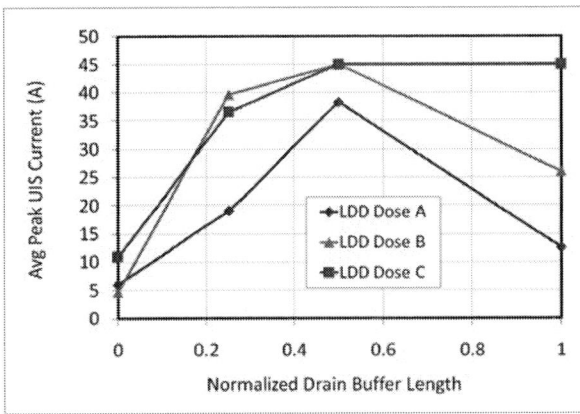

Figure 8. Peak UIS current vs. drain buffer length for processes with different LDD doses. The LDD dose determines whether the buffer is fully or partially depleted when the device enters avalanche. Self driven current was limited to 45A in this series of tests.

REFERENCES

[1] S. Merchant, E. Arnold, H. Baumgart, S. Mukhejee, H. Pein, and R. Pinker, "Realization of high breakdown voltage (>700 V) in thin SOI devices", ISPSD 1991, pp. 31-35.

[2] S. Merchant, E. Arnold, H. Baumgart, R. Egloff, T. Letavic, S. Mukherjee, and H. Pein, "Dependence of Breakdown Voltage on Drift Length and Buried Oxide Thickness in SOI RESURF LDMOS Transistors", ISPSD 1993, pp. 124-128.

[3] Il-Jung Kim, Satoshi Matsumoto, Tatsuo Sakai, and Toshiaki Yachi, "Analytical Approach to Breakdown Voltages in Thin-Film SOI Power MOSFETs", Journal of Solid-State Electronics Vol. 39, No. I, pp. 95-100, 1996.

[4] Steve Merchant, "Analytical Model for the Electric Field Distribution in SOI RESURF and TMBS Structures", IEEE Transactions on Electron Devices, Vol. 46, No. 6, June 1999.

[5] Sameer Pendharkar, Taylor Efland, and Chin-yu Tsaj, "Analysis of High Current Breakdown and UIS Behavior of Resurf LDMOS (RLDMOS) Devices", ISPSD 1998, pp. 419-422.

[6] P. Hower, J. Lin, S. Merchant and S. Paiva, "Using "Adaptive Resurf" to Improve the SOA of Ldmos Transistors", ISPSD 2000, pp. 345-348.

[7] V. Parthasarathy, V. Khemka, R. Zhu, J. whifield, R. Ida and A. Bose, "Drain Profile Engineering of RESURF LDMOS Devices for ESD Ruggedness", ISPSD 2002, pp. 265-268.

[8] Ettore Napoli and Florin Udrea, "Substrate Deep Depletion: An Innovative Design Concept to Improve the Voltage Rating of SOI Power Devices", ISPSD 2006.

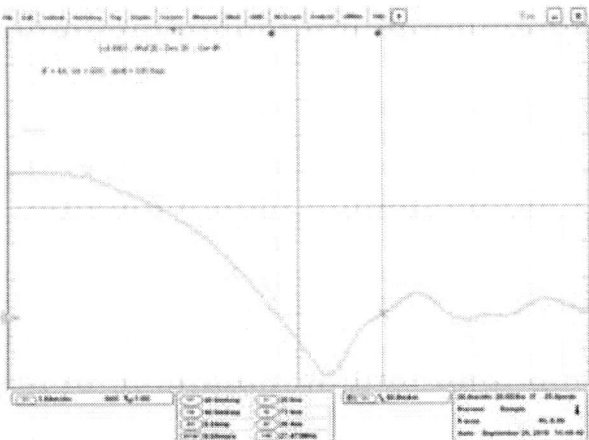

Figure 7. Measured reverse recovery current waveform at IF = 4 A, Vrr = 50 V, di/dt = 100 A/μs. Trr = 36 ns, Qrr = 32 nC.

IV. CONCLUSION

In this work, we report a new class of SOI lateral power MOSFET intended for MHz frequency power switching applications. The new device incorporates features to improve avalanche ruggedness and switching efficiency, both of which are primary concerns for high frequency power switching. These SOI LDMOS demonstrate improved performance over state of the art commercially available power MOSFETs. The discrete power MOSFETs were designed into a low profile, low inductance chip-scale package. These lateral devices were fabricated in a CMOS foundry and offer the possibility of monolithic integration into power integrated circuits.

978-1-4244-8425-6/11 $26.00 © 2011 IEEE

Proceedings of the 23rd International Symposium on Power Semiconductor Devices & IC's
May 23-26, 2011 San Diego, CA

P-type Isolated GGNMOS with a Deep Current Path for ESD Protection

Jae-Hyun Yoo, Jongmin Kim, Joong-Hyeok Byeon, Young-Sang Son, Jaeyoung Park and Won-Young Jung
Device Technology Team, Dongbu Hitek, Inc.
222-1, Dodang-Dong, Wonmi-Gu, Bucheon, Gyeonggi-Do, 420-712 Korea.
jaehyun.yoo@dongbu.com

Abstract— **In this paper, we propose a P-type Isolated GGNMOS (PI-GGNMOS) with a deep current path to improve holding voltage (Vh) of Electro-Static Discharge (ESD) protection device. In order to make the deep current path under the channel, the proposed ESD protection device has a p-type stud between source and the channel, compared to the conventional GGNMOS (Gate-Grounded NMOS). To verify the performance of the proposed structure, we simulated and measured the test structure that is fabricated in a 0.35μm Bipolar-CMOS-DMOS (BCD) process. We found that the proposed structure improves the holding voltage from 6.4V to 8.48V for 5V GGNMOS at 5.3μm pitch. In case of conventional 7V GGNMOS at 7.0μm pitch, the holding voltage is 8.7V. Therefore, we can use 5V PI-GGNMOS as a 7V ESD protection device with 32 % pitch reduction compared to conventional 7V ESD device without any additional process. The actual size of ESD cell is saved by 42.3%, considering It2. This improvement is attributed to the p-type stud which reduces gain and extends effective base width of parasitic NPN in GGNMOS. Consequently, the PI-GGNMOS can apply for upper range ESD protection at same cost.**

I. INTRODUCTION

The market share of PMIC in the semiconductor industry is growing gradually. These days, BCD technology has been used widely for various applications and required the several logic components to implement a lot of power management utilities. In the similar context, these demands should be accompanied with the guarantees of ESD protection. As an example, the 7V IO applications are usually used to secure the operating margin such as a Low-Drop-Output (LDO), thereby designers need the ESD protection device for 7V IO applications. However, because most BCD processes have generally 5V CMOS logic device, it is hard to provide 7V ESD protection device [1].

The alternate solution to satisfy these needs is to change the process for the purpose of application or is to add some process to cover the additional operating level such as 7V CMOS. However, the development of the ESD protection device for wide operating voltage levels is not simple and spends a lot of resource like cost, time, and man-power. The second solution for the requirements is that design layout as

like gate-length (Lg) is modified. This method is very simple and don't need a lot of development resource. However, it's not easy to obtain the ESD capability at ESD stress and the off-current (Ioff) at direct-current (DC) status without additional process. Even if these specific items are satisfied, the device area for the ESD protection become much larger than expected. Finally, it's also difficult to satisfy the ESD capability within limited ESD device size in conventional BCD process having broad operating voltage levels.

In this paper, we developed 7V ESD protection device to improve the ESD capability within limited ESD device size without adding or modifying the process in not only general but also Ultra-High-Voltage (UHV) 0.35μm BCD process with 5V logic components. For 7V ESD device, we transform the conventional 5V ESD protection device to new PI-GGNMOS by inserting the p-type stud. To test the feasibilities of the proposed PI-GGNMOS, TCAD tools such as Tsupream4 and Medici is used. Through the experiment, we could obtain the 7V ESD protection device Therefore, we propose the PI-GGNMOS to support 7V ESD protection in smaller pitch-size than that of 5V ESD protection.

II. DEVICE STRUCTURE AND SIMULATION

Firstly, in order to make 7V ESD protection device, we modify the structure of the conventional 5V GGNMOS. We add a p-type stud to the source region of it. Fig. 1-(a) and (b) show the design layouts for the conventional 5V ESD protection having 0.8μm gate-length and the PI-GGNMOS having 0.5μm gate-length respectively. The p-type stud in fig. 1-(b) is formed by using p-type LDD (pLDD) mask and p-type source-drain (pSD) mask. Then, we try to shrink the gate length of the PI-GGNMOS from 0.8μm to 0.5μm.

Generally, in the case of GGNMOS for 5V ESD protection, the gate-length is bigger than minimum design rule to guarantee the process instability. However, the PI-GGNMOS can overcome the process instability in spit of using the minimum gate-length of 0.5μm because the p-type stud between source and the channel removes perfectly this cause of the process instability associated with the surface channel leakage.

978-1-4244-8425-6/11 $26.00 © 2011 IEEE   380

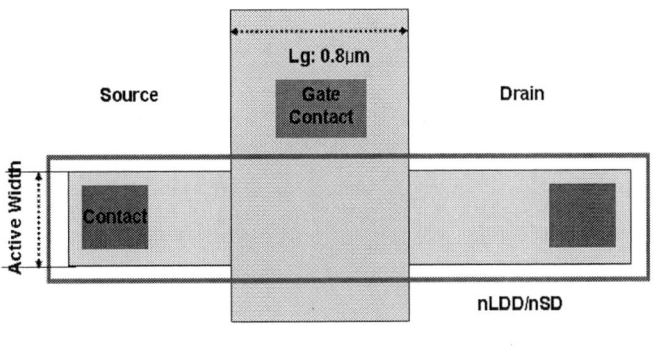

(a)

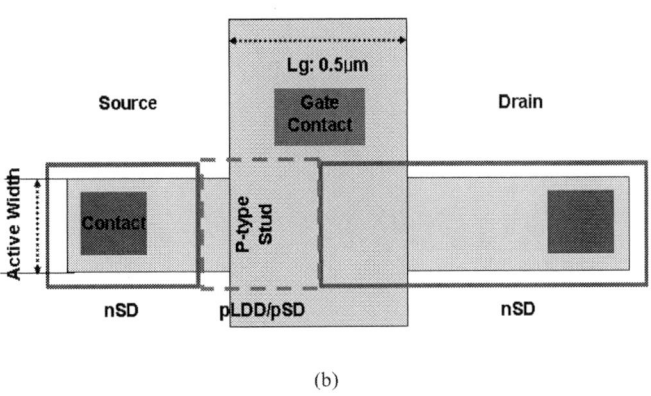

(b)

Figure 1.   Layout top views of *(a)* conventional GGNMOS and *(b)* PI-GGNMOS

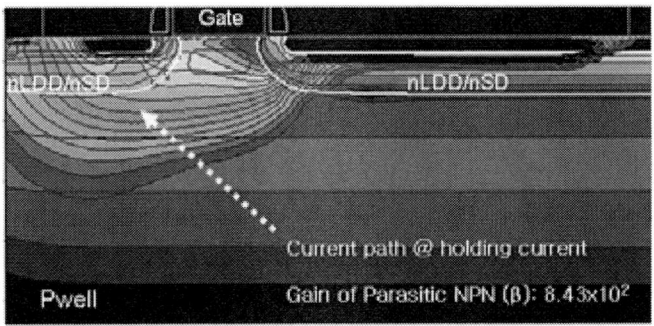

(a)

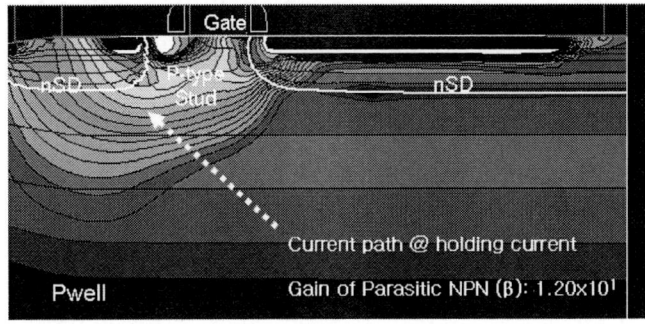

(b)

Figure 2.   Cross-sections beta gains of *(a)* conventional GGNMOS and *(b)* PI-GGNMOS

In the meanwhile, the LDD implant condition optimized for the performance of 5V NMOS leads to other leakage mechanism, band to band tunneling (BTBT) current, when used in the drain of 7V GGNMOS device. Therefore, for lower DC leakage level in the 7V PI-GGNMOS, the n-type LDD (nLDD) mask in drain region of PI-GGNMOS is blocked as shown in fig. 1-(b). The PI-GGNMOS without nLDD can avoid the leakage induced by BTBT. The PI-GGNMOS without nLDD can avoid the leakage induced by BTBT

The holding voltage, current density, ACS (TLP) curve and DC leakage of the conventional one and the proposed one are analyzed through 2D simulation. Simulated TLP curves were performed with ACS (average current slope) method [2].Fig. 2-(a) and (b) present the 2D simulation results of the conventional 5V ESD protection and the PI-GGNMOS respectively. From 2D simulation results, it is realized that the deep current path is formed under the channel compared to the conventional one as the p-type stud of high concentrated impurity leads it in fig. 2-(b). The p-type stud also reduces the beta of parasitic NPN from 843 of the conventional one to 12 of the proposed one due to the extension of effective base width in fig. 2-(a) and (b). Generally, The Vh is given by equation (1)

$$V_{sb} \approx BVdss(1 - \alpha) \approx \frac{BVdss}{(2)^{1/n}}(L_{eff}/L_d) \qquad (1)$$

Where n is constant, Leff is effective channel length, α is common base current gain, Ld is diffusion length [3]. In the case of PIGGNMOS, the isolation region with p-type impurity lead to change the direction of the current path, thereby Leff is extended and α is reduced.

Fig. 3-(a) and (b) illustrate the simulated TLP curves and DC leakage curves in each structure. The Vh increases from 6.3V of the conventional GGNMOS to 7.5V of PI-GGNMOS in simulation while The Vt1 increases from 10.8V to 12.2V.

In fig. 3-(b), DC leakage consists of surface channel leakage, junction leakage and BTBT leakage. Surface channel leakage and junction leakage under operation voltage is major while the tunneling leakage is dominant above the operating voltage. In the proposed PIGGNMOS, the surface channel leakage disappears as the p-type stud disconnects the current path between source and channel, and junction leakage only remains. Therefore, the DC leakage improves from 2.3×10^{-12} A to 1.0×10^{-13} A under operating voltage, from 7.0×10^{-9} A to 5.0×10^{-13} A above operating voltage. The DC leakage level improves more than one order in the measurement condition of 7.7V, which is normally defined 110 % of operating voltage. The reduction of surface channel leakage allows the gate-length to shrink in the PIGGNMOS.

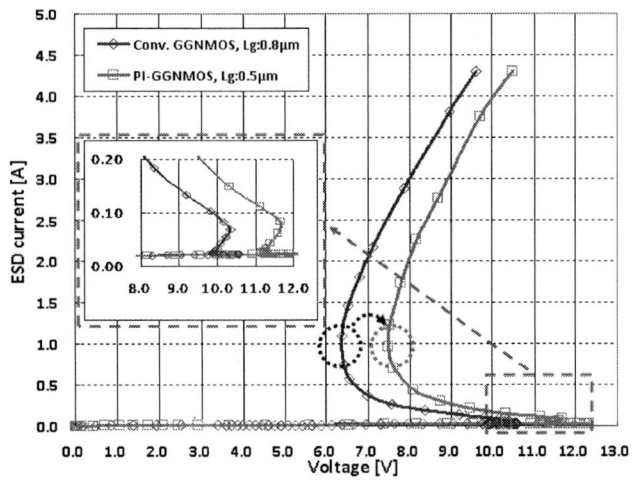

(a)

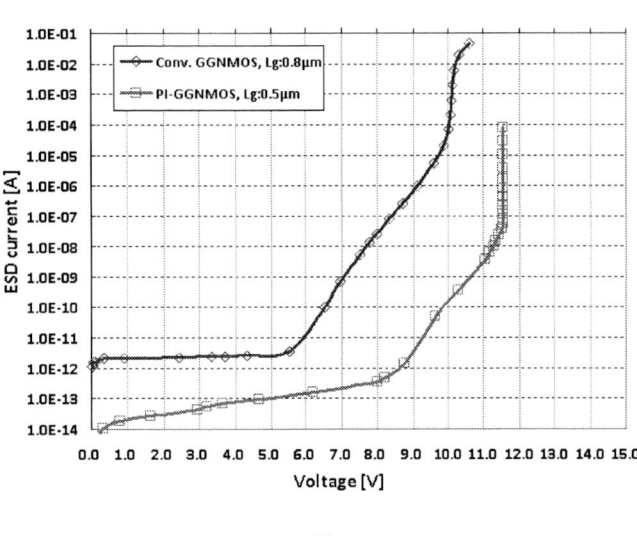

(b)

Figure 3.   Simulation results: (a) ACS (TLP) curves and (b) DC leakage curves of conventional GGNMOS and PI-GGNMOS

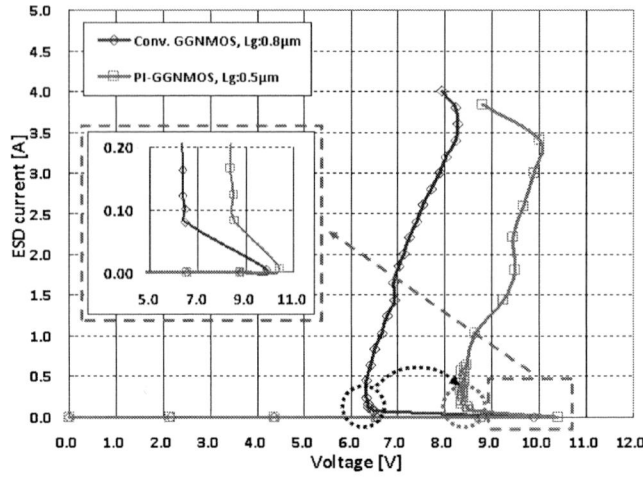

(a)

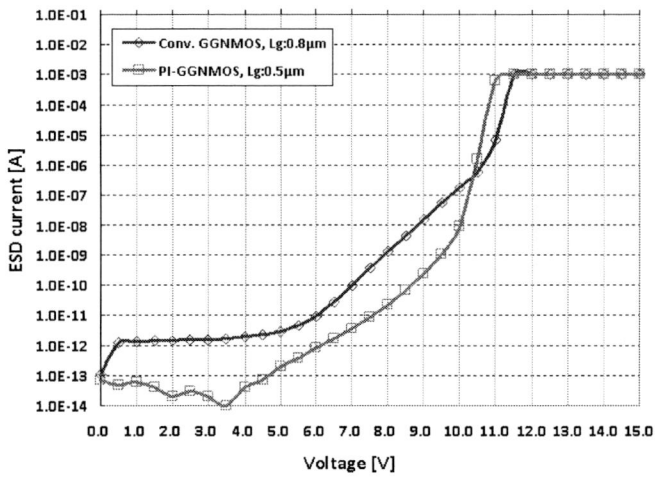

(b)

Figure 4.   Experiment results: *(a)* TLP curve and *(b)* DC leakage curve of conventional GGNMOS and PI-GGNMOS

III.   MEASUREMENT RESULTS

To verify the proposed structure, we fabricate the PI-GGNMOS in 0.35μm Bipolar-CMOS-DMOS (BCD) process. Fig. 4-(a) and (b) show the TLP curves and DC leakage curves in each structure. The Vh increases from 6.4V for conventional GGNMOS to 8.48V for PI-GGNMOS in experiment while the variation of Vt1 is slight. The DC leakage trend of both experiment and simulation looks similar. Therefore, we can sure that the p-type stud acts as a blocker of surface channel leakage and a former of deep current path as previously described.

TABLE I.   VT1, VH, AND IT2 OF CONVENTIONAL GGNMOS VS. GATE LENGTH AND VT1, VH, AND IT2 OF PI-GGNMOS

Device Type	Monitoring Factor			
	Lg $[\mu m]$	$Vt1$ $[V]$	Vh $[V]$	$It2$ $[mA/um]$
Conventional GGNMOS	0.50	8.60	4.80	9.40
	0.60	9.60	6.10	9.80
	0.80	9.75	6.40	8.90
	2.50	10.40	8.70	5.40
PI-GGNMOS	0.50	10.50	8.48	7.35

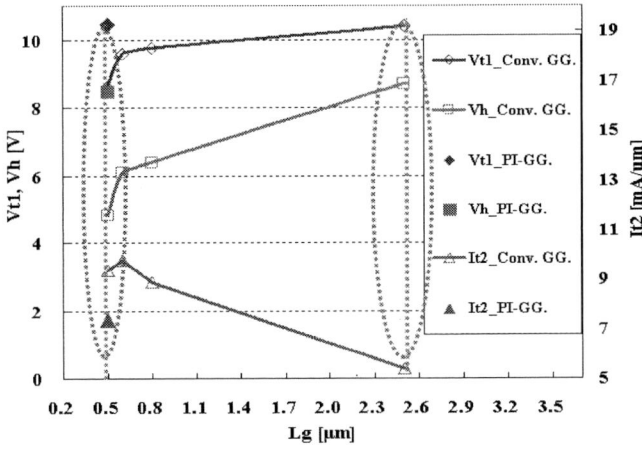

Figure 5. Vt1, Vh, and It2 of conventional GGNMOS vs. gate length and Vt1, Vh, and It2 of PI-GGNMOS

Table 1 and Fig. 5 compare the ESD characteristics - Triggering voltage (Vt1), Holding voltage (Vh) and second breakdown current (It2) - of the proposed device with those of the conventional one. Usually, the holding voltage in the conventional one becomes higher as the increasing of gate-length [4]. From our experimental results, we could determine that the ESD protection for 7V IO need to extend the gate-length about 2.5μm in conventional GGNMOS, whereas the novel structures obtain the ESD characteristics for 7V IO without any addition of process or extension of pitch size. The proposed one has 2.0V higher holding voltage, and 2.0mA/um² higher It2. In a pitch-size point of view, the proposed one reduces 32% over conventional one. The actual size of ESD cell is saved by 42.3 %, considering It2.

IV. CONCLUSION

We propose a novel PI-GGNMOS with a deep current path and prove that the PI-GGNMOS is applicable for upper range ESD protection without any additional process through TCAD simulation and experiment not in only general but also UHV 0.35μm BCD process. In this paper, we obtain the 7V ESD protection experimentally with competitive cost and size of ESD area. Finally, we increase the holding voltage from 6.4V to 8.4V while the triggering voltage is remained for 7V ESD protection. The actual size of ESD cell is saved by 42.3 %, compared to conventional 7V ESD device.

ACKNOWLEDGMENT

This work was supported by the IT R&D program of the Korea Ministry of the Knowledge. The authors would like to thank to B.-K Jun, K.-D Yoo, T.-S. Kim and L. N. Hutter, both with the Dongbu Hitek Company Ltd., Bucheon, Korea, for supporting and encouraging this work.

REFERENCES

[1] Xu Gong, " A Wide Range High Power Supply Rejection Ratio And Transient Enhanced Low Drop-out Regulator", 6th International Conference on Wireless Comunications Networking and Mobile Computing. pp.1-4, 2010.

[2] Ph. Galy, "Numerical study of a gg-nMOS protection transistor under ionizaing radiation and electrostatic discharge", 6th European Conference on Radiation and Its Effects on Components and Systems. pp. 300 – 304. 2001.

[3] Kueing-Long Chen. "Effects of Interconnect Process and Snapback Voltage on the ESD Failure Threshold of NMOS Transistor" Proc. EOS/ESD Symp. pp. 212 - 218. 1988.

[4] K. Bock, "Influence of gate length on ESD-performance for deep sub micron CMOS technology", Proc. EOS/ESD Symp. pp. 95 - 104. 1999.

AUTHOR INDEX

A

Abbate, C.	144
Abiko, Y.	308
Agarwal, Anant	227
Aida, Satoshi	316
Aizawa, Junichi	32
Alagi, F.	340
Allard, P.	172
Antoniou, M.	336
Aoki, Hirofumi	28
Arai, Taiga	32, 48
Aresu, Stefano	20
Arnould, Jean-Daniel	360
Asano, K.	296
Avenas, Yvan	332
Aykroyd, Craig	192

B

Baburske, R.	104, 108
Baccarani, G.	152
Baele, J.	304
Bahat-Treidel, E.	239
Bang, Sun-Kyung	24
Bauer, J.G.	104
Bauwens, F.	200, 312
Beckedahl, Peter	324
Belletti, F.	36
Bellini, M.	56
Bhalla, Anup	256
Biermann, J.	104
Blondel, Gael	368
Bogman, F.	304
Bonera, E.	36
Boos, P.W.M.	348
Borella, F.	340
Bou, Rachana	72
Bourennane, A.	140
Böving, Heike	64
Braspenning-Girault, V.	348

Breglio, G.	124
Brewer, Forrest	164
Brunner, F.	239
Bu, Jiankang	72, 320
Busatto, G.	144
Byeon, Joong-Hyeok	380

C

Caballero, Mar	300
Cai, Jun	208
Callanan, Robert	227
Cammarata, M.	88
Capell, Craig	227
Castellazzi, Alberto	268
Celaya, José R.	160
Cha, Jaehan	211
Chai, Francis K.	300
Challa, A.	156
Chan, Chien-Ling	188
Chang, Hsueh-Rong	72, 320
Chang, Kuo-Cheng	188
Chatterjee, Amitabh	164
Chen, Hongwei	276
Chen, Kevin J.	276
Chen, Max	80
Cheng, Chih-Chang	208
Cheng, Lin	227
Cho, Cheol-Ho	24
Cho, E.	239
Choi, Hong Goo	231
Choi, Yong-Keon	219
Chou, H.L.	208
Chow, T. Paul	243
Chu, F.Y.	208
Cirrone, G.A.P.	144
Constantin, Delphine	204
Cortés, I.	112
Corvasce, C.	88
Crébier, Jean-Christophe	204, 328, 332, 356, 360
Croce, G.	340

D

De Vleeschouwer, H.	304
Denison, M.	152
Dhar, S.	227, 348
Dhyani, Mahesh H.	280

Donlon, John F.	132
Duvvury, Charvaka	164

E

Ebiike, Y.	288
Eguchi, Hiroomi	28
Eguchi, S.	308
Enea, V.	144
Escoffier, Rene	368

F

Feilchenfeld, Natalie	215
Feldmann, Uwe	148
Felsl, H.P.	100, 104, 108
Fields, Terri	300
Fomani, A.A.	364
Fu, D.P.	76
Fujii, Hidenori	132
Fujii, Hiroki	372
Fujishima, N.	52
Fukada, Yusuke	132
Furukawa, A.	288
Furuya, K.	344

G

Galbiati, P.	36
Gallerano, Antonio	128
Gassot, P.	200
Gattari, P.	340
Gaude, Victor	328
Gebhardt, Karl-Heinz	20
Geil, Bruce	227
Gendron, Amaury	192
Ghandi, Reza	10
Gill, Chai	192
Glaser, Ulrich	20
Gnani, E.	152
Gnudi, A.	152
Göbl, Christian	324
Goebel, Kai	160
Grivna, G.	304
Grummel, Brian	260
Gu, Sung-Mo	24

H

Ha, Min-Woo **231**
Hahn, Cheol-Koo **231**
Hamada, Kimimori **28**
Han, Min-Koo **235, 247, 251**
Hao, Yang **336**
Hara, Kenji **32**
Harada, Takashi **48**
Harada, Tatsuo **132**
Haraguchi, Yuki **68**
Hatade, K. **136**
Hatasako, K. **344**
Hayashi, M. **40**
Haynie, S. **172**
Hefner, Allen R. **260**
Hella, Mona **243**
Henlser, Alexander **264**
Heringa, A. **348**
Herold, Christian **264**
Herzer, R. **196**
Hilt, O. **239**
Hinata, Yuichiro **272**
Hoelke, A. **336**
Honda, Hironobu **32**
Honda, Norihiro **28**
Honda, Shigeto **68**
Hori, Motohito **272**
Horio, Masafumi **272**
Hsu, Wesley Chih-Wei **80**
Hu, Xi **16, 180**
Huang, Alex Q. **184**
Huang, Guangzuo **180**
Huang, Yong **180**
Huesken, H. **120**
Hui, S.Y. (Ron) **352**
Hutter, Lou **24, 219**
Hwang, Shinwhan **223**

I

Iannuzzo, F. **144**
Ichikawa, Toshihiko **372**
Iizuka, Yuji **272**
Ikeda, Nariaki **284**
Ikeda, Yoshinari **272**
Ikegami, Y. **308**
Imai, Toshinori **44**
Imaizumi, M. **288**

Imbernon, E.	140
Irace, A.	124
Irène, Pheng	328
Ishibashi, Kohsuke	48
Ishikawa, M.	308
Ishizawa, Shinichi	132
Ito, A.	40
Izumisawa, Masaru	316

J

Jakobi, Waldemar	64
Jeon, Woochul	223
Jiang, Lingli	180
Jonas, Charlotte	227
Jong, Y.C.	208
Jordà, X.	112
Jung, Hyung-Gyun	24
Jung, Won-Young	380

K

Kagawa, Y.	288
Kambayashi, Hiroshi	284
Kaminski, N.	56
Kanai, H.	308
Kaneda, M.	136
Kang, Sun-Kyoung	24
Kao, Tzu-Cheng	188
Kato, Sadahiro	284
Katzenberger, G.	196
Kawai, Hiroji	280
Kho Ching Tee, E.	336
Kijima, Shinya	28
Kim, Han-Geon	24
Kim, Jongmin	380
Kim, Juho	211
Kim, Minki	247
Kim, Min-Seok	24
Kim, Min-Woo	24
Kim, Mi-Young	219
Kim, Nam-Joo	219
Kim, S.	156
Kim, Sungoo	211
Kim, Young-Shil	235, 247, 251
Kimoto, Tsunenobu	292
Kimura, Yoshinobu	168
Kinouchi, S.	288

Knaipp, Martin	215
Knauer, A.	239
Ko, Choul-Joo	24
Koduki, Shigeo	316
Kokawa, Takuya	284
Komatsu, Shinichi	372
Kong, George	72, 320
Kong, Soon Tat	176
Königsmann, G.	196
Kopta, A.	56, 88
Kuepper, Paul	20
Kuhn-Heinrich, Barbara	20
Kumagai, Yukihiro	168
Kuroi, T.	344
Kusunoki, S.	136
Kwon, T.	172

L

Labayen, Ricky	320
Lai, Li	180
Laska, Thomas	64
Lee, Chris	176
Lee, Hee-Bae	24
Lee, Jian-Hsing	188
Lee, Jun Ho	231
Lee, Junghee	223
Lee, Kyungho	211
Lee, Taejong	211
Lee, Yong-Jun	24
Lehmann, J.	196
Lei, L.	76
Lei, T.F.	76
Lelis, Aivars	227
Lesaint, Olivier	356
Levy, Max	215
Li, Zhaoji	16, 76, 180
Liang, Tao	180
Liao, Hong	180
Lim, Hyun-Chol	219
Lim, Jiyong	235, 247, 251
Lin, Pai-Li	80
Lin, Y.C.	208
Lin, Yih-Yin	80
Liou, R.S.	208
Litaudon, Stephane	332
Loechelt, G.	304
Lu, David Hongfei	116

Luo, Bo	16
Luo, Xiao	16
Luo, Xiaorong	76, 180
Luther-King, N.	92
Lutz, J.	104
Lutz, Josef	108, 264

M

Maegawa, S.	344
Mallinger, Mike	300
Malzer, Thomas	324
Marchesi, G.	340
Masuoka, Fumihito	96
Masuoka, Hiroki	148
Matsumoto, Yasuaki	84
Mattausch, Hans Juergen	148
Matthias, S.	88
Maxwell, Ed	300
Mayerhofer, Alevtina	20
McDonald, J.	312
Merlini, D.	36
Millán, J.	112
Minato, T.	136
Minixhofer, Rainer	215
Miura, N.	288
Miura-Mattausch, Mitiko	148
Miyake, Hiroki	292
Miyake, Masataka	148
Miyakoshi, Kenji	44
Miyamoto, Masafumi	168
Miyanagi, Y.	296
Miyoshi, T.	40
Mizuno, Y.	124
Moens, P.	200, 304, 312
Molfese, A.	340
Mori, Mutsuhiro	48
Mouhoubi, S.	200
Mukherjee, Satyen	1
Mustain, Habib A.	260

N

Nakajima, Akira	280
Nakajima, Tsunehiro	116
Nakamura, Katsumi	96
Nakano, H.	52
Nakao, Y.	288

Nakashima, Junichi 148
Nakata, Kazunari 68
Nakatake, H. 288
Nakayama, K. 296
Nakazawa, Haruo 116
Nakazawa, Y. 308
Napoli, E. 124
Narazaki, Atsushi 68
Ng, Daniel 256
Ng, W.T. 364
Niedernostheide, F.-J. 100, 104, 108, 120
Nishii, Akito 96
Nishio, H. 364
Nitta, T. 344
Nomura, Takehiko 284

O

O'Riain, Lincoln 20
Oda, Tetsuo 48
Odekirk, Bruce 300
Ogata, S. 296
Ogawa, Takaoki 148
Ogino, Masaaki 116
Omichi, A. 344
Omura, Ichiro 60, 84
Ong, Michaelina 176
Onozawa, Y. 52
Oomori, T. 288
Oshima, T. 40, 44
Östling, Mikael 10
Ozaki, D. 52

P

Pal, D.K. 336
Pala, Vipindas 243
Palmour, John 227
Pan, Ji 256
Park, Il-Yong 219
Park, Jaeyoung 380
Park, Kiyeol 223
Park, Namkyu 211
Park, Younghwan 223
Parsey, J. 304
Pendharkar, S. 152, 164
Perpiñà, X. 112
Pfaffenlehner, Manfred 108

Pfirsch, Frank 100, 108
Phelps, Rick 215
Pilkington, S. 336
Pizzo, G. 340
Poli, S. 152
Pont, L. 140
Probst, D. 156
Pugatschow, Anton 64

Q

Qiao, M. 16, 76, 180
Quddus, T. 304

R

Rahimo, M. 56
Raker, Thomas 100
Rebollo, J. 112
Reggiani, S. 152
Riccio, M. 124
Roh, Cheong Hyun 231
Roig, J. 200, 312
Ronsisvalle, C. 144
Roßberg, M. 196
Rouger, Nicolas 328, 356, 360
Rouleau, Blaise 368
Rudolf, Ralf 20
Ryu, Sei-Hyung 227

S

Sadovnikov, A. 172
Saha, Sankalita 160
Saito, Katsuaki 48
Saito, Wataru 316
Sakano, Junichi 32, 44
Sambi, M. 36
Sanchez, J.-L. 140
Sankara Narayanan, E.M. 280
Sanseverino, A. 144
Sapp, S. 156
Sarrabayrouse, G. 140
Sato, Masaharu 372
Sato, Yoshihiro 284
Saxena, Abhinav 160
Schaeffer, Christian 204, 332
Schulze, H.-J. 104, 120
Schulze, Hans-Joachim 100, 108

Scozzie, Charles	227
Sdrulla, Dumitru	300
Seok, Ogyun	231, 235, 247
Seok, O-Gyun	251
Shankar Narayanan, E.M.	92
Shea, Patrick M.	376
Shen, Z. John	260, 376
Shi, Yun	215
Shibkov, Andrei	128
Shimamoto, Satoshi	44
Shirakawa, Shinji	44
Shorten, A.	364
Simonot, Timothé	328, 360
Sin, Johnny K.O.	352
Solomon, Adane	268
Son, Young-Sang	380
Song, Hong Joo	231
Spirito, P.	124
Stecher, Matthias	20
Stefanov, Evgueniy N.	368
Stockmeier, Thomas	324
Stokes, R.	156
Storasta, L.	56
Strachan, A.	172
Strasser, Marc	20
Stribley, Paul	176
Strout, J.	172
Su, Hung-Der	188
Su, Jin-Lian	188
Su, Yi	256
Suda, Jun	292
Sugii, Nobuyuki	168
Sumida, Yasunobu	280
Sumitani, H.	288
Sun, Zhen	180
Suzuki, Kenji	132
Swanenberg, M.J.	348
Sweet, M.	92

T

Tack, M.	200, 304, 312
Tadokoro, Chihiro	136
Tahir, H.	140
Takahashi, Tetsuo	132
Takahashi, Y.	116, 272, 364
Takano, K.	136
Tamaki, T.	308

Tamura, Ryosuke 284
Tanaka, A. 296
Tanimura, Takuya 60
Terasaki, Yoshiaki 68
Terashima, Tomohide 96, 132
Thoben, Markus 264
Tominari, T. 40
Tomita, Hidemoto 28
Toyoda, Yasushi 48
Tsai, C.L. 208
Tsuda, Motohiro 84
Tuan, H.C. 208

U

Udrea, F. 336
Udrea, Florin 80
Ueno, Masaya 148
Ueno, S. 40
Ueta, Takashi 148
Urresti-Ibañez, J. 112
Uzuka, Tetsuo 6

V

van Dalen, R. 348
van der Wal, A.B. 348
Vanmeerbeek, P. 312
Vashchenko, Vladislav 128, 160
Vemulapati, U.R. 56
Vladimirova, Kremena 204, 332
Vobecky, J. 88
von Ehrenwall, Andreas 20
von Ehrenwall, Birgit 20
Voss, S. 120

W

Wada, S. 40
Wada, Shinichiro 44
Wagner, Cajetan 20
Wakimoto, E. 308
Wakimoto, Hiroki 116
Wang, Fei 256
Wang, Meng 16, 180
Wang, Xiaopeng 184
Wang, Y.G. 76
Wang, Zhuo 16, 180
Watanabe, So 48

Wen, Hengjuan	16
Wingert, Daniel	264
Wise, R.	152
Wu, K.M.	208
Wu, Rongxiang	352
Wu, Y.	304
Würfl, J.	239

X

Xue, Jonathan	256

Y

Yahiro, J.	136
Yamada, Tetsuya	28
Yamashita, Junichi	132
Yamawaki, Hideo	28
Yamazaki, T.	52
Yanagi, S.	344
Yanagida, Yohei	44
Yao, G.L.	76
Yasuda, T.	308
Yaw, Kee Kia	336
Yedinak, J.	156
Yoo, Jae-Hyun	380
Yoo, Kwang-Dong	24, 219
Yoon, Chul-Jin	219
Yoshihisa, Y.	344
Yoshinaga, M.	40
Yuan, Li	276
Yuasa, Kazufumi	60

Z

Zdebel, P.	304
Zetterling, Carl-Mikael	10
Zhan, Carol	192
Zhang, Bo	16, 76, 180
Zhao, Yuanyuan	180
Zhou, Chunhua	276
Zhou, Qi	276
Zhuang, Xiang	180
Ziad, H.	304

9781424484256